T0236951

Springer Series in the Data Sciences

Series Editors

David Banks, Duke University, Durham, NC, USA

Jianqing Fan, Department of Financial Engineering, Princeton University, Princeton, NJ, USA

Michael Jordan, University of California, Berkeley, CA, USA

Ravi Kannan, Microsoft Research Labs, Bangalore, India

Yurii Nesterov, CORE, Universite Catholique de Louvain, Louvain-la-Neuve, Belgium

Christopher Ré, Department of Computer Science, Stanford University, Stanford, USA

Ryan J. Tibshirani, Department of Statistics, Carnegie Melon University, Pittsburgh, PA, USA

Larry Wasserman, Department of Statistics, Carnegie Mellon University, Pittsburgh, PA, USA

Springer Series in the Data Sciences focuses primarily on monographs and graduate level textbooks. The target audience includes students and researchers working in and across the fields of mathematics, theoretical computer science, and statistics. Data Analysis and Interpretation is a broad field encompassing some of the fastest-growing subjects in interdisciplinary statistics, mathematics and computer science. It encompasses a process of inspecting, cleaning, transforming, and modeling data with the goal of discovering useful information, suggesting conclusions, and supporting decision making. Data analysis has multiple facets and approaches, including diverse techniques under a variety of names, in different business, science, and social science domains. Springer Series in the Data Sciences addresses the needs of a broad spectrum of scientists and students who are utilizing quantitative methods in their daily research. The series is broad but structured, including topics within all core areas of the data sciences. The breadth of the series reflects the variation of scholarly projects currently underway in the field of machine learning.

More information about this series at http://www.springer.com/series/13852

Ovidiu Calin

Deep Learning Architectures

A Mathematical Approach

 Springer

Ovidiu Calin
Department of Mathematics & Statistics
Eastern Michigan University
Ypsilanti, MI, USA

ISSN 2365-5674   ISSN 2365-5682   (electronic)
Springer Series in the Data Sciences
ISBN 978-3-030-36723-7   ISBN 978-3-030-36721-3   (eBook)
https://doi.org/10.1007/978-3-030-36721-3

Mathematics Subject Classification (2010): 68T05, 68T10, 68T15, 68T30, 68T45, 68T99

© Springer Nature Switzerland AG 2020
This work is subject to copyright. All rights are reserved by the Publisher, whether the whole or part
of the material is concerned, specifically the rights of translation, reprinting, reuse of illustrations,
recitation, broadcasting, reproduction on microfilms or in any other physical way, and transmission
or information storage and retrieval, electronic adaptation, computer software, or by similar or dissimilar
methodology now known or hereafter developed.
The use of general descriptive names, registered names, trademarks, service marks, etc. in this
publication does not imply, even in the absence of a specific statement, that such names are exempt from
the relevant protective laws and regulations and therefore free for general use.
The publisher, the authors and the editors are safe to assume that the advice and information in this
book are believed to be true and accurate at the date of publication. Neither the publisher nor the
authors or the editors give a warranty, expressed or implied, with respect to the material contained
herein or for any errors or omissions that may have been made. The publisher remains neutral with regard
to jurisdictional claims in published maps and institutional affiliations.

This Springer imprint is published by the registered company Springer Nature Switzerland AG
The registered company address is: Gewerbestrasse 11, 6330 Cham, Switzerland

Foreword

The multiple commercial applications of neural networks have been highly profitable. Neural networks are imbedded into novel technologies, which are now used successfully by top companies such as Google, Microsoft, Facebook, IBM, Apple, Adobe, Netflix, NVIDIA, and Baidu.

A neural network is a collection of computing units, which are connected together, called neurons, each producing a real-valued outcome, called activation. Input neurons get activated from the sensors that perceive the environment, while the other neurons get activated from the previous neuron activations. This structure allows neurons to send messages among themselves and consequently, to straighten those connections that lead to success in solving a problem and diminishing those which are leading to failure.

This book describes how neural networks operate from the mathematical perspective, having in mind that the success of the neural networks methods should not be determined by trial-and-error or luck, but by a clear mathematical analysis. The main goal of the present work is to write the ideas and concepts of neural networks, which are used nowadays at an intuitive level, into a precise modern mathematical language. The book is a mixture of old good classical mathematics and modern concepts of deep learning. The main focus is on the mathematical side, since in today's developing trend many mathematical aspects are kept silent and most papers underline only the computer science details and practical applications.

Keeping the emphasis on the mathematics behind the scenes of neural networks, the book also makes connections with Lagrangian Mechanics, Functional Analysis, Calculus, Differential Geometry, Equations, Group Theory, and Physics.

The book is not supposed to be exhaustive in methods and neural network type description, neither a cookbook of machine learning algorithms. However, it concentrates on the basic mathematical principles, which govern this branch of science, providing a clear theoretical motivation for why do neural nets work that well in so many applications, providing students the tools needed for the analytical understanding of this field. This is neither a programming book, since it contains no code. Most computer software for neural nets become easily obsolete in this age of technological revolution. The book treats the mathematical concepts that lay behind neuronal networks algorithms, without explicitly implementing code in any specific programming language.

The readers interested in practical aspects of neural networks including the programming point of view are referred to several recent books on the subject, which implement machine learning algorithms into different programming languages, such as TensorFlow, Python, or R. A few examples of books like that are: Deep Learning with Python by F. Chollet, Machine Learning with Python Cookbook: Practical Solutions from Preprocessing to Deep Learning by C. Albon, Python Machine Learning: Machine Learning and Deep Learning with Python, Scikit-learn, and TensorFlow by S. Raschka and V. Mirjalili, Introduction to Deep Learning Using R: A Step-by-Step Guide to Learning and Implementing Deep Learning Models Using R by T. Beysolow, Deep Learning: A Practitioners Approach by J. Patterson and A. Gibson, Hands-On Machine Learning with Scikit-Learn and TensorFlow: Concepts, Tools and Techniques to Build Intelligent Systems by A. Geron, and Deep Learning for beginners: A Practical Guide with Python and Tensor Flow by F. Duval.

The targeted audience for this text are graduate students in mathematics, electrical engineering, statistics, theoretical computer science, and econometrics, who would like to get a theoretical understanding of neural networks. To this end, the book is more useful for researchers rather than practitioners. In order to preserve completeness, we have included in the appendix-relevant material on measure theory, probability, and linear algebra, as well as real and functional analysis. Depending on their background, different readers will find useful different parts of the appendix. For instance, most graduate students in mathematics, who typically take basic functional analysis and real analysis, will find accessible most of the book, while statistics students, who take a minimal amount of measure theory and probability, will find useful to read the analysis part of the appendix. Finally, computer science students, who usually do not have a strong mathematical background, will find useful most of the appendix. The minimal prerequisites to read this book are: Calculus I, Linear Algebra with vectors and matrices, and elements of Probability Theory.

However, we recommend this book to students who have already had an introductory course in machine learning and are further interested to deepen their understanding of the machine learning material from the mathematical point of view. The book can be used as a one-semester, or a two-semester graduate course, covering parts I–II and, respectively, I–V.

The book contains a fair number of exercises, mostly solved or containing hints. The proposed exercises range from very simple to rather difficult, many of them being some applications or extensions of the presented results. The references to the literature are not exhaustive. Anything even approaching completeness was out of the question.

Despite the large number of neural network books published in the last few years, very few of them are of a theoretical nature, which discuss the "mathematical concepts" underlying neural networks. One of the well-known books in the field, providing a comprehensive review of deep learning literature, is Goodfellow et al. [46], which can be used both as a textbook and as a reference. Another book of great inspiration is the online book of Nielsen [92]. Also, the older book of Roja [101] presents a classical introduction to neural networks. A notable book containing a discussion of statistical machine learning algorithms such as "artificial neural networks" is Hastie et al. [53]. The mathematical concepts underlying neural networks are also discussed in the books of Mohri et al. [88] and Shalev-Shwartz et al. [110].

Even if the previous books cover important aspects related to statistical learning and mathematical statistics of deep learning, or the mathematics relevant to the computational complexity of deep learning, there is still a niche in the literature, which this book attempts to address. This book attempts to provide a useful introductory material discussion of what types of functions can be represented by deep learning neural networks.

Overview

The book is structured into four main parts, from simple to complex topics. The first one is an accessible introduction to the theory of neural networks; the second part treats neural networks as universal approximators; the third regards neural nets as information processors. The last part deals with the theory behind several topics of interest such as pooling, convolution, CNNs, RNNs, GANs, Boltzmann machines, and classification. For the sake of completeness, the book contains an Appendix with topics collected from Measure Theory, Probability Theory, Linear Algebra, and Functional Analysis. Each chapter ends with a summary and a set of end-of-chapter exercises. Full solutions or just hints for the proposed exercises are provided at the end of the book; also a comprehensive index is included.

Part I

This is the most elementary part of the book. It contains classical topics such as activation functions, cost functions, and learning algorithms in neural networks.

Chapter 1 introduces a few daily life problems, which lead toward the concept of abstract neuron. These topics introduce the reader to the process of adjusting a rate, a flow, or a current that feeds a tank, cell, fund, transistor, etc., which triggers a certain activation function. The optimization of the process involves the minimization of a cost function, such as volume, energy, potential, etc. More neural units can work together as a neural network. As an example, we provide the relation between linear regression and neural networks.

In order to learn a nonlinear target function, a neural network needs to use activation functions that are nonlinear. The choice of these activation functions defines different types of networks. **Chapter 2** contains a systematic presentation of the zoo of activation functions that can be found in the literature. They are classified into three main classes: sigmoid type (logistic, hyperbolic tangent, softsign, arctangent), hockey-stick type (ReLU, PReLU, ELU, SELU), and bumped type (Gaussian, double exponential).

During the learning process a neural network has to adjust parameters such that a certain objective function gets minimized. This function is also known under the names of cost function, error function, or loss function. **Chapter 3** describes some of the most familiar cost functions used in neural networks. They include the following: the supremum error function, L^2-error function, mean square error function, cross-entropy, Kullback-Leibler divergence, Hellinger distance, and others. Some of these cost functions are suited for learning random variables, while others for learning deterministic functions.

Chapter 4 presents a series of classical minimization algorithms. They are needed for the minimization of the associated cost function. The most used is the Gradient Descent Algorithm, which is presented in full detail. Other algorithms contained in the chapter are the linear search method, momentum method, simulated annealing, AdaGrad, Adam, AdaMax, Hessian, and Newton's methods.

Chapter 5 introduces the concept of abstract neuron and presents a few classical types of neurons, such as: the perceptron, sigmoid neuron, linear neuron, and the neuron with a continuum input. Also, some applications to logistic regression and classification are included.

The study of networks of neurons is done in **Chap. 6**. The architecture of a network as well as the backpropagation method used for training the network is explained in detail.

Part II

The main idea of this part is that neural networks are universal approximators, that is, the output of a neural network can approximate a large number of types of targets, such as continuous function, square integrable, or integrable functions, as well as measurable functions.

Chapter 7 introduces the reader to a number of classical approximation theorems of analytic flavor. This powerful tool with applications to learning contains Dini's Theorem, Arzela-Ascoli's Theorem, Stone-Weierstrass' Theorem, Wiener's Tauberian Theorems, and the Contraction Principle.

Chapter 8 deals with the case when the input variable is bounded and 1-dimensional. Besides its simplicity, this case provides an elementary treatment of learning and has a constructive nature. Both cases of multi-perceptron and sigmoid neural networks with one hidden layer are covered. Two sections are also dedicated to learning with ReLU and Softplus functions.

Chapter 9 answers the question of what kind of functions can be learned by one-hidden layer neural networks. It is based on the approximation theory results developed by Funahashi, Hornik, Stinchcombe, White, and Cybenko in late 1980s and early 1990s. The chapter provides mathematical proofs of analytic flavor that one-hidden layer neural networks can learn continuous, L^1 and L^2-integrable functions, as well as measurable functions on a compact set.

Chapter 10 deals with the case of exact learning, namely, with the case of a network that can reproduce exactly the desired target function. The chapter contains results regarding the exact learning of finite support functions, max functions, and piecewise linear functions. It also contains Kolmogorov-Arnold-Sprecher Theorem, Irie and Miyake's integral formula, as well as exact learning using integral kernels in the case of a continuum number of neurons.

Part III

The idea of this part is that neural networks can be regarded as information processors. The input to a neural net contains information, which is processed by the multiple hidden layers of the network, until certain characteristic features are selected. The output layer contains the compressed information needed for certain classification purposes. The job of a neural network can be compared with the one done by an artist in the process of making a sculpture. In the first stage the artist prepares the marble block, and then he cuts rough shapes. Then he continues with sketching general features and ends with refined details. Similarly with the artist, who removes the right amount of material at each step, a neural network neglects at each layer a certain amount of irrelevant information, such that at the end only the desired features are left.

Chapter 11 models the information processed by a neural net from the point of view of sigma-fields. There is some information lost in each layer of a feedforward neural net. Necessary conditions for the layers with trivial lost information are provided, and an information description of several types of neurons is given. For instance, a perceptron learns the information provided by half-planes, while a multi-perceptron is able to learn more sophisticated structures. Compressible layers are also studied from the point of view of information compression.

Chapter 12 is a continuation of the previous one. It deals with a quantitative assessment of how the flow of information propagates through the layers of a neural network. The tools used for this activity are entropy, conditional entropy, and mutual information. As an application, we present a quantitative approach of network's capacity and information bottleneck. Some applications for the information processing with the MNIST data set are provided at the end of the chapter.

Part IV

This part contains elements of geometric theory applied to neural networks. Each neural network is visualized as a manifold endowed with a metric, which can be induced from the target space or can be the Fisher information metric.

Since the output of any feedforward neural net can be parametrized by its weights and biases, in **Chap.** 13 we shall consider them as coordinate systems for the network. Consequently, a manifold is associated with each given network, and this can be endowed with a Riemannian metric, which describes the intrinsic geometry of the network. Each learning algorithm, which involves a change of parameters with respect to time, corresponds to a curve on this manifold. The most efficient learning corresponds to the shortest curve, which is a geodesic. The embedded curvature of this manifold into the target space can be used for regularization purposes, namely, the flatter the manifold, the least overfitting to the training data.

Chapter 14 is similar with the previous one, the difference being that neurons are allowed to be noisy. In this case the neural manifold is a manifold of probability densities and the intrinsic distance between two networks is measured using the Fisher information metric. The smallest variance of the estimation of optimal density parameters has a lower bound given by the inverse of the Fisher metric. The estimators which reach this lower bound are called Fisher-efficient. The parameters estimated by online learning or by batch learning are efficient in an asymptotic way.

Part V

This part deals with a few distinguished neural network architectures, such as CNNs, RNNs, GANs, Boltzmann machines, etc.

Pooling and convolution are two machine learning procedures described in **Chaps.** 15 and 16, respectively. Pooling compresses information with irremediable loss, while convolution compresses information by smoothing using a convolution with a kernel. The equivariance theory is also presented in terms of convolution layers.

Chapter 17 deals with recurrent neural networks, which are specialized in processing sequential data such as audio and video. We describe their training by backpropagation through time and discuss the vanishing gradient and exploding gradient problems.

Chapter 18 deals with the process of classification of a neural network. It includes the linear and nonlinear separability of clusters, and it studies decision maps and their learning properties.

Chapter 19 treats generative networks, of which GANs play a central role. The roles of discriminator and generator networks are described, and their optimality is discussed in the common setup of a zero-sum game, or a mini-max problem.

Chapter 20 contains a presentation of stochastic neurons, Boltzmann machines, and Hopfield networks as well as their applications. We also present the equilibrium distribution of a Boltzmann machine, its entropy, and the associated Fisher information metric.

Bibliographical Remarks

In the following we shall make some bibliographical remarks that will place the subject of this book in a proper historical perspective. The presentation is not by far exhaustive, the reader who is interested in details being referred to the paper of Jurgen Schmidhuber [109], which presents a chronological review of deep learning and contains plenty of good references.

First and foremost, neural networks, as variants of linear or nonlinear regression, have been around for about two centuries, being rooted in the work of Gauss and Legendre. The concept of estimating parameters using basic functions, such as Fourier Analysis methods, can be also viewed as a basic mechanism of multilayer perceptrons operating in a time domain, which has been around for a while.

Prior to 1980s all neural networks architectures were "shallow", i.e., they had just very few layers. The first attempt to model a biologic neuron was done by Warren S. McCulloch and Walter Pitts in 1943 [82]. Since this model had serious learning limitations, Frank Rosenblatt introduced the multilayer perceptron in 1959, endowed with better learning capabilities. Rosenblatt's perceptron had an input layer of sensory units, hidden units called association units, and output units called response units. In fact, the perceptron was intended to be a pattern recognition device, and the association units correspond to feature or pattern detectors. The theory was published in his 1961 book, "Principles of neurodynamics: Perceptrons and the theory of brain mechanism", but it was haunted by the lack of appreciation at the time due to its limitations, as pointed out in the book of Minsky and Papert [87].

Learning in Rosenblatt's multilayer perceptron is guaranteed by a convergence theorem, which assures learning in finite time. However, in most cases this convergence is too slow, which represents a serious limitation of the method. This is the reason why the next learning algorithms involved a procedure introduced by Hadamard in as early as 1908, called the gradient descent method. Shunichi-Amari in his 1967 paper [3] describes the use of gradient descent methods for adaptive learning machine. The early use of the gradient descent method has been also employed by Kelley [62], Bryson [18], Dreyfus [33], etc., in the same time period. The previously described period lasting between 1940 and 1960 and dealing with theories inspired by biological learning is called *cybernetics*.

However, the first artificial neural network that was "deep" indeed was Fukishima's "neocognitron" (1979), which was based on the principle of convolutional neural networks and inspired by the mammalian visual cortex. The difference from the later contest-winning modern architectures was that the weights were not found by a backpropagation algorithm, but by using a local, winner-take-all based learning algorithm. The backpropagation algorithm was used first time, without any reference to neural networks, by Linnainmaa in 1970s [77]. The first reference on backpropagation specific related to neural networks is contained in Werbos [123, 124].

The next period, from 1990 to 1995, is called *connectionism* and is characterized by training deep neural networks using backpropagation. The method of backpropagation was described first in mid-1980s by Parker [93], LeCun [73], and Rumelhart, Hinton, and Williams [108], and it is still the most important procedure to train neural networks.

In late 1980s several representation theorems dealing with neural networks as universal approximators were found, see Hornik et al. [89] and Cybenko [30]. The results state that a feedforward neural network with one hidden layer and linear output layer can approximate with any accuracy any continuous, integrable, or measurable function on a compact set, provided there are enough units in the hidden layer. Other theoretically comforting results state the existence of an exact representation of continuous functions, see Sprecher [115], Kolmogorov [64], and Irie [60]. One of the main points of the present book is to present this type of representation theorems.

During 1990s important advances have taken place in modeling sequences with recurrent neural networks with later application to natural language processing. These can be mostly attributed to the introduction of the long short-term memory (LSTM) by Hochreiter and Schmidhuber [56] in 1997.

Other novel advances occurred in computer vision, being inspired by the mammalian visual cortex mechanism, and can be credited to the use of convolutional neural networks, see LeCun et al. [74, 75]. This marked the beginning of the *deep learning* period, which will briefly present below.

The deep learning pioneering started with LeNet (1998), a 7-level convolutional network, which was used for handwritten digit classification that was applied by several banks to recognize handwritten numbers on checks. After this the models became deeper and more complex, with an enhanced ability to process higher resolution images.

The first famous deep network in this direction was AlexNet (2012), which competed in the ImageNet Large Scale Visual Recognition Challenge (ILSVRC), significantly outperforming all the prior competitors and achieving an error of 15.3%. Just a year later, in 2013, the ILSVRC winner network was also a convolutional network, called ZFNet, which achieved an error rate of 14.8%. This network maintained a similar structure with the AlexNet and tweaked its hyperparameters.

The winner of the ILSVRC 2014 competition was GoogLeNet from Google, a 22-layer deep network using a novel inception module, which achieved an error rate of 6.67%, being quite close to the human-level performance. Another 2014 ILSVRC top 5 competition winner was a network with 16 convolutional layers, which is known as VGGNet (standing for Visual Geometry Group at University of Oxford). It is similar to AlexNet but has some new standards (all its filters have size 3×3, max-poolings are placed after each 2 convolutions, and the number of filters is doubled after each max-pooling). Nowadays it is the most preferred choice for extracting features from images.

ResNet (Residual Neural Network) was introduced in 2015. Its novel 152-layer architecture has a strong similarity with RNNs and a novel feature of skipping connections, while using a lower complexity than VGGNet. It had achieved an error rate of 3.57%, which beats human-level performance on that data set, which is of about 5.1%.

In conclusion, the idea of teaching computers how to learn a certain task from raw data is not new. Even if the concepts of neuron, neural network, learning algorithm, etc. have been created decades ago, it is only around 2012 when Deep Learning became extremely fashionable. It is not a coincidence that this occurred during the age of Big Data, because these models are very hungry for data, and this requirement could be satisfied only recently when the GPU technology was developed enough.

Besides the presentation, this book contains a few more novel features. For instance, the increasing resolution method in sect 4.10, stochastic search in section 4.13, continuum input neuron in sect 5.8, continuum deep networks in sects 10.6 and 10.9, and information representation in a neural network as a sigma-algebra in Part III, to enumerate only a few, are all original contributions.

Acknowledgements

The work on this book has started while the author was a Visiting Professor at Princeton University, during the 2016–2017 academic year, and then continued during 2018–2019, while the author was supported by a sabbatical leave from the Eastern Michigan University. I henceforth express my gratitude to both Princeton University, where I first got exposed to this fascinating subject, and Eastern Michigan University, whose excellent conditions of research are truly appreciated. Several chapters of the book have been presented during the Machine Learning of Ann Arbor Group meetings, during 2017–2018.

This books owes much of its clarity and quality to the many useful comments of several unknown reviewers, whose time and patience in dealing with the manuscript is very much appreciated. Finally, I wish to express many thanks to Springer and its editors, especially Donna Chernyk for making this endeavor a reality.

Michigan, USA   Ann Arbor
September 2019

Chapters Diagram

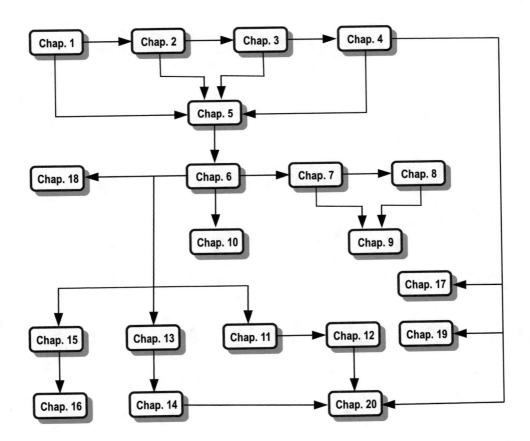

Notations and Symbols

The following notations have been frequently used in the text.

Calculus

H_f	Hessian of f
$\frac{\partial F}{\partial x}, J_F(x)$	Jacobian matrix of $F : \mathbb{R}^n \to \mathbb{R}^n$
$\partial_{x_k}, \frac{\partial}{\partial x_k}$	Partial derivative with respect to x_k
$L^2[0, T]$	Squared integrable functions on $[0, T]$
$C(I^n)$	Continuous functions on I_n
$C^2(\mathbb{R}^n)$	Functions twice differentiable with second derivative continuous
$C_0^2(\mathbb{R}^n)$	Functions with compact support of class C^2
$\|x\|$	Euclidean norm $(= \sqrt{x_1^2 + \cdots + x_n^2})$
$\|f\|_{L^2}$	The L^2-norm $(= \sqrt{\int_a^b f(t)^2 \, dt})$
∇C	Gradient of the function C

Linear Algebra

e_i	Vector $(0, \ldots, 0, 1, 0, \ldots, 0)$
$\mathbb{R}^n$	n-dimensional Euclidean space
$\det A$	Determinant of matrix A
A^{-1}	Inverse of matrix A
$\|A\|$	Norm of matrix A
$\mathbb{I}$	Unitary matrix
$A * B$	Convolution of matrices A and B
$a \odot b$	Hadamard product of vectors a and b

Probability Theory

$(\Omega, \mathcal{F}, P)$	Probability space
Ω	Sample space
ω	Element of the sample space
X	Random variable
$\mathfrak{S}(X)$	Sigma-field generated by X
X_t	Stochastic process
$\mathcal{F}_t$	Filtration
W_t	Brownian motion
$\mathbb{E}(X)$	The mean of X
$\mathbb{E}[X_t \vert X_0 = x]$	Expectation of X_t, given $X_0 = x$
$\mathbb{E}(X \vert \mathcal{F})$	Conditional expectation of X, given $\mathcal{F}$
$Var(X)$	Variation of X
$\mathbb{E}^P[\cdot]$	Expectation operator in the distribution P
$Var(X)$	Variance of the random variable X
$cov(X, Y)$	Covariance of X and Y
$corr(X, Y)$	Correlation of X and Y
$\mathcal{N}(\mu, \sigma^2)$	Normally distributed with mean μ and variance σ^2

Measure Theory

1_A	The characteristic function of A
$\mathcal{F}$	Sigma-field
μ, ν	Measures
I_n	The n-dim hypercube
$M(I_n)$	Space of finite signed Baire measures on I_n
$\int_A p(x) d\nu(x)$	Integration of $p(x)$ in measure ν

Information Theory

$S(p, q)$	Cross-entropy between p and q
$D_{KL}(p \Vert q)$	Kullback-Leibler divergence between p and q
$H(p)$	Shannon entropy of probability density p
$H(X)$	Shannon entropy of random variable X
$H(\mathcal{A}, \mu)$	Entropy of partition $\mathcal{A}$ and measure μ
$H(X \vert Y)$	Conditional entropy of X given Y
$I(X \vert Y), I(X, Y)$	Mutual information of X and Y

Differential Geometry

$\mathcal{S}$	Neuromanifold
$dist(\mathbf{z}, \mathcal{S})$	Distance from point $\mathbf{z}$ to $\mathcal{S}$
U, V	Vector fields
g_{ij}	Coefficients of the first fundamental form
L_{ij}	Coefficients of the second fundamental form
$\Gamma_{ijk}, \Gamma_{ij}^{k}$	Christoffel coefficients of the first and second kind
$D_U V, \nabla_U V$	Derivation of vector field V in direction U
$\dot{c}(t)$	Velocity vector along curve $c(t)$
$\ddot{c}(t)$	Acceleration vector along curve $c(t)$

Neural Networks

$\phi(x)$	Activation function
$\sigma(x)$	Logistic sigmoid function
$\mathbf{t}(x)$	Hyperbolic tangent
$sp(x)$	Softplus function
$ReLU(x)$	Rectified linear units
$\mathbf{w}, \mathbf{b}$	Weights and biasses
$C(w, b)$	Cost function
δ_{x_0}	Dirac's distribution sitting at x_0
$f_{w,b}$	Input-output function
η, δ	Learning rates
$X^{(0)}$	Input layer of a net
$X^{(L)}, Y$	Output layer of a net
$X^{(\ell)}, Y$	Output of the ℓth layer
$W^{(\ell)}$	Weight matrix for the ℓth layer
$b^{(\ell)}$	Bias vector for the ℓth layer
$\delta_j^{(\ell)}$	Backpropagated delta error in the ℓth layer
$\mathcal{I}^{(\ell)}$	Information generated by the ℓth layer
$\mathcal{L}^{(\ell)}$	Lost information in the ℓth layer

Contents

Part II   Analytic Theory

Part I
Introduction to Neural Networks

Chapter 1

Introductory Problems

This chapter introduces a few daily life problems which lead to the concept of abstract neuron and neural network. They are all based on the process of adjusting a parameter, such as a rate, a flow, or a current that feeds a given unit (tank, transistor, etc.), which triggers a certain activation function. The adjustable parameters are optimized to minimize a certain error function. At the end of the section we shall provide some conclusions, which will pave the path to the definition of the abstract neuron and neural networks.

1.1   Water in a Sink

A pipe is supplied with water at a given pressure P, see Fig. 1.1 **a**. The knob K can adjust the water pressure, providing a variable outgoing water flow at rate r. The knob influences the rate by introducing a nonnegative control w such that $P = r/w$.

A certain number of pipes of this type are used to pour water into a tank, see Fig. 1.1 **b**. Simultaneously, the tank is drained at a rate R. The goal of this problem is stated in the following:

Given the water pressure supplies $P_1, \ldots, P_n$, how can one adjust the knobs $K_1, \ldots, K_n$ such that after an a priori fixed interval of time t there is a predetermined volume V of water in the tank?

Denote by $r_1, \ldots, r_n$ the inflow rates adjusted by the knobs $K_1, \ldots, K_n$. If the outflow rate exceeds the total inflow rate, i.e., if $R > \sum_{i=1}^{n} r_i$, then no water will accumulate in the tank, i.e., $V = 0$ at any future time. Otherwise, if the total inflow rate is larger than the outflow rate, i.e., if $\sum_{i=1}^{n} r_i > R$, then the water accumulates at the rate given by the difference of the rates,

© Springer Nature Switzerland AG 2020
O. Calin, *Deep Learning Architectures*, Springer Series in the Data Sciences,
https://doi.org/10.1007/978-3-030-36721-3_1

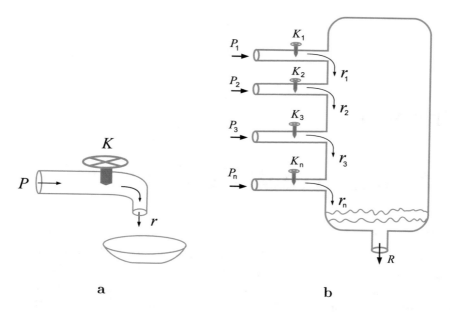

Figure 1.1: **a.** *The knob K adjusts the pressure P through a pipe to an outflow rate r.* **b.** *The n knobs $K_1, \ldots, K_n$ adjust pressures through pipes that pour water into a tank with a sink hole having an outflow rate R.*

$\sum_{i=1}^{n} r_i - R$, accumulating over time t an amount of water given by $V = (\sum_{i=1}^{n} r_i - R)t$. The resulting water amount can be written as a piecewise function

$$V = \begin{cases} 0, & \text{if } R > \sum_{i=1}^{n} r_i \\ (\sum_{i=1}^{n} r_i - R)t, & \text{otherwise.} \end{cases}$$

We shall write this function in an equivalent way. First, let w_i denote the control provided by the knob K_i, so the ith pipe supplies water at a rate $r_i = P_i w_i$. Consider also the *activation function*[1]

$$\varphi_t(x) = \begin{cases} 0, & \text{if } x < 0 \\ xt, & \text{otherwise,} \end{cases}$$

which depends on the time parameter $t > 0$. Now, the volume of water, V, can be written in terms of the pressures P_i, controls w_i, and function φ_t as

$$V = \varphi_t \left(\sum_{i=1}^{n} P_i w_i - R \right). \tag{1.1.1}$$

---

[1] As we shall see in a future chapter, this type of activation function is known under the name of ReLU.

The rate R produces a horizontal shift for the graph of φ_t and in the neural networks language it is called *bias*. The problem reduces now to find a solution of the equation (1.1.1), having the unknowns w_i and R. In practice it would suffice to obtain an approximation of the solution by choosing the controls w_i and the bias R such that the following proximity function

$$L(w_1, \ldots, w_n, R) = \frac{1}{2}\left(\varphi_t\left(\sum_{i=1}^{n} P_i w_i - R\right) - V\right)^2$$

is minimized. Other proximity functions can be also considered, for instance,

$$\tilde{L}(w_1, \ldots, w_n, R) = \left|\varphi_t\left(\sum_{i=1}^{n} P_i w_i - R\right) - V\right|,$$

the inconvenience being that it is not smooth. Both functions L and $\tilde{L}$ are nonnegative and vanish along the solution. They are also sometimes called error functions.

In practice we would like to learn from available data. To this purpose, we are provided with N training data consisting in n-tuples of pressures and corresponding volumes, $(\mathbf{P}^k, V_k)$, $1 \leq k \leq N$, with $\mathbf{P}^{k^T} = (P_1^k, \ldots, P_n^k)$. Using these data, we should be able to forecast the volume V for any other given pressure vector $\mathbf{P}$. The forecast is given by formula (1.1.1), provided the controls and bias are known. These are obtained by minimizing the following cost function:

$$C(w_1, \ldots, w_n, R) = \frac{1}{2}\sum_{k=1}^{N}\left(\varphi_t\left(\mathbf{P}^{k^T}\mathbf{w} - R\right) - V_k\right)^2,$$

where $\mathbf{w}^T = (w_1, \ldots, w_n)$.

1.2 An Electronic Circuit

The next example deals with an electronic circuit. It is worthwhile to mention that a problem regarding pressure, P, rates, r, and knobs can be "translated" into electronics language, using voltage, V, current intensity, I, and resistors, R, respectively. The following correspondence table will be useful:

Fluids	pressure	rate	knob
Electronics	voltage	intensity	resistivity

Consider the electronic circuit provided in Fig. 1.2. Voltages $x_1, \ldots, x_n$ applied at the input lines are transmitted through the variable resistors

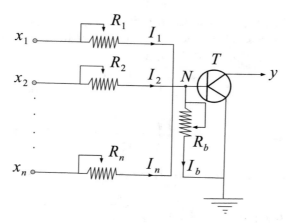

Figure 1.2: *The neuron implemented as an electronic circuit.*

$R_1, \ldots, R_n$. The emerging currents I_k are given by the Ohm's law

$$I_k = \frac{x_k}{R_k}, \qquad k = 1, \ldots, n.$$

For simplicity reasons, define the weights $w_k = \dfrac{1}{R_k}$ as the inverse of resistances. Then the previous relation becomes $I_k = x_k w_k$.

The n currents converge at the node N. Another variable resistor, R_b, connects the node N with the base. The current through this wire is denoted by I_b. By Kirchhoff's second law, the sum of the incoming currents at the node N is equal to the sum of the outgoing currents at the same node. Therefore, the current I that enters the transistor T is given by

$$I = \sum_{k=1}^{n} I_k - I_b = \sum_{k=1}^{n} x_k w_k - I_b.$$

The transistor is an electronic component that acts as a switch. More precisely, if the ingoing current I is lower than the transistor's threshold θ, then no current gets through. Once the current I exceeds the threshold θ, the current gets through, eventually magnified by a certain factor. The output current y of the transistor T can be modeled as

$$y = \begin{cases} 0, & \text{if } I < \theta \\ k, & \text{if } I \geq \theta, \end{cases}$$

for some constant $k > 0$. In this case the activation function is the step function

$$\varphi_\theta(x) = \begin{cases} 0, & \text{if } x < \theta \\ 1, & \text{if } x \geq \theta \end{cases} \tag{1.2.2}$$

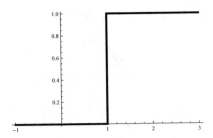

Figure 1.3: *The unit step function $\varphi_\theta(x)$ with $\theta = 1$.*

and the output can be written in the familiar form

$$y = k\varphi_\theta(I) = k\varphi_0(I - \theta) = k\varphi_0\left(\sum_{k=1}^{n} I_k - I_b - \theta\right) = k\varphi_0\left(\sum_{k=1}^{n} x_k w_k - \beta\right),$$

where $\beta = I_b + \theta$ is considered as a bias. Note that $\varphi_0(x)$ is the Heaviside function, or the unit step function, centered at the origin. For a Heaviside function centered at $\theta = 1$, see Fig. 1.3.

The learning problem can be formulated as: *Given a current $z = z(x_1, \ldots, x_n)$ depending on the input voltages $x_1, \ldots, x_n$, adjust the resistances $R_1, \ldots, R_n$, R_b such that the output y approximates z in the least squares way, i.e., such that $\frac{1}{2}(z - y)^2$ is minimum.*

Remark 1.2.1 Attempts to build special hardware for artificial neural networks go back to the 1950s. We refer the reader to the work of Minsky [86], who implemented adaptive weights using potentiometers, Steinbuch, who built associative memories using resistance networks [117], as well as Widrow and Hoff [125], who developed an adaptive system specialized for signal processing, called the Adaline neuron.

1.3 The Eight Rooks Problem

A *rook* is a chess figure which is allowed to move both vertically and horizontally any number of chessboard squares. The problem requires to place eight distinct rooks in distinct squares of an 8×8 chessboard. Assuming that all possible replacements are equally likely, we need to find the probability that all the rooks are safe from one another. Even if this problem can be solved in closed form by combinatorial techniques, we shall present in the following a computational approach (Fig. 1.4).

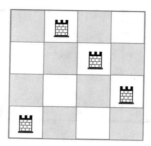

Figure 1.4: *A solution for the rooks problem on a* 4×4 *chess board.*

Let x_{ij} be the state of the (i, j)th square, which is

$$x_{ij} = \begin{cases} 1, & \text{if there is a rook at position } (i, j); \\ 0, & \text{if there is no rook at that place.} \end{cases}$$

These states are used in the construction of the objective function as in the following. Since there is only one rook placed in the jth column, we have $\sum_{i=1}^{8} x_{ij} = 1$. The fact that this property holds for all columns can be stated equivalently in a variational way by requiring the function

$$F_1(x_{11}, \ldots, x_{88}) = \sum_{j=1}^{8} \left(\sum_{i=1}^{8} x_{ij} - 1 \right)^2.$$

to be minimized. The minimum of F_1 is reached when all squared expressions vanish and hence there is only one rook in each column. Applying a similarly procedure for the rows, we obtain that the function

$$F_2(x_{11}, \ldots, x_{88}) = \sum_{i=1}^{8} \left(\sum_{j=1}^{8} x_{ij} - 1 \right)^2.$$

must also be minimized.

The rooks problem has many solutions, each of them corresponding to a global minimum of the objective function

$$E(x_{11}, \ldots, x_{88}) = F_1(x_{11}, \ldots, x_{88}) + F_2(x_{11}, \ldots, x_{88}).$$

The minima of the previous function can be obtained using a neural network, called *Hopfield network*. This application will be accomplished in Chapter 20.

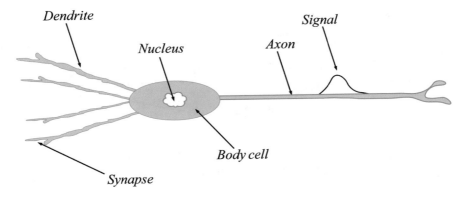

Figure 1.5: *A neuron cell.*

1.4 Biological Neuron

A neuron is a cell which consists of the following parts: *dendrites, axon,* and *body-cell*, see Fig. 1.5. The *synapse* is the connection between the axon of one neuron and the dendrite of another. The functions of each part is briefly described below:

- Dendrites are transmission channels that collect information from the axons of other neurons. This is described in Fig. 1.6 **a**. The signal traveling through an axon reaches its terminal end and produces some chemicals x_i which are liberated in the synaptic gap. These chemicals are acting on the dendrites of the next neuron either in a strong or a weak way. The connection strength is described by the weight system w_i, see Fig. 1.6 **b**.
- The body-cell collects all signals from dendrites. Here the dendrites activity adds up into a total potential and if a certain threshold is reached, the neuron fires a signal through the axon. The threshold depends on the sensitivity of the neuron and measures how easy is to get the neuron to fire.
- The axon is the channel for signal propagation. The signal consists in the movement of ions from the body-cell towards the end of the axon. The signal is transmitted electrochemically to the dendrites of the next neuron, see Fig. 1.5.

A schematic model of the neuron is provided in Fig. 1.6 **b**. The input information coming from the axons of other neurons is denoted by x_i. These get multiplied by the weights w_i modeling the synaptic strength. Each dendrite supplies the body-cell by a potential $x_i w_i$. The body cell is the unit which collects and adds these potentials into $\sum_{i=1}^{n} x_i w_i$ and then compares it with the threshold b. The outcome is as follows:

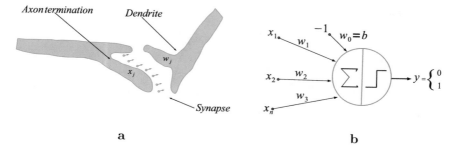

Figure 1.6: **a.** *Synaptic gap.* **b.** *Schematic model of a neuron.*

- if $\sum_{i=1}^{n} x_i w_i > b$, the neuron fires a signal through the axon, $y = 1$.
- if $\sum_{i=1}^{n} x_i w_i \leq b$, the neuron does not fire any signal, i.e., $y = 0$.

This can be written equivalently as

$$y = \varphi_b \left(\sum_{i=1}^{n} x_i w_i \right) = \varphi_0 \left(\sum_{i=1}^{n} x_i w_i - b \right),$$

where the activation function φ_b is the unit step function centered at b, see equation (1.2.2) and φ_0 is the Heaviside function.

Bias elimination In order to get rid of the bias, a smart trick is used. The idea is to introduce an extra weight, $w_0 = b$, and an extra constant signal, $x_0 = -1$. Then the previous outcome can be written as

$$y = \varphi_0 \left(\sum_{i=0}^{n} x_i w_i \right).$$

The incoming signals x_i and the synaptic weights w_i can be seen in Fig. 1.6 **b**. The neuron body-cell is represented by the main circle, while its functions are depicted by the two symbols, Σ (to suggest the summation of incoming signals) and the unit step function (to emphasize the activation function) of the neuron. The outcome signal of the axon is represented by an arrow with the outcome y, which is either 0 or 1.

The neuron learns an output function by updating the synaptic weights w_i. Only one neuron cannot learn complicated functions, but it can learn a simple function like, for instance, the piecewise function

$$z(x_1, x_2) = \begin{cases} 0, & \text{if } x_2 \leq 0 \\ 1, & \text{if } x_2 > 0, \end{cases}$$

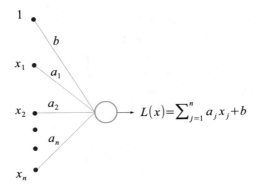

Figure 1.7: *A linear approximator neuron.*

which takes the value 1 on the upper half-plane and 0 otherwise. This function can be learned by choosing the weights $w_1 = 0$, $w_2 = 1$, and bias $b = 0$. The neuron fires when the inequality $x_1 w_1 + x_2 w_2 > b$ is satisfied. This is equivalent to $x_2 > 0$, which is the equation of the upper half-plane. Hence, one neuron can learn the upper-half plane. By this we mean that the neuron can distinguish whether a point (x_1, x_2) belongs to the upper-half plane. Similarly, one can show that a neuron can learn any half-plane.

However, the neuron cannot learn a more complicated function, such as

$$\zeta(x_1, x_2) = \begin{cases} 0, & \text{if } x_1^2 + x_2^2 \geq 1 \\ 1, & \text{if } x_1^2 + x_2^2 < 1, \end{cases}$$

i.e., it cannot learn to decide whether a point belongs to the interior of the unit circle.

1.5 Linear Regression

Neural networks are not a brand new concept. They have first appeared under the notion of linear regression in the work of Gauss and Legendre around the 1800s. The next couple of sections show that a linear regression model can be naturally interpreted as a simple "neural network".

The case of a real-valued function Consider a continuous function $f : K \to \mathbb{R}^m$, with $K \subset \mathbb{R}^n$ compact set. We would like to construct a linear function that approximates $f(x)$ in the L^2-sense using a neuron with a linear activation. The input is the continuous variable $X = (x_1, \ldots, x_m) \in K$ and the output is the linear function $L(X) = \sum_{j=1}^{n} a_j x_j + b$, where a_j are the weights and b is the neuron bias, see Fig. 1.7.

For the sake of simplicity we assume the approximation is performed near the origin, i.e., $L(0) = f(0)$, fact that implies the bias value $b = f(0)$. Since using a vertical shift we can always assume $f(0) = 0$, then it suffices to take $b = 0$, so the linear function takes the form $L = \sum_{j=1}^{n} a_j x_j$.

Consider the cost function that measures the L^2-distance between the neuron output and target function

$$C(a) = \frac{1}{2} \int_K (L(x) - f(x))^2 \, dx_1 \ldots dx_n = \frac{1}{2} \int_K \left(\sum_{j=1}^{n} a_j x_j - f(x) \right)^2 dx_1 \ldots dx_n$$

and compute its gradient

$$\frac{\partial C}{\partial a_k} = \int_K x_k \left(\sum_{j=1}^{n} a_j x_j - f(x) \right) dx_1 \ldots dx_n$$

$$= \sum_{j=1}^{n} a_j \rho_{jk} - m_k, \qquad 1 \leq k \leq n,$$

where

$$\rho_{jk} = \int_K x_j x_k \, dx_1 \ldots dx_n, \qquad m_k = \int_K x_k f(x) \, dx_1 \ldots dx_n.$$

In the equivalent matrix form this becomes

$$\nabla_a C = \rho a - m,$$

where $\nabla_a C = \left(\frac{\partial C}{\partial a_1}, \ldots, \frac{\partial C}{\partial a_n} \right)^T$, $\rho = \rho^T = \rho_{jk}$ and $m = (m_1, \ldots, m_n)^T$. The optimal value of weights are obtained solving the Euler equation $\nabla_a C = 0$. We obtain $a = \rho^{-1} m$, provided the matrix ρ is invertible, see Exercise 1.9.4. This means

$$a_j = \sum_{k=1}^{n} g^{jk} m_k, \qquad 1 \leq j \leq n,$$

where $g^{-1} = (g^{jk})$.

Remark 1.5.1 In order to avoid computing the inverse ρ^{-1}, we can employ the gradient descent method, which will be developed in Chapter 4. Let $a_k(0)$ be the weight initialization (for instance, either uniform or normally distributed with mean zero and a small standard distribution, as it shall be described in section 6.3). Then it can be shown that the sequence of approximations of the weight vector a is defined recursively by

$$a_k(t+1) = a_k(t) - \lambda \frac{\partial C}{\partial a_k} = a_k(t) - \lambda \left(\sum_{j=1}^{n} a_j(t) \rho_{jk} - m_k \right), \qquad 1 \leq k \leq n,$$

where $\lambda > 0$ is the learning rate.

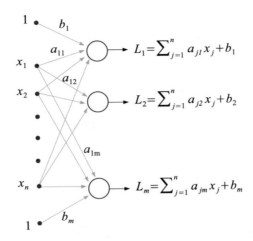

Figure 1.8: *A linear approximator neural net.*

The case of a multivalued function This section presents a neural network which learns the linear approximation of a multivalued function. Consider a continuous function $F : \mathbb{R}^n \to \mathbb{R}^m$ and we would like to approximate it around the origin by a linear function of type $L(X) = AX + b$, where A is an $m \times n$ matrix and $b \in \mathbb{R}^m$ is a vector.

We start by noting that $L(X)$ is the output of a 2-layer neural network having linear activations on each output neuron. The inputs are $X = (x_1, \ldots, x_m)$, the weight matrix is $A = (a_{ij})$, and the bias vector is given by b. Since the approximation is applied at $X = 0$, it follows that $b = F(0)$, which fixes the bias vector. The output of the kth neuron is given by the real-valued linear function $L_k(X) = \sum_{j=1}^n a_{jk}x_j + b_k$, see Fig. 1.8.

We assume the input $X \in \mathbb{R}^n$ be a continuous variable taking values in the compact set $K \subset \mathbb{R}^n$. In this case the cost function of the network is given by

$$C(A) = \int_K \|L(X) - F(X)\|^2 \, dx_1 \ldots dx_m = \sum_{k=1}^m C_k(A),$$

where

$$C_k(A) = \frac{1}{2} \int_K (L_k(X) - F_k(X))^2 \, dx_1 \ldots dx_n,$$

with $F = (F_1, \ldots, F_m)$. The minimization of the cost function $C(A)$ is equivalent with the simultaneous minimization of all cost functions $C_k(A)$, $1 \leq k \leq m$. Each of these cost functions is relative to a neuron and we have shown their optimization procedure in the previous case.

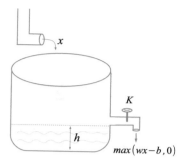

Figure 1.9: *A water tank with a knob and outflow $\phi(x) = \max(wx - b, 0)$.*

The next two examples will introduce the idea of "deep learning" network, which is the subject of this book.

1.6   The Cocktail Factory Network

Assume there is a water inflow at rate r into a tank, which has a knob K that controls the outflow from the tank, see Fig. 1.9. The knob is situated at a distance h above the bottom of the tank. After a given time, t, the amount of water that flowed in the tank is $x = rt$. Denote by w the parameter that controls the outflow from the knob. If the level of the water inflow is below h, no water will flow out. If the level of the water inflow is above h, a fraction w of the overflow will flow out of the tank. If A is the tank base area, then $x - Ah$ represents the overflow and the amount of $w(x - Ah) = wx - b$ will flow through the knob, where $b = wAh$. Then the tank outflow is modeled by the hockey-stick function

$$\phi(x) = \begin{cases} 0, & \text{if } x \le b/w \\ wx - b, & \text{otherwise.} \end{cases}$$

In the following we shall organize our tanks into a network of six tanks that mix three ingredients into a cocktail, see Fig. 1.10. Sugar, water, and wine are poured into tanks A, B, and C. A mixture of contents of tanks A and B goes into tank D as sweet water, and a mixture of contents of tanks B and C flows into tank E, as diluted wine. Mixing the outflows of tanks D and E provides the content of container F. The knob of container F controls the final cocktail production. This represents an example of a feedforward neural network with two hidden layers, formed by containers (A, B, C) and (D, E). The input to the network is made through three pipes and the output through the knob of container F.

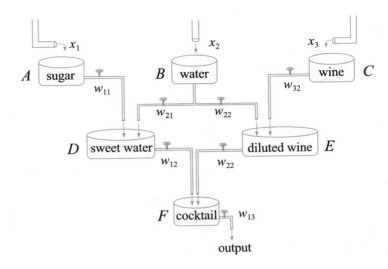

Figure 1.10: *A cocktail factory as a neural network with two hidden layers.*

The final production of cocktail is the result of a complex combination of the input ingredients, sugar, water, and wine, poured into quantities x_1, x_2, and x_3. The knobs' control the proportion of solution allowed to pass to the next container, and are used to adjust the quality of the cocktail production. After tasting the final outcome, a cocktail expert will adjust the knobs such that the cocktail will improve its quality as much as possible. The knobs' adjusting process corresponds to learning of how to make cocktail from the given ingredients.

1.7 An Electronic Network

We can stack together electronic devices like the ones described in Fig. 1.2 to construct more complex electronic networks. An example of this type of network is given in Fig. 1.11. It contains six transistors, three in the first layer, two in the second, and one in the output layer. There are n current inputs, $x_1, \ldots, x_n$, and $3n + 14$ variable resistors.

Since there is a transistor, T, in the last layer, the output y_{out} is Boolean, i.e., it is 1 if the current through T is larger than the transistor's threshold, and 0 otherwise. This type of electronic network can be used to learn to classify the inputs $x_1, \ldots, x_n$ into 2 classes after a proper tuning of the variable resistors. If the inputs $x_1, \ldots, x_n$ are connected to n photocells, the network can be used for image classification with two classes. A similar idea using

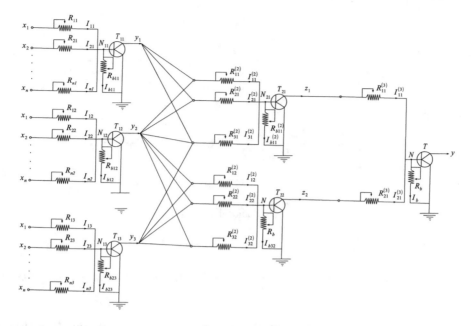

Figure 1.11: *Neural network with two hidden layers and 6 transistors. Each transistor implements a logic gate.*

custom-built hardware for image recognition was used in the late 50s by Rosenblatt in the construction of his perceptron machine, see [102].

The last two examples provided a brief idea about what a "deep neural network" is. Next we shall bring up some of the "questions" regarding deep neural networks, which will be addressed mathematically later in this book.

(i) What classes of functions can be represented by a deep neural network? It will be shown in Part II that this class is quite rich, given the network has at least one hidden layer. The class contains, for instance, any continuous functions, or integrable functions, provided the input data belongs to a compact set. This the reason why neural networks deserve the name of "universal function approximators".

(ii) How are the optimal parameters of a deep neural network estimated? Tuning the network parameters (the knobs or the variable resistors in our previous examples) in order to approximate the desired function corresponds to a learning process. This is accomplished by minimizing a certain distance or error function, which measures a proximity between the desired function and the network output. One of the most used learning method is the gradient

descent method, which changes parameters in the direction where the error function decreases the most. Other learning methods will be studied in Part I, Chapter 4.

(iii) How can we measure the amount of information which is lost or compressed in a deep learning network?
Neural networks are also information processors. The information contained in the input is compressed by the hidden layers until the output layer produces the desired information. For instance, if a deep network is used to classify whether an animal is a cat or a dog, the network has to process the input information until extracts the desired information, which corresponds to 1 bit (cat or dog). This type of study is done in Part III.

1.8 Summary

We shall discuss here all the common features of previous examples. They will lead to new abstract concept, the perceptron, and the neural network.

All examples start with some incoming information provided by the input variables $x_1, \ldots, x_n$. The way they have been considered in the previous examples are as deterministic variables, i.e., one can measure clearly their values, and these do not change from one observation to another. The analysis also makes sense in the case when the input variables x_i are random, i.e., they have different values for distinct observations, and these values are distributed according to a certain density law. Even more general, the input information can be supplied by a stochastic process $x_i(t)$, i.e., a sequence of n-uples of random variables parametrized over time t.

Another common feature is that the inputs x_i are multiplied by a weight w_i and then the products are added up to form the sum $\sum_{i=1}^{n} x_i w_i$, which is the inner product between the input vector $\mathbf{x}$ and weight vector $\mathbf{w}$. We note that in most examples a bias b is subtracted from the aforementioned inner product. If a new weight $w_0 = b$ and a new input $x_0 = -1$ are introduced, this leads to $\sum_{i=1}^{n} x_i w_i - b = \sum_{i=0}^{n} x_i w_i$, which is still the inner product of two vectors that extend the previous input and weight vectors.

Another common ingredient is the activation function, denoted by $\varphi(x)$. We have seen in the previous sections several examples of activation functions: linear, unit step functions, and piecewise-linear; each of them describes a specific mode of action of the neuronal unit and suits a certain type of problem.

The output function y is obtained applying the activation function φ to the previous inner product, leading to the expression $y = \varphi\left(\sum_{i=0}^{n} x_i w_i\right)$. If the input variables x_i are deterministic, then the output y is also deterministic.

However, if the inputs are random variables, the output is also a random variable.

The goal of all the presented examples is to adjust the system of weights $\{w_i\}_{i=0,n}$ such that the outcome y approximates a certain given function as good as possible. Even if this can be sometimes an exact match, in most cases it is just an approximation. This is achieved by choosing the weights w_i such that a certain proximity function is minimized. The proximity function is a distance-like function between the realized output and the desired output. For the sake of simplicity we had considered only Euclidean distances, but there are also other possibilities.

Most examples considered before contain the following ingredients: inputs $\{x_i\}$, weights, $\{w_i\}$, bias, b, an activation function, $\varphi(x)$, and an error function. These parts are used to construct a function approximator, by the process of tuning weights, called *learning* from observational data. An abstract concept, which enjoys all the aforementioned properties and contains all these parts, will be introduced and studied from the abstract point of view in Chapter 5.

A few examples use a collection of neurons structured on layers. The neurons in the input layer get activated from the sensors that perceive the environment. The other neurons get activated from the weight connections from the previous neuron activations. Like in the case of a single neuron, we would like to minimize the proximity between the network output and a desired result. Neural networks will be introduced and studied from their learning perspective in Chapter 6.

1.9   Exercises

Exercise 1.9.1 A factory has n suppliers that produce quantities $x_1, \ldots, x_n$ per day. The factory is connected with suppliers by a system of roads, which can be used at variable capacities $c_1, \ldots, c_n$, so that the factory is supplied daily the amount $x = c_1 x_1 + \cdots + c_n x_n$.
(a) Given that the factory production process starts when the supply reaches the critical daily level b, write a formula for the daily factory revenue, y.
(b) Formulate the problem as a learning problem.

Exercise 1.9.2 A number n of financial institutions, each having a wealth x_i, deposit amounts of money in a fund, at some adjustable rates of deposit w_i, so the money in the fund is given by $x = x_1 w_1 + \cdots + x_n w_n$. The fund is set up to function as in the following: as long as the fund has less than a certain reserve fund M, the fund manager does not invest. Only the money exceeding the reserve fund M is invested. Let $k = e^{rt}$, where r and t denote the investment rate of return and time of investment, respectively.

(a) Find the formula for the investment.

(b) State the associate learning problem.

Exercise 1.9.3 (a) Given a continuous function $f : [0,1] \to \mathbb{R}$, find a linear function $L(x) = ax + b$ with $L(0) = f(0)$ and such that the quadratic error function $\frac{1}{2} \int_0^1 (L(x) - f(x))^2 \, dx$ is minimized.

(b) Given a continuous function $f : [0,1] \times [0,1] \to \mathbb{R}$, find a linear function $L(x,y) = ax + by + c$ with $L(0,0) = f(0,0)$ and such that the error $\frac{1}{2} \int_{[0,1]^2} (L(x,y) - f(x,y))^2 \, dxdy$ is minimized.

Exercise 1.9.4 For any compact set $K \subset \mathbb{R}^n$ we associate the symmetric matrix

$$\rho_{ij} = \int_K x_i x_j \, dx_1 \ldots dx_n.$$

The invertibility of the matrix (ρ_{ij}) depends both on the shape of K and dimension n.

(a) Show that if $n = 2$ then $\det \rho_{ij} \neq 0$ for any compact set $K \subset \mathbb{R}^2$.

(b) Assume $K = [0,1]^n$. Show that $\det \rho_{ij} \neq 0$, for any $n \geq 1$.

Chapter 2

Activation Functions

In order to learn a nonlinear target function, a neural network uses activation functions which are nonlinear. The choice of each specific activation function defines different types of neural networks. Even if some of them have already been introduced in Chapter 1, we shall consider here a more systematic presentation. Among the zoo of activation functions that can be found in the literature, three main types are used more often: linear, step functions, sigmoid, and hockey-stick-type functions. Each of them has their own advantages (they are differentiable, bounded, etc.) and disadvantages (they are discontinuous, unbounded, etc.) which will rise in future computations.

2.1 Examples of Activation Functions

Linear functions The slope of a given line can be used to model the firing rate of a neuron. A linear activation function is mostly used in a multilayer network for the output layer. We note that computing with a neuron having a linear activation function is equivalent with a linear regression. We have

- *Linear Function* This is the simplest activation function. It just multiplies the argument by a positive constant, i.e., $f(x) = kx$, $k > 0$ constant, see Fig. 2.1 **a**. In this case the firing rate is constant, $f'(x) = k$.
- *Identity Function* For $k = 1$ the previous example becomes the identity function $f(x) = x$, which was used in the linear neuron, see Fig. 2.1 **b**.

Step functions This biologically inspired type of activation functions exhibit an upward jump that simplistically models a neuron activation. Two of them are described in the following.

© Springer Nature Switzerland AG 2020
O. Calin, *Deep Learning Architectures*, Springer Series in the Data Sciences,
https://doi.org/10.1007/978-3-030-36721-3_2

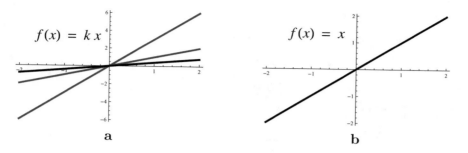

Figure 2.1: *Linear activation functions:* **a.** f(x) = kx **b.** *The identity function* $f(x) = x$.

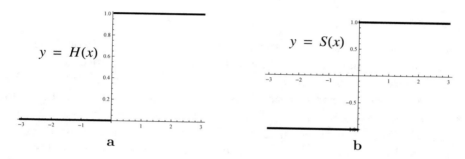

Figure 2.2: *Step activation functions:* **a.** *Heaviside function* **b.** *Signum function.*

- *Threshold step function* This function is also known under the equivalent names of binary step, unit step function, or Heaviside function, see Fig. 2.2 **a**. It is given by

$$H(x) = \begin{cases} 0, & \text{if } x < 0 \\ 1, & \text{if } x \geq 0 \end{cases}$$

This activation function will be used later to describe a neuron that fires only for values $x \geq 0$. It is worth noting that the function is not differentiable at $x = 0$. However, in terms of generalized functions[1] its derivative is given by the Dirac's function, $H'(x) = \delta(x)$, where $\delta(x)$ is the probability distribution measure for a unit mass centered at the origin. Roughly speaking, this is

$$\delta(x) = \begin{cases} 0, & \text{if } x \neq 0 \\ +\infty, & \text{if } x = 0 \end{cases}$$

having unit mass $\int_{\mathbb{R}} \delta(x)\, dx = 1$.
- *Bipolar Step Function* This function is also known under the name of signum function, see Fig. 2.2 **b**, and is defined as in the following:

---

[1]This follows Schwartz's distribution theory.

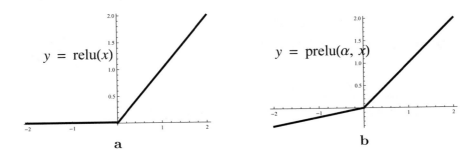

Figure 2.3: *Hockey-stick functions:* **a.** $y = ReLU(x)$ **b.** $y = PReLU(\alpha, x)$.

$$S(x) = \begin{cases} -1, & \text{if } x < 0 \\ 1, & \text{if } x \geq 0 \end{cases}$$

It is not hard to see that $S(x)$ relates to the Heaviside function by the linear relation

$$S(x) = 2H(x) - 1.$$

Consequently, its derivative equals twice the Dirac's function,

$$S'(x) = 2H'(x) = 2\delta(x).$$

Hockey-stick functions This class of functions has graphs that resemble a hockey stick or an increasing L-shaped curve. A few of them are discussed in the following.

• *Rectified Linear Unit (ReLU)* In this case the activation is linear only for $x \geq 0$, see Fig. 2.3 **a**, and it is given by

$$ReLU(x) = xH(x) = \max\{x, 0\} = \begin{cases} 0, & \text{if } x < 0 \\ x, & \text{if } x \geq 0. \end{cases}$$

It is worth realizing that its derivative is the Heaviside function, $ReLU'(x) = H(x)$. The fact that ReLU activation function does not saturate, as the sigmoid functions do, was proved useful in recent work on image recognition. Neural networks with ReLU activation function tend to learn several times faster than a similar network with saturating activation functions, such as logistic or hyperbolic tangent, see [68]. A generalization of the rectified linear units has been proposed in [48] under the name of *maxout*.

• *Parametric Rectified Linear Unit (PReLU)* In this case the activation is piecewise linear, having different firing rates for $x < 0$ and $x > 0$, see Fig. 2.3 **b**:

$$PReLU(\alpha, x) = \begin{cases} \alpha x, & \text{if } x < 0 \\ x, & \text{if } x \geq 0, \end{cases} \qquad \alpha > 0.$$

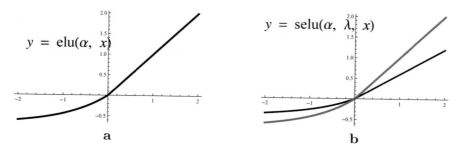

Figure 2.4: *Hockey-stick functions:* **a.** $y(x) = ELU(\alpha, x)$ **b.** $y(x) = SELU(\alpha, \lambda, x)$.

It is worth realizing that its derivative is the Heaviside function, $ReLU'(x) = H(x)$.

 • *Exponential Linear Units (ELU)* This activation function is positive and linear for $x > 4$, and negative and exponential otherwise, see Fig. 2.4 **a:**

$$ELU(\alpha, x) = \begin{cases} x, & \text{if } x > 0 \\ \alpha(e^x - 1), & \text{if } x \le 0 \end{cases}$$

Note that the function is differentiable at $x = 0$ only for the value $\alpha = 1$; in this case the derivative equals the value 1.

 • *Scaled Exponential Linear Units (SELU)* This activation function is a scaling of the previous function, see Fig. 2.4 **b:**

$$SELU(\alpha, \lambda, x) = \lambda \begin{cases} x, & \text{if } x > 0 \\ \alpha(e^x - 1), & \text{if } x \le 0. \end{cases}$$

 • *Sigmoid Linear Units (SLU)* This one is obtained as the product between a linear function and a sigmoid

$$\phi(x) = x\sigma(x) = \frac{x}{1 + e^{-x}}.$$

Unlike the other activation functions of hockey-stick type, this activation function is not monotonic, having a minimum, see Fig. 2.5 **a.** Its derivative can be computed using product rule as

$$\begin{aligned} \phi'(x) &= \sigma(x) + x\sigma'(x) = \sigma(x) + x\sigma(x)\big(1 - \sigma(x)\big) \\ &= \sigma(x)\big(1 + x\sigma(-x)\big). \end{aligned}$$

The critical point is located at $x_0 = -u \approx -1.27$, where u is the positive solution of the equation $\sigma(u) = \dfrac{1}{u}$, see Fig. 2.5 **b.** The SLU might perform slightly better than the ReLU activation functions in certain cases.

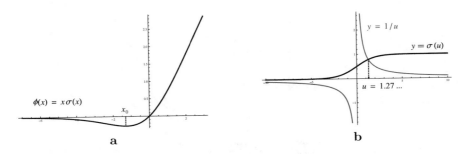

Figure 2.5: **a**. *The activation function $\phi(x) = x\sigma(x)$ has a minimum at the point $x_0 \approx -1.27$.* **b**. *The graphical solution of the equation $\sigma(x) = 1/x$.*

This activation function reminds of a profile usually observed in nature, for instance, in heart beating. Before the heart emits the pressure wave, there is a slight decrease in the pressure, which corresponds to the dip in Fig. 2.5. A neuron might exhibit a similar behavior, fact that explains the good results obtained using this activation function.

The size and the location of the dip can be controlled by a parameter $c > 0$ as

$$\phi_c(x) = x\sigma(cx) = \frac{x}{1 + e^{-cx}},$$

see Fig. 2.6.

- *Softplus* The softplus function is an increasing positive function that maps the real line into $(0, \infty)$, which is given by

$$sp(x) = \ln(1 + e^x).$$

Its name comes from the fact that it is a smoothed version of the hockey-stick function $x^+ = \max\{0, x\}$. Its graph is represented in Fig. 2.7 **a**. The decomposition of x into its positive and negative part, via the identity $x = x^+ - x^-$, where $x^- = \max\{0, -x\}$ has in the case of the softplus function the following analog:

$$sp(x) - sp(-x) = x.$$

This follows from the next algebraic computation that uses properties of the logarithm

$$
\begin{aligned}
sp(x) - sp(-x) &= \ln(1 + e^x) - \ln(1 + e^{-x}) = \ln \frac{1 + e^x}{1 + e^{-x}} \\
&= \ln \frac{e^x(1 + e^x)}{e^x(1 + e^{-x})} = \ln \frac{e^x(1 + e^x)}{1 + e^x} = \ln e^x = x.
\end{aligned}
$$

The graph of the function $sp(-x)$ is obtained from the graph of $sp(x)$ by symmetry with respect to the y-axis and represents a softened version of the function x^-.

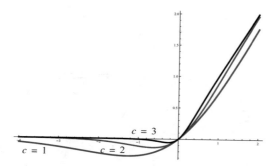

Figure 2.6: *The activation function $\phi_c(x) = x\sigma(cx) = \frac{x}{1+e^{-cx}}$ for $c = 1, 2, 3$.*

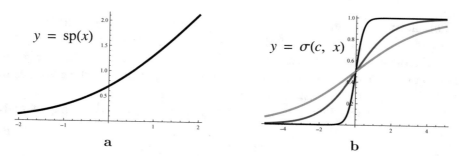

Figure 2.7: **a.** *Softplus function $sp(x) = \ln(1 + e^x)$* **b.** *Family of logistic functions approximating the step function.*

It is also worthy mentioning in the virtue of formula (2.1.1) the following relation with the logistic function:

$$sp'(x) = \frac{1}{1 + e^{-x}} > 0, \tag{2.1.1}$$

which means that softplus is an antiderivative of the logistic function. Consequently, the softplus is invertible, with the inverse given by

$$sp^{-1}(x) = \log(e^x - 1), \qquad x > 0.$$

Sigmoid functions These types of activation functions have the advantage that they are smooth and can approximate a step function to any degree of accuracy. In the case when their range is $[0, 1]$, its value can be interpreted as a probability. We encounter the following types:

    • *Logistic function with parameter $c > 0$* Also called the soft step function, this sigmoid function is defined by

$$\sigma_c(x) = \sigma(c, x) = \frac{1}{1 + e^{-cx}},$$

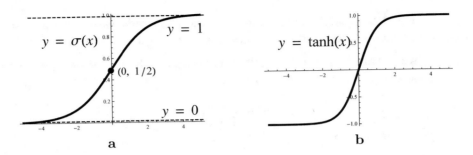

Figure 2.8: **a.** *Logistic function* $\sigma(x) = \frac{1}{1+e^{-x}}$ *and its horizontal asymptotes,* $y = 0$ *and* $y = -1$. **b.** *Hyperbolic tangent function.*

where the parameter $c > 0$ controls the firing rate of the neuron, in the sense that large values of c correspond to a fast change of values from 0 to 1. In fact, the sigmoid function $\sigma_c(x)$ approximates the Heaviside step function $H(x)$ as $c \to +\infty$, see Fig. 2.7 **b.** This follows from the observation that

$$\lim_{c \to \infty} e^{-cx} = \begin{cases} 0, & \text{if } x > 0 \\ \infty, & \text{if } x < 0, \end{cases}$$

which leads to

$$\lim_{c \to \infty} \sigma_c(x) \quad = \quad \lim_{c \to \infty} \frac{1}{1 + e^{-cx}} = \begin{cases} 1, & \text{if } x > 0 \\ 0, & \text{if } x < 0 \end{cases} = H(x), \qquad x \neq 0.$$

At $x = 0$ the value of the sigmoid is independent of c, since $\sigma_c(0) = \frac{1}{2}$, and hence the limit at this point will be equal to $\lim\limits_{c \to \infty} \sigma_c(0) = \dfrac{1}{2}$.

The logistic function σ_c maps monotonically the entire real line into the unit interval $(0, 1)$. Since $\sigma_c(0) = \dfrac{1}{2}$, the graph is symmetric with respect to the point $(0, 1/2)$, see Fig. 2.8 **a.**

The sigmoid function satisfies a certain differential equation, which will be useful later in the backpropagation algorithm. Since its derivative can be computed using quotient's rule as

$$\sigma_c'(x) \quad = \quad -\frac{1}{(1 + e^{-cx})^2}(1 + e^{-cx})' = \frac{c(1 + e^{-cx})}{1 + e^{-cx}}$$
$$= \quad c\big(\sigma_c(x) - \sigma_c^2(x)\big) = c\sigma_c(x)\big(1 - \sigma_c(x)\big),$$

it follows that the rate of change of the sigmoid function σ_c can be represented in terms of σ_c as

$$\sigma_c' = c\sigma_c(1 - \sigma_c).$$

It is worth noting that the expression on the right side vanishes for $\sigma_c = 0$ or $\sigma_c = 1$, which correspond to the horizontal asymptotes of the sigmoid.

If $c = 1$, we obtain the standard logistic function, which will be used in the construction of the sigmoid neuron

$$\sigma(x) = \frac{1}{1 + e^{-x}}.$$

A straightforward algebraic manipulation shows the following symmetry property of the sigmoid:

$$\sigma_c(-x) = 1 - \sigma_c(x).$$

It is worth noting that the logistic function can be generalized to a 2-parameter sigmoid function by the relation

$$\sigma_{c,\alpha}(x) = \frac{1}{(1 + e^{-cx})^\alpha}, \qquad c, \alpha > 0,$$

to allow for more sophisticated models.

Sometimes the inverse of the sigmoid function is needed. It is called the *logit* function and its expression is given by

$$\sigma^{-1}(x) = \log\left(\frac{x}{1 - x}\right), \qquad x \in (0, 1).$$

Note that the softplus function can be recovered from the logistic function as

$$sp(x) = \int_{-\infty}^{x} \sigma(x)\,dx. \qquad (2.1.2)$$

This relation can be verified directly by integration, or by checking the following antiderivative conditions:

$$\lim_{x \to -\infty} sp(x) = \lim_{x \to -\infty} \ln(1 + e^x) = \ln 1 = 0$$

$$sp'(x) = \left(\ln(1 + e^x)\right)' = \frac{e^x}{1 + e^x} = \sigma(x).$$

- *Hyperbolic tangent* This is sometimes called the bipolar sigmoidal function. It maps the real numbers line into the interval $(-1, 1)$, having horizontal asymptotes at $y = \pm 1$, see Fig. 2.8 **b**. The hyperbolic tangent is defined by

$$\mathbf{t}(x) = \tanh x = \frac{e^x - e^{-x}}{e^x + e^{-x}}.$$

An algebraic computation shows the following linear relation with the logistic function:

$$\mathbf{t}(x) = \frac{1 - e^{-2x}}{1 + e^{2x}} = 2\sigma_2(x) - 1.$$

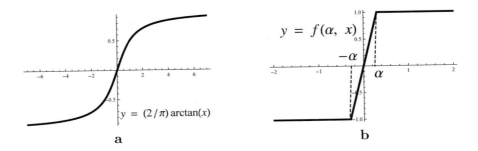

a

b

Figure 2.9: **a.** *Arctangent function* **b.** *Piecewise linear function with parame-ter* α.

Differentiating yields

$$\mathbf{t}'(x) = 2\sigma_2'(x) = 4\sigma_2(x)(1 - \sigma_2(x))$$
$$= (1 + t(x))(1 - t(x)) = 1 - t^2(x),$$

we arrive at the well-known differential equation

$$\mathbf{t}'(x) = 1 - \mathbf{t}^2(x),$$

which will be useful in the backpropagation algorithm. We also make the remark that $\mathbf{t}(0) = 0$ and the graph of $\mathbf{t}(x)$ is symmetric with respect to the origin. The fact that $\mathbf{t}(x)$ is centered at the origin constitutes an advantage of using hyperbolic tangent over the logistic function $\sigma(x)$.

- *Arctangent function* Another sigmoid function, which maps the entire line $(-\infty, \infty)$ into $(0, 1)$, is the following arctangent function, see Fig. 2.9 **a:**

$$h(x) = \frac{2}{\pi} \tanh^{-1}(x), \qquad x \in \mathbb{R}.$$

- *Softsign function* Also known as Hahn's function, softsign is an increasing sigmoid function, mapping the entire real line into interval $(-1, 1)$. Its definition relation is given by

$$so(x) = \frac{x}{1 + |x|}, \qquad x \in \mathbb{R}.$$

Despite the absolute value term, the function is differentiable; however, it is not twice differentiable at $x = 0$. In spite of its similarities with the hyperbolic function, softsign has tails that are quadratic polynomials rather than exponential, and hence it approaches asymptotes at a slower rate. Softsign was proposed as an activation function by Bergstra in 2009, see [16]. It has been shown that the advantage of using softsign over the hyperbolic tangent is the fact that it saturates slower than the latter, see [44].

• *Piecewise Linear* This is a sigmoid function, depending on a parameter $\alpha > 0$, which is defined as

$$f_\alpha(x) = f(\alpha, x) = \begin{cases} -1, & \text{if } x \leq -\alpha \\ x/\alpha, & \text{if } -\alpha < x < \alpha \\ 1, & \text{if } x \geq \alpha. \end{cases}$$

Note that as $\alpha \to 0$ then $f(\alpha, x) \to S(x)$, for $x \neq 0$ and that the graph has always two corner points, see Fig. 2.9 **b**.

Most aforementioned functions have similar sigmoid graphs, with two horizontal asymptotes, to $\pm\infty$. However, the functions distinguish themselves by the derivative at $x = 0$ as in the following:

$$\sigma'_c(0) = \frac{c}{4} \qquad \sigma'(0) = \frac{1}{4} \qquad so'(0) = 1$$

$$\mathbf{t}'(0) = 1 \qquad h'(0) = 1 \qquad f'_\alpha(0) = \frac{1}{\alpha}.$$

Roughly speaking, the larger the derivative at zero, the closer to a step function the sigmoid function is.

The logistic function $\sigma(x)$ encountered before satisfies the following three properties:

1. nondecreasing, differentiable, with $\sigma'(x) \geq 0$;
2. has horizontal asymptote for $x \to -\infty$ with value $\sigma(-\infty) = 0$;
3. has horizontal asymptote for $x \to +\infty$ with $\sigma(+\infty) = 1$.

These properties are typical for a cumulative distribution function of a random variable that takes values in $(-\infty, +\infty)$. Consequently, the integral $\sigma(x) = \int_{-\infty}^{x} p(u)\, du$ of any continuous, nonnegative, bumped-shaped function, $p(x)$, is a sigmoidal function satisfying the previous properties. For instance, if we take $p(x) = \frac{1}{\sqrt{2\pi}} e^{-x^2}$, then $F(x) = \frac{1}{2\pi} \int_{-\infty}^{x} e^{-t^2}\, dt$ is a sigmoid function.

Bumped-type functions These types of activation functions are used when the neuron gets activated to a maximum for a certain value of the action potential; it can occur in a smooth bumped way, see Fig. 2.10 **a**, or with a bump of a thorn type, see Fig. 2.10 **b**. These are given by

• *Gaussian* It maps the real line into $(0, 1]$ and is given by

$$g(x) = e^{-x^2}, \qquad x \in \mathbb{R}.$$

• *Double exponential* It maps the real line into $(0, 1]$ and is defined as

$$f(x) = e^{-\lambda|x|}, \qquad x \in \mathbb{R},\ \lambda > 0.$$

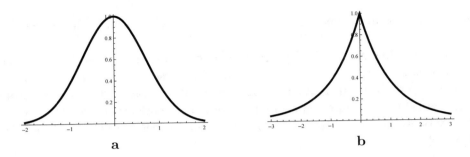

Figure 2.10: **a.** *Gaussian function* $g(x) = e^{-x^2}$ **b.** *Double exponential function* $f(x) = e^{-|x|}$.

Classification functions A *one-hot vector* with n components is a coordinate vector in $\mathbb{R}^n$ of the form

$$e_i = (0, \ldots, 0, 1, 0, \ldots, 0)$$

with a 1 on the ith slot and zero in rest. These vectors are useful in classification problems. If one would like to have the network output as close as possible to this type, then the activation function of the last layer should be a *softmax* function. If $x \in \mathbb{R}^n$ is a vector, then define the n-valued function $softmax(y) = z$, with

$$z_j = \frac{e^{x_j}}{\|e^x\|}, \qquad j = 1, \ldots, n.$$

Usually, the L^1-norm is considered, $\|e^x\| = \sum_{i=1}^n e^{x_i}$, but also other norms can be applied.

We can also define the softmax function depending on a parameter $c > 0$ by

$$softmax_c(x) = \left(\frac{e^{cx_1}}{\sum_{i=1}^n e^{cx_i}}, \ldots, \frac{e^{cx_n}}{\sum_{i=1}^n e^{cx_i}} \right).$$

We note that the usual softmax function is obtained for $c = 1$. However, if $c \to \infty$, we obtain

$$\lim_{c \to \infty} softmax_c(x) = (0, \ldots, 1, \ldots, 0) = e_k,$$

where $x_k = \max\{x_1, \ldots, x_n\}$. We shall verify this for the case $n = 2$ only. The general case can be treated similarly. Assume $x_1 > x_2$. Then

$$
\begin{aligned}
\lim_{c \to \infty} softmax_c(x) &= \lim_{c \to \infty} \left(\frac{e^{cx_1}}{e^{cx_1} + e^{cx_2}}, \frac{e^{cx_2}}{e^{cx_1} + e^{cx_2}} \right) \\
&= \lim_{c \to \infty} \left(\frac{1}{1 + e^{c(x_2-x_1)}}, \frac{1}{e^{c(x_1-x_2)} + 1} \right) = (1, 0) = e_1.
\end{aligned}
$$

Hence, the softmax function is a smooth softened version of the maximum function.

In the following we shall present two generic classes of activation functions: *sigmoidal* and *squashing functions*. These notions are more advanced and can be omitted at a first reading. They will be useful later when dealing with neural nets as function approximators. The reader can find the measure theory background needed for the next section in sections C.3 and C.4 of the Appendix.

2.2　Sigmoidal Functions

This section introduces and studies the properties of a useful family of activation functions. The results of this section will be useful in the approximation theorems presented in Chapter 9.

Definition 2.2.1 *A function* $\sigma : \mathbb{R} \to [0, 1]$ *is called sigmoidal if*

$$\lim_{x \to -\infty} \sigma(x) = 0, \qquad \lim_{x \to +\infty} \sigma(x) = 1.$$

We note that the previous definition does not require monotonicity; however, it states that it suffices to have two horizontal asymptotes. The prototype example for sigmoidal functions is the logistic function.

Recall that a measure μ can be regarded as a system of beliefs used to asses information, so $d\mu(x) = \mu(dx)$ represents an evaluation of the information cast into x. Consequently, the integral $\int f(x)\, d\mu(x)$ represents the evaluation of the function $f(x)$ under the system of beliefs μ.

The next definition introduces the notion of *discriminatory function*. This is characterized by the following property: if the evaluation of the neuron output, $\sigma(w^T x + \theta)$, over all possible inputs x, under the belief μ vanishes for any threshold θ and any system of weights, w, then μ must vanish, i.e., it is a void belief.

The next definition uses the concept of a signed Baire measure. The reader can peek in sections C.3 and C.9 of the Appendix to refresh the definition of Baire measures and signed measures. The main reason for using the Baire measure concept is its good behavior with respect to compactly supported functions (all compactly supported continuous functions are Baire-measurable and any compactly supported continuous function on a Baire space is integrable with respect to any finite Baire measure).

More precisely, if we denote by $I_n = [0, 1]^n = [0, 1] \times \cdots \times [0, 1]$ the n-dimensional unit cube in $\mathbb{R}^n$ and by $M(I_n)$, the space of finite, signed regular Baire measures on I_n, we have:

Definition 2.2.2 *Let $\mu \in M(I_n)$. A function σ is called discriminatory for the measure μ if:*

$$\int_{I_n} \sigma(w^T x + \theta) d\mu(x) = 0, \quad \forall w \in \mathbb{R}^n, \forall \theta \in \mathbb{R} \implies \mu = 0.$$

We note that the function σ in the previous definition is not necessarily a logistic sigmoid function, but any function satisfying the required property.

We shall denote by $\mathcal{P}_{w,\theta} = \{x; w^T x + \theta = 0\}$ the hyperplane with normal vector w and $(n+1)$-intercept θ. Consider also the open half-spaces

$$\begin{aligned}
\mathcal{H}_{w,\theta}^+ &= \mathcal{H}_{w,\theta} = \{x; w^T x + \theta > 0\} \\
\mathcal{H}_{w,\theta}^- &= \{x; w^T x + \theta < 0\},
\end{aligned}$$

which form a partition of the space as $\mathbb{R}^n = \mathcal{H}_{w,\theta}^+ \cup \mathcal{H}_{w,\theta}^- \cup \mathcal{P}_{w,\theta}$. The following lemma is useful whenever we want to show that a function σ is discriminatory.

Lemma 2.2.3 *Let $\mu \in M(I_n)$. If μ vanishes on all hyperplanes and open half-spaces in $\mathbb{R}^n$, then μ is zero. More precisely, if*

$$\mu(\mathcal{P}_{w,\theta}) = 0, \quad \mu(\mathcal{H}_{w,\theta}) = 0, \quad \forall w \in \mathbb{R}^n, \theta \in \mathbb{R},$$

then $\mu = 0$.

Proof: Let $w \in \mathbb{R}^n$ be fixed. Consider the linear functional $F : L^\infty(\mathbb{R}) \to \mathbb{R}$ defined by

$$F(h) = \int_{I_n} h(w^T x) \, d\mu(x),$$

where $L^\infty(\mathbb{R})$ denotes the space of almost everywhere bounded functions on $\mathbb{R}$.

The fact that F is a bounded functional follows from the inequality

$$|F(h)| = \left| \int_{I_n} h(w^T x) \, d\mu(x) \right| \leq \|h\|_\infty \left| \int_{I_n} d\mu(x) \right| = |\mu(I_n)| \cdot \|h\|_\infty,$$

where we used that μ is a finite measure.

Consider now $h = 1_{[\theta,\infty)}$, i.e., h is the indicator function of the interval $[\theta, \infty)$. Then

$$F(h) = \int_{I_n} h(w^T x) \, d\mu(s) = \int_{\{w^T x \geq \theta\}} d\mu(x) = \mu(\mathcal{P}_{w,-\theta}) + \mu(\mathcal{H}_{w,-\theta}) = 0.$$

If h is the indicator function of the open interval (θ, ∞), i.e., $h = 1_{(\theta, \infty)}$, a similar computation shows

$$F(h) = \int_{I_n} h(w^T x) \, d\mu(x) = \int_{\{w^T x > \theta\}} d\mu(x) = \mu(\mathcal{H}_{w, -\theta}) = 0.$$

Using that the indicator of any interval can be written in terms of the aforementioned indicator functions, i.e.,

$$1_{[a,b]} = 1_{[a,\infty)} - 1_{(b,\infty)}, \quad 1_{(a,b)} = 1_{(a,\infty)} - 1_{[b,\infty)}, \quad etc.,$$

it follows from linearity that F vanishes on any indicator function. Applying the linearity again, we obtain that F vanishes on simple functions

$$F\left(\sum_{i=1}^{N} \alpha_i 1_{J_i}\right) = \sum_{i=1}^{N} \alpha_i F(1_{J_i}) = 0,$$

for any $\alpha_j \in \mathbb{R}$ and J_i intervals. Since simple functions are dense in $L^\infty(\mathbb{R})$, it follows that $F = 0$. More precisely, for any fixed $f \in L^\infty(\mathbb{R})$, by density reasons, there is a sequence of simple functions $(s_n)_n$ such that $s_n \to f$, as $n \to \infty$. Since F is bounded, then it is continuous, and then we have

$$L(f) = L(\lim_n s_n) = \lim_n L(s_n) = 0.$$

Now we compute the Fourier transform of the measure μ as

$$\begin{aligned}
\widehat{\mu}(w) &= \int_{I_n} e^{iw^T x} \, d\mu(x) = \int_{I_n} \cos(w^T x) \, d\mu(x) + i \int_{I_n} \sin(w^T x) \, d\mu(x) \\
&= F(\cos(\cdot)) + iF(\sin(\cdot)) = 0, \qquad \forall w \in \mathbb{R}^n,
\end{aligned}$$

since F vanishes on the bounded functions sine and cosine. From the injectivity of the Fourier transform, it follows that $\mu = 0$. The reader can find the the definition of the Fourier transform in the section D.6.4 of the Appendix. ∎

The next result presents a class of discriminatory functions, see [30].

Proposition 2.2.4 *Any continuous sigmoidal function is discriminatory for all measures $\mu \in M(I_n)$.*

Proof: Let $\mu \in M(I_n)$ be a fixed measure. Choose a continuous sigmoidal σ that satisfies

$$\int_{I_n} \sigma(w^T x + \theta) d\mu(x) = 0, \qquad \forall w \in \mathbb{R}^n, \theta \in \mathbb{R}. \tag{2.2.3}$$

We need to show that $\mu = 0$. First, construct the continuous function

$$\sigma_\lambda(x) = \sigma\left(\lambda(w^T x + \theta) + \phi\right)$$

for given w, θ and θ, and use the definition of a sigmoidal to note that

$$\lim_{\lambda \to \infty} \sigma_\lambda(x) = \begin{cases} 1, & \text{if } w^T x + \theta > 0 \\ 0, & \text{if } w^T x + \theta < 0 \\ \sigma(\phi), & \text{if } w^T x + \theta = 0. \end{cases}$$

Define the bounded function

$$\gamma(x) = \begin{cases} 1, & \text{if } x \in \mathcal{H}_{w,\theta}^+ \\ 0, & \text{if } x \in \mathcal{H}_{w,\theta}^- \\ \sigma(\phi), & \text{if } x \in \mathcal{P}_{w,\theta} \end{cases}$$

and notice that $\sigma_\lambda(x) \to \gamma(x)$ pointwise on $\mathbb{R}$, as $\lambda \to \infty$. The Bounded Convergence Theorem allows switching the limit with the integral, obtaining

$$\lim_{\lambda \to \infty} \int_{I_n} \sigma_\lambda(x)\, d\mu(x) = \int_{I_n} \gamma(x)\, d\mu(x)$$

$$= \int_{\mathcal{H}_{w,\theta}^+} \gamma(x)\, d\mu(x) + \int_{\mathcal{H}_{w,\theta}^-} \gamma(x)\, d\mu(x) + \int_{\mathcal{P}_{w,\theta}} \gamma(x)\, d\mu(x)$$

$$= \mu\left(\mathcal{H}_{w,\theta}^+\right) + \sigma(\phi)\mu\left(\mathcal{P}_{w,\theta}\right).$$

Equation (2.2.3) implies that $\int_{I_n} \sigma_\lambda(x)\, d\mu(x) = 0$, and hence the limit in the previous left term vanishes. Consequently, the right term must also vanish, fact that can be written as

$$\mu\left(\mathcal{H}_{w,\theta}^+\right) + \sigma(\phi)\mu\left(\mathcal{P}_{w,\theta}\right) = 0.$$

Since this relation holds for any value of ϕ, taking $\phi \to +\infty$ and using the properties of σ, yields

$$\mu\left(\mathcal{H}_{w,\theta}^+\right) + \mu\left(\mathcal{P}_{w,\theta}\right) = 0.$$

Similarly, taking $\phi \to -\infty$, implies

$$\mu\left(\mathcal{H}_{w,\theta}^+\right) = 0, \qquad \forall w \in \mathbb{R}^n, \theta \in \mathbb{R}. \tag{2.2.4}$$

Note that, as a consequence of the last two relations, we also have $\mu\left(\mathcal{P}_{w,\theta}\right) = 0$. Since $\mathcal{H}_{w,\theta}^+ = \mathcal{H}_{-w,-\theta}^-$, relation (2.2.4) states that the measure μ vanishes on all half-spaces of $\mathbb{R}^n$. Lemma 2.2.3 states that a measure with such property is necessarily the zero measure, $\mu = 0$. Therefore, σ is discriminatory. ∎

Remark 2.2.5 1. The conclusion still holds true if in the hypothesis of the previous proposition we replace "continuous" by "bounded measurable".
2. A discriminatory function is not necessarily monotonic.

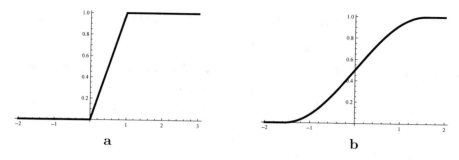

Figure 2.11: *Squashing functions:* **a.** *Ramp function* **b.** *Cosine squasher function.*

2.3   Squashing Functions

Another slightly different type of activation function used in the universal approximators of Chapter 9 is given in the following.

Definition 2.3.1 *A function* $\varphi : \mathbb{R} \to [0, 1]$ *is a squashing function if:*
(i) it is nondecreasing, i.e., for $x_1, x_2 \in \mathbb{R}$ *with* $x_1 < x_2$, *then* $\varphi(x_1) \leq \varphi(x_2)$
(ii) $\lim\limits_{x \to -\infty} \varphi(x) = 0$, $\lim\limits_{x \to \infty} \varphi(x) = 1$.

A few examples of squashing functions are:
1. The step function $H(x) = 1_{[0,\infty)}(x)$.
2. The ramp function $\varphi(x) = x1_{[0,1]}(x) + 1_{(1,\infty)}(x)$, see Fig. 2.11 **a**.
3. All monotonic sigmoidal functions are squashing functions, the prototype being the logistic function.
4. The *cosine squasher* of Gallant and White (1988)

$$\varphi(x) = \frac{1}{2}\Big(1 + \cos(x + \frac{3\pi}{2})\Big)1_{[-\frac{\pi}{2},\frac{\pi}{2}]}(x) + 1_{(\frac{\pi}{2},\infty)}(x).$$

In the previous relation, the first term provides a function, which is nonzero on $(-\frac{\pi}{2}, \frac{\pi}{2})$, while the second term is a step function equal to 1 on the right side of $\frac{\pi}{2}$. Their sum is a continuous function, see Fig. 2.11 **b**.

Being monotonic, squashing functions have at most a countable number of discontinuities, which are of jump type. In particular, squashing functions are measurable with respect to the usual Borel σ-algebra.

The next result states that we can always select a sequence of convergent squashing functions.

Lemma 2.3.2 *Let* $\mathcal{S}$ *be an infinite family of squashing functions. Then we can choose a sequence* $(\varphi_n)_n$ *in* $\mathcal{S}$ *such that* φ_n *converges at each point* x *in* $[0, 1]$. *The limit is also a squashing function.*

Proof: The family $\mathcal{S}$ is uniformly bounded with $\sup_{x \in [a,b]} |f(x)| \leq 1$. Since squashing functions are nondecreasing, they have finite total variation given by $V_{-\infty}^{\infty}(\varphi) = \varphi(+\infty) - \varphi(-\infty) = 1$, for all $\varphi \in \mathcal{S}$. Applying Helley's theorem, see Theorem E.5.4 in the Appendix, there is a sequence $(\varphi_n)_n$ of functions in $\mathcal{S}$ that is convergent at each point $x \in [a,b]$. Consider the limit function $\varphi(x) = \lim_{n \to \infty} \varphi_n(x)$ and show that φ satisfies the properties of a squashing function.

Let $x_1 < x_2$. Then $\varphi_n(x_1) \leq \varphi_n(x_2)$ for all $n \geq 1$. Taking the limit yields

$$\varphi(x_1) = \lim_{n \to \infty} \varphi_n(x_1) \leq \lim_{n \to \infty} \varphi_n(x_2) = \varphi(x_2),$$

which means that φ is nondecreasing. Interchanging limits, we also have

$$\varphi(-\infty) = \lim_{x \to -\infty} \varphi(x) = \lim_{x \to -\infty} \lim_{n \to \infty} \varphi_n(x) = \lim_{n \to \infty} \lim_{x \to -\infty} \varphi_n(x) = 0$$
$$\varphi(+\infty) = \lim_{x \to +\infty} \varphi(x) = \lim_{x \to +\infty} \lim_{n \to \infty} \varphi_n(x) = \lim_{n \to \infty} \lim_{x \to +\infty} \varphi_n(x) = 1,$$

where we used that φ_n are squashing functions. ∎

The next consequence states that the set of squashing functions is sequentially compact:

Corollary 2.3.3 *Any infinite sequence of squashing functions $(\varphi_n)_n$ contains a convergent subsequence (φ_{n_k}).*

Example 2.3.1 Consider the family of squashing functions

$$\mathcal{S} = \left\{ \frac{1}{1 + e^{-cx}} ; c \in \mathbb{R}_+ \right\}.$$

Then the sequence $(\varphi_n)_n$, $\varphi_n = \frac{1}{1 + e^{-nx}}$ is convergent to the squashing function $H(x) = 1_{[0,\infty)}(x)$.

2.4   Summary

Each unit of a neural net uses a function that processes the incoming signal, called the activation function. The importance of activation functions is to introduce nonlinearities into the network. Linear activations can produce only linear outputs and can be used only for linear classification or linear regression problems, from this point of view being very restrictive. Nonlinear activation functions fix this problem, the network being able to deal with more complex decision boundaries or to approximate nonlinear functions.

There are many types of activation functions that have been proposed over time for different purposes, having diverse shapes, such as step functions, linear, sigmoid, hockey stick, bumped-type, etc. The threshold step function was used as an activation function for the McCulloch-Pitts neuron in their 1943 paper [82]. Due to its serious limitations including the lack of good training algorithms, discontinuous activation functions are not used that often, and therefore, nonlinear activations functions have been proposed.

The most common nonlinear activation functions used for neural networks are the standard logistic sigmoid and the hyperbolic tangent. Their differentiability and the fact that their derivatives can be represented in terms of the functions themselves make these activation functions useful when applying the backpropagation algorithm. It has been noticed that when training multiple-layer neural networks the hyperbolic tangent provides better results than the logistic function.

The idea of using a ReLU as an activation function was biologically inspired by the observation that a cortical neuron is rarely in its maximum saturation regime and its activation increases proportionally with the signal, see [19] and [32]. The ReLU is, in this case, an approximation of the sigmoid. It has been shown that rectifier linear units, despite their non-differentiability at zero, are doing a better job than regular sigmoid activation functions for image-related tasks [90] and [45].

Other activation functions (sigmoidal activations and squashing functions) were included for the theoretical purposes of the universal approximator theorems of Chapter 9.

2.5   Exercises

Exercise 2.5.1 (a) Show that the logistic function σ satisfies the inequality $0 < \sigma'(x) \leq \frac{1}{4}$, for all $x \in \mathbb{R}$.

(b) How does the inequality change in the case of the function σ_c?

Exercise 2.5.2 Let $S(x)$ and $H(x)$ denote the bipolar step function and the Heaviside function, respectively. Show that

(a) $S(x) = 2H(x) - 1$;

(b) $ReLU(x) = \frac{1}{2}x(S(x) + 1)$.

Exercise 2.5.3 Show that the softplus function, $sp(x)$, satisfies the following properties:

(a) $sp'(x) = \sigma(x)$, where $\sigma(x) = \frac{1}{1+e^{-x}}$;

(b) Show that $sp(x)$ is invertible with the inverse $sp^{-1}(x) = \ln(e^x - 1)$;

(c) Use the softplus function to show the formula $\sigma(x) = 1 - \sigma(-x)$.

Exercise 2.5.4 Show that $\tanh(x) = 2\sigma(2x) - 1$.

Exercise 2.5.5 Show that the softsign function, $so(x)$, satisfies the following properties:

(a) It is strictly increasing;

(b) It is onto on $(-1, 1)$, with the inverse $so^{-1}(x) = \frac{x}{1-|x|}$, for $|x| < 1$;

(c) $so(|x|)$ is subaddtive, i.e., $so(|x + y|) \leq so(|x|) + so(|y|)$.

Exercise 2.5.6 Show that the softmax function is invariant with respect to addition of constant vectors $\mathbf{c} = (c, \ldots, c)^T$, i.e.,

$$softmax(y + \mathbf{c}) = softmax(y).$$

This property is used in practice by replacing $\mathbf{c} = -\max_i y_i$, fact that leads to a more stable numerically variant of this function.

Exercise 2.5.7 Let $\rho : \mathbb{R}^n \to \mathbb{R}^n$ defined by $\rho(y) \in \mathbb{R}^n$, with $\rho(y)_i = \frac{y_i^2}{\|y\|_2^2}$. Show that

(a) $0 \leq \rho(y)_i \leq 1$ and $\sum_i \rho(y)_i = 1$;

(b) The function ρ is invariant with respect to multiplication by nonzero constants, i.e., $\rho(\lambda y) = \rho(\lambda y)$ for any $\lambda \in \mathbb{R} \backslash \{0\}$. Taking $\lambda = \frac{1}{\max_i y_i}$ leads in practice to a more stable variant of this function.

Exercise 2.5.8 (cosine squasher) Show that the function

$$\varphi(x) = \frac{1}{2}\left(1 + \cos(x + \frac{3\pi}{2})\right)1_{[-\frac{\pi}{2}, \frac{\pi}{2}]}(x) + 1_{(\frac{\pi}{2}, \infty)}(x)$$

is a squashing function.

Exercise 2.5.9 (a) Show that any squashing function is a sigmoidal function;

(b) Give an example of a sigmoidal function which is not a squashing function.

Chapter 3

Cost Functions

In the learning process, the parameters of a neural network are subject to minimize a certain objective function, which represents a measure of proximity between the prediction of the network and the associated target. This is also known under the equivalent names of *cost function, loss function,* , or *error function.* In the following we shall describe some of the most familiar cost functions used in neural networks.

3.1 Input, Output, and Target

The *input* of a neural network is obtained from given data or from sensors that perceive the environment. The input is a variable that is fed into the network. It can be a one-dimensional variable $x \in \mathbb{R}$, or a vector $\mathbf{x} \in \mathbb{R}^n$, a matrix, or a random variable X. In general, the input can be a tensor, see section B of Appendix.

The network acts as a function, i.e., modifies the input variable in a certain way and provides an output, which can be, again, one-dimensional, $y \in \mathbb{R}$, or a vector, $\mathbf{y} \in \mathbb{R}^n$, or a random variable, Y, or a tensor. The law by which the input is modified into the output is done by the *input-output mapping*, $f_{w,b}$. The index (w, b) suggests that the internal network parameters, while making this assignment, are set to (w, b). Following the previous notations, we may have either $y = f_{w,b}(x)$, or $\mathbf{y} = f_{w,b}(\mathbf{x})$, or $Y = f_{w,b}(X)$.

The *target function* is the desired relation which the network tries to approximate. This function is independent of parameters w and will be denoted either by $z = \phi(x)$, in the one-dimensional case, or by $\mathbf{z} = \phi(\mathbf{x})$, in the vector case, or $Z = \phi(X)$, for random variables. Also some mixtures

© Springer Nature Switzerland AG 2020
O. Calin, *Deep Learning Architectures*, Springer Series in the Data Sciences,
https://doi.org/10.1007/978-3-030-36721-3_3

may occur, for instance, $z = f(\mathbf{x})$, in the case when the input is a vector and the output is one-dimensional.

The neural network will tune the parameters (w, b) until the output variable y will be in a proximity of the target variable z. This proximity can be measured in a variety of ways, involving different types of cost functions, $C(w) = dist(y, z)$, which are parameterized by (w, b), with the distance to be specified in the following. The optimal parameters of the network are given by $(w^*, b^*) = \arg\min_{w,b} C(w, b)$. The process by which the parameters (w, b) are tuned into (w^*, b^*) is called *learning*. Equivalently, we also say that the network learns the target function $\phi(x)$. This learning process involves the minimization of a cost function. We shall investigate next a few types of cost functions.

3.2   The Supremum Error Function

Assume that a neural network, which takes inputs $x \in [0, 1]$, is supposed to learn a given continuous function $\phi : [0, 1] \to \mathbb{R}$. If $f_{w,b}$ is the input-output mapping of the network then the associate cost function is

$$C(w, b) = \sup_{x \in [0,1]} |f_{w,b}(x) - \phi(x)|.$$

For all practical purposes, when the target function is known at n points

$$z_1 = \phi(x_1), z_2 = \phi(x_2), \ldots, z_n = \phi(x_n)$$

the aforementioned cost function becomes

$$C(w, b) = \max_{1 \le i \le n} |f_{w,b}(x_i) - z_i|.$$

3.3   The L^2-Error Function

Assume the input of the network is $x \in [0, 1]$, and the target function $\phi : [0, 1] \to \mathbb{R}$ is square integrable. If $f_{w,b}$ is the input-output mapping, the associate cost function measures the distance in the L^2-norm between the output and the target

$$C(w, b) = \int_0^1 \left(f_{w,b}(x) - \phi(x) \right)^2 dx.$$

If the target function is known at only n points

$$z_1 = \phi(x_1), z_2 = \phi(x_2), \ldots, z_n = \phi(x_n),$$

then the appropriate associated cost function is the square of the Euclidean distance in $\mathbb{R}^n$ between $f_{w,b}(\mathbf{x})$ and $\mathbf{z}$

$$C(w, b) = \sum_{i=1}^n (f_{w,b}(x_i) - z_i)^2 = \|f_{w,b}(\mathbf{x}) - \mathbf{z}\|^2, \tag{3.3.1}$$

where $\mathbf{x} = (x_1, \ldots, x_n)$ and $\mathbf{z} = (z_1, \ldots, z_n)$. For $\mathbf{x}, \mathbf{z} \in \mathbb{R}^n$ fixed, $C(w, b)$ is a smooth function of (w, b), which can be minimized using the gradient descent method.

The geometric interpretation is given in the following. For any fixed $\mathbf{x}$, the mapping $(w, b) \to f_{w,b}(\mathbf{x})$ represents a hypersurface in $\mathbb{R}^n$ parametrized by (w, b). The cost function, $C(w, b)$, represents the Euclidean distance between the given point $\mathbf{z}$ and a point on this hypersurface. The cost is minimized when the distance is the shortest. This is realized when the point $f_{w,b}(\mathbf{x})$ is the orthonormal projection of $\mathbf{z}$ onto the hypersurface. The optimal value of parameters are

$$(w^*, b^*) = \arg\min_{w,b} C(w, b) = \arg\min_{w,b} \|f_{w,b}(\mathbf{x}) - \mathbf{z}\| = \arg\min_{w,b} \operatorname{dist}(\mathbf{z}, f_{w,b}(\mathbf{x})).$$

The fact that the vector $\mathbf{z} - f_{w^*,b^*}(\mathbf{x})$ is perpendicular to the hypersurface is equivalent to the fact that the vector is perpendicular to the tangent plane, which is generated by $\partial_{w_k} f_{w^*}(\mathbf{x})$ and $\partial_{b_j} f_{b^*}(\mathbf{x})$. Writing the vanishing inner products, we obtain the *normal equation*

$$\sum_{i=1}^{n} (z_i - f_{w^*,b^*}(x_i)) \partial_{w_k} f_{w^*,b^*}(x_i) = 0$$

$$\sum_{i=1}^{n} (z_i - f_{w^*,b^*}(x_i)) \partial_b f_{w^*,b^*}(x_i) = 0.$$

In general, this system cannot be solved in closed form, and hence the need of numerical methods, such as the gradient descent method or its variants.

3.4 Mean Square Error Function

Consider a neural network whose input is a random variable X, and its output is the random variable $Y = f_{w,b}(X)$, where $f_{w,b}$ denotes the input-output mapping of the network, which depends on the parameter w and bias b. Assume the network is used to approximate the target random variable Z. The error function in this case measures a proximity between the output and target random variables Y and Z. A good candidate is given by the expectation of their squared difference

$$C(w, b) = \mathbb{E}[(Y - Z)^2] = \mathbb{E}[(f_{w,b}(X) - Z)^2], \qquad (3.4.2)$$

where $\mathbb{E}$ denotes the expectation operator, see section D.3 of Appendix. We search for the pair (w, b), which achieves the minimum of the cost function, i.e., we look for $(w^*, b^*) = \arg\min C(w, b)$. This is supposed to be obtained by one of the minimization algorithms (such as the steepest descent method) presented in Chapter 4.

We shall discuss next a few reasons regarding the popularity of the previous error function.

1. It is known that the set of all square-integrable random variables on a probability space forms a Hilbert space with the inner product given by $\langle X, Y \rangle = \mathbb{E}[XY]$. This defines the norm $\|X\|^2 = \mathbb{E}[X^2]$, which induces the distance $d(X, Y) = \|X - Y\|$. Hence, the cost function (3.4.2) represents the square of a distance, $C(w, b) = d(Y, Z)^2$. Minimizing the cost is equivalent to finding the parameters (w, b) that minimize the distance between the output Y and target Z. For the definitions of norm and Hilbert spaces, see section D.3 in Appendix.

2. It is worth emphasizing the relation between the aforementioned cost function and the conditional expectation. The neural network transforms the input random variable X into an output random variable $Y = f_{w,b}(X)$, which is parameterized by w and b. The information generated by the random variable $f_{w,b}(X)$ is the sigma-field $\mathcal{E}_{w,b} = \mathfrak{S}(f_{w,b}(X))$, i.e., the smallest sigma-field with respect to which $f_{w,b}(X)$ is measurable. All these sigma-fields generate the output information, $\mathcal{E} = \bigvee_{w,b} \mathcal{E}_{w,b}$, which is the sigma-field generated by the union $\bigcup_{w,b} \mathcal{E}_{w,b}$. (The notation $\mathcal{E}$ comes from from *exit*).

In general, the target random variable, Z, is not determined by the information $\mathcal{E}$. The problem now can be stated as in the following: *Given the output information $\mathcal{E}$, find the best prediction of Z based on the information $\mathcal{E}$.* This is a random variable, denoted by $Y = \mathbb{E}[Z|\mathcal{E}]$, called *the conditional expectation* of Z given $\mathcal{E}$. The best predictor, Y, is determined by $\mathcal{E}$ (i.e., it is $\mathcal{E}$-measurable) and is situated at the smallest possible distance from Z, i.e., we have (with notations of part 1)

$$d(Y, Z) \leq d(U, Z)$$

for any $\mathcal{E}$-measurable random variable U. This is equivalent to

$$\mathbb{E}[(Y - Z)^2] \leq \mathbb{E}[(U - Z)^2], \tag{3.4.3}$$

which means that

$$Y = \arg \min_U \mathbb{E}[(U - Z)^2].$$

If $U = f_{w,b}(X)$, then the right side of (3.4.3) is the cost function $C(w, b)$. Then $Y = f_{w^*,b^*}(X)$ minimizes the cost function, and hence

$$(w^*, b^*) = \arg \min_{w,b} \mathbb{E}[(f_{w,b}(X) - Z)^2] = \arg \min_{w,b} C(w, b).$$

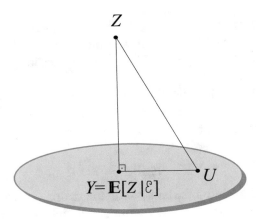

Figure 3.1: *The best predictor of the target Z, given the information $\mathcal{E}$, is the orthogonal projection of Z on the space of $\mathcal{E}$-measurable functions.*

We note that Y is the orthogonal projection of Z onto the space of $\mathcal{E}$-measurable functions, see Fig. 3.1. The orthogonality is considered in the sense introduced in the previous part 1, and Z is considered to belong to the Hilbert space of square integrable functions.

Another plastic way to present the problem is as in the following. Assume that the humans' goal is the ultimate knowledge of the world. The inputs to our brain are provided by the stimuli from our surroundings that our senses are able to collect and evaluate. This information is interpreted by the brain's neural network and an output information is obtained, which represents our knowledge about the world. This is a subset of the total information, which we might never be able to comprehend. However, given the limited information we are able to acquire, what is the best picture of the real world one can obtain?

3. The definition of the cost function can be easily extended to the case when the random variables are known by measurements. Consider n measurements of random variables (X, Z), which are given by $(x_1, z_1), (x_2, z_2), \ldots, (x_n, z_n)$. This forms the so-called training set for the neural network. Then in this case the cost function is defined by the following average

$$\tilde{C}(w, b) = \frac{1}{n} \sum_{j=1}^{n} \left(f_{w,b}(x_j) - z_j \right)^2,$$

which can be considered as the empirical mean of the square difference of Y and Z.

For the sake of completeness, we shall discuss also the case when the input variable X and the target variable Z are independent. In this case the neuronal network is trained on pairs of independent variables. What is the best estimator, Y, in this case?

Since X and Z are independent, then also $f_\theta(X)$ and Z will be independent for all values of the parameter $\theta = (w, b)$. Then the output information, $\mathcal{E}$, which is generated by the random variables $f_\theta(X)$ will be independent of Z. The best estimator is

$$Y = \mathbb{E}[Z|\mathcal{E}] = \mathbb{E}[Z],$$

where we used that an independent condition drops out from the conditional expectation. The best estimator is a number, which is the mean of the target variable. This will be expressed more plastically in the following.

Let's say that an experienced witch pretends to read the future of a gullible customer using the coffee traces on her mug. Then the coffee traces represent the input X and the customer future is the random variable Z. The witch-predicted future is the best estimator, Y, which is the mean of all possible futures, and does not depend on the coffee traces. Hence, the witch is an impostor.

The next few cost functions measure the proximity between probability densities of random variables.

3.5 Cross-entropy

We shall present the definitions and properties of the cross-entropy in the one-dimensional case. Later, we shall use them in the 2-dimensional case. For more details on this topic the reader can consult [81] and [22].

Let p and q be two densities on $\mathbb{R}$ (or any other interval). It is known that the negative likelihood function, $-\ell_q(x) = -\ln q(x)$, measures the information given by $q(x)$.[1] The *cross-entropy* of p with respect to q is defined by the expectation with respect to p of the negative likelihood function as

$$S(p, q) = \mathbb{E}^p[-\ell_q] = -\int_{\mathbb{R}} p(x) \ln q(x)\, dx.$$

This represents the information given by $q(x)$ assessed from the point of view of distribution $p(x)$, which is supposed to be given. It is obvious that for discrete densities we have $S(p, q) \geq 0$. Actually, a sharper inequality holds:

---

[1]This is compatible with the following properties: (*i*) non-negativity: $-\ell_q(x) \geq 0$; (*ii*) we have $-\ell_{q_1 q_2}(x) = -\ell_{q_1}(x) - \ell_{q_2}(x)$ for any two independent distributions q_1 and q_2; (*iii*) the information increases for rare events, with $\lim_{q(x) \to 0}(-\ell_q(x)) = \infty$.

Proposition 3.5.1 *We have $S(p,q) \geq H(p)$, where $H(p)$ is the Shannon entropy, see [111]*

$$H(p) = -\mathbb{E}^p[\ell_p] = -\int_{\mathbb{R}} p(x) \ln p(x) \, dx.$$

Proof: Evaluate the difference, using the inequality $\ln u \leq u - 1$ for $u > 0$:

$$
\begin{aligned}
S(p,q) - H(p) &= -\int_{\mathbb{R}} p(x) \ln q(x) \, dx + \int_{\mathbb{R}} p(x) \ln p(x) \, dx \\
&= -\int_{\mathbb{R}} p(x) \ln \frac{q(x)}{p(x)} \, dx \geq \int_{\mathbb{R}} p(x) \left(\frac{q(x)}{p(x)} - 1 \right) dx \\
&= \int q(x) \, dx - \int p(x) \, dx = 1 - 1 = 0.
\end{aligned}
$$

Hence, $S(p,q) - H(p) \geq 0$, with $S(p,q) = H(p)$ if and only if $p = q$. ∎

The previous result states that for a given density p, the minimum of the cross-entropy $S(p,q)$ occurs for $q = p$; this minimum is equal to the Shannon entropy of the given density p. It is worth noting that in the case of a continuous density the entropy $H(p)$ can be finite (positive or negative) or infinite, while in the case of a discrete distribution it is always finite and positive (and hence the cross-entropy becomes positive).

The Shannon entropy $H(p)$ is a measure of the information contained within the probability distribution $p(x)$. Therefore, the previous proposition states that the information given by the distribution $q(x)$ assessed from the point of view of distribution $p(x)$ is always larger than the information defined by the distribution $p(x)$.

Corollary 3.5.2 *Let p be the density of a one-dimensional random variable X on $\mathbb{R}$. Then*

$$H(p) \leq \frac{1}{2} \ln(2\pi e Var(X)).$$

Proof: Let $\mu = \mathbb{E}[X]$ and $\sigma^2 = Var(X)$, and consider the normal density $q(x) = \frac{1}{\sqrt{2\pi}\sigma} e^{-\frac{(x-\mu)^2}{2\sigma^2}}$, which has the same mean and variance as $p(x)$. The cross-entropy of p with respect to q can be computed as

$$
\begin{aligned}
S(p,q) &= -\int_{\mathbb{R}} p(x) \ln q(x) \, dx = \frac{1}{2} \ln(2\pi\sigma^2) + \frac{1}{2\sigma^2} \int_{\mathbb{R}} p(x)(x-\mu)^2 \, dx \\
&= \frac{1}{2} \ln(2\pi\sigma^2) + \frac{1}{2} = \frac{1}{2} \ln(2\pi e \sigma^2).
\end{aligned}
$$

By Proposition 3.5.1 we have $H(p) \leq S(p,q) = \frac{1}{2}\ln(2\pi e\sigma^2)$. The equality is reached for random variables X that are normally distributed. ∎

Remark 3.5.3 Cross-entropy is often used in classification problems, while the mean squared error is usually useful for regression problems.

3.6   Kullback-Leibler Divergence

The difference between the cross-entropy and the Shannon entropy is the *Kullback-Leibler divergence* (see [69–71])

$$D_{KL}(p||q) = S(p,q) - H(p).$$

Equivalently, this is given by

$$D_{KL}(p||q) = -\int_{\mathbb{R}} p(x) \ln \frac{q(x)}{p(x)} \, dx.$$

By the previous result, $D_{KL}(p||q) \geq 0$. However, this is not a distance, since it is neither symmetric, nor it satisfies the triangle inequality.

Both the cross-entropy and the Kullback-Leibler divergence can be considered as cost functions for neuronal networks. This will be discussed in the following.

Let X be the input random variable for a given neural network, and let $Y = f_\theta(X, \xi)$ be its output, where $\theta = (w, b)$ and ξ is a random variable denoting the noise in the network. The target random variable is denoted by Z. Another way to match the output Y to target Z is using probability density functions. The conditional density of Y, given the input X, is denoted by $p_\theta(y|x)$, and is called the *conditional model density function*. The joint density of (X, Z) is denoted by $p_{X,Z}(x, z)$, and it is regarded as the *training distribution*. There are several ways to compare the densities, as it will be shown next.

One way is to tune the parameter θ, such that for the given training distribution $p = p_{X,Z}(x, z)$, we obtain a conditional model distribution $q = p_\theta(\cdot|x)$ for which the cross-entropy of p and q is as small as possible. The value

$$\theta^* = \arg\min_\theta S(p_{X,Z}, p_\theta(Z|X))$$

is the minimum for the cost function

$$C(\theta) = S(p_{X,Z}, p_\theta(Z|X)).$$

In the best-case scenario, in the virtue of Proposition 3.5.1, the aforementioned minimum equals the Shannon entropy of the training distribution, $H(p_{X,Z})$.

Let $p_X(x)$ be the density of the input variable X. Using the properties of conditional densities, the previous cross-entropy can be written equivalently as

$$
\begin{aligned}
C(\theta) &= S(p_{X,Z}, p_\theta(Z|X)) = -\iint p_{X,Z}(x,z) \ln p_\theta(z|x)\, dx dz \\
&= -\iint p_{X,Z}(x,z) \ln \left(\frac{p_\theta(x,z)}{p_X(x)}\right) dx dz \\
&= -\iint p_{X,Z}(x,z) \ln p_\theta(x,z)\, dx dz + \iint p_{X,Z}(x,z) \ln p_X(x)\, dx dz \\
&= S(p_{X,Z}, p_\theta(X,Z)) + \int \left(\int p_{X,Z}(x,z)\, dz\right) \ln p_X(x)\, dx \\
&= S(p_{X,Z}, p_\theta(X,Z)) + \int p_X(x) \ln p_X(x)\, dx \\
&= S(p_{X,Z}, p_\theta(X,Z)) - H(p_X),
\end{aligned}
$$

where $H(p_X)$ is the *input entropy*, i.e., the Shannon entropy of the input variable X. Since $H(p_X)$ is independent of the model parameter θ, the new cost function

$$
\bar{C}(\theta) = S(p_{X,Z}, p_\theta(X,Z)),
$$

which is the cross-entropy of the training density with the model density, reaches its minimum for the same parameter θ as $C(\theta)$

$$
\theta^* = \arg\min_\theta C(\theta) = \arg\min_\theta \bar{C}(\theta).
$$

In conclusion, given a training density, $p_{X,Z}$, and either a model density, $p_\theta(X,Y)$, or a conditional model density, $p_\theta(Y|X)$, we search for the parameter value θ for which either the cost $\bar{C}(\theta)$ or $C(\theta)$, respectively, is minimum.

In practical applications, the random variables (X,Z) are known through n measurements

$$
(x_1, z_1), (x_2, z_2), \ldots, (x_n, z_n).
$$

In this case, we assume that the joint density of the pair (X,Z) is approximated by its empirical training distribution $\hat{p}_{X,Z}(x,z)$. The new cost function is the cross-entropy between the empirical training distribution defined by the training set and probability distribution defined by the model. Approximating the expectation with an average, we have

$$
\tilde{C}(\theta) = S(\hat{p}_{X,Z}, p_\theta(Z|X)) = \mathbb{E}^{\hat{p}_{X,Z}}[-\ln p_\theta(Z|X)] = -\frac{1}{n}\sum_{j=1}^n \ln p_\theta(z_j|x_j).
$$

A similar measurement-based error can be defined as

$$
\hat{C}(\theta) = S(\hat{p}_{X,Z}, p_\theta(X,Z)) = \mathbb{E}^{\hat{p}_{X,Z}}[-\ln p_\theta(X,Z)] = -\frac{1}{n}\sum_{j=1}^n \ln p_\theta(x_j, z_j).
$$

There is an equivalent description to the previous one, when the cost function

$$C(\theta) = D_{KL}(p_{X,Z} \| p_\theta(X, Z))$$

is given by the Kullback-Leibler divergence of the training density with the model density. Since Shannon entropy, $H(p_{X,Z})$, is independent of the parameter θ, we have

$$\theta^* = \arg\min_\theta D_{KL}(p_{X,Z} \| p_\theta(X, Z)) = \arg\min_\theta S(p_{X,Z}, p_\theta(X, Z)).$$

In the best-case scenario, when the training and the model distributions coincide, the previous minimum is equal to zero.

In the case when a training set is provided

$$(x_1, z_1), (x_2, z_2), \dots, (x_n, z_n),$$

the cost function is written using the empirical density, $\hat{p}_{X,Z}$, as

$$C(\theta) = \mathbb{E}^{\hat{p}_{X,Z}}\left[-\ln\frac{p_\theta(X, Z)}{\hat{p}_{X,Z}}\right] = -\frac{1}{n}\sum_{j=1}^n \left(\ln p_\theta(x_i, z_i) - \ln \hat{p}(x_i, z_i)\right).$$

Maximum Likelihood The minimum of the aforementioned empirical cost function

$$\widehat{C}(\theta) = \mathbb{E}^{\hat{p}_{X,Z}}[-\ln p_\theta(X, Z)] = -\frac{1}{n}\sum_{j=1}^n \ln p_\theta(x_j, z_j),$$

which is $\theta^* = \arg\min_\theta \widehat{C}(\theta)$, has the distinguished statistical property that is the *maximum likelihood estimator* of θ, given n independent measurements

$$(x_1, z_1), (x_2, z_2), \dots, (x_n, z_n).$$

This follows from the next computation, which uses properties of logarithms:

$$
\begin{aligned}
\theta^* &= \arg\min_\theta \widehat{C}(\theta) = \arg\max_\theta \frac{1}{n}\sum_{j=1}^n \ln p_\theta(x_j, z_j) \\
&= \arg\max_\theta \sum_{j=1}^n \ln p_\theta(x_j, z_j) = \arg\max_\theta \ln\left(\prod_{j=1}^n p_\theta(x_j, z_j)\right) \\
&= \arg\max_\theta \prod_{j=1}^n p_\theta(x_j, z_j) = \arg\max_\theta p_\theta(X = \mathbf{x}, Z = \mathbf{z}) \\
&= \theta_{ML}.
\end{aligned}
$$

The popularity of the empirical cross-entropy and Kullback-Leibler divergence as cost functions is due to this relationship with the maximum likelihood method. Furthermore, it is the hope that using these cost functions in a neural network will lead to cost surfaces with less plateaus than in the case of sum of squares cost function, fact that improves the network training time.

3.7 Jensen-Shannon Divergence

Another proximity measure between two probability distributions p and q is given by the *Jensen-Shannon divergence*

$$D_{JS}(p||q) = \frac{1}{2}\Big(D_{KL}(p||m) + D_{KL}(q||m)\Big), \qquad (3.7.4)$$

where $m = \frac{1}{2}(p+q)$ and D_{KL} denotes the Kullback-Leibler divergence. This will be useful in optimizing GANs in Chapter 19.

Proposition 3.7.1 *The Jensen-Shannon divergence has the following properties:*

(i) $D_{JS}(p||q) \geq 0$ *(non-negative);*

(ii) $D_{JS}(p||q) = 0 \Leftrightarrow p = q$ *(non-degenerate);*

(iii) $D_{JS}(p||q) = D_{JS}(q||p)$ *(symmetric).*

Proof: *(i)* It follows from the fact that the Kullback-Leibler divergence is nonnegative, namely $D_{KL}(p||m) \geq 0$, $D_{KL}(q||m) \geq 0$.
(ii) If $p = q$, then $p = q = m$ and then $D_{KL}(p||m) = D_{KL}(q||m) = 0$, which implies $D_{JS}(p||q) = 0$. Conversely, if $D_{JS}(p||q) = 0$, then $D_{KL}(p||m) + D_{KL}(q||m) = 0$, and since the Kullback-Leibler divergence is nonnegative, it follows that $D_{KL}(p||m) = 0$ and $D_{KL}(q||m) = 0$. This implies $p = m$ and $q = m$. Multiplying by 2 we obtain $2p = p + q$, which is equivalent to $p = q$.
(iii) The symmetry follows from the addition commutativity of the right-side terms in formula (3.7.4). ∎

The Jensen-Shannon divergence will be useful in Chapter 19 in the study of GANs.

3.8 Maximum Mean Discrepancy

If X is a continuous random variable with probability density $p(x)$ over the space $\in \mathcal{X}$, then for any vector function $\phi : \mathcal{X} \to \mathbb{R}^N$ we define the ϕ-*moment* of X by

$$\mu_\phi(X) = \mathbb{E}[\phi(X)] = \int_{\mathcal{X}} \phi(x)p(x)\,dx.$$

For instance, if $\phi(x) = x$, then $\mu_\phi(X)$ is the mean, or the first moment of X. If $\phi(x) = (x, x^2, \ldots, x^N)^T$, then $\mu_\phi(X)$ is an N-dimensional vector containing the first N moments of the random variable X.

Now, we shall consider two continuous random variables, X and Y on the same space $\mathcal{X}$ and having probability densities $p(x)$ and $q(y)$, respectively.

For a fixed function $\phi : \mathcal{X} \to \mathbb{R}^N$, the proximity between p and q can be measured by the Euclidean distance between the ϕ-moments of X and Y as

$$d_{MMD}(p, q) = d_{Eu}(\mu_\phi(X), \mu_\phi(Y)) = \|\mu_\phi(X) - \mu_\phi(Y)\|_{Eu}. \qquad (3.8.5)$$

The number $d_{MMD}(p, q)$ is called the *maximum mean discrepancy* of p and q. Formula (3.8.5) can be also written in the more convenient integral form

$$d_{MMD}(p, q) = \left\| \int_{\mathcal{X}} (p(u) - q(u)) \phi(u) \, du \right\|_{Eu}.$$

We note that this is a generalization of the L^1-distance, which is obtained in the particular case $\phi(x) = 1$. We shall emphasize in the following the relation with kernels.

Using that the length of any vector v can be written as $\|v\|^2 = v^T v$, we have

$$\begin{aligned}
d_{MMD}(p, q)^2 &= \|\mathbb{E}_p[\phi(X)] - \mathbb{E}_q[\phi(Y)]\|_{Eu}^2 \\
&= \Big(\mathbb{E}_p[\phi(X)] - \mathbb{E}_q[\phi(Y)]\Big)^T \Big(\mathbb{E}_p[\phi(X)] - \mathbb{E}_q[\phi(Y)]\Big) \\
&= \mathbb{E}_p[\phi(X)^T]\mathbb{E}_p[\phi(X)] + \mathbb{E}_q[\phi(Y)^T]\mathbb{E}_q[\phi(Y)] \\
&\quad - \mathbb{E}_p[\phi(X)^T]\mathbb{E}_q[\phi(Y)] - \mathbb{E}_q[\phi(Y)^T]\mathbb{E}_p[\phi(X)].
\end{aligned}$$

We shall compute each term using Fubini's theorem and show that the last two are equal as in the following:

$$\begin{aligned}
\mathbb{E}_p[\phi(X)^T]\mathbb{E}_p[\phi(X)] &= \int \phi(x)^T p(x) \, dx \int \phi(x') p(x') \, dx' \\
&= \iint \phi(x)^T \phi(x') p(x) p(x') \, dx dx' \\
\mathbb{E}_q[\phi(Y)^T]\mathbb{E}_q[\phi(Y)] &= \iint \phi(y)^T \phi(y') p(y) p(y') \, dy dy' \\
\mathbb{E}_p[\phi(X)^T]\mathbb{E}_q[\phi(Y)] &= \int \phi(x)^T p(x) \, dx \int \phi(y) q(y) \, dy \\
&= \int \phi(x)^T \phi(y) p(x) q(y) \, dx dy.
\end{aligned}$$

Since $\phi(x)^T \phi(y) = \phi(y)^T \phi(x)$, we obtain

$$\mathbb{E}_p[\phi(X)^T]\mathbb{E}_q[\phi(Y)] = \mathbb{E}_q[\phi(Y)^T]\mathbb{E}_p[\phi(X)].$$

Substituting back, yields

$$d_{MMD}(p,q)^2 = \iint \phi(x)^T \phi(x')p(x)p(x')\,dxdx' + \iint \phi(y)^T \phi(y')p(y)p(y')\,dydy'$$
$$- 2\int \phi(x)^T \phi(y)p(x)q(y)\,dxdy.$$

Consider the kernel

$$K(u,v) = \phi(u)^T \phi(v) = \sum_{j=1}^{N} \phi_j(u)\phi_j(v).$$

This kernel is symmetric, $K(u,v) = K(v,u)$, and is nonnegative definite, namely for any real-valued function $\alpha(x)$ we have

$$\iint K(u,v)\alpha(u)\alpha(v)\,dudv \geq 0.$$

The last inequality follows from the use of Fubini's theorem and the definition of the kernel $K(u,v)$ as

$$\iint K(u,v)\alpha(u)\alpha(v)\,dudv = \sum_{j=1}^{N} \iint \phi_j(u)\alpha(u)\phi_j(v)\alpha(v)\,dudv$$
$$= \sum_{j=1}^{N} \left(\int \phi_j(u)\alpha(u) \right)^2 \geq 0.$$

Then the aforementioned expression for the maximum mean discrepancy can be written in terms of the kernel as

$$d_{MMD}(p,q)^2 = \iint K(x,x')p(x)p(x')\,dxdx' + \iint K(y,y')p(y)p(y')\,dydy'$$
$$- 2\int K(x,y)p(x)q(y)\,dxdy$$
$$= \iint K(u,v)\Big(p(u)p(v) + q(u)q(v) - 2p(u)q(v) \Big)\,dudv$$
$$= \iint K(u,v)(p(u) - q(u))(p(v) - q(v))\,dudv,$$

where we changed variables and grouped the integrals under one integral. In the case of discrete random variables the previous formula becomes

$$d_{MMD}(p,q)^2 = \sum_{i,j} K_{ij}(p_i - q_i)(p_j - q_j),$$

where $K_{ij} = K_{ji} = K(u_i, u_j)$ is a symmetric, nonnegative definite matrix defined by

$$K_{ij} = \phi(u_i)^T \phi(u_j) = \sum_{k=1}^{N} \phi_k(u_i)^T \phi_k(u_j),$$

and $p_i = p(u_i)$, $q_j = q(u_j)$, where $\{u_1, \ldots, u_k\}$ denotes the sample space.

Maximum Mean Discrepancy For all practical purposes, the random variable X is known from a sample of n observations, $x_1, \ldots, x_n$, which are drawn from the distribution $p(x)$. Similarly, the random variable Y is known from a sample of m observations, $y_1, \ldots, y_m$, drawn from the distribution $q(y)$. The means are estimated as averages by

$$\mathbb{E}_{X \sim p}[\phi(X)] = \frac{1}{n} \sum_{i=1}^{n} \phi(x_i),$$

$$\mathbb{E}_{Y \sim q}[\phi(Y)] = \frac{1}{m} \sum_{i=1}^{m} \phi(y_i),$$

and the maximum mean discrepancy between p and q can be estimated using the previous two samples as

$$
\begin{aligned}
d_{MMD}(p, q) &= \left\| \frac{1}{n} \sum_{i=1}^{n} \phi(x_i) - \frac{1}{m} \sum_{i=1}^{m} \phi(y_i) \right\|^2 \\
&= \left(\frac{1}{n} \sum_{i=1}^{n} \phi(x_i) - \frac{1}{m} \sum_{i=1}^{m} \phi(y_i) \right)^T \left(\frac{1}{n} \sum_{i=1}^{n} \phi(x_i) - \frac{1}{m} \sum_{i=1}^{m} \phi(y_i) \right) \\
&= \frac{1}{n^2} \sum_{i,j} K(x_i, x_j) + \frac{1}{m^2} \sum_{i,j} K(y_i, y_j) - \frac{2}{mn} \sum_{i,j} K(x_i, y_j),
\end{aligned}
$$

(3.8.6)

where we used the kernel notation $K(x, y) = \phi(x)^T \phi(y)$.

The maximum mean discrepancy will be used in Chapter 19 in the study of generative moment matching networks.

3.9 Other Cost Functions

Other possibilities of forming cost functions by comparing the model density, $p_\theta(x, z)$, to the training density, $p_{X,Z}(x, z)$, are glanced in the following. For more details, the reader is referred to [22].

L^1**-distance** Assuming the densities are integrable, the distance between them is measured by $D_1(p_\theta, p_{X,Z}) = \iint |p_{X,Z}(x, z) - p_\theta(x, z)| \, dx dz$. The minimum of D_1 is zero and it is reached for identical distributions.

L^2-distance Assuming the densities are square integrable, the distance between them is measured by $D_2(p_\theta, p_{X,Z}) = \iint (p_{X,z}(x,z) - p_\theta(x,z))^2 \, dxdz$. For identical distributions this distance vanishes.

Hellinger distance Another variant to measure the distance is

$$H^2(p_\theta, p_{X,z}) = 2 \iint \left[\sqrt{p_\theta(x,z)} - \sqrt{p_{X,z}(x,z)} \right]^2 dxdz.$$

Jeffrey distance This is given by

$$J(p_\theta, p_{X,z}) = \frac{1}{2} \iint (p_\theta(x,z) - p_{X,z}(x,z)) (\ln p_\theta(x,z) - \ln p_{X,z}(x,z)) \, dxdz.$$

Renyi entropy For any $\alpha > 0$, $\alpha \neq 1$ define the *Renyi entropy* by

$$H_\alpha(p) = \frac{1}{\alpha} \ln \int p(x)^\alpha \, dx.$$

This generalizes the Shannon entropy, which is obtained as a limit, $H(p) = \lim_{\alpha \to 1} H_\alpha(p)$, see Exercise 3.15.9. A distinguished role is played by the *quadratic Renyi entropy*, which is obtained for $\alpha = 2$

$$H_2(p) = - \ln \int p(x)^2 \, dx.$$

3.10   Sample Estimation of Cost Functions

The practical utility of the aforementioned cost functions in machine learning resides in the ability of approximating them from a data sample, $\{(x_1, z_1), \ldots, (x_N, z_N)\}$. In the following we shall present a few of these estimations.

Mean squared error The expectation of the squared difference of the target, Z and the network outcome, $Y = f(X; \theta)$, can be estimated as the average

$$\mathbb{E}[(Z - f(X; \theta))^2] \approx \frac{1}{N} \sum_{j=1}^{N} (z_j - f(x_j; \theta))^2.$$

Quadratic Renyi entropy This estimation will use the Parzen window method [94]. We replace first the density $p(x)$ by a sample-based density using an window W_σ as

$$\hat{p}(x) = \frac{1}{N} \sum_{k=1}^{K} W_\sigma(x, x_k).$$

For simplicity reasons we assume the window as an one-dimensional Gaussian

$$W_\sigma(x, x_k) = \frac{1}{\sqrt{2\pi}\sigma} e^{-\frac{1}{2\sigma^2}|x-x_k|^2}.$$

Consider the quadratic potential energy $U(p) = \int p(x)^2\, dx$. Since the quadratic Renyi entropy is $H_2(p) = -\ln \int \hat{p}(x)^2\, dx = -\ln U(p)$, then it suffices to estimate $U(p)$. The estimation is given by

$$
\begin{aligned}
U(\hat{p}) &= \int \hat{p}(x)^2\, dx = \int \hat{p}(x)\hat{p}(x)\, dx \\
&= \int \frac{1}{N}\sum_{k=1}^{N} W_\sigma(x, x_k) \frac{1}{N}\sum_{j=1}^{N} W_\sigma(x, x_j)\, dx \\
&= \frac{1}{N^2}\sum_{k=1}^{N}\sum_{j=1}^{N} \int W_\sigma(x, x_k) W_\sigma(x, x_j)\, dx.
\end{aligned}
$$

In the case when the window is Gaussian, $W_\sigma(x, x') = \phi_\sigma(x - x')$, with $\phi_\sigma(t) = \frac{1}{\sqrt{2\pi}\sigma} e^{-\frac{t^2}{2\sigma^2}}$, the previous integral can be computed explicitly by changing the variable and transforming it into a convolution

$$
\begin{aligned}
\int W_\sigma(x, x_k) W_\sigma(x, x_j)\, dx &= \int \phi_\sigma(x - x_k)\phi_\sigma(x - x_j)\, dx \\
&= \int \phi_\sigma(t)\phi_\sigma(t - (x_j - x_k))\, dt \\
&= (\phi_\sigma * \phi_\sigma)(x_j - x_k) = \phi_{\sigma\sqrt{2}}(x_j - x_k) \\
&= W_{\sigma\sqrt{2}}(x_j, x_k).
\end{aligned}
$$

In the last equality we have used that the convolution of a Gaussian with itself is a scaled Gaussian, see Exercise 3.15.10. Substituting back into the quadratic potential energy, we arrive to the following estimation:

$$U(\hat{p}) = \frac{1}{N^2}\sum_{k=1}^{N}\sum_{j=1}^{N} W_{\sigma\sqrt{2}}(x_j, x_k).$$

Consequently, an estimation for the quadratic Renyi entropy is given by

$$H_2(\hat{p}) = -\ln\left(\frac{1}{N^2}\sum_{k=1}^{N}\sum_{j=1}^{N} W_{\sigma\sqrt{2}}(x_j, x_k)\right). \tag{3.10.7}$$

Integrated squared error If p_Z and p_Y represent the target and outcome densities, the cost function

$$C(p_Z, p_Y) = \int |p_Z(u) - p_Y(u)|^2 \, du$$

can be written using the quadratic potential energy as

$$C(p_Z, p_Y) = U(p_Z) + U(p_Y) - 2 \int p_Z(u) p_Y(u) \, du,$$

where $U(p)$ stands for the quadratic potential energy of density p defined before. Then the estimation takes the form

$$C(\hat{p}_Z, \hat{p}_Y) = U(\hat{p}_Z) + U(\hat{p}_Y) - 2 \int \hat{p}_Z(u) \hat{p}_Y(u) \, du.$$

The first two terms have been previously computed. It suffices to deal only with the integral term, called also the *Renyi cross-entropy*. We have

$$
\begin{aligned}
\int \hat{p}_Z(u) \hat{p}_Y(u) \, du &= \int \frac{1}{N} \sum_{j=1}^{N} W_\sigma(u, z_j) \frac{1}{N'} \sum_{k=1}^{N'} W_\sigma(u, y_j) \, du \\
&= \frac{1}{NN'} \sum_{j=1}^{N} \sum_{k=1}^{N'} \int W_\sigma(u, z_j) W_\sigma(u, y_j) \, du \\
&= \frac{1}{NN'} \sum_{j=1}^{N} \sum_{k=1}^{N'} W_{\sigma\sqrt{2}}(z_j, y_j) \\
&= \frac{1}{NN'} \sum_{j=1}^{N} \sum_{k=1}^{N'} W_{\sigma\sqrt{2}}(z_j, f(x_j; \theta)),
\end{aligned}
$$

where $y = f(x; \theta)$ is the input-output mapping of the neural net. Therefore, we obtain the following estimation

$$
\begin{aligned}
C(\hat{p}_Z, \hat{p}_Y) &= \frac{1}{N^2} \sum_{j=1}^{N} \sum_{k=1}^{N} W_{\sigma\sqrt{2}}(z_j, z_k) + \frac{1}{N'^2} \sum_{j=1}^{N'} \sum_{k=1}^{N'} W_{\sigma\sqrt{2}}(f(x_j; \theta), f(x_k; \theta)) \\
&\quad - \frac{2}{NN'} \sum_{j=1}^{N} \sum_{k=1}^{N'} W_{\sigma\sqrt{2}}(z_j, f(x_j; \theta)).
\end{aligned}
$$

Maximum Mean Discrepancy For all practical purposes, the random variable X is known from a sample of n observations, $x_1, \ldots, x_n$, which are drawn from the distribution $p(x)$. Similarly, the random variable Y is known from a sample of m observations, $y_1, \ldots, y_m$, drawn from the distribution $q(y)$. The means are estimated as averages by

$$\mathbb{E}_{X \sim p}[\phi(X)] = \frac{1}{n} \sum_{i=1}^{n} \phi(x_i),$$

$$\mathbb{E}_{Y \sim q}[\phi(Y)] = \frac{1}{m} \sum_{i=1}^{m} \phi(y_i),$$

and the maximum mean discrepancy between p and q can be estimated using the previous two samples as

$$
\begin{aligned}
d_{MMD}(p, q) &= \left\| \frac{1}{n} \sum_{i=1}^{n} \phi(x_i) - \frac{1}{m} \sum_{i=1}^{m} \phi(y_i) \right\|^2 \\
&= \left(\frac{1}{n} \sum_{i=1}^{n} \phi(x_i) - \frac{1}{m} \sum_{i=1}^{m} \phi(y_i) \right)^T \left(\frac{1}{n} \sum_{i=1}^{n} \phi(x_i) - \frac{1}{m} \sum_{i=1}^{m} \phi(y_i) \right) \\
&= \frac{1}{n^2} \sum_{i,j} K(x_i, x_j) + \frac{1}{m^2} \sum_{i,j} K(y_i, y_j) - \frac{2}{mn} \sum_{i,j} K(x_i, y_j),
\end{aligned}
$$

(3.10.8)

where we used the kernel notation $K(x, y) = \phi(x)^T \phi(y)$.

3.11   Cost Functions and Regularization

In order to avoid overfitting to the training data, the cost functions may include additional terms. It was noticed that the model overfits the training data if the parameters are allowed to take arbitrarily large values. However, if they are constrained to have bounded values, that would impede the model capability to pass through most of the training data points, and hence to prevent overfitting. Hence, the parameter values have to be kept small about the zero value. In order to minimize a cost function subject to small values of parameters, regularization terms of types L^1 or L^2 are usually used.

L^2-**regularization** Consider the weights parameter $w \in \mathbb{R}^n$. Its L^2-norm, $\|w\|_2$, is given by $\|w\|_2^2 = \sum_{i=1}^{n} w_i^2$. The cost function with L^2-regularization is obtained by adding the L^2-norm to the initial cost function

$$L_2(w) = C(w) + \lambda \|w\|_2^2,$$

where λ is a positive Lagrange multiplier, which controls the trade-off between the size of the weights and the minimum of $C(w)$. A large value of λ means a smaller value of the weights, and a larger value of $C(w)$. Similarly, a value of λ

closed to zero allows for large values of the weights and a smaller value of the cost $C(w)$, case that is prone to overfitting. The value of the hyperparameter λ should be selected such that the overfitting effect is minimized.

L^1-**regularization** The L^1-norm of $w \in \mathbb{R}^n$ is defined by $\|w\|_1 = \sum_{i=1}^{n} |w_i|$. The cost function with L^1-regularization becomes

$$L_1(w) = C(w) + \lambda \|w\|_1,$$

where λ is a Lagrange multiplier, with $\lambda > 0$, which controls the strength of our preference for small weights. Since $\|w\|_1$ is not differentiable at zero, the application of the usual gradient method might not work properly in this case. This disadvantage is not present in the case of the L^2-regularization.

Potential regularization This is a generalization of the previous two regularization procedures. Consider a function $U : \mathbb{R}^n \to \mathbb{R}_+$ satisfying

(*i*) $U(x) = 0$ if and only if $x = 0$

(*ii*) U has a global minimum at $x = 0$.

In the case when U is smoothly differentiable, condition (*ii*) is implied by the derivative conditions $U'(0) = 0$ and $U''(0) > 0$. The *potential function U* is a generalization of the aforementioned L^1 and L^2 norms.

The regularized cost function is defined now as

$$G(w) = C(w) + \lambda U(w), \qquad \lambda > 0.$$

Choosing the optimum potential with the best regularization properties for a certain cost function is done by verifying its performance on the test error (see for definition section 3.12). The test error has to decrease significantly when the term $U(w)$ is added to the initial cost function $C(w)$. More clearly, if $\epsilon_{test,1}$, $\epsilon_{test,2}$ are the test errors corresponding to the cost $C(w)$ and to the regularized cost $G(w)$, respectively, then U should be chosen such that

$$\epsilon_{test,2} < \epsilon_{test,1}.$$

In this case we say that the neural network generalizes well to new unseen data. We shall deal in more detail with this type of errors in the next section.

3.12   Training and Test Errors

One main feature that makes machine learning different than a regular optimization problem, which minimizes only one error, is the double optimization problem of two types of errors, which will be discussed in this section. It is the difference between these two errors that will determine how well a machine learning algorithm will perform.

In a machine learning approach the available data $\{(x_i, z_i)\}$ is divided into three parts: *training set*, *testing set* and *validation set*. It is assumed that all these sets are identically distributed, being generated by a common probability distribution; another assumption is that each of the aforementioned data sets are independent of each other. Size-wise, the largest of these is the training set $\mathcal{T}$ (about 70%), followed by the test set T (about 20%) and then by the validation set $\mathcal{V}$ (about 10%).

The cost function evaluated on the input data given by the training set is called *training error*. Similarly, the cost function evaluated on the input data given by the test set is called *test error* or *generalization error*. For instance, the errors given by

$$C_{\mathcal{T}}(w, b) = \frac{1}{m} \sum_{j=1}^{m} \big(f_{w,b}(x_j) - z_j\big)^2, \quad (x_j, z_j) \in \mathcal{T}$$

$$C_{\mathrm{T}}(w, b) = \frac{1}{k} \sum_{j=1}^{k} \big(f_{w,b}(x_j) - z_j\big)^2, \quad (x_j, z_j) \in \mathrm{T},$$

with $m = \mathrm{card}(\mathcal{T})$ and $k = \mathrm{card}(\mathrm{T})$ are the training error and the test error, respectively, associated with the average of the sum of squares. Similar training errors can be constructed for the other cost functions mentioned before.

In the first stage, an optimization procedure (such as the gradient descent) is used to minimize the training error $C_{\mathcal{T}}(w, b)$ by tuning parameters (w, b). Let's denote by (w^*, b^*) their optimal values. This procedure is called *training*.

In the second stage, we evaluate the test error at the previous optimal parameter values, obtaining the testing error $C_{\mathrm{T}}(w^*, b^*)$. In general, the following inequality is expected to hold:

$$C_{\mathcal{T}}(w^*, b^*) \leq C_{\mathrm{T}}(w^*, b^*).$$

The following few variants are possible:

(i) Both error values, $C_{\mathcal{T}}(w^*, b^*)$ and $C_{\mathrm{T}}(w^*, b^*)$, are small. This means that the network generalizes well, i.e., performs well not only on the training set, but also on new, unseen yet inputs. This would be the desired scenario of any machine learning algorithm.

(ii) The training error $C_{\mathcal{T}}(w^*, b^*)$ is small, while the test error $C_{\mathrm{T}}(w^*, b^*)$ is still large. In this case the network does not generalize well. It actually

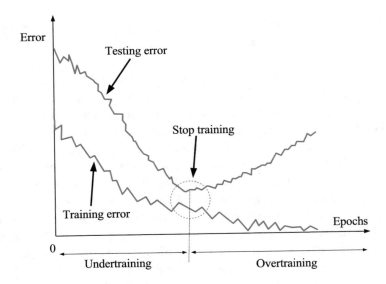

Figure 3.2: *Overtraining and undertraining regions for a neural network. Since the training and testing errors are stochastic the optimal stopping is a random variable which is mostly contained in the dotted circle.*

overfits the training set. If the training error is zero, then the network is "memorizing" the training data into the system of weights and biases. In this case a regularization technique needs to be applied. If even after this the test error does not get smaller, probably the network architecture has to be revised, one way being to decrease the number of parameters. In general, over-parametrization of the network usually leads to overfitting.

(*iii*) Both error values, $C_{\mathcal{T}}(w^*, b^*)$ and $C_{\mathrm{T}}(w^*, b^*)$, are large. In this case the neural network underfits the training data. To fix this issue, we need to increase the capacity of the network by changing its architecture into a network with more parameters.

The validation set is used to tune hyperparameters (learning rate, network depth, etc.) such that the smallest validation error is obtained. Finding the optimal hyperparameters is more like an art rather than science, depending on the scientist's experience, and we shall not deal with it here.

One other important issue to discuss here is *overtraining* and *optimal stopping* from training. By training for a long enough time (involving a large enough number of epochs) the training error can be made, in general, as small as possible (provided the network capacity is large enough). In the beginning of training period the testing error also decreases until a point, after which it starts increasing slowly. The optimal stopping time from training is when the testing error reaches its minimum. After that instance the gap between the training error and testing error gets larger, fact that leads to an overfit, namely the network performs very well on the training data but less well on the testing data, see Fig. 3.2.

In fact, the training and testing errors are not deterministic in practice. The use of the stochastic gradient descent method (or its variants) assures that errors are stochastic. If (w_t, b_t) are the parameter values at the time step t, then we consider the stochastic process $C_t = C_{\mathcal{T}}(w_t, b_t)$, which is the sequence of the training errors as a function of time steps. It has been determined experimentally that this process has a decreasing trend, approaching zero, with a nonvanishing variance, see Fig. 3.2. One simple way to model this error is to assume the recurrence[2] of first order $C_{t+1} = \phi C_t + \alpha \epsilon_t$, with $0 < \phi < 1$, and $\alpha \epsilon_t$ an white noise term controlled by the parameter α. This says that the value of the error at step $t+1$ is obtained from the value of the error at step t by shrinking it by a factor of ϕ, and then add some noise.

The process can be transformed into a difference model as $C_{t+1} - C_t = -(1-\phi)C_t dt + \alpha \epsilon_t$. Considering the time step small, this becomes a stochastic differential equation, $dC_t = -rC_t + \alpha dW_t$, where $r = 1 - \phi > 0$ and the white noise ϵ_t was replaced by increments of a Brownian motion, dW_t, which are normally distributed, see section D.8 in the Appendix. The solution to this equation is the *Ornstein-Uhlenbeck* process

$$C_t = C_0 e^{-rt} + \alpha \int_0^t e^{-r(t-u)} \, dW_u.$$

This is the sum of two terms. The first is the mean of the process, which shows an exponentially decreasing trend. The second is a Wiener integral,[3] which is a random variable normally distributed, with mean zero and variance $\frac{\alpha^2}{2r}(1 - e^{-2rt})$. The model parameters r and α depend on the network hyperparameters, such as learning rate, batch size, number of epochs, etc.

Example 3.12.1 There are several data sets on which machine learning algorithms are usually tested for checking their efficiency. A few of these examples are glanced in the following.
(*i*) The MNIST data set of handwritten digits (from 0 to 9) consists of 60,000 training and 10,000 test examples. Each image has 28×28 pixels. Usually 5,000 images from the training examples are used as validation set. Feedforward and convolutional networks can be trained on this data.

---

[2]In statistics this is called an autoregresive AR(1) model.

[3]The reader can think of the integral $I_t = \int_0^t f(u) \, dW_u$ as a random variable I_t obtained as the mean square limit of the partial sums $S_n = \sum_{i=0}^{n-1} f(u_i)(W_{u_{i+1}} - W_{u_i})$. Its distribution is normal, given by $I_t \sim \mathcal{N}(0, \int_0^t f(u)^2 \, du)$.

(*ii*) The CIFAR-10 data set contains 50,000 training and 10,000 test images and has 10 categories of images (airplanes, cars, birds, cats, deer, dogs, frogs, horses, ships, and trucks). Each one is a 32×32 color image. Usually 5,000 images are kept for validation. State-of-the-art results on the CIFAR-10 dataset have been achieved by convolutional deep belief networks, densely connected convolutional networks and others.

(*iii*) The CIFAR-100 data set is similar with the CIFAR-10, the only difference being that it has 100 image categories.

(*iv*) The Street View House Numbers (SVHN) is a real-world image data set containing approximately 600,000 training images and 26,000 test images. 6,000 examples out of the training set are used as validation set. SVHN is obtained from house numbers in Google Street View images. This data can be processed using convolutional networks, sparse autoencoders, recurrent convolutional networks, etc.

3.13 Geometric Significance

If $C(w,b)$ and $U(w)$ are both smooth, then the regularized cost function $G(w,b) = C(w,b) + \lambda U(w)$ is also smooth, and its minimum is realized for

$$(w^*, b^*) = \arg\min(C(w,b) + \lambda U(w)).$$

At this minimum point we have the vanishing gradient condition satisfied, $\nabla G(w^*, b^*) = 0$, which easily implies $\nabla_w C(w^*, b^*) = -\lambda \nabla_w U(w^*)$. This means that the normal vectors to the level surfaces of $C(w,b)$ and $U(w)$ are collinear. This occurs when the level surfaces are tangent, see Fig. 3.3. Since the normal vectors to the previous level surfaces are collinear and of opposite directions, it follows that $\lambda > 0$.

The significance of λ In order to understand better the role of the multiplier λ, we shall assume for simplicity that $U(w) = \|w\|_2^2$. Let w^* be a contact point between the level curves $\{C(w,b) = k\}$ and $\{U(w) = c\}$, see Fig. 3.3. Then the equation $\nabla_w C(w^*, b^*) = -\lambda \nabla_w U(w^*)$ becomes $\nabla_w C(w^*, b^*) = -2\lambda w^*$. This implies

$$\|\nabla_w C(w^*, b^*)\| = 2\lambda\sqrt{c},$$

that is, the magnitude of the normal vector to the level surface $\{C(w,b) = k\}$ at w^* depends on λ and c. The following remarks follow:

(*i*) Assume $\nabla_w C(w^*, b^*) \neq 0$. Then small (large) values of c correspond to large (small) values of λ and vice versa. Equivalently, small values of λ correspond to large values of weights w, and large values of λ correspond to small values of w.

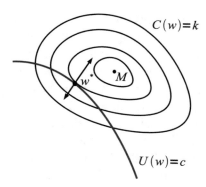

Figure 3.3: *The cost function $C(w, b)$ has a minimum at point M; its level surfaces are given by $\mathcal{S}_k = \{C(w, b) = k\}$. One of these level surfaces is tangent to the surface $\{U(w) = c\}$ at a point where the gradients ∇C and ∇U are collinear.*

(*ii*) Assume $\nabla_w C(w^*, b^*) = 0$. Then either $\lambda = 0$, or $c = 0$. The condition $c = 0$ is equivalent to $w = 0$, which implies that $w^* = 0$, namely the cost function $C(w, b)$ has a (global) minimum at $w = 0$. Since most cost functions do not have this property, we conclude that in this case $\lambda = 0$.

The role of the constraint Let $w^* = \arg\min_{U(w) \leq c} C(w, b)$ and assume $w^* \neq 0$. There are two cases:

(**a**). $w^* \in \{w; U(w) \leq c\}$. This corresponds to the case when c is large enough such that the minimum of $C(w, b)$ is contained into the interior of the level surface $\{U(w) = c\}$. See Fig. 3.4 **a**. In this case we choose $\lambda = 0$.

(**b**). $w^* \notin \{w; U(w) \leq c\}$. This describes the case when c is small enough such that the minimum of $C(w, b)$ is outside of the closed domain $D_c = \{U(w) \leq c\}$. In this case the minimum of $C(w, b)$ over the domain D_c is realized on the domain boundary, along the level surface $\{U(w) = c\}$, at a point w^{**} through which passes a level curve of $C(w, b)$ tangent to $\{U(w) = c\}$, see Fig. 3.4 **b**. The unknowns, $w^{**} = (w_1, \ldots, w_n)^T$ and λ, satisfy the equations

$$\partial_{w_k} C(w^{**}) = -\lambda \partial_{w_k} U(w^{**}), \quad U(w^{**}) = c.$$

In order to obtain an approximation for the solution of the previous equation, we minimize $G(w, b) = C(w, b) - \lambda U(w)$ using the gradient descend method. The hyperparameter λ is tuned such that the minimum of $G(w)$ is the lowest.

Overfitting removal We have seen that in the case when the training error $C_\mathcal{T}(w, b)$ is small while the test error $C_\mathrm{T}(w, b)$ is large, we deal with an overfit.

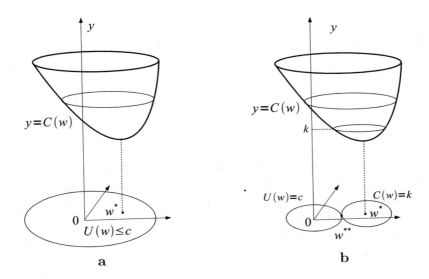

Figure 3.4: *The cost function $C(w,b)$ in two distinguished possibilities:* **a.** $w^* \in \{w; U(w) \leq c\}$ **b.** $w^* \notin \{w; U(w) \leq c\}$.

One way to try to fix the situation is to require the weight vector w to be small. In this case both test and training errors are close to each other. The argument follows from the fact that if we let $w \to 0$, then the output variable is $Y = f_{w,b}(X) \approx f_{0,b}(X) = \phi(b^{(L)})$, where ϕ is the activation function and $b^{(L)}$ is the bias vector of the neurons on the last layer. We note that $\phi(b^{(L)})$ is independent of the input X, either if it is of test or of training type. If the cost function is given by the cross-entropy between the densities of the model output Y and target Z, then the test and the training errors have approximately the same value because

$$C_{\mathcal{T}}(0,b) = S_{\mathcal{T}}(p_Y, p_Z) \approx S_{\mathrm{T}}(p_Y, p_Z) = C_{\mathrm{T}}(0,b).$$

We have also used that (X, Z) for both test and training samples are drawn from the same underlying distribution. We can formalize this mathematically by minimizing the cost function $C(w,b)$ subject to the constraint $U(w) \leq c$, with $c > 0$ small enough, as explained in the previous sections.

3.14 Summary

During the learning process, neural networks try to match their outputs to given targets. This approximation procedure involves the use of cost functions, which measure the difference between what it is predicted (output) and what it is desired (target). These proximities are of several types, depending of what the network tries to learn: a function, a random variable, a probability density, etc. Some of these cost functions are real distance functions (i.e., they are nonnegative, symmetric, and satisfy the triangle inequality), while others are not, such as the KL-divergence or the cross-entropy. However, they all measure the departure of the output from the target by a nondegenerate nonnegative function.

Sometimes, for overfitting removal purposes, the cost functions are augmented with regularization terms. These terms depend on the weights and have the task of minimizing the cost function subject to weights of small magnitude. The coefficient of the regularization term is a hyperparameter, which controls the trade-off between the size of the weights and the minimum of the cost function. The value of this hyperparameter is obtained using a validation set. This set is independent of the training and test sets.

The available data is split into three parts that are used for the following purposes: training, testing, and validation. When the cost function is computed using data from the training set and gets optimized, the training error is obtained. When testing data is used, the test error is obtained. A small training error and a large test error is a sign of overfitting. A large training error signals an underfit. The purpose of regularization is to obtain a lower test error.

3.15   Exercises

Exercise 3.15.1 Let p, p_i, q, q_i be density functions on $\mathbb{R}$ and $\alpha \in \mathbb{R}$. Show that the cross-entropy satisfies the following properties:

(a) $S(p_1 + p_2, q) = S(p_1, q) + S(p_2, q)$;

(b) $S(\alpha p, q) = \alpha S(p, q) = S(p, q^\alpha)$;

(c) $S(p, q_1 q_2) = S(p, q_1) + S(p, q_2)$.

Exercise 3.15.2 Show that the cross-entropy satisfies the following inequality:

$$S(p, q) \geq 1 - \int p(x) q(x) \, dx.$$

Exercise 3.15.3 Let p be a fixed density. Show that the symmetric relative entropy

$$D_{KL}(p\|q) + D_{KL}(q\|p)$$

reaches its minimum for $p = q$, and the minimum is equal to zero.

Exercise 3.15.4 Consider two exponential densities, $p_1(x) = \xi^1 e^{-\xi^1 x}$ and $p_2(x) = \xi^2 e^{-\xi^2 x}$, $x \geq 0$.

(a) Show that $D_{KL}(p_1||p_2) = \frac{\xi^2}{\xi_1} - \ln \frac{\xi^2}{\xi_1} - 1$.

(b) Verify the nonsymmetry relation $D_{KL}(p_1||p_2) \neq D_{KL}(p_2||p_1)$.

(c) Show that the triangle inequality for D_{KL} does not hold for three arbitrary densities.

Exercise 3.15.5 Let X be a discrete random variable. Show the inequality $H(X) \geq 0$.

Exercise 3.15.6 Prove that if p and q are the densities of two discrete random variables, then $D_{KL}(p||q) \leq S(p, q)$.

Exercise 3.15.7 We assume the target variable Z is $\mathcal{E}$-measurable. What is the mean squared error function value in this case?

Exercise 3.15.8 Assume that a neural network has an input-output function $f_{w,b}$ linear in w and b. Show that the cost function (3.3.1) reaches its minimum for a unique pair of parameters, (w^*, b^*), which can be computed explicitly.

Exercise 3.15.9 Show that the Shannon entropy can be retrieved from the Renyi entropy as

$$H(p) = \lim_{\alpha \to 1} H_\alpha(p).$$

Exercise 3.15.10 Let $\phi_\sigma(t) = \frac{1}{\sqrt{2\pi}\sigma} e^{-\frac{t^2}{2\sigma^2}}$. Consider the convolution operation $(f * g)(u) = \int f(t)g(t - u) \, dt$.

(a) Prove that $\phi_\sigma * \phi_\sigma = \phi_{\sigma\sqrt{2}}$;

(b) Find $\phi_\sigma * \phi_{\sigma'}$ in the case $\sigma \neq \sigma'$.

Exercise 3.15.11 Consider two probability densities, $p(x)$ and $q(x)$. The Cauchy-Schwartz divergence is defined by

$$D_{CS}(p, q) = -\ln \left(\frac{\int p(x)q(x) \, dx}{\sqrt{\int p(x)^2 \, dx \int q(x)^2 \, dx}} \right).$$

Show the following:

(a) $D_{CS}(p, q) = 0$ if and only if $p = q$;

(b) $D_{CS}(p, q) \geq 0$;

(c) $D_{CS}(p, q) = D_{CS}(q, p)$;

(d) $D_{CS}(p, q) = -\ln \int pq - \frac{1}{2}H_2(p) - \frac{1}{2}H_2(q)$, where $H_2(\cdot)$ denotes the quadratic Renyi entropy.

Exercise 3.15.12 (a) Show that for any function $f \in L^1[0,1]$ we have the inequality $\|\tanh f\|_1 \leq \|f\|_1$.

(b) Show that for any function $f \in L^2[0,1]$ we have the inequality $\|\tanh\|_2 \leq \|f\|_2$.

Exercise 3.15.13 Consider two distributions on the sample space $\mathcal{X} = \{x_1, x_2\}$ given by

$$p = \begin{pmatrix} x_1 & x_2 \\ \frac{1}{2} & \frac{1}{2} \end{pmatrix}, \qquad q = \begin{pmatrix} x_1 & x_2 \\ \frac{1}{3} & \frac{2}{3} \end{pmatrix}.$$

Consider the function $\phi : \mathcal{X} \to \mathbb{R}^2$ defined by $\phi(x_1) = (0,1)$ and $\phi(x_2) = (1,0)$. Find the maximum mean discrepancy between p and q.

Chapter 4

Finding Minima Algorithms

The learning process in supervised learning consists of tuning the network parameters (weights and biases) until a certain cost function is minimized. Since the number of parameters is quite large (they can easily be into thousands), a robust minimization algorithm is needed. This chapter presents a number of minimization algorithms of different flavors, and emphasizes their advantages and disadvantages.

4.1 General Properties of Minima

This section reviews basic concepts regarding minima of functions having a real variable or several real variables. These theoretically feasible techniques are efficient in practice only if the number of variables is not too large. However, in machine learning the number of variables is into thousands or more, so these classical theoretical methods for finding minima are not lucrative for these applications. We include them here just for completeness, and to have a basis to build on the next more sophisticated methods.

4.1.1 Functions of a real variable

A well-known calculus result states that any real-valued continuous function $f : [a, b] \to \mathbb{R}$, defined on a compact interval is bounded and achieves its bounds within the interval $[a, b]$; then there is (at least) a value $c \in [a, b]$ such that $f(c) = \min_{x \in [a,b]} f(x)$. This is a *global minimum* for $f(x)$. However, the function might also have local minima. Furthermore, if the (local or global) minimum value is reached inside the interval, i.e., $c \in (a, b)$, and if the function is differentiable, then Fermat's theorem states that $f'(c) = 0$, i.e., the

© Springer Nature Switzerland AG 2020
O. Calin, *Deep Learning Architectures*, Springer Series in the Data Sciences,
https://doi.org/10.1007/978-3-030-36721-3_4

derivative vanishes at that value. Geometrically, this means that the tangent line to the graph of $y = f(x)$ at the point $(c, f(c))$ is horizontal. However, this condition is necessary but not sufficient. If the function has the additional property to be convex (i.e. to satisfy $f''(x) \geq 0$), the aforementioned condition becomes also sufficient.

4.1.2   Functions of several real variables

Assume K is a compact set in $\mathbb{R}^n$, i.e., it is a closed and bounded set. For instance, $K = [a_1, b_1] \times \cdots \times [a_n, b_n]$, with $a_i, b_i \in \mathbb{R}$, or $K = \{x; \|x\| \leq R\}$, with $R > 0$. If $f : K \to \mathbb{R}$ is a continuous function, then there is a point $c \in K$ such that $f(c) = \min_{x \in K} f(x)$, i.e., the function has a global minimum. Similarly with the one-dimensional case, if the (global or local) minimum c is in the interior of K (i.e., if there is possibility to center a ball with small enough radius at c, which is contained in K), then the following system of partial differential equations holds:[1]

$$\frac{\partial f}{\partial x_i}(c) = 0, \qquad i = 1, \ldots, n. \tag{4.1.1}$$

This can be written equivalently in the gradient notation as $\nabla f(c) = 0$, where $\nabla f = \sum_{i=1}^n \frac{\partial f}{\partial x_i} e_i$ is a vector with components given by the partial derivatives. We have denoted $e_i = (0, \ldots, 1, \ldots, 0)^T$, the unit vector pointing the ith Cartesian direction (we used ^{T} for the transpose notation). The system (4.1.1) is equivalent with the condition that the tangent plane at $(c, f(c))$ to the surface $z = f(x)$ is horizontal, i.e., parallel to the x-hyperplane.

The second-order Taylor approximation of $f(x)$ in a neighborhood of $x = c$ is given by

$$f(x) = f(c) + \sum_i \frac{\partial f}{\partial x_i}(c)(x_i - c_i)$$

$$+ \frac{1}{2} \sum_{i,j} \frac{\partial^2 f}{\partial x_j \partial x_k}(c)(x_j - c_j)(x_k - c_k) + o(\|x - c\|^2)$$

$$= f(c) + (x - c)^T \nabla f(c) + \frac{1}{2}(x - c)^T H_f(c)(x - c) + o(\|x - c\|^2),$$

where $H_f = \dfrac{\partial^2 f}{\partial x_j \partial x_k}$ is the Hessian matrix of f. Assume the following:

(*i*) c satisfies the equation $\nabla f(c) = 0$;
(*ii*) The Hessian is positive definite and nondegenerate, i.e.,

$$v^T H_f v > 0, \quad \forall v \in \mathbb{R}^n \backslash \{0\}.$$

---

[1]This is sometimes called the Euler system of equations.

Heuristically speaking, if H_f is positive definite, then for x close enough to the critical point c, we may neglect the quadratic term in the Taylor expansion and obtain $f(x) > f(c)$, for $x \neq c$. This means that c is a local minimum for f. In the following we shall supply a more formal argument for supporting this statement.

Proposition 4.1.1 *Let c be a solution of $\nabla f(c) = 0$ and assume that H_f is positive definite in a neighborhood of c. Then c is a local minimum of f.*

Proof: Let $x(t)$ be an arbitrary fixed curve in the domain of f with $x(0) = c$, and consider the composite function $g(t) = f(x(t))$. To show that c is a local minimum for $f(x)$ is equivalent to prove that $t = 0$ is a local minimum for $g(t)$, for any curve $x(t)$ with the aforementioned properties.

Let $v = x'(0)$ be the velocity vector along the curve $x(t)$ at $t = 0$. Then the directional derivative of f in the direction u is given by

$$D_u f(c) = \lim_{t \to 0} \frac{f(x(t)) - f(c)}{t} = \lim_{t \to 0} \frac{g(t) - g(0)}{t} = g'(0).$$

On the other side, an application of the chain rule yields

$$D_u f(c) = \sum \frac{\partial f}{\partial x_k}(x(t)) x'_k(t)_{|t=0} = \langle \nabla f(c), v \rangle = v^T \nabla f(c) = 0.$$

From the last two relations it follows that $g'(0) = 0$. In order to show that $t = 0$ is a local minimum for $g(t)$, it suffices to prove the inequality $g''(0) > 0$ and then apply the second derivative test to the function $g(t)$ at $t = 0$. The desired result will follow from the positive definiteness of the Hessian H_f and an application of the formula $g''(0) = u^T H_f(c) u$, which will be proved next.

Iterating the formula $D_u f = u^T \nabla f$, we have

$$
\begin{aligned}
g''(0) &= D_u^2 f(c) = D_u(D_u)(c) = D_u(u^T \nabla f)(c) = D_u\left(\sum_k \partial_{x_k} f \, u_k\right)(c) \\
&= \sum_k \partial_{x_k}(D_u f)(c) \, u_k = \sum_k \partial_{x_k}\left(\sum_j \partial_{x_j} f \, u_j\right)(c) \, u_k \\
&= \sum_{j,k} (\partial_{x_j x_k} f)(c) u_j u_k = u^t H_f(c) u.
\end{aligned}
$$

Therefore, $g'(0) = 0$ and $g''(0) > 0$, and hence c is a local minimum of f. ∎

Example 4.1.2 (Positive definite Hessian in two dimensions) Consider a twice differentiable function, f, with continuous derivatives on $\mathbb{R}^2$. Its Hessian is given by the 2×2 matrix

$$H_f(x, y) = \begin{pmatrix} f_{xx} & f_{xy} \\ f_{yx} & f_{yy} \end{pmatrix},$$

where we used notation $f_{xx} = \dfrac{\partial^2 f}{\partial x^2}$ to denote the double partial derivative in x. For any vector $u = (a, b) \in \mathbb{R}^2$, a straightforward matrix multiplication provides the quadratic form

$$u^T H_f u = f_{xx}a^2 + 2f_{xy}ab + f_{yy}b^2,$$

which after completing the square, becomes

$$\begin{aligned} u^T H_f u &= f_{xx}\left(a + \frac{f_{xy}}{f_{xx}}b\right)^2 + \frac{b^2}{f_{xx}}(f_{xx}f_{yy} - f_{xy}^2) \\ &= f_{xx}\left(a + \frac{f_{xy}}{f_{xx}}b\right)^2 + \frac{b^2}{f_{xx}}\det H_f. \end{aligned}$$

The following concluding remarks follow from the previous computation:

(*i*) If $f_{xx} > 0$ and $\det H_f > 0$, then $u^T H_f u > 0$ for any $u \in \mathbb{R}^2$, i.e., the Hessian H_f is positive definite.

(*ii*) If $f_{xx} < 0$ and $\det H_f > 0$, then $u^T H_f u < 0$ for any $u \in \mathbb{R}^2$, i.e., the Hessian H_f is negative definite.

Example 4.1.3 (Quadratic functions) Let A be a symmetric, positive definite, and nondegenerate $n \times n$ matrix and consider the following real-valued quadratic function of n variables

$$f(x) = x^T A x - 2b^T x + d, \qquad x \in \mathbb{R}^n,$$

with $b \in \mathbb{R}^n$, $d \in \mathbb{R}$, where $x^T A x = \sum_{i,j} a_{ij}x_i x_j$ and $b^T x = \sum_k b_k x_k$. We have the gradient $\nabla f(x) = 2Ax - 2b$ and the Hessian $H_f = 2A$, so the function is convex. The solution of $\nabla f(x) = 0$, which is $c = A^{-1}b$, is a local minimum for f. Since the solution is unique, it follows that it is actually a global minimum. The invertibility of A follows from the properties of the Hessian, whose determinant is nonzero.

If the previous quadratic function $f(x)$ is defined just on a compact set K, sometimes the solution $c = A^{-1}b$ might not belong to K, and hence we cannot obtain all minima of f by solving the associated Euler system. On the other side, we know that f achieves a global minimum on the compact domain K. Hence, this minimum must belong to the boundary of K, and finding it requires a boundary search for the minimum.

Example 4.1.4 (Harmonic functions) A function $f(x)$ is called *harmonic* on the domain $D \subset \mathbb{R}^n$ if $\Delta_x f = 0$, $\forall x \in D$, where $\Delta_x f = \sum_i \dfrac{\partial^2 f(x)}{\partial x_i^2}$ is the Laplacian of f. The minimum (maximum) property of the Laplacian states that a harmonic function achieves its minima (maxima) on the boundary of the domain D. In other words, the extrema of a harmonic function are reached always on the boundary of the domain.

Since any affine function $f(x) = b^T x + d$, $b \in \mathbb{R}^n$, $d \in \mathbb{R}$ is harmonic, the minima (maxima) of f are reached on the boundary of D. If D is the interior of a convex polygon, then the minima (maxima) are achieved at the polygon vertices. Looking for the solution vertex is the basic idea of the simplex algorithm.

We have seen that minima cannot be always found by solving the system $\nabla f(x) = 0$. And even if it were possible, this analytical way of finding the minima is not always feasible in practice due to the large number of variables involved. In the following we shall present some robust iterative methods used to approximate the minimum of a given function.

4.2 Gradient Descent Algorithm

The gradient descent algorithm is a procedure of finding the minimum of a function by navigating through the associated level sets into the direction of maximum cost decrease. We shall present first the ingredient of level sets. The reader who is not interested in daunting details can skip directly to section 4.2.2.

4.2.1 Level sets

Consider the function $z = f(x)$, $x \in D \subset \mathbb{R}^n$, with $n \geq 2$ and define the set $\mathcal{S}_c = f^{-1}(\{c\}) = \{x \in \mathbb{R}^n; f(x) = c\}$. Assume the function f is differentiable with nonzero gradient, $\nabla f(x) \neq 0$, $x \in \mathcal{S}_c$. Under this condition, $\mathcal{S}_c$ becomes an $(n-1)$-dimensional hypersurface in $\mathbb{R}^n$. The family $\{\mathcal{S}_c\}_c$ is called the *level hypersurfaces* of the function f. For $n = 2$ they are known under the name of *level curves*. Geometrically, the level hypersurfaces are obtained intersecting the graph of $z = f(x)$ with horizontal planes $\{z = c\}$, see Fig. 4.1.

Proposition 4.2.1 *The gradient ∇f is normal to $\mathcal{S}_c$.*

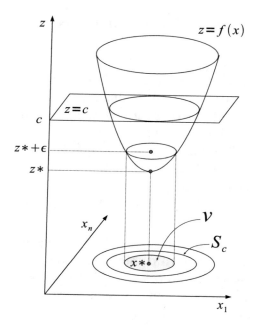

Figure 4.1: *Level sets for the function* $z = f(z)$.

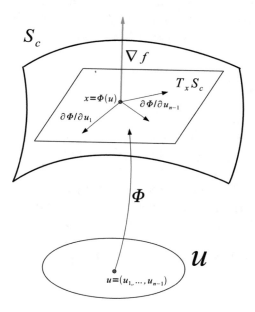

Figure 4.2: *Local parametrization* $\Phi : \mathcal{U} \to \mathbb{R}^n$ *of the hypersurface* $\mathcal{S}_c$.

Proof: It suffices to show the relation locally, in a coordinates chart. Consider a local parametrization of $\mathcal{S}_c$ given by

$$x^1 = \Phi^1(u_1, \ldots, u_{n-1})$$
$$\cdots = \cdots$$
$$x^n = \Phi^n(u_1, \ldots, u_{n-1}),$$

which depends on $n-1$ local parameters $u = (u_1, \ldots, u_{n-1}) \in \mathcal{U} \subset \mathbb{R}^{n-1}$, with $\Phi = (\Phi^1, \ldots, \Phi^n) : \mathcal{U} \subset \mathbb{R}^{n-1} \to \mathbb{R}^n$, having the Jacobian of rank $n-1$. The tangent vector fields to $\mathcal{S}_c$ are given by $\{\frac{\partial \Phi}{\partial u_1}, \ldots, \frac{\partial \Phi}{\partial u_{n-1}}\}$. The rank condition on Φ implies that they are linearly independent, and hence they span the tangent space at each point of $\mathcal{S}_c$. We shall show that the gradient vector ∇f is normal to each of these tangent vectors. This follows from taking the derivative with respect to u_i in the relation

$$f\big(\Phi(u_1, \ldots, u_{n-1})\big) = c, \quad \forall u \in \mathcal{U},$$

and applying the chain rule

$$0 = \frac{\partial}{\partial u_i} f(\Phi(u)) = \sum_k \frac{\partial f}{\partial x_k}\bigg|_{\Phi(u)} \frac{\partial \Phi^k(u)}{\partial u_i} = \langle \nabla f|_{\Phi(u)}, \frac{\partial \Phi}{\partial u_i} \rangle,$$

where $\langle \, , \, \rangle$ denotes the Euclidean inner product. This shows that the gradient ∇f is normal to the tangent space at each point $\Phi(u)$. $\blacksquare$

In equivalent notations, for each $x \in \mathcal{S}_c$, the vector $\nabla f(x)$ is normal to the tangent plane, $T_x \mathcal{S}_c$, of $\mathcal{S}_c$ at x, see Fig. 4.2. The orientation of the hypersurface is chosen such that ∇f points into the outward direction. The tangent plane, $T_x \mathcal{S}_c$, acts as an infinitesimal separator for the points about x which are inside and outside of $\mathcal{S}_c$, see Fig. 4.3 a.

Assume now that the function $z = f(x)$ has a (local) minimum at $x^* \in D$, i.e., $f(x^*) < f(x)$, for all $x \in \mathcal{V} \backslash \{x^*\}$, with $\mathcal{V}$ neighborhood of x^* (we may assume that $\mathcal{V}$ is a ball centered at x^*). Denote by $z^* = f(x^*)$ the local minimum value of f at x^*. Then there is an $\epsilon > 0$ such that $\mathcal{S}_c \subset \mathcal{V}$ for any $c \in [z^*, z^* + \epsilon)$, see Fig. 4.1. For $c = z^*$ the hypersurface degenerates to a point, $\mathcal{S}_c = \{x^*\}$. For small enough ϵ the family $\{\mathcal{S}_c\}_{c \in [z^*, z^* + \epsilon)}$ is nested, i.e., if $c_1 < c_2$ then $\mathcal{S}_{c_1} \subset \mathrm{Int}(\mathcal{S}_{c_2})$.

The next result states the existence of curves of arbitrary initial direction, emanating from x^*, which evolve normal to the family $\mathcal{S}_c$, see Fig. 4.3 **b**. Recall that a function $\phi(x)$ is called *Lipschitz continuous* if there is a constant $K > 0$ such that $|\phi(x) - \phi(y)| \le K\|x - y\|$, for all x and y in the domain of ϕ. The following two existence results will use this assumption.

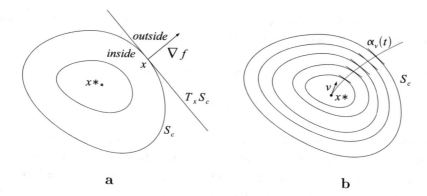

Figure 4.3: **a.** *Gradient ∇f normal to $\mathcal{S}_c$.* **b.** *The curve $\alpha_v(t)$ starting at x^* with initial velocity v.*

Lemma 4.2.2 *Assume ∇f is Lipschitz continuous. For any vector $v \in \mathbb{R}^n$, there is $\delta > 0$ and a differentiable curve $\alpha : [0, \delta) \to \mathbb{R}^n$ such that:*

(i) $\alpha(0) = x^*$;

(ii) $\dot{\alpha}(0) = v$;

(iii) $\dot{\alpha}(t)$ is normal to $\mathcal{S}_{f(\alpha(t))}$, for all $t \in [0, \delta)$.

Proof: Let $x^* = (x_1^*, \ldots, x_n^*)$, $v^T = (v_1, \ldots, v_n)$, and $\phi^k(x) = \frac{\partial f(x)}{\partial x_k}$. It suffices to show the existence of the curve components. Picard-Lindelöf theorem of existence and uniqueness of solutions states that the nonlinear system with initial conditions

$$
\begin{aligned}
\dot{\alpha}^k(t) &= \phi^k(\alpha(t)) \\
\alpha^k(0) &= x_k^* \\
\dot{\alpha}^k(0) &= v^k.
\end{aligned}
$$

can be solved locally, with the solution $\alpha^k : [0, \delta_k) \to \mathbb{R}$. Set $\delta = \min_k \delta_k$ and consider the curve $\alpha(t) = (\alpha^1(t), \ldots, \alpha^n(t))$, with $0 \leq t < \delta$. The curve obviously satisfies conditions (i) and (ii). Since $\dot{\alpha}(t)$ has the direction of the gradient, ∇f, it follows that $\alpha(t)$ is normal to the family $\mathcal{S}_{f(\alpha(t))}$, i.e., condition (iii) holds. ∎

For future reference, when we would like to indicate also the initial direction v, the curve constructed as a solution of the aforementioned ODEs system will be denoted by $\alpha_v(t)$.

It is worth noting that the curve is unique up to a reparametrization, i.e., the curve can change speed, while its geometric image remains the same.

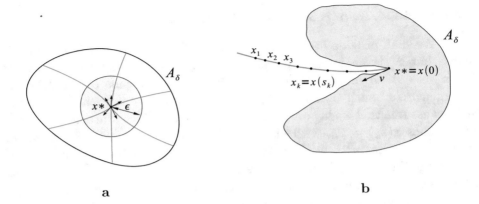

Figure 4.4: **a.** *There is a ball $B(x^*, \epsilon)$ included in A_δ.* **b.** *The sequence $(x_k)_k$ tends to x^* along the curve $x(s)$.*

One distinguished parametrization is over the level difference parameter $\tau = c - z^*$. This follows from the fact that for small values of t we have $\dot{c}(t) > 0$. If the curve in the new parametrization is denoted by $\beta(\tau)$, then $\beta(0) = x^* = \alpha(0)$ and $\beta'(\tau) = \dot{\alpha}(t(\tau))t'(\tau)$, so $\beta'(0) = t'(0)v$. We also have the convenient incidence relation $\beta(\tau) = \beta(c - z^*) \in \mathcal{S}_c$, which can be written also as $f(\beta(\tau)) = c$.

Now we state and prove the following local connectivity result, which will be used later in the gradient descent method:

Theorem 4.2.3 *Assume ∇f is Lipschitz continuous. For any point x^0 close enough to x^* there is a differentiable curve $\gamma : [0, \delta] \to \mathbb{R}^n$ such that*

(i) $\gamma(0) = x^0$;

(ii) $\gamma(\delta) = x^*$;

(iii) $\dot{\gamma}(s)$ is normal to $\mathcal{S}_{f(\gamma(s))}$, for all $s \in [0, \delta)$.

We shall provide a nonconstructive proof. We prepare first the ground with a few notations. Let $\alpha_v : [0, \delta_v) \to \mathbb{R}^n$ be the curve provided by Lemma 4.2.2. Assume the initial vector sub-unitary, $\|v\| \leq 1$. Since the end value δ_v is continuous with respect to v, it reaches its minimum on the unitary ball as $\delta = \min_{\|v\| \leq 1} \delta_v$. Denote now by $A_\delta = \{\alpha_v(t); t \in [0, \delta), \forall v \in \mathbb{R}^n, \|v\| \leq 1\}$. The set A_δ denotes the solid domain swapped by the curves $\alpha_v(t)$, which emanate from x^* into all directions v, until time δ. The value of δ was chosen such that the definition of A_δ makes sense. The set A_δ is nonempty, since obviously, $x^* \in A_\delta$. The next result states that in fact A_δ contains a nonempty ball centered at x^*, see Fig. 4.4 **a**. This is actually a result equivalent to the statement of Theorem 4.2.3.

Theorem 4.2.4 *There is an $\epsilon > 0$ such that $B(x^*, \epsilon) \subset A_\delta$.*

Proof: We shall provide first an empirical proof. The fact that $B(x^*, \epsilon) \subset A_\delta$ means that x^* is an interior point of A_δ. If, by contradiction, we assume that x^* is not an interior point, then there is a sequence of points $(x_k)_k$ convergent to x^* such that $x_k \notin A_\delta$. Here is where the empirical assumption is made: assume that x_k lay on a smooth curve $x(s)$, which starts at $x(0) = x^*$ and satisfies $x(s_k) = x_k$, with s_k decreasing sequence of negative numbers, see Fig. 4.4 **b**. Let $v^0 = x'(0)$ be the direction under which the curve $x(s)$ approaches the point x^*. Lemma 4.2.2 produces a curve α_{v^0} starting into this direction, which will coincide with $x(s)$ on a neighborhood. This follows from the fact that both curves $x(s)$ and α_{v^0} have the same initial points and velocities. Since in this case we would have $x(s) = \alpha_{v^0}(-s) \in A_\delta$, for $-s < \delta$, this leads to a contradiction. ∎

We make two remarks:

1. All points that can be joined to x^* by a curve γ satisfying the afore-mentioned properties form the *basin of attraction* of x^*. The theorem can be stated equivalently by saying that the basin of attraction contains a ball centered at x^*.

2. The direction $v^0 = x'(0)$ is a degenerate direction. The previous proof shows that there are no degenerate directions. A formal proof of this fact can be done using the Inverse Function Theorem (see Theorem F.1 in Appendix) as in the following.

Proof: Consider $\mathbb{B}^n = \{v \in \mathbb{R}^n; \|v\| \leq 1\}$ and define the function $F : \mathbb{B}^n \to A_\delta$ by $F(v) = \alpha_v(\delta)$. The existence of degenerate directions is equivalent to the fact that the Jacobian $\dfrac{\partial F}{\partial v}$ is degenerate. This follows from the Inverse Function Theorem, which states that if the Jacobian $\dfrac{\partial F}{\partial v}$ is nondegenerate, then F is a local diffeomorphism, i.e., there is a neighborhood $\mathcal{V}$ of $F(0) = x^*$ in A_δ and an $0 < \rho < 1$ such that $F_{|\mathbb{B}(0,\rho)} : \mathbb{B}(0, \rho) \to \mathcal{V}$ is a diffeomorphism.[2] For details, see section F.1 of Appendix. Consequently, $x^* \in F(\mathbb{B}(0,\rho)) \subset \mathcal{V} \subset A_\delta$. We may choose a ball centered at x^* of radius ϵ such that $\mathbb{B}(x^*, \epsilon) \subset (\mathbb{B}(0, \rho))$, which ends the proof.

So, it suffices to show that the Jacobian of F is nondegenerate. Note that $\dfrac{\partial F}{\partial v}$ is an $n \times n$ squared matrix. The degeneracy is equivalent with the existence of a nonzero vector $w = v - v^0$ such that

$$\frac{\partial F}{\partial v}(v - v^0) = 0.$$

---

[2]This means that $F_{|\mathbb{B}(0,\rho)}$ is bijective with both F and its inverse differentiable.

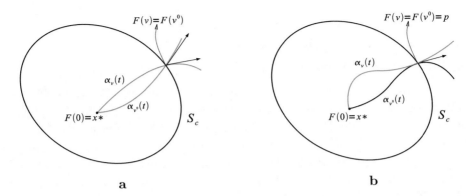

Figure 4.5: *The existence of a cut-point:* **a.** *The curves* α_v *and* α_{v^0} *are transversal.* **b.** *The curves* α_v *and* α_{v^0} *are tangent.*

Using the linear approximation

$$F(v) = F(v^0) + \frac{\partial F}{\partial v}(v - v^0) + o(\|v - v^0\|^2)$$

we obtain that $F(v) = F(v^0) + o(\|v - v^0\|^2)$, i.e. we have $\alpha_v(\delta) = \alpha_{v^0}(\delta) + o(\|v - v^0\|^2)$. Neglecting the quadratic term, we shall assume for simplicity that $\alpha_v(\delta) = \alpha_{v^0}(\delta)$, i.e., there is a "cut-point" where two curves with distinct initial velocities, v and v^0, intersect again. We need to show that the flow $\{\alpha_v\}_v$ is free of cut points on a neighborhood of x^*. By contradiction, assume there are cut points in any neighborhood, $\mathcal{N}$, of x^*. Let p be one of them, so $p = \alpha_v(\delta) = \alpha_{v^0}(\delta) \in \mathcal{S}_c$, with $c = f(p)$. Assume the curves are transversal at the cut-point p, see Fig. 4.5 **a**. Since $\dot{\alpha}_v(\delta)$ and $\dot{\alpha}_{v^0}(\delta)$ are both normal to $\mathcal{S}_c$, the curves have to be tangent at p. Now, by a similar procedure as in Lemma 4.2.2, we have two curves intersecting at the same point, p, and having the same direction, see Fig. 4.5 **b**, so they have to coincide on a neighborhood $\mathcal{N}_2$. Choosing $\mathcal{N}_1 \subset \mathcal{N}_2$, it follows that the curves coincide, which is contradictory. ■

Proof of Theorem 4.2.3. The proof of Theorem 4.2.3 follows by choosing x^0 in the ball $B(x^*, \epsilon)$ provided by Theorem 4.2.4. ■

The previous nonconstructive proof provides the existence of a curve of *steepest descent*, γ, from x^0 to x^*, which intersects normally the level sets family $\mathcal{S}_c$. However, this continuous result is not useful when it comes to computer implementations. In order to implement the curve construction we need to approximate the curve γ by a polygonal line $\mathfrak{P}_m = [x^0 x^1 \ldots x^m]$ satisfying the following properties:

(i) $x^k \in \mathcal{S}_{c_k}$.

(ii) $c_{k+1} < c_k$, for all $k = 0, \ldots, m - 1$;

(iii) the line $x^j x^{j+1}$ is normal to $\mathcal{S}_{c_j}$.

The construction algorithm goes as in the following. We start from the point x^0 and go for a distance η along the normal line at $\mathcal{S}_{c_0}$ into the inward direction, or equivalently, in the direction of $-\nabla f$ at point x^0. Thus, we obtain the point $x^1 \in \mathcal{S}_{c_1}$. We continue by going again a distance η along the normal line at $\mathcal{S}_{c_1}$ into the inwards direction, obtaining the point x^2. After m steps we obtain the point x^m, which we hope to be in a proximity of x^*, and hence a good approximation of it.

But how do we choose m, or in other words, how do we know when to stop the procedure? The algorithm continues as long as $c_{k+1} < c_k$, i.e., the landing hypersurfaces are nested. For any a priori fixed $\eta > 0$ there is a smallest m with properties (i)–(iii). This means that we stop when x^{m+1} lands on a hypersurface $\mathcal{S}_{c_{m+1}}$ that is not nested inside of the previous hypersurface $\mathcal{S}_{c_m}$. The smaller the step η, the larger the stopping order m, and the closer to x^* it is expected to get. When $\eta \to 0$, the polygonal line $\mathfrak{P}_m$ tends toward the curve γ.

If the algorithm stops after m steps, the upper and lower error bounds are given by

$$\|x^0 - x^*\| - m\eta \leq \|x^m - x^*\| \leq \mathrm{dia}(\mathcal{S}_{c_m}), \qquad (4.2.2)$$

where $\mathrm{dia}(\mathcal{S}_c)$ denotes the diameter of $\mathcal{S}_c$, i.e., the largest distance between any two elements of $\mathcal{S}_c$. It also makes sense to consider a bounded number of steps with $m < \dfrac{\|x^0 - x^*\|}{\eta}$.

The left inequality follows from the fact that any polygonal line is longer than the line segment joining its end points

$$
\begin{aligned}
\|x^0 - x^*\| &\leq \|x^0 - x^1\| + \|x^1 - x^2\| + \cdots + \|x^{m-1} - x^m\| + \|x^m - x^*\| \\
&= m\eta + \|x^m - x^*\|.
\end{aligned}
$$

This becomes identity if the polygonal line is a straight line, a case which occurs if the hypersurfaces are hyperspheres centered at x^*.

The inequality on the right of (4.2.2) follows from the construction of x^m, which belongs to $\mathcal{S}_{c_m}$. Therefore, we have the estimations

$$\|x^* - x^m\| \leq \max_{y \in \mathcal{S}_{c_m}} \|x^* - y\| \leq \sup_{x,y \in \mathcal{S}_{c_m}} \|x - y\| = \mathrm{dia}(\mathcal{S}_{c_m}).$$

The shrinking condition $\mathcal{S}_{c_m} \to \{x^*\}$ allows for diameters $\mathrm{dia}(\mathcal{S}_{c_m})$ as small as possible, provided η is small enough.

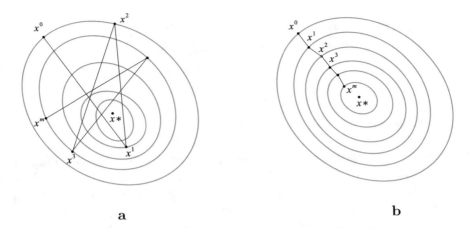

a   **b**

Figure 4.6: *The polygonal line* $\mathfrak{P}_m = [x^0, \ldots, x^m]$ *in two cases:* **a.** *Step* η *large.* **b.** *Step* η *small.*

We have a few comments regarding the size of η:

(i) if η is large, the algorithm stops too early, before reaching a good approximation of the point x^*, see Fig. 4.6 **a**;

(ii) if η is too small, the stopping order m is large and it might not be time effective in the case of a computer implementation, see Fig. 4.6 **b**.

In practical applications, the size of the step η is a trade-off between the margin of error and time effectiveness of a running application. We shall formalize this idea further in section 4.2.4.

4.2.2 Directional derivative

Another concept used later is the *directional derivative*, which measures the instantaneously the rate of change of a function at a point in a given direction. More precisely, let v be a unitary vector in $\mathbb{R}^n$ and consider the differentiable function $f : \mathcal{U} \subset \mathbb{R}^n \to \mathbb{R}$. The directional derivative of f at the point $x^0 \in \mathcal{U}$ is defined by

$$\frac{\partial f}{\partial v}(x^0) = \lim_{t \searrow 0} \frac{f(x^0 + tv) - f(x^0)}{t}.$$

Note that partial derivatives with respect to coordinates, $\frac{\partial f}{\partial x_k}$, are directional derivatives with respect to the coordinate vectors $v = (0, \ldots, 1, \ldots, 0)^T$. An application of chain rule provides a computation of the directional derivative as a scalar product:

$$\frac{\partial f}{\partial v}(x^0) = \frac{d}{dt} f(x^0 + tv)\Big|_{t=0+} \quad = \sum_{k+1}^{n} \frac{\partial f}{\partial x_k}(x^0 + tv)\, v^k \Big|_{t=0}$$

$$= v^T \nabla f(x^0) = \langle \nabla f(x^0), v \rangle.$$

4.2.3   Method of Steepest Descent

The *method of steepest descent* (or, *gradient descent method*) is a numerical method based on a greedy algorithm by which a minimum of a function is searched by directing a given step into the direction that decreases the most of the value of the function. One can picture the method by considering a blindfolded tourist who would like to get down a hill in the fastest possible fashion. At each point the tourist is checking the proximity to find the direction with the steepest descent and then make one step in that direction. Then the procedure repeats, until the tourist will eventually reach the bottom of the valley (or, get stuck in a local minimum, if his step is too small).

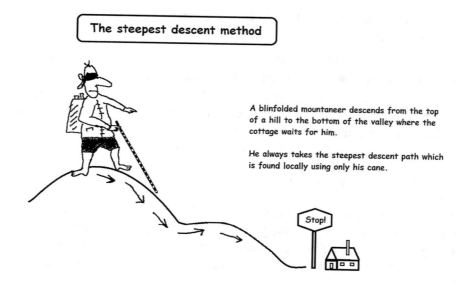

Cartoon 1: A blindfolded man gets off a mountain using the steepest descent method by taking advantage of the local geometry of the environment.

In order to apply this method, we are interested in finding the unitary direction v, in which the function f decreases as much as possible within a given small step size η. The change of the function f between the value at the initial point x^0 and the value after a step of size η in the direction v is written using the linear approximation as

$$f(x^0 + \eta v) - f(x^0) = \sum_{k=1}^{n} \frac{\partial f}{\partial x_k}(x^0)\eta v^k + o(\eta^2)$$

$$= \eta \langle \nabla f(x^0), v \rangle + o(\eta^2).$$

In the following, since η is small, we shall neglect the effect of the quadratic term $o(\eta^2)$. Hence, in order to obtain v such that the change in the function has the largest negative value, we shall use the Cauchy inequality for the scalar product

$$-\|\nabla f(x^0)\| \, \|v\| \leq \langle \nabla f(x^0), v \rangle \leq \|\nabla f(x^0)\| \, \|v\|.$$

It is known that the inequality on the left is reached for vectors that are negative proportional.[3] Since $\|v\| = 1$, the minimum occurs for

$$v = -\frac{\nabla f(x^0)}{\|\nabla f(x^0)\|}.$$

Then the largest change in the function is approximately equal to

$$f(x^0 + \eta v) - f(x^0) = \eta \langle \nabla f(x^0), v \rangle = -\eta \|\nabla f(x^0)\|.$$

The constant η is called the *learning rate*. From the previous relation, the change in the function after each step is proportional to the magnitude of the gradient as well as to the learning rate.

The algorithm consists of the following iteration that constructs the following sequence (x^n):

(*i*) Choose an initial point x^0 in the basin of attraction of the global minimum x^*.

(*ii*) Construct the sequence $(x^n)_n$ using the iteration

$$x^{n+1} = x^n - \eta \frac{\nabla f(x^n)}{\|\nabla f(x^n)\|}. \tag{4.2.3}$$

This guarantees a negative change of the objective function, which is given by $f(x^{n+1}) - f(x^n) = -\eta \|\nabla f(x^n)\| < 0$.

We note that the line $x^n x^{n+1}$ is normal to the level hypersurface $\mathcal{S}_{f(x^n)}$. Hence, we obtain the polygonal line $\mathfrak{P}_m = [x^0, \ldots, x^m]$ from the previous section, which is an approximation of the curve γ provided by Theorem 4.2.3.

---

[3]This is more transparent in the case of $\mathbb{R}^3$, when $\langle \nabla f(x^0), v \rangle = \|\nabla f(x^0)\| \, \|v\| \cos\theta$. The minimum is realized for $\theta = \pi$, i.e. when the vectors have opposite directions.

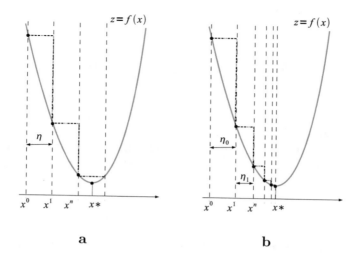

Figure 4.7: **a.** *The use of a fixed learning rate η leads to missing the minimum x^*.* **b.** *An adjustable learning rate η_n provides a much better approximation of the minimum. In this case the descent amount is proportional with the slope of the curve at the respective points; these slopes become smaller as we approach the critical point x^*.*

However, this construction has a drawback, which will be fixed shortly. Since $\|x^{n+1} - x^n\| = \eta > 0$, the approximation sequence $(x^n)_n$ does not converge, so it is easy to miss the minimum point x^*, see Fig. 4.7 **a**. To overcome this problem, we shall assume that the learning rate η is adjustable, in the sense that it becomes smaller as the function changes slower (when the gradient is small), see Fig. 4.7 **b**. We assume now there is a positive constant $\delta > 0$ such that the learning rate in the nth iteration is proportional with the gradient, $\eta_n = \delta \|\nabla f(x^n)\|$. Then the iteration (4.2.3) changes into

$$x^{n+1} = x^n - \delta \nabla f(x^n). \qquad (4.2.4)$$

Proposition 4.2.5 *The sequence $(x^n)_n$ defined by (4.2.4) is convergent if and only if the sequence of gradients converges to zero, $\nabla f(x^n) \to 0$, $n \to \infty$.*

Proof: " $\Longrightarrow$ " Since the sequence is convergent,

$$0 = \lim_{n \to \infty} \|x^{n+1} - x^n\| = \delta \lim_{n \to \infty} \|\nabla f(x^n)\|,$$

which leads to the desired result.

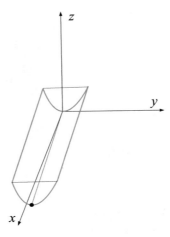

Figure 4.8: *The graph of $z = \frac{1}{2}y^2 - x$.*

" $\Longleftarrow$ " In order to show that $(x^n)_n$ is convergent, it suffices to prove that x^n is a Cauchy sequence. Let $p \geq 1$. An application of triangle inequality yields

$$\|x^{n+p} - x^n\| \leq \|x^{n+p} - x^{n+p-1}\| + \cdots + \|x^{n+1} - x^n\| = \delta \sum_{j=0}^{p} \|\nabla f(x^{n+j})\|.$$

Keeping p fixed, we have

$$\lim_{n\to\infty} \|x^{n+p} - x^n\| \leq \delta \sum_{j=0}^{p} \lim_{n\to\infty} \|\nabla f(x^{n+j})\| = 0.$$

∎

It is worth noting that if f is continuously differentiable, since $x^n \to x^*$, then $\nabla f(x^n) \to \nabla f(x^*)$. This agrees with the condition $\nabla f(x^*) = 0$.

Example 4.2.1 Consider the function $f : (0,1) \times (-2,2) \to \mathbb{R}$, given by $f(x,y) = \frac{1}{2}y^2 - x$. Its graph has a canyon-type shape, with the minimum at the point $(1,0)$, see Fig. 4.8. Let (x^0, y^0) be a fixed point in the function domain. Since the gradient is given by $\nabla f(x,y)^T = (\frac{\partial f}{\partial x}, \frac{\partial f}{\partial y}) = (-1, y)$, the equation (4.2.4) writes as

$$
\begin{aligned}
x^{n+1} &= x^n + \delta \\
y^{n+1} &= (1 - \delta)y^n.
\end{aligned}
$$

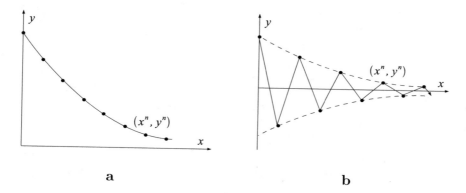

Figure 4.9: *Iterations:* **a.** *Case* $0 < \delta < 1$. **b.** *Case* $1 < \delta < 2$.

This iteration can be solved explicitly in terms of the initial point (x^0, y^0). We get

$$
\begin{aligned}
x^n &= n\delta + x^0 \\
y^n &= (1 - \delta)^n y^0.
\end{aligned}
$$

The sequence $(y^n)_n$ converges for $|1 - \delta| < 1$, i.e., for $0 < \delta < 2$. There are two distinguished cases, which lead to two distinct behaviors of the iteration:

(i) If $0 < \delta < 1$, the sequence $(y^n)_n$ converges to 0 keeping a constant sign (the sign of y^0). The sequence (x^n) is an arithmetic progression with the step equal to the learning rate δ. The iteration stops at the value $n = \lfloor \frac{1-x^0}{\delta} \rfloor$, where $\lfloor x \rfloor$ denotes the floor of x, i.e., the largest integer smaller or equal to x. See Fig. 4.9 **a**.

(ii) If $1 < \delta < 2$, the sequence $(y^n)_n$ converges to 0 in an oscillatory manner. This corresponds to the situation when the iteration ascends the canyon walls overshooting the bottom of the canyon, see Fig. 4.9 **b**.

4.2.4   Line Search Method

This is a variant of the method of steepest descent with an adjustable learning rate η. The rate is chosen as in the following. Starting from an initial point x^0, consider the normal direction on the level hypersurface $\mathcal{S}_{f(x^0)}$ given by the gradient $\nabla f(x^0)$. We need to choose a point, x^1, on the line given by this direction at which the objective function f reaches a minimum. This is equivalent with choosing the value $\eta_0 > 0$ such that

$$
\eta_0 = \arg\min_{\eta} f(x^0 - \eta \nabla f(x^0)). \tag{4.2.5}
$$

The procedure continues with the next starting point $x^1 = x^0 - \eta_0 \nabla f(x^0)$. By an iteration we obtain the sequence of points (x^n) and the sequence of learning rates (η_n) defined recursively by

$$\eta_n = \arg\min f(x^n - \eta \nabla f(x^n))$$
$$x^{n+1} = x^n - \eta_n \nabla f(x^n).$$

The method of line search just described has the following geometric significance. Consider the function

$$g(\eta) = f(x^0 - \eta \nabla f(x^0)),$$

and differentiate it to get

$$g'(\eta) = -\langle \nabla f(x^0 + \eta \nabla f(x^0)), \nabla f(x^0) \rangle. \tag{4.2.6}$$

If η_0 is chosen to realize the minimum (4.2.5), then $g'(\eta_0) = 0$, which implies via (4.2.6) that $\nabla f(x^1)$ and $\nabla f(x^0)$ are normal vectors. This occurs when the point x^1 is obtained as the tangent contact between the line $\{x^0 - \eta \nabla f(x^0)\}$ and the level hypersurface $\mathcal{S}_{f(x^1)}$, see Fig. 4.10.

In general, the algorithm continues as in the following: consider the normal line at x^n to the hypersurface $\mathcal{S}_{f(x^n)}$ and pick x^{n+1} to be the point where this line is tangent to a level surface. This algorithm produces a sequence which converges to x^* much faster than in the case of the steepest descent method. Note that the polygonal line $[x^0, x^1, x^2, \dots]$ is infinite and has right angles.

Before getting any further, we shall provide some examples.

Example 4.2.6 Consider the objective function $f(x) = \frac{1}{2}(ax - b)^2$ of a real variable $x \in \mathbb{R}$, with real coefficients $a \neq 0$ and b. It is obvious that its minimum is given by the exact formula $x^* = b/a$. We shall show that we arrive to the same expression applying the steepest descent method.

The gradient in this case is just the derivative $f'(x) = a^2 x - ab$. Starting from an initial point x^0, we construct the approximation sequence

$$x^{n+1} = x^n - \delta f'(x^n) = (1 - \delta a^2)x^n + \delta ab.$$

Denote $\alpha = 1 - \delta a^2$ and $\beta = \delta ab$. The linear recurrence $x^{n+1} = \alpha x^n + \beta$ can be solved explicitly in terms of x^0 as

$$x^n = \alpha^n x^0 + \frac{1 - \alpha^n}{1 - \alpha}\beta.$$

For δ small enough, $0 < \delta < \frac{1}{a^2}$, we have $0 < \alpha < 1$, which implies that $\alpha^n \to 0$, as $n \to \infty$ and hence the sequence x^n is convergent with the aforementioned desired limit

$$x^* = \lim_{n \to \infty} x^n = \frac{\beta}{1 - \alpha} = \frac{b}{a}.$$

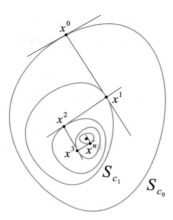

Figure 4.10: *The approximation of the minimum x^* by the method of line search.*

Example 4.2.7 This is an extension of the previous example to several variables. Consider the function $f : \mathbb{R}^k \to \mathbb{R}$ defined by $f(x) = \frac{1}{2}\|Ax - b\|^2$, where $b \in \mathbb{R}^m$ and A is an $m \times k$ matrix of rank k, and $\|\cdot\|$ denotes the Euclidean norm.

We note that the minimum x^* satisfies the linear system $Ax^* = b$, since $f(x) \geq 0$ and $f(x^*) = 0$. Multiplying to the left by A^T and then inverting, we obtain the exact form solution $x^* = (A^T A)^{-1} A^T b$. The existence of the previous inverse is provided by the fact that the $k \times k$ square matrix $A^T A$ has a maximum rank, since $\text{rank}(A^T A) = \text{rank} A = k$. The solution x^* is called the *Moore-Penrose pseudoinverse* and some of its algebraic and geometric significance can be found in section G.2 of the Appendix.

We shall apply next the method of steepest descent and show that we obtain the same previously stated exact solution, i.e., the Moore-Penrose pseudoinverse. Since

$$
\begin{aligned}
f(x) &= \frac{1}{2}\|Ax - b\|^2 = \frac{1}{2}(Ax - b)^T (Ax - b) \\
&= \frac{1}{2}\Big(\langle A^T Ax, x\rangle - 2\langle A^T b, x\rangle + \|b\|^2\Big),
\end{aligned}
$$

its gradient is given by $\nabla f(x) = A^T Ax - A^T b$. The approximation sequence can be written as

$$
\begin{aligned}
x^{n+1} &= x^n - \delta \nabla f(x^n) \\
&= x^n - \delta (A^T A x^n - A^T b) \\
&= (\mathbb{I}_k - \delta A^T A) x^n + \delta A^T b \\
&= M x^n + \delta A^T b,
\end{aligned}
$$

where $M = \mathbb{I}_k - \delta A^T A$, and $\mathbb{I}_k$ denotes the unitary matrix. Since the matrix $\mathbb{I}_k - M = \delta A^T A$ is invertible, an iteration provides

$$
\begin{aligned}
x^n &= M^n x^0 + (M^{n-1} + M^{n-2} + \cdots + M + \mathbb{I}_k) \delta A^T b \\
&= M^n x^0 + (\mathbb{I}_k - M^n)(\mathbb{I}_k - M)^{-1} \delta A^T b \\
&= M^n \big(x^0 - (A^T A)^{-1} A^T b\big) + (A^T A)^{-1} A^T b.
\end{aligned}
$$

This implies the limit

$$
x^* = \lim_{n \to \infty} x^n = (A^T A)^{-1} A^T b,
$$

provided we are able to show $\lim_{n \to \infty} M^n = 0$. For this to occur, it suffices to show that all the eigenvalues $\{\lambda_i\}$ of M are real and bounded, with $|\lambda_i| < 1$. This can be shown as in the following. Since $M = M^T$, all its eigenvalues are real and we have the decomposition $M = VDV^{-1}$, with D diagonal matrix, having the eigenvalues $\{\lambda_i\}$ along the diagonal. The nth power of M, given by $M^n = VD^nV^{-1}$, converges to the zero matrix, provided $D^n \to 0$, which occurs if $|\lambda_i| < 1$. We shall show that δ can be tuned such that this condition holds for all eigenvalues. This will be done in two steps:

Step 1. Show that $\lambda_i < 1$.
As an eigenvalue of M, λ_i satisfies the equation $\det(M - \lambda_i \mathbb{I}_k) = 0$. Substituting for M, this becomes $\det \big((1 - \lambda_i)\mathbb{I}_k - \delta A^T A\big) = 0$, which is equivalent to $\det \big(A^T A - \frac{1 - \lambda_i}{\delta} \mathbb{I}_k\big) = 0$. This means that $\eta_i = \frac{1 - \lambda_i}{\delta}$ is an eigenvalue of matrix $A^T A$. Since $A^T A$ is positive definite and nondegenerate, it follows that $\eta_i > 0$, which implies that $\lambda_i < 1$.

Step 2. Show that for δ small enough we have $\lambda_i \geq 0$.
The operator $F : \mathbb{R}^n \to \mathbb{R}^m$, defined by $F(w) = Aw$ is linear and continuous, and hence, it is bounded. Therefore, there is a constant $K > 0$ such that $\|Aw\| \leq K\|w\|$, for all $w \in \mathbb{R}^n$. Choosing $\delta \leq \frac{1}{K^2}$, we have

$$
\delta \|Aw\|^2 \leq \|w\|^2, \quad \forall w \in \mathbb{R}^n.
$$

This can be written as $\delta w^T A^T A w \leq w^T w$, or equivalently, $w^T(\mathbb{I}_k - \delta A^T A)w \geq 0$. This means $\langle Mw, w \rangle \geq 0$ for all $w \in \mathbb{R}^n$, i.e., the matrix M is nonnegative definite, and hence $\lambda_i \geq 0$.

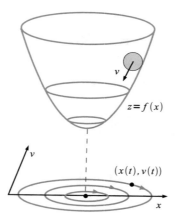

Figure 4.11: *A ball rolling downhill without friction and its trajectory representation in the phase space.*

4.3   Kinematic Interpretation

This section deals with the kinematic interpretation of the method of steepest descent. Consider a ball of mass $m = 1$ which rolls downhill toward the bottom of a valley, without friction. The state of the ball is described by the pair $s = (x, v)$, where x and v denote the coordinate and the velocity of the ball, respectively. The space of states s is called the *phase space*. The dynamics of the ball can be traced in the phase space as a curve $s(t) = (x(t), v(t))$, where $x(t)$ is its coordinate and $v(t) = \dot{x}(t)$ is its velocity at time t, see Fig. 4.11.

     The geometry of the valley is modeled by a convex function $z = f(x)$. We consider two types of energy acting on the ball:

- the kinetic energy due to movement: $E_k = \frac{1}{2}\|v\|^2$;
- the potential energy due to altitude: $E_p = f(x)$.

     The ball's dynamics is described by the classical Lagrangian, which is the difference between the kinetic and potential energies:

$$L(x, \dot{x}) = E_c - E_p = \frac{1}{2}\|\dot{x}\|^2 - f(x).$$

The equation of motion is given by the Euler-Lagrange variational equation $\frac{d}{dt}\left(\frac{\partial L}{\partial \dot{x}}\right) = \frac{\partial L}{\partial x}$, which in the present case becomes

$$\ddot{x}(t) = -\nabla f(x(t)). \tag{4.3.7}$$

This can be regarded as the Newton law of motion of a unit mass ball under a force derived from the potential $f(x)$.

The *total energy* of the ball at time t is defined as the sum between the kinetic and potential energies

$$E_{tot}(t) = E_k(t) + E_p(t) = \frac{1}{2}\|\dot{x}(t)\|^2 + f(x(t)). \qquad (4.3.8)$$

Applying (4.3.7) yields

$$
\begin{aligned}
\frac{d}{dt}E_{tot}(t) &= \frac{d}{dt}\left(\frac{1}{2}\dot{x}(t)^T\dot{x}(t) + f(x(t))\right) \\
&= \dot{x}(t)^T\ddot{x}(t) + \dot{x}(t)^T\nabla f(x) \\
&= \dot{x}(t)^T\left(\ddot{x}(t) + \nabla f(x(t))\right) = 0,
\end{aligned}
$$

i.e., the total energy is preserved along the trajectory. We can use this observation to construct level sets as in the following. Consider a function defined on the phase space by

$$H(x, v) = \frac{1}{2}\|v\|^2 + f(x), \qquad (4.3.9)$$

called *Hamiltonian function*. Define the energy levels given by the hypersurfaces $\mathcal{S}_c$ associated with the function H by

$$\mathcal{S}_c = H^{-1}(\{c\}) = \{(x, v); H(x, v) = c\}.$$

The previous computation shows that any solution trajectory in the phase space, $(x(t), v(t))$, belongs to one of these level hypersurfaces.

An equivalent way to look at equation (4.3.7) is to write it as a first-order system of ODEs

$$
\begin{aligned}
\dot{x}(t) &= v(t) & (4.3.10) \\
\dot{v}(t) &= -\nabla f(x(t)). & (4.3.11)
\end{aligned}
$$

Given some initial conditions, $x(0) = x^0$, $v(0) = v^0$, standard ODE results provide the existence of a unique local solution $(x(t), v(t))$ starting at the point (x^0, v^0).

In order to find the minimum of the potential function $z = f(x)$, it suffices to find the stable equilibrium point of the ball, which is achieved at the bottom of the valley. If the ball is placed at this point, with zero velocity, it will stay there forever. This point is an *equilibrium point* for the ODE system (4.3.10)–(4.3.11) and corresponds in the phase space to an energy level which is degenerated to a point. Setting $\dot{x}(t) = 0$ and $\dot{v}(t) = 0$ in the previous ODE system provides the equilibrium point (x^*, v^*) given by

$$v^* = 0, \quad \nabla f(x^*) = 0. \qquad (4.3.12)$$

Finding the equilibrium state $s^* = (x^*, v^*)$ can be achieved by applying the method of steepest descent to the energy level hypersurfaces $\mathcal{S}_c$ in the phase space. Theorem 4.2.3 provides the existence of a steepest descent curve joining the initial state of the ball $s_0 = (x^0, v^0)$ (initial position and velocity) to the equilibrium state $s^* = (x^*, v^*)$, provided s_0 is close enough to s^*.[4] The iteration (4.2.4) becomes

- Set an initial state $s_0 = (x^0, v^0)$;
- Construct recursively the sequence of states

$$s_{n+1} = s_n - \delta \nabla_s H(s_n), \qquad \forall n \geq 0. \tag{4.3.13}$$

Using the expression of the Hamiltonian function (4.3.9), its gradient becomes

$$\nabla_s H(s) = \left(\frac{\partial H(x, v)}{\partial x}, \frac{\partial H(x, v)}{\partial v} \right) = (\nabla f(x), v).$$

Hence, the expression (4.3.13) can now be written on components as

$$\begin{aligned} x^{n+1} &= x^n - \delta \nabla f(x^n) \\ v^{n+1} &= (1 - \delta) v^n. \end{aligned}$$

We have obtained two separated equations, which can be solved independently. The second one has the closed-form solution $v^n = (1 - \delta)^n v^0$, and for δ small, we have $v^n \to v^* = 0$, $n \to \infty$. The first equation is nothing but the iteration (4.2.4). Hence, besides a nice kinematic interpretation, this approach does not provide any improvement over the method described in section 4.2.3. In order to make an improvement we need to introduce a friction term, fact that will be done in the next section. The qualitative difference between these two approaches will be discussed in the following.

The solutions flow, $(x(t), v(t))$, satisfies the system (4.3.10)–(4.3.11). This can be written in terms of the Hamiltonian function, equivalently, as

$$\dot{x}(t) = \frac{\partial H}{\partial v} \tag{4.3.14}$$

$$\dot{v}(t) = -\frac{\partial H}{\partial x}. \tag{4.3.15}$$

The tangent vector field to the solutions flow is defined by $X = \dot{x}(t) \frac{\partial}{\partial x} + \dot{v}(t) \frac{\partial}{\partial v}$. Using (4.3.14)–(4.3.15) we compute its divergence

$$\operatorname{div} X = \frac{\partial}{\partial x} \frac{\partial H}{\partial v} - \frac{\partial}{\partial v} \frac{\partial H}{\partial x} = 0.$$

---

[4] This proximity condition can be waived if f is a convex function.

Since the divergence of a vector field represents the rate at which the volume evolves along the flow curves, the previous relation can be interpreted by saying that the solutions flow is incompressible, i.e., any given volume of particles preserves its volume during the evolution of the dynamical system, see Fig. 4.12 **a**. In fact, all Hamiltonian flows (solutions of the system (4.3.14)–(4.3.15) for any smooth function H) have zero divergence. Consequently, if the ball starts rolling from some initial neighboring states, then at any time during the system evolution the states are in the same volume proximity, without the possibility of converging to any equilibrium state. A ball which rolls downhill without friction in a convex cup will never stop at the bottom of the cup; it will continue to bounce back and forth on the cup wall passing, without stopping, near the equilibrium point infinitely many times.

4.4 Momentum Method

In order to avoid getting stuck in a local minimum of the cost function, several methods have been designed. The basic idea is that shaking the system by adding extra velocity or energy will make the system to pass over the energy barrier and move into a state of lower energy.[5]

We have seen that the gradient descent method can be understood by considering the physical model of a ball rolling down into a cup. The position of the ball is updated at all times into the negative direction of the gradient by a given step, which is the learning rate.

The *momentum method* modifies the gradient descent by introducing a velocity variable and having the gradient modify the velocity rather than the position. It is the change in velocity that will affect the position. Besides the learning rate, this technique uses an extra hyperparameter, which models the friction, which reduces gradually the velocity and has the ball rolling toward a stable equilibrium, which is the bottom of the cup. The role of this method is to accelerate the gradient descent method while performing the minimization of the cost function.

The classical momentum method (Polyak, 1964, see [98]) provides the following simultaneous updates for the position and velocity

$$x^{n+1} = x^n + v^{n+1} \tag{4.4.16}$$
$$v^{n+1} = \mu v^n - \eta \nabla f(x^n), \tag{4.4.17}$$

where $\eta > 0$ is the learning rate and $\mu \in (0, 1]$ is the momentum coefficient.

---

[5]For instance, shaking a basket filled with potatoes of different sizes will bring the large ones to the bottom of the basket and the small ones to the top – this corresponds to the state of the system with the smallest gravitational energy.

It is worth noting that for $\mu \to 0$, the previous model recovers the familiar model of the gradient descent, $x^{n+1} = x^n - \eta \nabla f(x^n)$.

4.4.1 Kinematic Interpretation

We picture again the model of a ball rolling into a cup, whose equation is given by $y = f(x)$, where f is the objective function subject to minimization. We shall denote by F_f the friction force between the ball and the cup walls. This force is proportional to speed and has its direction opposite to velocity, $F_f = -\rho \dot{x}(t)$, for some damping coefficient $\rho > 0$. Therefore, Newton's law of motion is written as

$$\ddot{x}(t) = -\rho \dot{x}(t) - \nabla f(x(t)). \tag{4.4.18}$$

The left side is the acceleration of a unit mass ball and the right side is the total force acting of the ball, which is the sum between the friction force and the force provided by the potential f. This equation can be used to show that in this case the ball's total energy (4.3.8) is not preserved along the solution. Since

$$
\begin{aligned}
\frac{d}{dt} E_{tot}(t) &= \frac{d}{dt}\left(\frac{1}{2}\dot{x}(t)^T \dot{x}(t) + f(x(t))\right) = \dot{x}(t)^T \ddot{x}(t) + \dot{x}(t)^T \nabla f(x) \\
&= \dot{x}(t)^T (\ddot{x}(t) + \nabla f(x(t))) = -\rho \dot{x}(t)^T \dot{x}(t) = -\rho \|\dot{x}(t)\|^2 \\
&= -\rho \|v(t)\|^2 < 0,
\end{aligned}
$$

it follows that the total energy decreases at a rate proportional to the square of the speed. Hence, $E_{tot}(t)$ is a decreasing function, which reaches its minimum at the equilibrium point of the system.

The equation (4.4.18) can be written equivalently as a first-order ODEs system as

$$
\begin{aligned}
\dot{x}(t) &= v(t) & (4.4.19) \\
\dot{v}(t) &= -\rho v(t) - \nabla f(x(t)), & (4.4.20)
\end{aligned}
$$

where $v(t)$ represents the velocity of the ball at time t. The tangent vector field to the solution flow, $(\dot{x}(t), \dot{v}(t))$, has the divergence equal to

$$\text{div}(\dot{x}, \dot{v}) = -\rho < 0.$$

This implies that the solution flow is contracting, the solution trajectories getting closer together, converging eventually to the equilibrium point. This point is obtained by equating to zero the right terms of the previous system

$$
\begin{aligned}
v &= 0 & (4.4.21) \\
-\rho v - \nabla f(x) &= 0. & (4.4.22)
\end{aligned}
$$

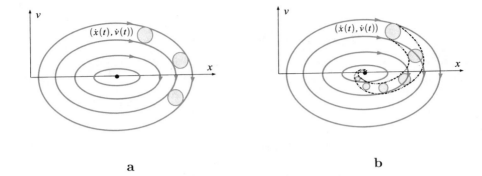

Figure 4.12: *Solution in phase space:* **a.** *The divergence of the tangent flow is zero and the volumes are preserved along the flow. The solution oscillates forever around the equilibrium point.* **b.** *When friction forces are present the tangent flow shrinks and the trajectory evolves to lower levels of energy. The oscillation is amortized in time.*

The solution (v^*, x^*) satisfies $v^* = 0$, $\nabla f(x^*) = 0$, which is the same equilibrium point as in the case of no friction case described by (4.3.12). The only difference in this case is that, due to energy loss, the solutions in the phase space move from higher energy levels to lower energy levels, spiraling down toward the equilibrium point, see Fig. 4.12 **b**.

In order to obtain an algorithm that can be implemented on a computer, we shall transform the ODE system (4.4.19)–(4.4.20) into a finite- difference system. Consider the equidistant time division $0 = t_0 < t_1 < \cdots < t_n < \infty$ and let $\Delta t = t_{n+1} - t_n$ be a constant time step. Denote the nth state of the system by $(x^n v^n) = \big(x(t_n), v(t_n)\big)$. The system (4.4.19)–(4.4.20) becomes

$$
\begin{aligned}
x^{n+1} - x^n &= v^n \, \Delta t \\
v^{n+1} - v^n &= -\rho v^n \, \Delta t - \nabla f(x^n) \, \Delta t.
\end{aligned}
$$

Substituting $\epsilon = \Delta t$ and $\mu = 1 - \rho\epsilon$, with $\mu < 1$, we obtain the finite-difference system

$$
\begin{aligned}
x^{n+1} &= x^n + \epsilon \, v^n & (4.4.23) \\
v^{n+1} &= \mu v^n - \epsilon \, \nabla f(x^n). & (4.4.24)
\end{aligned}
$$

Rescaling the velocity into $\tilde{v} = \epsilon v$ (physically feasible by changing the units of measure), the system (4.4.23)–(4.4.24) becomes

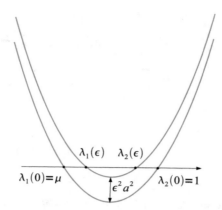

Figure 4.13: *For small upward shifts the roots* λ_i *remain real and situated between* μ *and* 1.

$$
\begin{aligned}
x^{n+1} &= x^n + \tilde{v}^n & (4.4.25) \\
\tilde{v}^{n+1} &= \mu \tilde{v}^n - \eta \nabla f(x^n), & (4.4.26)
\end{aligned}
$$

where $\eta = \epsilon^2$ is the learning rate (the time step is given by $\Delta t = \sqrt{\eta}$).

We notice the velocity index difference between the equations (4.4.25) and (4.4.16), or Polyak's classical momentum method. This can be fixed by replacing in our analysis the backward equation $x^{n+1} - x^n = v^n \Delta t$ by the forward equation $x^{n+1} - x^n = v^{n+1} \Delta t$.

Note that when the function f is quadratic, its gradient, ∇f, is linear and hence the equations (4.4.23)–(4.4.24) form a linear system that can be solved explicitly. We shall do this in the next example.

Example 4.4.1 Consider the quadratic function of a real variable $f(x) = \frac{1}{2}(ax + b)^2$. Since the gradient is $f'(x) = a(ax - b)$, the system (4.4.23)–(4.4.24) is linear:

$$
\begin{aligned}
x^{n+1} &= x^n + \epsilon v^n \\
v^{n+1} &= -\epsilon a^2 x^n + \mu v^n + \epsilon ab.
\end{aligned}
$$

In matrix notation this writes more simply as $s_{n+1} = M s_n + \beta$, with

$$
s_n = \begin{pmatrix} x^n \\ v^n \end{pmatrix}, \qquad
M = \begin{pmatrix} 1 & \epsilon \\ -\epsilon a^2 & \mu \end{pmatrix}, \qquad
\beta = \begin{pmatrix} 0 \\ \epsilon ab \end{pmatrix}.
$$

Inductively, we can express the state s_n in terms of the initial state s_0 as

$$s_n = M^n s_0 + (\mathbb{I}_2 - M^n)(\mathbb{I}_2 - M)^{-1}\beta. \tag{4.4.27}$$

A computation provides the inverse

$$(\mathbb{I}_2 - M)^{-1} = \frac{1}{\epsilon^2 a^2}\begin{pmatrix} 1-\mu & \epsilon \\ -\epsilon\,a^2 & 0 \end{pmatrix}.$$

Next we shall show that the eigenvalues of M are between 0 and 1. To denote the dependence of ϵ, we write $M = M(\epsilon)$. Then

$$M(0) = \begin{pmatrix} 1 & 0 \\ 0 & \mu \end{pmatrix}$$

is a diagonal matrix having eigenvalues $\lambda_1(0) = \mu$ and $\lambda_2(0) = 1$. The characteristic equation of M is the quadratic equation

$$\lambda^2 - (\mu+1)\lambda + (\mu + \epsilon^2 a^2) = 0,$$

with solutions $\lambda_1(\epsilon)$ and $\lambda_2(\epsilon)$. Instead of computing explicitly these eigenvalues we shall rather show that they are between 0 and 1 in the following qualitative way. Note that the characteristic equation depends additively on ϵ. When $\epsilon = 0$, there are two intersections between the parabola and the x-axis given by $\lambda_1(0) = \mu$ and $\lambda_2(0) = 1$. For small enough $\epsilon > 0$, the parabola shifts up by the small amount $\epsilon^2 a^2$, and by continuity reasons, there are still two intersections with the x-axis, situated in between μ and 1, see Fig. 4.13. Hence, $0 < \mu < \lambda_1(\epsilon) < \lambda_2(\epsilon) < 1$. Consequently, $M^n \to 0$, as $n \to \infty$ by Proposition G.1.2 of the Appendix.

Taking the limit in relation (4.4.27) yields the equilibrium state

$$s^* = \lim_{n\to\infty} s_n = (\mathbb{I}_2 - M)^{-1}\beta = \frac{1}{\epsilon^2 a^2}\begin{pmatrix} 1-\mu & \epsilon \\ -\epsilon\,a^2 & 0 \end{pmatrix}\begin{pmatrix} 0 \\ \epsilon a b \end{pmatrix} = \begin{pmatrix} \frac{b}{a} \\ 0 \end{pmatrix},$$

which retrieves the minimum $x^* = \dfrac{b}{a}$, as expected.

Remark 4.4.2 The name of "momentum method" is used in an improper way here. A momentum to a particle is usually a push forward, while in this case we damp the particle using a friction force. We do this in the effort of avoiding the particle to overshoot the equilibrium point.

However, sometimes we would like the opposite effect: to make the particle avoid getting stuck in a local minimum. In this case, the friction factor is replaced by a momentum factor meant to give a bust to the particle velocity to overshoot the local minimum. This is easily fulfilled by asking the condition $\mu > 1$ in equation (4.4.24).

4.4.2   Convergence conditions

In this section we are interested in studying the convergence of sequences x^n and v^n defined by the momentum method equations (4.4.16)–(4.4.17). To accomplish this task we shall find exact formulas for the sequences.

Iterating equation (4.4.16)

$$
\begin{aligned}
x^n &= x^{n-1} + v^n \\
x^{n-1} &= x^{n-2} + v^{n-1} \\
\cdots &= \cdots \\
x^1 &= x^0 + v^1
\end{aligned}
$$

and then adding, yields the expression of the position in terms of velocities

$$x^n = x^0 + \sum_{k=1}^{n} v^k. \tag{4.4.28}$$

Denote for simplicity $b_n = \nabla f(x^n)$. Iterating equation (4.4.17), we obtain

$$
\begin{aligned}
v^n &= \mu v^{n-1} - \eta b_{n-1} \\
&= \mu(\mu v^{n-2} - \eta b_{n-2}) - \eta b_{n-1} \\
&= \mu^2 v^{n-2} - \mu\eta b_{n-2} - \eta b_{n-1} \\
&= \mu^2(\mu v^{n-3} - \eta b_{n-3}) - \mu\eta b_{n-2} - \eta b_{n-1} \\
&= \mu^3 v^{n-3} - \mu^2 \eta b_{n-3} - \mu\eta b_{n-2} - \eta b_{n-1}.
\end{aligned}
$$

It can be shown by induction that

$$v^n = \mu^n v^0 - \eta\Big(\mu^{n-1} b_0 + \mu^{n-2} b_1 + \cdots + \mu b_{n-2} + b_{n-1}\Big).$$

In the following it is more convenient to shift indices and use the summation convention to obtain the expression

$$v^{n+1} = \mu^{n+1} v^0 - \eta \sum_{i=0}^{n} \mu^{n-i} b_i. \tag{4.4.29}$$

In order to understand the behavior of v^{n+1} we need to introduce the following notion.

Definition 4.4.3 The convolution series of two numerical series $\sum_{n\geq 0} a_n$ and $\sum_{n\geq 0} b_n$ is the series $\sum_{n\geq 0} c_n$ with the general term $c_n = \sum_{i=0}^{n} a_i b_{n-i} = \sum_{i=0}^{n} a_{n-i} b_i$.

The following two results will be used in the convergence analysis of sequences $(x^n)_n$ and $(v^n)_n$.

Proposition 4.4.4 *Let $(a_n)_n$ be a sequence of real numbers convergent to 0 and $\sum_{n \geq 0} b_n$ be an absolute convergent numerical series. Then*

$$\lim_{n \to \infty} \sum_{i=0}^{n} a_i b_{n-i} = 0.$$

Theorem 4.4.5 *Consider two numerical series $\sum_{n \geq 0} a_n$ and $\sum_{n \geq 0} b_n$, one convergent and the other absolute convergent. Then their convolution series is convergent and its sum is equal to the product of the sums of the given series, i.e.*

$$\sum_{n \geq 0} \left(\sum_{i=0}^{n} a_i b_{n-i} \right) = \left(\sum_{n \geq 0} a_n \right) \left(\sum_{n \geq 0} b_n \right).$$

The following result provides a characterization for the convergence of sequences $(x^n)_n$ and $(v^n)_n$.

Proposition 4.4.6 *(a) If $\nabla f(x^n)$ converges to 0, then the sequence $(v^n)_n$ is also convergent to 0, as $n \to \infty$.*
(b) Assume the series $\sum_{n \geq 0} \|\nabla f(x^n)\|$ is convergent. Then both sequences $(x^n)_n$ and $(v^n)_n$ are convergent.

Proof: (a) Recall that $0 < \mu < 1$. Since (b_n) converges to 0 and the geometric series $\sum_{n \geq 0} \mu^n$ is absolute convergent, Proposition 4.4.4 implies $\lim_{n \to \infty} \sum_{i=0}^{n} \mu^{n-i} b_i = 0$. Since μ^n is convergent to 0, taking the limit in (4.4.29) we obtain

$$v^* = \lim_{n \to \infty} v^{n+1} = v^0 \lim_{n \to \infty} \mu^{n+1} - \eta \lim_{n \to \infty} \sum_{i=0}^{n} \mu^{n-i} b_i = 0.$$

(b) Since $\sum_{n \geq 0} \|\nabla f(x^n)\|$ is convergent, then $\nabla f(x^n)$ converges to 0. From part (a) it follows that v^n is convergent to zero.

To show the convergence of x^n, we use relations (4.4.28) and (4.4.29) and manipulate the expressions algebraically as follows:

$$
\begin{aligned}
x^{n+1} &= x^0 + \sum_{k=1}^{n+1} v^k = x^0 + \sum_{k=0}^{n} v^{k+1} \\
&= x^0 + \sum_{k=0}^{n} \left[\mu^{k+1} v^0 - \eta \sum_{i=0}^{k} \mu^{k-i} b_i \right] \\
&= x^0 + v^0 \frac{\mu(1 - \mu^{n+1})}{1 - \mu} - \eta \sum_{k=0}^{n} \sum_{i=0}^{k} \mu^{k-i} b_i.
\end{aligned}
$$

This provides a closed-form expression for x^{n+1}. Taking the limit using Theorem 4.4.5 we obtain

$$
\begin{aligned}
x^* &= \lim_{n\to\infty} x^{n+1} = x^0 + v^0 \frac{\mu}{1-\mu} - \eta \Big(\sum_{n\geq 0} \mu^n\Big)\Big(\sum_{n\geq 0} b_n\Big) \\
&= x^0 + v^0 \frac{\mu}{1-\mu} - \frac{\eta}{1-\mu}\sum_{n\geq 0} \|\nabla f(x^n)\|.
\end{aligned}
$$

∎

Remark 4.4.7 An improvement of the classical momentum method has been proposed by Nesterov [91]. This is obtained by modifying the momentum method the argument of the gradient; instead of computing it at the current position x^n, it is evaluated at the corrected value $x^n + \mu v^n$:

$$
\begin{aligned}
x^{n+1} &= x^n + v^{n+1} & (4.4.30)\\
v^{n+1} &= \mu v^n - \eta \nabla f(x^n + \mu v^n). & (4.4.31)
\end{aligned}
$$

The Nesterov Accelerated Gradient (abbreviated as NAG) is a first-order optimization method with better convergence rate than the gradient descent. Compared with the momentum method, NAG changes velocity v in a more responsive way, fact that makes the method more stable, especially for larger value of μ.

4.5 AdaGrad

A modified stochastic gradient descent method with an adaptive learning rate was published in 2011 under the name of *AdaGrad* (Adaptive Gradient), see Duchi et al. [34]. If $C(x)$ denotes the cost function, which is subject to minimization, with $x \in \mathbb{R}^N$, then the gradient vector evaluated at step t is denoted by $g_t = \nabla C(x(t))$. We consider the $N \times N$ matrix

$$
G_t = \sum_{\tau=1}^{t} g_\tau g_\tau^T
$$

and consider the update

$$
x(t+1) = x(t) - \eta G_t^{-1/2} g_t,
$$

where $\eta > 0$ is the learning rate. For discrete time steps this can be written equivalently as

$$
x^{n+1} = x^n - \eta G_n^{-1/2} g_n.
$$

Since $G_t^{-1/2}$ is computationally impractical in high dimensions, the update can be done using only the diagonal elements of the matrix

$$x(t+1) = x(t) - \eta \operatorname{diag}(G_t)^{-1/2} g_t.$$

The diagonal elements of G_t can be calculated by

$$(G_t)_{jj} = \sum_{\tau=1}^{t} (g_{\tau,1})^2,$$

where we use that

$$g_\tau g_\tau^T = \begin{pmatrix} g_{\tau,1} \\ \vdots \\ g_{\tau,N} \end{pmatrix} (g_{\tau,1}, \ldots, g_{\tau,N}) = \begin{pmatrix} (g_{\tau,1})^2 & \cdots & \cdots \\ \cdots & \cdots & \cdots \\ \cdots & \cdots & (g_{\tau,N})^2 \end{pmatrix}.$$

4.6 RMSProp

The *Root Mean Square Propagation*, or RMSProp, is a variant of the gradient descent method with adaptive learning rate, which is obtained if the gradient is divided by a running average of its magnitude, see Tieleman and Hinton [118], 2012 .

If $C(x)$ denotes the cost function, and $g_t = \nabla C(x(t))$ is its gradient evaluated at time step t, then the running average is defined recursively by

$$v(t) = \gamma v(t-1) + (1-\gamma) g_{t-1}^2,$$

where $\gamma \in (0,1)$ is the forgetting factor, which controls the exponential decay rate and the vector g_{t-1}^2 denotes the elementwise square of the gradient g_{t-1}.[6] It can be shown inductively that the exponential moving average of the squared gradient $v(t)$ is given by

$$v(t) = \gamma^t v(0) + (1-\gamma) \sum_{i=1}^{t} \gamma^{t-i} g_i^2.$$

Since the coefficients sum up to 1, namely

$$\gamma^t + (1-\gamma) \sum_{i=1}^{t} \gamma^{t-i} = 1,$$

---

[6]For discrete time steps this can be written equivalently as

$$v^n = \gamma v^{n-1} + (1-\gamma)(g_{n-1})^2.$$

it follows that $v(t)$ is a weighted mean of $v(0)$ (which is usually taken equal to zero) and all the squared gradients until step t, giving larger weights to more recent gradients.

The minimum of the cost function, $x^* = \arg\min_x C(x)$, is obtained by the approximation sequence $(x(t))_{t \geq 1}$ defined recursively by the updating rule

$$x(t+1) = x(t) - \eta \frac{g_t}{\sqrt{|v(t)|}}, \qquad (4.6.32)$$

where $\eta > 0$ is a learning rate. We note that $v(t)$ can be interpreted as an estimation of the second moment (uncentered variance) of the gradient. The equation (4.6.32) can be seen as a gradient descent variant where the gradient g_t is scaled by its standard deviation estimation, $\sqrt{|v(t)|}$.

Example 4.6.1 We shall track mathematically the minimum of the real-valued function $C(x) = \frac{1}{2}x^2$ using the RMSProp method. The gradient in this case is $g_t = x(t)$ and the moving average is given by

$$v(t) = (1 - \gamma) \sum_{i=1}^{t} \gamma^{t-i} x(i)^2,$$

where we considered $v(0) = 0$. The sequence that estimates the minimum $x^* = \arg\min_x C(x)$ is given recursively by

$$x(t+1) = x(t) - \eta \frac{x(t)}{\sqrt{|v(t)|}} = x(t)\left[1 - \eta \frac{1}{\sqrt{|v(t)|}}\right].$$

We denote $\rho_t = 1 - \eta/\sqrt{|v(t)|}$ and assume that $0 < \rho_t < \rho < 1$. Then the relation $x(t+1) = x(t)\rho_t$ together with the initial condition $x(0) > 0$ implies that the sequence $x(t)$ is decreasing and bounded from below by 0, satisfying the inequality

$$0 < x(t) < x(0)\rho^t.$$

Talking the limit and using the Squeeze Theorem yields $x^* = \lim_{t \to \infty} x(t) = 0$, namely the minimum of $C(x) = \frac{1}{2}x^2$ is reached for $x = 0$.

Now, we go back and show the double inequality $0 < \rho_t < \rho < 1$. The first inequality, $0 < \rho_t$, is equivalent to $\frac{\eta}{\sqrt{v(t)}} < 1$. In order to show this inequality, we shall assume that $x(t)$ does not converge to 0 (because otherwise we already arrived to the conclusion), namely there is $\epsilon > 0$ such that $x(t)^2 > \epsilon$, for all $t \geq 1$. This implies that there is $N > 1$ such that $v(t) > \frac{\epsilon}{2}$, for all $t > N$. This fact follows from the computation

$$
\begin{aligned}
v(t) &= (1 - \gamma) \sum_{i=1}^{t} \gamma^{t-i} x(i)^2 > \epsilon(1 - \gamma) \sum_{i=1}^{t} \gamma^{t-i} \\
&= \epsilon(1 - \gamma)\frac{1 - \gamma^t}{1 - \gamma} = \epsilon(1 - \gamma^t) > \frac{\epsilon}{2},
\end{aligned}
$$

where we choose N such that $\gamma^t < \frac{1}{2}$ for $t > N$. The previous inequality implies

$$\frac{\eta}{v(t)} < \frac{2\eta}{\epsilon} < 1, \qquad \forall t > N,$$

where we chose the learning rate satisfying $\eta < \frac{\epsilon}{2}$.

The second inequality, $\rho_t < \rho < 1$, is equivalent to $v(t) < \left(\frac{\eta}{1-\rho}\right)^2$, i.e., to the boundness of the sequence $v(t)$. A sufficient condition for this is the boundness of the approximation sequence, $|x(t)| < M$, for all t. This can be shown as in the following

$$
\begin{aligned}
v(t) &= (1-\gamma)\sum_{i=1}^{t}\gamma^{t-i}x(i)^2 < M^2(1-\gamma)\sum_{i=1}^{t}\gamma^{t-i} \\
&= M^2(1-\gamma^t) < M^2.
\end{aligned}
$$

4.7 Adam

This adaptive learning method was inspired by the previous AdaGrad and RMSProp methods and was introduced in 2014 by Diederik and Ba [31]. The method uses the estimation of the first and second moment of the gradient by exponential moving averages and then apply some bias corrections.

The cost function, $C(x)$, subject to minimization, may have some stochasticity build in and is assumed to be differentiable with respect to x. We are interested in the minimum $x^* = \arg\min_x \mathbb{E}[C(x)]$.

For this, we denote the gradient of the cost function at the time step t by $g_t = \nabla C(x(t))$. We consider two exponential decay rates for the moment estimates, $\beta_1, \beta_2 \in [0,1)$, fix an initial vector $x(0) = x_0$, initialize the moments by $m(0) = 0$ and $v(0) = 0$, and consider the moments updates

$$
\begin{aligned}
m(t) &= \beta_1 m(t-1) + (1-\beta_1)g_t \\
v(t) &= \beta_2 v(t-1) + (1-\beta_2)(g_t)^2,
\end{aligned}
$$

where $(g_t)^2$ denotes the elementwise square of the vector g_t. The moments $m(t)$ and $v(t)$ can be interpreted as biased estimates of the first moment (mean) and second moment (uncentered variance) of the gradient given by exponential moving averages. The bias can be corrected as in the following. We can write inductively

$$
\begin{aligned}
m(t) &= (1-\beta_1)\sum_{i=1}^{t}\beta_1^{t-i}g_i \\
v(t) &= (1-\beta_2)\sum_{i=1}^{t}\beta_2^{t-i}(g_i)^2.
\end{aligned}
$$

Applying the expectation operator and assuming that the first and second moments are stationary, we obtain

$$\mathbb{E}[m(t)] = (1 - \beta_1) \sum_{i=1}^{t} \beta_1^{t-i} \mathbb{E}[g_i] = (1 - \beta_1^t) \mathbb{E}[g_t]$$

$$\mathbb{E}[v(t)] = (1 - \beta_2) \sum_{i=1}^{t} \beta_2^{t-i} \mathbb{E}[(g_i)^2] = (1 - \beta_2^t) \mathbb{E}[(g_t)^2].$$

Therefore, the bias-corrected moments are

$$\widehat{m}(t) = m(t)/(1 - \beta_1^t)$$
$$\widehat{v}(t) = v(t)/(1 - \beta_2^t).$$

The final recursive formula is given by

$$x(t + 1) = x(t) - \eta \frac{\widehat{m}(t)}{\sqrt{|\widehat{v}(t)|} + \epsilon},$$

with $\epsilon > 0$ a small scalar used to prevent division by zero. Some default settings for the aforementioned hyperparameters used in [31] are $\eta = 0.001$, $\beta_1 = 0.9$, $\beta_2 = 0.99$, and $\epsilon = 10^{-8}$.

4.8 AdaMax

Adaptive Maximum method, or AdaMax, is a variant of Adam based on the infinity norm, [31]. The model is defined by the following set of iterations:

$$m(t) = \beta_1 m(t - 1) + (1 - \beta_1)g_t$$
$$u(t) = \max(\beta_2 u(t - 1), |g(t)|)$$
$$x(t) = x(t - 1) - \frac{\eta}{1 - \beta_1^t} \frac{m(t)}{u(t - 1)},$$

with moment initializations $m(0) = u(0) = 0$. We note that in this case there is no correction needed for the bias.

4.9 Simulated Annealing Method

Another method for finding global minima of a given function is the *simulated annealing* (SA) introduced to neural networks in 1983 by Kirkpatrick et al. [63]. The method is inspired by a metallurgical process called annealing. During this process, the metal is tempered, i.e., is overheated and after

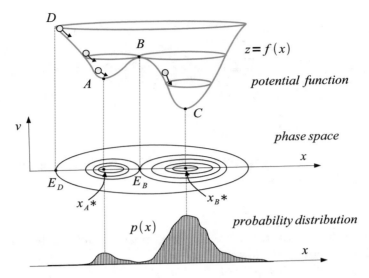

Figure 4.14: *Potential function, phase space representation and final distribution of particles.*

that is slowly cooled. By this procedure the crystalline structure of the metal reaches a global minimum energy, the method being used to eliminate eventual defects in the metal, making it stronger.

Since the kinetic energy of a molecule is proportional to the absolute temperature, see Einstein [38], then the metal molecules get excited at high temperatures, and then, during the cooling process, they loose energy and will finally end up at positions of global minimum energy in the crystal lattice. We shall discuss this behavior in the following from both the kinematics and thermodynamics points of view.

4.9.1 Kinematic Approach for SA

Consider the energy potential $z = f(x)$, which has some local minima and also a global minimum. Assume that the metal molecules have a dynamics induced by this potential. For a fixed temperature, T, each particle has constant total energy. Consequently, it has to move along constant energy level curves in the phase space, see section 4.3. When the temperature increases, the kinetic energy of particles increases too, and hence they will translate up to higher energy levels.

When the temperature is slowly decreased, the kinetic energy of particles decreases too, the effect on particles movement being similar to the effect of a damping factor. This brings the particle to a lower level of energy, hopefully

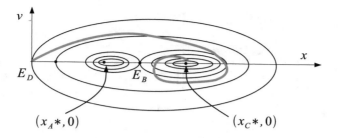

Figure 4.15: *Most of damped trajectories will spiral in approaching the equilibrium point $(x_C^*, 0)$, which has a larger basin of attraction.*

even lower than the initial one. In the end, most particles will be situated at the global minimum state in the phase space.

We shall discuss this behavior using Fig. 4.14. Consider the potential function $z = f(x)$, having a local minimum at the point A and a global minimum at C. Assume that a particle is in the proximity of the local minimum A at the initial temperature T_0. Then its associated trajectory in the phase space will be a small closed loop around the equilibrium point $(x_A^*, 0)$. We say that the particle got stuck in a local minimum, and without any extra energy the particle cannot escape from the basin of attraction of point A.

Now, we increase the temperature to $T_{\max}$ such that the particle has a kinetic energy that allows it to go as high as point D. Then, at this stage, the associated trajectory in the phase space corresponds to the energy level E_D, which is the largest loop in the figure. Then we start decreasing the temperature slowly. During this process the particle has a damped movement in the phase space, and the energy level E_D will get down to the energy level E_B, which has a number eight shape. Hence, after lowering the temperature even more, the trajectory in the phase space will revolve around the global minimum $(x_C^*, 0)$. When the temperature gets down to zero, the damping will bring the trajectory toward the global minimum $(x_C^*, 0)$, see Fig. 4.15.

If we consider now a large number of particles to which we apply the previous reasoning, then most of them will end up in a proximity of the equilibrium point $(x_C^*, 0)$, and only a few will not be able to escape from the basin of attraction of point A. This explains the double-peaked distribution of particles, $p(x)$, in the lower part of figure Fig. 4.14. The slower the cooling process, the smaller the peak above x_A^*. If the cooling would be infinitely long, the distribution will be just single peaked, fact that corresponds to the case when all particles have a minimum potential energy. It is worth noting that when $T = T_{\max}$ the distribution of particles is uniform.

4.9.2   Thermodynamic Interpretation for SA

We assume now that all particles form a thermodynamic system whose internal energy has to be minimized. The following concept of Boltzmann probability distribution will be useful. It is known that *a system in thermal equilibrium at a temperature T has a distribution of energy given by*

$$p(E) = e^{-\frac{E}{kT}},$$

where E is the energy level, T is the temperature, and k denotes the Boltzmann constant (which for the sake of simplicity we shall take it from now on equal to 1), see Fig. 4.16 **a**. In fact, in order to make the previous formula mathematically valid as a probability density, we need to specify an interval, $0 \le E \le E_{\max}$, and a normalization constant in front of the exponential. For our purposes we do not need to worry about it for the time being. The previous distribution is used as in the following. If a particle is in an initial state, s_0, with energy E_0, the probability to jump to a state s_1 of a higher energy, E_1, is

$$p(\Delta E) = e^{-\frac{\Delta E}{T}},$$

where $\Delta E = E_1 - E_0$ is the difference in the energies of the states. This model has a few consequences:

(i) Consider the temperature T fixed. Then the higher the energy jump, ΔE, the lower the probability of the jump, $p(\Delta E)$.

(ii) For a given energy jump ΔE, the higher the temperature, the larger the probability of the jump. Consequently, at low temperatures, the probability for a given jump size in energy is lower than the probability for the same jump size at a higher temperature.

(iii) For very large T, the jump energies are distributed uniformly.

These can be stated equivalently by saying that a thermodynamic system is more stable at low energies and low temperatures. These observations will have an uttermost impact on the following global minimum search algorithm.

Consider the function $E = f(x)$, which needs to be minimized globally. Here, x plays the role of a state of the system, while E denotes its internal energy. We start from a point x^0, which is the initial state of the system, and set the temperature parameter T to very high. A second point x^1 is created randomly in a neighborhood of x^0. Then we compute the energy

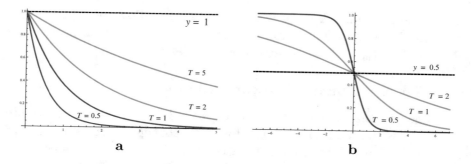

Figure 4.16: **a**. *The Boltzmann distribution* $p(x) = e^{-x/T}$. **b**. *The reflected sigmoid functions* $f(x) = \frac{1}{1+e^{x/T}}$. *Both functions are represented for a few different values of* T.

jump between the states x^0 and x^1 as $\Delta E = f(x^1) - f(x^0)$. There are two cases that can occur:

1) If $\Delta E < 0$, which means $f(x^1) < f(x^0)$, then we accept the new state x^1 with probability 1, and we shall continue the next steps in the search from x^1.

2) If $\Delta E \geq 0$, then we do not discard the state x^1 right away (as a greedy algorithm would do). We *do accept* x^1 with probability $p(\Delta E)$, where p denotes the Boltzmann distribution.

We note that the acceptance probability used in the above two steps can be written in only one formula as $P(\Delta E; T) = \min\left\{1, e^{-\frac{\Delta E}{T}}\right\}$.

In the next step we reduce the temperature parameter and choose randomly a new point in the neighborhood of the last accepted point; then continue with the above steps.

By this procedure, when T is high, the probability to accept states that increase the energy is large. In fact, for $T \to \infty$, any point can be accepted with probability 1. Setting T high in the beginning facilitates the system to get out of an eventual local minimum.

When T gets closer to 0, there is a small probability to accept points that increase the energy and a much higher probability to choose points that decrease the energy.

We conclude next with the following pseudocode for the simulated annealing:

1. Fix $\epsilon > 0$. Choose x^0 and set T high.
2. Choose a random point x^1 in a neighborhood of x^0.
3. Compute $\Delta E = f(x^1) - f(x^0)$.
   - if $\Delta E < 0$, set $x^2 = x^1$.
   - if $\Delta E \geq 0$, choose a random number $r \in (0, 1)$.

        ◦ if $r \leq e^{-\frac{\Delta E}{T}}$, set $x^2 = x^1$.

        ◦ if $r > e^{-\frac{\Delta E}{T}}$, then go to step 2.

4. If $\|x^{n+1} - x^n\| < \epsilon$ then stop; else reduce T and go to step 2.

We note that decreasing the temperature too fast leads to premature convergence, which might not reach the global minimum.[7] Decreasing the temperature too slowly leads to a slow convergence of the algorithm. In this case the search algorithm constructs a sequence $(x^n)_n$ which covers the states space pretty well and a global minimum cannot be missed. It can be proved that using the simulated annealing method the global minimum of the energy function is approached asymptotically, see [1].

One way to set the initial temperature is to choose it equal to the average of the function f, as $T_{\max} = Ave(f) = \frac{1}{vol(D)} \int_D f(x)\, dx$, where D denotes the states space.

Remark 4.9.1 (i) Boltzmann distribution was chosen for convenience and for the sake of similarity with thermodynamics. There are other acceptance probabilities which work as well as the previous one, for instance, the reflected sigmoid function $f(x) = \frac{1}{1+e^{x/T}}$, see [101]. When $T \to \infty$, the function tends to the horizontal line $y = 0.5$ and its behavior is similar to Boltzmann's distribution, see Fig. 4.16 **b**. This distribution is of paramount importance in the study of Boltzmann machines in Chapter 20.

(ii) A situation similar to the simulated annealing method occurs in the case of a roulette wheel. When the ball is vigorously spun along the wheel, it does not land in any of the numbered slots immediately. These slots act as local minima for the potential energy function, and as long as the ball has enough energy, it can easily get in and out of these pockets. It is only after the ball's energy decreases sufficiently, due to friction, that it lands in one of the pockets, and this occurs with equal probability for all numbers. If we assume that one of the slots is deeper than the others, the ball will have the tendency of landing in this specific pocket with a higher probability.

4.10 Increasing Resolution Method

In this section we shall present a non-stochastic method for finding global minima of functions that have plenty of local minima. This approach can also be performed in several variables, but for the sake of simplicity we shall present it here just for the case of a one-dimensional signal, $y = f(x)$, depending of a real variable, x. The method works for signals which are integrable

---

[7]From the Physics point of view, in this case the substance does not reach to the state of a crystalline structure, but rather to an amorphous structure.

and not necessarily differentiable. This technique can be considered as a version of the simulated annealing method, as we shall see shortly.

The idea is to blur the signal $y = f(x)$ using a Gaussian filter, $G_\sigma(x) = \frac{1}{\sqrt{2\pi}\sigma}e^{-\frac{x^2}{2\sigma^2}}$, obtaining the signal $f_\sigma(x)$, which is given by the convolution

$$f_\sigma(x) = (f * G_\sigma)(x) = \int_{\mathbb{R}} f(u)G_\sigma(x - u)\, du = \int_{\mathbb{R}} f(x - u)G_\sigma(u)\, du.$$

The positive parameter σ controls the signal resolution. A large value of σ provides a signal $f_\sigma(x)$, which retains a rough shape of $f(x)$. On the other side, a small σ means to retain details. Actually using the fact that the Gaussian tends to the Dirac measure, $\lim_{\sigma \searrow 0} G_\sigma(x) = \delta(x)$, we have

$$\lim_{\sigma \searrow 0} f_\sigma(x) = \lim_{\sigma \searrow 0} (f * G_\sigma)(x) = (f * \delta)(x) = f(x).$$

This means that for small values of σ the filtered signal is very close to the initial signal, fact which agrees with the common sense.

If one tries to find the global minimum of $y = f(x)$ directly, using the gradient descent method, two things would stand in the way:
1. The function f might not be differentiable and the algorithm can't be applied. For instance, neural nets of perceptrons (multilayer perceptrons) belong to this category;
2. The function f is differentiable, the algorithm applies, but it gets stuck in a local minimum. One deficient way to fix this issue is to increase the learning rate to allow skipping over local minima. However, this will lead to an oscillation around the minimum, which is not a precise approximation.

The next method avoids the previous two inconveniences. We first note that the smoothed function $f_\sigma(x)$ is integrable and differentiable with respect to x (even if f might not be), satisfying

$$\begin{aligned} f'_\sigma(x) &= (f * G'_\sigma)(x) \\ \|f_\sigma\|_1 &\leq \|f\|_1. \end{aligned}$$

The first identity follows from a change of variables and the interchangeability of the derivative with the integral, due to the exponential decay of the Gaussian. For the second inequality, see Exercise 4.17.5.

The algorithm Since the signal resolution increases as σ decreases, we shall first find an approximative location of the global minimum, and then increasing the resolution we look for more accurate approximations of the global minimum. This way, we look for the global minimum in more and more accurate proximity of the global minimum.

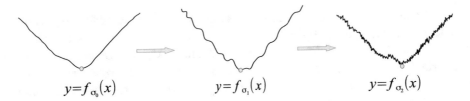

$$y=f_{\sigma_0}(x) \qquad y=f_{\sigma_1}(x) \qquad y=f_{\sigma_2}(x)$$

Figure 4.17: *The increase in the signal resolution. The minimum on each resolution profile is represented by a little circle.*

Consider the resolution schedule $\sigma_0 > \sigma_1 > \cdots > \sigma_k > 0$. Then $f_{\sigma_0}, \ldots, f_{\sigma_k}$ are different resolutions levels of the signal $y = f(x)$, see Fig. 4.17. We shall start with a large value of σ_0. Then the signal $y = f_{\sigma_0}(x)$ looks like a rough shape of $y = f(x)$, and it does not have any local minima if σ_0 is large enough. Let

$$x_0 = \arg\min_x f_{\sigma_0}(x).$$

Since f_{σ_0} is smooth and x_0 is the only minimum, it can be achieved using the gradient descent method.

We consider the next resolution signal $y = f_{\sigma_1}(x)$ and apply the gradient descent method, starting from x_0, to obtain the lowest value in a neighborhood $\mathcal{V}_0$ of x_0 at

$$x_1 = \arg\min_{x \in \mathcal{V}_0} f_{\sigma_1}(x).$$

Next, we apply the gradient descent on a neighborhood $\mathcal{V}_1$ of x_1 for the signal f_{σ_2} and obtain

$$x_2 = \arg\min_{x \in \mathcal{V}_1} f_{\sigma_2}(x).$$

We continue the procedure until we reach

$$x_k = \arg\min_{x \in \mathcal{V}_{k-1}} f_{\sigma_k}(x).$$

Due to the increase in resolution we have the descending sequence of neighborhoods $\mathcal{V}_0 \supset \cdots \supset \mathcal{V}_{k-1}$. Each one contains an approximation of the global minimum. If the schedule is correctly chosen, then x_k is a good approximation of the global minimum of $f(x)$.

Remark 4.10.1 1. The parameter σ plays a role similar to the temperature in the simulated annealing method.

2. This method is similar to the process of finding a small store located next to a certain street intersection in a given city, using Google Maps. First, we map the country in a chart $\mathcal{V}_0$; then we increase the resolution to map the

city in a chart V_1; increasing the resolution further, we look for the streets in a smaller chart V_2; magnifying the image even further, we look for the store at the prescribed streets' intersection in a chart V_3. We obtain the obvious inclusion $V_3 \subset V_2 \subset V_3 \subset V_0$, which corresponds to the resolution schedule store, streets, city, country. Neglecting only one resolution level will make the search almost impossible.

Blurring schedules If the signal $y = f(x)$ is blurred using the Gaussian filter $G_{\sigma_1}(x)$, we obtain the signal $f_{\sigma_1} = f * G_{\sigma_1}$. Next, we blur f_{σ_1} using the Gaussian filter $G_{\sigma_1}(x)$ and obtain $f_{\sigma_1,\sigma_2} = f_{\sigma_1} * G_{\sigma_2}$. We can also blur f directly using the Gaussian $G_{\sigma_1+\sigma_2}$ and obtain the signal $f_{\sigma_1+\sigma_2} = f * G_{\sigma_1+\sigma_2}$.

We shall show in the following that the signal f_{σ_1,σ_2} has more resolution than the signal $f_{\sigma_1+\sigma_2}$. Using the associativity of convolution as well as Exercise 4.17.6, we obtain

$$f_{\sigma_1,\sigma_2} = f_{\sigma_1} * G_{\sigma_2} = (f * G_{\sigma_1}) * G_{\sigma_2} = f * (G_{\sigma_1} * G_{\sigma_2}) = f * G_\sigma = f_\sigma,$$

where $\sigma = \sqrt{\sigma_1^2 + \sigma_2^2}$. Since $\sigma < \sigma_1 + \sigma_2$, it follows that f_σ contains more details than $f_{\sigma_1+\sigma_2}$. Similarly, if we blur the signal n times, we obtain

$$f * G_{\sigma_1} * \cdots * G_{\sigma_n} = f * G_\sigma,$$

with $\sigma = \sqrt{\sigma_1^2 + \cdots + \sigma_n^2} < \sigma_1 + \cdots + \sigma_n$. If consider the constraint

$$\sigma_1 + \cdots + \sigma_n = s,$$

which means that the blurring schedule has given length, s, then by Exercise 4.17.7, the minimum of σ is reached for the case

$$\sigma_1 = \cdots = \sigma_n = s/n.$$

Therefore, blurring the signal in smaller steps preserves more details of the initial signal, while blurring it in larger steps loses information.

It is worth noting the existence of infinite length schedules that provide a signal with finite resolution. For instance, the blurring schedule

$$\sigma_1 = 1, \sigma_2 = \frac{1}{2}, \ldots, \sigma_k = \frac{1}{k}, \ldots$$

satisfies $s = \sum_{k \geq 1} \sigma_j = \infty$, while $\sigma^2 = \sum_{j \geq 1} \frac{1}{k^2} = \frac{\pi^2}{6}$.

4.11 Hessian Method

The gradient methods presented in the previous sections do not take into account the curvature of the surface $z = f(x)$, which is described by the matrix of second partial derivatives

$$H_f(x) = \left(\frac{\partial^2 f(x)}{\partial x_i \partial x_j} \right)_{i,j},$$

called *Hessian*. The eigenvalues of the matrix H_f will be denoted by $\{\lambda_i\}$, $1 \leq i \leq n$. The corresponding eigenvectors, $\{\xi_j\}_j$, satisfy $H_f \xi_i = \lambda_i \xi_i$, with $\|\xi_i\| = 1$.

A few properties of the Hessian that will be used later are given in the following.

Proposition 4.11.1 *Let f be a real-valued function, twice continuous differentiable. The following hold:*

(a) The Hessian matrix is symmetric;

(b) For any x, the eigenvalues of $H_f(x)$ are real;

(c) Assume H has distinct eigenvalues. Then for any nonzero vector v, we have

$$\frac{\langle H_f(x)v, v \rangle}{\|v\|^2} = \sum_{i=1}^{n} w_j \lambda_j,$$

where $\{\lambda_j\}$ are the eigenvalues of $H_f(x)$ and w_j are weights, with $w_j \geq 0$ and $\sum_{j=1}^{n} w_j = 1$.

(d) Let $\lambda_{\min}$ and $\lambda_{\max}$ be the largest and the smallest eigenvalue of $H_f(x)$, respectively. Then

$$\lambda_{\min} \|v\|^2 \leq \langle H_f(x)v, v \rangle \leq \lambda_{\max} \|v\|^2.$$

Proof: (a) It is implied by the regularity of the function f and commutativity of derivatives

$$\frac{\partial^2 f(x)}{\partial x_i \partial x_j} = \frac{\partial^2 f(x)}{\partial x_j \partial x_i}.$$

(b) It follows from the fact than any symmetric real matrix has real eigenvalues.

(c) Expanding the vector v in the eigenvector orthonormal basis $\{\xi_i\}$ as $v = \sum_{i=1}^{n} v^i \xi_i$ and using inner product bilinearity, we have

$$\begin{aligned}
\langle Hv, v \rangle &= \sum_i \sum_j v^i v^j \langle H\xi_i, \xi_j \rangle = \sum_i \sum_j v^i v^j \lambda_i \langle \xi_i, \xi_j \rangle \\
&= \sum_i \sum_j v^i v^j \lambda_i \delta_{ij} = \sum_i (v^i)^2 \lambda_i.
\end{aligned}$$

Therefore, $\dfrac{\langle Hv, v \rangle}{\|v\|^2} = \sum_i w_i \lambda_i$, with $w_i = \dfrac{(v^i)^2}{\|v\|^2}$.

(d) Using that

$$\lambda_{\min} \leq \sum_i w_i \lambda_i \leq \lambda_{\max},$$

multiplying by $\|v\|^2$ yields the desired result.   ∎

Formula (4.2.4) provides the recurrence in the case of gradient descent method

$$x^{n+1} = x^n - \delta \nabla f(x^n). \tag{4.11.33}$$

The linear approximation

$$
\begin{aligned}
f(x^{n+1}) &\approx f(x^n) + (x^{n+1} - x^n)^T \nabla f(x^n) \\
&= f(x^n) - \delta \|\nabla f(x^n)\|^2 < f(x^n)
\end{aligned}
$$

produces a value less than the previous value, $f(x^n)$. But this is just an approximation, and if quadratic terms are being taking into account, then the previous inequality might not hold any more. We shall investigate this in the following. The second-order Taylor approximation provides

$$f(x^{n+1}) \approx f(x^n) + (x^{n+1} - x^n)^T \nabla f(x^n) + \frac{1}{2}(x^{n+1} - x^n)^T H_f(x^n)(x^{n+1} - x^n)$$

$$= f(x^n) - \delta \|\nabla f(x^n)\|^2 + \frac{1}{2}\delta^2 \langle H_f(x^n)\nabla f(x^n), \nabla f(x^n)\rangle.$$

The last term on the right is the correction term due to the curvature of f. There are two distinguished cases to look into:

1. The Hessian H_f is negative definite (all eigenvalues are negative). In this case the correction term is negative

$$\frac{1}{2}\delta^2 \langle H_f(x^n)\nabla f(x^n), \nabla f(x^n)\rangle < 0.$$

The previous Taylor approximation implies the inequality

$$f(x^{n+1}) < f(x^n) - \delta \|\nabla f(x^n)\|^2 < f(x^n),$$

i.e., for any learning rate $\delta > 0$ each iteration provides a lower value for f than the previous one.

2. The Hessian H_f is positive definite (all eigenvalues are positive). In this case the correction term is positive

$$\frac{1}{2}\delta^2 \langle H_f(x^n)\nabla f(x^n), \nabla f(x^n)\rangle > 0.$$

Proposition 4.11.1, part (d), provides the following margins of error for the correction term

$$0 < \frac{1}{2}\delta^2\lambda_{\min}\|\nabla f(x^n)\|^2 \leq \frac{1}{2}\delta^2\langle H_f(x^n)\nabla f(x^n), \nabla f(x^n)\rangle \leq \frac{1}{2}\delta^2\lambda_{\max}\|\nabla f(x^n)\|^2.$$

The idea is that if the correction term is too large, then the inequality $f(x^{n+1}) < f(x^n)$ might not hold. Considering the worst-case scenario, when the term is maximum, the following inequality has to be satisfied:

$$f(x^n) - \delta\|\nabla f(x^n)\|^2 + \frac{1}{2}\delta^2\lambda_{\max}\|\nabla f(x^n)\|^2 < f(x^n).$$

This is equivalent with the following condition on the learning rate

$$\delta < \frac{2}{\lambda_{\max}}. \tag{4.11.34}$$

Hence, if the learning rate is bounded as in (4.11.34), then the quadratic approximation provides for $f(x^{n+1})$ a value lower than $f(x^n)$. Formula (4.11.33) provides a minimization sequence for f, given the aforementioned constraint for the learning rate.

4.12 Newton's Method

Assume $f : \mathbb{R}^n \to \mathbb{R}$ is a function of class C^2, and let $x^0 \in \mathbb{R}^n$ be an initial guess for the minimum of f. Assume f is convex on a neighborhood of x^0. We approximate the function f about the point x^0 by a quadratic function as

$$f(x) \approx F(x) = f(x^0) + (x - x^0)^T\nabla f(x^0) + \frac{1}{2}(x - x^0)^T H_f(x^0)(x - x^0),$$

where $H_f(x^0)$ is the Hessian matrix of f evaluated at x^0. Since f is convex, its Hessian is positive definite. Therefore, the quadratic function F has a minimum at the critical point x^*, which satisfies $\nabla F(x) = 0$. Using that $\nabla F(x) = \nabla f(x^0) + H_f(x^0)(x - x^0)$, the equation $\nabla F(x) = 0$ becomes

$$H_f(x^0)x = -\nabla f(x^0) + H_f(x^0)x^0.$$

Assuming the Hessian does not have any zero eigenvalues, then it is invertible and we obtain the following critical point:

$$x^* = x^0 - H_f^{-1}(x^0)\nabla f(x^0),$$

which is the minimum of $F(x)$.

Newton's method consists of using this formula iteratively constructing the sequence $(x^n)_{n \geq 1}$ defined recursively by

$$x^{n+1} = x^n - H_f^{-1}(x^n)\nabla f(x^n). \tag{4.12.35}$$

This method converges to the minimum of f faster than the gradient descent method. However, despite its powerful value, it has a major weakness, which is the need to compute the inverse of the Hessian of f at each approximation.

Remark 4.12.1 It is worth noting that in the case $n = 1$ the previous iterative formula becomes

$$x^{n+1} = x^n - \frac{f'(x^n)}{f''(x^n)}.$$

If we denote $h(x) = f'(x)$, then looking for the critical point of f is equivalent to searching for the zero of $h(x)$. Convexity of f implies that h is increasing, which guarantees the uniqueness of the zero. The previous iteration formula is written as

$$x^{n+1} = x^n - \frac{h(x^n)}{h'(x^n)}.$$

We have arrived now to the familiar Newton-Raphson method for searching a zero of h using successive tangent lines.

4.13   Stochastic Search

The gradient descent method does not provide concluding results in all cases. This can be corrected by developing certain variants of the method, such as the momentum method, which is used to avoid getting stuck in a local minimum or to avoid overshooting the minimum. But sometimes there are other reasons, such as getting lost in a *plateau*, from which it takes a long time to get out, if ever. A plateau is a region where the gradient of the objective function ∇f is very small (or zero) corresponding to a relatively flat region of the surface. This section presents a method of stochastic flavor, which overcomes this major problem. The minimum is searched along a diffusion process rather than along a deterministic path. We shall present it in comparison with its deterministic counterpart, which we include first.

4.13.1   Deterministic variant

If a vector field in $\mathbb{R}^n$ is given by

$$b(x) = \sum_{k=1}^{n} b^k(x)\frac{\partial}{\partial x_k},$$

then its integral curve, $x(t)$, passing through the point x_0 is the solution of the differential equation

$$
\begin{aligned}
dx(t) &= b(x(t))\, dt \\
x(0) &= x_0.
\end{aligned}
$$

The vector field $b(x(t)) = \dot{x}(t)$ represents the field of velocities along the integral curves.

Consider the objective function $f : \mathbb{R}^n \to \mathbb{R}$, subject to minimization. In order to achieve this, we look for a vector field $b(x)$ for which the function $\varphi(t) = f(x(t))$ is decreasing. Equivalently, we look for a flow $x(t)$ along which the objective function f decreases its value.

Using the derivative given by the chain rule

$$
\varphi'(t) = \sum_{k=1}^{n} \frac{\partial f}{\partial x_k}_{|x(t)} \dot{x}_k(t) = \langle \nabla f, b \rangle_{|x(t)},
$$

a linear approximation provides

$$
\varphi(t + dt) = \varphi(t) + \varphi'(t)\, dt + o(dt^2).
$$

Hence, to ensure that the difference $\Delta\varphi(t) = \varphi(t + dt) - \varphi(t)$ is as negative as possible, we chose the vector field $b(x)$ such that

$$
b = \operatorname{argmin}\langle \nabla f, b \rangle,
$$

which is achieved for $b(x) = -\lambda \nabla f(x)$, for $\lambda > 0$, i.e., the vector field b is chosen to point to the opposite direction of the gradient of f. This method is equivalent to the method of steepest descent. This method fails if ∇f is very small or equal to zero on an entire region. The next variant takes care of this peculiarity.

4.13.2 Stochastic variant

The idea of this approach is to replace the deterministic trajectory $x(t)$ by a stochastic process X_t as in the following. In order to drive the iteration out of the plateau, we shall superpose some random noise term on the deterministic trajectory $x(t)$. This method uses the concepts of Brownian motion, Ito diffusion, Dynkin's formula, and infinitesimal generator operator, which the reader can give a glance in section D.8 of the Appendix.

After introducing the noise we obtain an Ito diffusion process X_t, starting at x_0, given by

$$
\begin{aligned}
dX_t &= b(X_t)dt + \sigma(X_t)dW_t \\
X_0 &= x_0,
\end{aligned}
$$

where $W_t = \big(W_1(t), \ldots, W_m(t)\big)$ is an m-dimensional Brownian motion. The coefficients $b(x)^T = (b_1(x), \ldots, b_n(x))$ and $\sigma(x) = (\sigma_{ij}(x)) \in \mathbb{R}^n \times \mathbb{R}^m$ represent, respectively, the *drift* and the *dispersion* of the process. They have to be selected such that the conditional expectation function

$$\varphi(t) = \mathbb{E}[f(X_t)|X_0 = x_0]$$

is decreasing as fast as possible. The value of the objective function along the diffusion, $f(X_t)$, is a random variable dependent on t, and we would like to minimize its expectation, given the starting point $X_0 = x_0$. This will be accomplished using Dynkin's formula. Before doing this, we need to recall the expression of the infinitesimal operator, $\mathcal{A}$, associated with the Ito process X_t given by the second-order differential operator

$$\mathcal{A} = \frac{1}{2} \sum_{i,j} (\sigma\sigma^T)_{ij} \frac{\partial^2}{\partial x_i \partial x_j} + \sum_k b_k \frac{\partial}{\partial x_k} = \frac{1}{2} \sigma\sigma^T \nabla^2 + \langle \nabla, b \rangle.$$

The matrix $\sigma\sigma^T$ is called *diffusion matrix*, and we shall get back to its form shortly.

Applying Dynkin's formula twice, we have

$$\varphi(t) = f(x_0) + \int_0^t \mathbb{E}[\mathcal{A}(f(X_s))|X_0 = x_0]\, ds$$

$$\varphi(t + dt) = f(x_0) + \int_0^{t+dt} \mathbb{E}[\mathcal{A}(f(X_s))|X_0 = x_0]\, ds.$$

Subtracting and using a linear approximation yields

$$\Delta\varphi(t) = \varphi(t + dt) - \varphi(t) = \int_t^{t+dt} \mathbb{E}[\mathcal{A}(f(X_s))|X_0 = x_0]\, ds$$

$$= \mathbb{E}[\mathcal{A}(f(X_t))|X_0 = x_0] + o(dt^2).$$

Following the same idea as in the method of steepest descent, we are interested in the process X_t for which the change $\Delta\varphi(t)$ is as negative as possible. Consequently, we require that

$$\mathbb{E}[\mathcal{A}(f(X_t))|X_0 = x_0] < 0. \qquad (4.13.36)$$

To construct the process X_t with this property it suffices to provide $\sigma(x)$ and $b(x)$. In a plateau the objective function is flat, and hence the Hessian is small. In order to get out of a plateau, we adopt the requirement that the "diffusion is large" when the "Hessian is small". This will steer the process X_t away from plateaus. Consequently, we choose the diffusion matrix $\sigma\sigma^T$ proportional to the inverse of the Hessian of the objective function, $H_f = \left(\frac{\partial^2 f}{\partial x_i \partial x_j}\right)_{ij}$, i.e.,

$$(\sigma\sigma^T)_{ij} = \lambda(H_f)_{ij}^{-1}, \tag{4.13.37}$$

with $\lambda > 0$ constant. In order to investigate the existence of solutions σ for equation (4.13.37), we assume the objective function satisfies the following properties:

(i) f is of class C^2;

(ii) f is strictly convex.

Under these two conditions the Hessian H_f is a real, symmetric, positive definite, and nondegenerate matrix. Hence, it is invertible, with the inverse also symmetric and positive definite. By the Cholesky decomposition, there is a lower triangular matrix σ satisfying (4.13.37). Asking for the strictly convexity of f is maybe a too strong requirement, but this condition guarantees the existence of a dispersion matrix σ. Under this choice, we can conveniently compute $\mathcal{A}f$ as

$$
\begin{aligned}
\mathcal{A}f(x) &= \frac{1}{2}\lambda\sum_{i,j}(H_f^{-1})_{ij}(H_f)_{ij} + \sum_k b_k(x)\frac{\partial f}{\partial x_k} \\
&= \frac{n}{2}\lambda + \langle \nabla f(x), b(x)\rangle
\end{aligned}
$$

In order to minimize $\mathcal{A}f(x)$, the drift term b is chosen such that it minimizes the second term, $\langle \nabla f(x), b(x)\rangle$, which comes with the choice $b(x) = -\eta\nabla f(x)$, with $\eta > 0$, constant. Consequently,

$$\mathcal{A}f(x) = \frac{n}{2}\lambda - \eta\|\nabla f(x)\|^2.$$

Requiring $\mathcal{A}f(x) < 0$ implies that the learning rates λ and η satisfy the inequality

$$\frac{\lambda}{\eta} < \frac{2}{n}\|\nabla f(x)\|^2.$$

If $L = \inf_x \|\nabla f(x)\|^2 \neq 0$, then it suffices to choose λ and η such that $\lambda = \frac{2L}{n}\eta$. Under this condition the inequality (4.13.36) always holds. The process X_t previously constructed represents a stochastic search of the minimum which has the advantage of scattering away the search from plateau regions.

For a better understanding we shall consider the following simple and explicit example.

Example 4.13.1 We shall consider the one-dimensional case given by the objective function $f(x) = \frac{1}{2}x^2$, $x \in (a, b)$, with $a > 0$. The diffusion $dX_t = b(X_t)dt + \sigma(X_t)dW_t$, $X_0 = x_0 \in (a, b)$ is one-dimensional, with $b(x)$ and $\sigma(x)$ continuous functions. Since the Hessian is $f''(x) = 1$, the equation $\sigma^2 = \lambda \frac{1}{f''(x)}$ implies $\sigma = \sqrt{\lambda}$. Also, $b(x) = -\eta f'(x) = -\eta x$. Since $L = \inf_{a<x<b} f'(x)^2 = a^2$, then $\lambda = 2a^2\eta$. The stochastic differential equation becomes the *Langevin's equation*

$$dX_t = -\eta X_t dt + \sqrt{\lambda}dW_t, \qquad X_0 = x_0.$$

The solution is given by the following Orstein-Uhlenbeck process

$$X_t = x_0 e^{-\eta t} + \sqrt{\lambda} \int_0^t e^{-\eta(t-s)}\, dW_s,$$

which is the sum between a deterministic function and a Wiener integral. Consequently, the process X_t is normally distributed with mean $x_0 e^{-\eta t}$ and variance $\frac{\lambda}{2\eta}(1 - e^{-2\eta t})$.

An ant searches for food in a stochastic way. If the food is located in the valley, the ant will eventually find it.

Ant food

Cartoon 2: An ant in a stochastic search for food.

Therefore, the search of the minimum is done along an Orstein-Uhlenbeck process. Its mean represents the expected direction of movement, while the integral is the white noise part. The previous relation between learning rates, $\lambda = 2a^2\eta$, can be seen as a condition stating that *the noise does not dominate the mean direction of movement.*

One can picture this as an ant that would like to get off a mountain and it tries to do so by searching its way in all directions, such that, on average, it lowers its altitude position.

The expected altitude $\varphi(t)$ at time t can be actually computed explicitly as in the following

$$
\begin{aligned}
\varphi(t) &= \mathbb{E}[f(X_t)|X_0 = x_0] = \frac{1}{2}\mathbb{E}[(X_t)^2|X_0 = x_0] \\
&= \frac{1}{2}\mathbb{E}\left[x_0^2 e^{-2\eta t} + 2x_0\sqrt{\lambda}e^{-\eta t}\int_0^t e^{-\eta(t-s)}\,dW_s + \lambda\left(\int_0^t e^{-\eta(t-s)}\,dW_s\right)^2\right] \\
&= \frac{1}{2}x_0^2 e^{-2\eta t} + \frac{1}{2}\lambda\int_0^t e^{-2\eta(t-s)}\,ds = \frac{1}{2}\left(x_0^2 e^{-2\eta t} + \frac{\lambda}{2\eta}(1 - e^{-2\eta t})\right) \\
&= \frac{1}{2}\left(e^{-2\eta t}\left(x_0^2 - \frac{\lambda}{2\eta}\right) + \frac{\lambda}{2\eta}\right).
\end{aligned}
$$

Since $\dfrac{\lambda}{2\eta} = a^2 < x_0^2$, it follows that $\varphi(t)$ is decreasing with the minimum

$$
\lim_{t \searrow 0} \varphi(t) = \frac{1}{2}\frac{\lambda}{2\eta} = \frac{1}{2}a^2.
$$

This is actually the expected result which can be also obtained using standard Calculus techniques to minimize $f(x) = \frac{1}{2}x^2$ on (a, b).

4.14 Neighborhood Search

The method presented in this section searches for a minimum on a neighborhood of the current point of evaluation and then makes a step into the direction of the smallest value. The difference from the gradient descent method is that we do not have an a priori knowledge of which will the direction of greatest descent be. In this method we search along a sphere of radius equal to the learning step, η. In the case when the dimension is $n = 1, 2$, the method can be implemented exactly, while for $n \geq 3$ it is computationally convenient to consider a stochastic search.

4.14.1 Left and Right Search

This search algorithm is applicable just for the one- dimensional case. Given that most practical examples of cost functions in machine learning have a lot more than one variable, this example is not applicable in real-life applications. However, we include it here for completeness reasons and for building a discussion basis for the multidimensional case.

Consider a smooth function $f(x)$ of a real variable x. We start the search for its minimum from the initial point x_0. We fix $\eta > 0$ small enough and evaluate the function to the left, $f(x_0 - \eta)$, and to the right, $f(x_0 + \eta)$, of x_0.

- If both neighbor values are larger than the initial value, $f(x_0 \pm \eta) > f(x_0)$, then x_0 is already a minimum point. In the case of the first iteration it is indicated to choose a different initial value x_0 and start the procedure again.
- If one neighbor value is smaller and the other one is larger, then make a step in the direction of the smaller value; for instance, if $f(x_0 - \eta) < f(x_0) < f(x_0 + \eta)$, then choose $x_1 = x_0 - \eta$.
- If both neighbor values are smaller than the current value, make a step into the direction of the smallest value (this is a greedy type algorithm). For instance, if $f(x_0 - \eta) < f(x_0 + \eta) < f(x_0)$, then choose $x_1 = x_0 - \eta$.
- If in any of the previous bullet points we encounter identities rather than inequalities, readjust the size of η such that the identities disappear.
- Starting the search procedure from x_1 we construct x_2, and so on. The sequence $(x_n)_n$ thus constructed will approximate the minimum of the function $f(x)$.

By its very construction, the sequence $(x_n)_n$ is stationary, i.e., there is a rank $N > 1$ such that $x_n = x_N$, for all $x \geq N$. In this case, N represents the optimum number of training epochs.

The method can yet be improved. After the approximation sequence $(x_n)_n$ has reached its stationary value, then decreasing the learning rate η changes the sequence into a better approximation sequence. Decreasing the learning rate can be applied until the desired degree of accuracy for the minimum is obtained.

4.14.2   Circular Search

In the two-dimensional case we shall look for a minimum about (x_0, y_0) using a search along the circle of radius η, whose equation is

$$(x, y) = (x_0, y_0) + \eta e^{it}, \qquad 0 \leq t < 2\pi.$$

Then we move into the direction of the smallest value and repeat the procedure, see Fig. 4.18. In order to do this, we choose n equidistant points, (x_0^k, y_0^k), along the previous circle, whose coordinates are given by

$$x_0^k = x_0 + \eta \cos \frac{2k\pi}{n}$$
$$y_0^k = y_0 + \eta \sin \frac{2k\pi}{n}, \qquad 0 \leq k \leq n - 1.$$

We evaluate the function f at the previous n points, (x_0^k, y_0^k), and pick the smallest value. Let

$$k_0 = \arg \min_k f(x_0^k, y_0^k)$$

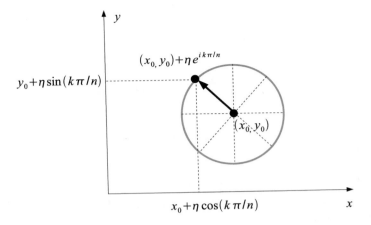

Figure 4.18: *The circle of center (x_0, y_0) and radius η is divided into n equal sectors. At each step the point of the smallest value is chosen and a step into that direction is made.*

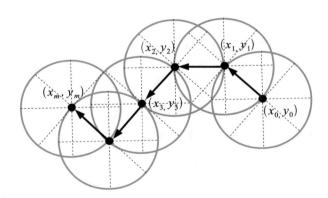

Figure 4.19: *The search of the minimum starts at (x_0, y_0) and follows the arrows. At each step the point of the smallest values is chosen and a step into that direction is made.*

and choose a new circle of radius η with the center at $(x_1, y_1) = (x_0^{k_0}, y_0^{k_0})$ and continue the previous procedure. This way we obtain the sequence $(x_m, y_m)_{m \geq 0}$, which approximates the minimum of $f(x, y)$. The recurrence relation is defined by $x_m = x_{m-1}^{k_{m-1}}$, $y_m = y_{m-1}^{k_{m-1}}$ with

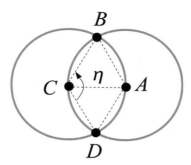

Figure 4.20: *Two equal circles intersect at points B and D and also pass through the centers of each other. Triangles $\triangle ABC$ and $\triangle ACD$ are equilateral, having all sides equal to η. Therefore the angle $\angle BCD = 120°$ and hence the arc BAD represents 1/3 of the total circle circumference, which means that the circle centered at C has 2/3 outside of the other circle.*

$$x_m^k = x_m + \eta \cos \frac{2k\pi}{n}$$
$$y_m^k = y_m + \eta \sin \frac{2k\pi}{n}, \qquad 0 \le k \le n-1,$$

where $k_m = \arg\min_k f(x_m^k, y_m^k)$. This means that the points (x_m^k, y_m^k) belong to the circle of center (x_m, y_m) and radius η, and (x_{m+1}, y_{m+1}) is roughly the point of the largest descent along this circles, see Fig. 4.19.

At each step it looks that the search of the minimum is done among n values. This number can be actually decreased. Assume that η is small enough so that we can assume that the minimum is not inside the previous circle. Then the search of the minimum should be applied only to the outside points of the next circle, see Fig. 4.20. The number of these points is about $2n/3$, since 1/3 of the points belong to the interior of the previous circle.

4.14.3   Stochastic Spherical Search

In the case when the dimension is large it is computationally expensive to construct equidistant points on a sphere of radius η. In this situation it is more convenient to choose n uniformly distributed points on the sphere.

Assume we have a smooth function $f(x)$ with $x \in \mathbb{R}^n$, which needs to be minimized. We start the search of the minimum from an initial point

x_0 by constructing n uniformly distributed points, $x_0^1, \ldots, x_0^n$, on the sphere $\mathbb{S}(x_0, \eta) = \{x; \|x - x_0\| = \eta\}$, where $\|\cdot\|$ denotes the Euclidean norm in $\mathbb{R}^n$. Choose $k^* = \arg\min_k f(x_0^k)$, and then consider the new sphere $\mathbb{S}(x_1, \eta)$, with $x_1 = x_0^{k^*}$. We continue the procedure by choosing n uniformly distributed points on the new sphere, and iterate the procedure. The sphere centers, x_m, form a sequence that tends toward the minimum point of the function f.

The only issue we still need to discuss is how to choose n random points on a sphere. We shall start with the two-dimensional case, when the sphere is a circle. We consider n uniformly distributed variables $\theta_1, \ldots, \theta_n \sim Unif(0, 2\pi)$. Then

$$(a + \eta \cos \theta_k, b + \eta \sin \theta_k), \qquad k = 1, \ldots n$$

are uniformly distributed on the circle centered at (a, b) and having radius η. The construction in the multidimensional case is similar. Consider the uniformly distributed random variables

$$\phi_1, \ldots, \phi_{n-2} \sim Unif(0, \pi), \quad \phi_{n-1} \sim Unif(0, 2\pi)$$

Then define $\mathbf{x} = (x_1, \ldots, x_n)$ on the m-dimensional sphere $\mathbb{S}(\mathbf{x}^0, \eta)$ by

$$
\begin{aligned}
x_1 &= x_1^0 + \eta \cos \phi_1 \\
x_2 &= x_2^0 + \eta \sin \phi_1, \cos \phi_2 \\
x_3 &= x_3^0 + \eta \sin \phi_1 \sin \phi_2 \cos \phi_3 \\
&\cdots \\
x_{m-1} &= x_{m-1}^0 + \eta \sin \phi_1 \cdots \sin \phi_{m-2} \cos \phi_{m-1} \\
x_m &= x_m^0 + \eta \sin \phi_1 \cdots \sin \phi_{m-2} \sin \phi_{m-1}.
\end{aligned}
$$

Repeating the procedure, we can create n instances of the above point $\mathbf{x}$.

As a variant of the previous algorithm, we may consider the n uniformly distributed points inside the ball $\mathbb{B}(x_0, \eta) = \{x; \|x - x_0\| \leq \eta\}$ instead of the sphere $\mathbb{S}(x_0, \eta)$.

Remark 4.14.1 Imagine you have to find an object hidden in the room. For each taken step you are provided with one of the hints "hot" or "cold", whether you get closer, or get farther from the hidden object, respectively. This is a stochastic search similar with the one described before. The indicator "hot" corresponds to the direction that minimizes the value of the function, and a step should be taken into that direction. Since the location of the hidden object is unknown, the steps are uniformly random and the one with the hottest indicator shall be considered.

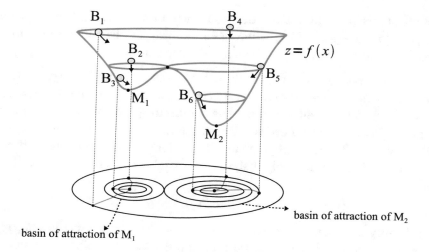

Figure 4.21: *In the search for the global minimum of the surface $z = f(x)$, several balls are left free to roll under the gravitation action. They will finally land to the local minima in their basins of attraction they initially belonged to.*

4.14.4   From Local to Global

Most search algorithms will produce an estimation of the local minimum which is closest to the initial search point. However, for machine learning purposes, we need a global minimum. In order to produce it, one should start the minimum search from several locations simultaneously. This way, we shall obtain several local minima points, among which the lowest one corresponds to the global minimum.

This method assumes the search domain is compact, so it can be divided into a finite partition. One initial search point will be chosen in each partition element, for instance, at its center, and a search is initiated there. If the partition is small enough, the number of local minima obtained is smaller than the number of partitions. All initial points situated in the basin of attraction of a local minimum will estimate that specific local minimum. At the end we should pick the one which is the smallest.

One way to imagine this is by enabling a certain number of balls to roll downhill on the surface $z = f(x)$, see Fig. 4.21. Balls B_1, B_2, and B_3 belong to the basin of attraction of the minimum M_1 and will roll toward M_1, while balls B_4, B_5, and B_6 will roll towards the minimum M_2. While only one ball will find only one local minimum, a relatively large number of balls, initially sparse enough distributed, will find all local minima of the function

$f(x)$. A final evaluation of f at these local minima points yield the global minimum.

4.15 Continuous Learning

This is learning with the gradient descent having an infinitesimal learning rate. In the classical version of gradient descent with learning rate $\eta > 0$, applied to the minimization of the cost function $f(x)$, with $x \in \mathbb{R}^n$, the update is given by

$$x(t_{n+1}) = x(t_n) - \eta \nabla f(x(t_n)).$$

We assume that $t_n = t_0 + n\Delta t$ is a sequence of equidistant instances of time and we consider the learning rate proportional to the time step, namely $\eta = \lambda \Delta t$, with $\lambda > 0$. Then the previous relation can be written as

$$\frac{x(t_n + \Delta t) - x(t_n)}{\Delta t} = -\lambda \nabla f(x(t_n)).$$

Taking the limit $\Delta t \to 0$ and replacing t_n by t, we obtain the differential equation $x'(t) = -\lambda \nabla f(x(t))$. This corresponds to *continuous learning*, which is learning with an infinitesimally small learning rate, $\eta = \lambda \, dt$. If the initial point is x_0, the continuous learning problem is looking for a differentiable curve $x(t)$ in $\mathbb{R}^n$ such that the following system of ODEs is satisfied

$$x'(t) = -\lambda \nabla f(x(t)) \tag{4.15.38}$$
$$x(0) = x_0. \tag{4.15.39}$$

In fact, $x(t)$ is an integral curve of the vector field $V = -\lambda \nabla f(x)$, namely the gradient vector $V(x(t))$ is the velocity vector field along the curve $x(t)$.

The curve $x(t)$ represents the curve along which the cost function has the lowest rate of decrease. If $u(t)$ is another curve, then

$$\frac{d}{dt} f(u(t)) = \langle \nabla f(u(t)), u'(t) \rangle$$

and from Cauchy's inequality

$$-\|\nabla f(u(t))\| \, \|u'(t)\| \le \frac{d}{dt} f(u(t)) \le \|\nabla f(u(t))\| \, \|u'(t)\|.$$

The lowest rate of decrease is obtained when the left inequality becomes identity

$$-\|\nabla f(u(t))\| \, \|u'(t)\| = \frac{d}{dt} f(u(t))$$

and this is achieved when $u'(t) = -\lambda \nabla f(u(t))$, with $\lambda > 0$ constant.

If $u(0) = x_0$, then by the local uniqueness of solutions property of linear systems of differential equations, it follows that $u(t) = x(t)$ for small enough values of the parameter t. In this case, the rate of change of the cost function along the gradient descent curve $x(t)$ is given by

$$\frac{d}{dt} f(x(t)) = \langle \nabla f(x(t)), x'(t) \rangle = -\frac{1}{\lambda} \|x'(t)\|^2 < 0.$$

Then the larger the magnitude of the velocity vector $x'(t)$, the faster the cost function decreases. If the minimum $x^* = \arg\min_x f(x)$ is realized at $x^* = \lim_{t \to T} x(t)$, with $T \in (0, \infty]$, then $\nabla f(x^*) = 0$, or equivalently, $\lim_{t \to T} \|x'(t)\| = 0$.

We shall provide an example of continuous learning following Example 4.2.7.

Example 4.15.1 Let $f : \mathbb{R}^n \to \mathbb{R}$ defined by $f(x) = \frac{1}{2}\|Ax - b\|^2$, where $b \in \mathbb{R}^m$ and A is an $m \times n$ matrix of rank n, and $\|\cdot\|$ denotes the Euclidean norm. Since

$$f(x) = \frac{1}{2}\|Ax\|^2 - x^T A^T b + \frac{1}{2}\|b\|^2,$$

the gradient is given by $\nabla f(x) = A^T A x - A^T b$. The continuous learning system (4.15.38)–(4.15.39) in this case becomes

$$\begin{aligned} x'(t) &= -\lambda A^T A x(t) + \lambda A^T b \\ x(0) &= x_0. \end{aligned}$$

Multiplying by the integrating factor $e^{\lambda A^T A t}$, we reduce it to the following exact equation

$$\frac{d}{dt}\left(e^{\lambda A^T A t} x(t)\right) = e^{\lambda A^T A t} \lambda A^T b.$$

Integrating, solving for $x(t)$, and using the initial condition, yields

$$x(t) = e^{-\lambda A^T A t} \int_0^t e^{\lambda A^T A s}\, ds \cdot \lambda A^T b + e^{-\lambda A^T A t} x_0. \tag{4.15.40}$$

The integral can be evaluated as in the following

$$\int_0^t e^{\lambda A^T A s}\, ds = \frac{1}{\lambda}(A^T A)^{-1}\left[e^{\lambda A^T A t} - \mathbb{I}_n\right].$$

Then substituting in (4.15.40) we obtain

$$x(t) = (A^T A)^{-1} A^T b - e^{-\lambda A^T A t}\left[(A^T A)^{-1} A^T b - x_0\right].$$

Denoting the Moore-Penrose pseudoinverse by $x^* = (A^T A)^{-1} A^T b$, see section G.2 of the Appendix, we can write the gradient descent curve as

$$x(t) = x^* - e^{-\lambda A^T A t} \left[x^* - x_0 \right].$$

Since $\lim_{t \to \infty} e^{-\lambda A^T A t} = \mathbb{O}_n$, see Proposition G.2.1 of Appendix, in the long run the curve $x(t)$ approaches the Moore-Penrose pseudoinverse, $\lim_{t \to \infty} x(t) = x^*$, which is the minimum of the function $f(x) = \frac{1}{2} \|Ax - b\|^2$.

4.16 Summary

Finding the global minimum of a given cost function is an important ingredient of any machine learning algorithm. The learning process is associated with the procedure of tuning the cost function variables into their optimal values. Some methods are of first order, i.e., involve just the first derivative (such as the gradient descent, line search, momentum method, etc.), fact that makes them run relatively fast. Others are second-order methods, involving the second derivative, or curvature, of the cost function (such as the Hessian method, Newton's method, etc.). Other methods have a topological or stochastic flavor.

The most well-known method is the gradient descent. The minimum is searched in the direction of the steepest descent of the function, which is indicated by the negative direction of the gradient. The method is easy to implement but exhibits difficulties for nonconvex functions with multiple local minima or functions with plateaus. The method has been improved in a number of ways. In order to avoid to get stuck in a local minimum or overshooting the minimum, the momentum method (and its variants) has been developed. The line search algorithm looks for a minimum along a line normal to the level surfaces and implements an adjustable rate.

Several adaptive optimization techniques derived from the gradient descent method, like AdaGrad, Adam, AdaMax, and RMSProp, are presented. They share the common feature that involve moments approximations computed by exponentially moving averages.

The Hessian method and Newton's method involve the computation of the second-order approximation of the cost function. Even if theoretically they provide good solutions, there are computationally expensive.

The simulated annealing is a probabilistic method for computing minima of objective functions which are interpreted as the internal energy of a thermodynamical system. Increasing the heat parameter of the system and then cooling it down slowly enables the system to reach its lowest energy level.

Stochastic search proposes an algorithm of driving the search out of plateaus. This involves looking for a minimum along a stochastic process rather than a deterministic path. This method is also computationally expensive, since it involves the computation of the inverse of the Hessian of the cost function.

Neighborhood search is a minimum search of a topological nature. The minimum is searched uniformly along spheres centered at the current search center.

All the above search methods produce local minima. In order to find a global minimum, certain methods have to be applied. One of them is applicable in the case of compact domains. In this case the domain is divided into a finite partition and a minimum is searched on each partition set. The lowest minimum approximates the global minimum.

4.17   Exercises

Exercise 4.17.1 Let $f(x_1, x_2) = e^{x_1} \sin x_2$, with $(x_1, x_2) \in (0, 1) \times (0, \frac{\pi}{2})$.
(a) Show that f is a harmonic function;
(b) Find $\|\nabla f\|$;
(c) Show that the equation $\nabla f = 0$ does not have any solutions;
(d) Find the maxima and minima for the function f.

Exercise 4.17.2 Consider the quadratic function $Q(x) = \frac{1}{2}\mathbf{x}^T A \mathbf{x} - b\mathbf{x}$, with A nonsingular square matrix of order n.
(a) Find the gradient ∇Q;
(b) Write the gradient descent iteration;
(c) Find the Hessian H_Q;
(d) Write the iteration given by Newton's formula and compute its limit.

Exercise 4.17.3 Let A be a nonsingular square matrix of order n and $b \in \mathbb{R}^n$ a given vector. Consider the linear system $A\mathbf{x} = b$. The solution of this system can be approximated using the following steps:
(a) Associate the cost function $C(\mathbf{x}) = \frac{1}{2}\|A\mathbf{x} - b\|^2$. Find its gradient, $\nabla C(\mathbf{x})$, and Hessian, $H_C(\mathbf{x})$;
(b) Write the gradient decent algorithm iteration which converges to the system solution $\mathbf{x}$ with the initial value $\mathbf{x}^0 = 0$;
(c) Write the Newton's iteration which converges to the system solution $\mathbf{x}$ with the initial value $\mathbf{x}^0 = 0$.

Exercise 4.17.4 (a) Let $(a_n)_n$ be a sequence with $a_0 > 0$ satisfying the inequality

$$a_{n+1} \leq \mu a_n + K, \forall n \geq 1,$$

with $0 < \mu < 1$ and $K > 0$. Show that the sequence $(a_n)_n$ is bounded from above.

(b) Consider the momentum method equations (4.4.16)–(4.4.17), and assume that the function f has a bounded gradient, $\|\nabla f(x)\| \leq M$. Show that the sequence of velocities, $(v^n)_n$, is bounded.

Exercise 4.17.5 (a) Let f and g be two integrable functions. Verify that

$$\int (f * g)(x)\, dx = \int f(x)\, dx \int g(x)\, dx;$$

(b) Show that $\|f * g\|_1 \leq \|f\|_1\, \|g\|_1$;

(c) Let $f_\sigma = f * G_\sigma$, where $G_\sigma(x) = \frac{1}{\sqrt{2\pi}\sigma} e^{-\frac{x^2}{2\sigma^2}}$. Prove that $\|f_\sigma\|_1 \leq \|f\|_1$ for any $\sigma > 0$.

Exercise 4.17.6 Show that the convolution of two Gaussians is also a Gaussian:

$$G_{\sigma_1} * G_{\sigma_2} = G_\sigma,$$

with $\sigma = \sqrt{\sigma_1^2 + \sigma_2^2}$.

Exercise 4.17.7 Show that if n numbers have the sum equal to s,

$$\sigma_1 + \cdots + \sigma_n = s,$$

then the numbers for which the sum of their squares, $\sum_{j=1}^n \sigma_j^2$, is minimum occurs for the case when all the numbers are equal to s/n.

Chapter 5

Abstract Neurons

The *abstract neuron* is the building block of any neural network. It is a unit that mimics a biological neuron, consisting of an input (incoming signal), weights (synaptic weights) and activation function (neuron firing model). This chapter introduces the most familiar types of neurons (perceptron, sigmoid neuron, etc.) and investigates their properties.

5.1 Definition and Properties

In the light of examples presented in Chapter 1, it makes sense to consider the following definition that formalizes our previously developed intuition:

Definition 5.1.1 *An abstract neuron is a quadruple* $(\mathbf{x}, \mathbf{w}, \varphi, y)$*, where* $\mathbf{x}^T = (x_0, x_1, \ldots, x_n)$ *is the input vector,* $\mathbf{w}^T = (w_0, w_1, \ldots, w_n)$ *is the weights vector, with* $x_0 = -1$ *and* $w_0 = b$*, the bias, and* φ *is an activation function that defines the outcome function* $y = \varphi(x^T w) = \varphi(\sum_{i=}^n w_i x_i)$*.*

The way the abstract neuron learns a desired target variable $\mathbf{z}$ is by tuning the weights vector $\mathbf{w}$ such that a certain function measuring the error between the desired variable $\mathbf{z}$ and the outcome y is minimized. Several possible error functions have been covered in Chapter 3 and their minimization algorithms were treated in Chapter 4.

The abstract neuron is represented in Fig. 5.1. The computing unit is divided into two parts: the first one contains the summation symbol, Σ, which indicates a summation of the inputs with the given weights, and the second contains the generic activation function φ used to define the output y.

© Springer Nature Switzerland AG 2020

O. Calin, *Deep Learning Architectures*, Springer Series in the Data Sciences,

https://doi.org/10.1007/978-3-030-36721-3_5

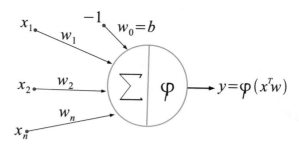

Figure 5.1: *Abstract neuron with bias b and activation function φ.*

It is worth noting that the threshold b is included in the weights system as a bias, $w_0 = b$, with constant input $x_0 = -1$. Equivalently, we may also consider the bias $w_0 = -b$ and the corresponding input $x_0 = 1$. In any of the cases, the expression of the inner product is given by

$$\mathbf{x}^T\mathbf{w} = \mathbf{w}^T\mathbf{x} = w_1x_1 + \cdots + w_nx_n - b.$$

The vectors $\mathbf{w}^T = (w_0, \ldots, w_n)$ and $\mathbf{x}^T = (x_0, \ldots, x_n)$ are the *extended weights* and *extended input* vectors. The output function $y = f(\mathbf{x}) = \varphi(\mathbf{w}^T\mathbf{x})$ is sometimes called the *primitive function* of the given computing unit.

The inputs $\{x_i\}_{i=1,n}$ are communicated through edges to the computing unit, after there are previously multiplied by a weight, see Fig. 5.1. The input data can be of several numerical types, such as:
- *binary* if $x_i \in \{0,1\}$ for $i = 1, n$;
- *signed* if $x_i \in \{-1,1\}$ for $i = 1, n$;
- *digital* if $x_i \in \{0,1,2,3,4,5,6,7,8,9\}$ for $i = 1, n$;
- *arbitrary real numbers* if $x_i \in (-\infty, \infty)$ for $i = 1, n$;
- *an interval of real numbers* if $x_i \in [0,1]$ for $i = 1, n$.

For instance, any handwritten digit can be transformed into binary data, see Fig. 5.2. The 4×4 matrix is read line by line and placed into a sequence of 0s and 1s with length 16. A value of 1 is assigned if the pixel is activated more than 50%. Otherwise, the value 0 is assigned. Coding characters this way is quite simplistic, some information regarding local shapes being lost in the process. However, there are better ways of encoding figures using convolution. In general, in the case of grayscale images each pixel has an activation which is a number between 0 and 1 (1 for full black activation and 0 for no activation, or white color).

Input efficiency Assume that one would like to implement a neuron circuit. One question which can be asked is which type of input is more efficient? Equivalently, how many switching states optimize the transmitted information from the circuit implementation cost point of view?

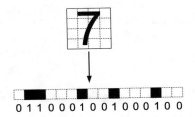

Figure 5.2: *Transformation of a 4 × 4 pixels character into a sequence of 0s and 1s.*

To answer this, let β denote the number of input signal states, for instance, $\beta = 2$ for the binary signals. The number of channels, or inputs, is provided by the number n. The implementation cost, F, is assumed proportional with both n and β, i.e., $F = cn\beta$, where $c > 0$ is a proportionality constant. Using n channels with β states, one can represent β^n numbers.[1] Now we assume the cost F is fixed and try to optimize the number of transmitted numbers as a function of β. Consider the constant $k = \frac{F}{c}$. Then $n = \frac{F}{c\beta} = \frac{k}{\beta}$. Therefore $f(\beta) = \beta^{\frac{k}{\beta}}$ numbers can be represented, for a given $\beta > 0$. This function reaches its maximum at the same point as its logarithm, $g(\beta) = \ln f(\beta) = \frac{k}{\beta} \ln \beta$. Since its derivative is $g'(\beta) = \frac{k}{\beta}(1 - \ln \beta)$, it follows that the maximum is achieved for $\beta = e \approx 2.718$. Taking the closest integer value, it follows that $\beta = 3$ is the optimal number of states for the input signals. However, this book deals with the theory of neuron models with any other number of input states.

In the following we shall present a few classical types of neurons specifying their input types and activation functions.

5.2 Perceptron Model

A *perceptron* is a neuron with the input either zero or one, i.e., $x_i \in \{0, 1\}$, and with the activation function given by the Heaviside function

$$\varphi(x) = \begin{cases} 0, & \text{if } x < 0 \\ 1, & \text{if } x \geq 0. \end{cases}$$

---

[1]This is the number of sequences of length n with elements chosen from the given β states. By a sequence we understand a function $h : \{1, \ldots, n\} \to \{1, \ldots, \beta\}$.

The output of the perceptron is given as a *threshold gate*

$$y = \varphi(x^T w - b) = \begin{cases} 0, & \text{if } \sum_{i=1}^n w_i x_i < b \\ 1, & \text{if } \sum_{i=1}^n w_i x_i \geq b. \end{cases}$$

This way, a perceptron is a rule that makes decisions by weighting up evidence supplied by the inputs x_i. The threshold b is a measure of how easy the perceptron's decision to get the output 1 is. For all general purposes, this is regarded as another weight, denoted by w_0, called bias.

We shall deal next with the geometric interpretation of a perceptron. Consider now an $(n-1)$-dimensional hyperplane in $\mathbb{R}^n$ defined by

$$\mathcal{H} = \{(x_1, \ldots, x_n); \sum_{i=1}^n w_i x_i = b\}.$$

Its normal vector N is given in terms of weights as

$$N^T = (w_1, \ldots, w_n),$$

where ^{T} denotes the transpose vector, as usual. The hyperplane passes through a point $\mathbf{p}$, which is related to the bias by relation $b = \mathbf{p}^T w$. Then the outcome y of a perceptron associates the value 0 to one of the half-spaces determined by the hyperplane $\mathcal{H}$, and 0 to the rest (the other half-space and $\mathcal{H}$). For the case of two inputs, $n = 2$, see Fig. 5.3. Roughly speaking, given a linear boundary between two countries, a perceptron can decide whether a selected point belongs to one of the countries, or not.

The perceptron can implement the logical gates *AND* ("$\wedge$") and *OR* ("$\vee$") as in the following. This property makes the perceptron important for logical computation.

Implementing *AND.* Consider the Boolean operation defined by the next table:

x_1	x_2	$y = x_1 \wedge x_2$
0	0	0
0	1	0
1	0	0
1	1	1

The same output function, $y = x_1 \wedge x_2$, can be generated by a perceptron with two inputs $x_1, x_2 \in \{0, 1\}$, weights $w_1 = w_2 = 1$, bias $b = -1.5$, and $x_0 = -1$, see Fig. 5.4 **a**. We have

$$y = \varphi(x_1 + x_2 - 1.5) = \begin{cases} 0, & \text{if } x_1 + x_2 < 1.5 \\ 1, & \text{if } x_1 + x_2 \geq 1.5 \end{cases}$$

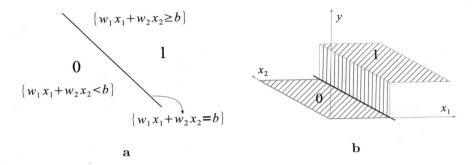

Figure 5.3: **a.** *The separation of the plane into half-planes.* **b.** *The graph of the activation function for a perceptron with two inputs, x_1 and x_2.*

The condition $\{x_1 + x_2 \geq 1.5\}$ is satisfied only for $x_1 = x_2 = 1$, while the second condition $\{x_1 + x_2 < 1.5\}$ holds for the rest of the combinations. It follows that this specific perceptron has the same output function as the one given by the previous table.

In the light of the previous perceptron's geometric interpretation, the line $x_1 + x_2 = 1.5$ splits the input data $\{(0,0), (0,1), (1,0), (1,1)\}$ into two classes, by associating the value 1 to the half-plane $\{x_1 + x_2 \geq 1.5\}$, and the value 0 to the open half-plane $\{x_1 + x_2 < 1.5\}$ see Fig. 5.5 **a**.

Implementing OR. This Boolean operation is defined by the table:

x_1	x_2	$y = x_1 \vee x_2$
0	0	0
0	1	1
1	0	1
1	1	1

The output function, $y = x_1 \vee x_2$, is generated by a perceptron with two inputs $x_1, x_2 \in \{0, 1\}$, weights $w_1 = w_2 = 1$, bias $b = -0.5$, and $x_0 = -1$, see Fig. 5.4 **b**. The output is written as

$$y = \varphi(x_1 + x_2 - 0.5) = \begin{cases} 0, & \text{if } x_1 + x_2 < 0.5 \\ 1, & \text{if } x_1 + x_2 \geq 0.5 \end{cases}$$

The conclusion follows from the observation that condition $\{x_1 + x_2 < 0.5\}$ is satisfied only for $x_1 = x_2 = 0$.

Figure 5.5 **b** represents a split of the input data $\{(0,0), (0,1), (1,0), (1,1)\}$ into two classes: the first is the data in the shaded half-plane $\{x_1 + x_2 \geq 0.5\}$ and the other is the data in the open half-plane $\{x_1 + x_2 < 0.5\}$. The value 1 is associated to the first class and the value 0 the the second class.

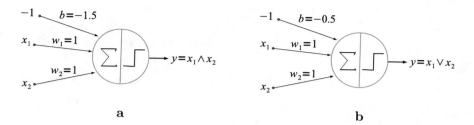

Figure 5.4: *Boolean function implementation using a perceptron:* **a.** *The output function AND.* **b.** *The output function OR.*

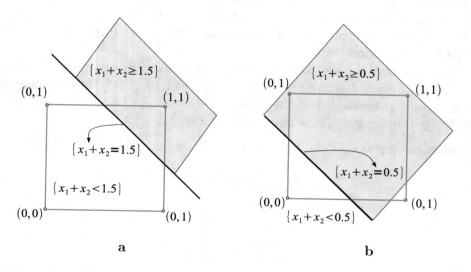

Figure 5.5: *Partition using a perceptron; the shaded half-plane is associated the value 1.* **a.** *The output function AND.* **b.** *The output function OR.*

However, a perceptron cannot implement all logical functions. For instance, it can't implement XOR (exclusive OR) function, which is an operation defined by the table:

x_1	x_2	y
0	0	0
0	1	1
1	0	1
1	1	0

The impossibility for a perceptron to implement XOR function follows from the fact that there is no line that separates symbols ⓪ and ① in Fig. 5.6 **a**. This can be shown in a couple of ways.

(*a*) One proof follows as a consequence of the separation properties of the plane. If one assumes, by contradiction, the existence of a separation line, then the plane is divided by the line in two half-planes, H_1 and H_2, which are convex sets[2], see Fig. 5.6 **b**. Since both ①-symbols belong to H_1, then by convexity the line segment joining them is entirely included into H_1. Similarly, the line segment joining both ⓪-symbols is contained in H_1. Since H_1 and H_2 do not have any points in common (as they form a partition of the plane), the line segments joining both ①-symbols and both ⓪-symbols are disjoint. However, it can be seen from Fig. 5.6 **a** that these two line segments do intersect, as diagonals of a a square. This leads to a contradiction, which shows that the assumption made is false, and hence there is no separation line.

(*b*) The second proof variant is based on algebraic reasons. Assume there is a separation line of the form $w_1 x_1 + w_2 x_2 = b$, which separates the points $\{(0,0),(1,1)\}$ from the points $\{(0,1),(1,0)\}$. Considering a choice, we have from testing the first pair of points that $0 < b$ and $w_1 + w_2 < b$ and for the second pair that $w_1 > b$, $w_2 > b$. The sum of the last two inequalities together with the second inequality implies the contradictory inequality $2b < b$, since b is positive.

The fact that one perceptron cannot learn the XOR function was pointed out in [87]. We shall see later in Remark 6.1.2 that a neural network of two perceptrons can successfully learn this Boolean function.

The next paragraph will present the perceptron as a linear classifier in more detail.

---

[2] A set is convex if for any two points that belong to the set the line segment defined by the points is included in the set.

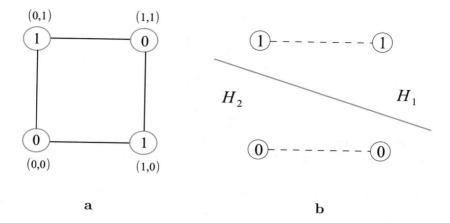

Figure 5.6: *Cluster classification using one perceptron:* **a.** *The output function XOR.* **b.** *A line divides the plane in two convex sets* H_1 *and* H_2.

Clusters splitting. Motivated by the previous application of splitting the input data into classes, we extend it to a more general problem of two-cluster classification. We allow now for the input data to be numbers taking values in the interval $[0, 1]$ and assume for the sake of simplicity that $n = 2$. Then each data is a pair of real numbers that can be represented as a point in the unit square. Assume now that we associate a double-valued label or characteristic with each point, given by "group 1" (or, red color, star shaped, etc.) and "group two" (or blue color, disk shaped, etc.). The question that arises is: *Can a perceptron decide whether a point belongs to one group or another?* The answer depends on the data distribution. If data can be separated by a line such that one group is contained in one half-plane and the other in the other half-plane (i.e., the groups are separated by a line), then the perceptron might eventually be able to decide which group is associated with what label after a proper tuning of the weights. In order to do this, we introduce the function

$$z(x_1, x_2) = \begin{cases} 0, & \text{if } (x_1, x_2) \in \mathcal{G}_1 \\ 1, & \text{if } (x_1, x_2) \in \mathcal{G}_2, \end{cases}$$

where notations $\mathcal{G}_1$ and $\mathcal{G}_2$ stand for "group 1" and "group 2", respectively. The percepton decides between the two groups if we are able to come up with two weights w_1, w_2, and a bias b such that the line $w_1 x_1 + w_2 x_2 - b = 0$ separates the groups $\mathcal{G}_1, \mathcal{G}_2$. If this is possible, then the output function

$$y = \varphi(w_1 x_1 + w_2 x_2 - b)$$

has the same expression as the aforementioned function z. More details about the learning algorithms of a perceptron will be provided in later chapter.

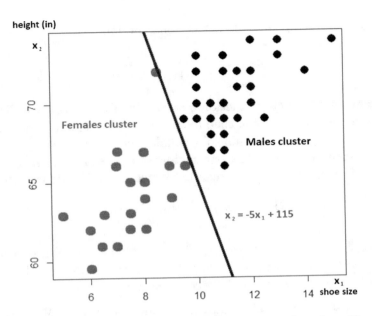

Figure 5.7: *Perceptron classifying groups of males and females. The line $x_2 = -5x_1 + 115$ is a decision boundary in the feature space $\{x_1, x_2\}$ between two linearly separable clusters.*

Example 5.2.1 Consider the input $x = (x_1, x_2)$, where x_1 denotes the shoe size and x_2 the height (in inches) of some individuals of a given population. We associate a label to each individual by considering the mapping $(x_1, x_2) \to \ell \in \{\text{man, woman}\}$. Given a feature pair (x_1, x_2), a perceptron is able to distinguish whether this corresponds to a man or to a woman if there is a separation line, $w_1 x_1 + w_2 x_2 = b$, between the two gender groups.

Data collected from a population of Michigan freshmen has the scatter plot shown in Fig. 5.7. We notice that the female group is linearly separable from the male group and a separation line is given by $x_2 = -5x_1 + 115$. Hence, given an individual with shoe size s and height h, the perceptron would classify it as "male" if $h > -5s + 115$ and as "female" otherwise.

This problem can be formulated in terms of neurons in a couple of ways, as follows.

(*i*) First, we shall associate numerical labels to data points. We shall label the points corresponding to males by $z = 1$ and the ones corresponding to females by $z = -1$. Consider a neuron with input $x = (x_1, x_2)$, weights w_1,

w_2, and bias b, having the step activation function

$$S(u) = \begin{cases} 1, & \text{if } u \geq 0 \\ -1, & \text{if } u < 0. \end{cases}$$

The neuron's output is the function $y = S(w_1x_1 + w_2x_2 - b)$, see Fig. 5.8 **a**. If (x_1, x_2) corresponds to a man, then $w_1x_1 + w_2x_2 - b > 0$, and hence the output is $y = 1$. If (x_1, x_2) corresponds to a female, then $w_1x_1 + w_2x_2 - b < 0$, and then the output is $y = -1$. The parameters of the separator line are obtained by minimizing the distance between the outcome and label vectors as

$$(w_1^*, w_2^*, b^*) = \arg\min_{w_i, b} \|y - z\|^2,$$

with $\|y - z\|^2 = \sum_{x \sim data} (y(x) - z(x))^2$, where $z(x)$ and $y(x)$ are, respectively, the label and the output of the data point x. This distance can be minimized using, for instance, the gradient descent method.

(ii) In this case we shall label data using one-hot vectors. We associate the vector $(1, 0)$ to males and $(0, 1)$ to females. We consider a neural network with input $x = (x_1, x_2)$, and two output neurons with a softmax activation function

$$(y_1, y_2) = softmax(u_1, u_2) = \left(\frac{e^{u_1}}{e^{u_1} + e^{u_2}}, \frac{e^{u_2}}{e^{u_1} + e^{u_2}} \right),$$

see Fig. 5.8 **b**. Here, u_i are the signals collected from the inputs as

$$\begin{aligned} u_1 &= w_{11}x_1 + w_{21}x_2 - b_1 \\ u_2 &= w_{21}x_1 + w_{22}x_2 - b_2. \end{aligned}$$

The cost function subject to be minimized in this case is

$$\sum_{x \sim data} \left(\|y_1(x) - z_1(x)\|^2 + \|y_2(x) - z_2(x)\|^2 \right).$$

5.3   The Sigmoid Neuron

We have seen that a perceptron works as a device that provides two decision states, 0 and 1, with no intermediate values. The situation can be improved if the neuron would take all the states in the interval $(0, 1)$; this way, outputs closer to 1 correspond to decisions that are more likely to occur, while the outputs closer to 0 represent decisions with small chances of occurrence. In this case, the neuron output acts as the likelihood of taking a certain weighted decision.

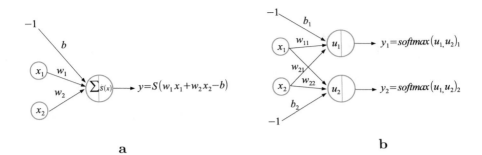

Figure 5.8: **a.** *The neuron in the case when the labels are $z \in \{-1, 1\}$.* **b.** *The neural networks in the case of one-hot-vector labels.*

A *sigmoid neuron* is a computing unit with the input $\mathbf{x} = (x_1, \ldots, x_n)$, and with the activation function given by the logistic function

$$\sigma(z) = \frac{1}{1 + e^{-z}}. \tag{5.3.1}$$

The vector $\mathbf{w}^T = (w_1, \ldots, w_n)$ denotes the weights and b is the neuron bias. The output of a sigmoid neuron is a number in $(0, 1)$, which is given by

$$y = \sigma(\mathbf{w}^T \mathbf{x} - b) = \frac{1}{1 + e^{-\mathbf{w}^T \mathbf{x} + b}}.$$

If the activation function is considered to be the scaled logistic function

$$\sigma_c(z) = \frac{1}{1 + e^{-cz}}, \qquad c > 0$$

then the output of the sigmoid neuron tends to the output of the perceptron for large values of c.

The advantage of this type of neurons is twofold: (i) it approximates perceptrons; (ii) the output function is continuous with respect to weights, i.e., small changes in the weights and threshold correspond to small changes in the output; this does not hold in the case of the perceptron, whose output has a jump discontinuity. This continuity property will be of uttermost importance in the learning algorithms in later chapters.

It is worth noting the way the weights w_i relate to the input x_i and the output y. For this we shall use the formula for the inverse of the logistic function, $\sigma^{-1}(x) = \ln \frac{x}{1 - x}$. Then the output equation $y = \sigma(\mathbf{w}^T \mathbf{x} - b)$ becomes $\ln \frac{y}{1 - y} = \mathbf{w}^T \mathbf{x} - b$. Now, we evaluate the expression at two distinct values of inputs, $x_i^* = x_i$ and $x_i^* = x_i + 1$:

$$\ln\left(\frac{y}{1-y}\bigg|x_i = x_i^*\right) = w_1 x_1 + \cdots + w_i x_i^* + \cdots + w_n x_n - b$$

$$\ln\left(\frac{y}{1-y}\bigg|x_i = x_i^* + 1\right) = w_1 x_1 + \cdots + w_i(x_i^* + 1) + \cdots + w_n x_n - b.$$

Subtracting, and using the properties of logarithmic functions, yields

$$w_i = \ln\left(\frac{\frac{y}{1-y}\big|x_i = x_i^* + 1}{\frac{y}{1-y}\big|x_i = x_i^*}\right), \tag{5.3.2}$$

which is an expression independent of the value x_i^*. As long as the ith input increases by 1 unit, formula (5.3.2) provides an expression for the ith weight of the neuron, w_i, in terms of the output y. In statistics, the expression $\frac{y}{1-y}$ is used under the name of *odds to favor*. This way, the weight w_i can be expressed in terms of the quotient of two odds to favor.

Different types of sigmoid neurons can be obtained considering activation functions given by other sigmoidal functions, such as: hyperbolic tangent, arctangent function, softsign, etc. The only difference between them is the learning speed due to different rates of saturation. However, from the mathematical point of view they will be treated in a similar way. The infinitesimal change in the output of a sigmoid neuron, $y = \varphi(\mathbf{w}^T \mathbf{x} - b)$, with activation function φ, in terms of the changes in weights and bias is given by the differential

$$\begin{aligned}
dy &= \sum_{i=1}^{n} \frac{\partial y}{\partial w_i} dw_i + \frac{\partial y}{\partial b} db \\
&= \sum_{i=1}^{n} \varphi'(\mathbf{w}^T \mathbf{x} - b) x_i dw_i + \varphi'(\mathbf{w}^T \mathbf{x} - b)(-1) db \\
&= \varphi'(\mathbf{w}^T \mathbf{x} - b)\left(\sum_{i=1}^{n} x_i dw_i - db\right).
\end{aligned}$$

We note that the change dy is proportional to the derivative of φ. The rate φ' depends on the sigmoidal function chosen, and in some cases can be easily computed. For instance, if $\varphi(z) = \sigma(z)$ is the logistic function, then

$$dy = \sigma(\mathbf{w}^T \mathbf{x} - b)\left(1 - \sigma(\mathbf{w}^T \mathbf{x} - b)\right)\left(\sum_{i=1}^{n} x_i dw_i - db\right).$$

In the case when $\varphi(z) = \mathbf{t}(z)$ is the hyperbolic tangent function, we have

$$dy = (1 - \mathbf{t}^2(\mathbf{w}^T \mathbf{x} - b))\left(\sum_{i=1}^{n} x_i dw_i - db\right).$$

The previous differentials are useful when computing the gradient of the output

$$\nabla y = (\nabla_w y, \partial_b y) = \left(\mathbf{x} \varphi'(\mathbf{w}^T \mathbf{x} - b), -\varphi'(\mathbf{w}^T \mathbf{x} - b) \right) = \varphi'(\mathbf{w}^T \mathbf{x} - b)(\mathbf{x}, -1).$$

This states that the gradient output, ∇y, is proportional to the extended input vector $(\mathbf{x}, -1)$, the proportionality factor being the derivative of φ. This will be useful in the back-propagation algorithm later.

5.4 Logistic Regression

In this section we shall present two applications of the sigmoid neuron which learns using *logistic regression*. The first example deals with forecasting the probability of default of a company, and the second one with a binary classification into clusters.

5.4.1 Default probability of a company

We are interested in predicting the probability of default of a company for a given time period $[0, T]$, using a sigmoid neuron. Assume the input $\mathbf{x}$ of the neuron is a vector which contains some information regarding the company, such as: cash reserves, revenue, costs, labor expenses, etc. The training set consists of n pairs $(\mathbf{x}_i, z_i)$, with $\mathbf{x}_i$ inputs as before and $z_i \in \{-1, 1\}$. A value $z_i = 1$ means the ith company has defaulted during $[0, T]$; a value $z_i = -1$ means that the ith company has not defaulted during $[0, T]$. The measurements $(\mathbf{x}_i, z_i)_{1 \leq i \leq n}$ represent empirically a pair of random variables (X, Z), where Z takes values ± 1 and X is a m-dimensional random vector. The conditional probability $P(Z|X)$ is given by the table

z	-1	1
$P(z|\mathbf{x})$	$1 - h(\mathbf{x})$	$h(\mathbf{x})$

for some positive function $h : \mathbb{R}^m \to [0, 1]$, where m is the dimension of the input X. One convenient way to choose $h(\mathbf{x})$ is using the sigmoid function

$$h(\mathbf{x}) = \sigma(\mathbf{w}^T \mathbf{x}).$$

The inner product $\mathbf{w}^T \mathbf{x}$ is a weighted sum of the inputs and can be interpreted as a "default score" of the company with the input $\mathbf{x}$. A high (positive) score $\mathbf{w}^T \mathbf{x}$ implies a value of $\sigma(\mathbf{w}^T \mathbf{x})$ close to 1, which can be interpreted as a high probability of default, provided that $\sigma(\mathbf{w}^T \mathbf{x})$ is considered as a probability. On the other side, a low (negative) score $\mathbf{w}^T \mathbf{x}$ implies a value of $\sigma(\mathbf{w}^T \mathbf{x})$ close to 0, which can be interpreted as a low probability of default. These two cases can be encapsulated in only one formula by writing the default probability as

$\sigma(z\mathbf{w}^T\mathbf{x})$, where $z \in \{-1, 1\}$. This can be shown as in the following. When $z = 1$, we recover the above formula, $h(\mathbf{x}) = \sigma(\mathbf{w}^T\mathbf{x})$, and when $z = -1$, using the property of the sigmoid, $\sigma(-x) = 1 - \sigma(x)$, we have

$$\sigma(z\mathbf{w}^T\mathbf{x}) = \sigma(-\mathbf{w}^T\mathbf{x}) = 1 - \sigma(\mathbf{w}^T\mathbf{x}) = 1 - h(\mathbf{x}).$$

Hence, the previous table can be written equivalently as

z	-1	1
$P(z\|x)$	$\sigma(-\mathbf{w}^T\mathbf{x})$	$\sigma(\mathbf{w}^T\mathbf{x})$

The probability distribution depends now on the parameter $\mathbf{w}$, and we shall consider it as a "model distribution", which is produced by the neuron. The next step is to find the weight $\mathbf{w}$ for which the model distribution approximates in "the best way" the training data $\{(\mathbf{x}_i, z_i)\}$, $1 \le i \le n$. This shall be accomplished by choosing $\mathbf{w}$ such that the likelihood of z_i being in a proximity of y_i is maximized. Assuming the training measurements are independent, we have

$$
\begin{aligned}
\mathbf{w}^* &= \arg\max_{\mathbf{w}} P(z_1, \ldots, z_n | \mathbf{x}_1, \ldots, \mathbf{x}_n) \\
&= \arg\max_{\mathbf{w}} \prod_{i=1}^{n} P(z_i | \mathbf{x}_i) = \arg\max_{\mathbf{w}} \ln\left(\prod_{i=1}^{n} P(z_i | \mathbf{x}_i)\right) \\
&= \arg\max_{\mathbf{w}} \sum_{i=1}^{n} \ln P(z_i | \mathbf{x}_i) = \arg\min_{\mathbf{w}}\left(-\sum_{i=1}^{n} \ln P(z_i | \mathbf{x}_i)\right) \\
&= \arg\min_{\mathbf{w}}\left(-\frac{1}{n}\sum_{i=1}^{n} \ln P(z_i | \mathbf{x}_i)\right) = \arg\min_{\mathbf{w}} \mathbb{E}^{\widehat{p}_X}\Big[-\ln P(Z|X)\Big],
\end{aligned}
$$

where we have used the properties of logarithmic function and the fact that a factor of $\frac{1}{n}$ does not affect the optimum, as well as the fact that a switch in the sign swaps the max into a min. The final expression involving the expectation with respect to $\widehat{p}_X$, is a the cross-entropy of the empirical probability $\widehat{p}_X$ with the conditional probability $P(Z|X)$. This expression of the cross-entropy can be computed explicitly using the previous formula for the probability and the expression (5.3.1) for the sigmoid function as in the following

$$
\begin{aligned}
\mathbb{E}^{\widehat{p}_X}\Big[-\ln P(Z|X)\Big] &= -\frac{1}{n}\sum_{i=1}^{n} \ln P(z_i | x_i) = -\frac{1}{n}\sum_{i=1}^{n} \ln \sigma(z_i \mathbf{w}^T \mathbf{x}_i) \\
&= -\frac{1}{n}\sum_{i=1}^{n} \ln \frac{1}{1 + e^{-z_i \mathbf{w}^T \mathbf{x}_i}} = \frac{1}{n}\sum_{i=1}^{n} \ln(1 + e^{-z_i \mathbf{w}^T \mathbf{x}_i}).
\end{aligned}
$$

The optimal weight is then given by

$$\mathbf{w}^* \;=\; \arg\min_{\mathbf{w}} \left(\frac{1}{n} \sum_{i=1}^{n} \ln(1 + e^{-z_i \mathbf{w}^T \mathbf{x}_i}) \right)$$

This can be written using the softplus function as

$$\mathbf{w}^* \;=\; \arg\min_{\mathbf{w}} \left(\frac{1}{n} \sum_{i=1}^{n} sp(-z_i \mathbf{w}^T \mathbf{x}_i) \right),$$

or, equivalently, as

$$\mathbf{w}^* \;=\; \arg\min_{\mathbf{w}} \left(\frac{1}{n} \sum_{i=1}^{n} sp(z_i \mathbf{w}^T \mathbf{x}_i) - \frac{1}{n} \sum_{i=1}^{n} z_i \mathbf{w}^T \mathbf{x}_i \right),$$

where we used that $sp(-x) = sp(x) - x$.

Gradient Descent There is no closed-form solution in this case for the optimum weight $\mathbf{w}^*$. We shall approximate the value $\mathbf{w}^*$ by a sequence $\mathbf{w}^{(j)}$ using the gradient descent method. The gradient of the previous cross-entropy error

$$F(\mathbf{w}) = \frac{1}{n} \sum_{i=1}^{n} \ln(1 + e^{-z_i \mathbf{w}^T \mathbf{x}_i})$$

is given by

$$\nabla_{\mathbf{w}} F \;=\; -\frac{1}{n} \sum_{i=1}^{n} \frac{z_i \mathbf{x}_i e^{-z_i \mathbf{w}^T \mathbf{x}_i}}{1 + e^{-z_i \mathbf{w}^T \mathbf{x}_i}} = -\frac{1}{n} \sum_{i=1}^{n} \frac{z_i \mathbf{x}_i}{1 + e^{z_i \mathbf{w}^T \mathbf{x}_i}}$$

$$= \frac{1}{n} \sum_{i=1}^{n} z_i \mathbf{x}_i \sigma(-z_i \mathbf{w}^T \mathbf{x}_i)$$

$$= \frac{1}{n} \sum_{i=1}^{n} z_i \mathbf{x}_i \sigma(z_i \mathbf{w}^T \mathbf{x}_i) - \frac{1}{n} \sum_{i=1}^{n} z_i \mathbf{x}_i.$$

Note that the last term is the empirical expectation of the product of random variables X and Z, i.e., $\frac{1}{n} \sum_{i=1}^{n} z_i \mathbf{x}_i = \widehat{\mathbb{E}}[XZ]$. The first term is a weighted sum of the entries $(\mathbf{x}_i, z_i)$ with the weights $\rho_i = \sigma(z_i \mathbf{w}^T \mathbf{x}_i) \in (0, 1)$.

To conclude, given an initial weight initialization $\mathbf{w}^{(0)}$, the approximation sequence that approaches the minimum argument, $\mathbf{w}^*$, of the error function $F(\mathbf{w})$ is given by

$$\mathbf{w}^{(j+1)} \;=\; \mathbf{w}^{(j)} - \eta \nabla_{\mathbf{w}} F(\mathbf{w}^{(j)})$$

$$= \mathbf{w}^{(j)} - \frac{\eta}{n} \sum_{i=1}^{n} z_i \mathbf{x}_i \, \sigma(-z_i \mathbf{w}^{(j)^T} \mathbf{x}_i),$$

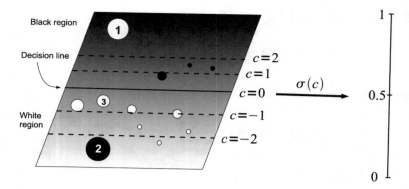

Figure 5.9: *The family of lines $w_1x_1 + w_2x_2 - b = c$, $c \in \mathbb{R}$. The sigmoid function σ maps each line to a point in $(0,1)$ regarded as a probability.*

where $\eta > 0$ is the learning rate.

How do we apply this technique in practice? Assume we have the training set $\{(\mathbf{x}_i, z_i)\}_{i \leq n}$ and consider a given input $\mathbf{x}$ describing the parameters of a certain company we would like to evaluate the probability of default. Then the desired probability is given by $h(\mathbf{x}) = \sigma(\mathbf{w}^{*T}x)$, where $\mathbf{w}^* = \lim\limits_{j \to \infty} \mathbf{w}^{(j)}$.

5.4.2   Binary Classifier

This section deals with the sigmoid neuron as a *binary classifier*. This means that the sigmoid neuron is trained to distinguish between two distinct clusters in the plane, the learning algorithm being the logistic regression.

Assume we have two groups of points in the plane: black and white. We would like to split the points in two groups: black points cluster, denoted by $\mathcal{G}_1$ and white points cluster, denoted by $\mathcal{G}_2$. If $x = (x_1, x_2)$ represents the coordinates of a point in the plane, consider the target given by the following decision function:

$$z(x_1, x_2) = \begin{cases} 1, & \text{if } x \in \mathcal{G}_1 \\ -1, & \text{if } x \in \mathcal{G}_2, \end{cases}$$

i.e., each black point is associated the label 1, while each white point the label -1. Assume there is a decision line in the plane, given by $w_1x_1 + w_2x_2 - b = 0$, which attempts to split the points into clusters $\mathcal{G}_1$ and $\mathcal{G}_2$; say $\mathcal{G}_1$ above the decision line and $\mathcal{G}_2$ below it, see Fig. 5.9. The job is to adjust parameters w_1, w_2, and b such that the aforementioned line represents "the best" split of data into clusters of identic color.

Consider the partition of the plane into lines parallel to the decision line

$$\{w_1 x_1 + w_2 x_2 - b = c;\ c \in \mathbb{R}\},$$

see Fig. 5.9. The lines with a positive value of c correspond to the black region, $\mathcal{G}_1$, while the lines with a negative value of c correspond to the white region, $\mathcal{G}_2$. The line corresponding to $c = 0$ is the decision line between the two clusters of distinct colors. We can think the plane as a gray scale from pure white, when $c = -\infty$, to pure black, for $c = +\infty$. Also the 50% black-white mixture is realized for $c = 0$.

Each point is more or less the same color to its neighboring background. If a given point has a color that is in large discrepancy with the background, then the "surprise" element is large, and this corresponds to a large amount of information. If the point color and the background do not differ much in tone, the amount of information provided by this point is small. We shall construct an information measure based on this color difference effect, subject to be minimized later. The information associated to a point of a given color with coordinate $\mathbf{x} = (x_1, x_2)$ is defined as

$$H(\mathbf{x}) = -\ln \sigma\big(z(\mathbf{x})(\mathbf{w}^T \mathbf{x} - b)\big),$$

where $\mathbf{w} = (w_1, w_2)$. We shall explain this construction for each of the following cases: unclassified points and correct classified points.

The case of unclassified points. Now, we pick an unclassified point of coordinate x, i.e. a point of a different color than its neighboring background. For instance, we pick the point with label "1" in Fig. 5.9. This is a white point in a very black region, so the information associated with this event is large. We represented this in the figure by considering a larger radius, so the information is associated with the area of the white disk. There is a positive constant $c > 0$ such that this point belongs to the line $w_1 x_1 + w_2 x_2 - b = c$. The information measure is constructed now as in the following. The sigmoid function σ maps the constant c into a value between 0 and 1, which can be considered as a probability, $P(x) = \sigma(c)$. The information associated with the point x is given by the negative log-likelihood function, $-\ln P(x) = -\ln \sigma(c)$. If c is large, then $\sigma(c)$ is close to 1 and hence the information $-\ln P(x)$ is close to 0, which does not make sense, since when c is large, the information should be also large (as a white point in very black region). To fix this problem, we shall define the information slightly different, using the complementary probability. Since the point is white, $x \in \mathcal{G}_2$, then $z(x) = -1$. Define the probability as $P(x) = \sigma(z(x)c) = \sigma(-c) = 1 - \sigma(c)$, so when c is large, the probability is close to 0 and hence the information $-\ln P(x) = -\ln \sigma(z(x)c) = -\ln \sigma(-c)$ is also large.

The same approach applies to the black point with the label "2" laying in the white region, see Fig. 5.9. In this case $c < 0$ and, since the point is white, $x \in \mathcal{G}_1$, we have $z(x) = 1$. The information is given again by $-\ln \sigma(z(x)c)$, and for c very negative the value of the information is large, as expected.

The case of correct classified points. We choose a correct classified point, for instance, the white point with label "3" situated in the white area, see Fig. 5.9. Let x denote its coordinate and $c < 0$ be the constant such that $w^T x - b = c$. The information associated with this point is small if the difference in the tone color with respect to the background is small. This occurs when the constant c is large and negative. We associate a probability with x, defined by $P(x) = \sigma(z(x)c)$. In this case $z(x) = -1$, since the point is white, $x \in \mathcal{G}_2$. Then, for c large and negative, the value of $\sigma(z(x)c)$ is close to 1, and hence its logarithm is close to 0. Therefore, it makes sense to define the information of this point as $-\ln \sigma(z(x)c)$.

Now we are in the situation of being able to find the best fitting line using an information minimization error. We shall see that this is equivalent to the maximum likelihood method. Consider n points in the plane with coordinates $\mathbf{x}_i$, and color type $z_i = z(\mathbf{x}_i)$, $1 \le i \le n$. The total information is defined as the sum of all individual points information (both correct and misclassified). This is

$$E(\mathbf{x}, z) = \sum_{i=1}^{n} H(\mathbf{x}_i) = -\sum_{i=1}^{n} \ln \sigma\big(z_i(\mathbf{w}^T \mathbf{x}_i - b)\big).$$

The "best decision line" is the line corresponding to the values of $\mathbf{w}$ and b that minimize the aforementioned information. The optimal values are given by

$$
\begin{aligned}
(\mathbf{w}^*, b^*) \;&=\; \arg\min_{\mathbf{w},b} E(\mathbf{x}, z) = \arg\max_{\mathbf{w},b} \sum_{i=1}^{n} \ln \sigma\big(z_i(\mathbf{w}^T \mathbf{x}_i - b)\big) \\
&=\; \arg\max_{\mathbf{w},b} \sum_{i=1}^{n} \ln \sigma\big(z_i(\mathbf{w}^T \mathbf{x}_i - b)\big) \\
&=\; \arg\max_{\mathbf{w},b} \ln \Big(\prod_{i=1}^{n} \sigma\big(z_i(\mathbf{w}^T \mathbf{x}_i - b)\big) \Big) \\
&=\; \arg\max_{\mathbf{w},b} \ln \Big(\prod_{i=1}^{n} P_{\mathbf{w},b}(z_i|\mathbf{x}_i) \Big),
\end{aligned}
$$

where $P_{\mathbf{w},b}(z|\mathbf{x})$ represents the probability that the color is of type z (black or white), given that the point coordinate is $\mathbf{x}$. We infer that finding the optimal parameters $(\mathbf{w}^*, b^*)$, which define the decision line $\mathbf{w}^{*T} \mathbf{x} - b^* = 0$, are obtained using a maximum likelihood method.

It is worth noting that the aforementioned error function, $E(\mathbf{x}, z)$, can be also regarded as a cross-entropy of the empirical probability of X with the conditional probability of Z, given X, as

$$E(\mathbf{x}, z) = \mathbb{E}^{\widehat{p}_X}[-\ln P(Z|X)],$$

with the model probability given by $P(z|X = \mathbf{x}) = -\ln \sigma\big(z(\mathbf{w}^T\mathbf{x}_i - b)\big)$.

Gradient descent In the absence of closed-form formulae for $\mathbf{w}^*$ and b^*, we shall employ the gradient descent method to obtain some appropriate approximation. The gradient has two components

$$\nabla E = (\nabla_{\mathbf{w}}E, \partial_b E).$$

Writing the error as

$$E(\mathbf{x}, z) = -\sum_{i=1}^n \ln \frac{1}{1 + e^{-\big(z_i(\mathbf{w}^T\mathbf{x}_i - b)\big)}} = \sum_{i=1}^n \ln(1 + e^{-(z_i(\mathbf{w}^T\mathbf{x}_i - b))}),$$

differentiating, we obtain

$$\nabla_{\mathbf{w}}E = -\sum_{i=1}^n \frac{z_i\mathbf{x}_i}{1 + e^{-(z_i\mathbf{w}^T\mathbf{x}_i - z_i b)}}$$

$$\partial_b E = \sum_{i=1}^n \frac{z_i}{1 + e^{-(z_i\mathbf{w}^T\mathbf{x}_i - z_i b)}}.$$

The approximation sequence is defined recursively by

$$\mathbf{w}^{(j+1)} = \mathbf{w}^{(j)} - \eta\,\nabla_{\mathbf{w}}E(\mathbf{w}^{(j)}, b^{(j)})$$

$$b^{(j+1)} = b^{(j)} - \eta\,\partial_b E(\mathbf{w}^{(j)}, b^{(j)}),$$

where $\eta > 0$ is the learning rate.

Remark 5.4.1 There are also other error functions one can consider. For instance, one simple error function is the number of misclassified points. If m_1 and m_{-1} represent the number of misclassified black and white points, respectively, then the error becomes $E = m_1 + m_{-1}$. The flaw of this error function is its discontinuity with respect to parameters w and b, and hence no differentiable method would work for it.

5.4.3 Learning with the square difference cost function

Consider an abstract neuron as in Fig.5.1. Its inputs are $\mathbf{x}^T = (x_1, \ldots, x_n)$, the output is $y = \sigma(\mathbf{w}^T\mathbf{x} - b)$, with weights $\mathbf{w}^T = (w_1, \ldots, w_n)$ and bias b. We shall assume the cost function is given by

$$C(\mathbf{w}, b) = \frac{1}{2}(y - z)^2,$$

where the target, z, is a number in $(0, 1)$. The minimum of $C(\mathbf{w}, b)$ is equal to zero and is achieved for $y = z$. This occurs when the weights and bias satisfy the linear equation $\mathbf{w}^T\mathbf{x} - b = \sigma^{-1}(z)$. Since this equation has a hyperplane of solutions, $(w_1, \ldots, w_n, b)$, the minimum of the cost function is not unique. Therefore, learning a real number, $z \in (0, 1)$, can be done exactly in multiple ways. This is no longer true for the case when the neuron learns a random variable. We shall deal with this case next.

Assume the input is an n-dimensional random vector $X = (X_1, \ldots, X_n)$ and the target Z a one-dimensional random variable taking values in $(0, 1)$. Consider m measurements of the variables (X, Z) given by the training set

$$\{(\mathbf{x}^{(1)}, z^{(1)}), \ldots, (\mathbf{x}^{(m)}, z^{(m)})\},$$

with $\mathbf{x}^{(j)} \in \mathbb{R}^n$ and $z^{(j)} \in \mathbb{R}$, $1 \le j \le m$.

The cost function is given by the empirical mean

$$C(w, b) = \mathbb{E}[(Y - Z)^2] = \frac{1}{2m} \sum_{j=1}^{m} \left(\sigma(\mathbf{w}^T\mathbf{x}^{(j)} - b) - z^{(j)}\right)^2.$$

If an exact learning would hold, i.e., if $C = 0$, then $\mathbf{w}^T\mathbf{x}^{(j)} - b = \sigma^{-1}(z^{(j)})$, $1 \le j \le m$. Therefore, (w, b) satisfies a linear system with m equations and $n + 1$ unknowns. Since in practice m is much larger than n (i.e., the number of observations is a lot larger than the input dimension), the previous linear system is incompatible, so exact learning does not hold.

In this case the minimum of the cost function is computed using the gradient descent method. For this it suffices to compute the partial derivatives with respect to weights and bias:

$$\frac{\partial C}{\partial w_k} = \frac{1}{m} \sum_{j=1}^{m} \left(\sigma(\mathbf{w}^T\mathbf{x}^{(j)} - b) - z^{(j)}\right)\sigma'(\mathbf{w}^T\mathbf{x}^{(j)} - b)x_k^{(j)}$$

$$\frac{\partial C}{\partial b} = -\frac{1}{m} \sum_{j=1}^{m} \left(\sigma(\mathbf{w}^T\mathbf{x}^{(j)} - b) - z^{(j)}\right)\sigma'(\mathbf{w}^T\mathbf{x}^{(j)} - b).$$

The approximation sequence for the optimal values of parameters is given by

$$(\mathbf{w}^{(k+1)}, b^{(k+1)}) = (\mathbf{w}^{(k)}, b^{(k)}) - \eta(\nabla_w C, \nabla_b C),$$

with η learning rate.

5.5   Linear Neuron

The *linear neuron* is a neuron with a linear activation function, $\varphi(x) = x$, and n random inputs. Its learning algorithm uses the least mean squares cost

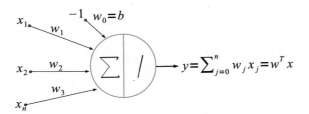

Figure 5.10: *Linear neuron with bias b and linear activation function.*

function. It was actually implemented as a physical device by Widrow and Hoff [125]. It can be used to recognize patterns, data filtering and, of course, to approximate linear functions.

The neuron has n inputs, which are random variables, $X_1, \ldots, X_n$. The weight for the input X_j is a number denoted by w_j. The bias is considered to be the weight $w_0 = b$, see Fig. 5.10. Consider $X_0 = -1$, constant. We shall adopt the vectorial notation

$$
X = \begin{pmatrix} X_0 \\ X_1 \\ \cdots \\ X_n \end{pmatrix}, \qquad \mathbf{w} = \begin{pmatrix} w_0 \\ w_1 \\ \cdots \\ w_n \end{pmatrix}
$$

Given that the activation function is the identity, the neuron output is given by the one-dimensional random variable

$$
Y = \sum_{j=0}^{n} w_j X_j = \mathbf{w}^T X = X^T \mathbf{w}.
$$

The desired output, or the target, is given by a one-dimensional random variable Z. The idea is to tune the parameter vector $\mathbf{w}$ such that Y learns Z; the learning algorithm in this case is to minimize the expectation of the squared difference, so the optimal parameter is

$$
\mathbf{w}^* = \arg \min_{\mathbf{w}} \mathbb{E}[(Z - Y)^2].
$$

We need to find $\mathbf{w}^*$, and we shall do this in a few ways. Some of these methods have more theoretical value, while others are more practically.

Exact solution A computation shows that the previous error function is quadratic in $\mathbf{w}$:

$$
\begin{aligned}
\mathbb{E}[(Z - Y)^2] &= \mathbb{E}[Z^2 - 2ZY + Y^2] = \mathbb{E}[Z^2 - 2ZX^T\mathbf{w} + (\mathbf{w}^T X)(X^T \mathbf{w})] \\
&= \mathbb{E}[Z^2] - 2\mathbb{E}[ZX^T]\mathbf{w} + \mathbf{w}^T \mathbb{E}[XX^T]\mathbf{w} \\
&= c - 2\mathbf{b}^T \mathbf{w} + \mathbf{w}^T A \mathbf{w}.
\end{aligned}
$$

The coefficients have the following meaning: $c = \mathbb{E}[Z^2]$ is the second centered moment of the target Z, $\mathbf{b} = \mathbb{E}[ZX]$ is a vector measuring the cross-correlation between the input and the output, and

$$A = \mathbb{E}[XX^T] = \begin{pmatrix} \mathbb{E}[X_0 X_0] & \mathbb{E}[X_0 X_1] & \cdots & \mathbb{E}[X_0 X_n] \\ \mathbb{E}[X_1 X_0] & \mathbb{E}[X_1 X_1] & \cdots & \mathbb{E}[X_1 X_n] \\ \cdots & \cdots & \cdots & \cdots \\ \mathbb{E}[X_n X_0] & \mathbb{E}[X_n X_1] & \cdots & \mathbb{E}[X_n X_n] \end{pmatrix}$$

is a matrix describing the autocorrelation of inputs. In the following analysis we shall assume that A is nondegenerate and positive definite (i.e., the inputs are coherent).

Denoting the aforementioned real-valued quadratic error function by

$$\xi(\mathbf{w}) = c - 2\mathbf{b}^T \mathbf{w} + \mathbf{w}^T A \mathbf{w},$$

its gradient is given by $\nabla_{\mathbf{w}} \xi(\mathbf{w}) = 2A\mathbf{w} - 2\mathbf{b}$, see Example 4.1.3. The optimal weight $\mathbf{w}^*$ is obtained as a solution of $\nabla_{\mathbf{w}} \xi(\mathbf{w}) = 0$, which becomes the linear system $A\mathbf{w} = \mathbf{b}$. Since A is nondegenerate, the system has the unique solution $\mathbf{w}^* = A^{-1}\mathbf{b}$. This is a minimum point, since the Hessian of the error is given by $H_\xi(\mathbf{w}) = 2A$, with A positive definite.

Gradient descent method In real life, when n is very large, it is computationally expensive to find the inverse A^{-1}. Hence, the need of a faster method to produce the optimal weight, $\mathbf{w}^*$, even if only as an approximation. This can be seen as a trade-off between the solution accuracy and the computer time spent to find the optimum.

In this case, the gradient descent method is more practical. The error function $\xi(\mathbf{w})$ is convex and has a minimum, see Fig. 5.11. We start from an arbitrary initial weight vector $\mathbf{w}^{(0)} = (w_0^{(0)}, \ldots, w_n^{(0)})^T \in \mathbb{R}^n$. Then construct the approximation sequence $(\mathbf{w}^{(j)})_j$ defined recursively by equation (4.2.4)

$$\begin{aligned} \mathbf{w}^{(j+1)} &= \mathbf{w}^{(j)} - \eta \nabla_{\mathbf{w}} \xi(\mathbf{w}^{(j)}) \\ &= \mathbf{w}^{(j)} - 2\eta(A\mathbf{w}^{(j)} - \mathbf{b}) \\ &= (\mathbb{I}_n - 2\eta A)\mathbf{w}^{(j)} + 2\eta\mathbf{b} \\ &= M\mathbf{w}^{(j)} + 2\eta\mathbf{b}, \end{aligned}$$

where $M = \mathbb{I}_n - 2\eta A$, with $\mathbb{I}_n$ denoting the unitary n-dimensional matrix. Iterating, we obtain

$$\begin{aligned} \mathbf{w}^{(j)} &= M^j \mathbf{w}^{(0)} + (M^{j-1} + M^{j-2} + \cdots + M + \mathbb{I}_n) 2\eta\mathbf{b} \\ &= M^j \mathbf{w}^{(0)} + (\mathbb{I}_n - M^j)(\mathbb{I}_n - M)^{-1} 2\eta\mathbf{b} \\ &= M^j \mathbf{w}^{(0)} + (\mathbb{I}_n - M^j)A^{-1}\mathbf{b}. \end{aligned} \tag{5.5.3}$$

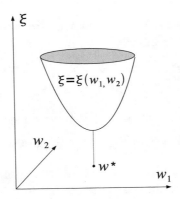

$$\xi = \xi(w_1, w_2)$$

w_2

w^*

w_1

Figure 5.11: *The quadratic error function has a global minimum* $\mathbf{w}^*$.

Assume the learning rate $\eta > 0$ is chosen such that $\lim_{j \to \infty} M^j = \mathbb{O}_n$. Then taking the limit in the previous formula yields $\mathbf{w}^* = \lim_{j \to \infty} \mathbf{w}^{(j)} = A^{-1}\mathbf{b}$, which recovers the previous result.

Now we return to the assumption $\lim_{j \to \infty} M^j = \mathbb{O}_n$. Since M is a symmetric matrix, its eigenvalues are all real. Similarly with the method applied in Example 4.2.7, this assumption is equivalent to showing that the eigenvalues $\{\lambda_i\}$ of M are bounded, with $|\lambda_i| < 1$. We do this in two steps:

Step 1. Show that $\lambda_i < 1$.
Let λ_i denote an eigenvalue of M. Then $\det(M - \lambda_i \mathbb{I}_n) = 0$. Substituting for M, this becomes $\det\left(A - \dfrac{1 - \lambda_i}{2\eta}\mathbb{I}_n\right) = 0$. This implies that $\alpha_i = \dfrac{1 - \lambda_i}{2\eta}$ is an eigenvalue of matrix A. Since A is positive definite and nondegenerate, it follows that $\alpha_i > 0$, which implies that $\lambda_i < 1$.

Step 2. Show that $\lambda_i > 0$ for η small enough.
The condition $\lambda_i > 0$ is equivalent to $\frac{1 - \lambda_i}{\alpha_i} < \frac{1}{\alpha_i}$, where we used that A has positive eigenvalues. This can be written in terms of η as $2\eta < \frac{1}{\alpha_i}$. Hence, the learning rate has to be chosen such that

$$0 < \eta < \min_i \frac{1}{2\alpha_i}. \tag{5.5.4}$$

The closed-form expression (5.5.3) is not of much practical use, since it contains the inverse A^{-1}. In real life we use the iterative formula

$$\mathbf{w}^{(j+1)} = M\mathbf{w}^{(j)} + 2\eta\mathbf{b},$$

where the learning rate η satisfies the inequality (5.5.4).

We shall estimate next the error at the jth iteration, $\epsilon_j = |\mathbf{w}^{(j)} - \mathbf{w}^*|$, using equation (5.5.3). We have:

$$\begin{aligned}
\mathbf{w}^{(j)} - \mathbf{w}^* &= M^j\mathbf{w}^{(0)} + (\mathbb{I}_n - M^j)\mathbf{w}^* - \mathbf{w}^* \\
&= M^j(\mathbf{w}^{(0)} - \mathbf{w}^*).
\end{aligned}$$

Using that $|M\mathbf{x}| \leq \|M\|\|\mathbf{x}|$ for all $\mathbf{x} \in \mathbb{R}^n$, we have

$$\epsilon_j = |\mathbf{w}^{(j)} - \mathbf{w}^*| = |M^j(\mathbf{w}^{(0)} - \mathbf{w}^*)| \leq \|M\|^j|(\mathbf{w}^{(0)} - \mathbf{w}^*)| = \mu^j d,$$

where $\mu = \|M\|$ is the norm of M (considering M as a linear operator), and $d = |(\mathbf{w}^{(0)} - \mathbf{w}^*)|$ is the distance from the initial value of the weight to the limit of the sequence. Since the norm of the matrix M is its largest eigenvalue, $\|M\| = \max_i \lambda_i$, and $\lambda_i < 1$, then $\mu \in (0, 1)$, and hence $\mu^j d \to 0$ as $j \to \infty$.

Gradient estimates For the sake of computer implementations, the error function $\xi(w) = \mathbb{E}[(Z-Y)^2]$ has to be estimated from measurements, $(\mathbf{x}^{(j)}, z^{(j)})$, where $\mathbf{x}^{(j)T} = (x_0^{(j)} \cdots x_n^{(j)})$. The empirical error is

$$\widehat{\xi}(w) = \frac{1}{m}\sum_{j=1}^{m}(z^{(j)} - \mathbf{w}^T\mathbf{x}^{(j)})^2 = \frac{1}{m}\sum_{j=1}^{m}\epsilon_j^2,$$

where $\epsilon_j = z^{(j)} - \mathbf{w}^T\mathbf{x}^{(j)}$ is the error between the desired value $z^{(j)}$ and the output value $\mathbf{w}^T\mathbf{x}^{(j)}$. The number m is regarded as the size of the mini-batch used for computing the empirical error.

In order to save computation time, a crude estimation is done, i.e., $m = 1$, which uses a single sample error, case in which the previous sum is replaced by only one term, $\widehat{\xi}(w) = \epsilon_j^2$. This use of a mini-batch of size of 1 is called *online learning*. The use of one training example at a time is prone to errors in the gradient estimation, but it turns out to be fine as long as it keeps the cost function decreasing. In this case the gradient is estimated as

$$\begin{aligned}
\nabla_w\widehat{\xi}(w) &= \nabla_\mathbf{w}\epsilon_j^2 = 2\epsilon_j\partial_\mathbf{w}\epsilon_j = 2\epsilon_j\partial_\mathbf{w}(z^{(j)} - \mathbf{w}^T\mathbf{x}^{(j)}) \\
&= -2\epsilon_j\mathbf{x}^{(j)}.
\end{aligned}$$

Applying now the gradient descent method, we obtain

$$\begin{aligned}
\mathbf{w}^{(j+1)} &= \mathbf{w}^{(j)} - \eta\nabla_\mathbf{w}\widehat{\xi} \\
&= \mathbf{w}^{(j)} - \eta(-2\epsilon_j\mathbf{x}^{(j)}) = \mathbf{w}^{(j)} + 2\eta\epsilon_j\mathbf{x}^{(j)}.
\end{aligned}$$

Substituting for $\epsilon_j = z^{(j)} - \mathbf{w}^T\mathbf{x}^{(j)}$ yields

$$\mathbf{w}^{(j+1)} = \mathbf{w}^{(j)} + 2\eta(z^{(j)} - \mathbf{w}^T\mathbf{x}^{(j)})\mathbf{x}^{(j)}, \tag{5.5.5}$$

where $\mathbf{x}^{(j)}$ is the jth measurement of the input X.

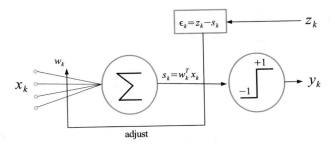

Figure 5.12: *Adaline neuron.*

Remark 5.5.1 If the inputs are random variables normally distributed, $X_j \sim \mathcal{N}(\mu_j, \sigma_j^2)$, $j = 1, \ldots, n$, then the output signal, $Y = \sum_{j=1}^{n} w_j X_j + b$, is a univariate Gaussian random variable, $Y \sim \mathcal{N}(\mathbf{w}^T \mu + b, \mathbf{w}^T A \mathbf{w})$. This follows from

$$
\begin{aligned}
\mathbb{E}[Y] &= \sum w_j \mathbb{E}[X_j] + b = \mathbf{w}^T \mu + b \\
Var(Y) &= Var(\sum w_j X_j) = Cov(\sum w_j X_j, \sum w_i X_i) \\
&= \sum_i \sum_j Cov(X_j, X_i) w_j w_i = \mathbf{w}^T A \mathbf{w},
\end{aligned}
$$

where $A_{ij} = Cov(X_i, X_j)$ and $\mu^T = (\mu_1, \ldots, \mu_n)$.

Remark 5.5.2 There is a case when the inverse matrix A^{-1} is inexpensive to compute. If the inputs X_i are independent and have zero mean, $\mathbb{E}[X_i] = 0$, then the inverse matrix A^{-1} is diagonal with $A_{ii}^{-1} = 1/\mathbb{E}[X_i^2]$.

5.6 Adaline

The Adaline (Adaptive Linear Neuron) is an early computing element with binary inputs $\{\pm 1\}$ and signum activation function $S(x) = sign(x)$ developed by Widrow and Hoff in 1960, see Fig. 5.12. It was built out of hardware and its weights were updated using electrically variable resistors. The difference between the Adaline and the standard perceptron is that during the learning phase the weights are adjusted according to the weighted sum of the inputs (while in the case of the perceptron the weights are adjusted in terms of the perceptron output).

The Adaline is trained by the α-LMS algorithm or Widrow-Hoff delta rule, see [127]. This is an example of algorithm that applies the *minimal disturbance principle.*[3] If w_k and x_k denote, respectively, the weight and input vector at step k, then the update rule is

$$w_{k+1} = w_k + \alpha \frac{\epsilon_k x_k}{|x_k|^2}, \qquad (5.6.6)$$

where $\alpha > 0$ and $\epsilon_k = z_k - w_k^T x_k$ denotes the error obtained as the difference between the target z_k and the linear output $s_k = w_k^T x_k$ before adaption. The aforementioned update rule changes the error as in the following

$$
\begin{aligned}
\Delta \epsilon_k &= \Delta(z_k - w_k^T x_k) = \Delta(z_k - x_k^T w_k) = -x_k^T \Delta w_k \\
&= -x_k^T(w_{k+1} - w_k) = -x_k^T \alpha \frac{\epsilon_k x_k}{|x_k|^2} = -\alpha \frac{\epsilon_k x_k^T x_k}{|x_k|^2} \\
&= -\alpha \epsilon_k.
\end{aligned}
$$

This computation shows that the error gets reduced by a factor of α, while the input vector x_k is kept fixed. A practical range for α is $0.1 < \alpha < 1$. The initial weight is chosen to be $w_0 = 0$ and the training continues until convergence.

It is worth noting that the if all input vectors x_k have equal length, then the α-LMS algorithm minimizes the mean-square error and the updating rule (5.6.6) becomes the gradient descent rule, see Exercise 5.10.9.

5.7 Madaline

Madaline (Multilayer adaptive linear element) was one of the first trainable neural networks used in pattern recognition research, see Widrow [126] and Hoff [80].

Madaline consists of a layer of Adaline neurons connected to a fixed logic gate (AND, OR, etc.), which produces an output, see Fig. 5.13 **a**. Only the weights associated with the Adaline units are adaptive, while the weights of the output logic gate are fixed.

For instance, with suitable chosen weights, the Madaline with two Adaline units and and AND output gate can implement the XNOR function given by

---

[3]Adapt to reduce the output error for the current training data, with minimal disturbance to responses already learned.

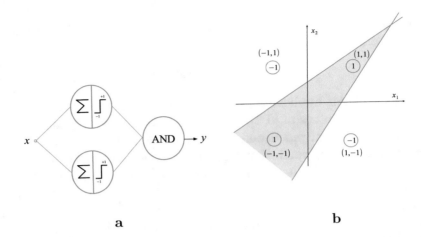

Figure 5.13: **a.** *Madaline with two Adaline units.* **b.** *Madaline implementation of XNOR function. The Madaline takes value 1 on the shaded area and value −1 in rest.*

x_1	x_2	XNOR
1	1	1
1	−1	−1
−1	−1	1
−1	1	−1

The separation boundaries are given by two lines, see Fig. 5.13 **b** and Exercise 5.10.10.

5.8 Continuum Input Neuron

This section introduces the concept of *continuum input neuron*, which is a neuron with an uncountable infinite number of weights. The concept will be used later in sections 7.3 and 10.9 for the construction of continuum neural networks. The learning in the context of continuum input neurons will be approached in section 10.6.

It is interesting to notice the relation with the Wilson-Cowan model [129] of excitatory and inhibitory interactions of model neurons, which also uses a continuous distribution of neural thresholds in a population of neurons.

A continuum input neuron has the inputs x continuum over the interval $[0, 1]$. The weight associated with the value x is given by $w(dx)$, where here w is a weighting measure on $[0, 1]$. The first half of the computing unit will integrate the inputs with respect to the measure w. The output function in this case takes the form

$$y = \sigma\left(\int_0^1 x\, dw(x)\right). \tag{5.8.7}$$

In particular, this is applicable to the situation when the input is a random variable X taking values in $[0, 1]$. If w represents the distribution measure of X, then the neuron output depends on its expectation, $y = \sigma\big(\mathbb{E}[X]\big)$.

Depending on the particular type of the weighting measure employed, formula (5.8.7) provides a general representation scheme for representing a neuron with either a finite number of inputs as previously discussed, a countably infinite number of inputs, or an uncountably infinite number of inputs. We shall provide next a few examples of neurons within the framework of some particular measures.

Example 5.8.1 (Neuron with a Dirac measure) Let $x_0 \in [0, 1]$ be fixed and consider the Dirac measure δ_{x_0} sitting at x_0, defined by

$$\delta_{x_0}(A) = \begin{cases} 1, & \text{if } x_0 \in A \\ 0, & \text{if } x_0 \notin A, \end{cases}$$

for any measurable set $A \in \mathcal{B}([0, 1])$. This corresponds to the case when the random variable X takes only the value x_0. The output function in this case is a constant:

$$y = \sigma\left(\int_0^1 x\, d\delta_{x_0}(x)\right) = \sigma(x_0).$$

Since there are no parameters to adjust, there is nothing to learn here.

In order to to avoid confusion, in the next examples we shall denote the weighting measure by μ and the weights at the points by w_i.

Example 5.8.2 (Neuron with discrete measure) Let $E = \{x_1, \ldots, x_n\}$ be a finite subset of $[0, 1]$ and consider the discrete measure

$$\mu(A) = \sum_{x_i \in E} w_i \delta_{x_i}(A), \quad \forall A \in \mathcal{B}([0, 1]),$$

where the positive number w_i is the mass attached to the point x_i. The output function is given by

$$y = \sigma\left(\int_0^1 x\, d\mu(x)\right) = \sigma\left(\sum_{j=1}^n w_j x_j\right),$$

which corresponds to the classical neuron model, see section 5.3. The learning algorithm in this case adjusts the weights w_j of the possible error function

$$F(w) = \frac{1}{2}\left(\sigma\left(\sum_{i=0}^n w_i x_i\right) - z\right)^2,$$

z being the desired value of the neuron, given the observation $x^T = (x_1, \dots, x_n)$.

In the case of m observations $\{(x^k, z^k)\}_{k=1,m}$ the error function takes the form

$$F(w) = \frac{1}{2} \sum_{k=1}^{m} \left(\sigma\left(\sum_{i=0}^{n} w_i x_i^k \right) - z^k \right)^2.$$

The optimal weights will be found, for instance, using the gradient descent algorithm.

Example 5.8.3 (Neuron with Lebesgue measure) Let μ be an absolute continuous measure with respect to the Lebesgue measure $dx_{|[0,1]}$. By Radon-Nikodym theorem there is a nonnegative measurable function p on $[0, 1]$ such that $d\mu(x) = p(x)dx$. Therefore, the output function

$$y = \sigma\left(\int_0^1 x \, d\mu(x) \right) = \sigma\left(\int_0^1 x p(x) \, dx \right)$$

depends on the weight function $p(x)$. The learning algorithm has to optimize the neuron by choosing the optimal function $p(x)$, which minimizes the error functional $F(p) = \frac{1}{2}\left(\sigma\left(\int_0^1 x p(x) \, dx \right) - z \right)^2$.

If μ is a probability measure, i.e., $\mu([0,1]) = 1$, then $p(x)$ is a density function. The optimality problem implies that $p(x)$ is a critical value for the the Lagrange multiplier functional

$$L(p) = \frac{1}{2}\left(\sigma\left(\int_0^1 x p(x) \, dx \right) - z \right)^2 - \lambda\left(\int_0^1 p(x) \, dx - 1 \right).$$

In order to find the optimal density, p, we consider the variation given by the family of density functions $(p_\epsilon)_{\epsilon > 0}$, with $p_\epsilon(x) = p(x) + \epsilon\varphi(x)$, where φ is a continuous function. We have $\frac{dp_\epsilon(x)}{d\epsilon} = \varphi(x)$, with $\int_0^1 \varphi(x) \, dx = 1$. The fact that $p(x)$ is a minimum for the functional $L(p)$ implies $\frac{dL(p_\epsilon(x))}{d\epsilon}\Big|_{\epsilon=0} = 0$. On the other side, an application of the chain rule provides

$$\frac{dL(p_\epsilon(x))}{d\epsilon} = \left(\sigma\left(\int_0^1 x p_\epsilon(x) \, dx \right) - z \right) \sigma'\left(\int_0^1 x p_\epsilon(x) \, dx \right) \int_0^1 x \frac{dp_\epsilon}{d\epsilon} \, dx$$
$$- \lambda \int_0^1 \frac{dp_\epsilon}{d\epsilon} \, dx,$$

and hence

$$\frac{dL(p_\epsilon(x))}{d\epsilon}\Big|_{\epsilon=0} = \left(\sigma\left(\int_0^1 x p(x) \, dx \right) - z \right) \sigma'\left(\int_0^1 x p(x) \, dx \right) \int_0^1 x \varphi(x) \, dx.$$

Since for $0 < x < 1$ we have $\int_0^1 x\varphi(x)\,dx < \int_0^1 \varphi(x)\,dx = 0$ and $\sigma'(u) = \sigma(u)(1 - \sigma(u)) \neq 0$, then the above variation vanishes if $p(x)$ satisfies

$$\int_0^1 xp(x)\,dx = \sigma^{-1}(z). \tag{5.8.8}$$

In the case of a repeatable experiment, the constant output z is replaced by the target random variable $Z : \Omega \to [0,1]$, where $(\Omega, \mathcal{H}, \mathbb{P})$ is a probability space. The problem asks for finding the random variable $X : \Omega \to [0,1]$ such that the amount

$$\mathbb{E}\big[\sigma\big(\mathbb{E}[X]\big) - Z\big]^2 = \int_\Omega \big[\sigma\big(\mathbb{E}[X]\big) - Z(\omega)\big]^2 d\mathbb{P}(\omega)$$

is minimized. It is known from the theory of random variables that a necessary condition for this to be achieved is that $\mathbb{E}[Z] = \sigma\big(\mathbb{E}[X]\big)$. Consequently, X is a random variables with the mean

$$\mathbb{E}[X] = \sigma^{-1}\big(\mathbb{E}[Z]\big),$$

which is equivalent to (5.8.8).

5.9 Summary

This chapter presented several types of neurons. Some of them are classical, such as the perceptron, the sigmoid neuron, or the linear neuron; others are of more theoretical value, such as neurons with a continuum input. Each neuron is characterized by an input, a system of weights, a bias, an activation function, and an output.

The perceptron is a neuron model with a step activation function. The neuron fires if the incoming signal exceeds a given threshold. Its output has a jump discontinuity. A perceptron can be used as a binary classifier to classify two linearly separable clusters in the plane. It can also be used to learn the Boolean functions AND and OR, but it cannot learn the function XOR.

The sigmoid neuron has a sigmoid activation function. Its output is continuous and can saturate if the signal is either too large or too small. The sigmoid neuron can learn using logistic regression, which is equivalent to the fact that the weights are also given by the maximum likelihood method. Some applications of sigmoid neurons are presented, such as binary classification and prediction of default probabilities.

Linear neurons have random variable inputs and the activation function is linear. Their optimal weights can be obtained by a closed-form solution. However, in practice it is easier to train them by the gradient descent method.

If the input to a neuron is a continuous variable in a given interval and the weights system is replaced by a measure on that interval, then we obtain a neuron with a continuum input. There are several types of neurons, corresponding to different types of measures, such as Dirac, discrete measures, or Lebesgue.

5.10 Exercises

Exercise 5.10.1 Recall that $\neg x$ is the negation of the Boolean variable x.

(a) Show that a single perceptron can learn the Boolean function $y = x_1 \wedge \neg x_2$, with $x_1, x_2 \in \{0, 1\}$.

(b) The same question as in part (a) for the Boolean function $y = x_1 \vee \neg x_2$, with $x_1, x_2 \in \{0, 1\}$.

(c) Show that a perceptron with one Boolean input, x, can learn the negation function $y = \neg x$. What about the linear neuron?

(d) Show that a perceptron with three Boolean inputs, x_1, x_2, x_3, can learn $x_1 \wedge x_2 \wedge x_3$. What about $x_1 \vee x_2 \vee x_3$?

Exercise 5.10.2 Show that two finite linearly separable sets A and B can be separated by a perceptron with rational weights.

Exercise 5.10.3 (a) Assume the inputs to a linear neuron are independent and normally distributed, $X_i \sim \mathcal{N}(0, \sigma_i^2)$, $i = 1, \ldots, n$. Find the optimal weights, w^*.

(b) A one-dimensional random variable with zero mean, Z, is learned by a linear neuron with input X. Assume the input, X, and the target, Z, are independent. Write the cost function and find the optimal parameters, w^*. Provide an interpretation for the result.

(c) Use Newton's method to obtain the optimal parameters of a linear neuron.

Exercise 5.10.4 Explain the equivalence between the linear regression algorithm and the learning of a linear neuron.

Exercise 5.10.5 Consider a neuron with a continuum input, whose output is $y = H\left(\int_0^1 x \, d\mu(x)\right)$. Find the output in the case when the measure is $\mu = \delta_{x_0}$.

Exercise 5.10.6 (Perceptron learning algorithm) Consider n points, $P_1, \ldots, P_n$, included in a half-circle, and denote by $\mathbf{x}_1, \ldots, \mathbf{x}_n$ their coordinate

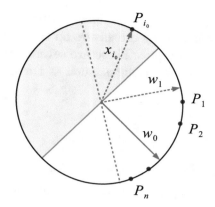

Figure 5.14: *For Exercise* 5.10.6.

vectors. A perceptron can learn the aforementioned half-circle by the following algorithm:

1. Start from an arbitrary half-circle determined by its diameter and a unit normal vector w_0. Then select an incorrectly classified point, P_{i_0}, i.e., a point for which $\langle w_0, \mathbf{x}_{i_0} \rangle < 0$, see Fig. 5.14.

2. Rotate the diameter such that the new normal is $w_1 = w_0 + \mathbf{x}_{i_0}$. Show that the point P_{i_0} is now correctly classified.

3. Repeating the previous two steps, we construct inductively the sequence of vectors $(w_m)_m$ such that $w_{m+1} = w_m + \mathbf{x}_{i_m}$, where P_{i_m} is a point misclassified at step m. Show that the process ends in a finite number of steps, i.e., there is $N > 1$ such that $\langle w_N, \mathbf{x}_j \rangle > 0$, $\forall 1 \le j \le n$. Find an estimation of the number N.

Exercise 5.10.7 Modify the perceptron learning algorithm given by Exercise 5.10.7 for the case when the points $P_1, \ldots, P_n$ are included in a half-plane.

Exercise 5.10.8 Let $1_A(x)$ denote the characteristic function of the set A, namely, $1_A(x) = 1$ if $x \in A$ and $1_A(x) = 0$ if $x \notin A$.

(a) Show that the function

$$\varphi(x_1, x_2) = 1_{\{x_2 > x_1 + 0.5\}}(x_1, x_2) + 1_{\{x_2 < x_1 - 0.5\}}(x_1, x_2)$$

implements XOR.

(b) Show that the XOR function can be implemented by a linear combinations of two perceptrons.

Exercise 5.10.9 Show that if all input vectors x_k have the same length, then the α-LMS algorithm minimizes the mean square error and in this case the updating rule (5.6.6) becomes the gradient descent rule.

Exercise 5.10.10 Find the weights of a Madaline with two Adaline units which implements the XNOR function.

Chapter 6

Neural Networks

We have discussed so far the case of a single neuron. In this section we shall consider multiple layers of neurons whose outputs are fed into other layers of neurons, forming *neural networks*. A layer of neurons is a processing step into a neural network and can be of different types, depending on the weights and activation function used in its neurons (fully-connected layer, convolution layer, pooling layer, etc.) The main part of this chapter will deal with training neural networks using the backpropagation algorithm.

Since the study of a neural network is heavily based on notation, we shall start with an warm-up example.

6.1 An Example of Neural Network

Consider two neurons having identical inputs, x_1, x_2 and (distinct) outputs y_1, y_2, respectively. The outputs are fed into the third neuron, with the output y, see Fig. 6.1. We assume the activation function for all neurons is the same and is denoted by ϕ. Each neuron has a bias that is considered as a weight of a "fake" input equal to -1. Now, we assemble these neurons together into an equivalently neural net, as given in Fig. 6.2. This forms a layer of two neurons in the middle, called a *hidden layer*.

The synapses between the neurons are depicted by edges and each of them is associated with a weight, denoted by $w_{ij}^{(\ell)}$. The indices have the following significance: the upper index, (ℓ), represents the index of the layer the synapse feds into. The index $\ell = 1$ is used for the weights that fed the hidden layer and the index $\ell = 2$ is used for the weights that enter the output layer, which consists of the last neuron. The lower index i indicates the input neuron. The

© Springer Nature Switzerland AG 2020

O. Calin, *Deep Learning Architectures*, Springer Series in the Data Sciences,

https://doi.org/10.1007/978-3-030-36721-3_6

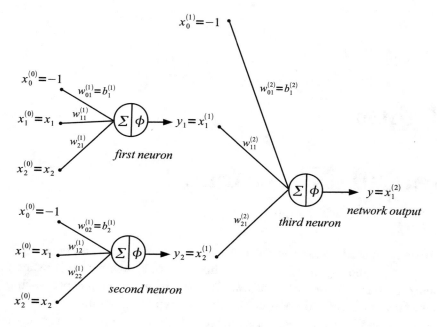

Figure 6.1: *Two neurons with identical inputs, x_1, x_2, and outputs y_1, y_2, are fed into a third neuron with the output y.*

index $i = 0$ is always used for the bias, while $i = 1, 2$ represent the index inputs. The second lower index, j, corresponds to the output neuron, i.e., the index of the neuron the synapse points to.

The inputs form the *first layer* of the network. They are denoted with a zero upper index: $x_0^{(0)} = -1$, $x_1^{(0)}$ and $x_2^{(0)}$. Note that the biasses of the neurons in the hidden layer are denoted by $b_1^{(1)} = w_{01}^{(1)}$ and $b_2^{(1)} = w_{02}^{(1)}$, while the bias for the neuron in the output layer is $b_1^{(2)} = w_{01}^{(2)}$.

The first neuron of the hidden layer collects the information from the input layer into the signal

$$s_1^{(1)} = w_{01}^{(1)} x_0^{(0)} + w_{11}^{(1)} x_1^{(0)} + w_{21}^{(1)} x_2^{(0)} = \sum_{i=1}^{2} w_{i1}^{(1)} x_i^{(0)} - b_1^{(1)}.$$

Then it applies the activation function ϕ on the signal to obtain the output

$$y_1 = x_1^{(1)} = \phi(s_1^{(1)}) = \phi\Big(\sum_{i=1}^{2} w_{i1}^{(1)} x_i^{(0)} - b_1^{(1)} \Big).$$

Similarly, the second neuron of the hidden layer collects the information into the signal

$$s_2^{(1)} = w_{02}^{(1)} x_0^{(0)} + w_{12}^{(1)} x_1^{(0)} + w_{22}^{(1)} x_2^{(0)} = \sum_{i=1}^{2} w_{i2}^{(1)} x_i^{(0)} - b_2^{(1)},$$

and then it applies the activation function ϕ to obtain the output

$$y_2 = x_2^{(1)} = \phi(s_2^{(1)}) = \phi\Big(\sum_{i=1}^{2} w_{i2}^{(1)} x_i^{(0)} - b_2^{(1)} \Big).$$

These relations can be manipulated easier if they are written in the matrix form. The signals in the hidden layer are related to the inputs by

$$\begin{pmatrix} s_1^{(1)} \\ s_2^{(1)} \end{pmatrix} = \begin{pmatrix} w_{11}^{(1)} & w_{21}^{(1)} \\ w_{12}^{(1)} & w_{22}^{(1)} \end{pmatrix} \begin{pmatrix} x_1^{(0)} \\ x_2^{(0)} \end{pmatrix} - \begin{pmatrix} b_1^{(1)} \\ b_2^{(1)} \end{pmatrix}.$$

This can be written in the condensed form as

$$s^{(1)} = W^{(1)^T} X^{(0)} - b^{(1)},$$

where $s^{(1)}$ represents the signal vector, $W^{(1)}$ is the weights matrix, $X^{(0)}$ is the input vector, and $b^{(1)}$ denotes the bias vector for the neurons in the hidden layer. The transpose notation follows from the way we wrote the indices.

The output of the hidden layer is the vector

$$X^{(1)} = \begin{pmatrix} y_1 \\ y_2 \end{pmatrix} = \begin{pmatrix} x_1^{(1)} \\ x_2^{(1)} \end{pmatrix} = \begin{pmatrix} \phi(s_1^{(1)}) \\ \phi(s_2^{(1)}) \end{pmatrix} = \phi(s^{(1)}),$$

where we adopt the convention that ϕ applied on a vector acts on its components.

The last two formulas imply the useful relation $X^{(1)} = \phi\Big(W^{(1)^T} X^{(0)} - b^{(1)} \Big)$, providing the output vector of the hidden layer as a function of inputs. The multiplication of $X^{(0)}$ by the matrix $W^{(1)^T}$ is a linear transformation. Subtracting $b^{(1)}$, this becomes an affine transformation. The activation function ϕ adds nonlinearity to the output.

The neuron in the last layer collects the incoming information into the signal

$$s^{(2)} = s_1^{(2)} = w_{01}^{(2)} x_0^{(1)} + w_{11}^{(2)} x_1^{(1)} + w_{21}^{(2)} x_2^{(1)} = \sum_{i=1}^{2} w_{i1}^{(2)} x_i^{(1)} - b_1^{(2)}.$$

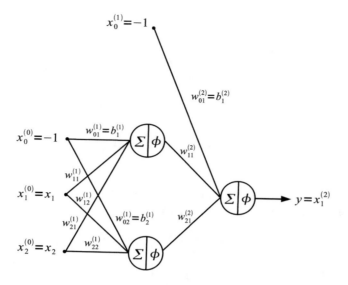

Figure 6.2: *Neural network with one hidden layer, having two inputs, x_1, x_2, and two neurons in the hidden layer.*

In the matrix form, this becomes

$$s^{(2)} = W^{(2)^T} X^{(1)} - b^{(2)},$$

where the matrix coefficients are given by

$$W^{(2)^T} = \left(w_{11}^{(2)} \quad w_{21}^{(2)} \right)$$

and the input vector is

$$X^{(1)} = \left(x_1^{(1)} \quad x_2^{(1)} \right)^T.$$

The signal coming out of the last neuron, $s^{(2)} = s_1^{(2)}$, is a one-dimensional matrix. The lower index, 1, indicates that there is only one neuron in the output layer. The network output is obtained applying the activation function on the signal $s^{(2)}$ as in the following:

$$Y = x_1^{(2)} = \phi(s^{(2)}) = \phi\left(W^{(2)^T} X^{(1)} - b^{(2)} \right).$$

The output was denoted by Y to be consistent with the notation used in previous sections and also by $x_1^{(2)}$; the upper script (2) indicates that this output is on the second layer (actually third, if we count the input layer) and the lower index, 1, means that there is only one neuron in the output layer.

Using the formula for $X^{(1)}$ obtained previously, we can express the final output in terms of the initial inputs as

$$Y = \phi\Big(W^{(2)^T}\phi(W^{(1)^T}X^{(0)} - b^{(1)}) - b^{(2)}\Big). \tag{6.1.1}$$

We notice that the affine transformations, consisting in multiplication by weight matrices and subtracting bias vector, alternate with nonlinear transformations, which are induced by the activation function ϕ. If consider the mapping $f_{w,b} : \mathbb{R}^2 \to \mathbb{R}$ defined by

$$f_{w,b}(X) = \phi\Big(W^{(2)^T}\phi(W^{(1)^T}X - b^{(1)}) - b^{(2)}\Big), \tag{6.1.2}$$

then we call $f_{w,b}$ the *input-output mapping* of the given neural network.

Remark 6.1.1 Assume all the neurons from the previous network are linear neurons, i.e., the activation function is $\phi(x) = x$. In this case, the output becomes

$$Y = (W^{(1)}W^{(2)})^T X^{(0)} - W^{(2)^T}b^{(1)} - b^{(2)} = W^T X^{(0)} - b.$$

This suggests that the entire neuronal network is equivalent with only one linear neuron having the weights matrix $W = W^{(1)}W^{(2)}$ and bias vector $b = W^{(2)^T}b^{(1)} - b^{(2)}$. Hence, the study of neural nets of linear neurons is reduced to the study of only one linear neuron.

Remark 6.1.2 Assume now that all the previous neurons are perceptrons, with the activation function $\phi(x) = H(x)$. The output of the network in this case is given by

$$Y = H\Big(w_1^{(2)}H\big(x_1 w_{11}^{(1)} + x_2 w_{21}^{(1)} - b_1^{(1)}\big) + w_2^{(2)}H\big(x_1 w_{12}^{(1)} + x_2 w_{22}^{(1)} - b_2^{(1)}\big) - b^{(2)}\Big).$$

It is worthy to note that this network can learn exactly the XOR function introduced in section 5.2. Choosing the weights such that the output becomes

$$Y = f(x_1, x_2) = H\Big(H(x + y - 0.5) - H(x + y - 1.5) - 0.5\Big),$$

we easily verify that $f(0, 0) = 0$, $f(1, 1) = 0$, $f(0, 1) = 1$, and $f(1, 0) = 1$. The graph of the function $f(x, y)$ is a square-shaped canyon with the bottom along the diagonal line $y = x$, having the bottom width equal to $\sqrt{2}$. The need of three perceptrons to learn the XOR function is explained heuristically in the following. It can be shown that that each of the functions $y_1 = x_1 \wedge \neg x_2$ and $y_2 = \neg x_1 \wedge x_2$ can be learned by one perceptron. It is needed one more perceptron to learn the function $y_1 \vee y_2$, which is the XOR function, $(x_1 \wedge \neg x_2) \vee (\neg x_1 \wedge x_2)$.

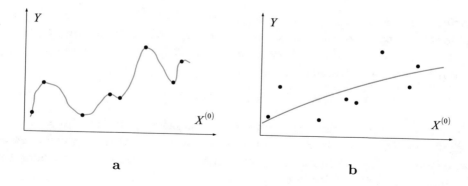

Figure 6.3: **a.** *The model overfits given data.* **b.** *The model generalizes well.*

6.1.1   Total variation and regularization

The generalization performance of a neural network depends on the output's sensitivity with respect to its input. Assume we have two networks, which have been trained on the same data: the one in Fig. 6.3 **a**, which overfits data, and the one in Fig. 6.3 **b**, which generalizes well. The geometric difference between these models consists in the fact that in the former case the total variation of the output is larger than in the latter. The total variation is related to the sensitivity of the output by a mathematical formula. If $Y = f(X^{(0)})$ denotes the input-output mapping of the network, the total variation of f represents the total cumulation effect of local absolute variations

$$V(f) = \int |df(X)|,$$

where the integral is taken on the input space. Therefore, $V(f)$ plays an important role in evaluating the generalization performance of a network. This is why a formula for the differential $dY = df(X)$ is needed and shall be computed in the following.

Formula (6.1.1) provides the net output in terms of its input. Assume that all weights and biasses are kept fixed. The changes in the output Y when small changes in the input $X^{(0)}$ occur can be computed by an iterative application of the chain rule, as in the following:

$$
\begin{aligned}
dY &= \frac{dY}{ds^{(2)}}\, ds^{(2)} = \frac{dY}{ds^{(2)}} \frac{ds^{(2)}}{dX^{(1)}}\, dX^{(1)} \\
&= \frac{dY}{ds^{(2)}} \frac{ds^{(2)}}{dX^{(1)}} \frac{dX^{(1)}}{ds^{(1)}}\, ds^{(1)} \\
&= \frac{dY}{ds^{(2)}} \frac{ds^{(2)}}{dX^{(1)}} \frac{dX^{(1)}}{ds^{(1)}} \frac{ds^{(1)}}{dX^{(0)}}\, dX^{(0)}
\end{aligned}
$$

Using that $\dfrac{dY}{ds^{(2)}} = \phi'(s^{(2)})$ and $\dfrac{dX^{(1)}}{ds^{(1)}} = \phi'(s^{(1)})$, as well as $\dfrac{ds^{(2)}}{dX^{(1)}} = W^{(2)^T}$ and $\dfrac{ds^{(1)}}{dX^{(0)}} = W^{(1)^T}$, we obtain

$$dY = \phi'(s^{(2)})W^{(2)^T}\phi'(s^{(1)})W^{(1)^T} dX^{(0)}. \tag{6.1.3}$$

It is important to remark the relation with regularization. We have noticed from the previous formula that the differential dY depends on the weights matrix $W^{(\ell)}$, $\ell \in \{1,2\}$. Keeping these weights small by requiring either an L^1 or L^2 norm constraint, $\|W^{(\ell)}\|_1 \leq 1$, or $\|W^{(\ell)}\|_2 \leq 1$, will decrease $|dY|$. This has a further diminishing effect on the total variation, $V(f)$, and hence this will improve the generalization performance.

In the case when the activation function is the logistic function, $\phi = \sigma$, using the properties of the logistic, formula (6.1.3) becomes

$$dY = \sigma(s^{(2)})(1 - \sigma(s^{(2)}))W^{(2)^T}\sigma(s^{(1)})(1 - \sigma(s^{(1)}))W^{(1)^T} dX^{(0)},$$

with the convention that σ applied to a vector acts on its components. The previous sensitivity formula will be used for the purpose of noise removal in Exercise 6.6.5.

6.1.2 Backpropagation

Consider a cost function $C(w, b)$, which measures the proximity between the network output, $Y = f_{w,b}(X)$, and the target Z. The cost is a smooth function of the weights, w, and biasses, b. Its gradient is given by

$$\nabla C = (\nabla_w C, \nabla_b C).$$

For the aforementioned example, this vector is nine-dimensional, depending on six weights and three biases. The gradient is needed for the gradient descent method. We shall start the computation backwards, from the last weights and biasses toward the first ones, by a procedure called *backpropagation*.

We compute the partial derivatives of the cost function with respect to the weights having the upper script $\ell = 2$. We shall also use that the cost C depends on Y, which depends on $w_{ij}^{(2)}$ through $s_1^{(2)}$ only, as $Y = \phi(s^{(2)})$. Hence, an application of the chain rule yields

$$\frac{\partial C}{\partial b_1^{(2)}} = \frac{\partial C}{\partial w_{01}^{(2)}} = \frac{\partial C}{\partial s_1^{(2)}}\frac{\partial s_1^{(2)}}{\partial w_{01}^{(2)}} = \delta_1^{(2)} x_0^{(1)} = -\delta_1^{(2)}$$

$$\frac{\partial C}{\partial w_{11}^{(2)}} = \frac{\partial C}{\partial s_1^{(2)}}\frac{\partial s_1^{(2)}}{\partial w_{11}^{(2)}} = \delta_1^{(2)} x_1^{(1)}$$

$$\frac{\partial C}{\partial w_{21}^{(2)}} = \frac{\partial C}{\partial s_1^{(2)}}\frac{\partial s_1^{(2)}}{\partial w_{21}^{(2)}} = \delta_1^{(2)} x_2^{(1)},$$

where we used the fact that $s^{(2)} = w_{01}^{(2)} x_0^{(1)} + w_{11}^{(2)} x_1^{(1)} + w_{21}^{(2)} x_2^{(1)}$, and employed the notation $\delta_1^{(2)} = \dfrac{\partial C}{\partial s_1^{(2)}}$.

Next, we compute the partial derivatives with respect to the weights and biases having the upper script $\ell = 1$. The cost C depends on Y, which depends on $w_{ij}^{(1)}$ through $s_j^{(1)}$, $j = 1, 2$, because $Y = \phi\left(W^{(2)^T} \phi(s^{(1)}) - b^{(2)}\right)$ by (6.1.1). If $j = 1$, then chain rule yields

$$\frac{\partial C}{\partial b_1^{(1)}} = \frac{\partial C}{\partial w_{01}^{(1)}} = \frac{\partial C}{\partial s_1^{(1)}} \frac{\partial s_1^{(1)}}{\partial w_{01}^{(1)}} = \delta_1^{(1)} x_0^{(0)} = -\delta_1^{(1)}$$

$$\frac{\partial C}{\partial w_{11}^{(1)}} = \frac{\partial C}{\partial s_1^{(1)}} \frac{\partial s_1^{(1)}}{\partial w_{11}^{(1)}} = \delta_1^{(1)} x_1^{(0)}$$

$$\frac{\partial C}{\partial w_{21}^{(1)}} = \frac{\partial C}{\partial s_1^{(1)}} \frac{\partial s_1^{(1)}}{\partial w_{21}^{(1)}} = \delta_1^{(1)} x_2^{(0)}$$

For $j = 2$, the chain rule implies

$$\frac{\partial C}{\partial b_2^{(1)}} = \frac{\partial C}{\partial w_{02}^{(1)}} = \frac{\partial C}{\partial s_2^{(1)}} \frac{\partial s_2^{(1)}}{\partial w_{02}^{(1)}} = \delta_2^{(1)} x_0^{(0)} = -\delta_2^{(1)}$$

$$\frac{\partial C}{\partial w_{12}^{(1)}} = \frac{\partial C}{\partial s_2^{(1)}} \frac{\partial s_2^{(1)}}{\partial w_{12}^{(1)}} = \delta_2^{(1)} x_1^{(0)}$$

$$\frac{\partial C}{\partial w_{22}^{(1)}} = \frac{\partial C}{\partial s_2^{(1)}} \frac{\partial s_2^{(1)}}{\partial w_{22}^{(1)}} = \delta_2^{(1)} x_2^{(0)},$$

where we used the fact that $s_j^{(1)} = \sum_{i=1}^{2} w_{ij}^{(1)} x_i^{(0)} - b_j^{(1)}$, and the notation $\delta_j^{(1)} = \dfrac{\partial C}{\partial s_j^{(1)}}$, $j = 1, 2$.

To conclude the previous computations, we write

$$\frac{\partial C}{\partial b_j^{(\ell)}} = -\delta_j^{(\ell)} \tag{6.1.4}$$

$$\frac{\partial C}{\partial w_{ij}^{(\ell)}} = \delta_j^{(\ell)} x_i^{(\ell-1)}. \tag{6.1.5}$$

It follows that the gradient of the cost function, ∇C, is known as long as we figure out the values of $\delta_j^{(k)}$. We shall show that the deltas of the hidden layer, $\delta_1^{(1)}$ and $\delta_2^{(1)}$, depend on the delta of the final layer, $\delta_1^{(2)}$, as in the

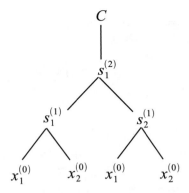

Figure 6.4: *The dependence tree used for applying chain rule.*

following. Using that $s_1^{(1)}$ affects the cost C through $s_1^{(2)}$ only, see Fig. 6.4, the chain rule provides

$$\delta_1^{(1)} = \frac{\partial C}{\partial s_1^{(1)}} = \frac{\partial C}{\partial s_1^{(2)}}\frac{\partial s_1^{(2)}}{\partial s_1^{(1)}} = \delta_1^{(2)}\frac{\partial s_1^{(2)}}{\partial s_1^{(1)}}. \qquad (6.1.6)$$

Since the signal $s_1^{(2)}$ can be written as

$$s_1^{(2)} = -w_{01}^{(2)} + w_{11}^{(2)}\phi(s_1^{(1)}) + w_{21}^{(2)}\phi(s_2^{(1)}),$$

it follows that

$$\frac{\partial s_1^{(2)}}{\partial s_1^{(1)}} = w_{11}^{(2)}\phi'(s_1^{(1)}).$$

Substituting in formula (6.1.6), we obtain

$$\delta_1^{(1)} = \delta_1^{(2)} w_{11}^{(2)} \phi'(s_1^{(1)}). \qquad (6.1.7)$$

Similarly, using the dependence tree given by Fig. 6.4, we have

$$\begin{aligned}\delta_2^{(1)} &= \frac{\partial C}{\partial s_2^{(1)}} = \frac{\partial C}{\partial s_1^{(2)}}\frac{\partial s_1^{(2)}}{\partial s_2^{(1)}} \\ &= \delta_1^{(2)}\frac{\partial}{\partial s_2^{(1)}}\Big(-w_{01}^{(2)} + w_{11}^{(2)}\phi(s_1^{(1)}) + w_{21}^{(2)}\phi(s_2^{(1)})\Big) \qquad (6.1.8) \\ &= \delta_1^{(2)} w_{21}^{(2)} \phi'(s_2^{(1)}). \qquad (6.1.9)\end{aligned}$$

The last two formulas can be written in the vector form as the following backpropagation formula for deltas:

$$
\begin{pmatrix} \delta_1^{(1)} \\ \delta_2^{(1)} \end{pmatrix} = \delta_1^{(2)} \begin{pmatrix} w_{11}^{(2)} \phi'(s_1^{(1)}) \\ w_{21}^{(2)} \phi'(s_2^{(1)}) \end{pmatrix}. \tag{6.1.10}
$$

Hence, in order to find the deltas of the hidden layer, it suffices to compute the delta of the outer layer, $\delta_1^{(2)}$. We shall compute this in the following.

The cost function, $C(w, b) = \text{dist}(Y, Z)$, depends on $s_1^{(2)}$ through the output, $Y = f_{w,b}(X^{(0)})$, so an application of the chain rule provides

$$
\delta_1^{(2)} = \frac{\partial C}{\partial s_1^{(2)}} = \frac{\partial C}{\partial Y} \frac{\partial Y}{\partial s_1^{(2)}} = \frac{\partial C}{\partial Y} \frac{\partial}{\partial s_1^{(2)}} \phi(s_1^{(2)}) = \frac{\partial C}{\partial Y} \phi'(s_1^{(2)}), \tag{6.1.11}
$$

where we used that the output is $Y = \phi(s_1^{(2)})$. Substituting in (6.1.10) yields

$$
\begin{pmatrix} \delta_1^{(1)} \\ \delta_2^{(1)} \end{pmatrix} = \frac{\partial C}{\partial Y} \phi'(s_1^{(2)}) \begin{pmatrix} w_{11}^{(2)} \phi'(s_1^{(1)}) \\ w_{21}^{(2)} \phi'(s_2^{(1)}) \end{pmatrix}. \tag{6.1.12}
$$

The factor $\dfrac{\partial C}{\partial Y}$ depends on the cost function considered. For instance, if $C = \frac{1}{2}(Y - Z)^2$, then $\dfrac{\partial C}{\partial Y} = Y - Z$, where Z is the target function that needs to be learned by the network. Up to this point, all deltas have been computed, their formulas being given by (6.1.11) and (6.1.12).

In order to minimize the cost function, C, we shall apply the gradient descent method. The initialization is given by some arbitrary values of weights and biasses, $(w_{ij}^{(\ell)}(0), b_j^{(\ell)}(0))$, and the approximation sequence $(w_{ij}^{(\ell)}(k), b_j^{(\ell)}(k))$, $k \geq 0$, is defined recursively by

$$
\begin{aligned}
b_j^{(\ell)}(k+1) &= b_j^{(\ell)}(k) - \eta \frac{\partial C}{\partial b_j^{(\ell)}} \left(w_{ij}^{(\ell)}(k), b_j^{(\ell)}(k) \right) \\
w_{ij}^{(\ell)}(k+1) &= w_{ij}^{(\ell)}(k) - \eta \frac{\partial C}{\partial w_{ij}^{(\ell)}} \left(w_{ij}^{(\ell)}(k), b_j^{(\ell)}(k) \right),
\end{aligned}
$$

where the argument k represents the number of iterations. In the virtue of formulas (6.1.4)–(6.1.5) the previous recursive system becomes

$$
\begin{aligned}
b_j^{(\ell)}(k+1) &= b_j^{(\ell)}(k) + \eta \delta_j^{(\ell)} \\
w_{ij}^{(\ell)}(k+1) &= w_{ij}^{(\ell)}(k) - \eta \delta_j^{(\ell)} x_i^{(\ell-1)}(k),
\end{aligned}
$$

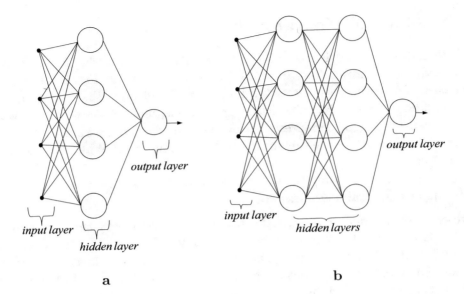

Figure 6.5: **a.** *One-hidden layer neural network;* **b.** *Two-hidden layer neural network.*

where $\eta > 0$ is the learning rate and $x_i^{(\ell-1)}(k)$ is the ith input to the layer ℓ, given the values of the weights and biasses at the kth run. The argument k used in the previous formulas represents the current number of iterations.

When all training examples have been used to update the network parameters we say one *epoch* has been processed. The process of parameters update continues until the learning system fits the training data. The gradient descent is used until the algorithm has converged to a locally optimal solution, fact that can be verified by checking whether the norm of the gradient has converged. If the model represents the training data perfectly, then the training loss can be made arbitrarily small. If the model underfits the training data, the training loss cannot decrease under a certain limit.

Remark 6.1.3 The idea of backpropagation dates since the 1960s when it appeared in the context of control theory [18], [62], [33]. Then it started to be used to neural networks in the mid-1970s, [77], [78], [123]. The first computer implementations with applications to neural networks with hidden layers have been developed in the mid-1980s in the classical paper [108]. After a break in popularity during the 2000s, the method returned in the literature and applications due to the computing power of the new GPU systems.

6.2   General Neural Networks

We shall discuss next the case of a *feedforward* neural network, i.e., a network where the information flows from the input to the output. Fig. 6.5 **a** depicts the case of a neural network with 3 layers: the input layer that has 4 inputs, the middle or hidden layer, which has 4 neurons, and the output layer with only 1 neuron. Fig. 6.5 **b** represents the case of a neural network with 4 layers, which looks similar to the previous network, but having 2 hidden layers, each with 4 neurons.

We start by presenting the notation regarding layers, weights, inputs, and outputs of a general neural network.

Layers We shall denote in the following the layer number by the upper script ℓ. We have $\ell = 0$ for the input layer, $\ell = 1$ for the first hidden layer, and $\ell = L$ for the output layer. Note that the number of hidden layers is equal to $L - 1$. The number of neurons in the layer ℓ is denoted by $d^{(\ell)}$. In particular, $d^{(0)}$ is the number of inputs and $d^{(L)}$ is the number of outputs.

Weights The system of weights is denoted by $w_{ij}^{(\ell)}$, with $1 \leq \ell \leq L$, $0 \leq i \leq d^{(\ell-1)}$, $1 \leq j \leq d^{(\ell)}$. The weight $w_{ij}^{(\ell)}$ is associated with the edge that joins the ith neuron in the layer $\ell - 1$ to the jth neuron in the layer ℓ. Note that the edges join neurons in consecutive layers only. The weights $w_{0j}^{(\ell)} = b_j^{(\ell)}$ are regarded as biasses. They are the weights corresponding to the fake input $x_0^{(\ell)} = -1$. The number of biases is equal to the number of neurons. The number of weights, without biasses, between the layers $\ell - 1$ and ℓ is given by $d^{(\ell-1)}d^{(\ell)}$, so the total number of weights in a feedforward network is $\sum_{\ell=1}^{L} d^{(\ell-1)}d^{(\ell)}$. For instance, the network in Fig. 6.5 **a** has $4 \times 4 + 4 \times 1 = 20$ weights, while the one in Fig. 6.5 **b** has $4 \times 4 + 4 \times 4 + 4 \times 1 = 36$ weights. The weights can be positive, negative, or zero. A positive weight encourages the neuron to fire, while a negative weight inhibits it from doing so.

Inputs and outputs The inputs to the network are denoted by $x_j^{(0)}$, with $1 \leq j \leq d^{(0)}$. We denote by $x_j^{(\ell)}$ the output of the jth neuron in the layer ℓ. Consequently, $x_j^{(L)}$ is the network output, with $1 \leq \ell \leq d^{(L)}$. The notation $x_0^{(\ell)} = -1$ is reserved for the fake input linked to the bias corresponding to the neurons in the layer $\ell + 1$.

Consider the jth neuron in the layer ℓ. In its first half, the neuron collects the information from the previous layer, as a weighted average, into the signal

$$s_j^{(\ell)} = \sum_{i=0}^{d^{(\ell-1)}} w_{ij}^{(\ell)} x_i^{(\ell-1)} = \sum_{i=1}^{d^{(\ell-1)}} w_{ij}^{(\ell)} x_i^{(\ell-1)} - b_j^{(\ell)}. \tag{6.2.13}$$

In its second half, the neuron applies an activation function ϕ to the previous signal and outputs the value

$$x_j^{(\ell)} = \phi\Big(\sum_{i=1}^{d^{(\ell-1)}} w_{ij}^{(\ell)} x_i^{(\ell-1)} - b_j^{(\ell)} \Big). \tag{6.2.14}$$

This can be written in the following equivalent matrix form:

$$X^{(\ell)} = \phi\Big(W^{(\ell)T} X^{(\ell-1)} - B^{(\ell)} \Big),$$

where we used the notations

$$X^{(\ell)} = (x_1^{(\ell)}, \ldots, x_{d^\ell}^{(\ell)})^T, \quad W^{(\ell)} = \Big(w_{ij}^{(\ell)} \Big)_{i,j}, \quad B^{(\ell)} = (b_1^{(\ell)}, \ldots, b_{d^\ell}^{(\ell)})^T,$$

with $1 \le i \le d^{(\ell-1)}$ and $1 \le j \le d^{(\ell)}$, and where we used the convention that the activation function applied on a vector acts on each of the components of the vector. We note that the input to the network is given by the vector $X^{(0)} = (x_1^{(0)}, \ldots, x_{d^0}^{(0)})^T$, while the network output is $X^{(L)} = (x_1^{(L)}, \ldots, x_{d^L}^{(L)})^T$.

6.2.1   Forward pass through the network

Given a system of weights and biasses, one can use equation (6.2.14) to find the outputs of each neuron in terms of their input. In particular, one finds the output of the network, which is the output of the last layer, $x^{(L)} = (x_1^{(L)}, \ldots, x_{d^{(L)}}^{(L)})$, where $d^{(L)}$ represents the number of network outputs (i.e. the number of neurons in the output layer). Going forward through the network means to find all the neuron outputs. This forward pass represents a target prediction step.

This prepares the ground for the second stage of the method, of going backwards through the network, to compute the gradient of the cost function in order to update parameters set by the gradient descent method. Passing forward and backward produces a sequence of parameters corresponding to an improved sequence of predictions of the target. This process ends when no cost improvements can be made anymore.

6.2.2   Going backwards through the network

Let $C(w, b)$ denote the cost function of the network, measuring a certain proximity between the target Z and the output $Y = x^{(L)}$. For the gradient descent purposes, we need to compute the gradient $\nabla C = (\nabla_w C, \nabla_b C)$.

Since the weight $w_{ij}^{(\ell)}$ joins the ith neuron in the layer $\ell - 1$ to the jth neuron in the layer ℓ, it will enter in the composition of the signal $s_j^{(\ell)}$ produced by the jth neuron of the layer ℓ. Furthermore, no other signal in the ℓth layer gets affected by this particular weight. Therefore, chain rule provides

$$\frac{\partial C}{\partial w_{ij}^{(\ell)}} = \frac{\partial C}{\partial s_j^{(\ell)}} \frac{\partial s_j^{(\ell)}}{\partial w_{ij}^{(\ell)}}. \tag{6.2.15}$$

with no summation over j. We use the following delta notation

$$\delta_j^{(\ell)} = \frac{\partial C}{\partial s_j^{(\ell)}}$$

to denote the sensitivity of the error with respect to the signal $s_j^{(\ell)}$. The second factor on the right side of (6.2.15) can be computed explicitly differentiating in relation (6.2.13) as

$$\frac{\partial s_j^{(\ell)}}{\partial w_{ij}^{(\ell)}} = x_i^{(\ell-1)}.$$

Substituting in (6.2.15) yields

$$\frac{\partial C}{\partial w_{ij}^{(\ell)}} = \delta_j^{(\ell)} x_i^{(\ell-1)}. \tag{6.2.16}$$

A similar analysis is carried out for the derivative of the cost with respect to the biasses. Since the cost C is affected by $b_j^{(\ell)}$ only through the signal $s_j^{(\ell)}$, we have

$$\frac{\partial C}{\partial b_j^{(\ell)}} = \frac{\partial C}{\partial s_j^{(\ell)}} \frac{\partial s_j^{(\ell)}}{\partial b_j^{(\ell)}}. \tag{6.2.17}$$

Differentiating in (6.2.13) yields

$$\frac{\partial s_j^{(\ell)}}{\partial b_j^{(\ell)}} = -1,$$

and hence (6.2.17) becomes

$$\frac{\partial C}{\partial b_j^{(\ell)}} = -\delta_j^{(\ell)}. \tag{6.2.18}$$

From relations (6.2.16) and (6.2.18) it follows that in order to find the gradient

$$\nabla C = (\nabla_w C, \nabla_b C) = \delta_j^{(\ell)}(x_i^{(\ell-1)}, -1), \tag{6.2.19}$$

it suffices to know the deltas $\delta_j^{(\ell)}$. The construction of deltas is done by an algorithm called *backpropagation* and will be presented in the next section.

6.2.3 Backpropagation of deltas

We start with the computation of deltas in the last layer, $\delta_j^{(L)}$, with $1 \leq j \leq d^{(L)}$. This essentially depends on the form of the cost function. We shall assume, for instance, that

$$C(w, b) = \frac{1}{2} \sum_{j=1}^{d^{(L)}} (x_j^{(L)} - z_j)^2 = \frac{1}{2} \|x^{(L)} - z\|^2,$$

where $z = (z_1, \ldots, z_{d^{(L)}})^T$ denotes the target variable. Using $x_j^{(L)} = \phi(s_j^{(L)})$, then chain rule implies

$$\delta_j^{(L)} = \frac{\partial C}{\partial s_j^{(L)}} = (x_j^{(L)} - z_j)\phi'(s_j^{(L)}). \tag{6.2.20}$$

In the case of some particular types of activation functions, this can be computed even further. If the activation function is the logistic function, $\phi = \sigma$, then

$$\delta_j^{(L)} = \frac{\partial C}{\partial s_j^{(L)}} = (x_j^{(L)} - z_j)\sigma(s_j^{(L)})\big(1 - \sigma(s_j^{(L)})\big).$$

Or, if the activation function is the hyperbolic tangent, $\mathbf{t}$, then

$$\delta_j^{(L)} = \frac{\partial C}{\partial s_j^{(L)}} = (x_j^{(L)} - z_j)(1 - \mathbf{t}^2(s_j^{(L)})).$$

The next task is to write the deltas of layer $\ell - 1$ in terms of the deltas of the layer ℓ. This will be achieved by a backpropagation formula. Chain formula provides

$$\delta_i^{(\ell-1)} = \frac{\partial C}{\partial s_i^{(\ell-1)}} = \sum_{j=1}^{d^{(\ell)}} \frac{\partial C}{\partial s_j^{(\ell)}} \frac{\partial s_j^{(\ell)}}{\partial s_i^{(\ell-1)}} = \sum_{j=1}^{d^{(\ell)}} \delta_j^{(\ell)} \frac{\partial s_j^{(\ell)}}{\partial s_i^{(\ell-1)}}. \tag{6.2.21}$$

The first equality follows from the definition of delta. The second one uses the fact that the signal $s_j^{(\ell-1)}$ in the layer $\ell - 1$ affects all signals, $s_j^{(\ell)}$, in the layer ℓ, with $1 \leq j \leq d^{(\ell)}$. Thus, the cost, C, is affected by $s_j^{(\ell-1)}$ through all signals in the layer ℓ, fact that explains the need of the summation over index j. The last identity uses again the definition of delta.

We shall compute next the partial derivative $\dfrac{\partial s_j^{(\ell)}}{\partial s_i^{(\ell-1)}}$, i.e., the sensitivity of a signal in the layer ℓ with respect to a signal in the previous layer. This can be achieved differentiating in formula (6.2.13) as in the following:

$$\frac{\partial s_j^{(\ell)}}{\partial s_i^{(\ell-1)}} = \frac{\partial}{\partial s_i^{(\ell-1)}} \Big(\sum_{i=1}^{d^{(\ell-1)}} w_{ij}^{(\ell)} x_i^{(\ell-1)} - b_j^{(\ell)} \Big)$$

$$= \frac{\partial}{\partial s_i^{(\ell-1)}} \Big(\sum_{i=1}^{d^{(\ell-1)}} w_{ij}^{(\ell)} \phi(s_i^{(\ell-1)}) - b_j^{(\ell)} \Big)$$

$$= w_{ij}^{(\ell)} \phi'(s_i^{(\ell-1)}).$$

Substituting in (6.2.21), we obtain the *backpropagation formula* for the deltas:

$$\delta_i^{(\ell-1)} = \phi'(s_i^{(\ell-1)}) \sum_{j=1}^{d^{(\ell)}} \delta_j^{(\ell)} w_{ij}^{(\ell)}. \tag{6.2.22}$$

This produces the deltas for the $(\ell-1)$th layer in terms of the deltas of the ℓth layer using a weighted sum.

If the activation function, ϕ, is the logistic function, σ, the previous formula becomes

$$\delta_i^{(\ell-1)} = \sigma(s_i^{(\ell-1)})\big(1 - \sigma(s_i^{(\ell-1)})\big) \sum_{j=1}^{d^{(\ell)}} \delta_j^{(\ell)} w_{ij}^{(\ell)}. \tag{6.2.23}$$

Or, if the activation function, ϕ, is the hyperbolic tangent, $\mathbf{t}$, the previous formula takes the form

$$\delta_i^{(\ell-1)} = \big(1 - \mathbf{t}^2(s_i^{(\ell-1)})\big) \sum_{j=1}^{d^{(\ell)}} \delta_j^{(\ell)} w_{ij}^{(\ell)}.$$

To conclude the last few sections, in order to find the gradient of the cost function given by (6.2.19), one uses the backpropagation formula of deltas (6.2.22) to compute $\delta_i^{(\ell)}$ and the forward propagation formula (6.2.14) to compute the outputs $x_i^{(\ell-1)}$.

6.2.4   Concluding relations

The computations done in the previous sections can be concluded with the following set of master equations

$$
x_j^{(\ell)} \;=\; \phi\Big(\sum_{i=1}^{d^{(\ell-1)}} w_{ij}^{(\ell)} x_i^{(\ell-1)} - b_j^{(\ell)} \Big), \qquad 1 \le j \le d^{(\ell)} \qquad (6.2.24)
$$

$$
\delta_j^{(L)} \;=\; (x_j^{(L)} - z_j)\phi'(s_j^{(L)}), \qquad 1 \le j \le d^{(L)} \qquad (6.2.25)
$$

$$
\delta_i^{(\ell-1)} \;=\; \phi'(s_i^{(\ell-1)}) \sum_{j=1}^{d^{(\ell)}} \delta_j^{(\ell)} w_{ij}^{(\ell)}, \qquad 1 \le i \le d^{(\ell-1)} \qquad (6.2.26)
$$

$$
\frac{\partial C}{\partial w_{ij}^{(\ell)}} \;=\; \delta_j^{(\ell)} x_i^{(\ell-1)} \qquad (6.2.27)
$$

$$
\frac{\partial C}{\partial b_j^{(\ell)}} \;=\; -\delta_j^{(\ell)}. \qquad (6.2.28)
$$

The first one provides a forward recursive formula of the outputs in terms of weights and biases. The second formula deals with the expression of the delta in the output layer (under the assumption that the cost function is a sum of squared errors). The third is the backpropagation formula for the deltas. The last two formulas provide the gradient components of C with respect to weights and biasses.

Equations (6.2.27)–(6.2.28) can be used to asses the sensitivity of the cost function, $C(w, b)$, with respect to small changes in weights and biasses. Using the formula for the differential of a function, yields

$$
\begin{aligned}
dC \;&=\; \sum \frac{\partial C}{\partial w_{ij}^{(\ell)}} \, dw_{ij}^{(\ell)} + \sum \frac{\partial C}{\partial b_j^{(\ell)}} db_j^{(\ell)} \\
&=\; \sum \delta_j^{(\ell)} x_i^{(\ell)} \, dw_{ij}^{(\ell)} - \sum \delta_j^{(\ell)} db_j^{(\ell)},
\end{aligned}
$$

where the summation is taken over i, j, ℓ, with $1 \le i \le d^{(\ell-1)}$, $1 \le j \le d^{(\ell)}$ and $1 \le \ell \le L$.

6.2.5   Matrix form

In order to avoid the complexity introduced by multiple indices, the aforementioned master equations can be written more compactly in matrix form.

First, we need to introduce a new product between vectors of the same dimension. If u and v are vectors in $\mathbb{R}^n$, then the *elementwise product* of u and v is a vector denoted by $u \odot v$ in $\mathbb{R}^n$ with components $(u \odot v)_j = u_j v_j$.

Sometimes, this is referred to as the *Hadamard product* of two vectors. Its use simplifies considerably the form of the equations.

Using the notations

$$X^{(\ell)} = (x_1^{(\ell)}, \ldots, x_{d^\ell}^{(\ell)})^T, \quad W^{(\ell)} = \left(w_{ij}^{(\ell)}\right)_{i,j}, \quad B^{(\ell)} = (b_1^{(\ell)}, \ldots, b_{d^\ell}^{(\ell)})^T,$$

$$\delta^{(\ell)} = (\delta_1^{(\ell)}, \ldots, \delta_{d^{(\ell)}}^{(\ell)})^T, \quad s^{(\ell)} = (s_1^{(\ell)}, \ldots, s_{d^{(\ell)}}^{(\ell)})^T, \quad x^{(L)} = (x_1^{(L)}, \ldots, x_{d^{(L)}}^{(l)})^T,$$

and the convention that ϕ acts on each component of its vector arguments, equations (6.2.24)–(6.2.28) take the following equivalent form:

$$X^{(\ell)} = \phi\left(W^{(\ell)T} X^{(\ell-1)} - B^{(\ell)}\right) \tag{6.2.29}$$

$$\delta^{(L)} = (x^{(L)} - z) \odot \phi'(s^{(L)}) \tag{6.2.30}$$

$$\delta^{(\ell-1)} = \left(W^{(\ell)}\delta^{(\ell)}\right) \odot \phi'(s^{(\ell-1)}) \tag{6.2.31}$$

$$\frac{\partial C}{\partial W^{(\ell)}} = X^{(\ell-1)}\delta^{(\ell)T} \tag{6.2.32}$$

$$\frac{\partial C}{\partial B^{(\ell)}} = -\delta^{(\ell)}. \tag{6.2.33}$$

It is worth noting that the right side of (6.2.32) is a multiplication of two matrices of vector type. More precisely, if $u = (u_1 \ldots u_n)^T$ and $v = (v_1 \ldots v_n)^T$ are two column vectors in $\mathbb{R}^n$, then $uv^T = (u_i v_j)_{i,j}$ is an $n \times n$ matrix. This should not be confused to $u^T v = \sum_i u_i v_i$, which is a number. We also note that in (6.2.31) the matrix $W^{(\ell)}$ does not have a transpose.

Remark 6.2.1 In the case of the linear neuron the previous formulae have a much simpler form. Since the activation function is $\phi(x) = x$, the backpropagation formula for the delta becomes $\delta^{(\ell-1)} = W^\ell \delta^{(\ell)}$. Iterating yields

$$\delta^{(\ell)} = W^{(\ell+1)} W^{(\ell+2)} \ldots W^{(L)} \delta^{(L)}.$$

Using (6.2.32) and (6.2.33) this implies

$$\frac{\partial C}{\partial w_{ij}^{(\ell)}} = x^{(\ell)} \left(W^{(\ell+1)} \ldots W^{(L)} \delta^{(L)}\right)^T$$

$$\frac{\partial C}{\partial b^{(\ell)}} = -W^{(\ell+1)} \ldots W^{(L)} \delta^{(L)},$$

formulas which provide a closed-form expression for the gradient of the cost function, $\nabla C = \left(\frac{\partial C}{\partial w_{ij}^{(\ell)}}, \frac{\partial C}{\partial b^{(\ell)}}\right)$.

Remark 6.2.2 Formula (6.2.30) is based on the particular assumption that the cost function is a sum of squares. If the cost function changes, *only* the formula for $\delta^{(L)}$ should be adjusted accordingly. This is based on the fact that deltas change backwards from $\ell = L$ to $\ell = 1$, see (6.2.31). In the following we shall provide formulas for $\delta^{(L)}$ for a couple of useful cost functions.

Example 6.2.3 Assume the cost function is given by the cross-entropy

$$C = -\sum_k z_k \ln x_k^{(L)} = -\sum_k z_k \ln \phi(s_k^{(L)}),$$

which represents the inefficiency of using the predicted output, $x^{(L)}$, rather than the true data z. The delta for the jth neuron in the output layer is now given by

$$
\begin{aligned}
\delta_j^{(L)} &= \frac{\partial C}{\partial s_j^{(L)}} = -\frac{\partial}{\partial s_j^{(L)}} \sum_k z_k \ln \phi(s_k^{(L)}) \\
&= -z_j \frac{\phi'(s_j^{(L)})}{\phi(s_j^{(L)})}.
\end{aligned}
$$

This expression can be further simplified if we consider the activation function to be the logistic function, $\phi(x) = \sigma(x)$. In this case, the computation can be continued as

$$
\begin{aligned}
\delta_j^{(L)} &= -z_j \frac{\sigma'(s_j^{(L)})}{\sigma(s_j^{(L)})} = -z_j \frac{\sigma(s_j^{(L)})\big(1 - \sigma(s_j^{(L)})\big)}{\sigma(s_j^{(L)})} \\
&= -z_j \big(1 - \sigma(s_j^{(L)})\big).
\end{aligned}
$$

Using the Hadamard product, this writes more compactly as

$$\delta^{(L)} = -z \odot \big(1 - \sigma(s(L))\big),$$

and can be regarded as a replacement for the relation (6.2.30) in the case when the cost function is given by the cross-entropy.

Example 6.2.4 Another cost function, which can be used in the case when the activation function (of the last layer) is the logistic function, $\phi(x) = \sigma(x)$, is the *regular cross-entropy function*,[1] where $z_k \in \{0, 1\}$.

$$C = -\sum_k z_k \ln x_k^{(L)} - \sum_k (1 - z_k) \ln(1 - x_k^{(L)}). \tag{6.2.34}$$

---

[1]This is the sum of binary entropy functions, see Exercise 12.13.15.

In this case the deltas of the output layer are computed as

$$\delta_j^{(L)} = \frac{\partial C}{\partial s_j^{(L)}} = -\frac{\partial}{\partial s_j^{(L)}}\left(\sum_k z_k \ln \sigma(s_k^{(L)}) + \sum_k (1 - z_k)\ln(1 - \sigma(s_k^{(L)}))\right)$$

$$= -z_j\frac{\sigma'(s_j^{(L)})}{\sigma(s_j^{(L)})} + (1 - z_j)\frac{\sigma'(s_j^{(L)})}{1 - \sigma(s_j^{(L)})}$$

$$= -z_j(1 - \sigma(s_j^{(L)})) + (1 - z_j)\sigma(s_j^{(L)})$$

$$= \sigma(s_j^{(L)}) - z_j = x_j^{(L)} - z_j,$$

where we have used the useful property $\sigma'(x) = \sigma(x)(1 - \sigma(x))$. Hence, in this case the deltas are equal to the differences between the model outcomes and target values, $\delta^{(L)} = x^{(L)} - z$.

6.2.6   Gradient descent algorithm

Consider a neural network with the initial system of weights and biasses given by $w_{ij}^{(\ell)}(0)$ and $b_j^{(\ell)}(0)$. The weights can be initialized efficiently by the procedure described in section 6.3. Let $\eta > 0$ be the learning rate. The optimum values of weights and biasses after the network is trained with the gradient descent algorithm are given by the approximation sequences defined recursively by

$$w_{ij}^{(\ell)}(n + 1) = w_{ij}^{(\ell)}(n) - \eta\delta_j^{(\ell)}(n)x_i^{(\ell-1)}(n)$$

$$b_j^{(\ell)}(n + 1) = b_j^{(\ell)}(n) + \eta\delta_j^{(\ell)}(n),$$

where the outputs $x_i^{(\ell)}(n)$ and deltas $\delta_j^{(\ell)}(n)$ depend of the weights, $w_{ij}^{(\ell)}(n)$, and biasses, $b_j^{(\ell)}(n)$, obtained at the nth step. We note that when writing the above equations we have used the aforementioned formulas for $\frac{\partial C}{\partial w_{ij}^{(\ell)}}$ and $\frac{\partial C}{\partial b_j^{(\ell)}}$.

6.2.7   Vanishing gradient problem

Assume we have a neural network where all neurons have the activation function given by the logistic function σ. If a signal is too large in absolute value, then the logistic function saturates, approaching either 0 or 1, fact that, via formula (6.2.23), implies that $\delta_i^{(\ell-1)}$ is equal, or close, to zero.

On the other side, since $\sigma' = \sigma(1 - \sigma)$, it is easy to show that $0 < \sigma' \leq 1/4$. This implies via (6.2.23) that backpropagation through a sigmoid layer reduces the gradient by a factor of at least 4. After propagating through several layers, see formulae (6.2.16) and (6.2.18), the resulting gradient becomes very small, fact that implies a slow, or a stopped learning. This phenomenon is called the *vanishing gradient problem*.

From the geometrical point of view, this means that the cost surface $(w, b) \to C(w, b)$ has flat regions, called *plateaus*. The plain-vanilla gradient descent method does not perform well on plateaus. In order to get the iteration out of a plateau, the method needs to be adjusted with a momentum term or a stochastic perturbation, see section 4.13. Another way to minimize the vanishing gradient problem is to use activation functions which do not saturate. Improved results can be obtained sometimes by using ReLU, or other hockey-type activation functions.

To conclude, the neuron learning process in the gradient descent method is affected by two ingredients: the learning rate, η, and the size of the cost function gradient, ∇C. If the learning rate is too large, the algorithm will not converge, the approximating sequence being oscillatory. On the other side, if it is too small, then the algorithm will take a long time to converge and might also get stuck in a point of local minimum.

The size of the gradient, ∇C, depends on two things: the type of activation function used, ϕ, and the choice of the cost function, C. The gradient ∇C depends on deltas $\delta_j^{(\ell)}$, which depend on the derivative of the learning function, ϕ'. Given the aforementioned saturation tendency of sigmoidal functions, it is recommended to used ReLU activation function for deep feedforward neural networks.

The choice of the cost function can also affect the learning efficiency. In the next table we provide three cost functions and their deltas in the output layers, $\delta_j^{(L)}$, which will be backpropagated to obtain the other deltas, $\delta_j^{(\ell)}$. The formulas are written for a sigmoid neuron, i.e., $\phi = \sigma$.

Cost formula	Deltas
$\frac{1}{2}\|x^{(L)} - z\|^2$	$\delta_j^{(L)} = (x_j^{(L)} - z_j)\sigma'(s_j^{(L)})$
$-\sum z_k \ln x_k^{(L)}$	$\delta_j^{(L)} = -z_j(1 - \sigma(s_j^{(L)}))$
$-\sum z_k \ln x_k^{(L)} - (1 - z_k)\ln(1 - x_k^{(L)})$	$\delta_j^{(L)} = x_j^{(L)} - z_j$

The first line, which corresponds to the quadratic cost, has a delta in the output layer that depends on σ', which by backpropagation influences the vanishing gradient tendency.

The second line corresponds to the cross-entropy function[2] and has a term

---

[2]This has the semantic interpretation that targets that take on the value zero are unobservable.

$(1 - \sigma)$ which may lead to vanishing gradient behavior, but not as much as in the case of the quadratic error.

The last line shows that the symmetric entropy delta is independent of the activation function, and hence, this produces the least vanishing gradient effect among all aforementioned cost functions.

There is one more factor that influences learning efficiency. This is the weights initialization. We shall deal with this important problem in section 6.3.

6.2.8 Batch training

Assume you are in an unknown city looking for the Science Museum. The only information you can get are directives from the people you meet in the street. It would be ideal to obtain directives from all people in town and then average their opinion, but this procedure is time consuming and hence not always feasible, if the town is large. Therefore, you are left with the options of asking one, or several individuals about the correct direction.

If you ask one individual "Where is the Science Museum?", you are given a directive, which may or may not be fully trustfully. However, you take the chance to go a certain distance in the suggested direction. Then ask again and walk another distance into the new suggested direction. This way, you are getting closer to the museum and eventually getting there.

There is another way to navigate toward the museum. Ask a group of 100 tourists about the museum direction and average out their suggested directions. This way, being more confident, you will walk for a larger distance into the average direction. Then repeat the procedure by asking again. In the latter case you expect to get to the museum sooner than in the former, since the learning rate is larger and you move into a more trustful direction.

A similar situation occurs when the gradient descent method is used. Each input $X^{(0)}$ produces a gradient direction $\nabla C(X^{(0)})$. Using full input data to compute the gradient of the cost function would provide the best result, $\nabla C(X^{(0)})$, but this procedure has the disadvantage that is time expensive. Therefore, a trade-off between accuracy and computation time is made by sampling a random minibatch and computing an estimation of the gradient from it. If consider a minibatch of N inputs, $\{X^{(0,1)}, \ldots, X^{(0,N)}\}$, the average direction is

$$\widetilde{\nabla C} = \frac{1}{N} \sum_{k=1}^{N} \nabla C(X^{(0,k)}).$$

By the Central Limit Theorem averages tend to have a smaller variance than each individual outcome. More precisely, if the inputs are independent random variables, then the average gradient $\widetilde{\nabla C}$ has the same mean as

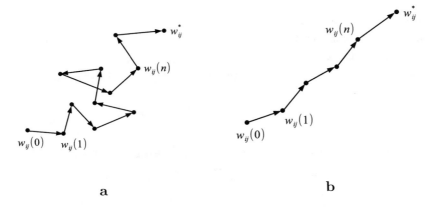

a   **b**

Figure 6.6: **a.** *Updating weights in the case of a small batch requires a large number of steps.* **b.** *Updating weights in the case of a large batch needs fewer steps.*

$\nabla C(X^{(0,k)})$ and the variance N times smaller. For instance, if the batch size is $N = 100$, the error of the mean direction $\widetilde{\nabla C}$ is one digit more accurate than the raw direction ∇C. Since the error is smaller, we are more confident in proceeding into this direction for finding the minimum, and hence we may choose a larger learning rate, at least in the beginning.

The procedure of updating the network parameters using the gradient descent method with a gradient estimated as an average of gradients measured on randomly selected training samples is called the *stochastic gradient descent* method.

It is worth noting the trade-off between the size of the batch and the training time, see Fig. 6.6. The larger the size of the batch, the more accurate the direction of the gradient, but the longer it takes to train. The fewer data fed into a batch, the faster the training but the less representative the result is. For instance, in the case of the MNIST data, the usual batch size used in practice is about 30.

6.2.9   Definition of FNN

After working with feedforward neural networks at a heuristic level, we shall provide in this section, for the sake of mathematical completeness, a formal definition of this concept.

First, we shall review all the ingredients used so far in the construction. We have seen that a feedforward neural network contains neurons arranged in $L + 1$ distinct layers. The first layer corresponding to the input data is obtained for $\ell = 0$ and the output layer is realized for $\ell = L$. The jth neuron in layer ℓ has the activation $x_j^{(\ell)}$. For the input layer, $x_j^{(0)}$, the activations come from the input data, while for the other layers the activations are computed in the terms of the previous layer activations by the forward pass formula

$$x_j^{(\ell)} = \phi^{(\ell)}\left(\sum_i w_{ij}^{(\ell)} x_i^{(\ell-1)} - b_j^{(\ell)}\right).$$

Here, $\phi^{(\ell)}$ is the activation function and $w_{ij}^{(\ell)}$ and $b_j^{(\ell)}$ are the weights and biasses of the neurons in the ℓth layer. We also note that $s_j^{(\ell)} = \sum_i w_{ij}^{(\ell)} x_i^{(\ell-1)} - b_j^{(\ell)}$ is an affine function of $x_i^{(\ell-1)}$ and the activation $x_j^{(\ell)}$ is obtained by applying the function ϕ on $s_j^{(\ell)}$.

Now we pursue a higher level of abstraction. We shall denote by $\mathcal{F}(U) = \{f : U \to \mathbb{R}\}$ the set of real-valued functions defined on the set U. Consider the set of indices $U_\ell = \{1, 2, \ldots, d^{(\ell)}\}$. Implement the affine function

$$\alpha_\ell : \mathcal{F}(U_{\ell-1}) \to \mathcal{F}(U_\ell)$$

by $\alpha_\ell(x^{(\ell-1)}) = s^{(\ell)}$. Then $x^{(\ell)} = (\phi^{(\ell)} \circ \alpha_\ell)(x^{(\ell-1)})$ is obtained composing the affine function on the previous layer activation with the nonlinearity ϕ. The formal definition of a feedforward neural network is as in the following.

Definition 6.2.5 *Let $U_\ell = \{1, 2, \ldots, d^{(\ell)}\}$, $0 \leq \ell \leq L$, and consider the sequence of affine functions $\alpha_1, \ldots, \alpha_L$*

$$\alpha_\ell : \mathcal{F}(U_{\ell-1}) \to \mathcal{F}(U_\ell)$$

and the sequence of activation functions $\phi^{(\ell)} : \mathbb{R} \to \mathbb{R}$. Then the corresponding feedforward neural network is the sequence of maps $f_0, f_1, \ldots, f_L$, where

$$f_\ell = \phi^{(\ell)} \circ \alpha_\ell \circ f_{\ell-1}, \quad 1 \leq \ell \leq L,$$

with f_0 given.

Hence, a deep feedforward neural network produces a sequence of progressively more abstract reparametrizations, f_ℓ, by mapping the input, f_0, through a series of parametrized functions, α_ℓ, and nonlinear activation functions, $\phi^{(\ell)}$. The network's output is given by

$$f_L = \phi^{(L)} \circ \alpha_L \circ \phi^{(L-1)} \circ \alpha_{L-1} \circ \cdots \circ \phi^{(1)} \circ \alpha_1.$$

The number L is referred as the depth of the network and $\max\{d^{(1)}, \ldots, d^{(L-1)}\}$ as the width. In the case of regression problems the last activation function is linear, $\phi^{(L)}(x) = x$. For classification problems, the the last activation function is a softmax, $\phi^{(L)}(x) = softmax(x)$, while in the case of logistic regression the activation function is a logistic sigmoid, $\phi^{(L)}(x) = \sigma(x)$.

6.3 Weights Initialization

In the case of networks with a small number of layers initializing all weights and biasses to zero, or sample them from a uniform or a Gaussian distribution of zero mean, usually provides satisfactory enough convergence results. However, in the case of deep neural networks, a correct initialization of weights makes a significant difference in the way the convergence of the optimality algorithm works.

The magnitude of weights plays an important role in avoiding as much as possible the vanishing and exploding gradient problems. This will be described in the following two marginal situations.

(i) *If the weights are too close to zero, then the variance of the input signal decreases as it passes through each layer of the network.*
This heuristic statement is based on the following computation. Let $X_j^{(\ell)}$ denote the output of the jth neuron in the ℓth layer. The outputs between two consecutive layers are given by the forward propagation equation

$$X_j^{(\ell)} = \phi\Big(\sum_i W_{ij}^{(\ell)} X_i^{(\ell-1)} - b_j^{(\ell)}\Big).$$

Using the Appendix formula (D.4.1)

$$Var(f(X)) \approx f'\big(\mathbb{E}[X]\big)^2 Var(X),$$

the aforementioned equation provides the variance of neuron activations in the ℓth layer in terms of variances of neuron activations in the previous layer as

$$Var(X_j^{(\ell)}) \approx \phi'\Big(\sum_i W_{ij}^{(\ell)}\mathbb{E}[X_i^{(\ell-1)}] - b_j^{(\ell)}\Big)^2\Big(\sum_i W_{ij}^{(\ell)} Var(X_i^{(\ell-1)})\Big). \quad (6.3.35)$$

By Cauchy's inequality, we have

$$\sum_i W_{ij}^{(\ell)} Var(X_i^{(\ell-1)}) \le \Big(\sum_i (W_{ij}^{(\ell)})^2\Big)^{1/2}\Big(\sum_i Var(X_i^{(\ell-1)})^2\Big)^{1/2}.$$

Making the assumption that the activation function has a bounded derivative, $(\phi')^2 < c$, then taking the sum over j and using Cauchy's inequality yields

$$\sum_j Var(X_j^{(\ell)})^2 < C^2 \sum_{i,j} (W_{ij}^{(\ell)})^2 \sum_i Var(X_i^{(\ell-1)})^2.$$

This shows that for small weights $W_{ij}^{(\ell)}$ the sum of squares of variances in the ℓth layer is smaller than the sum of squares of variances in the $(\ell-1)$th layer. Iterating the previous inequality, it follows that the signal's variance decreases to zero after passing through a certain number of layers. Therefore, after passing through a few layers, the signal becomes too low to be significant.

(ii) If the weights are too large, then either the variance of the signal tends to get amplified as it passes through the network layers, or the network approaches a vanishing gradient problem.

If the weights $W_{ij}^{(\ell)}$ are large, the sum $\sum_i W_{ij}^{(\ell)} X_i^{(\ell-1)}$ also gets large. If the activation function is linear, $\phi(x) = x$, then

$$Var(X_j^{(\ell)}) = \sum_i W_{ij}^{(\ell)} Var(X_i^{(\ell-1)}),$$

which implies that $Var(X_j^{(\ell)})$ is large.

If the activation function ϕ is of sigmoid type, then when the weights $W_{ij}^{(\ell)}$ are large the sum $\sum_i W_{ij}^{(\ell)} X_i^{(\ell-1)}$ tends to have large values, and hence the activation function ϕ becomes saturated, fact that leads to an approaching zero gradient problem.

Hence, neither choosing the weights too close to zero, or choosing them far too large is a feasible initialization, since in both cases the initialization is outside the right basin of attraction of the optimization procedure. We need to initialize the weights with values in a reasonable range before we start training the network. We shall deal next with the problem of finding this reasonable range.

The main idea is based on the fact that the propagation of the signal error through a deep neural network can be quantized by the signal variance, namely the variance of the neuron activations. In order to keep the variance under control, away from exploding or vanishing values, the simplest way is to find the weights values for which the variance remains roughly unchanged as the signal passes through each layer. Since assigning the initial weights is a random process, we assume the weights to be either uniform- or Gaussian-distributed random variables with zero mean.

To make notations simpler, after we fix a layer ℓ, we denote the incoming and outgoing vector signals from that layer by $X = X^{(\ell-1)}$ and $Y = X^{(\ell)}$, respectively. The number of the neurons into the layer ℓ will be denoted in this section by $N = d^{(\ell-1)}$. The jth component of Y is given by

$$Y_j = \phi\Big(\sum_i W_{ij} X_i + b_j \Big),$$

where W_{ij} are random variables with $\mathbb{E}[W_{ij}] = 0$, and X_i are independent, identically distributed random variables, with zero mean (for instance,

Gaussian distributed with zero mean). It makes sense to assume that W_{ij} and X_i are independent. Also, the bias b_j is a constant (usually initialized to zero). Using the Appendix formula (D.4.1), the variance of Y_j is approximately given by

$$
\begin{aligned}
Var(Y_j) &= \phi'\Big(\sum_i \mathbb{E}[W_{ij}X_i] + b_j\Big)^2 Var\Big(\sum_i W_{ij}X_i\Big) \\
&= \phi'(b_j)^2 \sum_i Var(W_{ij}X_i) \\
&= \phi'(b_j)^2 \sum_i Var(W_{ij})Var(X_i) \\
&= N\phi'(0)^2 Var(W_{ij})Var(X_i),
\end{aligned}
$$

where in the second identity we used $\mathbb{E}[W_{ij}X_i] = \mathbb{E}[W_{ij}]\mathbb{E}[X_i] = 0$ and the additivity of variance with respect to independent random variables. In the third identity we used the multiplicative property of variance as given by Goodman's formula, see Lemma D.4.1 from Appendix. In the last identity we used that both weights W_{ij} and inputs X_i are identically distributed and the number of incoming neurons is N. We also initialized here the bias $b_j = 0$.

Asking the condition that the variance is invariant under the ℓth layer, i.e., $Var(Y_j) = Var(X_i)$, we obtain

$$
N\phi'(0)^2 Var(W_{ij})Var(X_i) = Var(X_i).
$$

Simplifying by $Var(X_i)$, we obtain the variance of the weights

$$
Var(W_{ij}) = \frac{1}{N\phi'(0)^2}. \tag{6.3.36}
$$

We encounter two useful cases:
1. If W_{ij} are drawn from a normal distribution, then $W_{ij} \sim \mathcal{N}\big(0, \frac{1}{N\phi'(0)^2}\big)$. In the case of a linear activation, $\phi'(0) = 1$, we have $W_{ij} \sim \mathcal{N}\big(0, \frac{1}{N}\big)$.
2. If W_{ij} are drawn from a uniform distribution on $[-a, a]$, with zero mean, $W_{ij} \sim Unif[-a, a]$, then equating their variances, $\frac{1}{N\phi'(0)^2} = \frac{a^2}{3}$, yields $a = \frac{\sqrt{3}}{\phi'(0)\sqrt{N}}$. Hence, $W_{ij} \sim Unif\Big[-\frac{\sqrt{3}}{\phi'(0)\sqrt{N}}, \frac{\sqrt{3}}{\phi'(0)\sqrt{N}}\Big]$.

It is worth noting how the activation function nonlinearity at 0 influences the range of the initialization interval. The larger the slope at zero, the narrower the interval. One way to use this for obtaining better results is to set $\phi'(0)$ to be a new hyperparameter and tune it on a validation set.

Xavier initialization These weight initializations have taken into account only the number of the input neurons, N, into the layer ℓ, using a forward propagation point of view described by the invariance relation $Var(X^{(\ell)}) = Var(X^{(\ell+1)})$. For preserving the backpropagated signal as well, Glorot and Bengio [43] worked a formula involving also the outgoing number of neurons, called the *Xavier initialization*. This involves a backpropagation point of view, which is based on the assumption that the variances of the cost function gradient remain unchanged as the signal backpropagates through the ℓth layer, i.e.,

$$Var\Big[\frac{\partial C}{\partial W_{ij}^{(\ell-1)}}\Big] = Var\Big[\frac{\partial C}{\partial W_{ij}^{(\ell)}}\Big].$$

Using the chain rule equation (6.2.27) the previous equation becomes

$$Var\Big[\delta_j^{(\ell-1)} X_i^{(\ell-2)}\Big] = Var\Big[\delta_j^{(\ell)} X_i^{(\ell-1)}\Big],$$

where $\delta_j^{(\ell)} = \frac{\partial C}{\partial s_j^{(\ell)}}$. In order to make progress with the computation we shall assume that the activation function is linear, $\phi(x) = x$, i.e., the neuron is of linear type. Under this condition the equation (6.2.26) is written as

$$\delta_i^{(\ell-1)} = \phi'(s_i^{(\ell-1)}) \sum_{j=1}^{N'} \delta_j^{(\ell)} W_{ij}^{(\ell)} = \sum_{j=1}^{N'} \delta_j^{(\ell)} W_{ij}^{(\ell)}, \qquad (6.3.37)$$

where $N' = d^{(\ell)}$ is the number of neurons in the ℓth layer. Since $\delta_j^{(\ell)}$ and $W_{ij}^{(\ell)}$ are independent, see Exercise 6.6.7, then we have

$$\mathbb{E}\Big[\sum_{j=1}^{N'} \delta_j^{(\ell)} W_{ij}^{(\ell)}\Big] = \sum_{j=1}^{N'} \mathbb{E}[\delta_j^{(\ell)}]\mathbb{E}[W_{ij}^{(\ell)}] = 0,$$

where we used that $\mathbb{E}[W_{ij}^{(\ell)}] = 0$. Therefore, $\mathbb{E}[\delta_i^{(\ell-1)}] = 0$, and similarly, $\mathbb{E}[\delta_i^{(\ell)}] = 0$. Assuming that $\mathbb{E}[X_i^{(\ell-2)}] = \mathbb{E}[X_i^{(\ell-1)}] = 0$, and using the independence of $\delta_j^{(\ell)}$ and $x_i^{(\ell-1)}$, see Exercise 6.6.7, then Goodman's formula (Lemma D.4.1) yields

$$Var\big[\delta_j^{(\ell-1)}\big] Var\big[X_i^{(\ell-2)}\big] = Var\big[\delta_j^{(\ell)}\big] Var\big[X_i^{(\ell-1)}\big].$$

Using the forward propagation relation $Var\big[X_i^{(\ell-2)}\big] = Var\big[X_i^{(\ell-1)}\big]$, it follows that the deltas variance remains unchanged

$$Var\big[\delta_j^{(\ell-1)}\big] = Var\big[\delta_j^{(\ell)}\big].$$

Applying the variance to equation (6.3.37), using the independence and Goodman's formula, we obtain

$$Var(\delta_i^{(\ell-1)}) = \sum_{j=1}^{N'} Var(\delta_j^{(\ell)}) Var(W_{ij}^{(\ell)}) = N' Var(\delta_j^{(\ell)}) Var(W_{ij}^{(\ell)}),$$

where we also used the fact that the weights and deltas are identically distributed. Dividing the last two equations, we arrive at the relation $N'Var(W_{ij}^{(\ell)}) = 1$, or equivalently

$$Var(W_{ij}^{(\ell)}) = \frac{1}{N'}, \tag{6.3.38}$$

i.e., the variance of the weights in the ℓth layer is inversely proportional to the number of neurons in that layer. Under the assumption of the linear activation function, formula (6.3.36) becomes

$$Var(W_{ij}^{(\ell)}) = \frac{1}{N}, \tag{6.3.39}$$

where N is the number of neurons in the $(\ell-1)$th layer.

Now, the equations (6.3.38) and (6.3.39) are satisfied simultaneously only in the case $N = N'$, i.e., when the number of neurons is the same in any two consecutive layers. Since this condition is too restrictive, a lucrative compromise is to take the arithmetic average of the two, the case in which

$$Var(W_{ij}^{(\ell)}) = \frac{2}{N + N'}, \tag{6.3.40}$$

where $N = d^{(\ell-1)}$ and $N' = d^{(\ell)}$.

Again, we emphasize two cases of practical importance:

1. If W_{ij} are drawn from a normal distribution, then $W_{ij} \sim \mathcal{N}\left(0, \frac{2}{N+N'}\right)$.
2. If W_{ij} are drawn from a uniform distribution with zero mean, then $W_{ij} \sim Unif\left[-\frac{\sqrt{6}}{\sqrt{N+N'}}, \frac{\sqrt{6}}{\sqrt{N+N'}}\right]$, which is also known as the *normalized initialization*.

In conclusion, a reasonable initialization of weights, even it might not totally solve the vanishing and exploding gradient problems, it still improves significantly the backpropagation algorithm functionality.

6.4   Strong and Weak Priors

In the context of neural networks, a *prior* is a probability distribution over the parameters (weights and biasses) that encodes our initial beliefs before we have seen the data.

A prior is called *weak* if its entropy is high, i.e., if the initial belief about the parameters distribution is not very specific. A prior is called *strong* if its entropy is small, i.e., if the initial belief about parameters distribution is very specific.

This section will discuss the previous initializations in the aforementioned context. First, we recall from Chapter 12 that the entropy of a probability distribution p is given by

$$H(p) = -\int p(x) \ln p(x)\, dx.$$

We shall consider only those priors p with positive entropy only.

(i) In the case when $W_{ij} \sim Unif\left[-\dfrac{\sqrt{3}}{\phi'(0)\sqrt{N}}, \dfrac{\sqrt{3}}{\phi'(0)\sqrt{N}}\right]$, then by Exercise 6.6.10 we have the entropy

$$H(p_W) = \ln \frac{2\sqrt{3}}{\phi'(0)\sqrt{N}},$$

where p_W denotes the probability density of the weights W_{ij}. Consequently, the prior becomes strong if the number of neurons N is large.

(ii) In the case of Xavier initialization, $W_{ij} \sim \mathcal{N}\left(0, \frac{2}{N+N'}\right)$. Then by Exercise 6.6.11 we have

$$H(p_W) = \ln \frac{2\sqrt{e}}{\sqrt{N+N'}}.$$

Hence, a large number of incoming and outgoing neurons makes the prior to be strong.

6.5   Summary

A neural network is obtained by a concatenation of several layers of neurons. The first layer corresponds to the input while the last one to the output. The inner layers are called hidden layers. This chapter was concerned only with the feedforward neural networks. These are networks where the information flows forward, from the input to the output.

Neural networks are used for learning complicated features, which a single neuron can't do. They are trained by the gradient descent method. An important ingredient of this method is to compute the gradient of the cost function. The gradient is computed by a recursive method called backpropagation. The master equations for both forward and backward propagation are provided in detail.

Sometimes it is more lucrative to compute the gradient using an average over a training minibatch, which leads to the stochastic gradient descent method. The optimal size of the minibatch depends on the problem and can be considered as a trade-off between the training time and obtained accuracy.

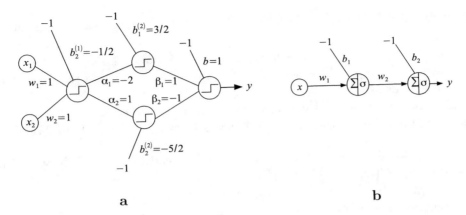

Figure 6.7: **a.** *Neural network used in Exercise* 6.6.3. **b.** *For Exercise* 6.6.6.

For the gradient descent method to work properly certain initialization for the weights and biases has to be done. If the network is shallow, usually a zero initialization is preferred. For deep networks there are some preferred initializations such as Xavier initialization, or the normalized initialization.

6.6   Exercises

Exercise 6.6.1 Draw the multi-perceptron neural network given by Remark 6.1.2.

Exercise 6.6.2 (a) Can a single perceptron learn the following mapping:

$$(0,0,0) \to 1, \quad (1,0,0) \to 0, \quad (0,1,0) \to 0, \quad (0,0,1) \to 0,$$

$$(0,1,1) \to 1, \quad (1,1,0) \to 0, \quad (1,0,1) \to 0, \quad (1,1,1) \to 1?$$

(b) Draw a multi-perceptron neural network that learns the previous mapping. State the weights and biases of all perceptrons in the network.

Exercise 6.6.3 Write an explicit formula for the output of the multi-perceptron neural network given in Fig. 6.7 **a**.

Exercise 6.6.4 Consider a sigmoid neuron with one-dimensional input x, weight w, bias b, and output $y = \sigma(wx+b)$. The target is the one-dimensional variable z. Consider the cost function $C(w,b) = \frac{1}{2}(y-z)^2$;

(a) Find $\nabla C(w,b)$ and show that $\|\nabla C\| < \frac{1}{4}\sqrt{1+x^2}\,(1+|z|)$;

(b) Write the gradient decent iteration for the sequence of weights (w_n, b_n).

Exercise 6.6.5 (noise reduction) Consider the neural network given by Fig. 6.2.

(a) Show that $\forall \epsilon > 0$, there is $\eta > 0$ such that if $\|dX\| < \eta$ then $\|dY\| < \epsilon$;

(b) Assume the input X is noisy. Discuss the effect of weights regularization on noise removal;

(c) Discuss the choice of the activation function in the noise removal. Which choice will denoise better: the logistic sigmoid or the tangent hyperbolic?

Exercise 6.6.6 Consider the neural network given by Fig. 6.7 **b**, with the cost function $C(w, b) = \frac{1}{2}(y - z)^2$.

(a) Compute the deltas $\delta^{(i)} = \frac{\partial C}{\partial s^{(i)}}$, $i \in \{1, 2\}$;

(b) Use backpropagation to find the gradient $\nabla C(w, b)$.

Exercise 6.6.7 Assume the activation function in the ℓth layer is linear and the weights and inputs are random variables.

(a) Show that $\delta_j^{(\ell)}$ and $W_{ij}^{(\ell)}$ are independent random variables;

(b) Show that $\delta_j^{(\ell)}$ and $X_i^{(\ell-1)}$ are independent random variables.

Exercise 6.6.8 Consider a sigmoid neuron with a random input normally distributed, $X \sim \mathcal{N}(0, 1)$, and the output $Y = \sigma(wX + b)$. Show that $Var Y \approx w^2 \sigma'(b)^2$. Note that the output variance decreases for small weights, w, or large values of the bias, b.

Exercise 6.6.9 Consider a one-hidden layer neural network with sigmoid neurons in the hidden layer. Given that the input is normally distributed, $X \sim \mathcal{N}(0, 1)$, and the output is $Y = \sum_{i=1}^{N} \alpha_i \sigma(w_i X + b_i)$, show that $Var Y \approx \sum_j \sigma'(b_i)^2 \alpha_i^2 w_i^2$.

Exercise 6.6.10 Let $p(x)$ be the uniform distribution over the interval $[a, b]$. Show that $H(p) = \ln(b - a)$.

Exercise 6.6.11 Let $p(x)$ be the one-dimensional normal distribution with mean μ and standard deviation σ. Show that its entropy is $H(p) = \ln(\sigma\sqrt{2\pi e})$.

Exercise 6.6.12 State a formula relating the batch size, N, the current iteration, k and the number of epochs, p, during a neural network training.

Part II
Analytic Theory

Chapter 7

Approximation Theorems

This chapter presents a few classical real analysis results with applications to learning continuous, integrable, or square-integrable functions. The approximation results included in this chapter contain Dini's theorem, Arzela-Ascoli's theorem, Stone-Weierstrass theorem, Wiener's Tauberian theorem, and the contraction principle. Some of their applications to learning will be provided within this chapter, while others will be given in later chapters.

7.1 Dini's Theorem

Learning a continuous function on a compact interval resumes to showing that the neural network produces a sequence of continuous functions, which converges uniformly to the continuous target function. In most cases, it is not difficult to construct outcomes that converge pointwise to the target function. However, showing the uniform convergence can be sometimes a complicated task. In some circumstances, this can be simplified by the use of certain approximations results, such as the following.

Theorem 7.1.1 (Dini) *Let $f_n : [a, b] \to \mathbb{R}$ be a sequence of continuous functions.*
(i) If $f_{n+1} \leq f_n$, for all $n \geq 1$, and $f_n(x) \to 0$, for any $x \in [a, b]$, then f_n converges uniformly to 0 on $[a, b]$.
(ii) Let $g \in C[a, b]$ such that $f_n(x) \searrow g(x)$ for any $x \in [a, b]$. Then f_n converges uniformly to g on $[a, b]$.

Proof: (i) Let $\epsilon > 0$ be arbitrary fixed and consider the sets

$$A_n = f_n^{-1}([\epsilon, \infty)) = \{x \in [a, b]; f_n(x) \geq \epsilon\},$$

© Springer Nature Switzerland AG 2020
O. Calin, *Deep Learning Architectures*, Springer Series in the Data Sciences,
https://doi.org/10.1007/978-3-030-36721-3_7

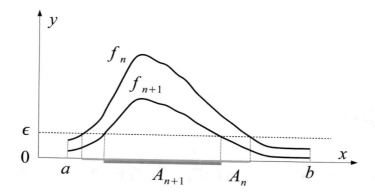

Figure 7.1: *The descending sequence of compact sets A_n.*

see Fig. 7.1. Since f_n are continuous functions, then A_n are closed. Furthermore, the sets A_n are compact, as closed subsets of a compact set, $A_n \subset [a, b]$. The sequence $(A_n)_n$ is descending, i.e., $A_{n+1} \subset A_n$, $\forall n \geq 1$. To show this, let $x \in A_{n+1}$ be arbitrary chosen. Since $f_n(x) \geq f_{n+1}(x) \geq \epsilon$, then $x \in A_n$, which implies the desired inclusion.

Assume now that $A_n \neq \emptyset$, for all $n \geq 1$. Then $\bigcap_{n \geq 1} A_n \neq \emptyset$, as an intersection of a descending sequence of nonempty compact sets.[1] Let $x_0 \in \bigcap_{n \geq 1} A_n$. Then $f_n(x_0) \geq \epsilon$, for all $n \geq 1$, which contradicts the fact that $f_n(x_0) \to 0$, as $n \to \infty$. Hence, the assumption made is false, that is, there is an $N \geq 1$ such that $A_N = \emptyset$. This means $f_N(x) < \epsilon$, $\forall x \in [a, b]$. And using that $(f_n)_n$ is decreasing, it follows that $f_n(x) < \epsilon$, $\forall n \geq N$ and $\forall x \in [a, b]$. This can be written as $\max_{x \in [a,b]} |f_n| \to 0$, as $n \to \infty$, namely, f_n converges uniformly to 0.

(ii) It follows from part (i) applied to the sequence $(f_n - g)_n$.   ∎

Remark 7.1.2 The result still holds if the compact interval $[a, b]$ is replaced with a compact complete metric space (S, d), and the convergence is considered in the sense of the metric d.

Example 7.1.1 1. Consider a continuous function $g : [0, 1] \to \mathbb{R}$ and the partition $\Delta_1 : 0 = x_0 < x_1 < \cdots < x_n = 1$, such that $(x_i)_i$ are inflection points for g, that is, $g_{|[x_i, x_{i+1}]}$ is either convex or concave, see Fig. 7.2. Denote by f_1 the piecewise linear function obtained by joining the points $(x_i, g(x_i))$.

---

[1]This result is known as the *Cantor's lemma.*

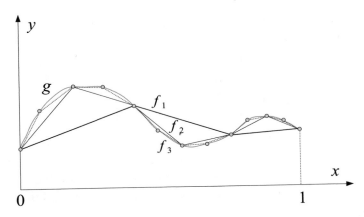

Figure 7.2: *A polygonal approximating sequence $(f_n)_n$ satisfying the inequality* $|g(x) - f_{n+1}(x)| \leq |g(x) - f_n(x)|$, *for all $x \in [0,1]$.*

Now, consider the partition Δ_2 obtained by enriching the partition Δ_1 with the segment midpoints $(x_i + x_{i+1})/2$. Denote the new obtained piecewise linear function by f_2. We continue the procedure, associating the function f_n to the partition Δ_n. The sequence of continuous functions $(f_n)_n$ satisfies the property

$$|g(x) - f_{n+1}(x)| \leq |g(x) - f_n(x)|, \qquad \forall x \in [0,1].$$

Dini's Theorem applied to the sequence $|g - f_n|$ implies that f_n converges to g, uniformly on $[0,1]$.

7.2   Arzela-Ascoli's Theorem

Another result of constructing a uniformly convergent sequence is to extract from an existing family of continuous functions a subsequence with this property. The *Arzela-Ascoli* theorem provides an existence result, rather than a constructive procedure, which is useful in theoretical investigations. First, we shall introduce a few definitions.

A family of functions $\mathcal{F}$ on a set A is called *uniformly bounded* if there is an $M > 0$ such that

$$|f(x)| \leq M, \quad \forall x \in A, \forall f \in \mathcal{F}.$$

This means that the functions in the family $\mathcal{F}$ are all bounded by the same bound M.

Example 7.2.1 Let $\mathcal{F} = \{\cos(ax + b); a, b \in \mathbb{R}\}$. Then the family $\mathcal{F}$ is uniformly bounded, with $M = 1$.

Example 7.2.2 Consider

$$\mathcal{F} = \Big\{ \sum_{j=1}^{N} \alpha_j \sigma(w_j x + b_j); \alpha_j, w_j, b_j \in \mathbb{R}, \sum_{j=1}^{N} \alpha_j^2 \leq 1 \Big\}$$

where $\sigma(x)$ is a sigmoid function satisfying $|\sigma(x)| \leq 1$ (such as the logistic function or the hyperbolic tangent). Then the family $\mathcal{F}$ is uniformly bounded, with $M = \sqrt{N}$. This follows from an application of Cauchy's inequality

$$\Big| \sum_{j=1}^{N} \alpha_j \sigma(w_j x + b_j) \Big|^2 \leq \sum_{j=1}^{N} \alpha_j^2 \sum_{j=1}^{N} \sigma(w_j x + b_j)^2 \leq 1 \cdot N = N.$$

A family of functions $\mathcal{F}$ on a set A is called *equicontinuous* if $\forall \epsilon > 0$, there is $\eta > 0$ such that $\forall f \in \mathcal{F}$ we have

$$|f(x) - f(x')| < \epsilon, \quad \forall x, x' \in A, \text{ with } |x - x'| < \eta.$$

Equivalently stated, the functions in the family $\mathcal{F}$ are all uniformly continuous with the same ϵ and η.

Example 7.2.3 Consider a family of continuous differentiable functions $\mathcal{F} \subset C^1[a, b]$, such that there is a constant $L > 0$ with the property

$$\sup_{x \in [a,b]} |f'(x)| < L, \quad \forall f \in \mathcal{F}.$$

Then $\mathcal{F}$ is equicontinuous.
To show this, using the Mean Value Theorem, we obtain the following estimation:

$$|f(x) - f(x')| \leq \sup_{x \in [a,b]} |f'(x)| \, |x - x'| \leq L|x - x'|, \quad \forall f \in \mathcal{F},$$

and then choose $\eta = \epsilon/L$.

Example 7.2.4 Consider the family of functions

$$\mathcal{F} = \Big\{ \sum_{j=1}^{N} \alpha_j \sigma(w_j x + b_j); \alpha_j, w_j, b_j \in \mathbb{R}, \sum_{j=1}^{N} (\alpha_j^2 + w_j^2) \leq 1 \Big\},$$

where σ stands for the logistic sigmoid. The family can be interpreted as outputs of a one-hidden layer neural network with exactly N units in the hidden layer. The weights constraint $\|\alpha\|^2 + \|w\|^2 \leq 1$ states that the weights are kept small, which is one of the ideas of regularization. Under this hypothesis, the family $\mathcal{F}$ is equicontinuous.

This can be shown as in the following. The maximum slope of $\sigma(x)$ is achieved at $x = 0$ and it is equal to

$$\max \sigma'(x) = \max \sigma(x)(1 - \sigma(x)) = \sigma(0)(1 - \sigma(0)) = \frac{1}{4}.$$

Then for any $f \in \mathcal{F}$, applying Cauchy's inequality, we have the derivative estimation

$$|f'(x)| = \Big| \sum_{j=1}^{N} \alpha_j w_j \sigma'(w_j x + b_j) \Big| \leq \frac{1}{4} \sum_{j=1}^{N} |\alpha_j| \, |w_j|$$

$$\leq \frac{1}{4} \Big(\sum_{j=1}^{N} \alpha_j^2 \Big)^{1/2} \Big(\sum_{j=1}^{N} w_j^2 \Big)^{1/2} \leq \frac{1}{4}.$$

Since $\sup_{x \in [a,b]} |f'(x)| \leq \frac{1}{4}$, $\forall f \in \mathcal{F}$, using Example 7.2.3, it follows that the family $\mathcal{F}$ is equicontinuous on $[a, b]$. It is worth noting that the conclusion holds in the more general setup of a sigmoid function σ satisfying $|\sigma'(x)| < \lambda < 1$. For instance, $\sigma(x) = \tanh(cx)$, or $\sigma(x) = \sin(cx + b)$, with $|c| < \lambda < 1$ and fixed $b \in \mathbb{R}$, satisfy the condition.

The proof of the following important result can be found in any advanced book of Real Analysis, such as [104].

Theorem 7.2.1 (Arzela-Ascoli) *Let $\mathcal{F} \subset C[a, b]$ be a family of continuous functions on $[a, b]$. Then the following statements are equivalent:*
(a) Any sequence $\forall (f_n)_n \subset \mathcal{F}$ contains a subsequence $(f_{n_k})_k$ that is uniform convergent;
(b) The family $\mathcal{F}$ is equicontinuous and uniformly bounded.

We make the remark that condition (a) is equivalent to the fact that $\mathcal{F}$ is a compact set in the metric space $(C[a, b], \| \cdot \|_\infty)$, see Appendix A for definitions.

In the following we state an existence result of an unsupervised learning. A three-layer neural net has capabilities of "potential learning" some continuous function, given constraint conditions on the size of the hidden layer, slope of the activation function, and magnitude of weights.

Theorem 7.2.2 *Let $N \geq 1$ be a fixed integer and consider a one-hidden layer neural network such that:*

(i) the network input is a real bounded variable, $x \in [a, b]$;
(ii) the output is a one-dimensional neuron with linear activation function and zero bias;
(iii) there are N neurons in the hidden layer with a differentiable activation function σ such that $|\sigma'| < \lambda < 1$;
(iv) the weights satisfy the regularization condition $\sum_{j=1}^{N}(\alpha_j^2 + w_j^2) \le 1$, where w_j are the input to hidden layer weights and α_j are the hidden layer to output weights;
Then there is a continuous function g on $[a, b]$ that can be approximately represented by the network.

Proof: The output function of the network is given by the sum

$$f_{\alpha,w,b}(x) = \sum_{j=1}^{N} \alpha_j \sigma(w_j x + b_j),$$

where w_j are the weights from the input to the hidden layer, and α_j are the weights from the hidden layer to the output. The biasses are denoted by b_j. Given the regularization constraint satisfied by the weights, Example 7.2.2 implies that the family $\mathcal{F} = \{f_{\alpha,w,b}; \alpha, w, b \in \mathbb{R}^N\}$ is uniformly bounded. Example 7.2.4 shows that $\mathcal{F}$ is equicontinuous. Then applying the Arzela-Ascoli Theorem, it follows that any sequence of outcomes, $(f_n)_n \subset \mathcal{F}$, given by

$$f_n(x) = \sum_{j=1}^{N} \alpha_j(n) \sigma(w_j(n)x + b_j(n)), \qquad n \ge 1$$

has a subsequence $(f_{n_k})_k$, which is uniformly convergent on $[a, b]$; its limit is a continuous function, $g \in C[a, b]$.

∎

Remark 7.2.3 In the previous result the continuous function g is not known a priori. The neural network learns "potentially" a continuous function, in the sense that among any infinite sequence of network trials, it can be extracted a subsequence along which the network approximates some continuous function. This is an existence result; finding an algorithm of constructing the convergent subsequence is a different problem and will be treated in a different chapter.

Example 7.2.5 (Sigmoid neuron) Consider a sigmoid neuron with the output

$$f_{w,b}(x) = \sigma(w^T x + b),$$

where σ is the logistic function, with weights $w \in \mathbb{R}^n$, bias $b \in \mathbb{R}$, and input $x \in I_n = [0, 1] \times \cdots \times [0, 1]$. Assume the weights are bounded, with $\|w\| \le 1$.

Then the family of continuous functions

$$\mathcal{F} = \{ f_{w,b}; \|w\| \le 1, b \in \mathbb{R} \}$$

is uniformly bounded (since $|f_{w,b}(x)| < 1$) and equicontinuous, because

$$
\begin{aligned}
|f_{w,b}(x') - f_{w,b}(x)| &= |\sigma(w^T x' + b) - \sigma(w^T x + b)| \\
&\le \max \|\sigma'\| \, \|w^T x' + b - w^t x - b\| \\
&= \frac{1}{4} \|w^T(x' - x)\| \le \frac{1}{4} \|w\| \, \|x' - x\| \le \frac{1}{4} \|x' - x\|.
\end{aligned}
$$

By the Arzela-Ascoli theorem, any sequence of outputs $\{ f_{w_k, b_k} \}_k$ contains a subsequence uniformly convergent to a continuous function g on I_n. It is worth to remark that no conditions have been asked on the bias b.

7.3 Application to Neural Networks

The previous notions can be applied to the study of the one-hidden layer neural networks with a continuum of neurons in the hidden layer, see also section 5.8.

Assume the input belongs to a compact interval, $x \in [a, b]$. It is well known that a one-hidden layer neural net, with N neurons in the hidden layer, has an output given by $g(x) = \sum_{j=1}^{N} \sigma(w_j x + b_j) \alpha_j$, where σ is a logistic sigmoid. Also, w_j are the weights between the input and the jth node, and α_j is the weight between the jth node and the output; b_j denotes the bias for the jth neuron in the hidden layer.

We want to generalize to the case where the number of hidden units is infinite. In particular, we want to consider the case where the number of hidden units is countably infinite. We assume the hidden neurons are continuously parametrized by t, taking values in the compact $[c, d]$. Hence, the parameter t will replace the index j. Furthermore, the summation weights α_j, will be replaced by a measure, which under the hypothesis of being absolute continuous with respect to the Lebesgue measure $dt_{[c,d]}$ will take the form $h(t)dt$. The output from the jth hidden neuron, $\sigma(w_j + b_j)$, is a continuous function of x and depends on j. This will be transformed into the kernel $K(x, t) = \sigma(w(t)x + b(t))$. The sum will become an integral with respect to the aforementioned measure.

Hence, the continuum analog of the output $g(x) = \sum_{j=1}^{N} \sigma(w_j x + b_j) \alpha_j$ is the integral $g(x) = \int_c^d K(x, t) h(t) \, dt$. If $\mathcal{K}$ is the integral transform with the kernel K, then $g = \mathcal{K}(h)$, i.e., the output is the integral transform of the measure density h. Assuming the regularization condition $\|h\|_2^2 \le M$, namely, the measure densities are L^2-uniformly bounded, then the set of outputs is equicontinuous and uniformly bounded, see Exercise 7.11.10. By Arzela-Ascoli Theorem, there is a sequence $(h_n)_n$ such that the sequence of outputs, $g_n = \mathcal{K}(h_n)$ is uniformly convergent on $[0, 1]$ to a continuous function g. In Chapter 9, it will be shown that the neural network can learn exactly any continuous function on $[0, 1]$.

It is worthy to note that we can consider an even more general case where the number of hidden units can be finite, countably infinite, or uncountably infinite. This can be accomplished by choosing the output $g(x) = \int_c^d K(x,t)\, d\mu(t)$, where the weighting measure μ is taken as in section 5.8.

7.4 Stone-Weierstrass' Theorem

In this section we denote by K a compact set in $\mathbb{R}^n$. For future applications, it suffices to consider $K = I_n = [0, 1] \times \cdots \times [0, 1]$, the unit hypercube, and in the case $n = 1$, we may consider $K = [a, b]$. The set of continuous real-valued functions on K is denoted by $C(K)$. The space $C(K)$ becomes a metric space if we set the metric

$$d(f,g) = \max_{x \in K} |f(x) - g(x)|, \qquad \forall f, g \in C(K).$$

$C(K)$ can also be considered as a normed linear space with the norm given by $\|f\| = \max_{x \in K} |f(x)|$.

We shall introduce next a central concept used in approximation theory and learning. The reader can find the definitions of distance and metric space in section E.1 of the Appendix.

Definition 7.4.1 *Let $(\mathcal{S}, d)$ be a metric space and $\mathcal{A}$ a subset of $\mathcal{S}$. Then $\mathcal{A}$ is called dense in $\mathcal{S}$ if for any element $g \in \mathcal{S}$ there is a sequence $(f_n)_n$ of elements in $\mathcal{A}$ such that $\lim_{n \to \infty} d(f_n, g) = 0$.*

Equivalently, this means that $\forall \epsilon > 0$, there is $N > 1$ such that

$$d(f_n, g) < \epsilon, \quad \forall n \geq N.$$

Example 7.4.2 Let $(\mathcal{S}, d) = (\mathbb{R}, |\ |)$ be the set of real numbers endowed with the distance given by the absolute value and $\mathcal{A} = \mathbb{Q} = \{\frac{p}{q}; p, q \text{ integers}, q \neq 0\}$, the set of rational numbers. Then $\mathbb{Q}$ is dense in $\mathbb{R}$, namely for any $x \in \mathbb{R}$, there is a sequence of rational numbers, $(r_n)_n$, which converges to x. Equivalently, any real number can be approximated by a rational number: given $x \in \mathbb{R}$, then $\forall \epsilon > 0$ there is $r \in \mathbb{Q}$ such that $|x - r| < \epsilon$.

Example 7.4.3 Consider $\mathcal{S} = C(K)$ endowed with the distance $d(f,g) = \max_{x \in K} |f(x) - g(x)|$. Then a subset $\mathcal{A}$ is dense in $C(K)$ if for any $g \in C(K)$ and any $\epsilon > 0$, there is a function $f \in \mathcal{A}$ such that

$$|g(x) - f(x)| < \epsilon, \qquad \forall x \in K.$$

The topological idea of density is that the elements of the subset $\mathcal{A}$ are as close as possible to the elements of the complementary set $\mathcal{S} \backslash \mathcal{A}$. Even if this concept is some sort of counterintuitive in the context of general spaces, the reader can think of it as in the familiar case of rational and real numbers.

A subset $\mathcal{A}$ of $C(K)$ is called an *algebra* if $\mathcal{A}$ is closed with respect to linear combinations and multiplication with reals:

(*i*) $\forall f, g \in \mathcal{A} \Rightarrow f + g \in \mathcal{A}$;

(*ii*) $\forall f \in \mathcal{A}, \forall c \in \mathbb{R} \Rightarrow cf \in \mathcal{A}$;

(*iii*) $\forall f, g \in \mathcal{A} \Rightarrow fg \in \mathcal{A}$;

We note that the first two conditions state that $\mathcal{A}$ is a linear subspace of $C(K)$.

Example 7.4.1 Let $\mathcal{A}$ be the set of all finite Fourier series on $[0, 2\pi]$

$$\mathcal{A} = \left\{ f(x) = c_0 + \sum_{k=1}^{N} (a_k \cos nx + b_k \sin nx); a_k, b_k \in \mathbb{R}, N = 0, 1, 2, \ldots \right\}.$$

It is obvious that $\mathcal{A}$ is closed with respect to linear combinations. Using formulas of transformations of products into sums, such as $\cos mx \cos nx = \frac{1}{2}[\cos(m+n)x + \cos(m-n)x]$, it follows that $\mathcal{A}$ is also closed to products. Hence, $\mathcal{A}$ is an algebra of $C[0, 2\pi]$.

The algebra $\mathcal{A}$ is said to *separate* points in K, if for any distinct $x, y \in K$, there is an f in $\mathcal{A}$ such that $f(x) \neq f(y)$.

We say that the algebra $\mathcal{A}$ contains the constant functions, if for any $c \in \mathbb{R}$, we have $c \in \mathcal{A}$. Given the algebra properties of $\mathcal{A}$, it only suffices to show that $1 \in \mathcal{A}$.

Example 7.4.2 Let $\mathcal{A}$ be the set of all polynomials defined on $[a, b]$

$$\mathcal{A} = \left\{ f(x) = \sum_{k=0}^{n} c_k x^k; x \in [a, b], c_k \in \mathbb{R}, n = 0, 1, 2, \ldots \right\}.$$

Then $\mathcal{A}$ is an algebra of $C[a, b]$ that separates points. To see this, let $x, y \in [a, b]$ and choose f to be the identity polynomial. Then $f(x) = x \neq y = f(y)$. Since $1 \in \mathcal{A}$, it follows that $\mathcal{A}$ contains the constant functions, too.

The proof of the following approximation theorem can be found, for instance, in [104].

Theorem 7.4.4 (Stone-Weierstrass) *Let K be a compact set in $\mathbb{R}^n$ and $\mathcal{A}$ an algebra of continuous real-valued functions on K satisfying the properties:*
(i) $\mathcal{A}$ separates the points of K;
(ii) $\mathcal{A}$ contains the constant functions.
Then $\mathcal{A}$ is a dense subset of $C(K)$.

Example 7.4.3 Let $\mathcal{A}$ be the set of all polynomials defined on $[a, b]$

$$\mathcal{A} = \left\{ f(x) = \sum_{k=0}^{n} c_k x^k; \; x \in [a, b], c_k \in \mathbb{R}, n = 0, 1, 2, \dots \right\}.$$

From Example 7.4.2 $\mathcal{A}$ is an algebra of $C[a, b]$ that separates points and contains constant functions. By Stone-Weierstrass theorem, for any continuous function $g : [a, b] \to \mathbb{R}$ and any $\epsilon > 0$, there is a polynomial function f such that $|g(x) - f(x)| < \epsilon, \forall x \in [a, b]$. If we consider a decreasing sequence $\epsilon_n \searrow 0$ and denote the associated polynomials by f_n, then the sequence $(f_n)_n$ tends uniformly to g on $[a, b]$. That is, $\max_{[a,b]} |f_n - g| \to 0$, as $n \to \infty$.

The following approximation result holds for continuous functions on product spaces:

Proposition 7.4.5 *Let $f : [a, b] \times [c, d] \to \mathbb{R}$ be a continuous function. Then for any $\epsilon > 0$, there is $N \geq 1$ and exists continuous functions $g_i \in C[a, b]$ and $h_i \in C[c, d]$, $i = 1, \dots, N$, such that*

$$\max_{x,y} \left| f(x, y) - \sum_{i=1}^{N} g_i(x) h_i(y) \right| < \epsilon.$$

Proof: Consider the set of functions

$$\mathcal{A} = \left\{ G(x, y) = \sum_{i=1}^{N} g_i(x) h_i(y); g_i \in C[a, b], h_i \in C[c, d], N = 1, 2, \dots \right\}.$$

It is easy to see that $\mathcal{A}$ is closed with respect to linear combinations and multiplications, i.e., $\mathcal{A}$ is an algebra of $C\big([a, b] \times [c, d]\big)$. It is obvious that $\mathcal{A}$ contains the constant functions. To show that $\mathcal{A}$ separates points, let $(x_1, y_1) \neq (x_2, y_2)$ be two distinct points. Then, either $x_1 \neq x_2$, or $y_1 \neq y_2$. In the former case the function $G(x, y) = x \cdot 1$ separates the points, while in the latter, $G(x, y) = 1 \cdot y$ does the same. Applying the Stone-Weierstrass theorem yields the desired result. ∎

Remark 7.4.6 The previous result can be easily extended to a finite product of compact intervals.

One of the applications of the Stone-Weierstrass Theorem is that neural networks with cosine (or sine) activation functions can learn any periodic function. The next section develops this idea.

7.5 Application to Neural Networks

Consider a one-hidden layer neural network with a real input, x, a one-dimensional output, y, and N neurons in the hidden layer. We assume the activation function is $\phi(x) = \cos(x)$. Since this is neither a sigmoidal, nor a squashing function, none of the results obtained in Chapter 9 can be useful to this case. The network's output is given by the sum

$$y = a_0 + \sum_{j=1}^{N} \alpha_j \cos(w_j x + b_j),$$

where w_j and α_j are the weights from the input to the hidden layer and from the hidden layer to the output, respectively. The biasses in the hidden layer are denoted by b_j, while a_0 denotes the bias of the neuron in the output layer. We note that the activation function in the hidden layer is the cosine function, while the activation function in the output neuron is linear. See Fig. 7.3 **a**.

We claim that the aforementioned network can represent continuous periodic functions. Consider a continuous function $f : \mathbb{R} \to \mathbb{R}$, which is periodic with period T, that is, $f(x + T) = f(x)$. Let $\nu = \frac{2\pi}{T}$ be the associated frequency. Consider the weights of the form $w_i = i\nu$. An application of trigonometric formulas provide

$$
\begin{aligned}
y &= a_0 + \sum_{j=1}^{N} \alpha_j \cos(j\nu x + b_j) \\
&= a_0 + \sum_{j=1}^{N} \alpha_j \cos(j\nu x)\cos(b_j) - \alpha_j \sin(j\nu x)\sin(b_j) \\
&= a_0 + \sum_{j=1}^{N} a_j \cos(j\nu x) - c_j \sin(j\nu x) \\
&= a_0 + \sum_{j=1}^{N} a_j \cos\left(\frac{2\pi j x}{T}\right) - c_j \sin\left(\frac{2\pi j x}{T}\right),
\end{aligned}
$$

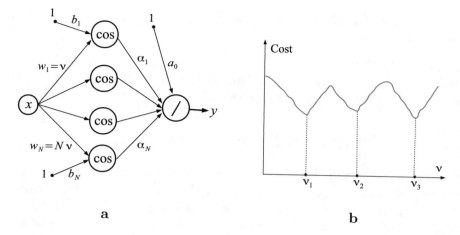

a   b

Figure 7.3: **a.** *One-hidden neural network with cosine activation functions.*
b. *The cost function reaches its relative minima for certain values of the*
hyperparameter ν.

with coefficients $a_j = \alpha_j \cos(b_j)$ and $c_j = \alpha_j \sin(b_j)$. By Exercise 7.11.7, these
trigonometric sums are dense in the set of continuous periodic functions on
$\mathbb{R}$ with period T. Therefore, $\forall \epsilon > 0$, there are $N \geq 1$, $\alpha_j, b_j, \nu \in \mathbb{R}$ such that

$$\left| a_0 + \sum_{j=1}^{N} \alpha_j \cos(j\nu x + b_j) - f(x) \right| < \epsilon.$$

Since cosine is an even function, it suffices to choose $\nu > 0$. Choosing the
decreasing sequence $\epsilon = \frac{1}{n}$, yields a sequence of frequencies ν_n, which approximates the proper frequency $\nu = \frac{2\pi}{T}$ of the target function f.

This method can be slightly modified to extract hidden frequencies from
a continuous signal. Consider the target signal $z = f(x)$, where x denotes
time. The weight ν from the previous computation is now taken as a hyperparameter. So, for each fixed value of ν, we tune the other parameters of
the neural network until the cost function is minimum. This can be done,
for instance, by the gradient descent method. Then, varying the value of ν,
we obtain different values of the associated cost function. We select those
values of ν for which the cost function achieves local minima, see Fig. 7.3 **b.**
They correspond to the proper frequencies contained in the signal. The global
minimum of the cost function is supposed to be obtained for the value of ν
obtained in the previous tuning, and this corresponds to the main frequency
contained in the signal.

This type of analysis can help in the study of stocks. The built-in frequencies can be used to trade stocks in a more efficient way.

Remark 7.5.1 The traditional way to find the frequencies contained into a signal is to use the Fourier transform, $\hat{f}(\nu) = \int e^{-i\nu x} f(x)\, dx$, which provides the amplitude of the signal in the frequency spectrum. High amplitudes correspond to proper frequencies of the signal $f(x)$.

7.6 Wiener's Tauberian Theorems

Sometimes, it is useful for a network to learn a time-dependent signal which is integrable, or square integrable on the entire real line. The following theorems can be used to prove the existence of these types of learning. The next two results are due to Wiener [128].

7.6.1 Learning signals in $L^1(\mathbb{R})$

Theorem 7.6.1 *Let $f \in L^1(\mathbb{R})$ and consider the translation functions $f_\theta(x) = f(x + \theta)$. Then the linear span of the family $\{f_\theta; \theta \in \mathbb{R}\}$ is dense in $L^1(\mathbb{R})$ if and only if the Fourier transform of f never takes on the value of zero on the real domain.*

The way to use this theorem is as in the following. Consider a neural network with an activation function given by an integrable function, f, with $\widehat{f}(\xi) \neq 0$. Then for any $g \in L^1(\mathbb{R})$ and any $\epsilon > 0$, there is a function

$$G(x) = \sum_{j=1}^{N} \alpha_j f(x + \theta_j), \quad \alpha_j, \theta_j \in \mathbb{R}, N = 1, 2, \ldots,$$

such that $\|g - G\|_1 = \int_{\mathbb{R}} |g(x) - G(x)|\, dx < \epsilon$. The function $G(x)$ is the output of a one-hidden layer neural network with the activation function f, weights α_j, and biasses θ_j.

We shall emphasize next some activation functions satisfying the above properties.

(i) *Double exponential* Let $f(x) = e^{-\lambda|x|}$, with $\lambda > 0$. Then

$$\int_{\mathbb{R}} |f(x)|\, dx = 2 \int_0^\infty e^{-\lambda x}\, dx = \frac{2}{\lambda} < \infty,$$

and hence $f \in L^1(\mathbb{R})$. The Fourier transform is given by

$$\widehat{f}(\omega) = \frac{2\lambda}{\lambda^2 + \omega^2} \neq 0.$$

(*ii*) *Laplace potential* Consider $f(x) = \dfrac{1}{a^2 + x^2}$, with some $a > 0$, fixed. Since

$$\int_{\mathbb{R}} |f(x)| \, dx = \frac{1}{a} \tan^{-1} \left(\frac{x}{a} \right) \Big|_{-\infty}^{\infty} = \frac{\pi}{2a} < \infty,$$

then $f \in L^1(\mathbb{R})$. Note that the Fourier transform is

$$\widehat{f}(\omega) = \frac{\pi}{a} e^{-a|\omega|} \neq 0.$$

(*iii*) *Gaussian* Let $f(x) = e^{-ax^2}$, with $a > 0$. We have $f \in L^1(\mathbb{R})$, because

$$\int_{\mathbb{R}} e^{-ax^2} \, dx = \sqrt{\frac{\pi}{a}}.$$

The Fourier transform, $\widehat{f}(\omega) = \sqrt{\dfrac{\pi}{a}} e^{-\frac{\omega^2}{4a}}$, never vanishes.

We conclude by stating that any one-hidden layer neural net with the input $x \in \mathbb{R}$ and an activation function such as the ones given by (*i*), (*ii*), or (*iii*), we can learn any given integrable function $g \in L^1(\mathbb{R})$, by tuning the weights α_j, biasses θ_j, and the number of units, N, in the hidden layer.

7.6.2   Learning signals in $L^2(\mathbb{R})$

Theorem 7.6.2 *Let $f \in L^2(\mathbb{R})$ and consider the translation functions $f_\theta(x) = f(x + \theta)$. Then the linear span of the family $\{f_\theta; \theta \in \mathbb{R}\}$ is dense in $L^2(\mathbb{R})$ if and only if the zero set of the Fourier transform of f is Lebesgue negligible.*

We use the theorem as in the following. Consider a squared integrable function, f, with the set $\{x; \widehat{f}(\xi) \neq 0\}$ Lebesgue negligible. In practice, it suffices to have a finite or countable number of zeros for $\widehat{f}(\xi)$. Then for any $g \in L^2(\mathbb{R})$ and any $\epsilon > 0$, there is a function

$$G(x) = \sum_{j=1}^{N} \alpha_j f(x + \theta_j), \quad \alpha_j, \theta_j \in \mathbb{R}, N = 1, 2, \ldots,$$

such that $\|g - G\|_2^2 = \int_{\mathbb{R}} (g(x) - G(x))^2 \, dx < \epsilon$. We note that the function $G(x)$ is the output of a one-hidden layer neural net with the activation function f, weights α_j, and biasses θ_j.

It is worth noting that the previously presented activation functions (*i*), (*ii*), and (*iii*) are also square integrable and do satisfy the zero Lebesgue measure condition.

Remark 7.6.3 This method has some limitations.

(*i*) One of them is that all the previous activation are bell-shaped. Activation functions of sigmoid type or hockey-stick shape, such as the logistic function or ReLU are not in $L^1(\mathbb{R})$, and hence, Wiener's theorem is not applicable in this case.

(*ii*) Another limitation is the fact that a closed-form expression for the Fourier transform is known only for a limited number of functions.

7.7 Contraction Principle

This section deals with an application of the contraction principle to deep learning.

If $(\mathcal{S}, d)$ is a metric space, an application $A : \mathcal{S} \to \mathcal{S}$ is called a *contraction* if there is a $0 < \lambda < 1$ such that

$$d(Ax, Ax') \leq \lambda\, d(x, x'), \qquad \forall x, x' \in \mathcal{S}.$$

Example 7.7.1 Let $\mathcal{S} = \mathbb{R}$ endowed with the metric $d(x, x') = |x - x'|$. Consider $A : \mathbb{R} \to \mathbb{R}$, given by $Ax = \dfrac{1}{2}\cos x$. By the mean value theorem, $\cos x - \cos x' = \sin(\xi)(x - x')$, with ξ between x and x'. Then

$$|Ax - Ax'| = \frac{1}{2}|\cos x - \cos x'| \leq \frac{1}{2}|x - x'|, \qquad \forall x, x' \in \mathbb{R},$$

and hence, the function A is a contraction.

A sequence $(x_n)_n$ in a metric space $(\mathcal{S}, d)$ is called a *Cauchy sequence* if the distance between any two elements, of a large enough index, can be made arbitrarily small. This is, $\forall \epsilon > 0$, there is $N \geq 1$ such that

$$d(x_n, x_m) < \epsilon, \qquad \forall n, m \geq N.$$

It is easy to notice that any convergent sequence is Cauchy. This is a consequence of the triangle identity. If x_n converges to x, then for any $\epsilon > 0$

$$d(x_n, x_m) \leq d(x_n, x) + d(x_m, x) < \frac{\epsilon}{2} + \frac{\epsilon}{2} = \epsilon, \qquad \forall n, m \geq N.$$

The converse statement is not necessarily true, so we need the following concept. A metric space $(\mathcal{S}, d)$ is called *complete* if any Cauchy sequence is convergent.

Example 7.7.2 The n-dimensional real space $\mathbb{R}^n$ is a complete metric space together with the Euclidean distance.

The element $x_0 \in \mathcal{S}$ is called a *fixed point* for the mapping $A : \mathcal{S} \to \mathcal{S}$ if $Ax_0 = x_0$. For instance, if $\mathcal{S} = \mathbb{R}$, the fixed points correspond to intersections between the graph of A and the line $y = x$. For several examples of fixed points the reader is referred to section E.6 in the Appendix.

The following main result will be useful shortly for the study of neural networks.

Theorem 7.7.1 (The Contraction Principle) *Let $(\mathcal{S}, d)$ be a complete metric space. Then any contraction $A : \mathcal{S} \to \mathcal{S}$ has an unique fixed point.*

The output of a one-hidden layer neural network, having the logistic activation function σ for all hidden neurons, is given by

$$A(x) = \sum_{j=1}^{N} \alpha_j \sigma(w_j x + b_j). \tag{7.7.1}$$

where we assumed that both the input and output are real numbers, i.e., $A : \mathbb{R} \to \mathbb{R}$.

Proposition 7.7.2 *Assume the following weights regularization conditions hold*

$$\sum_{j=1}^{N} w_j^2 \leq 1, \quad \sum_{j=1}^{N} \alpha_j^2 \leq 1. \tag{7.7.2}$$

Then the input-output function A, defined by (7.7.1), is a contraction with $\lambda = 1/4$.

Proof: By the mean value theorem, the logistic function satisfies

$$|\sigma(u) - \sigma(u')| \leq \max |\sigma'| \, |u - u'| \leq \frac{1}{4}|u - u'|, \qquad \forall u, u' \in \mathbb{R}.$$

Using the previous relation, Cauchy's inequality, and the regularization conditions, we have the estimate

$$
\begin{aligned}
|A(x) - A(x')| &\leq \left| \sum_{j=1}^{N} \alpha_j \Big(\sigma(w_j x + b_j) - \sigma(w_j x' + b_j) \Big) \right| \\
&\leq \sum_{j=1}^{N} |\alpha_j| \big| \sigma(w_j x + b_j) - \sigma(w_j x' + b_j) \big| \\
&\leq \frac{1}{4}|x - x'| \sum_{j=1}^{N} |\alpha_j| \, |w_j| \\
&\leq \frac{1}{4}|x - x'| \Big(\sum_{j=1}^{N} \alpha_j^2 \Big)^{1/2} \Big(\sum_{j=1}^{N} w_j^2 \Big)^{1/2} < \frac{1}{4}|x - x'|,
\end{aligned}
$$

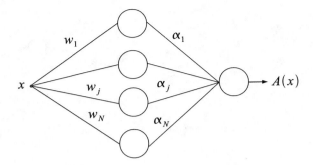

Figure 7.4: *A three-layer neural unit whose output is a contraction.*

which shows that A is a contraction.   ∎

Remark 7.7.3 In the case of an activation function with bounded slope, the regularization conditions can be weakened to

$$\sum_{j=1}^{N} w_j^2 \leq M, \quad \sum_{j=1}^{N} \alpha_j^2 \leq M,$$

with $M < \sup |\sigma'|$. Namely, the norm of the weights is bounded above by the maximum slope of the activation function.

Corollary 7.7.4 *There is a unique real value, $c \in \mathbb{R}$, which is invariant through the network, that is, $Ac = c$.*

Proof: From Proposition 7.7.2 the function A is a contraction, and since the space $\mathbb{R}$ is complete, the Contraction Principle, Theorem 7.7.1, implies that A has a unique fixed point, c.   ∎

Remark 7.7.5 The unique value c is related to the saturation level of a recurrent neural network as we shall see in the next section.

7.8   Application to Recurrent Neural Nets

In the following, a *neural unit* will stand for a one-hidden layer neural net, for which regularization conditions (7.7.2) hold, and have an input $x \in \mathbb{R}$ and an output $A(x) \in \mathbb{R}$ given by (7.7.1). A neural unit is depicted in Fig. 7.4.

We now concatenate neural units to construct a recurrent neural network as in Fig. 7.5. This means that the output of the nth neural unit is

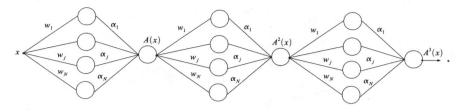

Figure 7.5: *A recurrent neural net constructed from three-layer neural units.*

the input of the $(n+1)$th neural unit, for $n \geq 1$. The output of the nth unit is $A^n(x) = A(\ldots A(A(x))\ldots)$. For the time being, all neural units are considered identical from the weights and biasses point of view. There is no update in the weights, which are considered constants. The learning in the network is unsupervised, i.e., there is no a priori specified target function.

Now, the question is: *What does this type of network learn?*

To be more precise, assume that for each input value, $x \in \mathbb{R}$, the nth outcome, $y_n = A^n(x)$, is a convergent sequence with the limit y, depending on x. If we define $y = \Lambda(x)$, the function learned by the network is $\Lambda(x)$. The next result deals with the existence, uniqueness, and the form of this function.

Proposition 7.8.1 *The recurrent network constructed in Fig. 7.5 learns a constant c, which is the unique fixed point of the input-output mapping $y = A(x)$.*

Proof: For each input value x, consider the sequence of real numbers $y_n = A^n(x)$. In order to prove that (y_n) is convergent, it suffices to show that it is Cauchy.

Using the contraction property of the function A, we estimate the difference between two consecutive terms in the sequence as

$$
\begin{aligned}
|y_{n+1} - y_n| &= |A^{n+1}(x) - A^n(x)| < \frac{1}{4}|A^n(x) - A^{n-1}(x)| < \cdots \\
&< \frac{1}{4^n}|A(x) - x| = \frac{1}{4^n}|y_1 - x|.
\end{aligned}
$$

Let $m = n + k$ and then estimate the difference between two distant terms in the sequence using the triangle inequality as

$$\begin{aligned}
|y_{n+k} - y_n| &\leq |y_{n+k} - y_{n+k-1}| + |y_{n+k-1} - y_{n+k-2}| + \cdots + |y_{n+1} - y_n| \\
&< \frac{1}{4^{n+k}}|y_1 - x| + \frac{1}{4^{n+k-1}}|y_1 - x| + \cdots + \frac{1}{4^n}|y_1 - x| \\
&< \frac{1}{4^n}\left(1 + \frac{1}{4} + \frac{1}{4^2}\cdots\right)|y_1 - x| \\
&= \frac{1}{4^n}\frac{1}{1 - \frac{1}{4}}|y_1 - x| = \frac{1}{3 \cdot 4^{n-1}}|y_1 - x|. \quad (7.8.3)
\end{aligned}$$

It is clear that the difference $|y_{n+k} - y_n|$ can be made arbitrarily small, if n is large enough. Hence, $(y_n)_n$ is a Cauchy sequence. Since $\mathbb{R}$ is complete, the sequence $(y_n)_n$ is convergent, with the limit $y = \lim_{n\to\infty} y_n$. Since y depends on x, we have $y = \Lambda(x)$.

In the following we shall prove that in fact y does not depend on the input x, and hence, $\Lambda(x)$ is constant. In order to show this, it suffices to consider two arbitrary fixed inputs $x, x' \in \mathbb{R}$ and show that for any $\epsilon > 0$ we have

$$|\Lambda(x) - \Lambda(x')| < \epsilon.$$

Let $y_n' = A^n(x')$, and y' be its limit. Triangle inequality provides

$$\begin{aligned}
|\Lambda(x) - \Lambda(x')| &\leq |\Lambda(x) - A^n(x)| + |A^n(x) - A^n(x')| + |A^n(x') - \Lambda(x')| \\
&\leq |y - y_n| + \frac{1}{4^n}|x - x'| + |y_n' - y'| \\
&\leq \frac{\epsilon}{3} + \frac{\epsilon}{3} + \frac{\epsilon}{3} = \epsilon,
\end{aligned}$$

since for n large enough, each of the terms can be made smaller than $\epsilon/3$. Hence, $y = \Lambda(x)$ is a constant. In order to determine its value, we estimate by triangle inequality

$$\begin{aligned}
|A(y) - y| &\leq |A(y) - y_n| + |y_n - y| = |A(y) - A(y_{n-1})| + |y_n - y| \\
&< \frac{1}{4}|y - y_{n-1}| + \frac{\epsilon}{2} \\
&< \frac{\epsilon}{2} + \frac{\epsilon}{2} = \epsilon,
\end{aligned}$$

for n large enough and any a priori fixed $\epsilon > 0$. It turns out that $A(y) = y = c$, i.e., y is the unique fixed point of the function A, i.e., $Ac = c$. In conclusion, the network learns the fixed point of the input-output mapping A. ∎

Remark 7.8.2 It is useful to measure the distance from an approximation point y_n to the fixed point c. For this, we take the limit $k \to \infty$ in relation (7.8.3) and obtain

$$|y_n - c| = |y_n - y| \leq \frac{1}{3 \cdot 4^{n-1}}|y_1 - x| = \frac{1}{3 \cdot 4^{n-1}}|A(x) - x|. \quad (7.8.4)$$

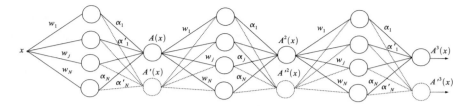

Figure 7.6: *A recurrent neural net constructed from two ϵ-close three-layer neural units.*

The integer n, which denotes the depth of the network, can be chosen such that the error becomes arbitrarily small.

Changing the weights of the network leads to a different learning constant c. We shall study next the continuous dependence of the fixed point c with respect to the weights α_j. First, we need to introduce a new concept.

Definition 7.8.3 *Two neural nets are called ϵ-close if for some $\epsilon > 0$ their input-output mappings satisfy $|A(x) - A'(x)| < \epsilon$, for all inputs $x \in \mathbb{R}$.*

Equivalently, this means that the output of one network belongs to an ϵ-neighborhood of the output produced by the other network, i.e.,

$$A'(x) \in \big(A(x) - \epsilon, A(x) + \epsilon\big).$$

Roughly stated, the error between the outputs of the networks is uniform with respect to the input x.

Lemma 7.8.4 *Consider two neural units with input-output mappings given by $A(x) = \sum_{j=1}^{N} \alpha_j \sigma(w_j x + b_j)$ and $A'(x) = \sum_{j=1}^{N} \alpha'_j \sigma(w_j x + b_j)$, such that $\sum_{j=1}^{N} |\alpha'_j - \alpha_j| < \epsilon$. Then the neural units are ϵ-close.*

Proof: Using that $|\sigma(x)| < 1$, a straightforward computation provides

$$|A'(x) - A(x)| = \Big| \sum_{j=1}^{N} (\alpha'_j - \alpha_j)\sigma(w_j x + b_j) \Big| \leq \sum_{j=1}^{N} |\alpha'_j - \alpha_j| < \epsilon,$$

for all $x \in \mathbb{R}$.   ∎

Lemma 7.8.5 *Consider two ϵ-close neural units as in Lemma 7.8.4, with input-output mappings $A(x)$ and $A'(x)$, and denote by c and c' their fixed points. Then $|c' - c| < \dfrac{4\epsilon}{3}$.*

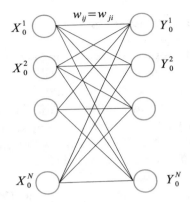

Figure 7.7: *A resonance network.*

Proof: Construct the sequence $y_n = A'^n(c)$, where $A(c) = c$. The sequence y_n converges to c', the fixed point of A'. If in relation (7.8.3) we take $k \to \infty$ and let $n = 0$, then we obtain a bound for the distance between the initial point and the limit

$$\lim_{n \to \infty} |y_n - y_0| < \frac{4}{3}|y_1 - y_0|,$$

or, equivalently,

$$|c' - c| < \frac{4}{3}|A'(c) - c| = \frac{4}{3}|A'(c) - A(c)| < \frac{4\epsilon}{3},$$

where we used that A and A' are ϵ-close. ∎

The continuous dependence of the fixed point of a neural unit with respect to the weights α_j can be formulated as in the following:

Proposition 7.8.6 *Consider two neural units as in Lemma 7.8.4. Then for any $\epsilon > 0$, there is an $\eta > 0$, such that if $\sum_{j=1}^{N} |\alpha'_j - \alpha_j| < \eta$, then $|c' - c| < \epsilon$.*

Proof: Let $\epsilon > 0$ be arbitrarily fixed and choose $\eta = \frac{3\epsilon}{4}$. By Lemma 7.8.4 the networks are η-closed. Then applying Lemma 7.8.5 leads to the desired result. ∎

A recurrent neural network obtained by using two ϵ-close neural units, which share the weights w_j and biasses b_j, is given in Fig. 7.6. The network learns the vector (c, c') of fixed points.

7.9   Resonance Networks

We consider a neural network with two layers, see Fig. 7.7. The information goes back and forth between the layers until it eventually converges to a steady state. Let N be the number of neurons in each layer. The initial state of the layer on the left is given by the vector X_0. This induces the state on the second layer $Y_0 = \phi(WX_0 + b)$, where the weights are assumed symmetric, $w_{ij} = w_{ji}$. This information is fed back into the first layer, which provides the output $X_1 = \phi(WY_0 + b)$. Here, ϕ is the activation function for both layers and b is the common bias vector.

This way, we obtain the sequences $(X_n)_n$ and $(Y_n)_n$ defined recursively by the system

$$\begin{aligned} Y_n &= \phi(WX_n + b) \\ X_n &= \phi(WY_{n-1} + b), \quad n \geq 1. \end{aligned}$$

This can be separated into two recursions as

$$\begin{aligned} Y_n &= F(Y_{n-1}) \\ X_n &= F(X_{n-1}), \end{aligned}$$

with $F : \mathbb{R}^n \to \mathbb{R}^n$, $F(U) = \phi\big(W\phi(WU + b) + b\big)$. The steady state of the network corresponds to the pair of fixed points, (X^*, Y^*), where $F(X^*) = X^*$ and $F(Y^*) = Y^*$.

We shall treat this problem using the Contraction Principle. First, we write the problem in an equivalent way as in the following. Denote $Z_0 = X_0$, $Z_1 = Y_0$, $Z_2 = X_1$, in general, $Z_{n+1} = G(Z_n)$, where $G(Z) = \phi(WZ + b)$. It suffices to show that Z_n is a sequence of vectors converging to some vector Z^*. Since $G : \mathbb{R}^n \to \mathbb{R}^n$, a computation shows

$$\|G(Z') - G(Z)\|_2 \leq \|\nabla G\| \, \|Z' - Z\|_2 \leq \sqrt{N} \|\phi'\|_\infty \|W\| \, \|Z' - Z\|_2, \quad \forall Z, Z' \in \mathbb{R}^n.$$

If the weights satisfy the inequality

$$\|W\| < \frac{1}{\sqrt{N} \|\phi'\|_\infty}$$

then G becomes a contraction mapping. Theorem E.6.5 assures that G has a unique fixed point, Z^*, which is approached by the sequence Z_n.

The conclusion is that if the weights w_{ij} are small enough, the state of the network tends to a stable state or resonant state.

Remark 7.9.1 If $N = 1$ there are only two neurons which change alternatively their states. If the activation function is the logistic function, the stability condition becomes $w_{12} = w_{21} < 4$.

Remark 7.9.2 An example of this type of architecture is Kosko's bidirectional associative memory (BAM) [67]. It can be also regarded as a special case of Hopfield (1984) [58] and is is worth noting the relation with the Cohen-Grossberg model [25].

7.10 Summary

This chapter contains some classical real analysis results useful for neural networks' representations.

For learning continuous functions $f \in C[a, b]$, the uniform convergence is required. The neural network produces a sequence of continuous functions, which converges pointwise to the target function f. In order to show that this convergence is uniform, we can use Dini's theorem.

Another technique for constructing a uniformly convergent sequence of functions on a compact set is by using Arzela-Ascoli's theorem. In this case one can extract a convergent subsequence among a set of outcomes which is uniformly bounded and equicontinuous. Consequently, one-hidden layer neural networks with sigmoid activation function can learn a continuous function on a compact set. Another approximation method of continuous functions on a compact set is using the Stone-Weierstrass theorem. This can be also used in neural networks to learn periodic functions.

Target functions from L^1 and L^2 can be learned using Wiener's Tauberian theorems. They apply mainly for bell-shaped activation functions.

The contraction principle, which states the existence of a fixed point, can be applied to input-output mappings of a one-hidden layer neural network as well as to resonance networks. Iterating the input-output mapping we obtain that the network learns a certain function.

7.11 Exercises

Exercise 7.11.1 Let $\mathcal{F} = \{\tanh(ax + b); a, b \in \mathbb{R}\}$. Show that the family $\mathcal{F}$ is uniformly bounded.

Exercise 7.11.2 Let $\mathcal{F} = \{ax + b; |a| + |b| < 1, x \in [0, 1]\}$ be the family of affine functions on $[0, 1]$ under a regularization constraint.
(a) Show that the family $\mathcal{F}$ is equicontinuous and uniformly bounded;
(b) What can be the application of part (a) for the Adaline neuron?

Exercise 7.11.3 Let $M > 0$ and consider the family of differentiable functions

$$\mathcal{F} = \{f : [a, b] \to \mathbb{R}; \int_a^b f'(x)^2 \, dx \leq M\}.$$

Show that $\mathcal{F}$ is equicontinuous.

Exercise 7.11.4 Denote by D a dense subset in (a, b) (such as the rational numbers between a and b). Let $\mathcal{F}$ be a family of functions such that
(i) it is equicontinuous on the interval (a, b);
(ii) for any $x_0 \in D$, the family $\mathcal{F}$ is uniformly bounded at x_0, i.e., there is $M > 0$ such that $f(x_0) \leq M$, $\forall f \in \mathcal{F}$.
Show that $\mathcal{F}$ is uniformly bounded at all points.

Exercise 7.11.5 Let $\mathcal{F}$ be equicontinuous and let $k \geq 1$ be a fixed integer and $\{w_1, \ldots, w_k\}$ a set of weights with $|w_j| < 1$. Show that the set remains equicontinuous if it is extended by including all linear combinations of the form $\sum_{j=1}^{k} w_j f_j$.

Exercise 7.11.6 A neural network using sigmoid neurons is designed to learn a continuous function $f \in C[a, b]$ by the gradient descent method. The network outcome obtained after the nth parameter update is denoted by $G_n(x) = G(x; w(n), b(n))$. Assume that each time the approximation is improved, i.e., $|f(x) - G_{n+1}(x)| < |f(x) - G_n(x)|$, for all $x \in [a, b]$ and any $n \geq 1$. Prove that G_n converges uniformly to $f(x)$ on $[a, b]$ (that is, the network learns any continuous function on $[a, b]$).

Exercise 7.11.7 Let $f : \mathbb{R} \to \mathbb{R}$ be a continuous periodic function with period T, that is $f(x + T) = f(x)$ for all $x \in \mathbb{R}$. Show that for any given $\epsilon > 0$, there is a function

$$F(x) = a_0 + \sum_{n=1}^{N} \left(a_n \cos \frac{2\pi n x}{T} + b_n \sin \frac{2\pi n x}{T} \right)$$

such that $|F(x) - f(x)| < \epsilon$ for all $x \in \mathbb{R}$.

We make the following remark. A function $f : \mathbb{R} \to \mathbb{R}$ that is an uniform limit of trigonometric polynomials on $\mathbb{R}$ is called *almost periodic*. The previous exercise states that any continuous periodic function on $\mathbb{R}$ is almost periodic. It is worth noting that $f : \mathbb{R} \to \mathbb{R}$ is almost periodic if and only if the following *Bohr condition* holds: for any $\epsilon > 0$, there is $\eta > 0$ such that for any real interval I, with $|I| < \epsilon$, there is $c \in I$ such that $|f(x) - f(x+c)| < \epsilon$, for all $x \in \mathbb{R}$, see [17].

Exercise 7.11.8 Proposition 7.7.2 has been proved under the hypothesis that σ is the logistic sigmoid. Does the result hold for other sigmoid activation functions?

Exercise 7.11.9 Let $\{f_j(x)\}_{j \geq 1}$ be a set of equicontinuous and uniformly bounded functions on $[a, b]$. Show that if $\int \lim_{j \to \infty} f_j^2(x)\, dx = 0$, then $\lim_{j \to \infty} f_j = 0$, uniformly.

Exercise 7.11.10 (The set of integral transforms of L^2-functions) Let $M > 0$ be a fixed constant and $K : [a, b] \times [c, d] \to \mathbb{R}$ a continuous function. Let $\mathcal{F}_M$ be the set of functions on $[a, b]$ given by

$$g(s) = \int_c^d K(s, t) h(t)\, dt,$$

with h satisfying $\int_c^d h(t)^2\, dt \leq M$. Show that

(a) $\mathcal{F}_M$ is equicontinuous;

(b) $\mathcal{F}_M$ is uniformly bounded.

Exercise 7.11.11 Let $a, b \in \mathbb{R}$, with $a < b$.

(a) Show that $\mathcal{A} = \{\sum_{i=1}^n \alpha_i e^{m_i x}; \alpha_i \in R, m_i \in \mathbb{Z}_+\}$ is a subalgebra of $C[a, b]$.

(b) Show that for any $\epsilon > 0$, there are $\alpha_1, \ldots, \alpha_n \in \mathbb{R}$ and nonnegative integers $m_1, \ldots, m_n \geq 0$ such that

$$\left| f(x) - \sum_{i=1}^n \alpha_i e^{m_i x} \right| \leq \epsilon, \quad \forall x \in [a, b].$$

(c) Formulate a machine learning interpretation of this result.

Chapter 8

Learning with One-dimensional Inputs

This chapter deals with the case of a neural network whose input is bounded and one-dimensional, $x \in [0, 1]$. Besides its simplicity, this case is important from a few points of view: it can be treated elementary, without the arsenal of functional analysis (as shall we do in Chapter 9) and, due to its constructive nature, it provides an explicit algorithm for finding the network weights.

Both cases of perceptron and sigmoid neural networks with one-hidden layer will be addressed. Learning with ReLU and Softplus is studied in detail.

8.1 Preliminary Results

In this section we shall use the notions of derivative in generalized sense, measure, integration, and Dirac's measure. The reader can find these notions in sections C.3, C.4, F.2, and F.3 of the Appendix.

1. We shall first show that any simple function can be written as a linear combination of translations of Heaviside step functions. Assume the support of the simple function is in $[0, 1]$. Then the function can be written as $c(x) = \sum_{i=0}^{N-1} \alpha_i 1_{[x_i, x_{i+1})}$, where $0 = x_0 < x_1 < \cdots < x_N = 1$ is a partition of the interval $[0, 1]$ and $1_{[x_i, x_{i+1})}$ is the indicator function of the interval $[x_i, x_{i+1})$, namely

$$1_{[x_i, x_{i+1})}(x) = \begin{cases} 1, & \text{if } x_i \leq x < x_{i+1} \\ 0, & \text{otherwise.} \end{cases}$$

© Springer Nature Switzerland AG 2020
O. Calin, *Deep Learning Architectures*, Springer Series in the Data Sciences,
https://doi.org/10.1007/978-3-030-36721-3_8

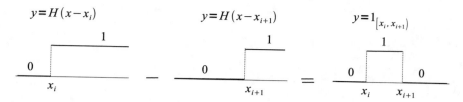

Figure 8.1: *The indicator function as a difference of two Heaviside functions,* $1_{[x_i,x_{i+1})}(x) = H(x - x_i) - H(x - x_{i+1}).$

Since any indicator function of an interval is a difference of two step functions

$$1_{[x_i,x_{i+1})}(x) = H(x - x_i) - H(x - x_{i+1}), \qquad \forall 0 \le i < N,$$

see Fig. 8.1, it follows that there are real numbers $c_0, \ldots, c_N$ such that

$$\sum_{i=0}^{N-1} \alpha_i 1_{[x_i,x_{i+1})}(x) = \sum_{i=0}^{N} c_i H(x - x_i), \qquad (8.1.1)$$

and hence a simple function could be written as a linear combination of Heaviside step functions. It is obvious now that the interval $[0,1]$ can be replaced by any compact interval.

2. As a consequence of the previous writing, it follows that the derivative of any simple function is a linear combination of Dirac measures

$$c'(x) = \frac{d}{dx} \sum_{i=0}^{N} c_i H(x - x_i) = \sum_{i=0}^{N} c_i \delta(x - x_i) = \sum_{i=0}^{N} c_i \delta_{x_i}(x),$$

where the relation $\frac{d}{dx} H(x - x_i) = \delta(x - x_i)$ holds in the generalized sense, see section F.2 of the Appendix. In other words, the sensitivity of the output with respect to the input, $c'(x)$ is given as the superposition of N shocks, $\delta(x - x_i)$, of magnitude c_i.

3. In this section it is shown that Dirac's measure can be approximated weakly by certain probability measures with "bell- shaped" densities. More precisely, we have:

Proposition 8.1.1 *Consider an activation function* $\varphi : \mathbb{R} \to \mathbb{R}$ *satisfying the following properties:*

(i) φ *is increasing;*

(ii) $\varphi(\infty) - \varphi(-\infty) = 1$;

(iii) φ is differentiable with $|\varphi'(x)|$ bounded.
Let $\varphi_\epsilon(x) = \varphi(\frac{x}{\epsilon})$, and consider μ_ϵ be the measure with density φ'_ϵ, namely $d\mu_\epsilon(x) = \varphi'_\epsilon(x)dx$. Then $\mu_\epsilon \to \delta$, as $\epsilon \searrow 0$, in the weak sense.

Proof: For any smooth function with compact support, $g \in C_0^\infty(\mathbb{R})$, we have by a change of variable

$$\int_{-\infty}^\infty \varphi'_\epsilon(x)g(x)\,dx = \int_{-\infty}^\infty \frac{1}{\epsilon}\varphi'\Big(\frac{x}{\epsilon}\Big)g(x)\,dx = \int_{-\infty}^\infty \varphi'(y)g(\epsilon y)\,dy.$$

The Bounded Convergence Theorem (Theorem C.4.3 in the Appendix) yields

$$\lim_{\epsilon \to 0}\int_{-\infty}^\infty \varphi'(y)g(\epsilon y)\,dy = g(0)\int_{-\infty}^\infty \varphi'(y)\,dy = g(0)(\varphi(\infty) - \varphi(-\infty)) = g(0).$$

Hence

$$\lim_{\epsilon \to 0}\int_{-\infty}^\infty g(x)\,d\mu_\epsilon(x) = \lim_{\epsilon \to 0}\int_{-\infty}^\infty \varphi'_\epsilon(x)g(x)\,dx = g(0) = \int_{-\infty}^\infty g(x)\delta(x)\,dx,$$

for all $g \in C_0^\infty(\mathbb{R})$, which means that $\mu_\epsilon \to \delta$, as $\epsilon \searrow 0$, in the weak sense. ∎

The function $\varphi'_\epsilon(x)$ has the shape of a "bump function", which tends to a "spike", as ϵ decreases to zero. An example of this type of activation function is the logistic function. In this case $\varphi_\epsilon(x) = \dfrac{1}{1 + e^{-\frac{x}{\epsilon}}}$.

Remark 8.1.2 Since φ is increasing, then $\varphi'_\epsilon \geq 0$. We also have

$$\int_{\mathbb{R}} d\mu_\epsilon(x)dx = \int_{\mathbb{R}} \varphi'_\epsilon(x)\,dx = \varphi_\epsilon(\infty) - \varphi_\epsilon(-\infty) = \varphi(\infty) - \varphi(-\infty) = 1.$$

This implies that μ_ϵ are probability measures on $(\mathbb{R}, \mathcal{B}(\mathbb{R}))$.

4. The convolution with the Dirac measure centered at a, namely $\delta_a(x) = \delta(x - a)$, is the same as a composition with a translation of size a:

$$(\delta_a * f)(x) = \int_{-\infty}^\infty \delta_a(y)f(x - y)\,dy = f(x - a).$$

This can be stated equivalently by saying that the signal $f(x)$ filtered through the Dirac measure δ_a is the same signal on which we have applied a phase shift.

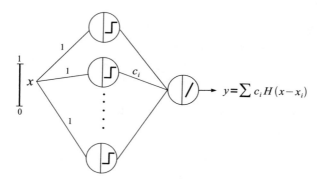

Figure 8.2: *A one-hidden layer perceptron network with input to hidden layer weights equal to 1 and hidden layer to output weights equal to c_i; the biasses are given by x_i.*

5. Any continuous real-valued function defined on a compact, $g : [a, b] \rightarrow \mathbb{R}$, is *uniformly continuous*, i.e., $\forall \epsilon > 0$, $\exists \delta > 0$ such that $\forall x, x' \in [a, b]$ with $|x - x'| < \delta$, then $|g(x) - g(x')| < \epsilon$. The proof of this classical statement is left to the reader in Exercise 8.8.3.

To understand this statement heuristically, we think of $g(x)$ as the speaker's volume of a radio set whose sliding button is situated in position x. This position can be anywhere in the interval $[a, b]$, where a stands for the minimum and b for the maximum possible positions. Then, for any ϵ-adjustment of the volume, there is a δ-control of the sliding button, such that moving from any position x into a δ-neighborhood position x', the corresponding volume has a net change less than ϵ.

8.2   One-Hidden Layer Perceptron Network

Consider a one-hidden layer neural network with the activation functions in the hidden layer given by the Heaviside function and a linear activation function in the output layer. We shall show that this network can potentially learn any continuous function $g \in C[0, 1]$. Equivalently stated, any continuous function on a compact interval can be approximated uniformly by simple functions. The network is represented in Fig. 8.2.

We first note that the right side of relation (8.1.1) is a particular output of this type of network, see Fig. 8.2. The following result shows that the network approximates continuous functions by stair functions.

Proposition 8.2.1 *For any function* $g \in C[0,1]$ *and any* $\epsilon > 0$, *there is a simple function* $c(x) = \sum_{i=0}^{N-1} c_i H(x - x_i)$ *such that*

$$|g(x) - c(x)| < \epsilon, \quad \forall x \in [0,1].$$

Proof: Fix $\epsilon > 0$ and let $\delta > 0$ be given by the uniform continuity property of the function g in the interval $[0,1]$. Consider an equidistant partition $0 = x_0 < x_1 < \cdots < x_N = 1$ with N large enough such that $\frac{1}{N} < \delta$. Choose u in $[0,1]$, fixed. Then there is a $k < N$ such that $u \in [x_k, x_{k+1})$. By the uniform continuity of g, we have

$$|g(u) - g(x_k)| \le \epsilon. \tag{8.2.2}$$

Construct the real numbers $c_0, c_1, \ldots, c_{N-1}$ such that

$$
\begin{aligned}
g(x_0) &= c_0 \\
g(x_1) &= c_0 + c_1 \\
g(x_2) &= c_0 + c_1 + c_2 \\
\cdots &= \cdots \\
g(x_{N-1}) &= c_0 + c_1 + \cdots + c_{N-1}.
\end{aligned}
$$

For the previous $u \in [x_k, x_{k+1})$ the Heaviside function evaluates as

$$
H(u - x_j) = \begin{cases} 1, & \text{if } j \le k \\ 0, & \text{if } j > k. \end{cases}
$$

The value at u of the simple function is

$$c(u) = \sum_{i=0}^{N-1} c_i H(u - x_i) = \sum_{j=0}^{k} c_j = g(x_k),$$

fact which implies

$$|g(x_k) - c(u)| = 0. \tag{8.2.3}$$

Now, triangle inequality together with relations (8.2.2) and (8.2.3) provide the inequality

$$|g(u) - c(u)| \le |g(u) - g(x_k)| + |g(x_k) - c(u)| \le \epsilon.$$

Since u was chosen arbitrary, the previous inequality holds for any $x \in [0,1]$, which leads to the desired conclusion. ∎

Remark 8.2.2 In the following we shall provide a few remarks:

1. The simple function $c(x)$ is the output function of a one-hidden layer neural network with N computing units in the hidden layer. Each unit is a classical perceptron with bias $\theta_i = -x_i$. The weight from the input to the hidden layer is $w = 1$. The weights from the hidden to the output layer are given by coefficients $\{c_i\}$.

2. The weights c_i can be constructed as in the following. Divide the interval $[0, 1]$ into an equidistant partition

$$0 = x_0 < x_1 < \cdots < x_N = 1$$

and define the weights

$$
\begin{aligned}
c_0 &= g(0) \\
c_1 &= g(x_1) - g(0) \\
c_2 &= g(x_2) - g(x_1) \\
\cdots &= \cdots \\
c_{N-1} &= g(x_{N-1}) - g(x_{N-2}).
\end{aligned}
$$

For N large enough (such that $\frac{1}{N} < \delta$) the aforementioned relations produce weights such that $|c(x) - g(x)| < \epsilon$.

3. From the uniform continuity of g, for any $\epsilon > 0$, there is a number N large enough such that $|g(x_{k+1}) - g(x_k)| < \epsilon$ for any k. Using the definition of c_k, this means that $|c_k| < \epsilon$ for any k, if the partition is small enough. Hence, the approximator $c(x) = \sum_{i=0}^{N-1} c_i H(x - x_i)$ given by the proposition can be assumed to satisfy $|c_k| < \epsilon$, $\forall k$, i.e., to have arbitrary small step variances. This property will be useful later. We note that in order to have the step variance small, the price paid is the large number of computing units in the hidden layer, N.

8.3   One-Hidden Layer Sigmoid Network

Consider a one-hidden layer neural network with a logistic sigmoid activation function, or more generally, with any activation function φ satisfying the hypothesis of Proposition 8.1.1, for the hidden layer and linear activation function for the output layer, see Fig. 8.3. We shall show that this network can potentially learn any continuous function $g \in C[0, 1]$.

The main result is given in the following.

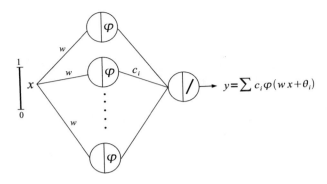

Figure 8.3: *A one-hidden layer sigmoid neural network, with input to hidden layer weights w and hidden layer to output weights c_i.*

Theorem 8.3.1 *Consider an activation function satisfying the hypothesis of Proposition 8.1.1 and let $g \in C[0,1]$. Then $\forall \epsilon > 0$ there are $c_i, w, \theta_i \in \mathbb{R}$, and $N \geq 0$ such that*

$$\left| g(x) - \sum_{i=1}^{N} c_i \varphi(wx + \theta_i) \right| < \epsilon, \quad \forall x \in [0,1].$$

Proof: Since the proof is relatively involved, for the sake of simplicity we shall divide it into several steps:

Step 1: Approximating the function g by a step function.
By Proposition 8.2.1 there is a step function $c(x)$ such that

$$| g(x) - c(x) | < \frac{\epsilon}{2}, \qquad \forall x \in [0,1]. \tag{8.3.4}$$

By remark 3 of Proposition 8.2.1 we may further assume that the coefficients of $c(x)$ satisfy $|c_k| < \frac{\epsilon}{4}$, for all k.

Step 2: Construct a "smoother" of the step function $c(x)$ using convolution. Define the function family $\varphi_\alpha(x) = \varphi(\frac{x}{\alpha})$, with $\alpha > 0$. Consider the bump function $\psi_\alpha(x) = \varphi'_\alpha(x) = \frac{1}{\alpha}\varphi'\left(\frac{x}{\alpha}\right)$. By Proposition 8.1.1 $\psi_\alpha(x)$ is the density of a probability measure, μ_α, which converges weakly to the Dirac measure $\delta(x)$, as $\alpha \to 0$. Define the "smoother" of $c(x)$ by the convolution

$$c_\alpha(x) = (c * \psi_\alpha)(x).$$

The geometrical flavor is depicted in Fig. 8.4.

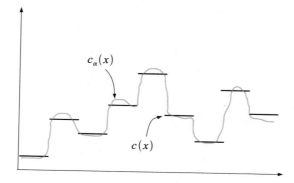

Figure 8.4: *The step function $c(x)$ and its "smoother" $c_\alpha(x) = (c * \psi_\alpha)(x)$.*

We shall show that the smoother $c_\alpha(x)$ is of the form

$$c_\alpha(x) = \sum_{i=1}^{N} c_i \varphi(wx + \theta_i)$$

for some $c_i, w, \theta_i \in \mathbb{R}$.

Using the convolution properties and the facts introduced in section 8.1, we have in the sense of derivatives in generalized sense:

$$
\begin{aligned}
c_\alpha(x) &= (c * \psi_\alpha)(x) = (c * \varphi'_\alpha)(x) = (c' * \varphi_\alpha)(x) \\
&= \frac{d}{dx}\Big(\sum_{i=1}^{N} c_i H(x - x_i)\Big) * \varphi_\alpha(x) \\
&= \Big(\sum_{i=1}^{N} c_i \delta(x - x_i)\Big) * \varphi_\alpha(x) = \sum_{i=1}^{N} c_i(\delta_{x_i} * \varphi_\alpha)(x) \\
&= \sum_{i=1}^{N} c_i \varphi_\alpha(x - x_i) = \sum_{i=1}^{N} c_i \varphi\Big(\frac{x - x_i}{\alpha}\Big) \\
&= \sum_{i=1}^{N} c_i \varphi(wx + \theta_i),
\end{aligned}
$$

with weight $w = \dfrac{1}{\alpha}$ and biasses $\theta_i = -\dfrac{x_i}{\alpha}$.

Step 3: Approximating the step function by a smoother.

We shall show that $\forall \epsilon > 0$ we have

$$|c(x) - c_\alpha(x)| < \frac{\epsilon}{2}, \qquad \forall x \in \mathbb{R} \qquad (8.3.5)$$

for α small enough. Actually, this states the continuity of the functional $\alpha \to c_\alpha$ at $\alpha = 0$. To prove this, we evaluate the difference and split it in a sum of three integrals

$$
\begin{aligned}
c(x) - c_\alpha(x) &= c(x) \int_{-\infty}^{\infty} \psi_\alpha(y)\, dy - \int_{-\infty}^{\infty} \psi_\alpha(y) c(x-y)\, dy \\
&= \int_{-\infty}^{\infty} \psi_\alpha(y) \big(c(x) - c(x-y) \big)\, dy \\
&= I_1 + I_2 + I_3, \qquad\qquad\qquad (8.3.6)
\end{aligned}
$$

where

$$
\begin{aligned}
I_1 &= \int_{-\infty}^{-\epsilon'} \psi_\alpha(y) \big(c(x) - c(x-y) \big)\, dy \\
I_2 &= \int_{-\epsilon'}^{\epsilon'} \psi_\alpha(y) \big(c(x) - c(x-y) \big)\, dy \\
I_3 &= \int_{\epsilon'}^{\infty} \psi_\alpha(y) \big(c(x) - c(x-y) \big)\, dy,
\end{aligned}
$$

for some $\epsilon' > 0$, subject to be specified later.
Note that $c(x)$ is bounded, with

$$|c(x)| = \Big| \sum_{i=1}^{N} c_i H(x - x_i) \Big| \le \sum_{i=1}^{N} |c_i| = M,$$

and hence $|c(x) - c(x-y)| \le |c(x)| + |c(x-y)| \le 2M$. Also, since ψ_α tends to $\delta(x)$ as $\alpha \to 0$, then ψ_α tends to zero on each of the fixed intervals $(-\infty, -\epsilon')$ and (ϵ', ∞). By the Dominated Convergence Theorem, see Theorem C.4.2, $I_1 \to 0$ and $I_3 \to 0$ as $\alpha \to 0$, so

$$I_1 + I_3 < \frac{\epsilon}{4},$$

for α small enough.
To evaluate I_2 we note that the graph of $c(x-y)$ is obtained from the graph of $c(x)$ by a horizontal shift of at most ϵ', see Fig. 8.5 **a**. For ϵ' small enough there is only one bump of the difference $c(x) - c(x-y)$ contained in the interval $(-\epsilon', \epsilon)$, see see Fig. 8.5 **b**. The height of the bump is equal to one of the coefficients c_k. Using the property of the coefficients developed in Step 1, we have the estimate

$$|c(x) - c(x-y)| \le |c_k| < \frac{\epsilon}{4},$$

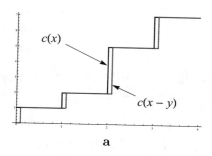 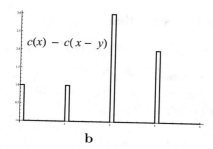

a   b

Figure 8.5: **a.** *The step functions $c(x)$ and $c(x-y)$ for $|y| < \epsilon'$.* **b.** *The graph of the difference $c(x) - c(x-y)$.*

for any $y \in (-\epsilon', \epsilon)$.

Using the properties of ψ_α we can now estimate the second integral:

$$
\begin{aligned}
I_2 &\le \int_{-\epsilon'}^{\epsilon'} \psi_\alpha(y)\,|c(x) - c(x-y)|\,dy < \frac{\epsilon}{4} \int_{-\epsilon'}^{\epsilon'} \psi_\alpha(y) \\
&\le \frac{\epsilon}{4} \int_{-\infty}^{\infty} \psi_\alpha(y) = \frac{\epsilon}{4}.
\end{aligned}
$$

To conclude Step 3, we have

$$
|I_1 + I_2 + I_2| \le \frac{\epsilon}{4} + \frac{\epsilon}{4} = \epsilon/2
$$

for α and ϵ' small enough. This implies inequality (8.3.5).

Step 4: Combining all the previous steps and finishing the proof. Triangle inequality, together with Step 1 and Step 2, provides

$$
\begin{aligned}
|g(x) - c_\alpha(x)| &\le |g(x) - c(x)| + |c(x) - c_\alpha(x)| \\
&\le \frac{\epsilon}{2} + \frac{\epsilon}{2} = \epsilon.
\end{aligned}
$$

By Step 2 the smoother $c_\alpha(x)$ is of the form $G(x) = \sum_{i=1}^{N} c_i \varphi(wx + \theta_i)$, which ends the proof.

∎

Remark 8.3.2 (*i*) It is worth noting that in the one-dimensional case the weight w does not depend on the index i.

(*ii*) The output of the one-hidden layer sigmoid neural network,

$$
y = \sum_{i=1}^{N} c_i \varphi(wx + \theta_i),
$$

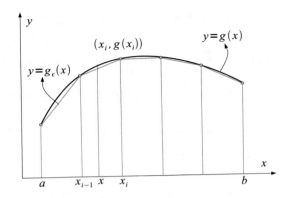

Figure 8.6: *The polygonal line $g_\epsilon(x)$ converges uniformly on $[a,b]$ to the continuous function $g(x)$, as $\epsilon \to 0$.*

is obtained by filtering the output of the one-hidden layer perceptron network, $y = \sum_{i=1}^{N} c_i H(x - x_i)$, through the filter given by the bump function $\varphi'_\alpha(x)$.

(*iii*) Both results proved in the last two sections state that any continuous function on $[0,1]$ can be learned in two ways: Proposition 8.2.1 assures that the learning can be done by simple functions, while Theorem 8.3.1 by smooth functions. However, the latter result is more useful, since the approximator is smooth, fact that makes possible the application of the backpropagation algorithm.

8.4   Learning with ReLU Functions

We start with a result regarding the polygonal approximation of continuous functions on a compact interval $[a,b]$, see Fig. 8.6.

Lemma 8.4.1 *Let $g : [a,b] \to \mathbb{R}$ be a continuous function. Then $\forall \epsilon > 0$ there is an equidistant partition*

$$a = x_0 < x_1 < \cdots < x_N = b$$

such that the piecewise linear function $g_\epsilon : [a,b] \to \mathbb{R}$, which passes through the points $\big(x_i, g(x_i)\big)$, $i = 0, \ldots, N$, satisfies

$$|g(x) - g_\epsilon(x)| < \epsilon, \quad \forall x \in [a,b].$$

Proof: We shall perform the proof in two steps.

Step 1. Define the partition and the piecewise linear function.

Let $\epsilon > 0$ be arbitrary fixed. Since g is uniform continuous on $[a, b]$, there is a $\delta > 0$ such that if $|x' - x''| < \delta$, then $|g(x') - g(x'')| < \epsilon/2$. Consider N large enough such that $\dfrac{b-a}{N} < \delta$, and define the equidistant partition $x_j = a + \dfrac{j(b-a)}{N}$, $j = 0, \ldots, N$. The piecewise polygonal function is given by

$$g_\epsilon(x) = g(x_{i-1}) + m_i(x - x_{i-1}), \quad \forall x \in [x_{i-1}, x_i],$$

with the slope $m_i = \dfrac{g(x_i) - g(x_{i-1})}{x_i - x_{i-1}}$.

Step 2. Finishing the proof by applying the triangle inequality.
Let $x \in [a, b]$ be fixed. Then there is a partition interval that contains x, i.e., $x \in [x_i, x_{i+1}]$. By the uniform continuity of g, we have

$$|g(x) - g(x_i)| < \frac{\epsilon}{2}. \tag{8.4.7}$$

Using that g_ϵ is affine on the interval $[x_i, x_{i+1}]$, we have the estimation

$$|g_\epsilon(x_i) - g_\epsilon(x)| < |g_\epsilon(x_i) - g_\epsilon(x_{i-1})| = |g(x_i) - g(x_{i-1})| < \frac{\epsilon}{2}, \tag{8.4.8}$$

where the last inequality used the uniform continuity of g. Now, triangle inequality together with relations (8.4.7) and (8.4.8) yield

$$
\begin{aligned}
|g(x) - g_\epsilon(x)| &\leq |g(x) - g(x_i)| + |g(x_i) - g_\epsilon(x)| \\
&= |g(x) - g(x_i)| + |g_\epsilon(x_i) - g_\epsilon(x)| \\
&< \frac{\epsilon}{2} + \frac{\epsilon}{2} = \epsilon.
\end{aligned}
$$

Since this inequality is satisfied by any $x \in [a, b]$, the lemma is proved.   ∎

Remark 8.4.2 1. The result can be reformulated by saying that any continuous function on a compact interval is the uniform limit of a sequence of piecewise linear functions.
2. A variant of this result in the case of a non-equidistant partition has been treated in Example 7.1.1.

Recall the definition of a ReLU function in terms of the Heaviside function as

$$ReLU(x) = xH(x) = \begin{cases} x, & \text{if } x \geq 0 \\ 0, & \text{if } x < 0. \end{cases} \tag{8.4.9}$$

Its generalized derivative is given by

$$ReLU'(x) = \big(xH(x)\big)' = x'H(x) + xH'(x) = H(x) + x\delta(x) = H(x),$$

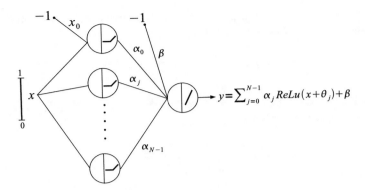

Figure 8.7: *One-hidden layer neural network with ReLU activation functions.*

since $x\delta(x) = 0$, as a product of functions with disjoint supports.[1] Another verification of this statement, directly from the definition, can be found in Example F.2.2 of the Appendix.

The next result shows that a neural network with a ReLU activation function can approximate continuous functions on a compact set, see Fig. 8.7.

Theorem 8.4.3 *Consider a one-hidden layer neural network with ReLU activation functions for the hidden layer neurons. Consider the output neuron having the identity activation function and the bias β. Let $g \in C[0, 1]$ be given. Then $\forall \epsilon > 0$ there are $\alpha_j, \theta_j \in \mathbb{R}$, and $N \geq 1$ such that*

$$\left| g(x) - \sum_{j=0}^{N-1} \alpha_j ReLU(x + \theta_j) - \beta \right| < \epsilon, \quad \forall x \in [0, 1].$$

Proof: Consider an equidistant partition $(x_i)_i$ of the interval $[0, 1]$

$$0 = x_0 < x_1 < \cdots < x_N = 1,$$

as given by Lemma 8.4.1. We need to show that parameters α_i, x_i, and β can be chosen such that the approximating function

$$G(x) = \sum_{j=0}^{N-1} \alpha_j ReLU(x + \theta_j) + \beta$$

---

[1]In general, the product between a function $f(x)$ and Dirac measure $\delta(x)$ is $f(x)\delta(x) = f(0)\delta(x)$.

becomes a piecewise linear function as in Lemma 8.4.1. We start by setting the biasses $\theta_j = -x_j$ and noting that

$$ReLU(x_k - x_j) = \begin{cases} 0, & \text{if } k \le j \\ \frac{k-j}{N}, & \text{if } k > j, \end{cases}$$

where we have taken into account formula (8.4.9). Therefore, the value of $G(x)$ at the partition point x_k is given by

$$G(x_k) = \sum_{j=0}^{N-1} \alpha_i ReLU(x_k - x_j) + \beta$$

$$= \frac{1}{N} \sum_{j<k} \alpha_j(k-j) + \beta.$$

The $N+1$ parameters, α_j and β, are uniquely obtained from the constraints $G(x_k) = g(x_k)$, $k = 0, \ldots, N$. Denoting for simplicity $y_k = g(x_k)$, the aforementioned constraints are written as

$$
\begin{aligned}
y_0 &= G(x_0) = G(0) = \beta \\
y_1 &= G(x_1) = \alpha_0(1-0)/N + \beta \\
y_2 &= G(x_2) = \alpha_0(2-0)/N + \alpha_1(2-1)/N + \beta \\
y_3 &= G(x_3) = \alpha_0(3-0)/N + \alpha_1(3-1)/N + \alpha_2(3-2)/N + \beta \\
\cdots &= \cdots\cdots\cdots\cdots\cdots\cdots\cdots\cdots\cdots\cdots \\
y_N &= G(x_N) = \alpha_0(N_0)/N + \cdots + \alpha_{N-1}/N + \beta.
\end{aligned}
$$

The system is equivalent to

$$
\begin{aligned}
y_0 &= \beta \\
N(y_1 - y_0) &= \alpha_0 \\
N(y_2 - y_0) &= 2\alpha_0 + \alpha_1 \\
N(y_3 - y_0) &= 3\alpha_0 + 2\alpha_1 + \alpha_2 \\
\cdots &= \cdots\cdots\cdots\cdots\cdots \\
N(y_N - y_0) &= N\alpha_0 + (N-1)\alpha_1 + \cdots + 2\alpha_{N-2} + \alpha_{N-1}.
\end{aligned}
$$

Subtracting consecutive equations yields

$$
\begin{aligned}
\beta &= y_0 \\
\alpha_0 &= N(y_1 - y_0) \\
\alpha_0 + \alpha_1 &= N(y_2 - y_1) \\
\alpha_0 + \alpha_1 + \alpha_2 &= N(y_3 - y_2) \\
\cdots\cdots\cdots &= \cdots\cdots\cdots\cdots \\
\alpha_0 + \alpha_1 + \cdots + \alpha_{N-1} &= N(y_N - y_{N-1}).
\end{aligned}
$$

The system has the unique solution

$$
\begin{aligned}
\beta &= y_0 \\
\alpha_0 &= N(y_1 - y_0) \\
\alpha_1 &= N(y_2 - 2y_1 + y_0) \\
\alpha_2 &= N(y_3 - 2y_2 + y_1) \\
\cdots &= \cdots\cdots\cdots\cdots \\
\alpha_{N-1} &= N(y_N - 2y_{N-1} + y_{N-2}).
\end{aligned}
$$

The approximator function

$$
G(x) = \sum_{j=0}^{N-1} \alpha_j ReLU(x - x_j) + \beta
$$

is piecewise linear, since its derivative

$$
G'(x) = \sum_{j=0}^{N-1} \alpha_j ReLU'(x - x_j) = \sum_{j=0}^{N-1} \alpha_j H(x - x_j)
$$

is constant on each interval (x_i, x_{i+1}), given by a sum of $\pm \alpha_j$. Applying Lemma 8.4.1 and choosing the polygonal line $G(x) = g_\epsilon(x)$, we obtain $|g(x) - G(x)| < \epsilon$ for all $x \in [0, 1]$, for any a priori arbitrary fixed $\epsilon > 0$. ∎

Example 8.4.4 (Hedging application) Let S denote the price of a stock and recall that the payoff of a European call with strike price K is given by $C(S) = ReLU(S - K)$. Then any portfolio value $P(S)$, which is a continuous function of S, for $0 \le S \le S_{max}$ can be learned by a one-hidden layer neural network, which has the output

$$
G(S) = \sum_{j=0}^{N-1} \alpha_j ReLU(S - K_j) + \beta = \sum_{j=0}^{N-1} \alpha_j C_j(S) + \beta.
$$

This can be constructed by buying α_j units of calls with strike price K_j, for $j = 0, \ldots, N-1$ and keeping an amount β in bonds. In order to hedge the portfolio position a trader should have to take the opposite market position, namely, to sell the new formed portfolio $G(S)$. The conclusion is that any portfolio of stocks can be replicated with a portfolio containing calls and bonds.

8.5   Learning with Softplus

The softplus activation function

$$\varphi(x) = sp(x) = \ln(1 + e^x)$$

is a smoothed version of the $ReLU(x)$ function, see Fig. 2.7 **a**, as the next result shows.

Proposition 8.5.1 *The softplus function is given by the convolution*

$$sp(x) = (ReLU * K)(x) = \int_{-\infty}^{\infty} ReLU(y)K(x-y)\,dy,$$

with the convolution kernel $K(x) = \dfrac{1}{(1+e^x)(1+e^{-x})}.$

Proof: We note that the kernel $K(x)$ decreases to zero exponentially fast as $x \to \pm\infty$, i.e., $\lim_{|x|\to\infty} K(x) = 0$. This condition suffices for the existence of the next improper integrals. We also note the following relation between the kernel and logistic function σ:

$$
\begin{aligned}
\sigma'(x) &= \sigma(x)(1 - \sigma(x)) = \frac{1}{1+e^{-x}}\left(1 - \frac{1}{1+e^{-x}}\right) \\
&= \frac{1}{(1+e^x)(1+e^{-x})} = K(x).
\end{aligned}
$$

The convolution can be now computed as

$$
\begin{aligned}
(ReLU * K)(x) &= \int_{-\infty}^{\infty} ReLU(y)K(x-y)\,dy = \int_{0}^{\infty} yK(x-y)\,dy \\
&= -\int_{x}^{-\infty}(x-t)K(t)\,dt = \int_{-\infty}^{x}(x-t)K(t)\,dt \\
&= x\int_{-\infty}^{x} K(t)\,dt - \int_{-\infty}^{x} tK(t)\,dt \\
&= x\int_{-\infty}^{x} \sigma'(t)\,dt - \int_{-\infty}^{x} t\sigma'(t)\,dt \\
&= x\sigma(x) - \left(t\sigma(t)\Big|_{-\infty}^{x} - \int_{-\infty}^{x} t'\sigma(t)\,dt\right) \\
&= \int_{-\infty}^{x} \sigma(t)\,dt = sp(x),
\end{aligned}
$$

which proves the desired relation. We note that in the last identity we used formula (2.1.2).   ∎

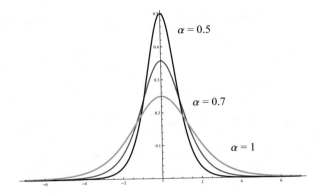

Figure 8.8: *The convolutional kernels $K_\alpha(x)$ converge to a "spike" centered at zero, as $\alpha \to 0$.*

Remark 8.5.2 An informal verification of the previous convolution formula is to compute the generalized derivative of the function

$$f(x) = (ReLU * K)(x) - sp(x).$$

Using convolution properties, as well as the derivation formulas

$$ReLU'(x) = H(x), \quad H'(x) = \delta(x), \quad sp'(x) = \sigma(x),$$

we obtain

$$
\begin{aligned}
f'(x) &= (ReLU * K)'(x) - sp'(x) = (ReLU' * K)(x) - \sigma(x) \\
&= (H * \sigma')(x) - \sigma(x) = (H' * \sigma)(x) - \sigma(x) \\
&= (\delta * \sigma)(x) - \sigma(x) = \sigma(x) - \sigma(x) = 0,
\end{aligned}
$$

where $H(x)$ and $\delta(x)$ stand for the Heaviside function and Dirac's measure. Hence, $f(x)$ is piecewise constant. Since it is continuous (as a difference of two continuous functions), it follows that $f(x)$ is constant. Then we make the point that $\lim_{x \to -\infty} f(x) = 0$, which shows that $f(x) = 0$ for all real numbers x.

The properties of the convolution kernel $K(x)$ are given in the next result, see Fig. 8.8.

Proposition 8.5.3 (*i*) *The kernel $K(x)$ is a symmetric probability density function on $\mathbb{R}$.*

(*ii*) *Consider the scaled perturbation $K_\alpha(x) = \dfrac{1}{\alpha} K\left(\dfrac{x}{\alpha}\right)$, with $\alpha > 0$, and define the measure μ_α by $d\mu_\alpha = K_\alpha(x)dx$. Then $\displaystyle\int_{-\infty}^{\infty} K_\alpha(x)\, dx = 1$ and $\mu_\alpha \to \delta$, as $\alpha \to 0$, in the weak sense.*

Proof: (i) It is obvious that $K(-x) = K(x)$ and $K(x) > 0$. We also have

$$\int_{-\infty}^{\infty} K(x)\, dx = \int_{-\infty}^{\infty} \sigma'(x)\, dx = \sigma(+\infty) - \sigma(-\infty) = 1.$$

(ii) Using the substitution $x = \alpha y$ and part (i), we have

$$\int_{-\infty}^{\infty} K_\alpha(x)\, dx = \frac{1}{\alpha}\int_{-\infty}^{\infty} K\left(\frac{x}{\alpha}\right) dx = \int_{-\infty}^{\infty} K(y)\, dy = 1.$$

For any function with compact support, $g \in C_0^\infty(\mathbb{R})$, the Bounded Convergence Theorem, Theorem C.4.3, yields

$$\begin{aligned}
\mu_\alpha(g) &= \int_{-\infty}^{\infty} g(x)\, d\mu_\alpha(x) = \int_{-\infty}^{\infty} K_\alpha(x)g(x)\, dx \\
&= \int_{-\infty}^{\infty} K(y)g(\alpha y)\, dy \to \int_{-\infty}^{\infty} K(y)g(0)\, dy = g(0) \\
&= \int_{-\infty}^{\infty} g(x)\delta(x)\, dx = \delta(g), \qquad \text{as } \alpha \to 0,
\end{aligned}$$

which means that $\mu_\alpha \to \delta$, as $\alpha \to 0$, in the weak sense.   ■

We consider now the *scaled softplus function*

$$\varphi_\alpha(x) = \alpha\varphi\left(\frac{x}{\alpha}\right) = \alpha\, sp\left(\frac{x}{\alpha}\right) = \alpha\, \ln(1 + e^{x/\alpha})$$

and note that

$$\varphi_\alpha''(x) = \frac{1}{\alpha} sp''\left(\frac{x}{\alpha}\right) = \frac{1}{\alpha}K\left(\frac{x}{\alpha}\right) = K_\alpha(x), \qquad (8.5.10)$$

since $sp'(x) = \sigma(x)$ and $\sigma'(x) = K(x)$.

It is worth noting that the scaled softplus functions $\varphi_\alpha(x)$ converge pointwise to $ReLU(x)$, in a decreasing way, see Fig. 8.9. This means

$$\varphi_\alpha(x) \searrow ReLU(x), \qquad \text{as } \alpha \searrow 0.$$

The following result provides a formula for the smoothed polygonal function $G(x)$ through the filter K_α, which is defined by (8.5.10).

Lemma 8.5.4 *Consider the convolution $G_\alpha = G * K_\alpha$, where*

$$G(x) = \sum_{j=0}^{N-1} \alpha_j ReLU(x - x_j) + \beta.$$

There are parameters c_j, w, and θ_j, depending on α_j and x_j, such that

$$G_\alpha(x) = \sum_{j=0}^{N-1} c_j sp(wx - \theta_j).$$

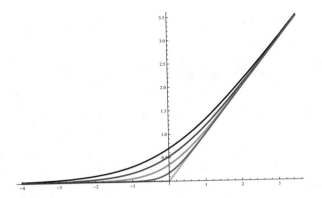

Figure 8.9: *The scaled softplus $\varphi_\alpha(x)$ for values $\alpha = 1, 0.75, 0.5, 0.25$.*

Proof: Using relation (8.5.10) and Proposition 8.5.3, part (ii), we have

$$
\begin{aligned}
G_\alpha(x) &= (G * K_\alpha)(x) = \sum_{j=0}^{N-1} \alpha_j ReLU(x - x_j) * K_\alpha(x) + \beta * K_\alpha(x) \\
&= \sum_{j=0}^{N-1} \alpha_j ReLU(x - x_j) * \varphi_\alpha''(x) + \beta \\
&= \sum_{j=0}^{N-1} \alpha_j H'(x - x_j) * \varphi_\alpha(x) + \beta \\
&= \sum_{j=0}^{N-1} \alpha_j \delta(x - x_j) * \varphi_\alpha(x) + \beta = \sum_{j=0}^{N-1} \alpha_j \varphi_\alpha(x - x_j) + \beta \\
&= \sum_{j=0}^{N-1} \alpha_j \alpha \varphi\Big(\frac{x - x_j}{\alpha}\Big) + \beta = \sum_{j=0}^{N-1} c_j sp(wx - \theta_j) + \beta,
\end{aligned}
$$

with $c_j = \alpha_j \alpha$, $w = 1/\alpha$, and $\theta_j = x_j/\alpha$. We also used that $\beta * K_\alpha(x) = \beta$, since K_α has the integral equal to 1. ∎

Remark 8.5.5 The function G_α is the output of a one-hidden layer neural network with softplus activation functions for the hidden layer and identity function for the output neuron. The bias for the output neuron is β.

The following result is an application of Dini's Theorem.

Lemma 8.5.6 *The sequence G_α converges uniformly to G on $[0,1]$, as $\alpha \searrow 0$, i.e., $\forall \epsilon > 0$ there is $\eta > 0$ such that for $\alpha < \eta$, we have*

$$
|G(x) - G_\alpha(x)| < \epsilon, \quad \forall x \in [0,1].
$$

Proof: From the previous construction, we have

$$G_\alpha(x) = \sum_{j=0}^{N-1} \alpha_j \alpha \varphi\left(\frac{x - x_i}{\alpha}\right) + \beta = \sum_{j=0}^{N-1} \alpha_j \varphi_\alpha (x - x_i) + \beta.$$

Since $\varphi_\alpha(x) \searrow ReLU(x)$ as $\alpha \searrow 0$, see Exercise 8.8.6, then $G_\alpha(x)$ converges pointwise and decreasing to $G(x)$, for any $x \in [0, 1]$. Since G_α and G are continuous on the compact interval $[0, 1]$, an application of Dini's Theorem (Theorem 7.1.1) implies that G_α converges to G, uniformly on $[0, 1]$. ∎

The next result is an analog of Theorem 8.4.3 for the case of the softplus function. The idea is to approximate a given continuous function on a compact interval with a polygonal line, and then to "blur" a little this approximation, using a convolution with a given kernel. This procedure leads to a differentiable approximation of the initial continuous function.

Theorem 8.5.7 *Consider a one-hidden layer neural network with softplus activation function for the hidden layer and identity activation function for the outcome neuron. Let $g \in C[0, 1]$. Then $\forall \epsilon > 0$ there are $c_j, w, \theta_j, \beta \in \mathbb{R}$ and $N \geq 1$ such that*

$$\left| g(x) - \sum_{j=0}^{N-1} c_j sp(wx - \theta_j) - \beta \right| < \epsilon, \quad \forall x \in [0, 1].$$

Proof: Let $\epsilon > 0$ be arbitrary fixed. By Theorem 8.4.3, there is a function $G(x) = \sum_{j=0}^{N-1} \alpha_j ReLU(x - x_j) + \beta$ such that

$$\left| g(x) - G(x) \right| < \frac{\epsilon}{2}, \quad \forall x \in [0, 1].$$

Lemma 8.5.6 implies that for any $\alpha > 0$, sufficiently small, we have the estimation

$$\left| G(x) - G_\alpha(x) \right| < \frac{\epsilon}{2}, \quad \forall x \in [0, 1].$$

The previous two estimations, together with the triangle inequality, yield

$$\begin{aligned} \left| g(x) - G_\alpha(x) \right| &\leq \left| g(x) - G(x) \right| + \left| G(x) - G_\alpha(x) \right| \\ &\leq \frac{\epsilon}{2} + \frac{\epsilon}{2} = \epsilon, \quad \forall x \in [0, 1]. \end{aligned}$$

Since Lemma 8.5.4 states that the function $G_\alpha(x)$ is the output of a one-hidden layer network with softplus activation functions, we obtain the desired relation.

∎

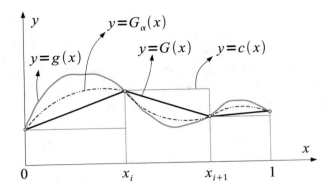

Figure 8.10: *The approximation of a continuous function $g \in C[0,1]$ by three approximators: $G(x) = \sum_{j=0}^{N-1} \alpha_j ReLU(x - x_j) + \beta$, $G_\alpha(x) \sum_{j=0}^{N-1} c_j sp(wx - \theta_j) + \beta$ and $c(x) = \sum_{i=0}^{N-1} c_i H(x - x_i)$.*

Remark 8.5.8 We have approximated continuous functions on $[0,1]$ using outputs of networks with different activation functions (Heaviside, ReLU, and Softplus). Now it is the time to ask the natural question: Which activation function is better? The following points will discuss this question.

(i) The outcome of a one-hidden neural network with Softplus activation function is obtained applying a convolution with kernel K_α to the outcome of a one-hidden neural network with ReLU activation function.

(ii) We note the following gain of smoothness: if the network outcome is continuous in the case of using ReLU, then in the case of using Softplus it becomes differentiable.

(iii) The approximation of a continuous function $g \in C[0,1]$ by different neural outcomes can be visualized in Fig. 8.10. The approximation using ReLU is a polygonal approximation denoted by $G(x)$. This is more accurate than the approximation using step functions, $c(x)$. However, the approximation using Softplus, denoted by $G_\alpha(x)$, is even better than $G(x)$. All errors in this case are measured in the sense of the $\| \cdot \|_\infty$-norm.

8.6 Discrete Data

In real life only some discrete data is provided. The ability of finding a continuous function rather than an arbitrary function that is arbitrarily close to the given set of data points says something about the learning machines ability to generalize of one-hidden layer neural networks.

We assume the novel input data consists in N data points, $(x_i, y_i) \in [0, 1] \times [a, b]$, $1 \le i \le N$. In order to move forward with our analysis we establish the following *ansatz*: y_i *are responses obtained from* x_i *applying a continuous function* g. In other words, we assume the existence of a continuous function $g : [0, 1] \to \mathbb{R}$ such that $g(x_i) = y_i$, $1 \le i \le N$. This mathematical assumption holds true in most cases when data is provided by measuring the cause and the effect in physical phenomena.

Let $\epsilon > 0$ be arbitrarily fixed. It can be inferred from the mathematical results proved in this section that the response generated by a one-hidden layer neural network, $f_\epsilon(x)$, has the maximal error from the given data of at most ϵ, namely

$$\max_i |f_\epsilon(x_i) - y_i| < \epsilon.$$

This procedure goes beyond lookup tables or memorizing data because it instantiated prior knowledge about how the learning machine should generalize. This means that if you have provided with some new data, then the network's output generates a response which is similar to a response from an input pattern and is close in a certain distance sense to the novel input pattern.

8.7 Summary

This chapter shows that one-hidden layer neural networks with one-dimensional input, $x \in [0, 1]$, can learn continuous functions in $C[0, 1]$. We prove this result in the cases when the activation function is either a step function, a sigmoid, a ReLU, or a Softplus function.

The outcome of a network having a sigmoid activation function is a smoothed version of a network having a step activation function. Similarly, learning with Softplus is a smoothed version of learning with a ReLU function. It is worth noting the mathematical reason behind the experimentally observed fact that the use of sigmoid activation function produces smaller errors than the use of step functions. Similarly, it is better off in many cases to use Softplus activation function rather than ReLU.

The practical importance of the method consists in the ability of the network of generalizing well when it is applied to discrete data. The network potentially learns a continuous function which underlies the given data.

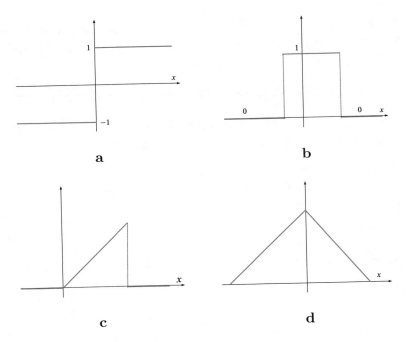

Figure 8.11: **a.** *Bipolar step function.* **b.** *Box function.* **c.** *Sawtooth function.* **d.** *Shark-tooth function.*

8.8 Exercises

Exercise 8.8.1 (generalized derivative) We say that $f^{(n)} = g$ in generalized sense, if for any compact supported function, $\phi \in C_0^\infty$, we have

$$(-1)^n \int f\phi^{(n)} = \int g\phi.$$

Show that the following relations hold in the generalized derivative sense:

(a) $H'(x - x_0) = \delta(x - x_0)$;

(b) $ReLU'(x) = H(x)$;

(c) $ReLU''(x) = \delta(x)$;

(d) $(xReLU(x))' = 2ReLU(x)$;

(e) $(xReLU(x))'' = 2H(x)$.

Exercise 8.8.2 Equation (8.1.1) shows that any simple function can be written as a linear combination of Heaviside functions. Find the coefficients c_i in terms of α_i.

Exercise 8.8.3 (uniform continuity) Consider the continuous function $g :$ $[a, b] \to \mathbb{R}$. Prove that $\forall \epsilon > 0$, $\exists \delta > 0$ such that $\forall x, x' \in [a, b]$ with $|x - x'| < \delta$, then $|g(x) - g(x')| < \epsilon$.

Exercise 8.8.4 Use Dini's theorem (Theorem 7.1.1) to prove Lemma 8.4.1.

Exercise 8.8.5 Let f be one of the functions depicted in Fig. 8.11. Which one has the property that the set of finite combinations

$$\left\{ \sum_{i=1}^{N} \alpha_i f(w_i x - b_i); N \geq 1, \alpha_i, w_i, b_i \mathbb{R} \right\}$$

is dense in $C[0, 1]$?

Exercise 8.8.6 Let $\varphi_\alpha(x) = \alpha \ln(1 + e^{x/\alpha})$. Show that $\varphi_\alpha(x) \searrow ReLU(x)$ as $\alpha \searrow 0$.

Chapter 9

Universal Approximators

The answer to the question *"Why neural networks work so well in practice?"* is certainly based on the fact that neural networks can approximate well a large family of real-life functions that depend on input variables. The goal of this chapter is to provide mathematical proofs of this behavior for different variants of targets.

The fact that neural networks behave as universal approximators translates by saying that their output is an accurate approximation to any desired degree of accuracy of functions in certain familiar spaces of functions. The process of obtaining an accurate approximation is interpreted as a learning process of a given target function.

The chapter deals with the existence of the learning process, but it does not provide any explicit construction of the network weights. The idea followed here is that it makes sense to know first that a solution exists before we should start looking for it.

The desired solution exists, provided that a sufficient number of hidden units are considered and the relation between the input variables and the target function is deterministic. In order to obtain the solution, an adequate learning algorithm should be used. This technique should assure the success of using neural networks in many practical applications.

9.1 Introductory Examples

A neural network has the capability of learning from data. Providing a target function of a certain type, $z = f(x)$, the network has to learn it by changing its internal parameters. Most of the time this learning is just an acceptable

© Springer Nature Switzerland AG 2020
O. Calin, *Deep Learning Architectures*, Springer Series in the Data Sciences,
https://doi.org/10.1007/978-3-030-36721-3_9

approximation. To facilitate understanding, in the following we provide a few examples.

Example 9.1.1 (Learning continuous functions) Assume we need to create a neural network which is meant to rediscover Newton's gravitational law. This is

$$f(m_1, m_2, d) = \frac{km_1m_2}{d^2},$$

i.e., the force between two bodies with masses m_1 and m_2 is inverse proportional with the square of the distance between them. We construct a network with inputs $(x_1, x_2, x_3) = (m_1, m_2, d)$ that can take continuous values in certain 3-dimensional region K. The target variable, which needs to be learned, is $z = f(m_1, m_2, d)$, with $f : K \to \mathbb{R}$ continuous function. The network has an output $y = g(m_1, m_2, d; w, b)$, which depends on both inputs and some parameters (weights w and bias b). The parameters have to be tuned such that the outcome function, g, can approximate to any degree of accuracy the continuous function f. Therefore, if one provides two bodies with given masses, situated at a known distance, then the network should be able to provide the gravitational force between them with an accuracy exceeding any a priori fixed error. The existence of such neural network is given by Theorem 9.3.6.

Example 9.1.2 (Learning finite energy functions) An audio signal needs to be learned by a neural network. It makes sense to consider signals of finite energy only, condition which will be formalized later by saying that the signal is a function in the space L^2. The input to the network is $x = (t, \nu)$ with $x_1 = t$, time, and $x_2 = \nu$, signal frequency. The target function, $z = f(x)$, provides the amplitude in terms of time and frequency. The network's output, $y = g(t, \nu; w, b)$, provides an approximation of the target, which can be adjusted by changing the values of the weights w and bias b. The existence of a neural network with this property is provided by Proposition 9.3.11.

9.2   General Setup

The concept of the neural network as an universal approximator goes as in the following. Let x be the input variable, z be the target, or desired variable, and denote the *target function* by $z = f(x)$, with f in a certain function space, $\mathcal{S}$, which is a metric space, endowed with distance d, i.e., $f \in (\mathcal{S}, d)$. The space of the neural network outcome functions, g, is denoted by $\mathcal{U}$ and is assumed to be a subset of $\mathcal{S}$. In fact, we choose the activation function of the network such that the outcome space is properly adjusted such that the previous inclusion is satisfied.

Now, we say that the neural network is an *universal approximator* for the space $(\mathcal{S}, d)$ if the space of outcomes $\mathcal{U}$ is d-dense in $\mathcal{S}$, i.e.,

$$\forall f \in \mathcal{S},\ \forall \epsilon > 0,\ \exists g \in \mathcal{U} \text{ such that } d(f - g) < \epsilon.$$

This means that for any function f in $\mathcal{S}$, there are functions in $\mathcal{U}$ situated in any proximity of f. The process by which the neural network approximates a target function $f \in \mathcal{S}$ by functions in $\mathcal{U}$ is called learning. $\mathcal{U}$ is the *approximation space* of the target space $\mathcal{S}$.

There are several types of target spaces, $(\mathcal{S}, d)$, which are useful in applications. The most familiar will be presented in the following examples. The reader can find more details regarding the normed vectorial spaces in the Appendix.

Example 9.2.1 Let $\mathcal{S} = C(K)$ be the space of real-valued continuous functions on a compact set K. The distance function is given by

$$d(f, g) = \sup_{x \in K} |f(x) - g(x)|, \qquad \forall f, g \in C(K),$$

which is the maximum absolute value difference between the values of two functions on the compact set. Since $|f - g|$ is continuous, the maximum is reached on K, and we may replace in the definition "sup" by "max". It is worth noting that this distance is induced by the norm $\|f\| = \sup_{x \in K} |f(x)|$, case in which $(C(K), \| \cdot \|)$ becomes a normed vectorial space.

Example 9.2.2 If $\mathcal{S} = L^1(K)$ is the space of real-valued integrable functions on a compact set K, the distance function in this case is

$$d(f, g) = \int_K |f(x) - g(x)|\, dx, \qquad \forall f, g \in L^1(K).$$

This can be interpreted as the total area between the graphs of the functions f and g over the set K. The corresponding norm is the L^1-norm defined by $\|f\|_1 = \int_K |f(x)|$. Consequently, $(L^1(K), \| \cdot \|_1)$ becomes a normed vectorial space.

Example 9.2.3 Consider $\mathcal{S} = L^2(K)$ be the space of real-valued squared integrable functions on a compact set K. The distance function is

$$d(f, g) = \left(\int_K |f(x) - g(x)|^2\, dx \right)^{1/2}, \qquad \forall f, g \in L^2(K).$$

The interpretation of this is the energy difference between two finite energy signals f and g over the set K. The corresponding norm is the L^2-norm defined by $\|f\|_2 = \left(\int_K |f(x)|^2\, dx \right)^{1/2}$, with respect to which $L^2(K)$ becomes a normed vector space.

Example 9.2.4 Let $K = \{x_1, \ldots, x_r\}$ be a finite set in $\mathbb{R}^n$ and consider $\mathcal{S} = \{f : K \to \mathbb{R}\}$ be the set of real-valued functions defined on a finite set. In this case the distance can be either the sum of squared errors

$$d_s(f, g) = \Big(\sum_{x_j \in K} |f(x_j) - g(x_j)|^2 \Big)^{1/2},$$

or the absolute deviation error

$$d_a(f, g) = \sum_{x_j \in K} |f(x_j) - g(x_j)|.$$

These distances are commonly used in regression analysis. Also, in the case of a classical perceptron, the inputs are finite, with $K = \{0, 1\}$. It is worth noting that $d_s(f, g)$ is the *Euclidean distance* in $\mathbb{R}^n$ between points $f(x)$ and $g(x)$, while $d_a(f, g)$ is the *taxi-cab distance* between the same points (this is the distance measured along the coordinate curves, similar to the distance a cab would follow on a grid of perpendicular streets, like the ones in New York city).

Example 9.2.5 In a very general setup, the target space can be considered to be the space of Borel-measurable functions from $\mathbb{R}^n$ to $\mathbb{R}$:

$$\mathcal{M}^n = \{f : \mathbb{R}^n \to \mathbb{R}; f \text{ Borel-measurable}\}.$$

This means that if $\mathcal{B}(\mathbb{R})$ and $\mathcal{B}(\mathbb{R}^n)$ denote the Borel $\mathfrak{S}$-fields on $\mathbb{R}$ and $\mathbb{R}^n$, respectively, then $f^{-1}(\mathcal{B}(\mathbb{R})) \subset \mathcal{B}(\mathbb{R}^n)$, or, equivalently

$$f^{-1}([a, b]) \in \mathcal{B}(\mathbb{R}^n), \qquad \forall a, b \in \mathbb{R}.$$

In order to introduce a distance function on $\mathcal{M}^n$, we shall consider first a probability measure μ on the measurable space $(\mathbb{R}^n, \mathcal{B}(\mathbb{R}))$, i.e., a measure $\mu : \mathcal{B}(\mathbb{R}) \to [0, 1]$ with $\mu(\mathbb{R}^n) = 1$. Define

$$d_\mu(f, g) = \inf\{\epsilon > 0; \ \mu(|f - g| > \epsilon) < \epsilon\}, \qquad \forall f, g \in \mathcal{M}^n.$$

Then $(\mathcal{M}^n, d_\mu)$ becomes a metric space. It is worth noting that the elements of $\mathcal{M}^n$ are determined almost everywhere, i.e., whenever $\mu\big(x; \ f(x) = g(x)\big) = 1$, the measurable functions f and g are identified.
It will be shown in section 9.3.4 that the convergence in the metric d_μ is equivalent to the convergence in probability.

We shall start the study of the learning process with the simple case of a deep network that has only one hidden layer.

9.3 Single Hidden Layer Networks

This section shows that a neural network with one hidden layer can be used as a function approximator for the spaces $C(I_n)$, $L^1(I_n)$, and $L^2(I_n)$, where I_n is the unit hypercube in $\mathbb{R}^n$. This means that given a function $f(x)$ in one of the aforementioned spaces, having the input variable $x \in I_n = [0,1]^n$, there is a tuning of the network weights such that the neural network output, $g(x)$, is "as close as possible" to the given function $f(x)$, in the sense of a certain distance.

9.3.1 Learning continuous functions $f \in C(I_n)$

Before the development of the main results, some preliminary results of functional analysis are needed. The first lemma states the existence of a separation functional, which vanishes on a subspace and has a nonzero value outside of the subspace. For the basic elements of functional analysis needed in this section, the reader is referred to section E in the Appendix.

Lemma 9.3.1 *Let $\mathcal{U}$ be a linear subspace of a normed linear space $\mathcal{X}$ and consider $x_0 \in \mathcal{X}$ such that*

$$dist(x_0, \mathcal{U}) \geq \delta,$$

for some $\delta > 0$, i.e., $\|x_0 - u\| \geq \delta$, $\forall u \in \mathcal{U}$. Then there is a bounded linear functional L on $\mathcal{X}$ such that:

(i) $\|L\| \leq 1$
(ii) $L(u) = 0, \quad \forall u \in \mathcal{U}$, i.e., $L_{|\mathcal{U}} = 0$
(iii) $L(x_0) = \delta$.

Proof: Consider the linear space $\mathcal{T}$ generated by $\mathcal{U}$ and x_0

$$\mathcal{T} = \{t;\ t = u + \lambda x_0, u \in \mathcal{U}, \lambda \in \mathbb{R}\}$$

and define the function $L : \mathcal{T} \to \mathbb{R}$ by

$$L(t) = L(u + \lambda x_0) = \lambda \delta.$$

Since for any $t_1, t_2 \in \mathcal{T}$ and $\alpha \in \mathbb{R}$ we have

$$
\begin{aligned}
L(t_1 + t_2) &= L(u_1 + u_2 + (\lambda_1 + \lambda_2)x_0) = (\lambda_1 + \lambda_2)\delta \\
&= L(t_1) + L(t_2) \\
L(\alpha t) &= \alpha L(t),
\end{aligned}
$$

it follows that L is a linear functional on $\mathcal{T}$.

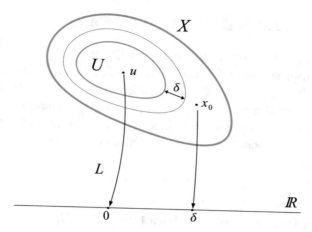

Figure 9.1: *The functional L that vanishes on subspace $\mathcal{U}$, see Lemma 9.3.1.*

Next, we show that $L(t) \leq \|t\|$, $\forall t \in \mathcal{T}$. First, notice that if $u \in \mathcal{U}$, then also $-\dfrac{u}{\lambda} \in \mathcal{U}$ and, using hypothesis, we have

$$\|x_0 + \frac{u}{\lambda}\| = \|x_0 - (-\frac{u}{\lambda})\| \geq \delta,$$

or, equivalently, $\dfrac{\delta}{\|x_0 + \frac{u}{\lambda}\|} \leq 1$. This is restated as $|\lambda|\delta \leq \|u + \lambda x_0\|$. Then

$$L(t) = L(u + \lambda x_0) = \lambda \delta \leq |\lambda|\delta \leq \|u + \lambda x_0\| = \|t\|.$$

Hence, $L(t) \leq \|t\|$, for all $t \in \mathcal{T}$.

Applying the Hahn-Banach Theorem, see Appendix, section E.3, with the norm $p(x) = \|x\|$, we obtain that L can be extended to a linear functional on $\mathcal{X}$, denoted also by L, such that $L(x) \leq \|x\|$, for all $x \in \mathcal{X}$, see Fig. 9.1. This implies $\|L\| \leq 1$, so the extension L is bounded. From the definition of L, we also have

$$\begin{aligned} L(u) &= L(u + 0 \cdot x_0) = 0 \cdot \delta = 0, &&\forall u \in \mathcal{U} \\ L(x_0) &= L(0 + 1 \cdot x_0) = 1 \cdot \delta = \delta > 0, \end{aligned}$$

which proves the lemma. ∎

The previous result can be reformulated using the language of dense subspaces as in the following. First, recall that a subspace $\mathcal{U}$ of $\mathcal{X}$ is *dense* in $\mathcal{X}$ with respect to the norm $\| \cdot \|$ if for any element $x \in \mathcal{X}$ there are elements $u \in \mathcal{U}$ as close as possible to x. Equivalently, $\forall x \in \mathcal{X}$ there is a sequence

$(u_n)_n$ in $\mathcal{U}$ such that $u_n \to x$, as $n \to \infty$. This can be restated as: $\forall x \in \mathcal{X}$, $\forall \epsilon > 0$, there is $u \in \mathcal{U}$ such that $\|u - x\| < \epsilon$.

Consequently, the fact that the subspace $\mathcal{U}$ is not dense in $\mathcal{X}$ can be described as: there are elements $x_0 \in \mathcal{X}$ such that no elements $u \in \mathcal{U}$ are close enough to x_0; or: there is a $\delta > 0$ such that $\forall u \in \mathcal{U}$ we have $\|u - x_0\| \geq \delta$. This is just the hypothesis condition of Lemma 9.3.1. Thus, we have the following reformulation:

Lemma 9.3.2 (reformulation of Lemma 9.3.1) *Let $\mathcal{U}$ be a linear, non-dense subspace of a normed linear space $\mathcal{X}$. Then there is a bounded linear functional L on $\mathcal{X}$ such that $L \neq 0$ on $\mathcal{X}$ and $L_{|\mathcal{U}} = 0$.*

Denote by $C(I_n)$ the linear space of continuous functions on the hypercube $I_n = [0,1]^n$, as a normed space with the norm

$$\|f\| = \sup_{x \in I_n} |f(x)|, \quad \forall f \in C(I_n).$$

Let $M(I_n)$ denote the space of finite signed Baire measures on I_n (see sections C.3 and C.9 of the Appendix for definitions). The reason for using Baire measures consists in their compatibility with compactly supported functions (which are Baire-measurable and integrable in any finite Baire measure). Even if in general Baire sets and Borel sets need not be the same, however, in the case of compact metric spaces, like I_n, the Borel sets and the Baire sets are the same, so Baire measures are the same as Borel measures that are finite on compact sets. These will be useful remarks when applying the representation theorem, Theorem E.5.6, of Appendix in the proof of the next result.

The next result states that there is always a signed measure that vanishes on a non-dense subspace of continuous functions.

Lemma 9.3.3 *Let $\mathcal{U}$ be a linear, non-dense subspace of $C(I_n)$. Then there is a measure $\mu \in M(I_n)$ such that*

$$\int_{I_n} h \, d\mu = 0, \quad \forall h \in \mathcal{U}.$$

Proof: Considering $\mathcal{X} = C(I_n)$ in Lemma 9.3.2, there is a bounded linear functional $L : C(I_n) \to \mathbb{R}$ such that $L \neq 0$ on $C(I_n)$ and $L_{|\mathcal{U}} = 0$. Applying the representation theorem of linear bounded functionals on $C(I_n)$, see Theorem E.5.6 from Appendix, there is a measure $\mu \in M(I_n)$ such that

$$L(f) = \int_{I_n} f \, d\mu, \quad \forall f \in C(I_n).$$

In particular, for any $h \in \mathcal{U}$, we obtain

$$L(h) = \int_{I_n} h \, d\mu = 0,$$

which is the desired result. ∎

Remark 9.3.4 It is worth noting that $L \neq 0$ implies $\mu \neq 0$.

The next result uses the concept of discriminatory function introduced in Chapter 2, Definition 2.2.2.

Proposition 9.3.5 *Let σ be any continuous discriminatory function. Then the finite sums of the form*

$$G(x) = \sum_{j=1}^{N} \alpha_j \sigma(w_j^T x + \theta_j), \qquad w_j \in \mathbb{R}^n, \alpha_j, \theta_j \in \mathbb{R}$$

are dense in $C(I_n)$.

Proof: Since σ is continuous, it follows that

$$\mathcal{U} = \{G; \ G(x) = \sum_{j=1}^{N} \alpha_j \sigma(w_j^T x + \theta_j)\}.$$

is a linear subspace of $C(I_n)$. We continue the proof adopting the contradiction method.

Assume that $\mathcal{U}$ is not dense in $C(I_n)$. By Lemma 9.3.3 there is a measure $\mu \in M(I_n)$ such that

$$\int_{I_n} h \, d\mu = 0, \quad \forall h \in \mathcal{U}.$$

This can also be written as

$$\sum_{j=1}^{N} \alpha_j \int_{I_n} \sigma(w_j^T x + \theta_j) \, d\mu = 0, \qquad \forall w_j \in \mathbb{R}^n, \alpha_j, \theta_j \in \mathbb{R}.$$

By choosing convenient coefficients α_j, we obtain

$$\int_{I_n} \sigma(w^T x + \theta) \, d\mu = 0, \qquad \forall w \in \mathbb{R}^n, \theta \in \mathbb{R}.$$

Using that σ is discriminatory, yields $\mu = 0$, which is a contradiction, see Remark 9.3.4. ∎

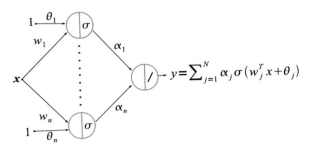

Figure 9.2: *The one-hidden layer neural network which approximates continuous functions.*

The previous result states that $\forall f \in C(I_n)$ and $\forall \epsilon > 0$, there is a sum $G(x)$ of the previous form such that

$$|G(x) - f(x)| < \epsilon, \qquad \forall x \in I_n.$$

Equivalently, the one-hidden layer neural network shown in Fig. 9.2 can learn any continuous function $f \in C(I_n)$ within an ϵ-error, by tuning its weights.

The next result states that, given no restrictions on the number of nodes and size of the weights, a one-hidden layer neural network is a continuous function approximator, see Fig. 9.2. Here the input is the vector $x^T = (x_1, \ldots, x_n) \in I_n$ and the weights from the input to the hidden layer are denoted by $(w_1, \ldots, w_N)$. Each weight is a vector given by $w_j^T = (w_{j1}, \ldots, w_{jn})$. The weights from the hidden layer to the output are $(\alpha_1, \ldots, \alpha_N)$, with each α_j real number. The number of neurons in the hidden layer is N, and their biasses are $(\theta_1, \ldots, \theta_N)$. Note that in the outer layer the activation function is linear, $\varphi(x) = x$, while in the hidden layer it is sigmoidal.

Theorem 9.3.6 (Cybenko, 1989) *Let σ be an arbitrary continuous sigmoidal function in the sense of Definition 2.2.1. Then the finite sums of the form*

$$G(x) = \sum_{j=1}^{N} \alpha_j \sigma(w_j^T x + \theta_j), \qquad w_j \in \mathbb{R}^n, \alpha_j, \theta_j \in \mathbb{R}$$

are dense in $C(I_n)$.

Proof: By Proposition 2.2.4, any continuous sigmoidal function is discriminatory. Then applying Proposition 9.3.5 yields the desired result. ∎

Remark 9.3.7 If the activation function is $\sigma = \frac{1}{1+e^{-x}}$, then all functions $G(x)$ are analytic in x, as finite sums of analytic functions. Since there are continuous functions in $C(I_n)$ that are not analytic, it follows that the approximation space $\mathcal{U}$ is a proper subspace of $C(I_n)$ (namely, $\mathcal{U} \neq C(I_n)$).

The next result is valid for any continuous activation function, see [59]. The price paid for this is to switch from the Σ-class networks, which approximates target functions using sums, to $\Sigma\Pi$-class networks, which involves sums of products.

Theorem 9.3.8 (Hornik, Stinchcombe, White, 1989) *Let $\varphi : \mathbb{R} \to \mathbb{R}$ be any continuous nonconstant activation function. Then the finite sums of finite products of the form*

$$G(x) = \sum_{k=1}^{M} \beta_k \prod_{j=1}^{N_k} \varphi(w_{jk}^T x + \theta_{jk}), \qquad (9.3.1)$$

with $w_{jk} \in \mathbb{R}^n, \beta_j, \theta_{jk} \in \mathbb{R}$, $x \in I_n$, $M, N_k = 1, 2, \ldots$, are dense in $C(I_n)$.

Proof: Consider the set $\mathcal{U}$ given by the finite sums of products of the type (9.3.1). Since φ is continuous, then for any $G, G_1, G_2 \in \mathcal{U}$, we obviously have

$$G_1 + G_2 \in \mathcal{U}, \qquad G_1 G_2 \in \mathcal{U}, \qquad \lambda G \in \mathcal{U}, \quad \forall \lambda \in \mathbb{R},$$

which shows that $\mathcal{U}$ is an algebra of real continuous functions on I_n. Next we shall verify the conditions in the hypothesis of Stone-Weierstrass theorem, see Theorem 7.4.4:

• $\mathcal{U}$ *separates points of I_n.* Let $x, y \in I_n$, $x \neq y$. We need to show that there is $G \in \mathcal{U}$ such that $G(x) \neq G(y)$.

Since φ is nonconstant, there are $a, b \in \mathbb{R}$, $a \neq b$, with $\varphi(a) \neq \varphi(b)$. Pick a point x and a point y in the hyperplanes $\{w^T x + \theta = a\}$ and $\{w^T x + \theta = b\}$, respectively. Then $G(u) = \varphi(w^T u + \theta)$ separates x and y:

$$\begin{aligned} G(x) &= \varphi(w^T x + \theta) = \varphi(a) \\ G(y) &= \varphi(w^T y + \theta) = \varphi(b) \neq \varphi(a). \end{aligned}$$

• $\mathcal{U}$ *contains nonzero constants.* Let θ be such that $\varphi(\theta) \neq 0$, and choose the vector $w^T = (0, \ldots, 0) \in \mathbb{R}^n$. Then $G(x) = \varphi(w^T x + \theta) = \varphi(\theta) \neq 0$ is a nonzero constant. Multiplying by any nonzero real number λ we obtain that all nonzero constants are contained in $\mathcal{U}$.

Applying now Stone-Weierstrass Theorem, it follows that $\mathcal{U}$ is dense in $C(I_n)$, which ends the proof of the theorem. ∎

Remark 9.3.9 1. Since Stone-Weierstrass Theorem holds in general on compact spaces, replacing the input space I_n by any compact set K in $\mathbb{R}^n$ (i.e., a bounded, closed set) does not affect the conclusion of the theorem.

2. This approach is shorter than the one using the Hahn-Banach Theorem used in the proof of Theorem 9.3.6, but it cannot be extended to Σ-class approximators.

9.3.2   Learning square integrable functions $f \in L^2(I_n)$

It makes sense to make the network learn signals f of finite energy on I_n, i.e., functions for which

$$\int_{I_n} f(x)^2 \, dx < \infty,$$

where dx denotes the Lebesgue measure on I_n. The space of functions with this property is denoted by $L^2(I_n)$. This is a normed linear space with the norm $\|f\|_2 = \left(\int_{I_n} f(x)^2 \, dx \right)^{1/2}$. The interaction between two signals $f, g \in L^2(I_n)$ is measured by their scalar product $\langle f, g \rangle = \int_{I_n} f(x)g(x) \, dx$. Two signals f, g, which do not interact have a zero scalar product, $\langle f, g \rangle = 0$, case in which they are called *orthogonal*; in this case we adopt the notation $f \perp g$.

Consider an activation function σ with the property $0 \le \sigma \le 1$. Then $h(x) = \sigma(w^T x + \theta) \in L^2(I_n)$, for any $w \in \mathbb{R}^n$ and $\theta \in \mathbb{R}$, since

$$\int_{I_n} h^2(x) \, dx = \int_{I_n} \sigma(w^T x + \theta)^2 \, dx \le \int_{I_n} dx = 1.$$

In the following we shall carry over the approximation theory from $C(I_n)$ as much as possible. Consider all finite sums of the type

$$\mathcal{U} = \{G; \ G(x) = \sum_{j=1}^{N} \alpha_j \sigma(w_j^T x + \theta_j), w_j \in \mathbb{R}^n, \alpha_j, \theta_j \in \mathbb{R}, N = 0, 1, 2, \dots \}$$

and notice that $\mathcal{U}$ is a linear subspace of $L^2(I_n)$. We are looking for a property of σ such that $\mathcal{U}$ becomes dense in $L^2(I_n)$.

The following discussion supports the next definition, Definition 9.3.10. We start by assuming that $\mathcal{U}$ is dense in $L^2(I_n)$, with respect to the L^2-norm. Consider $g \in L^2(I_n)$, such that $g \perp \mathcal{U}$ (i.e., g is orthogonal on all elements of the space $\mathcal{U}$). Then there is a sequence $(G_n)_n$ in $\mathcal{U}$ with $G_n \to g$ in L^2, as $n \to \infty$. Since $g \perp G_n$, then $\langle g, G_n \rangle = 0$, and using the continuity of the scalar product, taking the limit, yields $\langle g, g \rangle = \|g\|^2 = 0$. Hence $g = 0$, almost everywhere.

The relation $g \perp \mathcal{U}$ can be written equivalently as

$$\int_{I_n} \sigma(w^T x + \theta)g(x)\,dx = 0, \quad \forall w \in \mathbb{R}^n, \theta \in \mathbb{R}. \qquad (9.3.2)$$

We have shown that relation this implies $g = 0$ a.e. The conclusion of this discussion is the following: if a finite energy signal, g, does not interact with any of the neural outputs, $\sigma(w^T x + \theta)$ (i.e., relation (9.3.2) holds), then g must be an almost everywhere zero signal. We shall use this as our desired property for the activation function σ.

Definition 9.3.10 *The activation function σ is called discriminatory in L^2-sense if:*
(i) $0 \le \sigma \le 1$
(ii) if $g \in L^2(I_n)$ such that

$$\int_{I_n} \sigma(w^T x + \theta)g(x)\,dx = 0, \quad \forall w \in \mathbb{R}^n, \theta \in \mathbb{R},$$

then $g = 0$.

We shall provide next the analog L^2-version of Proposition 9.3.5.

Proposition 9.3.11 *Let σ be discriminatory function in L^2-sense. Then the finite sums of the form*

$$G(x) = \sum_{j=1}^{N} \alpha_j \sigma(w_j^T x + \theta_j), \qquad w_j \in \mathbb{R}^n, \alpha_j, \theta_j \in \mathbb{R}$$

are dense in $L^2(I_n)$.

Proof: By contradiction, assume the previously defined linear subspace $\mathcal{U}$ is not dense in $L^2(I_n)$. From Lemma 9.3.2 there is a bounded linear functional F on $L^2(I_n)$ such that $F \neq 0$ on $L^2(I_n)$ and $F_{|\mathcal{U}} = 0$. By Riesz theorem of representation, see Theorem E.5.1 in Appendix, there is $g \in L^2(I_n)$ such that

$$F(f) = \int_{I_n} f(x)g(x)\,dx,$$

with $\|F\| = \|g\|_2$. The condition $F_{|\mathcal{U}} = 0$ implies

$$\int_{I_n} G(x)g(x)\,dx = 0, \qquad \forall G \in \mathcal{U}.$$

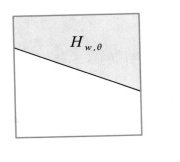

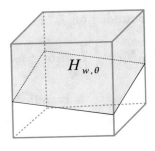

a   b

Figure 9.3: *The upper half-space for:* **a.** *the unit square I_2.* **b.** *the unit cube* I_3.

In particular,

$$\int_{I_n} \sigma(w^T x + \theta) g(x)\, dx = 0.$$

Using that σ is discriminatory in L^2-sense, it follows that $g = 0$. This triggers $\|F\| = \|g\|_2 = 0$, which implies that $F = 0$, contradiction. ∎
It follows that $\mathcal{U}$ is dense in $L^2(I_n)$.

The meaning of the previous proposition is: $\forall h \in L^2(I_n)$, $\forall \epsilon > 0$, there is G of the previous form such that

$$\int_{I_n} |G(x) - h(x)|^2\, dx < \epsilon.$$

Or, equivalently, any L^2-function with the input in the hypercube I_n can be learned by a one-hidden layer network within a mean square error as small as we desire.

The only concern left is whether the usual activation functions introduced in Chapter 2 are discriminatory in L^2-sense. We shall deal with this issue in the following. First, we develop a result, which is an L^2-analog of Lemma 2.2.3.

Lemma 9.3.12 *Let $g \in L^2(I_n)$ such that $\int_{\mathcal{H}_{w,\theta}} g(x)\, dx = 0$, for any half-space $\mathcal{H}_{w,\theta} = \{x;\ w^T x + \theta > 0\} \cap I_n$. Then $g = 0$ almost everywhere.*

Proof: Let $g \in L^2(I_n)$ and $w \in \mathbb{R}^2$ be fixed and consider the linear functional $F : L^\infty(\mathbb{R}) \to \mathbb{R}$ defined by

$$F(h) = \int_{I_n} h(w^T x) g(x)\, dx,$$

with h bounded on I_n. Since

$$|F(h)| = \left| \int_{I_n} h(w^T x) g(x) \, dx \right| \leq \left| \int_{I_n} g(x) \, dx \right| \|h\|_\infty \leq \|g\|_2 \|h\|_\infty,$$

it follows that F is bounded, with $\|F\| \leq \|g\|_2$.

Evaluating F on the indicator function $h(u) = 1_{(\theta,\infty)}(u)$, and using the hypothesis, yields

$$F(h) = \int_{I_n} 1_{(\theta,\infty)}(w^T x) g(x) \, dx = \int_{\mathcal{H}_{w,-\theta}} g(x) \, dx = 0.$$

Similarly, F vanishes on $h(u) = 1_{[\theta,\infty)}(u)$, since the hyperplane $\{w^T x - \theta = 0\}$ has zero measure. By linearity, F vanishes on combinations of indicator functions, such as 1_A, with A interval, and then vanishes on finite sums of these types of functions, i.e., on simple functions. Namely, if $s = \sum_{j=1}^{N} \alpha_j 1_{A_j}$, with $\{A_j\}_j$ disjoint intervals in $\mathbb{R}$, then $F(s) = 0$. Using that simple functions are dense in $L^\infty(\mathbb{R})$, then

$$F(h) = 0, \qquad \forall h \in L^\infty(\mathbb{R}).$$

This relation holds in particular for the bounded functions $s(u) = \sin u$ and $c(u) = \cos u$, i.e., $F(s) = 0$ and $F(c) = 0$. This way we are able to compute the Fourier transform of g

$$\begin{aligned} \widehat{g}(w) &= \int_{I_n} e^{iw^T x} g(x) \, dx = \int_{I_n} \cos(w^T x) g(x) \, dx + i \int_{I_n} \sin(w^T x) g(x) \, dx \\ &= F(c) + iF(s) = 0. \end{aligned}$$

Since the Fourier transform is one-to-one, it follows that $g = 0$, almost everywhere (For the definition of the Fourier transform, see Appendix, section D.6.4).

■

Remark 9.3.13 It is worth noting that by choosing a convenient value for the parameter θ, the upper half-space, see Fig. 9.3, may become the entire hypercube. Then the function g considered before has a zero integral, $\int_{I_n} g(x) \, dx = 0$.

Now we shall provide examples of familiar activation functions, which are discriminatory in L^2-sense. The one-hidden layer neural networks having this type of activation functions can learn finite energy signals.

Example 9.3.14 The step function $\varphi(x) = H(x) = \begin{cases} 1, & \text{if } x \geq 0 \\ 0, & x < 0 \end{cases}$ is discriminatory in L^2-sense.

The condition $0 \leq \varphi \leq 1$ is obvious. In order to show the rest, we shall use Lemma 9.3.12. Let $g \in L^2(I_n)$ and assume

$$\int_{I_n} \varphi(w^T x + \theta) g(x) \, dx = 0, \qquad \forall w \in \mathbb{R}^n, \theta \in \mathbb{R},$$

which becomes

$$\int_{\mathcal{H}_{w,-\theta}} g(x) \, dx = 0, \qquad \forall w \in \mathbb{R}^n, \theta \in \mathbb{R},$$

which implies $g = 0$, a.e., by Lemma 9.3.12. Hence, the aforementioned step function, $H(x)$, is discriminatory in L^2-sense.

Consequently, Proposition 9.3.11 implies that any finite energy function $g \in L^2(I_n)$ can be approximated by the output of a one-hidden layer perceptron, with enough perceptrons in the hidden layer.

Example 9.3.15 The logistic function $\sigma(x) = \dfrac{1}{1 + e^{-x}}$ is discriminatory in L^2-sense.

In order to show this, we set $g \in L^2(I_n)$ and assume

$$\int_{I_n} \sigma(w^T x + \theta) g(x) \, dx = 0, \qquad \forall w \in \mathbb{R}^n, \theta \in \mathbb{R}.$$

Let $\sigma_\lambda(x) = \sigma\big(\lambda(w^T x + \theta)\big)$. Then also

$$\int_{I_n} \sigma_\lambda(x) g(x) \, dx = 0, \qquad \forall \lambda \in \mathbb{R},$$

or, equivalently, $\langle \sigma_\lambda, g \rangle = 0$. Notice that $\lim_{\lambda \to \infty} \sigma_\lambda = \gamma$ pointwise, where

$$\gamma(x) = \begin{cases} 1, & \text{if } w^T x + \theta > 0 \\ 0, & \text{if } w^T x + \theta < 0 \\ \frac{1}{2}, & \text{if } w^T x + \theta = 0. \end{cases}$$

Assume for the time being that $\lim_{\lambda \to \infty} \sigma_\lambda = \gamma$ in L^2-sense. By the continuity of the scalar product we have

$$0 = \lim_{\lambda \to \infty} \langle \sigma_\lambda, g \rangle = \langle \gamma, g \rangle = \int_{I_n} \gamma(x) g(x) \, dx = \int_{\mathcal{H}_{w,\theta}} g(x) \, dx.$$

Since this relation holds for any w and θ, in the virtue of Lemma 9.3.12 it follows that $g = 0$ a.e., which proves that σ is discriminatory in L^2-sense.

We are still left to show that $\sigma_\lambda \to \gamma$ in L^2-sense, as $\lambda \to \infty$, which is equivalent to

$$\lim_{\lambda \to \infty} \int_{I_n} |\sigma_\lambda(x) - \gamma(x)|^2 \, dx = 0.$$

Computing the difference piecewise, the integrand becomes by straightforward computation

$$|\sigma_\lambda(x) - \gamma(x)|^2 = \begin{cases} \dfrac{1}{(1 + e^{\lambda(w^T x + \theta)})^2}, & \text{if } w^T x + \theta > 0 \\ \dfrac{1}{(1 + e^{-\lambda(w^T x + \theta)})^2}, & \text{if } w^T x + \theta < 0 \\ 0, & \text{if } w^T x + \theta = 0. \end{cases}$$

Then

$$\begin{aligned} \lim_{\lambda \to \infty} \int_{I_n} |\sigma_\lambda(x) - \gamma(x)|^2 \, dx &= \lim_{\lambda \to \infty} \int_{\mathcal{H}_{w,\theta}} \frac{1}{\left(1 + e^{\lambda(w^T x + \theta)}\right)^2} \, dx \\ &\quad + \lim_{\lambda \to \infty} \int_{\mathcal{H}_{-w,-\theta}} \frac{1}{\left(1 + e^{-\lambda(w^T x + \theta)}\right)^2} \, dx \\ &= 0, \end{aligned}$$

since the last two limits vanish as an application of the Dominated Convergence Theorem (Theorem C.4.2 in Appendix). Therefore, the logistic function is discriminatory in the L^2-sense.

Consequently, Proposition 9.3.11 implies that any finite energy function $g \in L^2(I_n)$ can be approximated by the output of a one-hidden layer neural network, with enough logistic sigmoid neurons in the hidden layer.

9.3.3 Learning integrable functions $f \in L^1(I_n)$

The previous theory can be carried over to approximating integrable functions $f \in L^1(I_n)$. The discriminatory property for the sigmoidal in this case is given by:

Definition 9.3.16 *The activation function σ is called discriminatory in L^1-sense if:*
(i) σ *is measurable and bounded;*
(ii) σ *is sigmoidal in the sense of Definition 2.2.1;*
(iii) if $g \in L^\infty(I_n)$ *such that*

$$\int_{I_n} \sigma(w^T x + \theta) g(x) \, dx = 0, \quad \forall w \in \mathbb{R}^n, \theta \in \mathbb{R},$$

then $g = 0$.

The L^1-version of Proposition 9.3.5 takes the following form:

Proposition 9.3.17 *Let σ be discriminatory function in L^1-sense. Then the finite sums of the form*

$$G(x) = \sum_{j=1}^{N} \alpha_j \sigma(w_j^T x + \theta_j), \qquad w_j \in \mathbb{R}^n, \alpha_j, \theta_j \in \mathbb{R}$$

are dense in $L^1(I_n)$.

Proof: The proof is similar with the one given for Proposition 9.3.5. We assume that the linear subspace of $L^1(I_n)$

$$\mathcal{U} = \{G;\ G(x) = \sum_{j=1}^{N} \alpha_j \sigma(w_j^T x + \theta_j), w_j \in \mathbb{R}^n, \alpha_j, \theta_j \in \mathbb{R}, N = 0, 1, 2, \dots \}$$

is not dense in $L^1(I_n)$. Applying Lemma 9.3.2, there is a bounded linear functional $F : L^1(I_n) \to \mathbb{R}$ such that $F \neq 0$ on $L^1(I_n)$ and $F_{|\mathcal{U}} = 0$. Using that $L^\infty(I_n)$ is the dual of $L^1(I_n)$, Riesz theorem (Theorem E.5.2 in Appendix) supplies the existence of $g \in L^\infty(I_n)$ such that

$$F(f) = \int_{I_n} f(x)g(x)\, dx,$$

with $\|F\| = \|g\|_1 \neq 0$. Now, the condition $F_{|\mathcal{U}} = 0$ writes as

$$\int_{I_n} G(x)g(x)\, dx = 0, \qquad \forall G \in \mathcal{U},$$

which implies

$$\int_{I_n} \sigma(w^T x + \theta)g(x)\, dx = 0.$$

Since σ is discriminatory in L^1-sense, we obtain $g = 0$, which contradicts $\|F\| \neq 0$. It follows that $\mathcal{U}$ is dense in $L^1(I_n)$. ∎

Since for any $g \in L^\infty(I_n)$ the measure $\mu(x) = g(x)dx$ belongs to the space of signed measures $M(I_n)$, using Remark 2.2.5 of Proposition 2.2.4 it follows that any measurable, bounded sigmoidal function is discriminatory in the L^1-sense. As a consequence of Proposition 9.3.17, we have:

Theorem 9.3.18 *Let σ be any measurable, bounded sigmoidal function (in the sense of Definition 2.2.1). Then the finite sums of the form*

$$G(x) = \sum_{j=1}^{N} \alpha_j \sigma(w_j^T x + \theta_j), \qquad w_j \in \mathbb{R}^n, \alpha_j, \theta_j \in \mathbb{R}$$

are dense in $L^1(I_n)$. This means that $\forall f \in L^1(I_n)$ and $\forall \epsilon > 0$, there is a function G of the previous form such that

$$\|G - f\|_{L^1} = \int_{I_n} |G(x) - f(x)| \, dx < \epsilon.$$

Remark 9.3.19 1. In particular, a one-hidden layer neural network with logistic activation function for the hidden layer and a linear neuron in the output layer can learn functions in $L^1(I_n)$.

2. Using the Wiener Tauberian Theorem and properties of Fourier transform, it can be shown that any function σ in $L^1(\mathbb{R})$ with a nonzero integral, $\int_{\mathbb{R}} \sigma(x) \, dx \neq 0$, is discriminatory in L^1-sense.

3. Consequently, the conclusion of Theorem 9.3.18 is still valid if the activation function σ is in $L^1(\mathbb{R})$ and has a nonzero integral.

9.3.4   Learning measurable functions $f \in \mathcal{M}(\mathbb{R}^n)$

A function $y = f(x)$ is *measurable* if it is the result of an observation that can be determined by a given body of information. This is usually the information that characterizes intervals in $\mathbb{R}$, squares in $\mathbb{R}^2$, cubes in $\mathbb{R}^3$, and so on; it is customarily denoted by $\mathcal{B}(\mathbb{R}^n)$ and is called the Borel $\mathfrak{S}$-field on $\mathbb{R}^n$. Consequently, all inputs x, which are mapped by f into an interval $[a-\epsilon, a+\epsilon]$, are known given the information $\mathcal{B}(\mathbb{R}^n)$. This fact can be written equivalently using the pre-image notation as

$$f^{-1}([a - \epsilon, a + \epsilon]) \subset \mathcal{B}(\mathbb{R}^n), \qquad \forall a \in \mathbb{R}, \epsilon > 0.$$

A measurable function is not necessarily continuous, nevertheless a continuous function is always Borel-measurable. The reader can find more details about measurable functions in the Appendix.

In this section the target space is the space of Borel-measurable functions from $\mathbb{R}^n$ to $\mathbb{R}$, denoted by $\mathcal{M}^n = \mathcal{M}(\mathbb{R}^n)$. Learning a measurable function will be done "almost everywhere" with respect to a certain measure, which will be introduced next.

Let μ be a probability measure on the measurable space $(\mathbb{R}^n, \mathcal{B}(\mathbb{R}))$. This can be regarded as the input pattern of the input variable x. It is natural not to distinguish between two measurements (measurable functions f and g) that differ on a negligible subset of the input pattern (For instance, looking at a circle which is missing a few points, we can still infer that it is a circle). Having this in mind, we say that two measurable functions, f and g, are *μ-equivalent* if

$$\mu\{x \in \mathbb{R}^n; f(x) \neq g(x)\} = 0.$$

Sometimes, we write $f = g$ a.e. (almost everywhere). For the sake of simplicity, identifying equivalent functions, the space of equivalence classes will be denoted also by $\mathcal{M}^n$. We make the remark that all measurable functions in this section are considered to be finite almost everywhere, i.e., $\mu\{x; |f(x)| = \infty\} = 0$.

The next metric will measure the distance between equivalence classes of measurable functions. More precisely, f and g are situated in a proximity if and only if it is a small probability that the differ significantly. The corresponding distance is defined by

$$d_\mu(f,g) = \inf\{\epsilon > 0; \mu(|f-g| > \epsilon) < \epsilon\}, \qquad \forall f,g \in \mathcal{M}^n. \tag{9.3.3}$$

The previous assumptions that f and g are finite a.e. make sense here, because the difference $f - g$ might be infinite only on a negligible set. We note that $d_\mu(f,g) = 0$ is equivalent to $f = g$ a.e.; the symmetry of d_μ is obvious and the triangle inequality is left as an exercise to the reader. Therefore, $(\mathcal{M}^n, d_\mu)$ becomes a metric space.

The next goal is to find a d_μ-dense subset, which can be represented by outputs of one-hidden layer neural networks. This will be done in a few stages. We shall start with a few equivalent convergence properties of the metric d_μ needed later in the density characterization. Recall the notation for the minimum of two numbers, $x \wedge y = \min(x, y)$.

Lemma 9.3.20 *Let $f_j, g \in \mathcal{M}^n$. The following statements are equivalent:*

(a) $d_\mu(f_j, f) \to 0$, $j \to \infty$;
(b) *For any* $\epsilon > 0$, $\mu\{x; |f_j(x) - f(x)| > \epsilon\} \to 0$, $j \to \infty$;
(c) $\int_{\mathbb{R}^n} |f_j(x) - f(x)| \wedge 1 \, d\mu(x) \to 0$, $j \to \infty$.

Proof: We shall show that $(a) \Leftrightarrow (b) \Leftrightarrow (c)$.

$(a) \Rightarrow (b)$ Assume that (a) holds. From the definition of d_μ, for any $\epsilon > 0$ that satisfies $\mu(|f_j - f| > \epsilon) < \epsilon$, we have $d_\mu(f_j, f) \leq \epsilon$. Since $d_\mu(f_j, f) \to 0$, we may choose a sequence $\epsilon = \frac{1}{n_j} \to 0$, such that

$$\mu\left(|f_j - f| > \frac{1}{n_j}\right) < \frac{1}{n_j}. \tag{9.3.4}$$

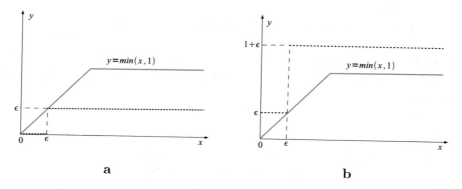

Figure 9.4: *Graphical interpretation for two inequalities:*
a. $\epsilon 1_{(\epsilon,+\infty)}(x) \leq x \wedge 1$. **b.** $x \wedge 1 \leq \epsilon + 1_{(\epsilon,+\infty)}(x)$.

By contradiction, assume that (b) does not hold, so there is an $\epsilon_0 > 0$ such that

$$\mu\{|f_j - f| > \epsilon\} > \epsilon_0.$$

Choosing $\epsilon = \frac{1}{n_j}$, we arrive at a relation that contradicts (9.3.4).

$(b) \Rightarrow (a)$ If (b) holds, then obviously (a) holds.

$(b) \Leftrightarrow (c)$. Let $\epsilon \in (0,1)$, fixed. The following double inequality can be easily inferred from Fig. 9.4 **a,b**:

$$\epsilon 1_{(\epsilon,+\infty)}(x) \leq x \wedge 1 \leq \epsilon + 1_{(\epsilon,+\infty)}(x), \qquad x \geq 0. \tag{9.3.5}$$

Replacing in (9.3.5) the variable x by $|f_j(x) - f(x)|$, integrating with respect to μ and using that

$$\int_{\mathbb{R}^n} 1_{|f_j(x)-f(x)|>\epsilon}\, d\mu(x) = \mu\{x; |f_j(x) - f(x)| > \epsilon\},$$

we obtain

$$\epsilon\,\mu\{x; |f_j(x) - f(x)| > \epsilon\} \leq \int_{\mathbb{R}^n} |f_j(x) - f(x)| \wedge 1\, d\mu(x) \tag{9.3.6}$$

$$\epsilon + \mu\{x; |f_j(x) - f(x)| > \epsilon\} \geq \int_{\mathbb{R}^n} |f_j(x) - f(x)| \wedge 1\, d\mu(x). \tag{9.3.7}$$

Now, if (c) holds, inequality (9.3.6) implies (b).
If (b) holds, then inequality (9.3.7) provides

$$\epsilon \geq \lim_j \int_{\mathbb{R}^n} |f_j(x) - f(x)| \wedge 1\, d\mu(x), \quad \forall \epsilon > 0,$$

which means that the limit on the right is zero; hence (c) holds. ∎

Next we shall make a few remarks.

1. Statement (a) represents the sequential convergence in the d_μ-metric.

2. Condition (b) states the convergence in probability of f_j to f. Assume one plays darts and $f_j(x)$ is the position shot in the jth trial given the initial condition (input) x. The set $\{x; |f_j(x) - f(x)| > \epsilon\}$ represents all inputs x which produce shots outside the disk of radius ϵ, centered at the target $f(x)$. Consequently, condition (b) states that the chances (probability measured using μ) that in the long run the dart eventually hits inside any circular target increase to 1.

3. Part (c) can be written using expectation as $\mathbb{E}[|f_j(x) - f(x)| \wedge 1] \to 0$. Since

$$\mathbb{E}[|f_j(x) - f(x)| \wedge 1] \le \mathbb{E}[|f_j(x) - f(x)|]$$

it follows that if $f_j \to f$ in L^1-sense, then (c) holds. Hence, the L^1-convergence implies convergence in probability. If we define

$$\rho_\mu(f, g) = \mathbb{E}[|f - g| \wedge 1] = \int_{\mathbb{R}^n} |f(x) - g(x)| \wedge 1 \, d\mu(x),$$

then ρ_μ is a distance on $\mathcal{M}^n$. The previous result states that the topologies defined by the distance functions d_μ and ρ_μ are equivalent. Consequently, a set is d_μ-dense in $\mathcal{M}^n$ if and only if it is ρ_μ-dense.

The uniform convergence and the d_μ-convergence are related by the following result:

Proposition 9.3.21 *Let $(f_j)_j$ be a sequence of functions in $\mathcal{M}^n$ that converges uniformly on compacta to f, i.e., $\forall K \subset \mathbb{R}^n$ compact,*

$$\sup_{x \in K} |f_j(x) - f(x)| \to 0, \; j \to \infty.$$

Then $d_\mu(f_j, f) \to 0$, as $j \to \infty$.

Proof: By Lemma 9.3.20, part (c), it suffices to shows that $\forall \epsilon > 0$, $\exists N \ge 1$ such that

$$\int_{\mathbb{R}^n} |f_j(x) - f(x)| \wedge 1 \, d\mu(x) < \epsilon, \quad \forall j \ge N. \tag{9.3.8}$$

This will be achieved by splitting the integral in two parts, each being smaller than $\epsilon/2$.

Step 1. We shall construct a compact set K such that $\mu(\mathbb{R}^n \backslash K) < \dfrac{\epsilon}{2}$.

Denote by $B(0, k)$ the closed Euclidean ball of radius k, centered at the origin. Then $\mathbb{R}^n$ can be written as a limit of an ascending sequence of compact sets as

$$\mathbb{R}^n = \bigcup_{k \geq 1} B(0, k) = \limsup_k B(0, k).$$

From the sequential continuity of the probability measure μ we get

$$1 = \mu(\mathbb{R}^n) = \lim_{k \to \infty} \mu(B(0, k)).$$

Therefore, for k large enough, $\mu(B(0, k)) > 1 - \epsilon/2$. Consider the compact $K = B(0, k)$, and note that $\mu(\mathbb{R}^n \backslash K) < \epsilon/2$.

Step 2. Using Step 1 we have the estimation:

$$\int_{\mathbb{R}^n \backslash K} |f_j(x) - f(x)| \wedge 1 \, d\mu(x) \leq \int_{\mathbb{R}^n \backslash K} 1 \, d\mu(x) = \mu(\mathbb{R}^n \backslash K) < \frac{\epsilon}{2}.$$

Since $\sup_{x \in K} |f_j(x) - f(x)| \to 0$, we can find $N \geq 1$ such that

$$\sup_{x \in K} |f_j(x) - f(x)| < \frac{\epsilon}{2}, \quad \forall j \geq N.$$

Therefore

$$\int_K |f_j(x) - f(x)| \wedge 1 \, d\mu(x) \leq \int_K \sup_{x \in K} |f_j(x) - f(x)| \, d\mu(x) \leq \frac{\epsilon}{2} \mu(K) < \frac{\epsilon}{2}.$$

Combining the last two inequalities yields

$$\int_{\mathbb{R}^n} |f_j(x) - f(x)| \wedge 1 \, d\mu(x) = \int_K |f_j(x) - f(x)| \wedge 1 \, d\mu(x)$$

$$+ \int_{\mathbb{R}^n \backslash K} |f_j(x) - f(x)| \wedge 1 \, d\mu(x)$$

$$\leq \frac{\epsilon}{2} + \frac{\epsilon}{2} = \epsilon, \quad \forall j \geq N,$$

which proves the desired relation (9.3.8).   ∎

The next result extends Theorem 9.3.6 and will be used in the proof of the main result shortly.

Theorem 9.3.22 *Let $\sigma : \mathbb{R} \to \mathbb{R}$ be any arbitrary continuous sigmoidal function (in the sense of Definition 2.2.1). Then the finite sums of the form*

$$G(x) = \sum_{j=1}^{N} \alpha_j \sigma(w_j^T x + \theta_j), \quad w_j \in \mathbb{R}^n, \alpha_j, \theta_j \in \mathbb{R}, N = 1, 2, \ldots$$

are uniformly dense on compacta in $C(\mathbb{R}^n)$. This means that for any given function $f \in C(\mathbb{R}^n)$, there is a sequence $(G_m)_m$ of functions of the previous type such that $G_m \to f$ uniformly on any compact set K in $\mathbb{R}^n$, as $m \to \infty$.

Proof: We shall follow a few steps.

Step 1: Show that if $f \in C(\mathbb{R}^n)$ and $K \subset \mathbb{R}^n$ is a compact set, then there is a sequence $(G_m^{(K)})_m$ of functions of the previous type that converges uniformly to f on K.

We note that Theorem 9.3.6 works also in the slightly general case when $C(I_n)$ is replaced by $C(K)$, with $K \subset \mathbb{R}^n$ compact set, the proof being similar. This means that for any fixed compact set K, the finite sums of the form

$$G(x) = \sum_{j=1}^N \alpha_j \sigma(w_j^T x + \theta_j), \qquad w_j \in \mathbb{R}^n, \alpha_j, \theta_j \in \mathbb{R}$$

are dense in $C(K)$. Since for any $f \in C(\mathbb{R}^n)$ we have $f_{|K} \in C(K)$, then there is a sequence $(G_m^{(K)})_m$ of functions of the previous type such that $G_m^{(K)} \to f$ uniformly on K.

Step 2: There is an ascending sequence of compact sets, $(K_j)_j$, such that $K_j \subset K_{j+1}$ and $K_j \nearrow \mathbb{R}^n$, $j \to \infty$.

It suffices to define the sequence as $K_j = \{x \in \mathbb{R}^n; \|x\| \leq j\}$.

Step 3: Use a diagonalization procedure to find the desired sequence.

For each compact set K_j defined by *Step 2*, we consider the sequence $(G_m^{(j)})_m$ defined by *Step 1*, so $G_m^{(j)} \to f$ on K_j, as $m \to \infty$. Applying this result on each compact K_j, we obtain the following table of sequences:

$$
\begin{array}{ccccccc}
G_1^{(1)} & G_2^{(1)} & G_3^{(1)} & G_4^{(1)} & \cdots & \to f \text{ on } K_1 \\
G_1^{(2)} & G_2^{(2)} & G_3^{(2)} & G_4^{(2)} & \cdots & \to f \text{ on } K_2 \\
G_1^{(3)} & G_2^{(3)} & G_3^{(3)} & G_4^{(3)} & \cdots & \to f \text{ on } K_3 \\
G_1^{(4)} & G_2^{(4)} & G_3^{(4)} & G_4^{(4)} & \cdots & \to f \text{ on } K_4 \\
\cdots & \cdots & \cdots & \cdots & \cdots & \cdots & \cdots
\end{array}
$$

The sequence in each row tends to f uniformly, on a given compact set. Since the compacts $(K_j)_j$ are nested, we have that in fact $G_m^{(p)} \to f$, uniformly on any K_j, with $j \leq p$. Using a diagonalization procedure, we construct the sequence $G_k = G_k^{(k)}$, $k \geq 1$. We shall show next that for any K compact we have $G_k \to f$, uniformly on K.

For this we select the smallest integer j_0 such that we have the inclusions $K \subset K_{j_0} \subset K_{j_0+1} \subset K_{j_0+2} \subset \cdots$. Using the previous property, the sequence $(G_k)_{k \geq j_0}$ converges to f, uniformly on K_{j_0}, and hence on K. ∎

The following result states that any measurable function can be approximated by a continuous function in the probability sense.

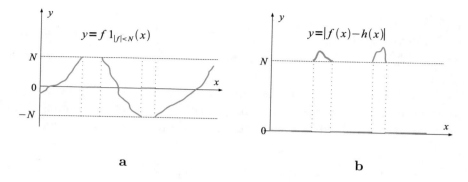

Figure 9.5: **a.** *The graph of $y = h(x)$.* **b.** *The graph of $y = |f(x) - h(x)|$.*

Proposition 9.3.23 *The set of continuous functions $C(\mathbb{R}^n)$ is d_μ-dense in $\mathcal{M}^n$.*

Proof: Let $f \in \mathcal{M}^n$ be a measurable function. Fix $\epsilon > 0$. We need to show that there is a continuous function $g \in C(\mathbb{R}^n)$ such that $d_\mu(f, g) < \epsilon$. This shall be achieved in a few steps:

Step 1. Show that there is an N large enough such that

$$\mu\{x; |f(x)| > N\} < \epsilon/2.$$

For this we consider $A_n = \{x; |f(x)| > n\}$. Since $A_{n+1} \subset A_n$ and $A_n = f^{-1}(n, +\infty) \cup f^{-1}(-\infty, -n) \in \mathcal{B}(\mathbb{R}^n)$, then $(A_n)_n$ is a descending sequence of measurable sets with limit $A_n \searrow \bigcap_n A_n = \{x; |f(x)| = \infty\}$. From the sequential continuity and the fact that f is finite a.e. we get $\mu(A_n) \to 0$, $n \to \infty$. Therefore, we can choose N large enough such that $\mu\{|f| > N\} < \epsilon/2$.

Step 2. Show that $\rho_\mu(f, h) < \dfrac{\epsilon}{2}$, where $h(x) = f(x)\,1_{|f|<N}(x)$.

This means

$$\int_{\mathbb{R}^n} |f(x) - h(x)| \wedge 1 \, d\mu(x) < \frac{\epsilon}{2}.$$

If $1_{|f|<N}$ denote the indicator function of the set $\{x; |f(x)| < N\}$, then

$$h(x) = f1_{|f|<N}(x) = \begin{cases} f(x), & \text{if } |f(x)| < N \\ 0, & \text{if } |f(x)| \geq N, \end{cases}$$

see Fig. 9.5 **a.** The difference becomes

$$|f(x) - h(x)| = \begin{cases} 0, & \text{if } |f(x)| < N \\ |f(x)|, & \text{if } |f(x)| \geq N, \end{cases}$$

see Fig. 9.5 **b**. Note that by Step 1, this is nonzero only on a set with measure arbitrary small. Integrating, using that $N > 1$ and the inequality obtained at Step 1, we obtain

$$\int_{\mathbb{R}^n} |f - h| \wedge 1 \, d\mu = \int_{|f|<N} |f - h| \wedge 1 \, d\mu + \int_{|f|\geq N} |f - h| \wedge 1 \, d\mu$$

$$= \int_{|f|<N} 0 \wedge 1 \, d\mu + \int_{|f|\geq N} |f| \wedge 1 \, d\mu$$

$$= \int_{|f|\geq N} 1 \, d\mu = \mu\{|f| > N\} < \frac{\epsilon}{2}.$$

Step 3. There is a continuous function, $g \in C(\mathbb{R}^n)$, with $\rho_\mu(h, g) < \frac{\epsilon}{2}$.

The function $h(x) = f1_{|f|<N}(x)$ is measurable and bounded, with $|h(x)| < N$. It is known that $h(x)$ is the limit of a sequence of simple functions $(s_k)_k$, see Appendix. We may further assume that $|s_k(x)| < N$, for k large enough. Since $\mu(\mathbb{R}^n) = 1$ and $s_k(x) \to h(x)$, the Bounded Convergence Theorem (Theorem C.4.3 in Appendix) implies

$$\int_{\mathbb{R}^n} |h(x) - s_k(x)| \wedge 1 \, d\mu(x) \to 0, \qquad k \to \infty.$$

In ϵ-language, there is a $k_0 > 1$ such that

$$\rho_\mu(h, s_k) < \frac{\epsilon}{4}, \qquad k > k_0. \tag{9.3.9}$$

Write the simple function as $s_k(x) = \sum_{i=1}^k \alpha_i 1_{A_i}(x)$ and consider compacta $K_i \subset A_i$ such that $\mu(A_i \backslash K_i) < \frac{\epsilon}{4kN}$ and the continuous functions

$$g_i(x) = \begin{cases} 1, & \text{if } x \in K_i \\ 0, & x \notin A_i. \end{cases}$$

Construct the continuous function $g(x) = \sum_{i=1}^k \alpha_i g_i(x) \in C(\mathbb{R}^n)$. We have the estimation

$$\int_{\mathbb{R}^n} |g(x) - s_k(x)| \wedge 1 \, d\mu(x) \leq \sum_{i=1}^k |\alpha_i| \int_{\mathbb{R}^n} |g_i(x) - 1_{A_i}(x)| \wedge 1 \, d\mu(x)$$

$$\leq \sum_{i=1}^k |\alpha_i| \int_{A_i \backslash K_i} 1 \, d\mu(x) = \sum_{i=1}^k |\alpha_i| \mu(A_i \backslash K_i)$$

$$\leq \frac{\epsilon}{4kN} \sum_{i=1}^k |\alpha_i| = \frac{\epsilon}{4},$$

or, equivalently,

$$\rho_\mu(g, s_k) < \frac{\epsilon}{4}, \qquad k > k_0. \tag{9.3.10}$$

Triangle inequality together with relations (9.3.9) and (9.3.10) now provides

$$\rho_\mu(h, g) \le \rho_\mu(h, s_k) + \rho_\mu(g, s_k) \le \frac{\epsilon}{4} + \frac{\epsilon}{4} = \frac{\epsilon}{2}.$$

Step 4. Finishing the proof.
Using the triangle inequality for the distance ρ_μ and the estimations proved in Step 2 and Step 3, we have

$$\rho_\mu(f, g) \le \rho_\mu(f, h) + \rho_\mu(h, g) \le \frac{\epsilon}{2} + \frac{\epsilon}{2} = \epsilon.$$

It follows that $C(\mathbb{R}^n)$ is ρ_μ-dense in $\mathcal{M}^n$. By Lemma 9.3.20, remark 3, $C(\mathbb{R}^n)$ is also d_μ-dense in $\mathcal{M}^n$, which is the desired conclusion.

∎

The following theorem is the main result of this section. It states that any measurable function on $\mathbb{R}^n$ can be learned to any desired accuracy by a one-hidden layer neural network (with a linear neuron in the output layer), regardless of the space dimension n and input probability measure μ.

Theorem 9.3.24 *Let* $\sigma : \mathbb{R} \to \mathbb{R}$ *be a continuous sigmoidal function (in the sense of Definition 2.2.1). Then for any integer n, and every probability density μ the finite sums of the form*

$$G(x) = \sum_{j=1}^{N} \alpha_j \sigma(w_j^T x + \theta_j), \qquad w_j \in \mathbb{R}^n, \alpha_j, \theta_j \in \mathbb{R}$$

are d_μ-dense in $\mathcal{M}^n$.

Proof: Let $f \in \mathcal{M}^n$ and fix $\epsilon > 0$. Consider the set of functions

$$\mathcal{U} = \{G(x) = \sum_{j=1}^{N} \alpha_j \sigma(w_j^T x + \theta_j), w_j \in \mathbb{R}^n, \alpha_j, \theta_j \in \mathbb{R}, N = 1, 2, \dots\}.$$

From Proposition 9.3.23 there is a continuous function $g \in C(\mathbb{R}^n)$ such that $d_\mu(f, g) < \epsilon/2$. By Theorem 9.3.22 there is a sequence $(G_k) \subset \mathcal{U}$ such that G_k converges uniformly on compacta to g. Applying Proposition 9.3.21 yields that $d_\mu(G_k, g) \to 0$, $k \to \infty$, so $d_\mu(G_{k_0}, g) < \epsilon/2$, for some large enough k_0. Triangle inequality now provides

$$d_\mu(f, G_{k_0}) \le d_\mu(f, g) + d_\mu(G_{k_0}, g) \le \frac{\epsilon}{2} + \frac{\epsilon}{2} = \epsilon,$$

which shows that $\mathcal{U}$ is d_μ-dense in $\mathcal{M}^n$.

∎

Remark 9.3.25 1. The previous result can be stated by saying that any one-hidden layer neural network, with sigmoidal neurons in the hidden layer and linear neuron in the output, can learn any measurable function.

2. With some tedious modifications it can be shown that the conclusion of Theorem 9.3.24 holds true even if the activation function is a squashing function, see Definition 2.3.1. Consequently, measurable functions can be learned by one-hidden layer neural networks of classical perceptrons.

3. The approximation of measurable functions will be used later when neural networks will learn random variables. In this case the measure μ is the distribution measure of the input random variable.

9.4   Error Bounds Estimation

We have seen that given a target function $f(x)$ in one of the classical spaces encountered before, $C(I_n)$, $L^1(I_n)$ or $L^2(I_n)$, there is an output function $G(x)$ produced by a one-hidden layer neural network, which is situated in a proximity of $f(x)$ with respect to a certain metric. It is expected that the larger the number of hidden units, the better the approximation of the target function. One natural question now is: *how does the number of the hidden units influence the approximation accuracy?*

The answer is given by the following result of Barron [12] that we shall discuss in the following. Assume the target function f has a Fourier representation of the form

$$f(x) = \int_{\mathbb{R}^n} e^{i\omega^T x} \widehat{f}(\omega)\, d\omega, \qquad x \in \mathbb{R}^n,$$

with $\omega \widehat{f}(\omega)$ integrable; this means the frequencies $\widehat{f}(\omega)$ decrease to zero fast enough for $\|\omega\|_1$ large, where $\|\omega\|_1 = \sum_k |\omega_k|$.

Denote $C_f = \int_{\mathbb{R}^n} \|\omega\|_1 |\widehat{f}(\omega)|\, d\omega$. The constant C_f measures the extent to which the function f oscillates. Since

$$|\partial_{x_j} f(x)| = \left| \int_{\mathbb{R}^n} \omega_j e^{i\omega^T x} \widehat{f}(\omega)\, d\omega \right| \le \int_{\mathbb{R}^n} |\omega_j|\, |\widehat{f}(\omega)|\, d\omega \le C_f,$$

then all functions f, with C_f finite, are continuously differentiable.

The accuracy of an approximation $G_k(x)$ to the target function $f(x)$ is measured in terms of the norm on $L^2(\mu, I_n)$, where μ is a probability measure on the hypercube, which describes the input pattern

$$\|f - G_k\|^2 = \int_{I_n} |f(x) - G_k(x)|^2\, d\mu(x).$$

This is also regarded as the integrated square error approximation. The rate of approximation is stated in the next result:

Theorem 9.4.1 *Given an arbitrary sigmoidal activation function (in the sense of Definition 2.2.1), a target function f with $C_f < \infty$, and a probability measure μ on I_n, then for any number of hidden units, $N \geq 1$, there is a one-hidden layer neural network with sigmoidal activation function in the hidden neurons and linear activation in the output neuron, having the output G_k such that*

$$\|f - G_k\| \leq \frac{C_f}{\sqrt{N}}.$$

Roughly speaking, for each 100-fold increase in the number of hidden units, the approximation earns an extra digit of accuracy. From this point of view, the approximation error resembles the error obtained when applying the Monte Carlo method.

9.5 Learning q-integrable functions, $f \in L^q(\mathbb{R}^n)$

For any $q \geq 1$, we define the norm $\| \cdot \|_q$ by

$$\|f\|_q = \int_{\mathbb{R}^n} |f(x)|^q \, dx.$$

We recall that the space $L^q(\mathbb{R}^n)$ is the set of Lebesgue integrable functions f on $\mathbb{R}^n$ for which $\|f\|_q < \infty$.

A function $f : \mathbb{R}^n \to \mathbb{R}$ is called *compactly supported* if there is a compact set $K \subset \mathbb{R}^n$ such that f vanishes outside of K, i.e.,

$$f(x) = 0, \quad \forall x \in \mathbb{R}^n \backslash K.$$

We shall denote the set of continuous, compactly supported functions on $\mathbb{R}^n$ by

$$C_0(\mathbb{R}^n) = \{f : \mathbb{R}^n \to \mathbb{R}; f \text{continuous, compactly suported}\}.$$

We first note the inclusion $C_0(\mathbb{R}^n) \subset L^q(\mathbb{R}^n)$. This follows from the following argument. Let K be a compact set such that $\{x; f(x) \neq 0\} \subset K$. Since f is continuous on K, then f is bounded, namely, there is a constant $M > 0$ such that $|f(x)| < M$, for all $x \in K$. The previous inclusion follows now from the estimation

$$\|f\|_q = \int_{\mathbb{R}^n} |f(x)|^q \, dx = \int_K |f(x)|^q \, dx \leq M^q vol(K) < \infty,$$

where we used that any compact set is bounded and hence has a finite volume.

In order to be able to approximate functions in $L^q(\mathbb{R}^n)$ by continuous piecewise linear functions we need two density results:

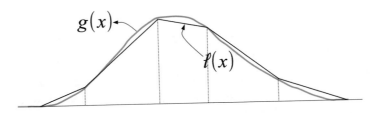

Figure 9.6: *A piecewise continuous function $\ell(x)$ approaches uniformly the compact supported continuous function $g(x)$.*

Proposition 9.5.1 *The family of compact supported continuous functions on $\mathbb{R}^n$ is dense in $L^q(\mathbb{R}^n)$.*

This means that given $f \in L^q(\mathbb{R}^n)$, then for any $\epsilon > 0$, there is $g \in C_0(\mathbb{R}^n)$ such that

$$\|f - g\|_q < \epsilon.$$

Even if there is no obvious proof for this result, we shall include next a rough argument. For more details, the reader is referred to the book [105].
If, by contradiction, $C_0(\mathbb{R}^n)$ is not dense in $L^q(\mathbb{R}^n)$, then there is a nonzero continuous linear functional $F : L^q(\mathbb{R}^n) \to \mathbb{R}$ such that $F_{|C_0(\mathbb{R}^n)} = 0$. By Riesz' theorem of representation, there is $g \in L^p(\mathbb{R}^n)$, with $1/p + 1/q = 1$, such that $F(f) = \int fg$, for all $f \in L^q(\mathbb{R}^n)$. Then $\int \phi g = 0$, for all $\phi \in C_0(\mathbb{R}^n)$, fact that implies $g = 0$, a.e. Therefore, $F = 0$ on $L^q(\mathbb{R}^n)$, which is a contradiction.
The second density result is:

Proposition 9.5.2 *The family of continuous, piecewise linear functions on $\mathbb{R}^n$ is dense in the set of compact supported continuous functions on $\mathbb{R}^n$.*

This means that given $g \in C_0(\mathbb{R}^n)$, then for any $\epsilon > 0$, there is a continuous piecewise linear function ℓ such that

$$\sup_{x \in K} |g(x) - \ell(x)| < \epsilon,$$

where K is the support of g, i.e., the closure of the set $\{x; g(x) \neq 0\}$. The proof considers a partition of K in a finite set of polyhedra and ℓ as an affine function over each polyhedron. When the partition gets refined the function ℓ gets closer to g. By a Dini-type theorem, Theorem 7.1.1, it follows the uniform convergence on K. This can be easily seen in the more simplistic one-dimensional case in Fig. 9.6.

The following result is taken from Arora et al. [10].

Theorem 9.5.3 *Any function in $L^q(\mathbb{R}^n)$, $(1 \leq q \leq \infty)$, can be arbitrarily well approximated in the $\|\cdot\|_q$ by a ReLU feedforward neural network with at most $L = 2(\lfloor \log_2 n \rfloor + 2)$ layers.*

This means that if $f \in L^q(\mathbb{R}^n)$, then for any $\epsilon > 0$ there is a ReLU-network, $\mathcal{N}_\epsilon$, such that $\|f - f_{\mathcal{N}_\epsilon}\|_q < \epsilon$.

Proof: Let $\epsilon > 0$ be fixed. By Proposition 9.5.1 there is a function $g \in C_0(\mathbb{R}^n)$ such that $\|f - g\|_q < \epsilon/2$. Let K be the support of the function g. Then by Proposition 9.5.2 there is a continuous piecewise linear function ℓ such that

$$\sup_K |g - \ell| < \left(\frac{\epsilon}{2 vol(K)} \right)^{1/q}.$$

Then integrating, we obtain $\|g - \ell\|_q < \epsilon/2$. By Proposition 10.2.7 there is a ReLU-network, $\mathcal{N}_\epsilon$, that can represent exactly the linear function ℓ, namely, $\ell = f_{\mathcal{N}_\epsilon}$. Applying the triangle inequality to the norm $\|\cdot\|_q$ we obtain

$$\|f - f_{\mathcal{N}_\epsilon}\|_q \;<\; \|f - g\|_q + \|g - \ell\|_q + \underbrace{\|\ell - f_{\mathcal{N}_\epsilon}\|_q}_{=0}$$

$$<\; \frac{\epsilon}{2} + \frac{\epsilon}{2} = \epsilon,$$

which ends the proof. ∎

Remark 9.5.4 The results of this section also hold true if the space $L^q(\mathbb{R}^n)$ is replaced by the space $L^q(K)$, with K compact set in $\mathbb{R}^n$.

9.6 Learning Solutions of ODEs

Neural networks can also learn solutions of ordinary differential equations, provided the equations are *well posed*, namely, if the following properties are satisfied:

(*i*) The solution exists;

(*ii*) The solution is unique;

(*iii*) The solution is smooth with respect to the initial data.

The case of the first-order ODEs is covered by equations of the type

$$\begin{aligned} y'(t) &= f(t, y(t)) \\ y(t_0) &= y_0, \qquad t \in [t_0, t_0 + \epsilon), \end{aligned}$$

for some $\epsilon > 0$. It is known that if $f(t, \cdot)$ is Lipschitz in the second variable, then the aforementioned ODE has a unique solution for ϵ small enough. For simplicity reasons, we assume that $f : [t_0, t_0 + \epsilon) \times \mathbb{R} \to \mathbb{R}$, namely, y is considered one-dimensional. Similar results hold if $y \in \mathbb{R}^m$, case in which the above equation becomes a system of ODEs.

We shall construct a feedforward neural network, whose input is the continuous variable $t \in [t_0, t_0 + \epsilon)$ and its output is a smooth function, $\phi_\theta(t)$, which approximates the solution $y(t)$ of the previous equation. The parameters of the network were denoted by θ.

Consider the cost function

$$
\begin{aligned}
C(\theta) &= \|\phi_\theta' - f(\cdot, \phi_\theta)\|_2^2 + \|\phi_\theta(t_0) - y_0\|_2^2 \\
&= \int_{t_0}^{t_0+\epsilon} [\phi_\theta'(t) - f(t, \phi_\theta(t))]^2 \, dt + (\phi_\theta(t_0) - y_0)^2, \quad (9.6.11)
\end{aligned}
$$

and let $\theta^* = \arg\min C(\theta)$. Then the network optimal output, ϕ_{θ^*}, starts with a value $\phi_{\theta^*}(t_0)$ close to y_0 and evolves close to the solution starting at y_0.

Since the previous cost function is difficult to implement, we shall consider the equidistant division $t_0 < t_1 < \cdots < t_n = t_0 + \epsilon$, with $\Delta t = \epsilon/n$, and associate the empirical cost function

$$
C(\theta; \{t_i\}) = \sum_{k=0}^{n-1} \left[\frac{\phi_\theta(t_{k+1}) - \phi_\theta(t_k)}{\Delta t} - f(t_k, \phi_\theta(t_k)) \right]^2 \Delta t + (\phi_w(t_0) - y_0)^2.
$$

One possible approach is to consider a neural network with one hidden layer, N hidden neurons, and linear activation in the output. Consequently, the output is given by

$$
\phi_\theta(t) = \sum_{j=1}^{N} \alpha_j \sigma(w_j t + b_j), \qquad \theta = (w, b, \alpha) \in \mathbb{R}^N \times \mathbb{R}^N \times \mathbb{R}^N. \quad (9.6.12)
$$

The differentiability of $\phi_\theta(t)$ is assured by that of the logistic function σ. Now we have two choices:

(*i*) Substitute relation (9.6.12) into the empirical cost function $C(\theta; \{t_i\})$ and minimize with respect to θ to obtain $\theta^* = \arg\min C(\theta; \{t_i\})$. Then $\phi_{\theta^*}(t)$ is an approximation of the solution $y(t)$.

(*ii*) Compute the gradient

$$
\begin{aligned}
\nabla_\theta C(\theta) &= 2 \int_{t_0}^{t_0+\epsilon} (\phi_\theta' - f(t, \phi_\theta))(\nabla_\theta \phi_\theta' - \nabla_y f(t, \phi_\theta)^T \nabla_\theta \phi_\theta) \, dt \\
&\quad + 2(\phi_\theta(t_0) - y_0)\nabla \phi_\theta(t_0)
\end{aligned}
$$

and substitute relation (9.6.12), see Exercise 9.8.11. Then apply the gradient descent to construct the approximation sequence

$$\theta_{j+1} = \theta_j - \lambda \nabla_w C(\theta),$$

for a certain learning rate $\lambda > 0$.

9.7 Summary

The chapter answers the question of what kind of functions can be approximated using multiple layer networks and is based on the approximation theory results developed by Funahashi, Hornik, Stinchcombe, White, and Cybenko in the late 1980s.

The results proved in this chapter show that a one-hidden layer neural network with sigmoid neurons in the hidden layer and linear activation in the output layer can learn continuous functions, integrable functions, and square integrable functions, as well as measurable functions. The quality of neural networks to be potential universal approximators is not the specific choice of the activation function, but rather the feedforward network architecture itself.

The price paid is that the number of neurons in the hidden layer does not have an a priori bound. However, there is a result stating that for each extra digit of accuracy gain in the target approximation, the number of hidden neurons should increase 100-fold.

The approximation results work also if the activation function is a ReLU or a Softplus. However, the proof is not a straightforward modification of the proofs provided in this chapter. The trade-off between depth, width and approximation error is still an active subject of research. For instance, in 2017, Lu et al. [79] proved an universal approximation theorem for width-bounded deep neural networks with ReLU activation functions that can approximate any Lebesgue integrable function on n-dimensional input space. Shortly after, Hanin [52] improved the result using ReLU-networks to approximate any continuous convex function of n-dimensional input variables.

Neural networks can also be used to solve numerically first-order differential equations with initial conditions.

9.8 Exercises

Exercise 9.8.1 It is known that the set of rational numbers, $\mathbb{Q}$, is dense in the set of real numbers, $\mathbb{R}$. Formulate the previous result in terms of machine learning terminology.

Exercise 9.8.2 A well-known approximation result is the Weierstrass Approximation Theorem: *Let f be a continuous real-valued function defined on the real interval $[a, b]$. Then there is a sequence of polynomials P_n such that* $\sup_{[a,b]} |f(x) - P_n(x)| \to 0$, *as $n \to \infty$.*
Formulate the previous result in terms of machine learning terminology.

Exercise 9.8.3 (separation of points) Let x_0, x_1 be two distinct, non-collinear vectors in the normed linear space $\mathcal{X}$. Show that there is a bounded linear functional L on $\mathcal{X}$ such that $L(x_0) = 1$ and $L(x_1) = 0$.

Exercise 9.8.4 Let x_0, x_1 be two distinct, non-collinear vectors in the normed linear space $\mathcal{X}$. Show that there is a bounded linear functional L on $\mathcal{X}$ such that $L(x_0) = L(x_1) = 1/2$, with $\|L\| \leq \frac{\delta_0 + \delta_1}{2\delta_0\delta_1}$, where δ_i is the distance from x_i to the line generated by the other vector.

Exercise 9.8.5 Given two finite numbers, a and b, show that there is a unique finite signed Borel measure on $[a, b]$ such that

$$\int_a^b \sin t \, d\mu(t) = \int_a^b \sin t \, d\mu(t).$$

Exercise 9.8.6 Let $\mathcal{P}([0, 1])$ be the space of polynomials on $[0, 1]$. For any $P \in \mathcal{P}([0, 1])$ define the functional

$$L(P) = a_0 + a_1 + \cdots + a_n,$$

where $P(x) = a_0 + a_1 x + \cdots + a_n x^n$, $a_i \in \mathbb{R}$.
(a) Show that L is a linear, bounded functional on $\mathcal{P}([0, 1])$.
(b) Prove that there is a finite, signed Baire measure, $\mu \in M([0, 1])$, such that

$$\int_0^1 P(x) \, d\mu(x) = a_0 + a_1 + \cdots + a_n, \qquad \forall P \in \mathcal{P}([0, 1]).$$

Exercise 9.8.7 (a) Is the tangent hyperbolic function discriminatory in the L^2-sense? What about in the L^1-sense?
(b) Show that the function $\phi(x) = e^{-x^2}$ is discriminatory in the L^1-sense on $\mathbb{R}$.

Exercise 9.8.8 (a) Write a formula for the output of a two-hidden layer FNN having N_1 and N_2 number of neurons in the hidden layers, respectively.
(b) Show that the outputs of all possible two-hidden layer FNNs with the same input form a linear space of functions.

Exercise 9.8.9 An activation function σ is called *strong discriminatory* for the measure μ if

$$\int_{I_n} (\sigma \circ f)(x)\, d\mu(x) = 0, \quad \forall f \in C(I_n) \quad \Longrightarrow \mu = 0.$$

(*a*) Show that if σ is strong discriminatory then it is discriminatory in the sense of Definition 2.2.2.

(*b*) Assume the activation functions of a two-layer FNN are continuous and strong discriminatory with respect to any signed measure. Show that this FNN can learn any continuous function in $C(I_n)$.

Exercise 9.8.10 Show that the metric d_μ defined by (9.3.3) satisfies the triangle inequality:

$$d_\mu(f, g) \le d_\mu(f, h) + d_\mu(h, g), \qquad \forall f, g, h \in \mathcal{M}(\mathbb{R}^n).$$

Exercise 9.8.11 Find $\phi'_\theta(t)$, $\nabla_w \phi_\theta(t)$, and $\nabla_w \phi'_\theta(t)$ for the expression of $\phi_\theta(t)$ given by (9.6.12).

Chapter 10

Exact Learning

By *exact learning* we mean the expressibility of a network to reproduce exactly the desired target function. For an exact learning the network weights do not need tuning; their values can be found exactly. Even if this is unlikely to occur in general, there are a few particular cases when this happens. These cases will be discussed in this chapter.

10.1 Learning Finite Support Functions

We shall start with the simplest case of finite support functions. The next result states that a one-hidden layer neural network, having the activation function in the hidden neurons given by the Heaviside function, $H(x)$, and one linear neuron in the output layer, has the capability to represent exactly any function with finite support on $\mathbb{R}^r$. More precisely, we have:

Proposition 10.1.1 *Let $g : \mathbb{R}^r \to \mathbb{R}$ be an arbitrary function and consider a set $S = \{x_1, \ldots, x_n\}$ of distinct points in $\mathbb{R}^r$. Then there is a function $G(x) = \sum_{i=1}^{n} \alpha_i H(w_i^T x - \theta_i)$, with $w_i \in \mathbb{R}^k$, $\alpha_i, \theta_i \in \mathbb{R}$, such that*

$$G(x_j) = g(x_j), \quad j = 1, \ldots, n.$$

Proof: The proof will be done in two steps.
Step 1: Assume $r = 1$. In this case, the points are real numbers, so we may assume the order $x_1 < x_2 < \cdots < x_n$. Choose thresholds θ_j such that $\theta_1 < x_1 < \theta_2 < x_2 < \theta_3 < x_3 < \cdots < x_{n-1} < \theta_n < x_n$. It is easy to see that

$$H(x_i - \theta_j) = \begin{cases} 1, & \text{if } \theta_j < x_i \text{ (or } j \leq i) \\ 0, & \text{otherwise.} \end{cases}$$

© Springer Nature Switzerland AG 2020
O. Calin, *Deep Learning Architectures*, Springer Series in the Data Sciences,
https://doi.org/10.1007/978-3-030-36721-3_10

Consider the function $G(x) = \sum_{i=1}^{n} \alpha_i H(x - \theta_i)$ and impose conditions $G(x_j) = g(x_j)$, $j = 1, \ldots, n$. This leads to the following diagonal linear system satisfied by $\alpha'_j s$:

$$
\begin{aligned}
\alpha_1 &= g(x_1) \\
\alpha_1 + \alpha_2 &= g(x_2) \\
\cdots &= \cdots \\
\alpha_1 + \cdots + \alpha_n &= g(x_n),
\end{aligned}
$$

with solution $\alpha_j = g(x_j) - g(x_{j-1})$, for $2 \leq j \leq n$.

Step 2: Assume $r \geq 2$. The idea is to reduce this case to Step 1. Consider the finite number of hyperplanes, $H_{ij} = \{p; p(x_i - x_j) = 0\}$, where $p(x_i - x_j)$ denotes the inner product between vectors p and $x_i - x_j$, and choose a vector, w, which does not belong to any of the hyperplanes

$$
w \in \mathbb{R}^k \setminus \bigcup_{i \neq j} H_{ij}.
$$

Then the inner products wx_i are distinct real numbers, which can be ordered, for instance, as

$$
wx_1 < wx_2 < \cdots < wx_n.
$$

Using Step 1 we can construct the weights $\{\alpha_j\}$, thresholds $\{\theta_j\}$ and consider the function

$$
G(x) = \sum_{i=1}^{n} \alpha_i H(wx - \theta_i),
$$

which via Step 1 satisfies $G(x_j) = g(x_j)$, for $1 \leq j \leq n$. ∎

Remark 10.1.2 We shall make a few remarks.

(i) We have seen in section 5.2 that a single perceptron cannot implement the function XOR, which is a function satisfying $g(0,0) = 0$, $g(0,1) = 1$, $g(1,0) = 1$, and $g(1,1) = 0$. The previous proposition infers that a neural with one hidden layer and 4 computing units in the hidden layer can implement exactly the XOR function. In fact, only 2 neurons in the hidden layer would suffice, see Exercise 10.11.1.

(ii) The number of neurons in the hidden layer is equal to the number of data points, n. Therefore, the more the data, the wider the network.

(iii) The weights and biasses are not unique. There are infinitely many good choices for the parameters such that the network learns the function g.

(iv) The proposition still holds true if the step function $H(\cdot)$ is replaced by a squashing function ψ (see section 2.3) that achieves 0 and 1. The proof idea is still the same.

(*v*) With some more effort the proof can be modified to obtain the result valid for arbitrary squashing functions. In particular, the result holds true if $H(x)$ is replaced by the logistic function, $\sigma(x)$. We make here the following relation with the message of section 8.6. In this case the network "memorizes" the values of the arbitrary function g, without having the capacity of generalizing to any underlying continuous model. From this point of view, the network behaves like a look-up table, and hence it is of not much interest for machine learning.

Replacing $g(x_i)$ by y_i, the result can be reformulated as in the following.

Corollary 10.1.3 *Given n data, $(x_i, y_i) \in \mathbb{R}^r \times \mathbb{R}$, $1 \leq i \leq n$, there is a function $G(x) = \sum_{i=1}^n \alpha_i H(w_i^T x - \theta_i)$, with $w_i \in \mathbb{R}^k$, $\alpha_i, \theta_i \in \mathbb{R}$, such that*

$$G(x_j) = y_j, \quad j = 1, \ldots, n.$$

This means that the neural network stores the data in its parameters. There are n weights α_j, n biasses θ_j and n matrices w_j of type $r \times r$, in total $2n + nr^2$ real number parameters, which can store the data, which is equivalent to a vector of length $n(r+1)$.

10.2 Learning with ReLU

The ReLU function can be extended to vectors $\mathbf{x} \in \mathbb{R}^n$ using an action on components as

$$ReLU(\mathbf{x}) = \big(ReLU(x_1), \ldots, ReLU(x_n)\big),$$

where $\mathbf{x}^T = (x_1, \ldots, x_n)$.

We also recall that an affine function $A : \mathbb{R}^n \to \mathbb{R}^m$ is given by

$$A(\mathbf{x}) = (A_1(\mathbf{x}), \ldots, A_m(\mathbf{x})),$$

where $A_j(\mathbf{x}) = \sum_{k=1}^n w_{kj} x_k + b_j$ is a real-valued affine function with real coefficients w_{kj} and b_j.

Definition 10.2.1 *A ReLU-feedforward neural network with input $\mathbf{x} \in \mathbb{R}^n$ and output $y \in \mathbb{R}^m$ is given by specifying a sequence of $L-1$ natural numbers, $\ell_1, \ldots, \ell_{L-1}$, representing the widths of the hidden layers, and a set of L affine functions $A_1 : \mathbb{R}^n \to \mathbb{R}^{\ell_1}$, $A_i : \mathbb{R}^{\ell_{i-1}} \to \mathbb{R}^{\ell_i}$, for $2 \leq i \leq L-1$, and $A_L : \mathbb{R}^{\ell_{L-1}} \to \mathbb{R}^m$. The output of the network is given by*

$$f = A_L \circ ReLU \circ A_{L-1} \circ ReLU \circ \cdots \circ A_2 \circ ReLU \circ A_1,$$

where $\circ$ denotes function composition.

The number L denotes the *depth* of the network. The *width* is defined by $w = \max\{\ell_1, \ldots, \ell_{L-1}\}$. The *size* of the network is $s = \ell_1 + \cdots + \ell_{L-1}$, i.e., the number of hidden neurons.

10.2.1 Representation of maxima

In this section we show that the function $f(x_1, \ldots, x_n) = \max\{x_1, \ldots, x_n\}$ can be represented exactly by a neural network having ReLU activation functions in the hidden layers and a linear activation function in the output layer. We also show that the depth of the network increases logarithmical with the input size n.

We start with a few notations and some elementary properties. Let $x^+ = \max\{x, 0\}$ and $x^- = \min\{x, 0\}$ denote the positive and the negative part of the real number x, respectively. We note that we have decompositions $x = x^+ + x^-$ and $|x| = x^+ - x^-$. We also have $x^+ = -(-x)^-$. Using the formula

$$\max\{x_1, x_2\} = \frac{1}{2}(x_1 + x_2 + |x_1 - x_2|)$$

and the fact that $ReLU(x) = x^+$, we can write the maximum of two numbers as a linear combinations of ReLU functions as in the following:

$$
\begin{aligned}
\max\{x_1, x_2\} &= \frac{1}{2}(x_1 + x_2 + |x_1 - x_2|) \\
&= \frac{1}{2}\left((x_1 + x_2)^+ + (x_1 + x_2)^- + (x_1 - x_2)^+ - (x_1 - x_2)^-\right) \\
&= \frac{1}{2}\left((x_1 + x_2)^+ - (-x_1 - x_2)^+ + (x_1 - x_2)^+ + (-x_1 + x_2)^+\right) \\
&= \frac{1}{2}ReLU(1 \cdot x_1 + 1 \cdot x_2) - \frac{1}{2}ReLU(-1 \cdot x_1 - 1 \cdot x_2) \\
&\quad + \frac{1}{2}ReLU(1 \cdot x_1 - 1 \cdot x_2) + \frac{1}{2}ReLU(-1 \cdot x_1 + 1 \cdot x_2).
\end{aligned}
$$

Therefore

$$\max\{x_1, x_2\} = \sum_{j=1}^{4} \lambda_j ReLU(w_{1j}x_1 + w_{2j}x_2), \tag{10.2.1}$$

which means that $\max\{x_1, x_2\}$ can be represented exactly by a feedforward neural network with one hidden layer. The activation in the 4 hidden neurons is given by the ReLU function, while the output neuron is linear, see Fig. 10.1. The weights from the input to first layer are

$$w_{11} = 1, \ w_{12} = -1, \ w_{13} = 1, \ w_{14} = -1$$

$$w_{21} = 1, \ w_{22} = -1, \ w_{23} = -1, \ w_{14} = 1,$$

and the weights from the first-to-second layer are given by

$$\lambda_1 = \lambda_3 = \lambda_4 = \frac{1}{2}, \ \lambda_2 = -\frac{1}{2}.$$

All biases are zero. In the following L denotes the depth of a network, which has $L - 1$ hidden layers.

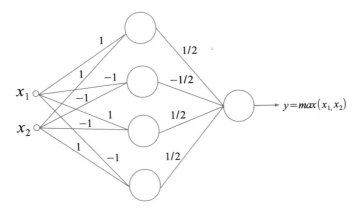

Figure 10.1: *A one-hidden layer network learning exactly the maximum of two numbers.*

Proposition 10.2.2 *The function $f(x_1, \ldots, x_n) = \max\{x_1, \ldots, x_n\}$ can be represented exactly by a ReLU-feedforward neural network (as in Definition 10.2.1) having*

$$L = \begin{cases} 2k, & \text{if } n = 2^k \\ 2(\lfloor \log_2 n \rfloor + 1), & \text{if } n \neq 2^k. \end{cases}$$

Proof: The proof will be done in a few steps.
Step 1: We prove the statement in the case $n = 2^k$, with $k \geq 1$. This will be done using induction over k.
The case $k = 1$ follows from the previous discussion, which shows that $\max\{x_1, x_2\}$ can be represented exactly by formula (10.2.1). In this case the network has depth $L = 2$.
In the case $k = 2$ we write

$$\max\{x_1, x_2, x_3, x_4\} = \max\{\max\{x_1, x_2\}, \max\{x_3, x_4\}\}.$$

Then $a = \max\{x_1, x_2\}$ and $b = \max\{x_3, x_4\}$ are represented by a neural network with one hidden layer each. We put these networks in parallel and add two more layers to compute $\max\{a, b\}$, see Fig. 10.2. The resulting network has $L = 2 + 2 = 4$, which verifies the given relation.
The induction hypothesis states that if $n = 2^k$ then

$$f(x_1, \ldots, x_{2^k}) = \max\{x_1, \ldots, x_{2^k}\}$$

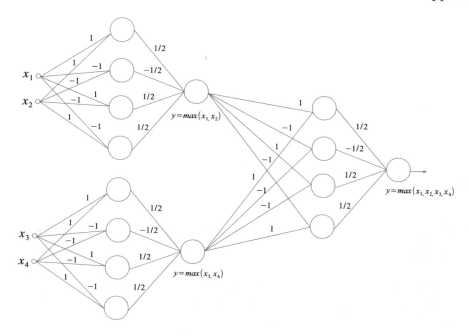

Figure 10.2: *A neural network learning exactly the maximum of 4 numbers. The 8 neurons in the first layer can be considered as only 4 overlaying neurons.*

can be represented by a network with $L_k = 2k$. Since

$$f(x_1, \ldots, x_{2^{k+1}}) = \max\{f(x_1, \ldots, x_{2^k}), f(x_{2^k+1}, \ldots, x_{2^k+2^k})\}.$$

then by the induction hypothesis $f(x_1, \ldots, x_{2^k})$ and $f(x_{2^k+1}, \ldots, x_{2^k+2^k})$ can be represented by a network each having $L_k + 1$ layers. We put these networks in parallel and add two more layers to compute $f(x_1, \ldots, x_{2^{k+1}})$. The resulting network has $L_{k+1} = L_k + 2 = 2k + 2 = 2(k + 1)$, which ends the induction.

Step 2: We prove the statement in the case $n \neq 2^k$, for $k \geq 1$.
In this case we have

$$2^k < n < 2^{k+1},$$

for $k = \lfloor \log_2 n \rfloor$. We complete the set $\{x_1, \ldots, x_n\}$ to a set with cardinal equal to the next power of 2 by appending $2^{k+1} - n$ numbers equal to x_n

$$x_n = x_{n+1} = x_{n+3} = \cdots = x_{2^{k+1}}.$$

Since

$$\max\{x_1, \ldots, x_n\} = \max\{x_1, \ldots, x_n, \ldots x_{2^{k+1}}\}$$

applying Step 1 we obtain a network of depth $L_{k+1} = 2(k+1) = 2(\lfloor \log_2 n \rfloor + 1)$, which learns the maximum in the right side. This is the relation provided in the conclusion. ∎

Remark 10.2.3 It is worth noting that a similar result holds for representing the minimum of n numbers. The proof is similar, using the formula

$$\min\{x_1, x_2\} = \frac{1}{2}\left(x_1 + x_2 - |x_1 - x_2|\right).$$

The following result shows that for representing $\max\{x_1, x_2\}$ the width of the hidden layer can be reduced to only 2 neurons if one of the input variables is nonnegative and a ReLU activation is applied to the output neuron.

Proposition 10.2.4 *There is a feedforward neural network with input of dimension 2, one hidden layer of width 2, output of dimension 1 and having ReLU activation functions for all neurons, which represents exactly the function*

$$f(x_1, x_2) = \max\{x_1, x_2\}, \quad \forall x_1 \in \mathbb{R}, \, x_2 \in \mathbb{R}_+.$$

Proof: We consider the neural network given by Fig. 10.3. All biasses are taken equal to zero and the weights are chosen to be

$$w_{11}^{(1)} = 1, \; w_{12}^{(1)} = 0, \; w_{21}^{(1)} = -1, \; w_{22}^{(1)} = 1, \; w_{11}^{(2)} = 1, \; w_{21}^{(2)} = 1.$$

Using these weights we construct the affine functions $A_1 : \mathbb{R}^2 \to \mathbb{R}^2$ and $A_2 : \mathbb{R}^2 \to \mathbb{R}$

$$A_1(x_1, x_2) = (x_1 - x_2, x_2), \quad A_2(u, v) = u + v.$$

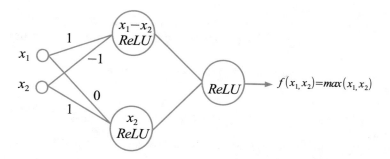

Figure 10.3: *A neural network learning exactly the maximum of 2 numbers.*

Since all neurons have a ReLU activation function, the input-output mapping of the network is computed as in the following:

$$
\begin{aligned}
f(x_1, x_2) &= (ReLU \circ A_2 \circ ReLU \circ A_1)(x_1, x_2) \\
&= (ReLU \circ A_2 \circ ReLU)(x_1 - x_2, x_2) \\
&= (ReLU \circ A_2)(ReLU(x_1 - x_2), x_2) \\
&= ReLU\big(ReLU(x_1 - x_2) + x_2, 0\big) \\
&= \max\{ReLU(x_1 - x_2) + x_2, 0\} \\
&= \max\{\max\{x_1 - x_2, 0\} + x_2, 0\} \\
&= \max\{\max\{x_1, x_2\}, 0\} = \max\{x_1, x_2\}, \quad \forall x_1 \in \mathbb{R},\ x_2 \in \mathbb{R}_+.
\end{aligned}
$$

The last identity holds since $\max\{x_1, x_2\} \geq x_2 \geq 0$. ∎

10.2.2   Width versus Depth

The next result deals with a trade-off between "shallow and wide" and "deep and narrow" for ReLU-nets, where by a ReLU-net we understand a feedforward neural network having ReLU activation functions for all its neurons (also including the output layer). More specifically, we have seen how a continuous function $f : [0, 1]^d \to \mathbb{R}$ can be learned by a one-hidden layer neural network with enough neurons in the hidden layer. We show here that in the case of a ReLU-neural network the width of the hidden layer can be reduced at the expense of the depth, namely, the same output is obtained by adding extra hidden layers of a certain bounded width, see Fig.10.4. The next result is taken from Hanin [52]:

Proposition 10.2.5 *Let $\mathcal{N}$ denote a ReLU-network (as in definition 10.2.1) with input dimension d, a single hidden layer of width n, and output*

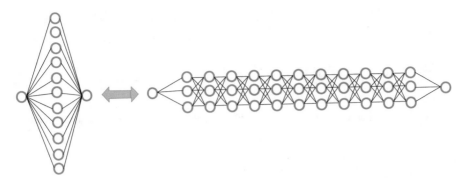

Figure 10.4: *Two ReLU-networks, one "shallow and wide" and the other, "deep and narrow" can compute the same function.*

dimension 1. Then there exists another ReLU-net, $\widetilde{\mathcal{N}}$, that computes the same function as $\mathcal{N}$ on $[0,1]^d$, having the same input dimension d, but having $n+2$ hidden layers, each of width $d+2$.

We note a decrease in the width from n to only $d+2$, while the number of hidden layers increased from 1 to $n+2$ (In general n is much larger than d).

Proof: Consider the affine functions computed by the hidden neurons

$$A_j(x) = \sum_{k=1}^{d} w_{kj} x_k + b_j, \qquad 1 \le j \le n, \ x \in [0,1]^d$$

Then the activation of the jth neuron in the hidden layer is given by $y_j = (ReLU \circ A_j)(x)$, $1 \le j \le n$. It follows that the input-output function of the network $\mathcal{N}$ is

$$f_{\mathcal{N}} = b + \sum_{j=1}^{n} \lambda_j ReLU(A_j(x)),$$

where b is the bias of the output neuron and λ_j are the hidden-to-output weights.

Next, we shall construct the network $\widetilde{\mathcal{N}}$, such that $f_{\widetilde{\mathcal{N}}}(x) = f_{\mathcal{N}}(x)$, for all $x \in [0,1]^d$. In order to specify $\widetilde{\mathcal{N}}$, it suffices to provide its weights and biasses, or equivalently, to specify the affine transformations $\tilde{A}_j$ computed by the jth hidden layer of $\widetilde{\mathcal{N}}$.

Since the function $x \to \sum_{j=1}^{n} \lambda_j ReLU(A_j(x))$ is continuous on the compact set $[0,1]^d$, then it is bounded, and hence there is a constant $T > 0$ such that

$$T + \sum_{j=1}^{n} \lambda_j ReLU(A_j(x)) > 0, \qquad \forall x \in [0,1]^d.$$

The affine transformations $\tilde{A}_1 : [0,1]^d \to \mathbb{R}^{d+2}$, $\tilde{A}_j : \mathbb{R}^{d+2} \to \mathbb{R}^{d+2}$, $2 \le j \le n+1$, and $\tilde{A}_{n+2} : \mathbb{R}^{d+2} \to \mathbb{R}$ are defined by

$$\begin{aligned}
\tilde{A}_1(x) &= (x, A_1(x), T) \\
\tilde{A}_j(x, y, z) &= (x, A_j(x), z + \lambda_{j-1} y), \quad 2 \le j \le n+1 \\
\tilde{A}_{n+2}(x, y, z) &= z - T + b.
\end{aligned}$$

We shall track the first few layer activations. The first layer activation is

$$\begin{aligned}
X^{(1)} &= ReLU \circ \tilde{A}_1(x) = ReLU(x, A_1(x), T) = (x, ReLU(A_1(x)), T) \\
X^{(2)} &= ReLU \circ \tilde{A}_2 \circ ReLU \circ \tilde{A}_1(x) = ReLU \circ \tilde{A}_2(x, ReLU(A_1(x)), T) \\
&= (x, ReLU(A_2(x)), T + \lambda_1 ReLU(A_1(x)) \\
X^{(3)} &= ReLU \circ \tilde{A}_3 \circ ReLU \circ \tilde{A}_2 \circ ReLU \circ \tilde{A}_1(x) \\
&= ReLU \circ \tilde{A}_3\big(x, ReLU(A_2(x)), T + \lambda_1 ReLU(A_1(x))\big) \\
&= ReLU\big(x, A_3(x), T + \lambda_1 ReLU(A_1(x)) + \lambda_2 ReLU(A_2(x))\big) \\
&= \big(x, ReLU(A_3(x)), T + \lambda_1 ReLU(A_1(x)) + \lambda_2 ReLU(A_2(x))\big).
\end{aligned}$$

Inductively, we obtain

$$X^{(p)} = \Big(x, ReLU(A_p(x)), T + \sum_{j=1}^{p-1} \lambda_j ReLU(A_j(x))\Big), \quad \forall p \le n+1.$$

The last layer activation of the network $\tilde{\mathcal{N}}$ is given by

$$\begin{aligned}
X^{(n+2)} &= \tilde{A}_{n+2} \circ X^{(n+1)} \\
&= \tilde{A}_{n+2}\Big(x, ReLU(A_{n+1}(x)), T + \sum_{j=1}^{n} \lambda_j ReLU(A_j(x))\Big) \\
&= b + \sum_{j=1}^{n} \lambda_j ReLU(A_j(x)) = f_{\mathcal{N}}(x), \qquad x \in [0,1]^d.
\end{aligned}$$

Hence, $f_{\tilde{\mathcal{N}}}(x) = f_{\mathcal{N}}(x)$, for all $x \in [0,1]^d$, which ends the proof. ∎

Lemma 10.2.6 *The input-output function of a ReLU-neural network is a continuous piecewise linear function.*

Proof: Let f denote the input-output function of the ReLU-network with L layers. Then

$$f(x) = (A_L \circ ReLU \circ \cdots \circ ReLU \circ A_2 \circ ReLU \circ A_1)(x).$$

Since f is a composition of continuous functions (both ReLU and the affine linear functions A_j are continuous), it follows that f is continuous. Because ReLU is piecewise, the aforementioned composition is also piecewise. Since

$$ReLU(ax - b) = \begin{cases} ax - b, & \text{if } x \geq b/a \\ 0, & \text{if } x < b/a, \end{cases}$$

the composition of ReLU and any of the affine functions A_j is also piecewise linear. The repetitive composition of these types of functions is also piecewise linear. ∎

The converse statement also holds true, see Arora et al. [10]. In the following we shall present a slightly different result where the network is allowed to have both ReLU and linear neurons.

Proposition 10.2.7 *Let $f : \mathbb{R}^n \to \mathbb{R}$ be a given continuous piecewise linear function. Then there is a feedforward neural network, having both ReLU and linear neurons, which can represent exactly the function f. The network has $L = 2(\lfloor \log_2 n \rfloor + 2)$ layers.*

Proof: The proof is based on a result of Wang and Sung [122], which states that any continuous piecewise linear function f can be represented as a linear combination of piecewise linear convex functions as

$$f = \sum_{j=1}^{p} s_j \max_{i \in S_j} \ell_i$$

where $\{\ell_1, \ldots, \ell_k\}$ is a finite set of affine linear functions and $S_i \subset \{1, 2, \ldots, k\}$, each set S_i having at most $n + 1$ elements and $s_j \in \{-1, 1\}$. The function $g_j = \max_{i \in S_j} \ell_i$ is a piecewise linear convex function having at most $n + 1$ pieces, since $|S_j| \leq n + 1$. Each affine function $\ell_i(x)$ is computed from the input x by a linear neuron (first hidden layer). Each value $g_j(x)$ is computed from inputs $\ell_i(x)$ by a ReLU-neural network given as in Proposition 10.2.2 (this are $2(\lfloor \log_2 n \rfloor + 1)$ hidden layers). An output linear neuron can compute the function $f = \sum_{j=1}^{p} s_j g_i$ (the output layer). The total number of layers is $L = 1 + 2(\lfloor \log_2 n \rfloor + 1) + 1 = 2(\lfloor \log_2 n \rfloor + 2)$. ∎

Remark 10.2.8 The previous result does not provide information about the width of the layers. Hanin [52] showed that in the case of a nonnegative continuous piecewise linear function, $f : \mathbb{R}^n \to \mathbb{R}+$, the hidden layers have width $n + 3$ and the ReLU-network has depth $2p$.

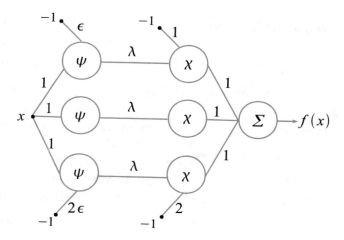

Figure 10.5: *Two-hidden layer exact representation of a continuous function f in the case $n = 1$.*

10.3 Kolmogorov-Arnold-Sprecher Theorem

A version of the famous Hilbert's thirteenth problem is that there are analytic functions of three variables which cannot be written as a finite superposition of continuous functions in only two arguments. The conjecture was proved false by Arnold and Kolmogorov in 1957, who proved the following representation theorem, see [9] and [64].

Theorem 10.3.1 (Kolmogorov) *Any continuous function $f(x_1, \ldots, x_n)$ defined on I_n, $n \geq 2$, can be written in the form*

$$f(x_1, \ldots, x_n) = \sum_{j=1}^{2n+1} \chi_j \Big(\sum_{i=1}^{n} \psi_{ij}(x_i) \Big),$$

where χ_j, ψ_{ij} are continuous functions of one variable and ψ_{ij} are monotone functions which are not dependent on f.

We note that for $n = 2$ the previous expression becomes

$$
\begin{aligned}
f(x_1, x_2) \;=\; & \chi_1\Big(\psi_{11}(x_1) + \psi_{21}(x_2) \Big) + \chi_2\Big(\psi_{12}(x_1) + \psi_{22}(x_2) \Big) \\
& + \cdots + \chi_5\Big(\psi_{15}(x_1) + \psi_{25}(x_2) \Big).
\end{aligned}
$$

Sprecher refined Kolmogorov's result to the following version, where the outer functions χ_j are replaced by a single function χ and the inner functions ψ_{ij} are of a special type and shifted by a translation:

Theorem 10.3.2 (Sprecher, 1964) *For each integer $n \geq 2$, there exists a real, monotone increasing function $\psi(x)$, with $\psi([0,1]) = [0,1]$, dependent on n and having the following property:*
For each preassigned number $\delta > 0$, there is a rational number ϵ, with $0 < \epsilon < \delta$, such that every real continuous function $f(x_1, \ldots, x_n)$, defined on I_n, can be represented as

$$f(x_1, \ldots, x_n) = \sum_{j=1}^{2n+1} \chi\left(\sum_{i=1}^{n} \lambda^i \psi(x_i + \epsilon(j-1)) + j - 1 \right),$$

where the function χ is real and continuous and λ is an independent constant of f.

The proof can be found in Sprecher paper [115]. This result is important to the field of neural networks because it states that any continuous function on I_n can be represented *exactly* by a neural network with two hidden layers. The activation function for the first hidden layer is ψ and for the second hidden layer is χ.

For clarity reasons we have presented the neural network for the case $n = 1$ in Fig. 10.5. In this case each hidden layer has three computing units. The weights from the first-to-second hidden layer are all equal to λ. All the other weights (input to first hidden layer and second hidden layer to output layer) are equal to 1. Note that the threshold ϵ can be made arbitrary small.

The case $n = 2$ is represented in Fig. 10.6, where for the sake of simplicity thresholds have been neglected and all edges without a weight have weight equal to 1 by default. The first hidden layer has 10 computing units while the second hidden layer has just 5 units.

Kolmogorov-Arnold-Sprecher's Theorem represents an interesting theoretical result regarding four-layer neural networks. However, it does not have much practical applicability due to the fact that the activation functions ψ and χ are not set a priori, and their construction is laborious. In fact, it is worthy to note the following trade-off property: in Kolmogorov's Theorem the number of intermediary units is given, but there is no control on the expression of ψ and χ. However, in some of the previous results the activation function was fixed a priori, while the number of hidden units were increased as needed until some desired level of approximation accuracy was reached.

10.4 Irie and Miyake's Integral Formula

Another result of theoretical importance is the integral formula of Irie and Miyake, see [60], which states that arbitrary functions of finite energy (that is

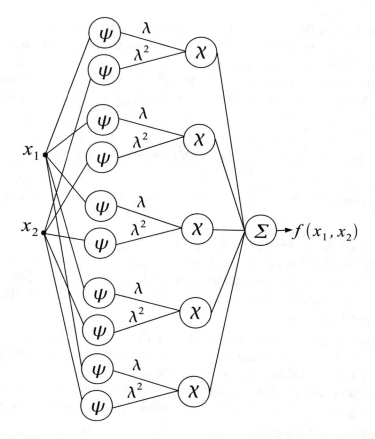

Figure 10.6: *Two-hidden layer exact representation of a continuous function f in the case n = 2.*

L^2-integrable) can be represented by a three-layer network with an continuum of computational units, see Fig. 10.7. More precisely, the results state the following:

Theorem 10.4.1 (Irie-Miyake, 1988) *Let* $f(x_1, \ldots, x_n) \in L^2(\mathbb{R}^n)$ *and* $\psi(x) \in L^1(\mathbb{R})$. *Let* $\Psi(\xi)$ *and* $F(w_1, \ldots, w_n)$ *be the Fourier transforms of* $\psi(x)$ *and* $f(x_1, \ldots, x_n)$, *respectively. If* $\Psi(1) \neq 0$, *then*

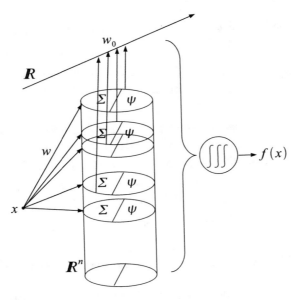

Figure 10.7: *A one-hidden layer neural network with a continuum infinite number of computational units.*

$$f(x_1, \ldots, x_n) = \int_{\mathbb{R}^{n+1}} \psi\Big(\sum_{i=1}^{n} x_i w_i - w_0\Big) \frac{1}{(2\pi)^n \Psi(1)} F(w_1, \ldots, w_n) e^{iw_0} \, d\mathbf{w},$$

where $d\mathbf{w} = dw_0 dw_1 \ldots dw_n$.

In the above formula w_0 corresponds to the bias, w_i to the connection weights, and ψ to the activation function in the hidden layer. The connection weights between the hidden layer and the output layer depend on the Fourier transform of f as

$$\lambda(w_0, w_1, \ldots, w_n) = \frac{1}{(2\pi)^n \Psi(1)} F(w_1, \ldots, w_n) e^{iw_0}.$$

The summation over a continuum infinite number of computational units is achieved by integrating with respect to the weights and thresholds. We have a few points to make.

1. Since the logistic function $\sigma(x) = \dfrac{1}{1+e^{-x}} \notin L^1(\mathbb{R})$, the conclusion of the theorem does not necessarily hold for neural networks with sigmoid neurons.

2. However, the Gaussian function $\psi(x) = e^{-x^2}$ satisfies both properties $\psi \in L^1(\mathbb{R})$ and $\Psi(1) \neq 0$, which makes the formula valid for neural networks with Gaussian activation function.

3. Another major drawback, which makes this formula not useful in practice, is that the function to be realized is given by an integral formula which assumes that the network has an continuum infinite number of units.

4. The idea of recovering a finite energy function of several variables from the values along hyperplanes and its associated frequency is similar to the principle of the computerized tomography.

10.5   Exact Learning Not Always Possible

The message of this section is that there are continuous functions on $[0,1]$, which cannot be represented exactly as outputs of neural networks. For this purpose, we consider a one-layer neural network having the activation in the hidden layer given by the logistic or hyperbolic tangent, and having linear activation in the output layer. If an exact representation would exist, then f could be written as a finite combination of analytic functions such as

$$f(x) = \sum_{j=1}^{N} \alpha_j \sigma(w_j^T x - \theta_j).$$

Since σ is an analytic function of x (as an algebraic combination of e^x), then the right side expression is analytic on $(0,1)$, while the continuous function f can be chosen not analytic, for instance,

$$f(x) = \begin{cases} 0, & \text{if } 0 \le x \le 0.5 \\ x - 0.5, & \text{if } 0.5 < x \le 1. \end{cases}$$

A similar lack of exact representations can be also found for L^1 and L^2 functions.

10.6   Continuum Number of Neurons

This section will extend the exact learning to a deep neural network with a continuum of neurons in each hidden layer. We assume the input belongs

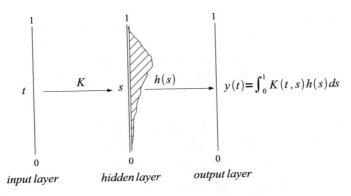

Figure 10.8: *A three-layer neural net with a continuum of neurons in the hidden layer.*

to a compact interval, $t \in [0,1]$. A one-hidden layer neural network, with N neurons in the hidden layer, has an output given by $y(t) = \sum_{j=1}^{N} v_j \sigma(w_j t + b_j)$, where σ is a sigmoid activation function, w_j are the weights between the input and the jth node, v_j is the weight between the jth node and the output, and b_j denotes the bias for the jth neuron in the hidden layer.

Now, we replace the finite number of neurons in the hidden layer, by a continuous infinite number of hidden neurons, which are parametrized by $s \in [0,1]$, see Fig. 10.8. Hence, the index j is replaced by the parameter s. Consequently, w_j, b_j become continuous functions $w(s)$, $b(s)$, respectively, while the weights v_j are replaced by the measure $h(s)\, ds$. The summation is transformed into an integral, the network output being given by the integral transform

$$y(t) = \int_0^1 K(t,s)h(s)\, ds, \qquad (10.6.2)$$

where the system of biasses and weights are encapsulated into the integral kernel $K(t,s) = \sigma\big(w(s)t + b(s)\big)$.

This section investigates the following problem: *Given a continuous function* $g \in C[0,1]$, *does the previous network learn exactly the function* g? *Namely, are there any continuous functions* $w, b, h \in C[0,1]$ *such that* $g(t) = \int_0^1 K(t,s)h(s)\, ds$ *for all* $t \in [0,1]$?

The provided answer is partial, holding for a certain subclass of continuous functions, whose Fourier coefficients tend to zero fast enough. While approaching this problem, we shall employ techniques of integral operators.

The theory of integral equations is well developed in the case of symmetric integral kernels. In order to understand better these types of kernels, the next lemma deals with the symmetry of kernels of the particular type $K(t, s) = \sigma(w(s)t+b(s))$. However, the rest of the section will assume symmetric kernels of a more general form.

Lemma 10.6.1 *Let $w, b \in C[0, 1]$ and consider $K(t, s) = \sigma(w(s)t + b(s))$, $\forall s, t \in [0, 1]$. The following conditions are equivalent:*

(i) K is symmetric: $K(s, t) = K(t, s)$, $\forall s, t \in [0, 1]$;

(ii) $w(s)$ and $b(s)$ are affine, given by

$$
\begin{aligned}
w(s) &= (w(1) - w(0))s + w(0) \\
b(s) &= w(0)s + b(0),
\end{aligned}
$$

with $w(0) = b(1) - b(0)$.

Proof: $(i) \implies (ii)$ Since σ is one-to-one, the symmetry of K is equivalent to

$$ w(s)t + b(s) = w(t)s + b(t), \qquad \forall s, t \in [0, 1]. \qquad (10.6.3) $$

Similarly, taking the limit $s \to 0$ and using the continuity of w and b yields

$$ b(t) = w(0)t + b(0), $$

which shows that $b(t)$ is affine. Taking the limit $t \to 1$ in (10.6.3), solving for $w(s)$ and using the previous relation for b, we have

$$
\begin{aligned}
w(s) &= w(1)s + b(1) - b(s) \\
&= (w(1) - w(0))s + b(1) - b(0).
\end{aligned}
$$

Then taking $s = 0$ provides $w(0) = b(1) - b(0)$.

$(ii) \implies (i)$ Using the expressions of $w(s)$ and $b(s)$, a computation shows that

$$ w(s)t + b(s) = (w(1) - w(0))st + w(0)(t + s) + b(0) = w(t)s + b(t). $$

This implies that (10.6.3) holds, and hence $K(t, s) = K(s, t)$. ∎

It is worth noting that the kernel in this case becomes

$$ K(t, s) = \sigma(\alpha st + \beta(s + t) + \gamma), $$

with $\alpha = w(1) - w(0)$, $\beta = w(0)$, and $\gamma = b(0)$. Then the symmetric kernel depends on the sum and the product of its variables, s and t.

So far, we have determined the functions $w(s)$ and $b(s)$. We still need to find the weight function $h(s)$. We shall determine it by solving the integral equation of the first kind

$$g(t) = \int_0^1 K(t,s)h(s)\,ds. \tag{10.6.4}$$

Since the kernel K is continuous, the integral operator on the right side improves the smoothness of the function h. Therefore, in general, the equation (10.6.4) with arbitrary continuous g cannot be solved by a continuous function h. Thus, we need to assume more restrictive properties on the function g.

Before getting any further, we shall recall a few properties of integral operators, which will be used later:

1. If there is a number λ and a function $e(\cdot)$ such that

$$\int_0^1 K(t,s)e(s)\,ds = \lambda e(t),$$

then λ is called an *eigenvalue* and $e(t)$ an *eigenfunction* corresponding to the eigenvalue λ.

2. If K is symmetric then all eigenvalues are real.

3. If K is symmetric and nondegenerate, there are infinite, countable number of eigenvalues and eigenvectors (The kernel K is called degenerate if $K(t,s) = \sum_{i=1}^p \alpha_i(t)\beta_i(s)$).

4. The eigenfunctions corresponding to nonzero eigenvalues are continuous.

5. The eigenfunctions corresponding to distinct eigenvalues are orthogonal in the L^2-sense.

6. If $f \in L^2[0,1]$, then $\sum_{n\geq 1} f_n^2 \leq \|f\|^2$, where $f_n = \int_0^1 f(t)e_n(t)\,dt$ and $\|f\|^2 = \int_0^1 f(t)^2\,dt$, where $e_n(t)$ denotes the nth eigenfunction. This is called the *Bessel inequality*.

We shall provide without proof three properties involving absolute continuity, which can be found in any analysis textbook and will be used later.

A useful test used to check for the uniform convergence of a series of functions is the following:

Proposition 10.6.2 (Cauchy) *Let $f_n : [0,1] \to \mathbb{R}$ be a sequence of functions. These are equivalent:*

(i) The series $\sum_{n\geq 1} f_n$ *is uniformly convergent on $[0,1]$;*

(ii) $\forall \epsilon > 0$, there is an integer $N \geq 1$ such that

$$\left| \sum_{i=n}^{n+p} f_i(t) \right| < \epsilon, \quad \forall t \in [0,1], \ \forall n \geq N, \ \forall p \geq 1.$$

Condition (*ii*) states that the sequence of partial sums is uniformly Cauchy. The next result states that the uniform convergence preserves continuity:

Proposition 10.6.3 (Weierstrass) *If each function f_n is continuous at $x_0 \in [0,1]$ and the series $\sum_{n \geq 1} f_n$ is uniformly convergent on $[0,1]$, then the series sum is continuous at x_0.*

The next result provides a necessary condition for uniform convergence:

Proposition 10.6.4 *Let $f, f_n : [0,1] \to [0,+\infty)$ be continuous functions, such that $f(t) = \sum_{n \geq 1} f_n(t)$ for all $t \in [0,1]$. Then the series $\sum_{n \geq 1} f_n$ is uniformly convergent.*

The next result states the existence of the weight function h, given some extra conditions on the function g. We note that the symmetric kernel K used in the next result is of general type, not of the particular type as the one given in Lemma 10.6.1.

Lemma 10.6.5 *Let $K : [0,1] \times [0,1] \to \mathbb{R}$ be a continuous and symmetric kernel. Consider $g \in C[0,1]$, satisfying the properties:*
(i) It has the representation $g(t) = \sum_{n \geq 1} g_n e_n(t)$, with $g_n = \int_0^1 g(t) e_n(t)\, dt$, where e_n is the nth eigenfunction of the kernel;
(ii) The series $\sum_{n \geq 1} \dfrac{g_n^2}{\lambda_n^4}$ is convergent, where λ_n denotes the eigenvalue corresponding to eigenfunction e_n.
Then, there is a function $h \in C[0,1]$ such that

$$g(t) = \int_0^1 K(t,s) h(s)\, ds.$$

Proof: We shall break the proof into several steps.
Step 1. *The series $\sum_{n \geq 1} \lambda_n^2 e_n^2(t)$ is absolutely convergent on $[0,1]$ and uniformly bounded in t.*
Applying Bessel's inequality for the function $f(s) = K(t,s)$ and using that e_n is an eigenfunction, yields

$$\int_0^1 K^2(t,s)\, ds \;\geq\; \sum_{n \geq 1} \left(\int_0^1 K(t,s) e_n(s)\, ds \right)^2 = \sum_{n \geq 1} \lambda_n^2 e_n^2(t).$$

The continuity of K implies $\int_0^1 K^2(t,s)\, ds < M$, for all $t \in [0,1]$. Thus, $\sum_{n \geq 1} \lambda_n^2 e_n^2(t)$ is convergent for any $t \in [0,1]$ and also uniformly bounded by M.

Step 2. *Let $h_n = g_n/\lambda_n$. The series $\sum_{n\geq 1} h_n e_n(t)$ is uniformly convergent on $[0,1]$.*

It suffices to show that condition (ii) of Proposition 10.6.2 holds. Let $\epsilon > 0$ be arbitrarily fixed. Using Cauchy's inequality we have

$$\left(\sum_{i=n}^{n+p} h_i e_i(t)\right)^2 = \left(\sum_{i=n}^{n+p} \frac{g_i}{\lambda_i^2}\lambda_i e_i(t)\right)^2 \leq \sum_{i=n}^{n+p}\left(\frac{g_i}{\lambda_i^2}\right)^2 \sum_{i=n}^{n+p}\lambda_i^2 e_i^2(t).$$

From Step 1 it follows that $\sum_{i=n}^{n+p}\lambda_i^2 e_i^2(t) < M$. From hypothesis condition (ii) we have that the series $\sum_{n\geq 1}\dfrac{g_n^2}{\lambda_n^4}$ is convergent, so for n large enough

$$\sum_{i=n}^{n+p}\left(\frac{g_i}{\lambda_i^2}\right)^2 < \frac{\epsilon^2}{M}.$$

Substituting in the previous inequality we obtain

$$\sum_{i=n}^{n+p} h_i e_i(t) < \epsilon, \qquad \forall t \in [0,1]$$

for n large enough and any $p \geq 1$. The assertion follows now from Proposition 10.6.2.

Step 3. *Check that $h(t) = \sum_{n\geq 1} h_n e_n(t)$ is a solution.*

This involves two things: $h \in C[0,1]$ and h satisfies the integral equation. By Step 2 and Proposition 10.6.3, using that $e_n(t)$ is continuous, it follows that $h(t) = \sum_{n\geq 1} h_n e_n(t)$ is continuous on $[0,1]$.

Since uniform convergence allows integration term by term, we have

$$\int_0^1 K(t,s)h(s)\,ds = \sum_{n\geq 1} h_n \int_0^1 K(t,s)e_n(s)\,ds = \sum_{n\geq 1} h_n \lambda_n e_n(t)$$
$$= \sum_{n\geq 1} g_n e_n(t) = g(t).$$

This ends the proof. ∎

Remark 10.6.6 1. The convergence condition $\sum_{n\geq 1}\frac{g_n^2}{\lambda_n^4} < \infty$ implies that the Fourier coefficients g_n tend to zero faster than eigenvalues λ_n do. This restricts the class of continuous functions learned by the network.

2. The fact that $\lim_{n \to \infty} \lambda_n = 0$ can be shown by integrating one more time in the inequality provided at Step 1

$$M = \int_0^1 \int_0^1 K^2(t,s)\,ds\,dt \geq \sum_{n \geq 1} \lambda_n^2 \int_0^1 e_n(t)^2\,dt = \sum_{n \geq 1} \lambda_n^2.$$

Since the series $\sum_{n \geq 1} \lambda_n^2$ is convergent, it follows that λ_n tends to zero.

3. There is a more general result of Picard, see [29] p.160, which states that a necessary and sufficient condition for the existence of a solution $h \in L^2[0,1]$ for the integral equation

$$g(t) = \int_0^1 K(t,s)h(s)\,ds$$

is the convergence of the series $\sum_{n \geq 1} \dfrac{g_n^2}{\lambda_n^2}$. It is worth noting that this result holds also for nonsymmetric kernels K.

The results of lemmas 10.6.1 and 10.6.5 can be combined, by considering a symmetric kernel of the type $K(t,s) = \sigma(\alpha st + \beta(s+t) + \gamma)$ and obtain the existence of a weight density $h(s)$ such that the network learns a continuous function g, whose Fourier coefficients decrease to zero faster than the square of the eigenvalues of the kernel.

10.7 Approximation by Degenerate Kernels

Another way of dealing effectively with kernels is to approximate them with degenerate kernels, i.e., with kernels that can be represented as a finite sum of separable functions in the underlying variables, namely, as finite sums of the type $A_n(t,s) = \sum_{i=1}^n \alpha_i(t)\beta_i(s)$. The next result approximates uniformly the continuous kernel $K(t,s)$ by degenerate kernels with α_i and β_i continuous.

Proposition 10.7.1 *Let $K : [0,1] \times [0,1] \to \mathbb{R}$ be a continuous kernel. Then $\forall \epsilon > 0$, there is an $n \geq 1$ and $\alpha_i, \beta_i \in C[0,1]$, $i = 1, \ldots, n$, such that*

$$\left| K(t,s) - \sum_{i=1}^n \alpha_i(t)\beta_j(s) \right| < \epsilon, \qquad \forall t, s \in [0,1].$$

Proof: This is a reformulation of Proposition 7.4.5 and a consequence of the Stone-Weierstrass Theorem. ∎

In the following we address the question of how is the network output affected if the kernel K is replaced by a nearby degenerate kernel.

Let $h \in L^2[0,1]$ be fixed and denote the outputs by

$$g(t) = \int_0^1 K(t,s)h(s)\,ds, \qquad g_n(t) = \int_0^1 A_n(t,s)h(s)\,ds,$$

with $A_n(t,s) = \sum_{i=1}^n \alpha_i(t)\beta_j(s)$ given by Proposition 10.7.1. Both functions g and g_n are continuous. In particular, g_n is a linear combination of continuous functions α_i

$$g_n(t) = \sum_{i=1}^n c_i \alpha_i(t),$$

with coefficients $c_i = \int_0^1 \beta_i(s)h(s)\,ds \le \|\beta_i\|\,\|h\|$.

The next result states that whatever the network can learn using arbitrary continuous kernels, it can also learn using degenerate kernels, and hence the sufficiency of using degenerate kernels.

Proposition 10.7.2 *The sequence g_n converges uniformly to g on $[0,1]$ as $n \to \infty$.*

Proof: Let $\epsilon > 0$ be arbitrarily fixed. By Proposition 10.7.1 there is an integer $n \ge 1$ such that

$$\left| K(t,s) - A_n(t,s) \right| < \frac{\epsilon}{\|h\|}, \qquad \forall t,s \in [0,1].$$

Using Cauchy's inequality, we have

$$
\begin{aligned}
|g(t) - g_n(t)|^2 &= \left| \int_0^1 \big(K(t,s) - A_n(t,s)\big)\,h(s)\,ds \right|^2 \\
&\le \int_0^1 |K(t,s) - A_n(t,s)|^2\,ds \int_0^1 h^2(s)\,ds \\
&\le \frac{\epsilon^2}{\|h\|^2}\|h\|^2 = \epsilon^2.
\end{aligned}
$$

Hence, for the previous n, we obtain

$$\max_{t \in [0,1]} |g(t) - g_n(t)| < \epsilon, \qquad \forall t \in [0,1],$$

which ends the proof. ∎

10.8   Examples of Degenerate Kernels

Since the functions α_i and β_i are provided by the Stone-Weierstrass Theorem, which is an existential result, it follows that an explicit form for the degenerate kernel, $A_n(t, s)$, is not known in general. If more restrictive conditions are required for the kernel, explicit formulas for the degenerate kernels can be constructed. We shall deal with this problem in the following.

The extra conditions imposed on K are to be symmetric and *nonnegative definite*. The symmetry is needed for the existence of real eigenvalues. Recall that a continuous, symmetric integral kernel K is called nonnegative definite if all its eigenvalues are nonnegative. Equivalently, this can be represented by the integral condition

$$\int_0^1 \int_0^1 K(t, s)u(t)u(s)\,dt\,ds \geq 0, \qquad \forall u \in C[0, 1],$$

which can be interpreted as a nonnegative integral quadratic form.[1]
The following characterization theorem of nonnegative definite kernels can be found in Mercer [83] and shall be used shortly:

Theorem 10.8.1 (Mercer) *Let K be a continuous, symmetric and nonnegative definite integral kernel on $[a, b]$. Denote by $\{e_n\}_{n\geq 1}$ an orthonormal basis for the space spanned by the eigenfunctions e_n corresponding to the nonzero eigenvalues λ_n. Then the kernel can be expanded as a series*

$$K(t, s) = \sum_{n\geq 1} \lambda_n e_n(t)e_n(s), \qquad \forall s, t \in [0, 1]$$

which converges absolutely, uniformly and in mean square.

This means that we may consider degenerate kernels of the type

$$A_n(t, s) = \sum_{i=1}^n \lambda_i e_i(t)e_i(s),$$

which converge uniformly to the kernel $K(t, s)$, by the previous result. In this case $g_n(t)$ can be written as a linear combination of eigenfunctions

$$g_n(t) = \int_0^1 A_n(t, s)h(s)\,ds = \sum_{i=1}^n \lambda_i h_i e_i(t), \qquad (10.8.5)$$

---

[1]This definition is based on the following result proved by Hilbert [54]: $\int_0^1 \int_0^1 K(t, s)u(t)u(s)\,dt\,ds = \sum_n \frac{1}{\lambda_n}\langle u, \psi_n \rangle$, where $\langle u, \psi_n \rangle = \int_0^1 u(s)\psi_n(s)\,ds$, where λ_n is the eigenvalue corresponding to the eigenfunction ψ_n.

with Fourier coefficients $h_i = \int_0^1 e_i(s)h(s)\,ds$. Consider

$$\mathbf{h}_n(t) = \sum_{i=1}^n h_i e_i(t)$$

be the projection of $h(t)$ on the space spanned by $\{e_1(t), \ldots, e_n(t)\}$. Since $e_i(t)$ are continuous, the function $\mathbf{h}_n(t)$ is continuous; these functions will be used to approximate the function $h \in L^2[0,1]$.

In order to simplify notations, we employ the integral operator notations

$$\mathcal{K}(h)(t) = \int_0^1 K(t,s)h(s)\,ds, \qquad \mathcal{A}_n(h)(t) = \int_0^1 A_n(t,s)h(s)\,ds.$$

From the expression of g_n given by (10.8.5), which uses only coefficients h_i with $i \le n$, we obtain

$$\int_0^1 A_n(t,s)h(s)\,ds = \int_0^1 A_n(t,s)\mathbf{h}_n(s)\,ds,$$

or, equivalently, $\mathcal{A}_n(h) = \mathcal{A}_n(\mathbf{h}_n)$.

Using that $e_j(t)$ are orthonormal, we obtain

$$
\begin{aligned}
\mathcal{K}(\mathbf{h}_n)(t) - \mathcal{A}_n(\mathbf{h}_n)(t) &= \int_0^1 [K(t,s) - A_n(t,s)]\mathbf{h}_n(s)\,ds \\
&= \int_0^1 \Big(\sum_{i \ge n+1} \lambda_i e_i(t)e_i(s) \sum_{j=1}^n h_j e_j(s) \Big)\,ds \\
&= \sum_{i \ge n+1} \sum_{j=1}^n \lambda_i e_i(t) h_j \int_0^1 e_i(s)e_j(s)\,ds = 0,
\end{aligned}
$$

since i and j are never equal. Hence $\mathcal{K}(\mathbf{h}_n) = \mathcal{A}_n(\mathbf{h}_n)$. Now, we put all parts together. Triangle inequality, together with Proposition 10.7.2 and previous two relations, yield

$$|\mathcal{K}(h) - \mathcal{K}(\mathbf{h}_n)| \le \underbrace{|\mathcal{K}(h) - \mathcal{A}_n(h)|}_{<\epsilon} + \underbrace{|\mathcal{A}_n(h) - \mathcal{A}_n(\mathbf{h}_n)|}_{=0} + \underbrace{|\mathcal{A}_n(\mathbf{h}_n) - \mathcal{K}(\mathbf{h}_n)|}_{=0}.$$

This means that for any $\epsilon > 0$, there is $n \ge 1$ such that

$$\max_{t \in [0,1]} |\mathcal{K}(h)(t) - \mathcal{K}(\mathbf{h}_n)(t)| < \epsilon.$$

The interpretation of this result follows. From part 3 of Remark 10.6.6, there is a solution $h \in L^2[0,1]$ of the equation $\mathcal{K}(h)(t) = g(t)$ if and only if the function g satisfies some restrictive condition in view of the increase in eigenvalues λ_n. The network can now learn the outcome $g(t)$ by continuous outcomes, $\mathcal{K}(\mathbf{h}_n)(t)$, produced by using continuous weight functions $\mathbf{h}_n(t)$.

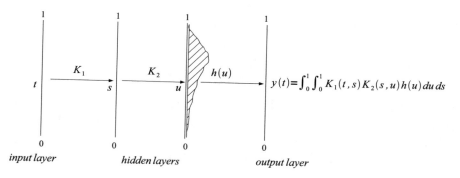

Figure 10.9: *A four-layer neural net with a continuum of neurons in each hidden layer.*

10.9 Deep Networks

Until now we have studied only the case of a network with one hidden layer having a continuum of neurons. Using iterated kernels, in this section we consider multiple hidden layers of the same type. For instance, Fig. 10.9 represents a two-hidden layer neural network of this form. The weights and biasses for the first hidden layer are encapsulated in the kernel $K_1(t, s)$, while the weights and biasses for the second layer are represented by the kernel $K_2(s, u)$. The weight function from the second hidden layer to the output is denoted by $h(u)$. The continuous parameters y, s, u represent the input and the parameters for the second and third layers of the network. The output of the network given in Fig. 10.9 is given by a double integral

$$y(t) = \int_0^1 \int_0^1 K_1(t, s) K_2(s, u) h(u) \, du ds.$$

If let

$$G(t, u) = \int_0^1 K_1(t, s) K_2(s, u) \, ds,$$

then the output can be represented equivalently as

$$y(t) = \int_0^1 G(t, u) h(u) \, du,$$

which looks like the output (10.6.2) but with a more complicated structure of the kernel K. If the kernels K_1 and K_2 are symmetric, then the kernel G is not necessarily symmetric. However, if we further assume that $K_1 = K_2 = K$, then the iterated kernel

$$K^{(2)}(t, u) = \int_0^1 K(t, s)K(s, u)\,ds$$

is symmetric. Hence, for preserving the symmetry of the network kernel, the layers have to share the same kernel, K. The same procedure can be applied to a network with any number of layers.

What is the novelty brought by these types of deep networks? We noticed that in the case of a one-hidden layer neural net, the expansion of the kernel as

$$K(t, s) = \sum_{n \geq 1} \lambda_n e_n(t)e_n(s)$$

is not valid in general. Mercer's Theorem fixed this problem by asking some restrictive assumptions on the kernel K regarding its definiteness. The gain of deeper networks is that the iterate kernel $K^{(2)}$ satisfies a similar expansion, provided the initial kernel K is only symmetric. The expansion is

$$K^{(2)}(t, u) = \sum_{n \geq 1} \lambda_n^2 e_n(t)e_n(u).$$

Similar expansions hold for iterated kernels associated with deeper networks. If we define the nth iteration recursively as

$$K^{(n)}(t, u) = \int_0^1 K^{(n-1)}(t, s)K(s, u)\,du,$$

then the expansion

$$K^{(n)}(t, u) = \sum_{i \geq 1} \lambda_i^n e_i(t)e_i(u),$$

converges absolutely and uniformly. The proof of this result can be found in Courant and Hilbert [29], p.138.

Therefore, all results of section 10.8, which are proved using Mercer's Theorem under the assumption that K is positive definite, are valid in the general case of a deep neural net with at least two hidden layers and the same kernels in each layer.

10.10   Summary

Sometimes a neural network can represent the target function exactly. We have discussed the case of exact learning of functions with finite support. This is equivalent with the fact that the network memorizes given data; the network can be replaced by a look-up table, having no ability to generalize to other new data.

Another exact representation result is Kolmogorov-Arnold-Sprecher's Theorem, which is related to the answer to the famous Hilbert's thirteenth problem. The theorem states that a two-hidden layer neural network can represent exactly any continuous function on the n-dimensional hypercube. This result was a joint effort of Arnold, Kolmogorov, and Sprecher around 1960s. This deep mathematical result has only a theoretical value for neural networks, the theorem being just existential and not constructive.

Irie and Miyake's integral formula is another theoretical result dealing with an integral formula, which states that arbitrary functions of finite energy can be represented by a three-layer network with an continuum of computational units.

The last part deals with the exact learning in the case of a continuum number of neurons with one and two hidden layers, and necessary conditions for learning continuous functions are developed. In this case the activations of the type $\sigma(w_j x + b_j)$ are replaced by integral kernels $K(s,t)$. The main result shows that the network can represent exactly continuous functions whose eigenvalues increase fast enough. Only the case of symmetric and nonnegative kernels can be fully treated, since in this case there are some mathematical tools already developed by Hilbert and Mercer. Deep neural network of continuum neurons is introduced, the case of shared kernels being the one with the nicest properties.

10.11   Exercises

Exercise 10.11.1 (a) Show that a multi-perceptron neural network with one hidden layer and 2 computing units in the hidden layer can implement exactly the XOR function, i.e., a function satisfying $g(0,0) = 0$, $g(0,1) = 1$, $g(1,0) = 1$, and $g(1,1) = 0$.
(b) Draw the network and state the weights and biases.

Exercise 10.11.2 (a) Show that there is a multi-perceptron neural neural network with one hidden layer that can implement exactly the following mapping:

$$(0,0,0) \to 1, \qquad (1,0,0) \to 0, \qquad (0,1,0) \to 0, \qquad (0,0,1) \to 0,$$

$$(0,1,1) \to 0, \qquad (1,1,0) \to 0, \qquad (1,0,1) \to 0, \qquad (1,1,1) \to 1?$$

(b) Draw the network and state the weights and biases of all perceptrons in the network.

Exercise 10.11.3 Write the Irie-Miyake formula for the exponential function $\psi(x) = e^{-x^2}$.

Exercise 10.11.4 Construct a continuous function $f \in C[0,1]$ that cannot be learned exactly by a neural network with a logistic sigmoid activation function.

Exercise 10.11.5 Prove a similar result to Lemma 10.6.5 for deep networks with n hidden layers.

Exercise 10.11.6 Let K be a symmetric kernel with eigenvalues λ_n. Show that the series $\sum_{i \geq 1} \lambda_i^n$ is convergent for any positive integer n.

Exercise 10.11.7 Consider the data points $\{(1,1), (3,3), (5,2), (7,1)\}$.
(a) Find a simple function $c(x)$, which learns exactly the given data;
(b) Find α_j and θ_j such that $G(x) = \sum_{j=1}^{4} \alpha_j H(x - \theta_j)$ learns exactly the given data.

Exercise 10.11.8 Consider $f_1 : \mathbb{R}^n \to R^m$ and $f_2 : \mathbb{R}^m \to \mathbb{R}^k$ be two functions that are represented exactly by two feedforward networks with activation function ϕ. Show that there is a neural network, having activation function ϕ, which can represent the composed function $f_2 \circ f_1 : \mathbb{R}^n \to \mathbb{R}^k$.

Part III
Information Processing

Chapter 11

Information Representation

This chapter deals with the information representation in neural networks and the description of the information content of several types of neurons and networks using the concept of sigma-algebra. The main idea is to describe the evolution of the information content through the layers of a network. The network's input is considered to be a random variable, being characterized by a certain information. Consequently, all network layer activations will be random variables carrying forward some subset of the input information, which are described by some sigma-fields. From this point of view, neural networks can be interpreted as information processors.

The neural network's ability to generalize consists in using only that part of the input information that is useful for the task at hand. This is a procedure by which most of the input information is discarded by a selective process of extracting only those useful characteristic features. The output of the network caries some information, which is a sub-sigma-algebra of the input information and contains the useful information for the generalization. For instance, if the inputs are pictures of males and females, we assume the network has to figure out the gender of the person. Knowing that the gender is a 1-bit of information, the network has to selectively decrease the information content from several bits of initial information of each input picture to only the useful needed information of 1 bit.

This process of compressing the information through the layers of a network involves a certain content of information that is lost at each layer of the network. We define the uncompressed layers by means of minimum lost information and we shall study their properties.

© Springer Nature Switzerland AG 2020
O. Calin, *Deep Learning Architectures*, Springer Series in the Data Sciences,
https://doi.org/10.1007/978-3-030-36721-3_11

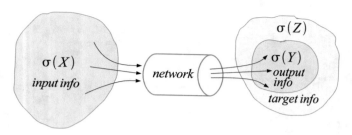

Figure 11.1: *Neural network as an information approximator: input, output and desired information fields.*

Before reading this chapter, the reader who is not very familiar with measure and probability theory notions is recommended to consult the Appendix.

11.1   Information Content

In this chapter we shall assume that the input, output, and target of a neural network are random variables. To be precise, let Ω denote the sample space of a probability space and $\mathcal{H}$ be a sigma-field on Ω. The input X is a random variable, $X : \Omega \to \mathbb{R}^n$, which maps each state $\omega \in \Omega$ into n real numbers $\big(X_1(\omega), \dots, X_n(\omega)\big) \in \mathbb{R}^n$. If g denotes the input-output mapping of the network, then the output $Y = g(X)$ is also a random variable. The network's target variable is considered to be a real-valued random variable, $Z : \Omega \to \mathbb{R}$.

We shall introduce next three bodies of information useful for the study of neural networks using the concept of sigma-algebra. They basically specify the set of possible events which can be potentially observed either at the input of the network, or at the output, or in the target. See Fig. 11.1 for a diagram of the information flows.

The *input information* represents the sigma-algebra generated by the input variable X and is denoted by $\mathcal{I}$. It contains all the events that can be observed from the given data.

The *output information* is the sigma-algebra generated by the output variable Y and is denoted by $\mathcal{E}$. This is the set of events that are observed at the output of the network. Since the output depends on the input in a deterministic way, namely, $Y = g(X)$ with g measurable function, then Proposition D.5.1 implies the inclusion $\mathfrak{S}(Y) \subset \mathfrak{S}(X)$, or $\mathcal{E} \subset \mathcal{I}$. This means the output information is always coarser than the input information.

The *target information* is the sigma-algebra generated by the target variable Z, namely, $\mathcal{Z} = \mathfrak{S}(Z)$. In general, it is expected to have $\mathcal{Z} \subset \mathcal{E}$, since we want to learn a finer set of events from a coarser set. However, there is no inclusion relation between the target and the input information, but if

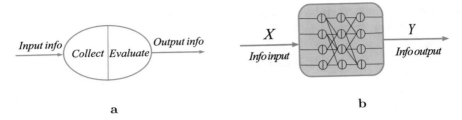

Figure 11.2: **a**. *Neuron as an information processing unit.* **b**. *Information flows in and out of a neural network.*

$\mathcal{Z} \subset \mathcal{I}$, we say that Z is *learnable* from X. This means that all events in $\mathcal{Z}$ are also events in $\mathcal{I}$. Equivalently, Z is determined by X in the sense that exists a measurable function f such that $Z = f(X)$.

The *error information* is the sigma-algebra generated by the events that can be observed in the target but not in the output information, namely, $\mathcal{R} = \mathfrak{S}(\mathcal{Z} \backslash \mathcal{E})$. Here, we assumed the inclusion $\mathcal{Z} \subset \mathcal{E}$ satisfied. In the particular case $\mathcal{Z} = \mathcal{E}$ the error information is minimal, $\mathcal{R} = \{\emptyset, \Omega\}$.

A neuron can be seen as an information processing unit consisting of two blocks, the first collecting the input information $\mathcal{I}$, and the second one evaluating the information using a system of weights and an activation function, see Fig. 11.2 **a**. Similarly, a neural network is a special case of a black box in which information $\mathcal{I}$ goes in and information $\mathcal{E}$ comes out, see Fig. 11.2 **b**. The hope is to be able to use the output information to learn the target information. That involves tuning the network parameters to obtain an ascending sequence of sigma-algebras, $\mathcal{E}_n$, which tends to $\mathcal{Z}$. This means $\mathcal{E}_n \subset \mathcal{E}_{n+1}$ with $\bigcup_n \mathcal{E}_n = \mathcal{Z}$. The errors of this approximation are given by the events that can be observed in the target but not in the output sequence. They generate the descending sequence of sigma-algebras $\mathcal{R}_n = \mathfrak{S}(\mathcal{Z} \backslash \mathcal{E}_n)$. The coarser the $\mathcal{R}_n$, the better the learning of Z is.

The pattern qualities of these information types are described by some distribution measures. We shall discuss them next.

The pattern of relative frequency of occurrence of X is described by the distribution measure, μ, of X. This can be induced from a probability measure $\mathbb{P}$ defined on the measurable space $(\Omega, \mathcal{H})$ as in the following

$$\mu(A) = \mathbb{P}(\omega \in X^{-1}(A)) = \mathbb{P}(\omega; X(\omega) \in A) = \mathbb{P}(X \in A), \qquad \forall A \in \mathcal{B}(\mathbb{R}^n),$$

i.e., $\mu(A)$ measures the chance that X takes values in the measurable set A. Using symbolic infinitesimal notation, we shall also write the previous relation informally as $d\mu(x) = \mu(dx) = \mathbb{P}(X \in dx)$. Since $\mu(\mathbb{R}^n) = \mathbb{P}(X \in \mathbb{R}^n) = 1$, it follows that μ is a probability measure on the measurable space $(\mathbb{R}^n, \mathcal{B}(\mathbb{R}^n))$.

The joint random variable of the input and target, $(X, Z) : \Omega \to \mathbb{R}^n \times \mathbb{R}$, defines *training evens*

$$\{\omega; X(\omega) \in A, Z(\omega) \in B\} \in \mathcal{H}$$

for any Borel sets $A \in \mathcal{B}(\mathbb{R}^n)$ and $B \in \mathcal{B}(\mathbb{R})$. The probability of each training event is given by

$$\rho(B, D) = \mathbb{P}(X \in A, Z \in B), \quad \forall A \in \mathcal{B}(\mathbb{R}^n), B \in \mathcal{B}(\mathbb{R}),$$

where ρ is the joint distribution measure of (X, Z), which describes the training pattern and is called the *training measure*. This can be equivalently written as[1]

$$d\rho(x, z) = \mathbb{P}(X \in dx, Z \in dz).$$

We shall present next a few examples concerned with characterizing the events that random variables can represent.

Example 11.1.1 We can make a comparison between learning of a neural network and understanding process of a human mind. The mental concepts are similar to the elements ω of the sample space Ω of a probability space. The information contained in the mind corresponds to a certain $\mathfrak{S}$-algebra $\mathcal{E}$ on Ω. The mind is able to understand an exterior body of information $\mathcal{Z}$ if its own information is finer than $\mathcal{Z}$. If the $\mathfrak{S}$-algebra $\mathcal{Z}$ is not contained into $\mathcal{E}$, then the mind can partially understand only the intersection information $\mathcal{E} \cap \mathcal{Z}$.

Example 11.1.2 As an extreme case, when $Z = c$, constant, then the target information $\mathcal{Z} = \mathfrak{S}(Z) = \{\emptyset, \Omega\}$ is the smallest possible $\mathfrak{S}$-field. This corresponds to the information with the least possible detail. This information is contained in any other body of information. For instance, we can think of Z as the random variable describing weather prediction at the North Pole. The information generated by Z in this case is minimal.

Example 11.1.3 If the target variable is an indicator function, $Z = 1_A$, with A measurable set, then the target information is given by $\mathcal{Z} = \mathfrak{S}(Z) = \{\emptyset, A, A^c, \Omega\}$. Consequently, the input information should contain A, i.e., $A \in \mathcal{I}$. The random variable

$$Z(\omega) = \begin{cases} 1, & \text{if } \omega \in A \\ 0, & \text{if } \omega \notin A \end{cases}$$

---

[1]In the case when ρ has a density, we write $d\rho(x, z) = p_{X,Z}(x, z)$, where $p_{X,Z}$ is the training density function.

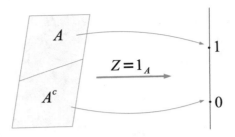

Figure 11.3: *Z acts as a space separator.*

acts as a separator between the set A and its complementary A^c, see Fig. 11.3. Therefore, learning Z is equivalent with learning the set A (or the set A^c). As an informal example, we can think of the random variable Z as classifying animals at the zoo into mammals and non-mammals.

Example 11.1.4 In general, if the target is a simple function $Z = \sum_{j=1}^{N} c_j 1_{A_j}$, with $\{A_j\}_j$ partition of Ω and $c_j \in \mathbb{R}$, then Z acts as a classifier, associating the number c_j as the label of the set A_j. The target information $\mathcal{Z}$ is the $\mathfrak{S}$-field generated by the sets $\{A_j\}_j$. In this case it is necessary that $A_j \in \mathcal{I}$.

11.2   Learnable Targets

This section will deal with the particular case of network target, Z, which is learnable from the input, X, namely, when the inclusion $\mathcal{Z} \subset \mathcal{I}$ holds.

   More precisely, if Z is learnable form X, then by Proposition D.5.1, there is a measurable function $f : \mathbb{R}^n \to \mathbb{R}$ such that $Z = f(X)$, see Fig. 11.4 **a**. Hence, to learn Z it suffices to approximate the measurable mapping f. Therefore, it suffices to show that a neural network behaves as an universal approximator for measurable functions $f \in \mathcal{M}^n$. This result was presented in Chapter 9, section 9.3.4 and will be used in the proof of the main result.

Before presenting the results, we prepare the ground with two summary statements:

(*i*) *What is given:* The input random variable X, its distribution measure ν, and the target random variable Z, as well as the training measure ρ.

(*ii*) *What is to be learned:* The function f, which satisfies $Z = f(X)$. In the learning process the function f is approximated by a measurable function g, where the random variable $Y = g(X)$ is the network output.

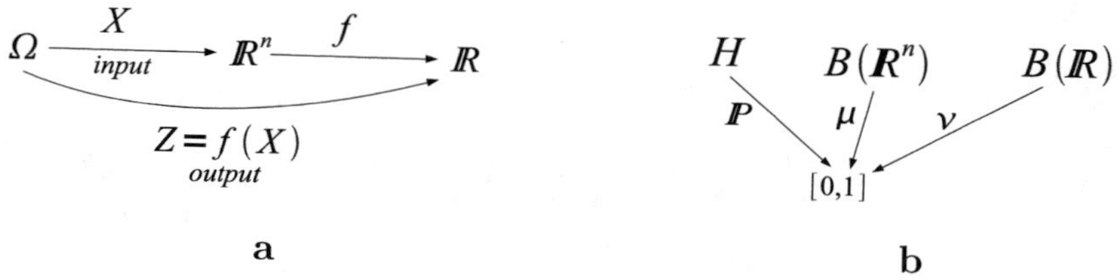

Figure 11.4: **a.** *The input and output variables.* **b.** *The measures* $\mathbb{P}$, μ, *and* ν.

We also note that learning the function f also implies learning the distribution measure of Z, denoted by ν, see Fig. 11.4 **b**. This follows from

$$
\begin{aligned}
\nu(B) &= \mathbb{P}(Z \in B) = \mathbb{P}(f(X) \in B) = \mathbb{P}(X \in f^{-1}(B)) \\
&= \mu\big(f^{-1}(B)\big) = \mu \circ f^{-1}(B), \qquad \forall B \in \mathcal{B}(\mathbb{R}),
\end{aligned}
$$

and the fact that μ and f^{-1} are either given or learnable. The relation with the training measure is given by

$$
\nu(B) = \mathbb{P}(Z \in B) = \mathbb{P}(X \in \mathbb{R}^n, Z \in B) = \rho(\mathbb{R}^n, B), \qquad \forall B \in \mathcal{B}(\mathbb{R}),
$$

which shows that the training measure determines the measure ν.

The first result of this section formulates an analog of Theorem 9.3.24 in terms of random variables. The theorem states that the target random variable, which are learnable, can be actually learned in the probability sense (for definition see Appendix, section D.6).

Theorem 11.2.1 *Consider a one-hidden layer neural network with a continuous sigmoidal activation function (in the sense of Definition 2.2.1) for the hidden layer and one linear neuron in the output layer. Given an n-dimensional input random variable, X, and an one-dimensional learnable target random variable, Z, there is a sequence of output random variables $(Y_k)_k$ produced by the network,*

$$
Y_k = \sum_{j=1}^{N_k} \alpha_j \varphi(w_j^T X + \theta_j), \qquad w_j \in \mathbb{R}^n, \alpha_j, \theta_j \in \mathbb{R},
$$

such that Y_k converges to Z in probability.

Proof: Since Z is learnable, by Proposition D.5.1 there is a measurable function $f \in \mathcal{M}^n$ such that $Z = f(X)$. From Theorem 9.3.24 there is a sequence of functions G_k of the type

$$
G_k(x) = \sum_{j=1}^{N_k} \alpha_j \varphi(w_j^T x + \theta_j), \qquad w_j \in \mathbb{R}^n, \alpha_j, \theta_j \in \mathbb{R}
$$

such that $d_\mu(G_k, f) \to 0$, as $k \to \infty$, where μ denotes the distribution measure of X.

Let $\epsilon > 0$, arbitrary fixed. Consider the set

$$A_k = \{x; |G_k(x) - f(x)| > \epsilon\} \subset \mathbb{R}^n.$$

The previous d_μ-convergence can be written as $\mu(A_k) \to 0$, $k \to \infty$. Note also that

$$\{\omega; X(\omega) \in A_k\} = \{\omega; |G_k(X(\omega)) - f(X(\omega))| > \epsilon\}.$$

Define $Y_k = G_k(X)$, see Fig. 11.5 **a**. Then

$$\begin{aligned}
\mathbb{P}(\omega; |Y_k(\omega) - Z(\omega)| > \epsilon) &= \mathbb{P}(\omega; |G_k(X(\omega)) - f(X(\omega))| > \epsilon) \\
&= \mathbb{P}(\omega; X(\omega) \in A_k) = \mathbb{P}(X^{-1}(A_k)) \\
&= \mu(A_k) \to 0, \quad k \to \infty.
\end{aligned}$$

Hence, Y_k converges in probability to Z. ∎

One possible interpretation of this theorem is given in the following. Assume that the neural network is a machine that can estimate the age $Z(\omega)$ of a person ω. The input $X(\omega)$ represents n features of the person ω (weight, height, fat body index, etc.), while the random variables $Y_k(\omega)$ are estimates of the person's age. In the beginning the estimates given by the machine will be very likely to be off, but after some practice the probability to estimate accurate the person's age will increase and in the long run will tend to 1.

But no matter how skilful the machine can be, there is always a small probability of making an error. If these errors are eliminated, we obtain a sequence that converges almost surely to the target variable, i.e., the age of the person. Next we present this consequence.

Corollary 11.2.2 *In the hypothesis of Theorem 11.2.1 there is a sequence of random variables $(Y_j)_j$ produced by the network such that $Y_j \to Z$ almost surely.*

Proof: By Theorem 11.2.1 the network can produce a sequence of random variables $(Y_k)_k$ such that Y_k converges to Z in probability. Since the estimation error, $|Y_k - Z|$, can be made arbitrary small with probability as close as possible to 1, fixing an error $\epsilon_j = \frac{1}{j}$, the probability of being off by more than ϵ_j decreases to zero as

$$\mathbb{P}(|Y_k - Z| > \epsilon_j) \to 0, \qquad k \to \infty.$$

This implies the existence of an integer $j_0 \geq 1$ and of a subsequence k_j such that

$$\mathbb{P}(|Y_{k_j} - Z| > \epsilon_j) < \frac{1}{2^j}, \qquad \forall j \geq j_0.$$

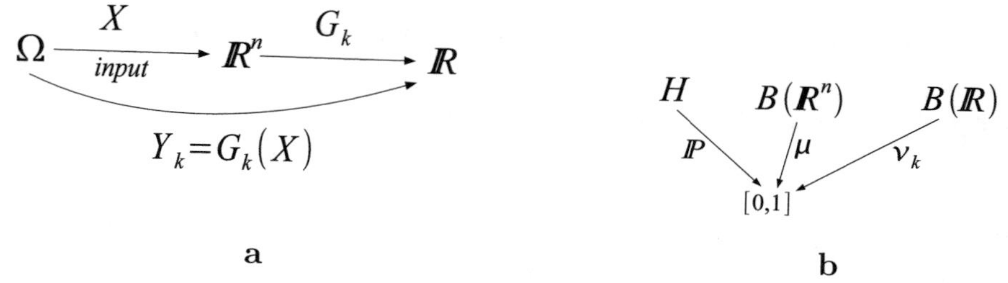

Figure 11.5: **a.** *The estimations* Y_k. **b.** *The distribution measures* ν_k *of* Y_k.

Taking the sum yields

$$\sum_{j \geq j_0} \mathbb{P}(|Y_{k_j} - Z| > \epsilon_j) \leq \sum_{j \geq j_0} \frac{1}{2^j} < 1.$$

Applying a variant of Borel-Cantelli lemma, see Appendix, Proposition D.6.2, implies that Y_{k_j} converges to Z almost surely. Renaming Y_{k_j} by Y_j and eliminating all the other terms from the sequence, we arrive at the desired sequence of estimations. ∎

In the input process we notice some frequency patterns for the person's features (weight, height, fat body index, etc.), which can be modeled by the distribution measure μ of X. We are interested in learning the distribution of the person's age, i.e., in learning the distribution measure ν of Z. The next result shows that we are able to learn this from the patterns induced by the estimations Y_1, Y_2, Y_3, etc. The distribution measure of Y_k is denoted by ν_k, with $\nu_k : \mathcal{B}(\mathbb{R}) \to \mathbb{R}_+$ and $\nu_k(B) = \mathbb{P}(Y_k \in B) = \mu \circ G_k^{-1}(B)$, for all $B \in \mathcal{B}(\mathbb{R})$, see Fig. 11.5 **b**. The concepts of weak convergence and convergence in distribution, which will be used shortly, can be found in section D.6 of the Appendix.

Theorem 11.2.3 *Let* $(Y_k)_k$ *be the sequence of approximations provided by Theorem 11.2.1. If* ν_k *and* ν *denote the distribution measures of the output* Y_k *and target* Z, *respectively, then the measures* $(\nu_k)_k$ *converge weakly to the measure* ν.

Proof: By Theorem 11.2.1 the sequence Y_k converges to Z in probability. Therefore, Y_k also converges to Z in distribution. This means that for any bounded continuous function $g : \mathbb{R} \to \mathbb{R}$ we have

$$\mathbb{E}(g \circ Y_k) \to \mathbb{E}(g \circ Z), \quad k \to \infty.$$

This can be written equivalently using measures as

$$\int_{\mathbb{R}} g(x) \, d\mu_k(x) \to \int_{\mathbb{R}} g(x) \, d\mu(x), \quad k \to \infty,$$

which means that $\nu_k \to \nu$ weakly, as $k \to \infty$.

■

11.3 Lost Information

We have seen that for any neural network the non-strict inclusion $\mathcal{E} \subset \mathcal{I}$ always holds, as a consequence of the fact that the output, Y, is determined by the input, X. Therefore, there might be events potentially observed in $\mathcal{I}$, which are not observed in $\mathcal{E}$. Roughly speaking, these events represent the *lost information* in the network. This concept is formalized in the next definition.

Definition 11.3.1 *The lost information in a neural network is the sigma-algebra generated by the sets that belong to the input information and do not belong to the output information,*

$$\mathcal{L} = \mathfrak{S}(\mathcal{I} \backslash \mathcal{E}).$$

The lost information describes the compressibility of the information in a neural network. The uncompressed case, given by $\mathcal{I} = \mathcal{E}$, implies $\mathcal{L} = \{\varnothing, \Omega\}$, which is the coarsest sigma-algebra. The total compression case occurs for $\mathcal{E} = \{\varnothing, \Omega\}$, which implies $\mathcal{L} = \mathcal{I}$, namely, the lost information is maximum (and then the network becomes useless).

Example 11.3.2 The input of a classical perceptron is given by a random variable X, which takes only the values 0 and 1. Let $A = X^{-1}(\{1\})$, so we have $X = 1_A$, i.e., X is the indicator function of the set A. The input information is generated by A and is given by

$$\mathcal{I} = \mathfrak{S}(X) = \{\varnothing, A, A^c, \Omega\}.$$

Consider the output variable $Y = f(X)$, with $f(1) = a$ and $f(0) = b$. Then the output information is $\mathcal{E} = \sigma(Y)$ and its content depends on the following two cases. If $a \neq b$, then $\mathcal{E} = \mathcal{I}$, and the lost information $\mathcal{L}$ is trivial. If $a = b$, then $\mathcal{E} = \{\varnothing, \Omega\}$, and then $\mathcal{L} = \mathcal{I}$, i.e., in this case all information is lost.

The next proposition provides an equivalent characterization of the lost information.

Proposition 11.3.3 *Let $\mathcal{I}$, $\mathcal{E}$, and $\mathcal{L}$ denote the input, output, and lost information of a neural network, respectively. The following hold:*
(a) The input information decomposes as

$$\mathcal{I} = \mathfrak{S}(\mathcal{L} \cup \mathcal{E}) = \mathcal{L} \vee \mathcal{E}.$$

(b) $\mathcal{L}$ is the largest field which satisfies (a).

Proof: (a) It will be shown by double inclusion. First, since $\mathcal{L} \subset \mathcal{I}$ and $\mathcal{E} \subset \mathcal{I}$, then $\mathcal{L} \cup \mathcal{E} \subset \mathcal{I}$. Using the monotonicity of the operator $\mathfrak{S}$ yields

$$\mathfrak{S}(\mathcal{L} \cup \mathcal{E}) \subset \mathfrak{S}(\mathcal{I}) = \mathcal{I}.$$

For the inverse inclusion, we start from $\mathfrak{S}(\mathcal{I} \backslash \mathcal{E}) = \mathcal{L}$, so $\mathcal{I} \backslash \mathcal{E} \subset \mathcal{L}$. Then taking in both sides the union with $\mathcal{E}$, we obtain

$$(\mathcal{I} \backslash \mathcal{E}) \cup \mathcal{E} \subset \mathcal{L} \cup \mathcal{E},$$

which is equivalent to $\mathcal{I} \subset \mathcal{L} \cup \mathcal{E}$. Taking the operator $\mathfrak{S}$ yields

$$\mathfrak{S}(\mathcal{I}) \subset \mathfrak{S}(\mathcal{L} \cup \mathcal{E}) = \mathcal{L} \vee \mathcal{E}.$$

Using that $\mathcal{I} = \mathfrak{S}(\mathcal{I})$ it follows that $\mathcal{I} \subset \mathfrak{S}(\mathcal{L} \cup \mathcal{E})$.
(b) Let $\mathcal{L}'$ be another $\mathfrak{S}$-field which verifies $\mathcal{I} = \mathfrak{S}(\mathcal{L}' \cup \mathcal{E})$. In order to show that $\mathcal{L}$ is the largest field which satisfies (a), it suffices to show that $\mathcal{L}' \subset \mathcal{L}$, i.e., $\mathcal{L}' \subset \mathfrak{S}(\mathcal{I} \backslash \mathcal{E})$. The right side can be written as

$$\mathfrak{S}(\mathcal{I} \backslash \mathcal{E}) = \mathfrak{S}(\mathfrak{S}(\mathcal{L}' \cup \mathcal{E}) \backslash \mathcal{E}).$$

Therefore, it suffices to show the inclusion

$$\mathcal{L}' \subset \mathfrak{S}(\mathfrak{S}(\mathcal{L}' \cup \mathcal{E}) \backslash \mathcal{E}).$$

This holds as a strict inclusion since the information $\mathcal{L}'$ is still contained into the right side, but after extracting $\mathcal{E}$ there might be some sets left that do not belong to $\mathcal{L}'$.

■

To see how the previous result works on a concrete case, we consider the following example.

Example 11.3.1 Let $A, B \in \mathcal{I}$ be two sets and consider the $\mathfrak{S}$-fields

$$\mathcal{L}' = \{\emptyset, A, A^c, \Omega\}, \quad \mathcal{E} = \{\emptyset, B, B^c, \Omega\}.$$

Then $\mathcal{L}' \cup \mathcal{E} = \{\emptyset, A, B, A^c, B^c, \Omega\}$, while $\mathfrak{S}(\mathcal{L}' \cup \mathcal{E})$ is the sigma-field generated by A and B and has 16 elements, see Exercise 11.10.3. Therefore, taking away the elements of $\mathcal{E}$, it follows that $\mathfrak{S}(\mathcal{L}' \cup \mathcal{E}) \backslash \mathcal{E}$ will have only 12 elements. Among these elements there are also the sets A and A^c. Then $\mathfrak{S}(\mathfrak{S}(\mathcal{L}' \cup \mathcal{E}) \backslash \mathcal{E})$ will contain, besides A, A^c, and others, also $\emptyset$ and Ω. Hence, $\mathcal{L}' \subset \mathfrak{S}(\mathfrak{S}(\mathcal{L}' \cup \mathcal{E}) \backslash \mathcal{E})$, with a strict inclusion, since sets such as $A \cap B$ do not belong to $\mathcal{L}'$ but belong to the $\mathfrak{S}$-field on the right side.

The relation with the target information is provided next.

Proposition 11.3.4 *Assume the target Z of a neural network is learnable from the input X. Then*

$$\mathcal{R} \subset \mathcal{L}, \tag{11.3.1}$$

where $\mathcal{R}$ is the error information and $\mathcal{L}$ is the lost information of the network.

Proof: Since Z is learnable from X, we have $\mathcal{Z} \subset \mathcal{I}$. Extracting the set $\mathcal{E}$ from both sides of the previous inclusion, we get $\mathcal{Z} \backslash \mathcal{E} \subset \mathcal{I} \backslash \mathcal{E}$. Using the monotonicity of the operator $\mathfrak{S}$ we obtain

$$\mathfrak{S}(\mathcal{Z} \backslash \mathcal{E}) \subset \mathfrak{S}(\mathcal{I} \backslash \mathcal{E}),$$

which is equivalent to relation (11.3.1). ∎

11.4 Recoverable Information

The concept of information recovery is related to the question of *how much information can be excluded from a $\mathfrak{S}$-field, such that it can be recovered from the remaining information by unions, intersections, and complement operations?*

The main result of this section states that the lost information of a neural network is non-recoverable. Therefore, in a sense, *lost* and *recovered* information have opposite connotations.

Definition 11.4.1 (*a*) *A $\mathfrak{S}$-algebra $\mathcal{F}$ in $\mathcal{I}$ is called recoverable if $\mathfrak{S}(\mathcal{I} \backslash \mathcal{F}) = \mathcal{I}$.*
(*b*) *A $\mathfrak{S}$-algebra $\mathcal{H}$ in $\mathcal{I}$ is called non-recoverable if $\mathfrak{S}(\mathcal{I} \backslash \mathcal{H}) \neq \mathcal{I}$.*

Two $\mathfrak{S}$-fields $\mathcal{A}$ and $\mathcal{B}$ are *comparable* if either $\mathcal{A} \subseteq \mathcal{B}$ or $\mathcal{B} \subseteq \mathcal{A}$. An ascending sequence of $\mathfrak{S}$-fields oriented by inclusion is called a *filtration*. For more details on filtrations, see the Appendix. Also, note that if any ascending sequence of $\mathfrak{S}$-fields is included in a given $\mathfrak{S}$-field $\mathcal{I}$, then by Zorn's lemma (see Lemma A.0.3 in the Appendix) there is a maximal element of the sequence. This is the existence basis of the next concepts.

A $\mathfrak{S}$-algebra $\mathcal{F}$ is *maximal recoverable* if it satisfies $\mathfrak{S}(\mathcal{I} \backslash \mathcal{F}) = \mathcal{I}$ and is maximal with respect to inclusion. Similarly, a $\mathfrak{S}$-algebra $\mathcal{H}$ is *minimal non-recoverable* if $\mathfrak{S}(\mathcal{I} \backslash \mathcal{H}) \neq \mathcal{I}$ and it is minimal with respect to inclusion.

The following two properties are straightforward:
1. Any maximal recoverable information field, $\mathcal{F}$, is included in a minimal non-recoverable information field, $\mathcal{H}$.
2. Any minimal non-recoverable information field, $\mathcal{H}$ contains a maximal recoverable information field, $\mathcal{F}$.

It will be clear from the next examples that the information fields $\mathcal{F}$ and $\mathcal{H}$ are not unique.

Example 11.4.2 Consider the $\mathfrak{S}$-algebra generated by one measurable set, A, given by $\mathcal{I} = \{\varnothing, A, A^c, \Omega\}$. It is easy to show that the maximal recoverable information is the trivial one, $\mathcal{F} = \{\varnothing, \Omega\}$. It is obvious that the minimal non-recoverable information is the entire space, $\mathcal{H} = \mathcal{I}$.

Example 11.4.3 Let A, B be two measurable sets in Ω and consider

$$\mathcal{I} = \{\varnothing, A, B, A^c, B^c, A \cap B, A \cup B, A^c \cap B^c, A^c \cup B^c, A^c \cap B, A \cap B^c,$$
$$A^c \cup B, A \cup B^c, (A \cap B^c) \cup (A^c \cap B), (A \cap B) \cap (A \cap B)^c, \Omega\}$$

be the $\mathfrak{S}$-field generated by A and B. Then

$$\mathcal{F}_A = \{\varnothing, A, A^c, \Omega\}$$

is a maximal recoverable information. In order to show that $\mathfrak{S}(\mathcal{I} \backslash \mathcal{F}_A) = \mathcal{I}$, it suffices to recover the set A from the set $\mathcal{I} \backslash \mathcal{F}_A$ by using union, intersection, and complementarity. This can be done as in the following:

$$A = (A \backslash B) \cup (A \cap B) = (B^c \cap A) \cup (A \cap B).$$

Then A^c is obtained by taking the complementary and applying de Morgan's formulas. The sets $\varnothing$ and Ω are obviously obtained from A and A^c by taking intersection and union, respectively.

It is worth to note that the maximal recoverable information fields are not unique, but they have the same cardinal. Other maximal recoverable information fields are $\mathcal{F}_B = \{\varnothing, B, B^c, \Omega\}$, $\mathcal{F}_C = \{\varnothing, C, C^c, \Omega\}$, $\mathcal{F}_{A \cap B} = \{\varnothing, A \cap B, (A \cap B)^c, \Omega\}$, $\mathcal{F}_{A \triangle B} = \{\varnothing, A \triangle B, (A \triangle B)^c, \Omega\}$, where $A \triangle B = (A \backslash B) \cup (B \backslash A)$ is the symmetric difference, etc.

Example 11.4.4 Let $\{E_j\}_{j=1,N}$ be a finite partition of the sample space Ω. It is known that the $\mathfrak{S}$-algebra generated by the partition consists of unions of elements of the partition

$$\mathcal{I} = \{\varnothing, \ E_i, \ E_i \cup E_j, \ E_i \cup E_j \cup E_k, \ \dots, \bigcup_{i=1}^{N} E_i = \Omega\}.$$

Pick an element of the partition, say E_1, and consider the $\mathfrak{S}$-algebra generated by it

$$\mathcal{F}_{E_1} = \{\varnothing, E_1, \bigcup_{i=2}^{N} E_i, \Omega\}.$$

Similar reasoning with the previous example shows that $\mathcal{F}$ is a recoverable information. For maximality, see Exercise 11.10.5.

The rest of the section deals with two results regarding the lost information.

Proposition 11.4.5 *Let $\mathcal{L}$ denote the lost information. We have:*
(a) The output information, $\mathcal{E}$, is included in a maximal recoverable information field if and only if the lost information is maximal, $\mathcal{I} = \mathcal{L}$.
(b) If the output information, $\mathcal{E}$, includes a minimal non-recoverable information field, then $\mathcal{L} \neq \mathcal{I}$.

Proof: (a) " $\Longrightarrow$ " Let $\mathcal{F}$ be a maximal recoverable information field and assume $\mathcal{E} \subseteq \mathcal{F}$. Then $\mathcal{I} \backslash \mathcal{F} \subseteq \mathcal{I} \backslash \mathcal{E}$, and using the monotonicity of the operator $\mathfrak{S}$ we get

$$\mathfrak{S}(\mathcal{I} \backslash \mathcal{F}) \subseteq \mathfrak{S}(\mathcal{I} \backslash \mathcal{E}).$$

Since $\mathcal{F}$ is recoverable, $\mathfrak{S}(\mathcal{I} \backslash \mathcal{F}) = \mathcal{I}$, and using that the right side is the lost information, we get $\mathcal{I} \subseteq \mathcal{L}$, which in fact implies equality, $\mathcal{I} = \mathcal{L}$.

" $\Longleftarrow$ " Assume $\mathcal{I} = \mathcal{L}$, so $\mathcal{I} = \mathfrak{S}(\mathcal{I} \backslash \mathcal{E})$. This means the output information $\mathcal{E}$ is recoverable. Since it can be included in a maximal recoverable field (Zorn's lemma), we obtain the desired result.
(b) Assume $\mathcal{H} \subseteq \mathcal{E}$, with $\mathcal{H}$ minimal non-recoverable information field. Then $\mathcal{I} \backslash \mathcal{E} \subseteq \mathcal{I} \backslash \mathcal{H}$. This implies $\mathfrak{S}(\mathcal{I} \backslash \mathcal{E}) \subseteq \mathfrak{S}(\mathcal{I} \backslash \mathcal{H})$. Therefore $\mathcal{L} \subseteq \mathfrak{S}(\mathcal{I} \backslash \mathcal{H}) \subsetneqq \mathcal{I}$, and hence $\mathcal{L} \neq \mathcal{I}$. ∎

We show next that in general the lost information is non-recoverable, i.e., $\mathfrak{S}(\mathcal{I} \backslash \mathcal{L}) \neq \mathcal{I}$.

Theorem 11.4.6 (Recoverable lost information) *If the lost information is recoverable, then it is trivial, $\mathcal{L} = \{\emptyset, \Omega\}$.*

Proof: By contradiction, assume the lost information $\mathcal{L}$ is recoverable, namely,

$$\mathfrak{S}(\mathcal{I} \backslash \mathcal{L}) = \mathcal{I}. \tag{11.4.2}$$

Taking the complement in the obvious inclusion $\mathcal{L} = \mathfrak{S}(\mathcal{I} \backslash \mathcal{E}) \supset \mathcal{I} \backslash \mathcal{E}$ implies

$$\mathcal{I} \backslash \mathcal{L} \subset \mathcal{I} \backslash (\mathcal{I} \backslash \mathcal{E}) = \mathcal{E}.$$

Then applying again the operator $\mathfrak{S}$ we obtain

$$\mathcal{I} = \mathfrak{S}(\mathcal{I} \backslash \mathcal{L}) \subset \mathfrak{S}(\mathcal{E}) = \mathcal{E},$$

where the first equality uses (11.4.2) and the last one uses that $\mathcal{E}$ is a $\mathfrak{S}$-field. By transitivity, $\mathcal{I} \subset \mathcal{E}$, which is in fact identity, $\mathcal{I} = \mathcal{E}$. This implies that the lost information $\mathcal{L} = \mathfrak{S}(\mathcal{I} \backslash \mathcal{E}) = \{\emptyset, \Omega\}$ is trivial. ∎

The previous result can be also restated equivalently by saying that no nontrivial information can be recovered from the lost information.

11.5   Information Representation Examples

This section deals with characterizing the events that a few types of neurons and some simple networks can represent. These examples suggest that the more complex the input-output mapping is, the richer its associated sigma-algebra tends to be.

11.5.1   Information for indicator functions

The simplest nontrivial case of sigma-algebra is represented by indicator functions and their combinations. The following examples will be useful in later sections.

Example 11.5.1 Let $A \subset \Omega$ be a measurable set and consider the indicator function $X = 1_A$. Since

$$X(\omega) = \begin{cases} 1, & \text{if } \omega \in A \\ 0, & \text{if } \omega \in A^c \end{cases}$$

then for any $k \in \mathbb{R}$ we obtain

$$X^{-1}(-\infty, k] = \{\omega; X(\omega) \le k\} = \begin{cases} \Omega, & \text{if } k \ge 1 \\ A^c, & \text{if } 0 \le k < 1 \\ \emptyset, & \text{if } k < 0. \end{cases}$$

The $\mathfrak{S}$-algebra generated by X is generated by the pre-images $\{X^{-1}(-\infty, k], k \in \mathbb{R}\}$ and is given by $\mathfrak{S}(X) = \{\emptyset, A, A^c, \Omega\}$.

Example 11.5.2 Let $A, B \subset \Omega$ be measurable and disjoint, $A \cap B = \emptyset$. Consider the linear combination of indicators $X = 1_A + 2\,1_B$. Then

$$X(\omega) = \begin{cases} 2, & \text{if } \omega \in B \\ 1, & \text{if } \omega \in A \\ 0, & \text{if } \omega \notin A \cup B. \end{cases}$$

For any real k, the half-lines pre-images through X are given by

$$X^{-1}(-\infty, k] = \{\omega; X(\omega) \le k\} = \begin{cases} \Omega, & \text{if } k \ge 2 \\ B^c, & \text{if } 1 \le k < 2 \\ (A \cup B)^c, & \text{if } 0 \le k < 1 \\ \emptyset, & \text{if } k < 0. \end{cases}$$

The $\mathfrak{S}$-algebra generated by the pre-images $\{X^{-1}(-\infty, k], k \in \mathbb{R}\}$ provides the information field generated by X:

$$\mathfrak{S}(X) = \{\emptyset, A, A^c, B, B^c, A \cup B, (A \cup B)^c, \Omega\}.$$

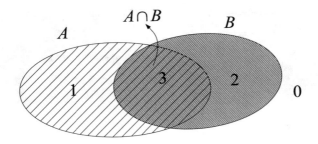

Figure 11.6: *Diagram for Example* 11.5.3.

Example 11.5.3 This is similar with Example 11.5.2, but under the hypothesis that the sets A and B are not disjoint. In this case we have

$$X(\omega) = 1_A(\omega) + 2\,1_B(\omega) = \begin{cases} 3, & \text{if } \omega \in A \cap B \\ 2, & \text{if } \omega \in B \cap A^c \\ 1, & \text{if } \omega \in A \cap B^c \\ 0, & \text{if } \omega \in (A \cup B)^c, \end{cases}$$

see Fig. 11.6. Then for any real k

$$X^{-1}(-\infty, k] = \begin{cases} \Omega, & \text{if } k \geq 3 \\ (A \cap B)^c, & \text{if } 2 \leq k < 3 \\ B^c, & \text{if } 1 \leq k < 2 \\ (A \cup B)^c, & \text{if } 0 \leq k < 1 \\ \varnothing, & \text{if } k < 0. \end{cases}$$

Since $(A \cup B)^c, (A \cap B)^c \in \mathfrak{S}(X)$, then $A \cup B, A \cap B \in \mathfrak{S}(X)$. Using the properties of intersection and union, we show next that A belongs to $\mathfrak{S}(X)$:

$$\begin{aligned} A &= A \cap \Omega = A \cap (B^c \cup B) \\ &= (A \cap B^c) \cup (A \cap B) \\ &= \underbrace{\left((A \cup B) \cap B^c\right)}_{\in \mathfrak{S}(X)} \cup \underbrace{(A \cap B)}_{\in \mathfrak{S}(X)} \in \mathfrak{S}(X). \end{aligned}$$

Since both A and B belong to the $\mathfrak{S}(X)$, it follows that $\mathfrak{S}(1_A, 1_B) \subset \mathfrak{S}(X)$. Since $X(\omega) = 1_A(\omega) + 2\,1_B(\omega)$ is $\mathfrak{S}(1_A, 1_B)$-measurable, then $\mathfrak{S}(X) \subset \mathfrak{S}(1_A, 1_B)$, from which we conclude $\mathfrak{S}(X) = \mathfrak{S}(1_A, 1_B)$.

We note that a similar analysis can be applied to the linear combination $X = \alpha 1_A + \beta 1_B$, with $\alpha \neq \beta$.

Example 11.5.4 Let $\{E_i\}_i$ be a countable partition of Ω and consider $X = \sum_i \alpha_i 1_{A_i}$, with distinct coefficients $\alpha_i \in \mathbb{R}$. It can be shown that $\mathfrak{S}(X)$ in this case is the set of countable unions of elements of the partition.

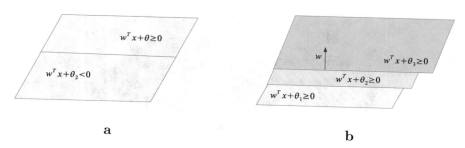

Figure 11.7: **a.** *Information generated by a perceptron.* **b.** *Information generated by a sigmoid or a linear neuron.*

11.5.2   Classical perceptron

In this case the input variable is $X : \Omega \to \mathbb{R}^n$, while the output is provided by $Y = H(w^T X + \theta)$, where H is the Heaviside step function and $w \in \mathbb{R}^n$ is a fixed system of weights and $\theta \in \mathbb{R}$ is a given threshold. Since

$$Y(\omega) = \begin{cases} 1, & \text{if } w^T X(\omega) + \theta \geq 0 \\ 0, & \text{if } w^T X(\omega) + \theta < 0 \end{cases} = \begin{cases} 1, & \text{if } X(\omega) \in \mathcal{H}^+_{w,\theta} \\ 0, & \text{if } X(\omega) \in \mathcal{H}^-_{w,\theta}, \end{cases}$$

then the output can be written as the indicator function $Y = 1_A$, with $A = X^{-1}(\mathcal{H}^+_{w,\theta})$, where $\mathcal{H}^+_{w,\theta} = \{x; w^T x + \theta \geq 0\}$ denotes a closed upper half-space in $\mathbb{R}^n$ with normal direction w.

Consequently, using Example 11.5.2, the $\mathfrak{S}$-algebra generated by the output Y is

$$\mathfrak{S}(Y) = \{\varnothing, A, A^c, \Omega\} = \{\varnothing, X^{-1}(\mathcal{H}^+_{w,\theta}), X^{-1}(\mathcal{H}^-_{w,\theta}), \Omega\},$$

where we used that

$$A^c = \big(X^{-1}(\mathcal{H}^+_{w,\theta})\big)^c = X^{-1}\big((\mathcal{H}^+_{w,\theta})^c\big) = X^{-1}(\mathcal{H}^-_{w,\theta}),$$

where

$$\mathcal{H}^-_{w,\theta} = \{x; w^T x + \theta < 0\}.$$

This can be further written as

$$\mathfrak{S}(Y) = X^{-1}\Big(\{\varnothing, \mathcal{H}^+_{w,\theta}, \mathcal{H}^-_{w,\theta}, \Omega\}\Big).$$

It is worth noting that the $\mathfrak{S}$-algebra $\{\varnothing, \mathcal{H}^+_{w,\theta}, \mathcal{H}^-_{w,\theta}, \Omega\}$ provides the information which classifies points above and below the hyperplane $w^T x + \theta = 0$. This is the reason why a classical perceptron can be used to classify two clusters, but no more than that, given the limited information it can accommodate, see Fig. 11.7 **a.**

11.5.3 Linear neuron

Consider the input $X : \Omega \to \mathbb{R}^n$ and the output $Y = w^T X + \theta$, where $w \in \mathbb{R}^n$ and $\theta \in \mathbb{R}$ are given. We are interested in finding the information field $\mathfrak{S}(Y)$. Since for any real number k we have

$$Y^{-1}(-\infty, k) = \{\omega; Y(\omega) < k\} = \{\omega; w^T X(\omega) + \theta < k\} = X^{-1}(\mathcal{H}^-_{w,\theta-k}),$$

then

$$\begin{aligned} \mathfrak{S}(Y) &= \mathfrak{S}\Big(Y^{-1}(-\infty, k); k \in \mathbb{R}\Big) = \mathfrak{S}\Big(X^{-1}(\mathcal{H}^-_{w,\theta-k}); k \in \mathbb{R}\Big) \\ &= X^{-1}\Big(\mathfrak{S}(\mathcal{H}^-_{w,\theta-k}; k \in \mathbb{R})\Big) = X^{-1}\Big(\mathfrak{S}(\mathcal{H}^-_{w,u}; u \in \mathbb{R})\Big) \\ &= X^{-1}\Big(\mathfrak{S}(\mathcal{H}^+_{w,u}; u \in \mathbb{R})\Big). \end{aligned}$$

The commutation between $\mathfrak{S}$ and X^{-1} follows from Proposition A.0.1 of the Appendix. Since u is a parameter, the hyperplanes $\{w^T x + u = 0\}$ are parallel. Hence, the upper half-spaces $\mathcal{H}^+_{w,u}$ are obtained by translating $\mathcal{H}^+_{w,0}$ a distance u along direction w, see Fig. 11.7 b. The field $\mathfrak{S}(\mathcal{H}^+_{w,u}; u \in \mathbb{R})$ accommodates a lot more information than the one corresponding to the classical perceptron described in section 11.5.2. Containing infinitely many parallel strips, it defines the information of a field of parallel hyperplanes in the space.

11.5.4 Sigmoid neuron

The input is given by the random variable $X : \Omega \to \mathbb{R}^n$ and the output by $Y = \sigma(w^T X + \theta)$, where σ is the logistic sigmoid activation function, and $w \in \mathbb{R}^n$, $\theta \in \mathbb{R}$ are fixed. Using the monotonicity and invertibility of the logistic function, for any $k \in (0, 1)$, we have

$$\begin{aligned} Y^{-1}(\infty, k) &= \{\omega; Y(\omega) < k\} = \{\omega; \sigma(w^T X(\omega) + \theta) < k\} \\ &= \{\omega; w^T X(\omega) + \theta < \sigma^{-1}(k)\} \\ &= \{\omega; X(\omega) \in \mathcal{H}_{w,\theta-\sigma^{-1}(k)}\} \\ &= X^{-1}(\mathcal{H}^-_{w,\theta-\sigma^{-1}(k)}) = X^{-1}(\mathcal{H}^-_{w,u}), \end{aligned}$$

with $u = \sigma^{-1}(k)$ real parameter. The $\mathfrak{S}$-field generated by Y is

$$\begin{aligned} \mathfrak{S}(Y) &= \mathfrak{S}\Big(Y^{-1}(\infty, k); k \in [0, 1]\Big) = \mathfrak{S}\Big(X^{-1}(\mathcal{H}^-_{w,u}); u \in \mathbb{R}\Big) \\ &= X^{-1}\Big(\mathfrak{S}(\mathcal{H}^-_{w,u}); u \in \mathbb{R}\Big) = X^{-1}\Big(\mathfrak{S}(\mathcal{H}^+_{w,u}); u \in \mathbb{R}\Big), \end{aligned}$$

where we used Proposition A.0.1. We have obtained the same $\mathfrak{S}$-field as in the case of the linear neuron given in section 11.5.3, so all conclusions mentioned there also apply here. Since this field is independent of the activation

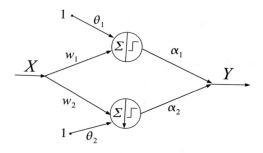

Figure 11.8: *Neural networks made of two perceptrons. Its output is given by* $Y = \alpha_1 H(w_1^T X + \theta_1) + \alpha_2 H(w_2^T X + \theta_2).$

function σ, we infer that all sigmoid neurons can learn the same amount of information as long as the activation function σ enjoys all previous used properties. Consequently, a sigmoidal neuron can learn a lot more than a classical perceptron. Note that the nonlinearity of the activation function σ does not have any impact on the information produced. We shall see that this will no longer hold for a network of this type of neurons, where the nonlinearity plays an important role.

11.5.5 Arctangent neuron

This is a particular case of a sigmoid neuron, which will be used again in section 11.5.8. The output variable is given by $Y = \tan^{-1}(w^T X + \theta)$ and takes values in $(-\pi/2, \pi/2)$. The pre-image of a half-line is

$$
\begin{aligned}
Y^{-1}(-\infty, k) = \{Y < k\} &= \{w^T X + \theta < \tan k\} = X^{-1}(\mathcal{H}^-_{w, \theta - \tan k}) \\
&= X^{-1}(\mathcal{H}^-_{w, u}),
\end{aligned}
$$

with $u \in \mathbb{R}$ real parameter. We have, like in the previous section,

$$
\mathfrak{S}(Y) = X^{-1}\Big(\mathfrak{S}\{\mathcal{H}^-_{w, u}; u \in \mathbb{R}\}\Big).
$$

11.5.6 Neural network of classical perceptrons

Consider a neural network with one hidden layer that has two units, each being a perceptron with a step activation function, see Fig. 11.8. There are two sets of weights: from the input to the hidden layer, $w_1, w_2 \in \mathbb{R}^n$, two thresholds, $\theta_1, \theta_2 \in \mathbb{R}$, and two weights from the hidden layer to the output layer, $\alpha_1, \alpha_2 \in \mathbb{R}$. We are interested to describe the information represented by this network.

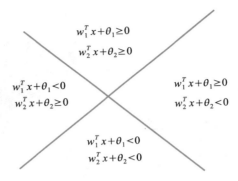

Figure 11.9: *Information provided by a network of two classical perceptrons.*

We shall write the outcome as a linear combination of two indicator functions:

$$
\begin{aligned}
Y(\omega) &= \alpha_1 H(w_1^T X(\omega) + \theta_1) + \alpha_2 H(w_2^T X(\omega) + \theta_2) \\
&= \alpha_1 1_{\{w_1^T X(\omega)+\theta_1 \geq 0\}} + \alpha_2 1_{\{w_2^T X(\omega)+\theta_2 \geq 0\}} \\
&= \alpha_1 1_{X^{-1}(\mathcal{H}^+_{w_1,\theta_1})} + \alpha_2 1_{X^{-1}(\mathcal{H}^+_{w_2,\theta_2})} \\
&= \alpha_1 1_{A_1} + \alpha_2 1_{A_2},
\end{aligned}
$$

where $A_i = X^{-1}(\mathcal{H}^+_{w_i,\theta_i}))$ is the pre-image of an upper half-space through X. Assume $\alpha_1 \neq \alpha_2$. Since we are in the conditions of Example 11.5.3, we have

$$
\mathfrak{S}(Y) = \mathfrak{S}(1_{A_1}, 1_{A_2}) = X^{-1}(\mathfrak{S}(1_{\mathcal{H}^+_{w_1,\theta_1}}, 1_{\mathcal{H}^+_{w_2,\theta_2}})).
$$

The sigma-algebra $\mathfrak{S}(1_{\mathcal{H}^+_{w_1,\theta_1}}, 1_{\mathcal{H}^+_{w_2,\theta_2}})$ contains the upper and lower half-spaces, $\mathcal{H}^+_{w_i,\theta_i}, \mathcal{H}^-_{w_i,\theta_i}, i = 1,2$, and all their combinations obtained by intersection and union, see Fig. 11.9. In particular, since it divides the space into at most four disjoint regions, the network can classify at most four clusters of points. This information is definitely richer than the one obtained in the case of a single classical perceptron.

When the number of units N in the hidden layer increases, the information structure gets more complex; this can be used to classify a given number of clusters. For a space of arbitrary dimension n this number might not be straightforward to obtain, but for $n = 2$, the maximal number of clusters that can be classified, i.e., the maximum number of regions obtained, is given by the formula $N(N + 1)/2 + 1$, as it can be shown by induction over N.

Before attempting to find the information produced by a network of sigmoid neurons, we need first the following preparation.

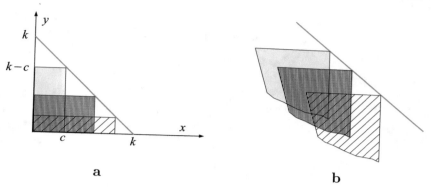

Figure 11.10: **a.** *Triangle as a union of rectangles.* **b.** *Half-space as a union of infinite rectangles.*

11.5.7 Triangle as a union of rectangles

Consider a triangle in the plane given by $T = \{x > 0, y > 0, x + y < k\}$, see Fig. 11.10 **a**. For any $c \in (0, k)$ the rectangle $R_c = [0, c] \times [0, k - c]$ is inscribed in the triangle. When the parameter c varies from 0 to k, the rectangle union covers the entire triangle $T = \bigcup_{0 < c < k} R_c$. By density reasons we may assume that c takes only rational values, $c \in \mathbb{Q}$, case in which the union becomes countable

$$T = \bigcup_{\substack{0 < c < k \\ c \in \mathbb{Q}}} R_c.$$

The results still holds even if conditions $x > 0$, $y > 0$ are dropped, case in which the half-plane can be written as a countable union of infinite rectangles, see Fig. 11.10 **b**:

$$\{x + y < k\} = \bigcup_{\substack{0 < c < k \\ c \in \mathbb{Q}}} \Big(\{x < c\} \cap \{y < k - c\} \Big).$$

11.5.8 Network of sigmoid neurons

Consider a neural network with one hidden layer that has two units, each being a sigmoid neuron with an activation function σ, see Fig. 11.11. The weights from the input to the hidden layer are denoted by $w_1, w_2 \in \mathbb{R}^n$, and the thresholds by $\theta_1, \theta_2 \in \mathbb{R}$. The two weights from the hidden layer to the output layer are $\alpha_1, \alpha_2 \in \mathbb{R}$. The output variable Y can be written in terms of the input random variable X as

$$Y = \alpha_1 \sigma(w_1^T X + \theta_1) + \alpha_2 \sigma(w_2^T X + \theta_2).$$

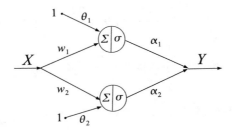

Figure 11.11: *Network of two sigmoid neurons.*

We are interested in finding the information generated by this network, which is the sigma-algebra $\mathfrak{S}(Y)$.

To get a glimpse of the complexity raised by the nonlinearity of σ, we shall produce an explicit computation for the case of an arctangent neuron with $\alpha_1 = \alpha_2 = 1$. The output in this case is given by

$$Y = \tan^{-1}(w_1^T X + \theta_1) + \tan^{-1}(w_2^T X + \theta_2) = u_1 + u_2.$$

Using a trigonometric formula, we write

$$\tan Y = \tan(u_1 + u_2) = \frac{\tan u_1 + \tan u_2}{1 - \tan u_1 \tan u_2} = \frac{(w_1^T x + \theta_1) + (w_2^T x + \theta_2)}{1 - (w_1^T x + \theta_1)(w_2^T x + \theta_2)}.$$

The output

$$Y = \tan^{-1}\left(\frac{(w_1 + w_2)^T x + \theta_1 + \theta_2}{1 - (w_1^T x + \theta_1)(w_2^T x + \theta_2)}\right)$$

is a lot more complex than the output of a single arctangent neuron, which is $\tan^{-1}(w^T x + \theta)$.

In the following we shall attempt to describe the sigma-field $\mathfrak{S}(Y)$. First we shall find a system of generators. Fix $\alpha_1, \alpha_2 > 0$. Using Example 11.5.7, we can write for any real k

$$\begin{aligned}
\{Y < k\} &= \{\alpha_1 \sigma(w_1^T X + \theta_1) + \alpha_2 \sigma(w_2^T X + \theta_2) < k\} \\
&= \bigcup \left(\{\alpha_1 \sigma(w_1^T X + \theta_1) < c\} \cap \{\alpha_2 \sigma(w_2^T X + \theta_2) < k - c\}\right)
\end{aligned}$$

$$= \bigcup_{c<k} \Big(\{ \sigma(w_1^T X + \theta_1) < c/\alpha_1 \} \cap \{ \sigma(w_2^T X + \theta_2) < (k-c)/\alpha_2 \} \Big)$$

$$= \bigcup_{c<k} \Big(\{ w_1^T X + \theta_1 < \sigma^{-1}\Big(\frac{c}{\alpha_1}\Big) \} \cap \{ w_2^T X + \theta_2 < \sigma^{-1}\Big(\frac{k-c}{\alpha_2}\Big) \} \Big)$$

$$= \bigcup_{c<k} \{ X \in \mathcal{H}^-_{w_1,\theta_1 - \sigma^{-1}(\frac{c}{\alpha_1})} \} \cap \{ X \in \mathcal{H}^-_{w_2,\theta_2 - \sigma^{-1}(\frac{k-c}{\alpha_2})} \}$$

$$= \bigcup_{\substack{c<k \\ c \in \mathbb{Q}}} X^{-1}(\mathcal{H}^-_{w_1,u_1(c)}) \cap X^{-1}(\mathcal{H}^-_{w_2,u_2(c)})$$

$$= X^{-1}\Big(\bigcup_{\substack{c<k \\ c \in \mathbb{Q}}} \mathcal{H}^-_{w_1,u_1(c)} \cap \mathcal{H}^-_{w_2,u_2(c)} \Big)$$

with $u_1(c) = \theta_1 - \sigma^{-1}(\frac{c}{\alpha_1})$ and $u_2(c) = \theta_2 - \sigma^{-1}(\frac{k-c}{\alpha_2})$. In the last identity we used Proposition A.0.1. Note that the intersection of half-spaces $\mathcal{H}^-_{w_1,u_1(c)} \cap \mathcal{H}^-_{w_2,u_2(c)}$ contains different sorts of strips and angle sectors. The information generated by the output Y is the sigma-algebra generated by the previous sets

$$\mathfrak{S}(Y) = \mathfrak{S}\{Y < k; k \in R\} = X^{-1}\Big(\mathfrak{S}\{ \bigcup_{\substack{c<k \\ c \in \mathbb{Q}}} \mathcal{H}^-_{w_1,u_1(c)} \cap \mathcal{H}^-_{w_2,u_2(c)} \} \Big).$$

This sigma-field contains a lot more information than all the other output information fields encountered so far. This behavior is due to a linear combination of nonlinear sigmoids. Furthermore, the information complexity depends on the weights w_i, α_i and thresholds θ_i, and therefore a maximum capacity learning algorithm should be able to tune these parameters such that the output information $\mathfrak{S}(Y)$ becomes maximal.

11.5.9   More remarks

A neural network with multiple hidden layers combines iteratively nonlinear combinations of sigmoids. The resulting output has an even more nonlinearity, fact that leads to a reacher output information $\mathfrak{S}(Y)$. However, there is an upper bound of all these improvements, since all these sigma-fields are contained into the input information, $\mathfrak{S}(Y) \subset \mathfrak{S}(X) = \mathcal{I}$. The role of a neural network is to adapt the input information $\mathcal{I}$ into an output information $\mathcal{E}$, hoping that eventually the target Z becomes $\mathcal{E}$-measurable, or, equivalently,

Z is determined by information $\mathcal{E}$. Since this reduces to $Z \subset \mathcal{E}$, the network weights have to be tuned to adjust $\mathcal{E}$ to the right maximal size. However, since sometimes Z is not a subset of $\mathcal{I}$, the best we can hope for is just the inclusion $Z \cap \mathcal{I} \subset \mathcal{E}$.

The massage of the input information $\mathcal{I}$ into the output information $\mathcal{E}$ is done composing the input X by a measurable function f to obtain the output $Y = f(X)$. The function f is highly nonlinear and depends on the network weights and biasses.

If the neural network is regarded as a human mind and the sigma-algebra $\mathcal{I}$ is the information collected by senses, the measurable function f plays the role of a thought in the brain, which structures the information into some understandable information $\mathcal{E}$.

Two minds supplied with the same input information might understand nonidentical things due to distinct thought functions f, which reduces to having different synaptic weights. There is some information lost, $\mathcal{L}$, which is the difference between of what we can see, $\mathcal{I}$, and what we can understand, $\mathcal{E}$. The learning process comes with producing more sophisticated thoughts, f, which achieve an expanded understanding, $\mathcal{E}$.

A neural network can be compared also with an engine, which is supplied some input energy and which produces some output work. At the output the engine produces always less energy than it is supplied with. A mechanic's job is to adjust the engine parameters to make it more efficient by increasing its capacity; in a similar way the weights are tuned in a neural network to match a certain target information.

11.6   Compressible Layers

In this section we discuss how the flow of information propagates through the layers of a deep feedforward neural network. In general, each internal layer does a certain classification job, by dividing the incoming information into pieces and reorganizing them into new information. The next layers perform a similar job, condensing the information into larger chunks.

Since there are presumably some events in the input data field, which are not relevant to classification, they can be dropped without consequence, while dropping the others has severe consequences. The selective process of removing the useless events and keeping only the necessary useful minimum cannot be done usually in only one layer. This process requires several layers, each layer dropping selectively certain events, which generate the lost information at that specific layer.

We consider now a deep feedforward neural network with $L - 1$ hidden layers and $d^{(\ell)}$ neurons in the ℓth layer. Recall that the network input is an n-dimensional random variable, $X = X^{(0)} = (x_1^{(0)}, \ldots, x_{d^{(0)}}^{(0)})^T$ and the activation of the ℓth layer is a $d^{(\ell)}$-dimensional random variable, denoted by $X^{(\ell)} = (x_1^{(\ell)}, \ldots, x_{d^{(\ell)}}^{(\ell)})^T$. The neural network's output is given by the random vector $Y = X^{(L)} = (x_1^{(L)}, \ldots, x_{d^{(L)}}^{(L)})^T$.

The information field generated by the activation of the ℓth layer is given by the $\mathfrak{S}$-algebra $\mathcal{I}^{(\ell)} = \mathfrak{S}(X^{(\ell)})$. In particular, the input information is $\mathcal{I}^{(0)} = \mathcal{I}$, and the output information of the network is $\mathcal{I}^{(L)} = \mathcal{E}$. We have seen that the inclusion $\mathcal{E} \subseteq \mathcal{I}$ holds always, and $\mathcal{L} = \mathfrak{S}(\mathcal{I} \backslash \mathcal{E})$ is the information lost in the network. We shall apply a similar approach to each layer of the network.

The first hidden layer is fed the information $\mathcal{I}^{(0)}$ and it produces the information $\mathcal{I}^{(1)}$, which is the $\mathfrak{S}$-field generated by the activation $X^{(1)}$. Since $X^{(1)}$ depends on the input $X^{(0)}$, then $\mathfrak{S}(X^{(1)}) \subseteq \mathfrak{S}(X^{(0)})$, i.e., $\mathcal{I}^{(1)} \subseteq \mathcal{I}^{(0)}$. The $\mathfrak{S}$-algebra $\mathcal{L}^{(1)} = \mathfrak{S}(\mathcal{I}^{(0)} \backslash \mathcal{I}^{(1)})$ is the lost information through the first hidden layer.

We shall apply the same construction for each layer. The information that goes into the ℓth layer is the information generated by the activation of the previous layer, $X^{(\ell-1)}$, and is given by $\mathcal{I}^{(\ell-1)}$. The information that comes out of the ℓth layer is generated by the activation $X^{(\ell)}$ and is given by $\mathcal{I}^{(\ell)}$. Since $X^{(\ell)}$ depends on input $X^{(\ell-1)}$ through the forward propagation relation (6.2.29)

$$X^{(\ell)} = \phi\Big(W^{(\ell)^T} X^{(\ell-1)} - B^{(\ell)}\Big),$$

then $\mathfrak{S}(X^\ell) \subseteq \mathfrak{S}(X^{(\ell-1)})$, or $\mathcal{I}^{(\ell)} \subseteq \mathcal{I}^{(\ell-1)}$. Hence, we obtain a descending sequence of sigma-algebras:

$$\mathcal{I}^{(L)} \subseteq \cdots \mathcal{I}^{(\ell)} \subseteq \mathcal{I}^{(\ell-1)} \subseteq \cdots \subseteq \mathcal{I}^{(0)}. \tag{11.6.3}$$

The difference $\mathcal{I}^{(\ell-1)} \backslash \mathcal{I}^{(\ell)}$ contains the events that have been dropped by the ℓth layer. This difference is not a sigma-algebra, but it generates one. The size of this sigma-algebra describes the compressibility of the ℓth layer.

Definition 11.6.1 (a) *The lost information in the ℓth layer is given by the sigma-algebra*

$$\mathcal{L}^{(\ell)} = \mathfrak{S}(\mathcal{I}^{(\ell-1)} \backslash \mathcal{I}^{(\ell)}).$$

(b) *The ℓth layer is called uncompressed if the lost information is trivial,* $\mathcal{L}^{(\ell)} = \{\emptyset, \Omega\}$.
(c) *The ℓth layer is called totally compressed if the lost information is maximal, i.e.,* $\mathcal{L}^{(\ell)} = \mathcal{I}^{(\ell-1)}$.
(d) *The ℓth layer is called partially compressed if* $\{\emptyset, \Omega\} \neq \mathcal{L}^{(\ell)} \neq \mathcal{I}^{(\ell-1)}$.

Remark 11.6.2 The definition implies:

(i) If the ℓth layer is uncompressed, then $\mathcal{I}^{(\ell-1)} = \mathcal{I}^{(\ell)}$, namely, the information field remains unchanged.

(ii) If the ℓth layer is totally compressed, then $I^{(\ell)} = \{\varnothing, \Omega\}$. The inclusion sequence (11.6.3) implies $I^{(p)} = \{\varnothing, \Omega\}$, for all $\ell \leq p \leq L$. This makes the rest of the layers, from $\ell + 1$ to L, useless.

We shall make in the following the relation between the information fields $\mathcal{I}$, $\mathcal{E}$, and $\mathcal{L}^{(\ell)}$. Proposition 11.3.3 provides the decomposition

$$\mathcal{I}^{(\ell-1)} = \mathfrak{S}(\mathcal{L}^{(\ell)} \cup \mathcal{I}^{(\ell)}) = \mathcal{L}^{(\ell)} \vee \mathcal{I}^{(\ell)}.$$

Iterating with $\ell = 1, 2, \ldots, L$, we have

$$\begin{aligned} \mathcal{I}^{(0)} &= \mathcal{L}^{(1)} \vee \mathcal{I}^{(1)} \\ \mathcal{I}^{(1)} &= \mathcal{L}^{(2)} \vee \mathcal{I}^{(2)} \\ \cdots &= \cdots \\ \mathcal{I}^{(L-1)} &= \mathcal{L}^{(L)} \vee \mathcal{I}^{(L)}. \end{aligned}$$

By backward substitution, we obtain a representation of the input information field in terms of all lost information in the layers and the output information as

$$\mathcal{I} = \bigvee_{\ell=1}^{L} \mathcal{L}^{(\ell)} \vee \mathcal{E}. \tag{11.6.4}$$

We note that we have applied the associativity property of the operator $\vee$, see Exercise 11.10.10. Comparing with $\mathcal{I} = \mathcal{L} \vee \mathcal{E}$, and using the maximality of $\mathcal{L}$, see Proposition 11.3.3 part (b), yields

$$\bigvee_{\ell=1}^{L} \mathcal{L}^{(\ell)} \subset \mathcal{L}. \tag{11.6.5}$$

This means that the total information lost in a network, $\mathcal{L}$, is at least as large as the cumulation of information lost in each of the network layers.

The next result is a consequence of the previous relation.

Proposition 11.6.3 *If the lost information in a neural network is trivial, then all of its layers are uncompressed.*

Proof: Since $\mathcal{L} = \{\varnothing, \Omega\}$. Formula (11.6.5) implies that $\bigvee_{\ell=1}^{L} \mathcal{L}^{(\ell)} = \{\varnothing, \Omega\}$, and hence for each ℓ we have $\mathcal{L}^{(\ell)} \subseteq \{\varnothing, \Omega\}$, which actually implies equality $\mathcal{L}^{(\ell)} = \{\varnothing, \Omega\}$. ∎

11.7   Layers Compressibility

We shall present necessary conditions for a layer to be uncompressed. We shall start with some background preparation.

We assume the ℓth layer is uncompressed, i.e., $\mathcal{L}^{(\ell)} = \{\varnothing, \Omega\}$. This means the incoming and outgoing information are equal, $\mathcal{I}^{(\ell-1)} = \mathcal{I}^{(\ell)}$, or in terms of sigma-algebras, $\mathfrak{S}(X^{(\ell-1)}) = \mathfrak{S}(X^{(\ell)})$. This identity occurs when $X^{(\ell)}$ and $X^{(\ell-1)}$ can be expressed in terms of each other, i.e., there are two measurable functions F and G such that $X^{(\ell)} = F(X^{(\ell-1)})$ and $X^{(\ell-1)} = G(X^{(\ell)})$, see Proposition D.5.1 in the Appendix. We shall investigate the existence of functions F and G.

The existence of F follows from the forward pass formula (6.2.29)

$$X^{(\ell)} = \phi\Big(W^{(\ell)^T} X^{(\ell-1)} - B^{(\ell)}\Big). \tag{11.7.6}$$

Therefore, it suffices to state necessary conditions for inverting this relation. This is done by the next result.

Theorem 11.7.1 (compressibility conditions I) *Assume the following conditions are satisfied:*
(i) The activation function ϕ is invertible;
(ii) There is the same number of incoming and outgoing variables for the ℓth layer, i.e., $d^{(\ell-1)} = d^{(\ell)}$;
(iii) $\det W^{(\ell)} \neq 0$.
Then the ℓth layer is uncompressed.

Proof: Since ϕ is invertible, relation (11.7.6) becomes

$$W^{(\ell)^T} X^{(\ell-1)} = \phi^{-1}(X^{(\ell)}) + B^{(\ell)}.$$

Since $d^{(\ell-1)} = d^{(\ell)}$, then W^{ℓ} is a square matrix of nonzero determinant, so it is invertible. Therefore

$$X^{(\ell-1)} = W^{(\ell)^T}{}^{-1}\Big(\phi^{-1}(X^{(\ell)}) + B^{(\ell)}\Big),$$

or $X^{(\ell-1)} = G(X^{(\ell)})$, with G measurable. Hence $\mathcal{I}^{(\ell-1)} \subseteq \mathcal{I}^{(\ell)}$. Since the converse inclusion is trivial, it turns out that $\mathcal{I}^{(\ell-1)} = \mathcal{I}^{(\ell)}$, i.e., the ℓth layer is efficient. ∎

Remark 11.7.2 Condition (i) is satisfied by all increasing activation functions, such as the logistic function, hyperbolic tangent, arctangent, softplus, etc. However, the invertibility condition is not satisfied by ReLU, unit step, or

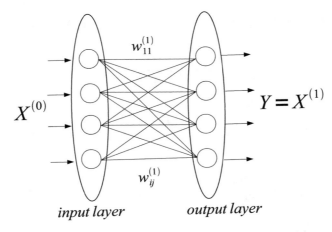

Figure 11.12: *A two-layer neural network with* $d^{(0)} = d^{(1)}$.

bipolar activation functions. Also, pooling functions are not invertible, since there are not one-to-one; therefore pooling provides compression, and hence, loss of information.

Example 11.7.1 (Two-layer network) Consider a neural network with no hidden layers, the only layers being the input and the output layers, see Fig. 11.12. Assume the dimension of the input equals the dimension of the output, i.e., $d^{(0)} = d^{(1)} = n$. Denote the weight matrix by $W^{(1)} = (w_{ij}^{(1)})_{1 \leq i,j \leq n}$ and the bias vector by $B^{(1)} = (b_1^{(1)}, \ldots, b_n^{(1)})^T$. We also assume that ϕ is increasing. The output is given by

$$X^{(1)} = \phi\big(W^{(1)^T} X^{(0)} - B^{(1)}\big).$$

The network is uncompressed as long as $\det W^{(1)} \neq 0$. This roughly states that the outputs are functional independent of the inputs. Next section deals with this topic in more detail.

11.8 Functional Independence

As usually, $X^{(\ell-1)} = (x_1^{(\ell-1)}, \ldots, x_n^{(\ell-1)})$ and $X_j^{(\ell)} = (x_1^{(\ell)}, \ldots, x_n^{(\ell)})$ denote the activations of layers $\ell - 1$ and ℓ, where we make the assumption that these layers have equal dimensions, $d^{(\ell-1)} = d^{(\ell)} = n$. Consider the determinant of the Jacobian

$$\Delta^{(\ell)} = \det\left(\frac{\partial x_j^{(\ell)}}{\partial x_i^{(\ell-1)}}\right)_{i,j}.$$

The activations $X^{(\ell-1)}$ and $X^{(\ell)}$ are called *functional independent* if $\Delta^{(\ell)} \neq 0$. Using the formula $s_j^{(\ell)} = \sum_{i=1}^{d^{(\ell-1)}} w_{ij}^{(\ell)} x_i^{(\ell-1)} - b_j^{(\ell)}$, the elements of the Jacobian can be computed as

$$\frac{\partial x_j^{(\ell)}}{\partial x_i^{(\ell-1)}} = \frac{\partial \phi(s_j^{(\ell)})}{\partial x_i^{(\ell-1)}} = \phi'(s_j^{(\ell)})\frac{\partial s_j^{(\ell)}}{\partial x_i^{(\ell-1)}} = \phi'(s_j^{(\ell)})w_{ij}^{(\ell)}.$$

Applying the determinant yields

$$\Delta^{(\ell)} = \det\left(\phi'(s_j^{(\ell)})w_{ij}^{(\ell)}\right) = \phi'(s_1^{(\ell)})\ldots\phi'(s_n^{(\ell)})\det W^{(\ell)}, \tag{11.8.7}$$

where we used the multilinearity property of determinants. If the activation function, ϕ, satisfies $\phi' \neq 0$, then $\Delta^{(\ell)} = 0$ if and only if $\det W^{(\ell)} = 0$.

Remark 11.8.1 The condition $\phi' \neq 0$ is satisfied by most sigmoid-type activation functions. For instance, if $\phi = \sigma$ is the logistic function, then

$$\Delta^{(\ell)} = \prod_{i=1}^{n} \sigma(s_i^{(\ell)})(1 - \sigma(s_i^{(\ell)}))\det W^{(\ell)}.$$

If $\phi = \mathbf{t}$ is the hyperbolic tangent, then

$$\Delta^{(\ell)} = \prod_{i=1}^{n} \left(1 - \mathbf{t}^2(s_i^{(\ell)})\right)\det W^{(\ell)}.$$

We notice that in these cases we have the inequality

$$|\Delta^{(\ell)}| < |\det W^{(\ell)}|.$$

The next result is a reformulation of Theorem 11.7.1:

Theorem 11.8.2 (compressibility conditions II) *Assume the following conditions hold:*
(i′) The activation function satisfies $\phi' > 0$;
(ii′) There is the same number of incoming and outgoing variables for the ℓth layer, i.e., $d^{(\ell-1)} = d^{(\ell)}$;
(iii′) $X^{(\ell)}$ is functional independent of $X^{(\ell-1)}$.
Then the ℓth layer is uncompressed.

Proof: Condition (i') implies that ϕ is both increasing and has nonvanishing derivative, $\phi' \neq 0$. From condition (iii') we infer $\Delta^{(\ell)} \neq 0$, and since $\phi' \neq 0$, it follows that $\det W^{(\ell)} \neq 0$. Applying Theorem 11.7.1 yields the desired result. ∎

A variant of the previous result covering the case of a *decoder layer*, i.e., the case when layers increase in size, is given next:

Theorem 11.8.3 (compressibility conditions III) *Assume the following conditions are satisfied:*
(i) The activation function ϕ is invertible;
(ii) The number of outgoing variables in the ℓth layer is greater than the number of incoming variables, i.e., $d^{(\ell)} > d^{(\ell-1)}$.
(iii) $W^{(\ell)}$ has maximal rank, i.e., $rank\, W^{(\ell)} = d^{(\ell-1)}$.
Then the lost information in the ℓth layer is trivial, i.e., the ℓth layer is uncompressed.

Proof: Using the invertibility of the activation function ϕ, relation (11.7.6) becomes

$$W^{(\ell)^T} X^{(\ell-1)} = \phi^{-1}(X^{(\ell)}) + B^{(\ell)}.$$

This can be considered as a linear system with $d^{(\ell)}$ equations and $d^{(\ell-1)}$ unknowns. Since $rank\, W^{(\ell)^T} = rank\, W^{(\ell)} = d^{(\ell)}$, by an eventual reindexing, we can solve for $X_1^{(\ell-1)}, X_2^{(\ell-1)}, \ldots, X_{d^{(\ell-1)}}^{(\ell-1)}$ in terms of $X_1^{(\ell)}, \ldots, X_{d^{(\ell-1)}}^{(\ell)}$. This implies

$$\mathcal{I}^{(\ell-1)} = \mathfrak{S}(X^{(\ell-1)}) \subset \mathfrak{S}(X_1^{(\ell)}, \ldots, X_{d^{(\ell-1)}}^{(\ell)}) \subset \mathfrak{S}(X^{(\ell)}) = \mathcal{I}^{(\ell)}, \qquad (11.8.8)$$

where the first and last of the previous identities result from the definition and the last inclusion is obvious. By transitivity it follows that $\mathcal{I}^{(\ell-1)} \subset \mathcal{I}^{(\ell)}$. Since for feedforward neural networks we always have $\mathcal{I}^{(\ell)} \subset \mathcal{I}^{(\ell-1)}$, the double inclusion implies the identity $\mathcal{I}^{(\ell-1)} = \mathcal{I}^{(\ell)}$. Therefore the ℓth layer is uncompressed. ∎

We shall discuss next the case of an *encoder layer*. This means $d^{(\ell-1)} > d^{(\ell)}$, i.e., the ℓth layer has fewer neurons than the previous layer. Then $X^{(\ell)}$ can be written in terms of $X^{(\ell-1)}$ as $X^{(\ell)} = G(X^{(\ell-1)})$, with $G : \mathbb{R}^{d^{(\ell-1)}} \to \mathbb{R}^{d^{(\ell)}}$ measurable, so $\mathcal{I}^{(\ell)} \subset \mathcal{I}^{(\ell-1)}$. Giving the inequality of dimensions it follows that G is non-invertible,[2] the previous inclusion being strict, fact

---

[2]For instance, the function $G : \mathbb{R}^3 \to \mathbb{R}^2$, $G(x_1, x_2, x_3) = (x_1, x_2)$, is not invertible.

that implies a compression in the ℓth layer. A neural network satisfying the sequence of inequalities

$$d^{(0)} > d^{(1)} > \cdots > d^{L-1} > d^{(L)}$$

exhibits compression effects in each of its layers.

11.9   Summary

This chapter formalizes in language of sigma-algebras the concept of information in a neural network, having the input and target given by random variables. We introduced the concepts of input, output, and target information fields and discussed their properties and relation to learning.

     The concept of lost information in a neural network describes the content of information that cannot be recovered when passing forward through the network. It can be also defined on individual layers and used to define compressed and uncompressed layers. We provided results regarding the compressibility of encoder and decoder layers. An encoder layer compresses the information, the lost information being nontrivial, while a decoder layer is always uncompressed, having a trivial lost information.

     We discussed the information representation for several examples involving neurons and simple neural networks, emphasizing the geometric interpretation. These examples include the perceptron, sigmoid neuron, linear neuron, and arctangent neuron.

11.10   Exercises

Exercise 11.10.1 Let c be a constant and X be a real-valued random variable. Show that:

(a) $\mathfrak{S}(c) = \{\emptyset, \Omega\}$;

(b) $\mathfrak{S}(c + X) = \mathfrak{S}(X)$;

(c) $\mathfrak{S}(cX) = \mathfrak{S}(X)$ for $c \neq 0$.

Exercise 11.10.2 Consider a sequence of random variables $(X_n)_n$ and define two more sequences, $(Y_n)_n$ and $(Z_n)_n$, by the following rules $Y_0 = 0$, $Y_n = \sum_{i=1}^{n} X_i$ and $Z_0 = 0$, $Z_n = \sum_{i=1}^{n} Y_i$.

(a) Show the strict inclusion $\mathfrak{S}(Y_n) \subset \mathfrak{S}(X_1, \ldots, X_n)$;

(b) Show the identity of sigma-fields $\mathfrak{S}(Y_1, \ldots, Y_n) = \mathfrak{S}(X_1, \ldots, X_n)$;

(c) Prove that $Z_n = nX_1 + (n-1)X_2 + \cdots + X_n$ and $\mathfrak{S}(Z_1, \ldots, Z_n) = \mathfrak{S}(X_1, \ldots, X_n)$.

Exercise 11.10.3 (The $\mathfrak{S}$-field generated by two sets) Show that the smallest $\mathfrak{S}$-field that contains two measurable sets, A and B has 16 elements and it is given by

$$\mathfrak{S}(A,B) = \{\emptyset, A, B, A^c, B^c, A\cap B, A\cup B, A^c\cap B^c, A^c\cup B^c, A^c\cap B, A\cap B^c,$$
$$A^c\cup B, A\cup B^c, (A\cap B^c)\cup(A^c\cap B), (A\cap B)\cap(A\cap B)^c, \Omega\}.$$

Exercise 11.10.4 How many elements does the $\mathfrak{S}$-field generated by three sets have?

Exercise 11.10.5 Let $\{E_j\}_{j=1,N}$, $N \geq 3$, be a finite partition of the sample space Ω. The $\mathfrak{S}$-field generated by the partition is given by

$$\mathcal{I} = \{\emptyset,\ E_i,\ E_i\cup E_j,\ E_i\cup E_j\cup E_k,\ \ldots, \bigcup_{i=1}^{N} E_i = \Omega\}.$$

Denote by $\mathcal{F}_{E_1}$, $\mathcal{F}_{E_1,E_2}$ the $\mathfrak{S}$-fields generated by E_1 and $\{E_1, E_2\}$, respectively. Show that:

(a) $\mathcal{F}_{E_1,E_2}$ is a recoverable information;

(b) $\mathcal{F}_{E_1}$ is not a maximal recoverable information.

Exercise 11.10.6 Consider the input of a neural network given by the n-dimensional random variable $X = (X_1, \ldots, X_n)$ on the measurable space $(\Omega, \mathcal{H})$. Prove that the input information $\mathfrak{S}(X)$ is the $\mathfrak{S}$-algebra generated by events of the form

$$\bigcap_{i=1}^{k}\{\omega; X_i(\omega) \leq x_i\}$$

with $x_1, \ldots, x_k \in \mathbb{R}$, $k \geq 1$.

Exercise 11.10.7 Assume that we drop one neuron from the last layer of a neural network. This means that we erase all synapses related to that particular neuron. Let Y be the output of the initial network and $\tilde{Y}$ the output after the neuron dropout has occurred. Let $\mathcal{E} = \mathfrak{S}(Y)$ and $\tilde{\mathcal{E}} = \mathfrak{S}(\tilde{Y})$ be the output information and $\mathcal{L}$ and $\tilde{\mathcal{L}}$ be the lost information in each case.
(a) Show that $\mathcal{L} \subset \tilde{\mathcal{L}}$.
(b) Formulate a similar result for the case when the neuron is dropped from the ℓth layer.

Exercise 11.10.8 Assume that we add an extra neuron to the last layer of a neural network. Let Y be the output of the initial network and $\tilde{Y}$ the output after the neuron addition has occurred. Let $\mathcal{E} = \mathfrak{S}(Y)$ and $\tilde{\mathcal{E}} = \mathfrak{S}(\tilde{Y})$ be the output information and $\mathcal{L}$ and $\tilde{\mathcal{L}}$ be the lost information in each case.

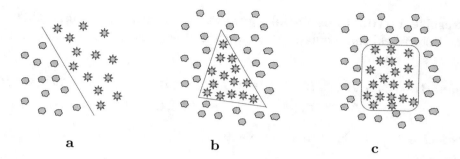

Figure 11.13: **a.** *The clusters can be separated by a line;* **b.** *One cluster is inside of a triangle and the other is outside;* **c.** *One cluster lies inside of a rectangle.*

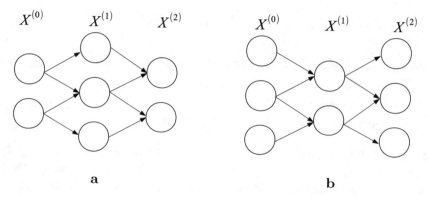

Figure 11.14: **a.** *For Exercise* 11.10.11. **b.** *For Exercise* 11.10.12.

(a) Show that $\tilde{\mathcal{L}} \subset \mathcal{L}$;

(b) Formulate a similar result for the case when the neuron is added to the ℓth layer.

Exercise 11.10.9 The clusters given in Fig. 11.13 **a**, **b**, and **c** are subject to be classified by a neural network with one hidden layer. What is the minimal number of neurons that need to be used in the hidden layer for each of the aforementioned cases?

Exercise 11.10.10 Show that operator $\vee$ on sigma-fields is associative, i.e., for any three sigma-fields, $\mathcal{F}, \mathcal{G}, \mathcal{H}$, we have

$$(\mathcal{F} \vee \mathcal{G}) \vee \mathcal{H} = \mathcal{F} \vee (\mathcal{G} \vee \mathcal{H}).$$

Exercise 11.10.11 Show that the hidden layer of the neural network given by Fig.11.14 **a** is efficient, i.e., $\mathcal{I}^{(0)} = \mathcal{I}^{(1)}$.

Exercise 11.10.12 Consider the one-hidden layer autoencoder given in Fig.11.14 **b**. Show that $\mathcal{I}^{(2)} = \mathcal{I}^{(1)}$.

Exercise 11.10.13 Let $\mathcal{I}$, $\mathcal{E}$, and $\mathcal{L}$ denote the input, output, and lost information in a network. Show that: $\mathcal{I} = \mathcal{E}$ if and only if $\mathcal{L} = \{\emptyset, \Omega\}$.

Exercise 11.10.14 Let $\mathcal{I}$ and $\mathcal{E}$ denote the input and the output information fields of a network. Prove that it is not possible to place $\mathcal{E}$ between a maximal recovery information field, $\mathcal{F}$, and a minimal non-recovery information field, $\mathcal{H}$, namely, we cannot have the double inclusion $\mathcal{H} \subseteq \mathcal{E} \subseteq \mathcal{F}$.

Exercise 11.10.15 Let $\mathcal{I}$ and $\mathcal{E}$ denote the input and the output information fields of a network. Show the following:

(i) If $\mathcal{E}$ is contained into a maximal recoverable information field, then $\mathcal{E}$ does not contain any minimal non-recoverable information fields;

(ii) If we assume that $\mathcal{E}$ contains a minimal non-recoverable information field, then $\mathcal{E}$ is not contained into a maximal recoverable information field.

Exercise 11.10.16 Show that any subfield of a recoverable information field is also recoverable.

Chapter 12

Information Capacity Assessment

This chapter deals with one of the main problems of Deep Learning, namely, *how can a neural network go from raw data to a more complex representation as the data flows through the network layers?* The organization of pixels into features can be assessed by some information measures, such as entropy, conditional entropy, and mutual information. These measures are used to describe the information evolution through the layers of a feedforward network.

If the previous chapter provided a qualitative description of the information content through the layers of a neural network, this chapter deals with a quantitative description of the evolution of the information in a neural network.

The reader interested in a more detailed presentation of these topics is referred to the book [11]. The applications to neural networks include definition and computation of network capacity and the information bottleneck method.

The previous chapter described the process of compressing the information through the layers of a network. In this chapter we shall introduce a measure of assessment of the compressibility of a layer using mutual information.

12.1 Entropy and Properties

The entropy describes the information organization of a system. It is maximum in the case when the system is completely uncertain. For instance, a picture with pixels represented by white noise will have maximum entropy, while a picture of a geometrical figure will have a smaller entropy. Raw data

© Springer Nature Switzerland AG 2020
O. Calin, *Deep Learning Architectures*, Springer Series in the Data Sciences,
https://doi.org/10.1007/978-3-030-36721-3_12

inputs of a neural network have a larger entropy than its output complex representation.

For instance, we shall consider the example of a neural network, which is fed data representing pictures of cats and dogs, and has to figure out the animal type. The entropy of the output layer is 1 bit, as the information needed to know is one out of two possible choices. However, each input picture has a larger entropy. The task of the neural network is to gradually decrease the entropy until the entropy of the output is just the correct one (namely, one bit), and hence the interest in how the entropy decreases along the layers of a neural network.

The next subsections introduce the concept of entropy and present some of its properties. Then, we shall approach the study of expressionless layers using the concept of entropy.

12.1.1 Entropy of a random variable

Let X be a discrete random variable taking values $\{x_1, x_2, \ldots, x_n\}$ with probabilities $p_k = P(X = x_k)$. The negative log-likelihood, $-\ln p_k$, represents the information that X takes the value x_k and its mathematical formulation is discussed in section D.7 of the Appendix. Then the entropy of X is defined by the sum

$$H(X) = -\sum_{k=1}^{n} p_k \ln p_k = \mathbb{E}^p[-\ln p], \tag{12.1.1}$$

where $\mathbb{E}^p$ represents the expectation operator with respect to the distribution p. Relation (12.1.1) is a weighted average of the negative log-likelihood function and represents the expected amount of information contained in the random variable X having the probability distribution $p = (p_1, \ldots, p_n)^T$. Since $p_k \in (0, 1)$, then $H(X) > 0$.

In the case when the random variable X is continuous, its distribution is given by the density $p(x)$. This is usually written heuristically as

$$P(x < X < x + dx) = p(x)dx,$$

i.e., the probability that X takes values in the infinitesimal interval $(x, x + dx)$ is proportional to its length, dx, and this proportionality function is the density $p(x)$. In this case the probability measure of X is absolutely continuous with respect to the Lebesgue measure, and is given by $\mu(dx) = p(x)dx$, see Appendix, section C.7. If the range of values of X is denoted by $\mathcal{X}$, the associated entropy is defined by

$$H(X) = H(p) = -\int_{\mathcal{X}} p(x) \ln p(x)\, dx = \mathbb{E}^p[-\ln p]. \tag{12.1.2}$$

Relation (12.1.1) defines the *discrete entropy* while (12.1.2) provides the *differential entropy*. One major difference between them is that the latter might be, sometimes, negative, or infinite. Some sufficient conditions for bounds on entropy are provided in Exercise 12.13.2.

Similarly, we can define the *joint entropy* of two continuous random variables X and Y as

$$H(X, Y) = -\int_{\mathbb{R}} \int_{\mathbb{R}} p(x, y) \ln p(x, y) \, dx dy.$$

This can be interpreted as the information contained in the pair of random variables (X, Y). The corresponding relation for the entropy in the case of two discrete random variables is

$$H(X, Y) = -\sum_{i,j} p_{ij} \ln p_{ij},$$

where $p_{ij} = P(X = x_i, Y = y_j)$ is the joint distribution of (X, Y).

12.1.2 Entropy under a change of coordinates

This section shows that the entropy is coordinate dependent. For a better understanding, we shall start with an example.

Exercise 12.1.1 We assume that X is a random variable describing the pixels activation of a gray-scaled picture with dimensions $m \times k$. Let p_{ij} denote the activation of the pixel with position (i, j), where $1 \leq i \leq m$, $1 \leq j \leq k$. We assume $0 \leq p_{ij} \leq 1$, with value 0 corresponding to white pixels and value 1 to black pixels. We can also assume the normality condition $\sum_{i,j} p_{ij} = 1$ satisfied, which can be obtained by dividing each p_{ij} by the sum. Thus, p_{ij} can be regarded as a probability distribution. The entropy of the image is given by

$$H(X) = -\sum_{i,j} p_{ij} \ln p_{ij}.$$

Now, we consider the picture obtained by applying a transformation, f, to the former picture, obtaining $Y = f(X)$. How does the entropy change? If the transformation is a rotation, a translation, or a flip the amount of information in the picture should not change, so $H(Y) = H(X)$. However, if the transformation is not rigid (but still smooth and invertible), the picture's information changes. For instance, you can deform continuously a picture of a circle into a picture of a square, changing the information, and hence, the entropy.

The next result provides a formula for the change in the entropy of a continuous random variable under a smooth coordinate transform.

Proposition 12.1.2 *Let* $\mathcal{X}, \mathcal{Y} \subset \mathbb{R}^n$ *be two domains. Let* X *and* Y *be continuous random variables on* $\mathcal{X}$ *and* $\mathcal{Y}$, *respectively, with* $Y = f(X)$, *where* $f : \mathcal{X} \to \mathcal{Y}$ *is a differentiable function having a nonsingular Jacobian,* $\det J_f(x) \neq 0$, *with*

$$J_f(x) = \left(\frac{\partial f^j(x)}{\partial x_k} \right)_{j,k}.$$

Then

$$H(Y) = H(X) - \mathbb{E}^{P_Y}[\ln |\det J_{f^{-1}}(Y)|]. \tag{12.1.3}$$

Proof: Let p_X and p_Y be the densities of the random variables X and Y, respectively. It is known that if $Y = f(X)$ then the relation between their densities is

$$p_X(x) = p_Y(f(x))|\det J_f(x)|.$$

This formula can be found in any book of elementary probabilities, such as in Wackerly et al. [120], Chapter 6. Using the change of variables $y = f(x)$, the formula $dy = |\det J_f(x)| \, dx$, and the fact that the transformation f is invertible (since the Jacobian is nonsingular), we have

$$
\begin{aligned}
H(X) &= -\int_{\mathcal{X}} p_X(x) \ln p_X(x) \, dx \\
&= -\int_{\mathcal{X}} p_Y(f(x))|\det J_f(x)| \ln \Big(p_Y(f(x))|\det J_f(x)| \Big) \, dx \\
&= -\int_{\mathcal{Y}} p_Y(y) \ln \Big(p_Y(y) \frac{1}{|\det J_{f^{-1}}(y)|} \Big) \, dy \\
&= -\int_{\mathcal{Y}} p_Y(y) \ln p_Y(y) \, dy + \int_{\mathcal{Y}} p_Y(y) \ln \big(|\det J_{f^{-1}}(y)| \big) \, dy \\
&= H(Y) + \mathbb{E}^{P_Y}[\ln |\det J_{f^{-1}}(Y)|],
\end{aligned}
$$

which ends the proof. ∎

The next consequence states that the entropy is invariant under rigid transforms of the plane. It follows from the fact that the determinant of these transforms is equal to ± 1.

Corollary 12.1.3 *Let* X *and* Y *be random variables on* $\mathbb{R}^2$. *If* $f : \mathbb{R}^2 \to \mathbb{R}^2$ *is a rotation, a flip into a line, or a vector translation, and* $Y = f(X)$, *then* $H(X) = H(Y)$.

In order to asses the change in the entropy we need to study the expectation term $\mathbb{E}^{P_Y}[\ln |\det J_{f^{-1}}(Y)|]$. We shall introduce first a few notions.

Definition 12.1.4 *A mapping* $f : \mathbb{R}^n \to \mathbb{R}^n$ *is called a* λ-*contraction if there is a number* $0 < \lambda < 1$ *such that for any compact domain* $K \subset \mathbb{R}^n$ *we have* $vol\big(f(K)\big) < \lambda vol(K)$, *where "vol" denotes the* n-*dimensional Lebesgue measure.*

Contractions often occur in neural networks when describing information compression, such as pooling, when pixels in a given square region are replaced by the pixel of maximum value. The geometric interpretation of the determinant of the Jacobian matrix in terms of contractions is given in the following result.

Proposition 12.1.5 *Let f be a mapping from $\mathbb{R}^n$ to $\mathbb{R}^n$ and $\lambda > 0$. There are equivalent:*
(i) $|\det J_f(x)| < \lambda$ for all $x \in \mathbb{R}^n$;
(ii) f is a λ-contraction.

Proof: $(i) \Longrightarrow (ii)$ It follows from an application of the change of variable formula

$$vol\big(f(K)\big) = \int_{f(K)} dy = \int_K |\det J_f(x)|\, dx < \lambda \int_K dx = \lambda\, vol(K).$$

$(ii) \Longrightarrow (i)$ Assume there is an x_0 such that $|\det J_f(x_0)| \geq \lambda$. By continuity reasons the inequality can be extended on a neighborhood of x_0. Let K be a compact set included in this neighborhood, so $|\det J_f(x)| \geq \lambda$, for all $x \in K$. Then

$$vol\big(f(K)\big) = \int_{f(K)} dy = \int_K |\det J_f(x)|\, dx \geq \lambda \int_K dx = \lambda\, vol(K),$$

which contradicts the definition of the λ-contraction. Therefore, the assumption was false and hence $|\det J_f(x_0)| < \lambda$ everywhere. ∎

The next proposition states that under a λ-contraction the entropy decreases.

Proposition 12.1.6 (Entropy change) *Let X and Y be given as in Proposition 12.1.2, and assume the mapping f is a λ-contraction, with $\lambda \in (0, 1)$. Then the entropy change, $H(X) - H(Y)$, is positive. A lower bound is given by*

$$H(X) - H(Y) > \ln \frac{1}{\lambda}.$$

Proof: Using Proposition 12.1.2 we have

$$H(X) - H(Y) = \mathbb{E}^{P_Y}[\ln |\det J_{f^{-1}}(Y)|].$$

Using Proposition 12.1.5 and flipping the inequality $0 < |\det J_f(x)| < \lambda < 1$ yields $|\det J_{f^{-1}}(y)| > \frac{1}{\lambda} > 1$, and hence $\ln |\det J_{f^{-1}}(y)| > 0$. Therefore, $\mathbb{E}^{P_Y}[\ln |\det J_{f^{-1}}(Y)|] > 0$. More precisely,

$$\mathbb{E}^{P_Y}[\ln |\det J_{f^{-1}}(Y)|] = \int p_Y(y) \ln |\det J_{f^{-1}}(y)|\, dy > \ln \frac{1}{\lambda} \int p_Y(y)\, dy$$

$$= \ln \frac{1}{\lambda}.$$

■

A positive entropy change corresponds to a loss of information occurred during this transformation. The previous proposition states that contraction mappings always cause information loss (the smaller the λ, the larger the loss). An example of this type of mapping is the max-pooling. The concept of entropy change will be further used when analyzing the entropy flow in neural networks.

12.2   Entropy Flow

We consider a feedforward neural network with L layers, whose activations are given by the random vector variables $X^{(\ell)}$, $0 \leq \ell \leq L$. The information field $\mathcal{I}^{(\ell)} = \mathfrak{S}(X^{(\ell)})$ can be assessed numerically using the entropy function of the ℓth layer activation, $H(X^{(\ell)})$. Assume the layers $\ell - 1$ and ℓ have the same number of neurons, $d^{(\ell-1)} = d^{(\ell)}$, and that $X^{(\ell)} = f(X^{(\ell-1)})$, with f deterministic smooth function. If let $\Delta^{(\ell)} = \det J_f(X^{(\ell-1)})$, then Proposition 12.1.2 provides

$$H(X^{(\ell)}) = H(X^{(\ell-1)}) + \mathbb{E}^{P_{X^{(\ell)}}}[\ln |\Delta^{(\ell)}|], \qquad 1 \leq \ell \leq L. \qquad (12.2.4)$$

Definition 12.2.1 *The entropy flow associated with a feedforward neural network with layer activations $X^{(\ell)}$ is the sequence $\{H(X^{(\ell)})\}_{0 \leq \ell \leq L}$ of entropies of the network layer activations.*

If the feedforward network performs a classification task, the entropy flow is expected to decrease to $\log_2 c$ bits, where c is the number of classes. This corresponds to the network's role of information organization and reduction of uncertainty. For example, if the network has to classify pictures with animals into mammals and non-mammals, the entropy of the last layer is $H(X^{(L)}) = 1$, even if the entropy of the input picture, $H(X^{(0)})$, is a lot larger.

The decrease in entropy is done gradually from layer to layer. For instance, in the case of a convolutional network applied on car images, the first layer determines small corners and edges; the next layer organizes the information obtained in the previous layer into some small parts, like wheel, window, bumper, etc. A later layer might be able to classify this information into higher level features, such as the car type, which has the smallest entropy.

The following concept is useful in the study of the entropy flow behavior. We define the *entropy leak* between the layers $\ell - 1$ and ℓ by the change in the entropies of the layer activations

$$\Lambda(\ell - 1, \ell) = H(X^{(\ell-1)}) - H(X^{(\ell)}). \qquad (12.2.5)$$

The next result states sufficient conditions for the information flow to be decreasing. This corresponds to a sequence of compressions of the information through the network.

Theorem 12.2.2 *Assume the following conditions are satisfied in a feedforward neural network:*
(i) The activation function is increasing, with $0 < \phi'(x) < 1$;
(ii) There is the same number of incoming and outgoing variables for the ℓth layer, i.e., $d^{(\ell-1)} = d^{(\ell)}$;
(iii) $0 < |\det W^{(\ell)}| < 1$.
Then there is a positive entropy leak between the layers $\ell - 1$ and ℓ, i.e.,

$$\Lambda(\ell - 1, \ell) > 0.$$

Proof: Conditions (i) and (iii) substituted into (11.8.7) imply $0 < |\Delta^{(\ell)}| < 1$. Using (12.2.4) and (12.2.5) we have

$$\Lambda(\ell - 1, \ell) = -\mathbb{E}^{P_{X^{(\ell)}}}[\ln |\Delta^{(\ell)}|] > 0.$$

$\blacksquare$

Remark 12.2.3 (i) If we assume $|\det W^{(\ell)}| < \lambda < 1$, then the following lower bound for the entropy leak holds:

$$\Lambda(\ell - 1, \ell) > -\ln \lambda > 0.$$

(ii) In the case when the activation function is the logistic function, $\phi(x) = \sigma(x)$, following Remark 11.8.1, we have the following explicit computation:

$$
\begin{aligned}
\ln |\Delta^{(\ell)}| &= \sum_i \ln \sigma(s_i^{(\ell)}) + \sum_i \ln(1 - \sigma(s_i^{(\ell)})) + \ln |\det W^{(\ell)}| \\
&= -\sum_i \ln(1 + e^{-s_i^{(\ell)}}) - \sum_i \ln(1 + e^{s_i^{(\ell)}}) + \ln |\det W^{(\ell)}| \\
&= -\sum_i sp(-s_i^{(\ell)}) - \sum_i sp(s_i^{(\ell)}) + \ln |\det W^{(\ell)}| \\
&= -2\sum_i sp(s_i^{(\ell)}) + \sum_i s_i^{(\ell)} + \ln |\det W^{(\ell)}|,
\end{aligned}
$$

where we used the softplus properties $sp(x) = \ln(1 + e^x)$ and $sp(x) - sp(-x) = x$ introduced in Chapter 2.

Furthermore, if there is a matrix A such that $W^{(\ell)} = e^A$, with positive determinant, then the last term becomes in the virtue of Liouville's formula $\ln \det W^{(\ell)} = \text{Trace} A$.

(*iii*) If $W^{(\ell)}$ is an orthonormal matrix, then the ℓth layer corresponds to a rigid transformation (i.e., a rotation composed with a translation induced by the bias vector $B^{(\ell)}$), then $\ln|\det W^{(\ell)}| = 0$, and hence, there is no information loss in this case, $\Lambda(\ell - 1, \ell) = 0$.

12.3   The Input Layer Entropy

Consider a feedforward neural network with the input $X = X^{(0)}$, having components $(X_1^{(0)}, \dots, X_n^{(0)})$. The entropy of the input, $H(X^{(0)})$, depends whether the input components are independent or not. In the case when the input variables $X_j^{(0)}$ are independent, the entropy of the input random variable is the sum of entropies of individual components.

Proposition 12.3.1 *Let $X = (X_1, \dots, X_n)^T$ be an n-dimensional random variable with independent components. Then*

$$H(X) = \sum_{j=1}^n H(X_j).$$

Proof: Since the independence implies $p_X(\mathbf{x}) = p_{X_1}(x_1) \cdots p_{X_n}(x_n)$, using Fubini's theorem, we have

$$
\begin{aligned}
H(X) &= -\int_{\mathbb{R}^n} p_X(\mathbf{x}) \ln p_X(\mathbf{x}) \, d\mathbf{x} = -\int_{\mathbb{R}^n} p_X(\mathbf{x}) \sum_j \ln p_{X_j}(x_j) \, d\mathbf{x} \\
&= -\sum_j \int_{\mathbb{R}^n} p_{X_1}(x_1) \cdots p_{X_n}(x_n) \ln p_{X_j}(x_j) \, d\mathbf{x} \\
&= -\sum_j \left(\int_{\mathbb{R}} p_{X_j}(x_j) \ln p_{X_j}(x_j) \, dx_j \right) \prod_{i \neq j} \underbrace{\left(\int_{\mathbb{R}} p_{X_i}(x_i) \, dx_i \right)}_{=1} \\
&= -\sum_j \left(\int_{\mathbb{R}} p_{X_j}(x_j) \ln p_{X_j}(x_j) \, dx_j \right) \\
&= \sum_j H(X_j).
\end{aligned}
$$

∎

Remark 12.3.2 If the components X_j are not independent, the incoming information is smaller, and hence we have the inequality $H(X) < \sum_{j=1}^n H(X_j)$.

Assume the components of the input are not independent. Then the entropy of the input, $H(X^{(0)})$, has an upper bound given by the determinant of its covariance matrix. This will be shown next.

Proposition 12.3.3 *Let* $X = (X_1, \ldots, X_n)^T$ *be an n-dimensional random variable with covariance matrix A. Assume* $\det A > 0$. *Then*

$$H(X) \le \frac{1}{2} \ln[(2\pi e)^n \det A].$$

Proof: Let $p(\mathbf{x})$ be the density of the random variable $X = (X_1, \ldots, X_n)^T$, with $\mathbf{x} = (x_1, \ldots, x_n)^T \in \mathbb{R}^n$. The covariance matrix of X is the $n \times n$ matrix $A = (A_{ij})_{i,j}$, with $A_{ij} = \mathbb{E}^p[(X_i - \mu_i)(X_j - \mu_j)]$, where $\mathbb{E}[X] = \mu = (\mu_1, \ldots, \mu_n)^T$ is the mean vector of X. By Proposition 3.5.1 we have the inequality

$$H(p) \le S(p, q), \tag{12.3.6}$$

for any density function $q(\mathbf{x})$ on $\mathbb{R}^n$. In particular, the inequality holds for the multivariate normal distribution

$$q(\mathbf{x}) = \frac{1}{\sqrt{(2\pi)^n \det A}} e^{-\frac{1}{2}(\mathbf{x} - \mu) A^{-1}(\mathbf{x} - \mu)}.$$

It suffices to compute the cross-entropy $S(p, q)$. In the following we shall use upper indices to denote the inverse of the covariance matrix, $A^{-1} = (A^{ij})_{i,j}$. Since the negative log-likelihood of q is given by

$$-\ln q(\mathbf{x}) = \frac{1}{2} \ln[(2\pi)^n \det A] + \frac{1}{2}(\mathbf{x} - \mu) A^{-1}(\mathbf{x} - \mu),$$

then

$$
\begin{aligned}
S(p, q) &= \mathbb{E}^p[-\ln q] = \frac{1}{2} \ln[(2\pi)^n \det A] + \frac{1}{2} \mathbb{E}^p[(\mathbf{x} - \mu) A^{-1}(\mathbf{x} - \mu)] \\
&= \frac{1}{2} \ln[(2\pi)^n \det A] + \frac{1}{2} \mathbb{E}^p\Big[\sum_{i,j=1}^{n} A^{ij}(x_i - \mu_i)(x_j - \mu_j) \Big] \\
&= \frac{1}{2} \ln[(2\pi)^n \det A] + \frac{1}{2} \sum_{i,j=1}^{n} A^{ij} \mathbb{E}^p\big[(x_i - \mu_i)(x_j - \mu_j)\big] \\
&= \frac{1}{2} \ln[(2\pi)^n \det A] + \frac{1}{2} \sum_{i,j=1}^{n} A^{ij} A_{ij} \\
&= \frac{1}{2} \ln[(2\pi)^n \det A] + \frac{1}{2} \sum_{i,j=1}^{n} \delta_{ij} = \frac{1}{2} \ln[(2\pi)^n \det A] + \frac{n}{2} \\
&= \frac{1}{2} \ln[(2\pi e)^n \det A].
\end{aligned}
$$

Using now inequality (12.3.6) yields the desired relation

$$H(X) = H(p) \le \frac{1}{2} \ln[(2\pi e)^n \det A].$$

Note that the identity is achieved when X is multivariate normal random variable with covariance matrix A. This means that among all random variables of given covariance matrix, the one with the maximal entropy is the multivariate normal.

∎

Corollary 12.3.4 *Let $X = (X_1, \ldots, X_n)^T$ be an n-dimensional random variable with independent components. Then*

$$H(X) \leq \frac{1}{2} \ln[(2\pi e)^n \prod_{i=1}^{n} Var(X_i)].$$

Proof: If X has independent components, then its covariance matrix has a diagonal form

$$A_{ij} = \begin{cases} Var(X_i), & \text{if } i = j \\ 0, & \text{if } i \neq j, \end{cases}$$

and hence $\det A = \prod_{i=1}^{n} Var(X_i)$. The result follows now from Proposition 12.3.3.

∎

As an exemplification, we shall apply next the concept of entropy flow to the linear neuron and to the autoencoder.

12.4   The Linear Neuron

Let $X = (X_1, \ldots, X_n)^T$ and $Y = (Y_1, \ldots, Y_n)^T$ be the input and the output vectors for a linear neuron, respectively, see section 5.5. We have assumed that the number of inputs is equal to the number of outputs. Since the activation function is linear, $\phi(x) = x$, the output is related to the input by $Y = f(X) = W^T X - B$, where $W = (w_{ij})_{i,j}$ is the weight matrix and $B = (b_1, \ldots, b_n)$ is the bias vector. The Jacobian of the input-output function f is

$$J_f(x) = \left(\frac{\partial f^j(x)}{\partial x_k}\right)_{j,k} = W^T.$$

By Proposition 12.1.2 the entropy of the output, Y, can be written in terms of the entropy of the input, X, as in the following:

$$\begin{aligned} H(Y) &= H(W^T X - B) = H(X) - \mathbb{E}^{P_Y}[\ln|\det(W^T)^{-1}|] \\ &= H(X) + \mathbb{E}^{P_Y}[\ln|\det W|] \\ &= H(X) + \ln|\det W|, \end{aligned}$$

where we used that $\det W = \det W^T$ and the fact that W is a matrix with constant entries.

By Theorem 11.7.1 the linear neuron is uncompressed if $\det W \neq 0$. In the particular case when W is an orthogonal matrix, $|\det W| = 1$, then there is no entropy leak, i.e., $H(Y) - H(X) = 0$. There is a positive entropy leak in the case $|\det W| < 1$. The condition $|\det W| < 1$ is implied by the use of small weights $w_{ij}^{(\ell)}$, see Exercise 12.13.19. Hence, the use of the regularization condition $\|W\|^2 < \alpha$, with α small enough, implies the previous inequality.

If one considers a neural network made of linear neurons with layers of the same dimension, then

$$H(X^{(\ell)}) = H(X^{(\ell-1)}) + \ln \det |W^{\ell}|, \qquad \ell = 1, \ldots, L.$$

Using the properties of logarithms, we obtain the relation

$$H(X^{(L)}) = H(X^{(0)}) + \ln |\det[W^{(1)} W^{(2)} \cdots W^{(L)}]|. \qquad (12.4.7)$$

Using (12.4.7) and Proposition 12.3.3 we obtain an upper bound for the entropy of the output layer of the neural network as

$$H(X^{(L)}) \leq \frac{1}{2} \ln[(2\pi e)^n \det A] + \ln |\det[W^{(1)} W^{(2)} \cdots W^{(L)}]|. \qquad (12.4.8)$$

12.5 The Autoencoder

We consider a feedforward neural network with 3 inputs, 3 outputs and one hidden layer having a single neuron, see Fig. 12.1. The first part of the network, including the neuron in the middle, encodes the input signal into

$$x_1^{(1)} = \phi(s_1^{(1)}) = \phi(w_{11}^{(1)} x_1^{(0)} + w_{21}^{(1)} x_2^{(0)} + w_{31}^{(1)} x_3^{(0)} - b_1^1),$$

while the second part decodes the signal into the output vector

$$x^{(2)} = \begin{pmatrix} x_1^{(2)} \\ x_2^{(2)} \\ x_3^{(2)} \end{pmatrix} = \phi(s^{(2)}) = \begin{pmatrix} \phi(w_{11}^{(2)} x_1^{(1)} - b_1^{(2)}) \\ \phi(w_{12}^{(2)} x_1^{(1)} - b_2^{(2)}) \\ \phi(w_{13}^{(2)} x_1^{(1)} - b_3^{(2)}) \end{pmatrix}.$$

This very simple neural net is an example of an *autoencoder*. Since the input and output dimensions are equal, $d^{(0)} = d^{(2)} = 3$, it makes sense to apply the information tools developed earlier. Denote the activation function in each neuron by ϕ.

The sensitivity of the ith output with respect to the jth input is given by

$$\begin{aligned} \frac{\partial x_i^{(2)}}{\partial x_j^{(0)}} &= \frac{\partial \phi(s_i^{(2)})}{\partial x_j^{(0)}} = \phi'(s_i^{(2)}) \frac{\partial s_i^{(2)}}{\partial x_j^{(0)}} \\ &= \phi'(s_i^{(2)}) w_{1i}^{(2)} \frac{\partial x_1^{(1)}}{\partial x_j^{(0)}} = \phi'(s_i^{(2)}) w_{1i}^{(2)} \phi'(s_1^{(1)}) w_{j1}^{(1)} \\ &= \phi'(s_1^{(1)}) \phi'(s_i^{(2)}) w_{1i}^{(2)} w_{j1}^{(1)}. \end{aligned}$$

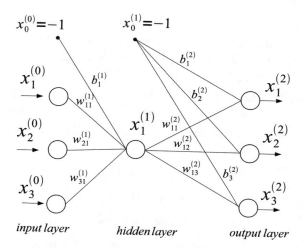

Figure 12.1: *Autoencoder with one hidden neuron.*

The determinant of the previous Jacobian can be computed using the determinant multilinearity property on rows and columns as in the following:

$$\det\left(\frac{\partial x_i^{(2)}}{\partial x_j^{(0)}}\right) \;=\; \phi'(s_1^{(1)})^3 \phi'(s_1^{(2)})\phi'(s_2^{(2)})\phi'(s_3^{(2)}) \det\left(w_{1i}^{(2)} w_{j1}^{(1)}\right).$$

If any of the weights $w_{1i}^{(2)}$ or $w_{j1}^{(1)}$ are zero, then the determinant vanishes, $\det\left(w_{1i}^{(2)} w_{j1}^{(1)}\right) = 0$. Assume now that none of the weights is vanishing. Then

$$\det\left(w_{1i}^{(2)} w_{j1}^{(1)}\right) \;=\; \begin{vmatrix} w_{11}^{(2)} w_{11}^{(1)} & w_{11}^{(2)} w_{21}^{(1)} & w_{11}^{(2)} w_{31}^{(1)} \\ w_{12}^{(2)} w_{11}^{(1)} & w_{12}^{(2)} w_{21}^{(1)} & w_{12}^{(2)} w_{31}^{(1)} \\ w_{13}^{(2)} w_{11}^{(1)} & w_{13}^{(2)} w_{21}^{(1)} & w_{13}^{(2)} w_{31}^{(1)} \end{vmatrix}$$

$$\;=\; w_{11}^{(1)} w_{21}^{(1)} w_{31}^{(1)} \begin{vmatrix} w_{11}^{(2)} & w_{11}^{(2)} & w_{11}^{(2)} \\ w_{12}^{(2)} & w_{12}^{(2)} & w_{12}^{(2)} \\ w_{13}^{(2)} & w_{13}^{(2)} & w_{13}^{(2)} \end{vmatrix} = 0.$$

Therefore, $\det\left(\dfrac{\partial x_i^{(2)}}{\partial x_j^{(0)}}\right) = 0$, which means that the autoencoder is a λ-contraction for any $\lambda > 0$, see Proposition 12.1.5. The information is compressed from the input to the hidden layer, $\mathcal{I}^{(1)} \subsetneq \mathcal{I}^{(0)}$. The inclusion is strict since we cannot solve for all input components, $x_j^{(0)}$, in terms of the component $x_1^{(1)}$.

If $w_{1j}^{(2)} \neq 0$, then rank $w_{1j}^{(2)} = 1$, and by Theorem 11.8.3 the output layer is uncompressed, $\mathcal{I}^{(1)} = \mathcal{I}^{(2)}$. Hence, the autoencoder behaves as an information compressor. The information is compressed in the encoder layer and it is preserved in the decoding layer.

12.6   Conditional Entropy

Another measure of information between two random variables, X and Y, is the *conditional entropy*, $H(Y|X)$. This evaluates the information contained in Y if the variable X is known.

The formal definition is introduced in the following. First, we shall assume that X and Y are discrete random variables with distributions

$$p(x_i) = P(X = x_i), \quad p(y_j) = P(Y = y_j), \quad i = 1, \ldots, N, \ j = 1, \ldots, M.$$

The *conditional entropy of* Y *given that* $X = x_i$ is defined by

$$H(Y|X = x_i) = -\sum_{j=1}^{M} p(y_j|x_i) \ln p(y_j|x_i),$$

where $p(y_j|x_i) = P(Y = y_j|X = x_i)$. The conditional entropy of Y given X is the weighted average of $H(Y|X = x_i)$, namely,

$$H(Y|X) = \sum_{i=1}^{N} p(x_i) H(Y|X = x_i).$$

Using the joint density relation $p(x_i)p(y_j|x_i) = p(x_i, y_j)$, the aforementioned relation becomes

$$H(Y|X) = -\sum_{i=1}^{N} \sum_{j=1}^{M} p(x_i, y_j) \ln p(y_j|x_i), \tag{12.6.9}$$

where $p(x_i, y_j) = P(X = x_i, Y = y_j)$ is the joint distribution of X and Y. The definition variant for the conditional entropy in case of continuous random variables is

$$H(Y|X) = -\int_{\mathbb{R}} \int_{\mathbb{R}} p(x, y) \ln p(y|x) \, dx dy, \tag{12.6.10}$$

where $p(x, y)$ is the joint probability density of (X, Y), and $p(y|x)$ is the conditional probability density of Y given X. The definition can be extended to the multivariate case. For instance,

$$H(Z|X, Y) = -\int\int\int p(x, y, z) \ln p(z|x, y) \, dx dy dz,$$

$$H(X,Y|Z) = -\iiint p(x,y,z)\ln p(x,y|z)\,dxdydz.$$

The first relation represents the conditional entropy of Z, given variables X and Y, while the latter is the conditional entropy of X and Y, given the variable Z.

Proposition 12.6.1 *Let X and Y be two random variables, such that Y depends deterministically on X, i.e., there is a function f such that $Y = f(X)$. Then $H(Y|X) = 0$.*

Proof: First we note that the entropy of a constant is equal to zero, $H(c) = 0$. This follows from the definition of the entropy as

$$H(c) - \sum_i p_i \ln p_i = -1\ln 1 = 0.$$

The entropy of Y conditioned by the event $\{X = x_i\}$ is

$$H(Y|X = x_i) = H(f(X)|X = x_i) = H(f(x_i)) = 0,$$

by the previous observation. Then

$$H(Y|X) = \sum_i p(x_i)H(Y|X = x_i) = 0.$$

The proof was done for discrete random variables, but with small changes it can also accommodate continuous random variables.

■

12.7 The Mutual Information

Heuristically, if X and Y are independent random variables, then $H(Y|X) = H(Y)$, i.e., the knowledge of X does not affect the information contained in Y. However, if X and Y are not independent, it will be shown that $H(Y|X) \leq H(Y)$, i.e., conditioning a random variable decreases its information. The difference

$$I(Y|X) = H(Y) - H(Y|X) \qquad (12.7.11)$$

represents the amount of *information conveyed by X about Y*. Examples include the information that face images provide about the names of the people portrayed, or the information that speech sounds provide about the words spoken.

This concept will be used later in the study of compression of information through the layers of a neural network. A few basic properties are presented in the following:

Proposition 12.7.1 *For any two random variables X and Y defined on the same sample space we have:*

(a) Nonnegativity: $I(X|Y) \geq 0$;

(b) Nondegeneracy: $I(X|Y) = 0 \Leftrightarrow X$ and Y are independent.

(c) Symmetry: $I(X|Y) = I(Y|X)$.

Proof: (a) Using the definition relation (12.7.11), it suffices to show that $H(Y) - H(Y|X) \geq 0$. We have

$$H(Y) - H(Y|X) = -\int p(y) \ln p(y) \, dy + \iint p(x,y) \ln p(y|x) \, dx dy$$

$$= -\iint p(x,y) \ln p(y) \, dx dy + \iint p(x,y) \ln p(y|x) \, dx dy$$

$$= \iint p(x,y) \ln \frac{p(y|x)}{p(y)} \, dx dy = \iint p(x,y) \ln \frac{p(x,y)}{p(x)p(y)} \, dx dy$$

$$= D_{KL}[p(x,y)||p(x)p(y)] \geq 0,$$

where we used the nonnegativity property of the Kullback-Leibler divergence D_{KL}, see section 3.6.

(b) Using the computation from part (a), we have

$$I(X|Y) = 0 \Leftrightarrow H(Y) - H(Y|X) = 0 \Leftrightarrow D_{KL}[p(x,y)||p(x)p(y)] = 0.$$

This occurs for the case $\ln \dfrac{p(x,y)}{p(x)p(y)} = 0$, namely, for $p(x,y) = p(x)p(y)$, which means that X and Y are independent random variables.

(c) The computation in part (a) shows that $I(X|Y) = D_{KL}[p(x,y)||p(x)p(y)]$. Since the expression on the right side is symmetric in x and y, it follows that $I(X|Y) = I(Y|X)$. ∎

The symmetry property enables us to write just $I(X,Y)$ instead of $I(X|Y)$ or $I(Y|X)$; we shall call it the *mutual information of X and Y*. This means that the amount of information contained in X about Y is the same as the amount of information carried in Y about X.

Corollary 12.7.2 *We have the following equivalent definitions for the mutual information:*

$$\begin{aligned} I(X,Y) &= H(Y) - H(Y|X) \\ &= H(X) - H(X|Y) \\ &= D_{KL}[p(x,y)||p(x)p(y)]. \end{aligned}$$

The mutual information can be also seen as the information by which the sum of separate information of X and Y exceeds the joint information of (X,Y).

Proposition 12.7.3 *The mutual information is given by*

$$I(X,Y) = H(X) + H(Y) - H(X,Y). \tag{12.7.12}$$

Proof: Using that $\ln p(y|x) = \ln \dfrac{p(x,y)}{p(x)} = \ln p(x,y) - \ln p(x)$, we have

$$
\begin{aligned}
I(X,Y) &= H(Y) - H(Y|X) = H(Y) + \iint p(x,y) \ln p(y|x)\, dxdy \\
&= H(Y) + \iint p(x,y) \ln p(x,y)\, dxdy - \iint p(x,y) \ln p(x)\, dxdy \\
&= H(Y) - H(X,Y) - \int p(x) \ln p(x)\, dx \\
&= H(Y) - H(X,Y) + H(Y).
\end{aligned}
$$

Since the mutual information is nonnegative, relation (12.7.12) implies that $H(X,Y) \le H(X) + H(Y)$.
∎

Remark 12.7.4 There is an interesting similarity between properties of entropy and properties of area. The area of intersection of two domains A and B, which given by

$$area(A \cap B) = area(A) + area(B) - area(A \cup B) = area(A) - area(A \backslash B)$$

is the analog of

$$I(X,Y) = H(X) + H(Y) - H(X,Y) = H(X) - H(X|Y).$$

This forms the basis of reasoning with entropies as if they were areas, like in Fig. 12.2. Thus, random variables can be represented as domains whose area corresponds to the information contained in the random variable. The area contained in domain X, which is not contained in domain Y, represents the conditional entropy, $H(X|Y)$, while the common area of domains X and Y is the mutual information $I(X,Y)$.

It is worth noting that the mutual information is not a distance on the space of random variables, since the triangle inequality does not hold.

One of the most important properties of the mutual information, which is relevant for deep neural networks, is the invariance property to invertible transformations, which is contained in the next result.

Proposition 12.7.5 (Invariance property) *Let X and Y be random variables taking values in $\mathbb{R}^n$. Then for any invertible and differentiable transformations $\phi, \psi : \mathbb{R}^n \to \mathbb{R}^n$, we have*

$$I(X,Y) = I(\phi(X), \psi(Y)). \tag{12.7.13}$$

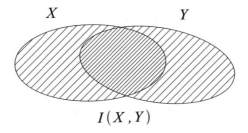

$$I(X,Y)$$

Figure 12.2: *Each random variable is represented by a domain; the area of intersection of domains represents the mutual information $I(X,Y)$.*

Proof: We shall use the formula of entropy change under a coordinate transformation given by Proposition 12.1.2. Let $X' = \phi(X)$ and $Y' = \psi(Y)$, and denote $F = (\phi, \psi) : \mathbb{R}^n_x \times \mathbb{R}^n_y \to \mathbb{R}^n_{x'} \times \mathbb{R}^n_{y'}$, i.e., $(x', y') = F(x,y) = (\phi(x), \psi(y))$. It follows that

$$J_F(x,y) = \frac{\partial(\phi, \psi)}{\partial(x,y)} = \begin{pmatrix} J_\phi(x) & 0 \\ 0 & J_\psi(y) \end{pmatrix},$$

and hence $\det J_F(x,y) = \det J_\phi(x) \det J_\psi(y)$. Inverting, we obtain a similar relation

$$\det J_{F^{-1}}(x', y') = \det J_{\phi^{-1}}(x') \det J_{\psi^{-1}}(y').$$

Taking the log function yields

$$\ln |\det J_{F^{-1}}(x', y')| = \ln |\det J_{\phi^{-1}}(x')| + \ln |\det J_{\psi^{-1}}(y')|. \qquad (12.7.14)$$

We then consider the expectation with respect to the joint distribution of (X', Y') and use formula (12.7.14)

$$\mathbb{E}^{P_{X'Y'}} [\ln |\det J_{F^{-1}}(X', Y')|] = \iint p(x', y') \ln |\det J_{F^{-1}}(x', y')| \, dx' dy'$$

$$= \iint p(x', y') \ln |\det J_{\phi^{-1}}(x')| \, dx' dy' + \iint p(x', y') \ln |\det J_{\psi^{-1}}(y')| \, dx' dy'$$

$$= \int p(x') \ln |\det J_{\phi^{-1}}(x')| \, dx' + \int p(y') \ln |\det J_{\psi^{-1}}(y')| \, dy'$$

$$= \mathbb{E}^{P_{X'}} [\ln |\det J_{\phi^{-1}}(X')|] + \mathbb{E}^{P_{Y'}} [\ln |\det J_{\psi^{-1}}(Y')|] \qquad (12.7.15)$$

Using Proposition 12.1.2 and formula (12.7.15), we obtain

$$
\begin{aligned}
I(\phi(X), \psi(Y)) &= I(X', Y') = H(X') + H(Y') - H(X', Y') \\
&= H(X) - \mathbb{E}^{P_{X'}}[\ln | \det J_{\phi^{-1}}(X')|] \\
&\quad + H(Y) - \mathbb{E}^{P_{Y'}}[\ln | \det J_{\psi^{-1}}(Y')|] \\
&\quad - H(X, Y) + \mathbb{E}^{P_{X'Y'}}[\ln | \det J_{F^{-1}}(X', Y')|] \\
&= H(X) + H(Y) - H(X, Y) \\
&= I(X, Y),
\end{aligned}
$$

which is the desired result.

$\blacksquare$

The following result states that the conditional entropy decreases if extra conditions are added:

Lemma 12.7.6 *For any three random variables, X, Y, and Z we have*

$$
H(X|Y, Z) \leq H(X|Z).
$$

Proof: Consider the difference $\Delta H = H(X|Z) - H(X|Y, Z)$. Using the definition of entropy and properties of log functions and Kullback-Leibler divergence, we have

$$
\begin{aligned}
\Delta H &= - \int p(x, z) \ln p(x|z)\, dx dz + \int p(x, y, z) \ln p(x|y, z)\, dx dy dz \\
&= - \int p(x, y, z) \ln p(x|z)\, dx dy dz + \int p(x, y, z) \ln p(x|y, z)\, dx dy dz \\
&= \int p(x, y, z) \ln \frac{p(x|y, z)}{p(x|z)}\, dx dy dz \\
&= \int p(x, y, z) \ln \frac{p(x, y, z)}{p(x|z)p(y, z)}\, dx dy dz \\
&= D_{KL}[p(x, y, z) || p(x|z)p(y, z)] \geq 0,
\end{aligned}
$$

with identity for $p(x, y, z) = p(x|z)p(y, z)$. Since $p(x, y, z) = p(x|y, z)p(y, z)$, the previous identity is equivalent to $p(x|z) = p(x|y, z)$, which shows a memoryless property of X, given Y, with respect to Y. $\blacksquare$

We say that three random variables X, Y, and Z form the *Markov chain* $X \to Y \to Z$, if each variable depends only on the previous variable, see Fig. 12.3. In terms of conditional probability densities, we have $p(Z|X, Y) = p(Z|Y)$, which represents the *memoryless* property of the chain. The relevance for neural nets is given by the fact that the layer activations $X^{(\ell)}$ of a feedforward neural network form a Markov chain.

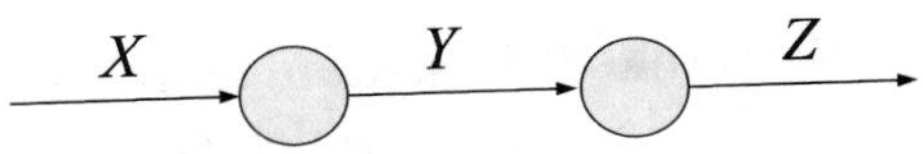

Figure 12.3: *A Markov chain* $X \to Y \to Z$ *satisfies the memoryless property* $p(Z|X,Y) = p(Z|Y)$.

Lemma 12.7.7 *For any three random variables, X, Y, and Z that form a Markov chain, $X \to Y \to Z$, we have*

(a) $H(Z|X,Y) = H(Z|Y)$;

(b) $H(X|Y,Z) = H(X|Y)$.

Proof: (a) Taking the log in the Markov property, $p(z|x,y) = p(z|y)$, and then integrating, yields

$$\int p(x,y,z) \ln p(z|x,y) \, dx dy dz = \int p(x,y,z) \ln p(z|y) \, dx dy dz.$$

The left side is equal to $-H(Z|X,Y)$ and the right side to $H(Z|Y)$, which proves the identity in part (a).

(b) From the Markov property, $p(z|x,y) = p(z|y)$. Then, using the law of conditional probabilities, we obtain

$$p(x|y,z) = \frac{p(x,y,z)}{p(y,z)} = \frac{p(x,y)p(z|x,y)}{p(y)p(z|y)} = \frac{p(x,y)}{p(y)} = p(x|y).$$

Taking the expectation in the previous formula and using the definition, we have

$$\int p(x,y,z) \ln p(x|y,z) \, dx dy dz = \int p(x,y,z) \ln p(x|y) \, dx dy dz,$$

which is equivalent to the desired identity $H(X|Y,Z) = H(X|Y)$. ∎

Remark 12.7.8 Identity (a) shows that the uncertainty of a variable in a Markov chain conditioned by the past is essentially conditioned only by the previous variable (or the nearest past value). Identity (b) states that the uncertainty of a variable in a Markov chain, conditioned by the future, is essentially conditioned by the next variable (or the nearest future value).

Another property of mutual information useful in the context of deep learning is given by the following result, which states that information cannot be increased by data processing. This will be used later in the study of neural nets, where we use the fact that lossy compression cannot convey more information than the original data.

Proposition 12.7.9 (Data processing inequalities) *For any three random variables X, Y, and Z, which form a Markov chain, $X \to Y \to Z$, we have:*

(a) $I(X,Y) \geq I(X,Z)$;

(b) $I(Y,Z) \geq I(X,Z)$.

Proof: (a) Subtracting the identities

$$I(X,Y) = H(X) - H(X|Y)$$

$$I(X,Z) = H(X) - H(X|Z),$$

and using Lemma 12.7.7, part (b), and then Lemma 12.7.6, yields

$$I(X,Y) - I(X,Z) = H(X|Z) - H(X|Y) = H(X|Z) - H(X|Y,Z) \geq 0.$$

(b) We have

$$I(Y,Z) = H(Z) - H(Z|Y)$$

$$I(X,Z) = H(Z) - H(Z|X).$$

Subtracting, using Lemma 12.7.7, part (a), and then Lemma 12.7.6, yields

$$I(Y,Z) - I(X,Z) = H(Z|X) - H(Z|Y) = H(Z|X) - H(Z|X,Y) \geq 0.$$

$\blacksquare$

The previous result can be restated by saying that in a Markov sequence near variables convey more information than distant variables.

Example 12.7.1 A group of children play the *wireless phone game*. One child whispers in the ear of a second child a word, the second one whispers the word to the third one, and so on. The last child says loudly to the group what he or she understood. The fun of the game is that the output is quite different than the input word.

Since each member whispers the message fast, this adds uncertainty to the process and fun to the game. This can be modeled as a Markov sequence, since the understanding of each child is conditioned by the previous child message. The uncertainty of the message is described by the entropy. Lemma 12.7.7 (a) states that the uncertainty of a given message is conditioned only by the previous message. The mutual information, $I(X,Y)$, represents the amount of common information between two whispers X and Y. Data processing inequalities (Proposition 12.7.9) state that the amount of common information of two consecutive whispers is larger than the common information of two distant ones.

Example 12.7.2 Consider the Markov chain $X^{(0)} \to X^{(1)} \to X^{(2)}$, where $X^{(0)}$ is the identity of a randomly picked card from a usual 52-card pack, $X^{(1)}$ represents the suit of the card (Spades, Hearts, Clubs, or Diamonds), and $X^{(2)}$ is the color of the card (Black or Red), see Fig. 12.4. We may consider $X^{(0)}$ as the input random variable to a one-hidden layer neural network. The hidden layer $X^{(1)}$ is pooling the suit of the card, while the output $X^{(2)}$ is a pooling layer that collects the color of the suit; $X^{(2)}$ can be also considered as the color classifier of a randomly chosen card.

We shall compute the entropies of each layer using their uniform distributions $P(X^{(0)} = x_i) = 1/52$, as well as

$X^{(1)}$	Hearts	Clubs	Spades	Diamonds
$P(s_i)$	1/4	1/4	1/4	1/4

$X^{(2)}$	Red	Black
$P(c_i)$	1/2	1/2

Using the definition of the entropy, we have

$$H(X^{(0)}) \;=\; -\sum_{i=1}^{52} p(x_i) \ln p(x_i) = -\ln(1/52) = \ln 52$$

$$H(X^{(1)}) \;=\; -\sum_{i=1}^{4} p(s_i) \ln p(s_i) = -\ln(1/4) = \ln 4$$

$$H(X^{(2)}) \;=\; -\sum_{i=1}^{2} p(c_i) \ln p(c_i) = -\ln(1/2) = \ln 2.$$

We notice the strictly decreasing flow of entropy

$$H(X^{(0)}) > H(X^{(1)}) > H(X^{(2)}).$$

Dividing by $\ln 2$, we obtain the entropy measured in bits. This way, $H(X^{(0)})$ is 5.726 bits, $H(X^{(1)})$ is 2 bits and $H(X^{(2)})$ is just 1 bit. This means that in order to determine a card randomly picked from a deck we need to ask on average 5.726 questions (with Yes/No answers), while to determine its color it's suffices to ask only 1 question (for instance, we may ask *Is the color Red?*).

If the card identity, $X^{(0)}$, is revealed, then its suit is determined, and hence, $X^{(1)}$ has no uncertainty, $H(X^{(1)}|X^{(0)}) = 0$. This can be formally shown by noting that

$$p(x_i^{(1)}|x_j^{(0)}) = P\big(X^{(1)} = x_i^{(1)}|X^{(0)} = x_j^{(0)}\big) = \begin{cases} 1, & \text{if card } x_j^{(0)} \text{ has suit } x_i^{(1)} \\ 0, & \text{otherwise,} \end{cases}$$

so that the conditional entropy becomes

$$
\begin{aligned}
H(X^{(1)}|X^{(0)}) &= -\sum_{j=1}^{52}\sum_{i=1}^{4} p(x_i^{(1)}, x_j^{(0)}) \ln p(x_i^{(1)}|x_j^{(0)}) \\
&= -\sum_{j=1}^{52}\sum_{i=1}^{4} p(x_i^{(1)}) p(x_i^{(1)}|x_j^{(0)}) \ln p(x_i^{(1)}|x_j^{(0)}) = 0.
\end{aligned}
$$

Similarly, we have $H(X^{(2)}|X^{(0)}) = 0$, since the color of a card is known once the card identity is revealed. We compute next the mutual information between the input layer and the other two layers:

$$
I(X^{(0)}, X^{(1)}) = H(X^{(1)}) - \underbrace{H(X^{(1)}|X^{(0)})}_{=0} = H(X^{(1)}) = \ln 4
$$

$$
I(X^{(0)}, X^{(2)}) = H(X^{(2)}) - \underbrace{H(X^{(2)}|X^{(0)})}_{=0} = H(X^{(2)}) = \ln 2.
$$

It follows that the following data processing inequality

$$
I(X^{(0)}, X^{(1)}) \geq I(X^{(0)}, X^{(2)})
$$

is verified strictly.

12.8   Applications to Deep Neural Networks

Before presenting the applications to deep neural networks, we recall some notations. Let $X^{(\ell-1)}$ and $X^{(\ell)}$ be the activations of the layers $\ell - 1$ and ℓ of a feedforward neural network. $X^{(\ell-1)}$ can be also seen as the input to the ℓth layer.

12.8.1   Entropy Flow

We have studied the entropy flow in the case when layer activations are continuous random variables. In this section we shall assume they are discrete.

Proposition 12.8.1 *If the layer activations of a feedforward neural network are discrete random variables, then the entropy flow is decreasing*

$$
H(X^{(0)}) \geq H(X^{(1)}) \geq \cdots \geq H(X^{(L)}) \geq 0.
$$

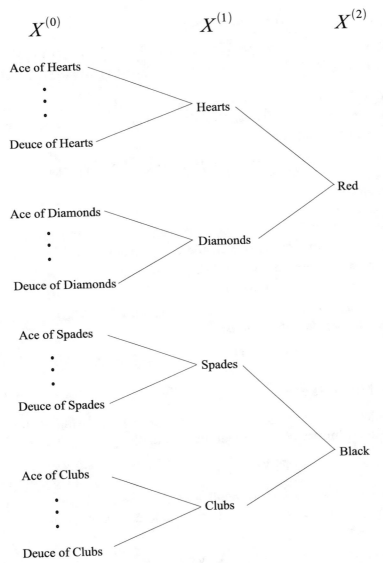

Figure 12.4: *The Markov chain $X^{(0)} \rightarrow X^{(1)} \rightarrow X^{(2)}$, where $X^{(0)}$ is the identity of a card, $X^{(1)}$ is the suit of a card, and $X^{(2)}$ is the color of a card.*

Proof: Since the activation of the ℓth layer depends deterministically on the activation of the previous layer, $X^{(\ell)} = F(X^{(\ell-1)})$, by Proposition 12.6.1 we have $H(X^{(\ell)}|X^{(\ell-1)}) = 0$. We shall compute the mutual information

$I(X^{(\ell)}, X^{(\ell-1)})$ in two different ways:

$$
\begin{aligned}
I(X^{(\ell)}, X^{(\ell-1)}) &= H(X^{(\ell)}) - H(X^{(\ell)}|X^{(\ell-1)}) = H(X^{(\ell)}) \\
I(X^{(\ell)}, X^{(\ell-1)}) &= H(X^{(\ell-1)}) - H(X^{(\ell-1)}|X^{(\ell)}) \le H(X^{(\ell-1)}),
\end{aligned}
$$

since $H(X^{(\ell-1)}|X^{(\ell)}) \ge 0$, as the variables are discrete. From the last two formulas we infer $H(X^{(\ell)}) \le H(X^{(\ell-1)})$. The equality holds for the case when F is invertible, since

$$
H(X^{(\ell-1)}|X^{(\ell)}) = H(F^{-1}(X^{(\ell)})|X^{(\ell)}) = 0,
$$

by Proposition 12.6.1. ∎

Corollary 12.8.2 *Assume the layer activations of a feedforward neural network are discrete random variables and the following conditions are satisfied:*
(i) The activation function is strict increasing;
(ii) The network has constant width: $d^{(0)} = d^{(1)} = \cdots = d^{(L)}$;
(iii) The weight matrices are nondegenerate, $\det W^{(\ell)} \ne 0$, for all $1 \le \ell \le L$. Then the entropy flow is constant

$$
H(X^{(0)}) = H(X^{(1)}) = \cdots = \cdots H(X^{(L)}).
$$

12.8.2 Noisy layers

In all feedforward neural networks used so far the activation of the ℓth layer depends deterministically on the activation of the previous layer

$$
X^{(\ell)} = F(X^{(\ell-1)}) = \phi(W^{(\ell)T}X^{(\ell-1)} - B^{(\ell)}).
$$

In the rest of this chapter we shall assume that the layer activations are perturbed by some noise, $\epsilon^{(\ell)}$, which is a random variable independent of the layer $X^{(\ell-1)}$, as

$$
X^{(\ell)} = F(X^{(\ell-1)}, \epsilon^{(\ell)}).
$$

One easy way to accomplish this from the computational perspective is to consider an *additive noise*

$$
X^{(\ell)} = F(X^{(\ell-1)}) + \epsilon^{(\ell)} = \phi(W^{(\ell)T}X^{(\ell-1)} - B^{(\ell)}) + \epsilon^{(\ell)},
$$

with $\mathbb{E}[\epsilon^{(\ell)}] = 0$ and some known positive entropy, $H(\epsilon^{(\ell)})$. In the case of an additive noise, a computation provides

$$
H(X^{(\ell)}|X^{(\ell-1)}) = H\big(F(X^{(\ell-1)}) + \epsilon^{(\ell)}|X^{(\ell-1)}\big) = H(\epsilon^{(\ell)}),
$$

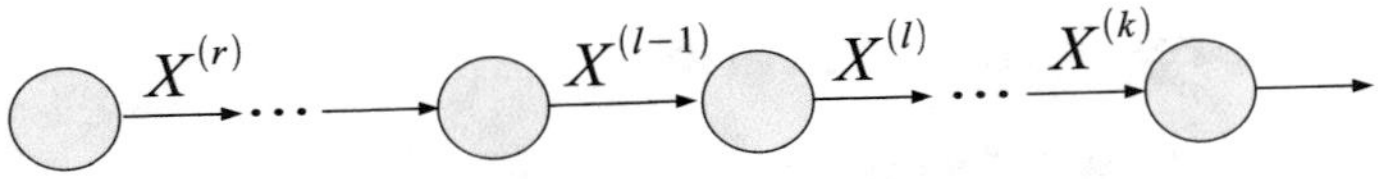

Figure 12.5: *If $X^{(\ell-1)}$ and $X^{(\ell)}$ are independent, then $X^{(\ell)}$ is independent of the previous layers of $X^{(\ell-1)}$, and $X^{(\ell-1)}$ is independent of the next layers of $X^{(\ell)}$.*

since the noise is independent of the $(\ell - 1)$th layer.

We shall present next some applications of mutual information to independence and information compression in a feedforward neural network with noisy layers.

12.8.3 Independent layers

The layers r and ℓ are called *independent* if $I(X^{(r)}, X^{(\ell)}) = 0$, i.e., if the layer activation $X^{(\ell)}$ does not contain any information from the layer activation $X^{(r)}$. Equivalently, the random variables $X^{(\ell)}$ and $X^{(r)}$ are independent by Proposition 12.7.1, part (b).

Proposition 12.8.3 (Separability by independence) *Assume the layers $\ell - 1$ and ℓ are independent. Then*

(a) *The layer ℓ is independent of any layer r, for any $r \leq \ell - 1$.*

(b) *The layer $\ell - 1$ is independent of any layer k, for any $k \geq \ell$.*

Proof: The proof is a consequence of the data processing inequalities, Proposition 12.7.9, and the nonnegativity of the mutual information, see Fig. 12.5.
(a) For any $r \leq \ell - 1$, we have

$$0 \leq I(X^{(r)}, X^{(\ell)}) \leq I(X^{(\ell-1)}, X^{(\ell)}) = 0,$$

from where $I(X^{(r)}, X^{(\ell)}) = 0$.
(b) For any $k \geq \ell$, we have

$$0 = I(X^{(\ell-1)}, X^{(\ell)}) \geq I(X^{(\ell-1)}, X^{(k)}) \geq 0,$$

which implies $I(X^{(\ell-1)}, X^{(k)}) = 0$. ∎

Roughly speaking, if two consecutive layers of a feedforward neural network are independent, then any layer before them is independent from any layer after them.

12.8.4   Compressionless layers

We consider a feedforward neural network with noisy layers. The ℓth layer is called *compressionless* if

$$I(X^{(0)}, X^{(\ell-1)}) = I(X^{(0)}, X^{(\ell)}),$$

i.e., the input $X^{(0)}$ conveys the same information about both $X^{(\ell-1)}$ and $X^{(\ell)}$.

Remark 12.8.4 In the absence of noise, both $X^{(\ell-1)}$ and $X^{(\ell)}$ depend deterministically on $X^{(0)}$ and in this case we have

$$
\begin{aligned}
I(X^{(0)}, X^{(\ell-1)}) &= H(X^{(\ell-1)}) - H(X^{(\ell-1)}|X^{(0)}) = H(X^{(\ell-1)}), \\
I(X^{(0)}, X^{(\ell)}) &= H(X^{(\ell)}) - H(X^{(\ell)}|X^{(0)}) = H(X^{(\ell)}),
\end{aligned}
$$

where we used Proposition 12.6.1. Therefore, the ℓth layer is compressionless in the sense of the previous definition if $H(X^{(\ell-1)}) = H(X^{(\ell)})$, namely, no entropy leak between these layers. This relation is implied by the following three conditions:

(i) the activation function satisfies $\phi' > 0$;
(ii) $d^{(\ell-1)} = d^{(\ell)}$;
(iii) the weight matrix $W^{(\ell)}$ is nonsingular.

We note that conditions (i)–(iii) are also necessary conditions for the layer $X^{(\ell)}$ to be uncompressed, see Proposition 11.6.3 and Theorem 11.7.1. These two are distinct concepts, as the former relates to the measure of information and the latter to the associated sigma-algebra. However, both concepts describe the fact that some sort of information invariance occurs between layers $\ell - 1$ and ℓ.

The next result deals with a few equivalent ways to show that a layer is compressionless. It is included here for completeness reasons.

Proposition 12.8.5 *Consider a feedforward neural network with noisy layers. The following are equivalent:*

(i) *The ℓth layer is compressionless;*

(ii) $I(X^{(0)}, X^{(\ell)}) = I(X^{(0)}, X^{(\ell+1)})$;

(iii) $H(X^{(0)}|X^{(\ell+1)}) = H(X^{(0)}|X^{(\ell)}, X^{(\ell+1)})$;

(iv) $p(x^{(0)}|x^{(\ell+1)}) = p(x^{(0)}|x^{(\ell)}, x^{(\ell+1)})$;

(v) $p(x^{(\ell)}, x^{(\ell+1)}|x^{(0)}) = p(x^{(\ell+1)}|x^{(0)})p(x^{(\ell)}|x^{(\ell+1)})$.

Proof: (i) $\Leftrightarrow$ (ii) It comes from the definition.

$(ii) \Leftrightarrow (iii)$ Writing the mutual information in terms of entropy, we have

$$
\begin{aligned}
I(X^{(0)}, X^{(\ell)}) &= I(X^{(0)}, X^{(\ell+1)}) \Leftrightarrow \\
H(X^{(0)}) - H(X^{(0)}|X^{(\ell)}) &= H(X^{(0)}) - H(X^{(0)}|X^{(\ell+1)}) \Leftrightarrow \\
H(X^{(0)}|X^{(\ell)}) &= H(X^{(0)}|X^{(\ell+1)}).
\end{aligned}
$$

$(iii) \Leftrightarrow (iv)$ Since $X^{(0)} \to X^{(\ell)} \to X^{(\ell+1)}$ is a Markov chain, Lemma 12.7.7, part (b) implies $H(X^{(0)}|X^{(\ell)}, X^{(\ell+1)}) = H(X^{(0)}|X^{(\ell)})$. On the other side, Lemma 12.7.6 yields $H(X^{(0)}|X^{(\ell+1)}) \geq H(X^{(0)}|X^{(\ell)}, X^{(\ell+1)})$. The last two expressions lead to the inequality

$$
H(X^{(0)}|X^{(\ell+1)}) \geq H(X^{(0)}|X^{(\ell)}).
$$

According to the proof of Lemma 12.7.7, this inequality becomes identity (as in (iii)) if and only if

$$
p(x^{(0)}|x^{(\ell+1)}) = p(x^{(0)}|x^{(\ell)}, x^{(\ell+1)}),
$$

which is (iv).

$(iv) \Leftrightarrow (v)$ It follows from a computation using conditional probability densities. We have

$$
\begin{aligned}
p(x^{(0)}|x^{(\ell+1)}) &= \frac{p(x^{(0)}, x^{(\ell+1)})}{p(x^{(\ell+1)})} = \frac{p(x^{(\ell+1)}|x^{(0)})p(x^{(0)})}{p(x^{(\ell+1)})} \\
p(x^{(0)}|x^{(\ell)}, x^{(\ell+1)}) &= \frac{p(x^{(0)}, x^{(\ell)}, x^{(\ell+1)})}{p(x^{(\ell)}, x^{(\ell+1)})} = \frac{p(x^{(\ell)}, x^{(\ell+1)}|x^{(0)})p(x^{(0)})}{p(x^{(\ell)}, x^{(\ell+1)})}.
\end{aligned}
$$

Equating $p(x^{(0)}|x^{(\ell+1)}) = p(x^{(0)}|x^{(\ell)}, x^{(\ell+1)})$ yields

$$
p(x^{(\ell)}, x^{(\ell+1)}|x^{(0)}) = p(x^{(\ell+1)}|x^{(0)}) \frac{p(x^{(\ell)}, x^{(\ell+1)})}{p(x^{(\ell+1)})}.
$$

Using $\frac{p(x^{(\ell)}, x^{(\ell+1)})}{p(x^{(\ell+1)})} = p(x^{(\ell)}|x^{(\ell+1)})$ yields equation (v). ∎

12.8.5 The number of features

In this section we discuss a useful interpretation of entropy and mutual information for estimating the size of the input and output layers of a feedforward neural network, as well as the number of described features.

Assume we have a discrete random variable, X, that takes n values, x_1, $x_2, \ldots, x_n$ with probabilities $p_1, p_2, \ldots, p_n$, respectively. Assume that the value of X is revealed to us by a person who can communicate only by means of the words "yes" and "no". Then we can always be able to arrive to the correct value of X by a finite sequence of "yes" and "no" answers. Then the entropy $H(X)$ is the average minimum number of "yes" and "no" answers needed to find the correct value of X. The content of this result is known in coding theory under the name of the *noiseless coding theorem*, see, for instance, the reference [11].

The number of values taken by the input data X can be estimated from its entropy as $n \approx 2^{H(X)}$, where $H(X)$ is measured in bits. Now, consider a neural network with input X and output Y. Each value, y_i, of Y is considered to be a certain feature of the input X. The conditional entropy of X, given the feature $Y = y_i$, is defined by

$$H(X|Y = y_i) = -\sum_{k=1}^{n} p(x_k|y_i) \ln p(x_k|y_i),$$

and represents the uncertainty of X given that X contains the feature y_i.

The size of the input data with feature y_i is approximated by $2^{H(X|Y=y_i)}$. This size depends on the feature, but we can consider their mean. The average size of the input data with a given feature is estimated by $2^{H(X|Y)}$, where

$$H(X|Y) = \sum_{j=1}^{m} p(y_j) H(X|Y = y_i)$$

is a weighted sum of the conditional expectations. The estimated number of subsets of the input data is obtained dividing the size of X to the average size of a subset of the same feature, i.e.,

$$\frac{2^{H(X)}}{2^{H(X|Y)}} = 2^{H(X)-H(X|Y)} = 2^{I(X,Y)}.$$

Since each subset of X is mapped one-to-one to a feature, then $2^{I(X,Y)}$ estimates the number of features of the output layer. The information $I(X,Y)$ is measured in bits.

For instance, in the case of the MNIST data classification, there are 10 classes of digits, or 10 features. In this case the network should be constructed such that the mutual information between the output and the input satisfies $I(X,Y) = \log_2 10$. However, if $I(X,Y) < \log_2 10$ the network will lead to an underfit, since the output information conveyed by the input is too small to classify 10 digits.

We shall verify these formulas on the following concrete example.

Example 12.8.6 We shall consider the example of selecting a random card from an ordinary 52-card deck. Then each card has the selection probability equal to $p_i = 1/52$. A first question, Q1, can be: *Is the suit color Red?* Assume the answer is "yes". Then the next question, Q2, can be: *Is the suit Diamond?* Assume the answer is "no". This means the correct suit is "Hearts". Since there are 13 cards, the next question, Q3, can be: *Is the card number larger than 6?* Let the answer be "no". This leaves us with only six choices: $1, 2, 3, 4, 5,$ and 6. We divide them into two sets and decide again the set the card belongs to by asking Q4: *Is the card number less than 4?* Assume the answer is "yes". This implies the card is either $1, 2,$ or 3. The next question, Q5, can be *Is the card number less than 2?* If the answer is "yes", it means that the card is the *Ace of Hearts*, and in this case 5 question would suffice to determine the card identity. However, if the last answer is "no", then the card is either a 2 or a 3. We determine it by asking Q6: *Is the card number 2?* An "yes" answer implies that the card is a *2 of Hearts*, while a "no" answer implies the card is a *3 of Hearts*. In conclusion, we need either 5 or 6 questions to determine exactly the card identity. If the experiment of finding the card is performed many times and then we consider the average of the number of questions needed, we obtain 5.7. This means that the knowledge of the identity of a randomly selected card from a 52-card deck contains an information of 5.7 bits.

On the other side, the entropy of the random variable of extracting randomly a card is

$$H(X) = -\sum_{i=1}^{52} p_i \ln(p_i) = -\ln(1/52) = 3.95.$$

Dividing by $\ln 2$ we obtain the information in bits as $3.95/\ln 2 = 5.7$, which agrees with the aforementioned value (If we do not divide by $\ln 2$ the information is measured in *nats* rather than bits).

Consider now the random variable Y, which represents the suit of a card randomly selected from the same deck. The variable Y takes only four values: *Spades, Hearts, Diamonds,* and *Clubs*. The conditional entropy given one of the features is

$$H(X|Y = Hearts) = -\sum_{k=1}^{13} \frac{1}{13} \ln \frac{1}{13} = \ln 13 \approx 2.56.$$

Considering the weighted sum with $p(x_i) = 1/4$, we obtain the conditional entropy of X, given features variable Y, as

$$H(X|Y) = \sum_{j=1}^{4} \frac{1}{4} H(X|Y = y_i) = \ln 13 \approx 2.56.$$

Then the mutual information of X and Y can be estimated as

$$I(X, Y) = H(X) - H(X|Y) \approx 3.95 - 2.56 = 1.39,$$

which, after dividing by $\ln 2$ we obtain 2 bits. Then $2^{I(X,Y)} = 4$, which corresponds to 4 classes, one for each suit.

12.8.6 Total Compression

The *compression factor* of the ℓth layer of a feedforward neural network with noisy layers is defined by the ratio of the following mutual information:

$$\rho_\ell = \frac{I(X^{(0)}, X^{(\ell)})}{I(X^{(0)}, X^{(\ell-1)})}.$$

From the data processing inequality, Proposition 12.7.9, part (a), we have $I(X^{(0)}, X^{(\ell-1)}) \geq I(X^{(0)}, X^{(\ell)})$. This fact implies $0 \leq \rho_\ell \leq 1$. The compression factor is 1 if the layer is compressionless and is equal to 0 if $X^{(0)}$ and $X^{(\ell)}$ are independent layers.

The product of all compression factors is independent of the hidden layers of the network as the following computation shows:

$$
\begin{aligned}
\rho_1 \rho_2 \cdots \rho_L &= \frac{I(X^{(0)}, X^{(1)})}{I(X^{(0)}, X^{(0)})} \frac{I(X^{(0)}, X^{(2)})}{I(X^{(0)}, X^{(1)})} \cdots \frac{I(X^{(0)}, X^{(L)})}{I(X^{(0)}, X^{(L-1)})} \\
&= \frac{I(X^{(0)}, X^{(L)})}{I(X^{(0)}, X^{(0)})} = \frac{I(X^{(0)}, X^{(L)})}{H(X^{(0)})},
\end{aligned}
$$

where we used $I(X^{(0)}, X^{(0)}) = H(X^{(0)}) - H(X^{(0)}|X^{(0)}) = H(X^{(0)})$. This suggests that the quotient

$$\rho = \frac{I(X^{(0)}, X^{(L)})}{H(X^{(0)})} \tag{12.8.11}$$

describes the *total compression* in a feedforward neural network. It represents the amount of information shared by the input and the output of the network, scaled by the amount of input information.

Remark 12.8.7 We make two remarks, which follow easily.
(i) The total compression $\rho = 1$ (no compression) if and only if all layers are compressionless.
(ii) The total compression $\rho = 0$ if and only if there is a layer independent from the input X^0.

We note that if the feedforward neural network has to classify the input data into c classes, then the last layer has to have the size $d^{(L)} = c$. If the input layer has the size $n = d^{(0)}$, then we shall estimate the total compression in terms of these two parameters.

Proposition 12.8.8 *The total compression of a feedforward neural network with n inputs that classifies data into c classes is given by $\rho = \log_n c$.*

Proof: Since the number of classes is $c = 2^{I(X^{(0)}, X^{(L)})}$, and the input size is $n = 2^{H(X^{(0)})}$, using relation (12.8.11) and the change of base formula for the logarithm, we obtain

$$\rho = \frac{I(X^{(0)}, X^{(L)})}{H(X^{(0)})} = \frac{\log_2 c}{\log_2 n} = \log_n c.$$

∎

For instance, in the case of the MNIST database classification of digits, the number of classes is $c = 10$ and the input size is $n = 28 \times 28$ (each picture is a matrix of 28×28 pixels), then the total compression needed for classification is $\rho = \log_{784} 10 = 0.345$.

Remark 12.8.9 We note that the total compression in the case of the absence of the noise from the layers is given by the quotient of the output and the input entropies

$$\rho = \frac{H(X^{(L)})}{H(X^{(0)})}.$$

12.9 Network Capacity

We have seen that a feedforward neural network can be interpreted as an information compressor. Now, this section addresses the following question:

How large is the information conveyed by the input pattern given you observe the network response?

In order to answer this question we need to define the concept of *network capacity*. This can be informally defined by stating that capacity is the ability of a neural network to fit a large variety of target data. A low capacity network struggles to fit the training data and may lead to an underfit, while a high capacity network memorizes the training set, leading to an overfit. The capacity depends on the network architecture, more specifically on the number of neurons, number of weights and biasses, learning algorithm adopted, etc.

In the following we shall further formalize this concept and also find an exact formula for the capacity in the case of feedforward neural networks with noisy layers given by discrete random variables.

12.9.1 Types of capacity

Consider a feedforward neural network, having $L - 1$ hidden layers and $d^{(\ell)}$ neurons in the ℓth layer, $0 \leq \ell \leq L$. Denote the input variable by $X = X^{(0)}$ and the output by $Y = X^{(L)}$, and the target variable by Z. If X is a random variable, then the output, Y, is also a random variable.

There are three mutual information of interest, $I(X, Y)$, $I(Y, Z)$, and $I(X, Z)$, which we shall address shortly.

1. The first one, $I(X, Y)$, represents the amount of information contained in X about Y, or, equivalently, the amount of information processed by the network. This depends on the input distribution $p(x)$, as well as on the system of weights, $W^{(\ell)}$, and biasses, $B^{(\ell)}$. Then, varying the distribution $p(x)$, keeping matrices $W^{(\ell)}$, $B^{(\ell)}$ fixed until the information processed reaches a maximum, we obtain the *network capacity* corresponding to the weight system (W, B) as

$$C(W, B) = \max_{p(x)} I(X, Y). \tag{12.9.12}$$

This represents the maximum amount of information a network can process if the weights and biasses are kept fixed. It is worth noting the similarity to the definition of *channel capacity*, which comes into the famous Channel Coding Theorem, see Chapter 8 of [131].

Varying the system of weights and biasses we obtain the maximum possible information processed by the network, called the *total network capacity*

$$C_{tot} = \max_{W,B} \{ C(W, B); \|W\|^2 + \|B\|^2 \leq 1 \}. \tag{12.9.13}$$

The regularization constraint $\|W\|^2 + \|B\|^2 \leq 1$ has been added in order to keep the variables (W, B) into a compact set and to assure for the existence of the maximum.

The maximum information processed by a network with a given bounded input entropy is measured by the *essential capacity*

$$C(M) = \max_{H(X) \leq M} I(X, Y). \tag{12.9.14}$$

2. The mutual information $I(Y, Z)$ represents the amount of information conveyed by the network output, Y, about the target variable Z. During the learning process we expect that $I(Y, Z)$ will tend to the target entropy, $H(Z)$. Assuming the random variables have discrete values, we have $H(Z|Y) > 0$. Then the inequality

$$I(Y, Z) = H(Z) - H(Z|Y) \leq H(Z)$$

shows that the mutual information $I(Y, Z)$ is bounded from above by $H(Z)$, and hence for a better learning of Z the mutual information $I(Y, Z)$ has to be as large as possible.

3. The third amount of interest, $I(X, Z)$, represents the mutual information contained in the training pair (X, Z). This can be computed explicitly since the training distribution p_{XZ} is given.

Before proving the existence of capacity, we shall introduce first some terminology.

12.9.2 The input distribution

The input layer has $d^{(0)} = n$ neurons and the input random variable is given by $X = (X_1, \ldots, X_n)$. Each component X_k is a real-valued random variable, which takes values in the finite set $\{x_k^1, x_k^2, \ldots, x_k^{r_k}\}$. The input probability is given by

$$P(X = \mathbf{x}) = P(X_1 = x_1^{i_1}, \ldots, X_n = x_n^{i_n}) = p(x_1^{i_1}, \ldots, x_n^{i_n}),$$

for $\mathbf{x} = (x_1^{i_1}, \ldots, x_n^{i_n})$.

In the particular case when there is only one neuron in the input layer, namely, $d^{(0)} = 1$ and $X = X_1$, we assume that X takes N values, $\{x_1, \ldots, x_N\}$, and each value is taken with a given probability

$$p(x_i) = P(X = x_i), \quad i = 1, \ldots, N,$$

which forms the input probability distribution, see Fig. 12.6.

12.9.3 The output distribution

The output layer has $d^{(L)} = m$ neurons and the output random variable is given by $Y = (Y_1, \ldots, Y_m)$, with each component Y_k real-valued random variable, taking values in the finite set $\{y_k^1, y_k^2, \ldots, y_k^{t_k}\}$. The output probability is given by

$$P(Y = \mathbf{y}) = P(Y_1 = y_1^{j_1}, \ldots, Y_m = y_m^{j_m}) = p(y_1^{j_1}, \ldots, y_m^{j_m}),$$

for $\mathbf{y} = (y_1^{j_1}, \ldots, y_m^{j_m})$.

In the particular case when the output layer has only one neuron, then $Y = Y_1$ takes values in $\{y_1, \ldots, y_M\}$ with probabilities

$$p(y_j) = P(Y = y_j), \quad j = 1, \ldots, M,$$

which forms the output distribution, see Fig. 12.6.

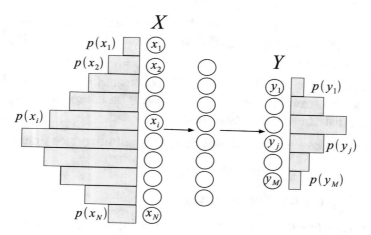

Figure 12.6: *The output density $p(y_j)$ in terms of the input density $p(x_i)$.*

12.9.4   The input-output tensor

We consider the multi-indices $\mathcal{I} = (i_1, \ldots, i_n)$ and $\mathcal{J} = (j_1, \ldots, j_m)$ and let

$$\mathbf{x}^{\mathcal{I}} = (x_1^{i_1}, \ldots, x_n^{i_n}), \qquad \mathbf{y}^{\mathcal{J}} = (y_1^{j_1}, \ldots, y_m^{j_m}).$$

The *input-output tensor* $q_{\mathcal{I},\mathcal{J}} = q_{i_1,\ldots,i_n;j_1,\ldots,j_m}$ is defined by the following conditional probability:

$$q_{\mathcal{I},\mathcal{J}} = P(Y = \mathbf{y}^{\mathcal{J}} | X = \mathbf{x}^{\mathcal{I}}) = p(y_1^{j_1}, \ldots, y_m^{j_m} | x_1^{i_1}, \ldots, x_n^{i_n}).$$

The tensor $q_{\mathcal{I},\mathcal{J}}$ depends on the architectural structure of the network (with fixed weights and biasses) and is independent of the input distribution $p(\mathbf{x})$. Since

$$P(Y = \mathbf{y}^{\mathcal{J}}) = \sum_{\mathcal{I}} P(Y = \mathbf{y}^{\mathcal{J}} | X = \mathbf{x}^{\mathcal{I}}) P(X = \mathbf{x}^{\mathcal{I}}),$$

the tensor $q_{\mathcal{I},\mathcal{J}}$ transforms the input distribution into the output distribution by the formula $p(\mathbf{y}^{\mathcal{I}}) = \sum_{\mathcal{I}} q_{\mathcal{I},\mathcal{J}} p(\mathbf{x}^{\mathcal{I}})$, or in the equivalent detailed form

$$p(y_1^{j_1}, \ldots, y_m^{j_m}) = \sum_{i_1,\ldots,i_n} q_{i_1,\ldots,i_n;j_1,\ldots,j_m} p(x_1^{i_1}, \ldots, x_n^{i_n}).$$

12.9.5   The input-output matrix

If $d^{(0)} = d^{(L)} = 1$, then the neural network is characterized by the following *input-output matrix* $q_{ij} = p(y_j | x_i)$, where

$$p(y_j | x_i) = P(Y = y_j | X = x_i), \quad 1 \leq i \leq N, 1 \leq j \leq M.$$

The role of the matrix $Q = q_{ij}$ is to transform the input distribution into the output distribution by the formula $p(y_j) = \sum_{i=1}^{N} p(x_i)p(y_j|x_i)$, or in an equivalent matrix form, $p(\mathbf{y}) = Q^T p(\mathbf{x})$. If $N = M$ and the matrix Q is nonsingular, given the output density $p(\mathbf{y})$, there is a unique input density $p(\mathbf{x})$ that is transformed into $p(\mathbf{y})$.

However, if $M < N$, then Q is not a square matrix, and then it does not make sense to consider its determinant. In this case it is useful to assume the maximal rank condition, $\text{rank}(Q) = \text{rank}(Q^T) = M$. Under this condition, there is at most one solution $p(\mathbf{x})$ for the aforementioned equation, see Exercise 12.13.4. The solution existence will be treated in the next section.

Since for a fixed value $X = x_i$, the random variable Y takes some value y_j, we have $\sum_{j=1}^{M} p(y_j|x_i) = 1$. This also writes as $\sum_{j=1}^{M} q_{ij} = 1$, i.e., the sum of the entries on each row of the input-output matrix Q is equal to 1. Since the entries of Q are nonnegative, $q_{ij} \geq 0$, it follows that Q is a *Toeplitz matrix*.[1] This property of the matrix q_{ij} will be used several times in the next section.

Remark 12.9.1 Consider a feedforward neural network with noiseless layers, such that the input-output mapping, f, is bijective, with $f(x_i) = y_i$ and $M = N$. The input-output matrix in this case is given by

$$
\begin{aligned}
q_{ij} &= P(Y = y_j | X = x_i) = P(f(X) = y_j | X = x_i) \\
&= P(f(x_i) = y_j) = \delta_{ij},
\end{aligned}
$$

since the event $\{f(x_i) = y_j\}$ is either sure or impossible. Hence, the input-output matrix is the identity, $Q = \mathbb{I}_N$.

If the injectivity of f is dropped, it is not hard to show that the entries of the matrix Q consist only in 0s and 1s, with only one 1 on each row (the rows are "hot-vectors"). This fact agrees with the Toeplitz property of Q.

12.9.6 The existence of network capacity

The fact that the definition of the network capacity makes sense reduces to the existence of the maximum of the mutual information $I(X, Y)$ under variations of the input distribution. For the sake of simplicity we shall treat the problem in the particular case $d^{(0)} = d^{(L)} = 1$, i.e., when there is only one input and one output neurons. We switch the indices into $n = N$ and $m = M$, and determine a formula for $I(X, Y)$ in terms of the input distri-

---

[1] Sometimes it is also called a *stochastic matrix*.

bution $p(x_i)$, using the definitions on the mutual information, entropy, and conditional entropy:

$$
\begin{aligned}
I(X,Y) &= H(Y) - H(Y|X) \\
&= -\sum_{j=1}^{m} p(y_j) \ln p(y_j) + \sum_{i=1}^{n} \sum_{j=1}^{m} p(x_i, y_j) \ln p(y_j|x_i) \\
&= -\sum_{j=1}^{m} \sum_{i=1}^{n} p(x_i) p(y_j|x_i) \ln p(y_j) \\
&\quad + \sum_{i=1}^{n} \sum_{j=1}^{m} p(x_i) p(y_j|x_i) \ln p(y_j|x_i) \\
&= \sum_{j=1}^{m} \sum_{i=1}^{n} p(x_i) p(y_j|x_i) \Big(\ln p(y_j|x_i) - \ln p(y_j) \Big) \\
&= \sum_{j=1}^{m} \sum_{i=1}^{n} p(x_i) p(y_j|x_i) \Big(\ln p(y_j|x_i) - \ln \sum_{i=1}^{n} p(x_i) p(y_j|x_i) \Big) \\
&= \sum_{j=1}^{m} \sum_{i=1}^{n} p(x_i) q_{ij} \Big(\ln q_{ij} - \ln \sum_{i=1}^{n} p(x_i) q_{ij} \Big).
\end{aligned}
$$

It follows that for a given input-output matrix, q_{ij}, the mutual information $I(X,Y)$ is a continuous function of n real numbers, $p(x_1), \ldots, p(x_n)$, which belong to the domain

$$
K = \{(p_1, \ldots, p_n); p_i \geq 0, \sum_{i=1}^{n} p_i = 1\}.
$$

Since $0 \leq p_i \leq 1$, then $K \subset B_{\mathbb{R}^n}(0,1)$, i.e., K is bounded. The set K is also closed, as an intersection of two closed sets, $K = S_n \cap \mathcal{H}_n$, the hyperplane

$$
\mathcal{H}_n = \{(p_1, \ldots, p_n); \sum_{i=1}^{n} p_i \leq 1\},
$$

and the first sector

$$
S_n = \{(p_1, \ldots, p_n); p_i \geq 0\}.
$$

Being bounded and closed, it follows that K is a compact set in $\mathbb{R}^n$. Since any continuous function on a compact set reaches its maxima on that set, using that $I(X,Y)$ is a continuous function of p_i on K, it follows that there is an input distribution $p_i^* = p(x_i)$ for which $I(X,Y)$ is maximum. The maximum distribution p^* can belong either to the boundary of K, case in which at least one of the components is zero, or to the interior of K. In the latter case, calculus methods will be employed to characterize the maximum capacity.

12.9.7   The Lagrange multiplier method

In this section we shall find the network capacity using a variational problem involving a Lagrange multiplier, λ, which is used to introduce the linear constraint $\sum_{i=1}^{n} p_i = 1$. The function to be maximized, $F : K \to \mathbb{R}$, is given by

$$F(p_1, \ldots, p_n) = I(X,Y) + \lambda \Big(\sum_{i=1}^{n} p_i - 1 \Big).$$

The function $F(p_1, \ldots, p_n)$ is concave, as a sum between a concave function, $I(X,Y)$, and the linear constraint function in p. Therefore, any critical point p^* of F, which belongs to the interior of set K, is a point where F reaches a relative maximum, i.e., where $\nabla_p F(p^*) = 0$. Furthermore, if this is unique, then the point corresponds to a global maximum. We shall compute next the variational equations $\dfrac{\partial F}{\partial p_k} = 0$, $1 \leq k \leq n$.

Using that

$$\frac{\partial p(y_j)}{\partial p_k} = \frac{\partial p(y_j)}{\partial p(x_k)} = p(y_j|x_k) = q_{kj}$$

$$\frac{\partial H(Y)}{\partial p(y_j)} = -\frac{\partial}{\partial p(y_j)} \sum_{r=1}^{m} p(y_r) \ln p(y_r) = -(1 + \ln p(y_j)),$$

then chain rule implies

$$\frac{\partial H(Y)}{\partial p_k} = \sum_{j=1}^{m} \frac{\partial H}{\partial p(y_j)} \frac{\partial p(y_j)}{\partial p_k} = -\sum_{j=1}^{m} (1 + \ln p(y_j)) q_{kj}$$

$$= -1 - \sum_{j=1}^{m} \ln p(y_j) q_{kj},$$

where we used the Toeplitz property of q_{kj}. We also have

$$\frac{\partial H(Y|X)}{\partial p_k} = -\frac{\partial}{\partial p_k} \sum_{i=1}^{n} \sum_{j=1}^{m} p(x_i) p(y_j|x_i) \ln p(y_j|x_i)$$

$$= -\frac{\partial}{\partial p_k} \sum_{i=1}^{n} \sum_{j=1}^{m} p_i q_{ij} \ln q_{ij} = -\sum_{j=1}^{m} q_{kj} \ln q_{kj}.$$

Assembling the parts, we have

$$\frac{\partial F}{\partial p_k} = \frac{\partial}{\partial p_k}\Big[I(X,Y) + \lambda\Big(\sum_{i=1}^{n} p_i - 1\Big)\Big]$$

$$= \frac{\partial H(Y)}{\partial p_k} - \frac{\partial H(Y|X)}{\partial p_k} + \lambda$$

$$= -1 - \sum_{j=1}^{m} \ln p(y_j) q_{kj} + \sum_{j=1}^{m} q_{kj} \ln q_{kj} + \lambda.$$

Hence, the equations $\dfrac{\partial F}{\partial p_k} = 0$ take the explicit form

$$1 - \lambda + \sum_{j=1}^{m} q_{kj} \ln p(y_j) = \sum_{j=1}^{m} q_{kj} \ln q_{kj}, \quad 1 \le k \le n. \tag{12.9.15}$$

These are n equations, which together with the constraint $\sum_{j=1}^{m} p(y_j) = 1$ forms a system of $n+1$ equations with $m+1$ unknowns, $p(y_1), \ldots, p(y_m)$ and λ. The right side of (12.9.15) is known, since the input-output matrix q_{kj} is given. We need to solve for $m+1$ unknowns, $p(y_1), \ldots, p(y_m)$ and λ, which are in the left side. Using the Toeplitz property of q_{kj}, the previous relation can be written as

$$\sum_{j=1}^{m} q_{kj}\big(1 - \lambda + \ln p(y_j)\big) = \sum_{j=1}^{m} q_{kj} \ln q_{kj}, \quad 1 \le k \le n.$$

We shall write this system of equations in a matrix form. First, we introduce a few notations

$$\ln p(\mathbf{y})^T = \big(\ln p(y_1), \ldots, \ln p(y_m)\big)$$

and $h^T = (h_1, \ldots, h_n)$, where $h_k = \sum_{j=1}^{m} q_{kj} \ln q_{kj}$. Then the previous system of equations becomes

$$Q(1 - \lambda + \ln p(\mathbf{y})) = h. \tag{12.9.16}$$

The input-output matrix $Q = (q_{ij})$ is of type $n \times m$, with $m < n$. The solution method uses the idea of Moore-Penrose pseudoinverse. We prior multiply the previous equation by Q^T and obtain

$$Q^T Q(1 - \lambda + \ln p(\mathbf{y})) = Q^T h.$$

We notice that $Q^T Q$ is a squared matrix of type $m \times m$. From the double inequality

$$m \le \operatorname{rank}(Q^T Q) \le \operatorname{rank}(Q) = m,$$

it follows that the matrix $Q^T Q$ is nonsingular. Therefore

$$1 - \lambda + \ln p(\mathbf{y}) = (Q^T Q)^{-1} Q^T h.$$

Using the Moore-Penrose pseudoinverse, $Q^+ = (Q^T Q)^{-1} Q^T$, see section G.2 in Appendix, we write

$$1 - \lambda + \ln p(\mathbf{y}) = Q^+ h.$$

Equating on components, we obtain

$$1 - \lambda + \ln p(y_j) = (Q^+ h)_j, \qquad 1 \le j \le m,$$

with

$$(Q^+ h)_j = \sum_{k=1}^{n} q_{jk}^+ h_k,$$

where $Q^+ = (q_{ij}^+)$ is the Moore-Penrose pseudoinverse of Q. Taking an exponential yields

$$e^{1-\lambda} p(y_j) = e^{(Q^+ h)_j}, \qquad 1 \le j \le m. \tag{12.9.17}$$

Summing over j, then using $\sum_{j=1}^{m} p(y_j) = 1$, and taking the log function solve for λ as

$$1 - \lambda = \ln \Big(\sum_{j=1}^{m} e^{(Q^+ h)_j} \Big). \tag{12.9.18}$$

This formula produces λ in terms of the input-output matrix. Substituting (12.9.18) back into (12.9.17) and solving for the output distribution yields

$$p(y_j) = e^{(Q^+ h)_j} / \Big(\sum_{k=1}^{m} e^{Q^+ h_k} \Big) = \frac{e^{Q^+ h}}{\|e^{Q^+ h}\|_1}.$$

Using the definition of the softmax function, the output distribution can be written as

$$p(\mathbf{y}) = \text{softmax}(Q^+ h), \tag{12.9.19}$$

where the right side of the previous expression depends only on the input-output matrix.

Our final goal was to find the input distribution, $p(\mathbf{x})$, which satisfies

$$Q^T p(\mathbf{x}) = p(\mathbf{y}). \tag{12.9.20}$$

If this equation has a solution $p^* = p(\mathbf{x})$, then by Exercise 12.13.4, it is unique. Furthermore, if all $p_i^* > 0$, for all $1 \le i \le n$, then this solution is in the interior of the definition domain and hence it is the point where the functional F achieves its maximum.

12.9.8   Finding the Capacity

Assume we have succeed in finding a maximum point p^* for $F(p)$. Then we can compute the network capacity substituting the solution p^* into the formula of mutual information $I(X, Y)$. We start by multiplying formula (12.9.15) by p_k

$$(1 - \lambda)p_k = \sum_{j=1}^{m} p_k q_{kj} \ln q_{kj} - \sum_{j=1}^{m} p_k q_{kj} \ln p(y_j)$$

and then sum over k to obtain

$$
\begin{aligned}
1 - \lambda &= \sum_{k=1}^{n}\sum_{j=1}^{m} p_k q_{kj} \ln q_{kj} - \sum_{k=1}^{n}\sum_{j=1}^{m} p_k q_{kj} \ln p(y_j) \\
&= \sum_{k=1}^{n}\sum_{j=1}^{m} p(x_k, y_j) \ln p(y_j | x_k) - \sum_{j=1}^{m} p(y_j) \ln p(y_j) \\
&= -H(Y|X) + H(Y) = I(X, Y).
\end{aligned}
$$

Hence, using (12.9.18), the capacity of a network with $d^{(0)} = d^{(L)} = 1$ and fixed weights W and biases b is given by

$$C(W, b) = 1 - \lambda = \ln \left(\sum_{j=1}^{m} e^{(Q^+ h)_j} \right), \qquad (12.9.21)$$

where Q^+ is the Moore-Penrose pseudoinverse of the input-output matrix Q and $h_j = \sum_{r=1}^{m} q_{jr} \ln q_{jr}$. It is worth noting that the capacity depends only on the input-output matrix $Q = q_{ij}$.

12.9.9   Perceptron Capacity

We have seen that a perceptron, as a computing unit with unit step activation function, is able to perform half-plane classifications. This section answers the question: *How much information a perceptron can process?*

We shall show that the capacity of a single perceptron is 1 bit. This means that the outcome of a perceptron carries 1 bit of information, and hence it can learn the decision function of whether a point belongs to a given half-plane, who carries an information also of 1 bit. In order to do this, it suffices to compute explicitly formula (12.9.21) in the case of a perceptron.

The perceptron input is given by a random variable, X, which takes only two values, $x_1 = 0$ and $x_2 = 1$, with probabilities α and $1 - \alpha$, respectively. The output variable, given by $Y = H(wx + b)$, has also two outcomes, $y_1 = 0$ and $y_2 = 1$, taken with probabilities β and $1 - \beta$. The weight, w, and bias, b, are fixed constants, with $w \neq 0$. If $q_{ij} = P(Y = y_j | X = x_i)$, $i, j \in \{1, 2\}$, denotes the input-output matrix, the relation (12.9.20) between the input and output distributions can be written as:

$$\begin{pmatrix} \beta \\ 1 - \beta \end{pmatrix} = \begin{pmatrix} q_{11} & q_{21} \\ q_{12} & q_{22} \end{pmatrix} \begin{pmatrix} \alpha \\ 1 - \alpha \end{pmatrix}. \tag{12.9.22}$$

We shall compute next the entries q_{ij}. We have

$$\begin{aligned} q_{11} &= P(Y = y_1 | X = x_1) = P(Y = 0 | X = 0) = P(H(b) = 0) \\ &= \begin{cases} 1, & \text{if } b < 0 \\ 0, & \text{if } b \geq 0 \end{cases} = 1_{\{b<0\}}, \end{aligned}$$

because the event $\{H(b) = 0\}$ is occurs surely for $b < 0$ and becomes impossible for $b \geq 0$. Using the same line of thinking, we compute the other entries as

$$q_{21} = P(Y = y_1 | X = x_2) = P(Y = 0 | X = 1) = P(H(w + b) = 0) = 1_{\{w+b<0\}},$$
$$q_{12} = P(Y = y_2 | X = x_1) = P(Y = 1 | X = 0) = P(H(b) = 1) = 1_{\{b\geq0\}},$$
$$q_{22} = P(Y = y_2 | X = x_2) = P(Y = 1 | X = 1) = P(H(w + b) = 1) = 1_{\{w+b\geq0\}}.$$

Therefore, the transpose of the input-output matrix is given by

$$Q^T = \begin{pmatrix} q_{11} & q_{21} \\ q_{12} & q_{22} \end{pmatrix} = \begin{pmatrix} 1_{\{b<0\}} & 1_{\{w+b<0\}} \\ 1_{\{b\geq0\}} & 1_{\{w+b\geq0\}} \end{pmatrix}. \tag{12.9.23}$$

We shall choose the weight, w, and bias, b, such that Q becomes nonsingular.

Proposition 12.9.2 *Let $w \neq 0$. If either $w > 0$, $b \in [-w, 0)$, or $w < 0$, $b \in [0, -w)$, then $\det Q \neq 0$. Furthermore, $Q = Q^{-1} = Q^+ = Q^T = (Q^T)^{-1}$.*

Proof: The entries of the matrix Q^T are either 0 or 1. Since the sum of the entries on each column of Q^T is equal to 1 (by the Toeplitz property), the only two cases in which Q^T is not singular are the following:

$$Q^T = \begin{pmatrix} 1 & 0 \\ 0 & 1 \end{pmatrix} \quad \text{and} \quad Q^T = \begin{pmatrix} 0 & 1 \\ 1 & 0 \end{pmatrix}.$$

Using (12.9.23), the first case corresponds to $b < 0$, $w + b \geq 0$, and the second to $b \geq 0$, $w + b < 0$, which are the hypothesis conditions. We note that the previous matrices are symmetric and are their own inverses. ∎

Since the entries of Q^T are either 0 or 1, we have

$$\begin{pmatrix} h_1 \\ h_2 \end{pmatrix} = \begin{pmatrix} q_{11} \ln q_{11} + q_{12} \ln q_{12} \\ q_{21} \ln q_{21} + q_{22} \ln q_{22} \end{pmatrix} = \begin{pmatrix} 0 \\ 0 \end{pmatrix},$$

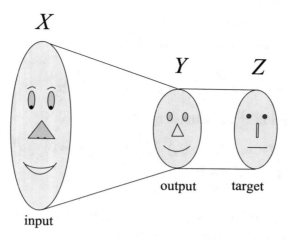

Figure 12.7: *Y is a compressed version of the input X, which preserves meaningful information about target Z. Y preserves the important features of the face given by X, which are meaningful for the idea of face required by the target Z.*

where we used $\lim_{x \searrow 0} x \ln x = 0$. Substituting in (12.9.21) we obtain the perceptron's capacity

$$C(W, b) \;\; = \;\; 1 - \lambda = \ln \left(\sum_{j=1}^{2} e^{(Q^{+}h)_j} \right) = \ln 2.$$

Dividing by $\ln 2$ we change the measure units from *nats* to *bits*. Hence, the capacity is of 1 bit.

We shall find next the input and output probability distributions. Since $h = 0$ and Q is invertible, formula (12.9.16) yields $Q(\ln 2 + \ln p(\mathbf{y})) = 0$, i.e., $p(\mathbf{y})) = 1/2$, or equivalently, $\beta = 1/2$. Solving for α from (12.9.22) provides $\alpha = 1/2$. Hence, both input and output probabilities are uniform when the maximum capacity is reached.

Remark 12.9.3 (i) In the case of a feedforward neural network with noiseless layers where the input-output mapping, f, is bijective, with $f(x_i) = y_i$ and $m = n$, the input-output matrix is the identity matrix, $Q = \mathbb{I}_n$, see Remark 12.9.1. Then the pseudoinverse is $Q^{+} = \mathbb{I}_n$ and $(Q^{+}h)_j = h_j = 0$. Then the network capacity is given by (12.9.21) as $C(w, b) = \ln \sum_{i=1}^{n} e^0 = \ln n$. Formulas (12.9.19) and (12.9.20) provide that both the output and input distributions which maximize the network capacity are uniform

$$p(x_j) = p(y_j) = \frac{1}{n}, \qquad 1 \le j \le n.$$

(*ii*) Capacity can be increased by supplying a larger input information. Increasing the number of inputs leads to an increase in the number of weights and biases of the network. The capacity depends on the matrix Q, which depends on the network activation function and parameters.

Capacity can be interpreted as the ability of a network to fit a large variety of target functions. Consequently, a network with a large capacity can lead to overfitting the training set by memorizing it. On the contrary, a low capacity network cannot process enough information and thus leads to an underfit of the training set. The *optimal capacity*, which is difficult to find theoretically, is somewhere in between these two limits, being close to the true complexity of the task subject to be performed by the network. A related problem to finding the optimal capacity is the "information bottleneck", which will be treated in the next section.

12.10　The Information Bottleneck

The information bottleneck method is a technique introduced by Naftali Tishby, Fernando C. Pereira, and William Bialek in [119]. We start illustrating the idea using a couple of suggestive examples.

Example 12.10.1 Assume that the input, X, represents face images and the target, Z, is the gender of the portrayed people. A face contains a lot of information, which needs to be squeezed as much as possible through a bottleneck that preserves the relevant information about the gender. Thus, the meaningful information about Z might contain features such as hairstyle, presence of makeup, clothes color, presence of a beard, etc. These features, which are contained in the initial data X, should also come up in the outcome Y and represent meaningful information for determining the gender of the portrayed person.

Example 12.10.2 A speaker has to deliver a talk on a certain topic in a foreign language in which his vocabulary is limited. He needs to squeeze the information of the talk, X, through a "bottleneck", Y, formed by the limited set of words available, such that the meaningful information about the talk topic, Z, is not affected.

Example 12.10.3 An employee has to write a narrative about his new proposed project. However, his busy boss imposes a 2-page limit per project. Therefore, the employee's challenge is to include a lot of information about the project in only a limited amount of space. The available project information, X, has to be compressed in a 2-page narrative, Y, the bottleneck, such that the important project features are not weaken too much.

We consider a feedforward neural network with input random variable X and output variable Y, which is subject to learn the target random variable Z. We assume that the training distribution, $p(x, z)$, which is the joint distribution of (X, Z) is provided. The input X is usually a high dimensional variable, while the target, Z, has a significantly lower dimension; thus, most entropy of X is not very informative about Z, and the information provided by X has to be compressed as much as possible into the output Y, such that there is still enough meaningful information left in Y about the target Z.

Since lossy compression of X into Y cannot convey more information about Z than the initial data X, we have the inequality

$$I(Y, Z) \leq I(X, Z).$$

Namely, the information conveyed about the target Z by the output Y does not exceed the information conveyed about Z by the initial data X. However, even if it is less than $I(X, Z)$, the information $I(Y, Z)$ should still be large enough such that Y still contains enough meaningful information about Z. Thus, we are looking for a network that keeps a fixed amount of meaningful information $I(Y, Z)$ about the target Z, while maximizing the compression of the input X into Y, i.e., minimizing the information $I(X, Y)$. This is represented in Fig.12.7. The amount of relevant information captured by the network is described by the ratio $\rho = I(Y, Z)/I(X, Z)$, with $\rho \in (0, 1)$.

The *bottleneck principle* states that we need to pass the information provided by X about Z through a "bottleneck" formed by the output variable Y, in the most efficient way. This means to minimize $I(X, Y)$ subject to a given fixed information $I(Y, Z)$. This can be formalized variationally by minimizing the functional with constraints

$$\mathcal{L}(p(y|x)) = I(X, Y) - \beta I(Y, Z), \qquad (12.10.24)$$

where β is a Lagrange multiplier. The first term denotes the *compressed information*, while the second term describes the *meaningful information*. Therefore, the multiplier β describes the trade-off between preserved meaningful information and compression. A value of β close to 0 emphasizes compression over meaningful information, in which case the representation becomes very sketchy. If β tends to ∞, then meaningful information prevails over compression, in which case the representation becomes very detailed. The argument of the functional is $p(y|x)$, or equivalently, the input-output matrix.

It is worth noting that this problem is distinct from the problem of finding the network capacity, where $I(X, Y)$ is maximized over all input patterns $p(x)$; here the information $I(X, Y)$ is minimized over $p(y|x)$, while the input distribution, $p(x)$, is kept fixed.

12.10.1 An exact formal solution

Assume the input distribution, $p(x)$, and the training distribution, $p(x, z)$, are given (and hence, the *posterior distribution*, $p(z|x) = p(x, z)/p(x)$, is given). The unknown distributions are $p(y)$, $p(y|x)$ and $p(z|y)$. The distribution $p(y|x)$ is called *encoder*, while the distribution $p(z|y)$ is called *decoder*. We need to find the output distribution as well as the encoder and decoder distributions such that the squeezing through the bottleneck procedure described before becomes optimal. The solution of the nonlinear variational problem (12.10.24) has an exact implicit solution, which can express $p(y|x)$, $p(z|y)$, and $p(y)$ in terms of themselves and also in terms of the known distributions $p(x)$ and $p(z|x)$ (prior and posterior distributions).

Theorem 12.10.1 *The optimal solution that minimizes functional (12.10.24) satisfies the following implicit equations:*

$$p(y|x) = \frac{p(y)}{Z(x, \beta)} e^{-\beta D_{KL}(p(z|x)\|p(z|y))}, \qquad (12.10.25)$$

$$p(z|y) = \frac{1}{p(y)} \sum_x p(z|x)p(y|x)p(x), \qquad (12.10.26)$$

$$p(y) = \sum_x p(y|x)p(x), \qquad (12.10.27)$$

where

$$D_{KL}(p(z|x)\|p(z|y)) = \sum_z p(z|x) \ln \frac{p(z|x)}{p(z|y)},$$

and $Z(x, \beta)$ is the normalization function.

Proof: We note that two of the unknown distributions, $p(y)$ and $p(z|x)$, can be written in terms of $p(x)$, $p(z|x)$ and $p(y|x)$ as follows:

$$p(y) = \sum_x p(y|x)p(x),$$

$$p(z|y) = \sum_x p(z|x)p(x|y) = \frac{1}{p(y)} \sum_x p(z|x)p(y|x)p(x),$$

where in the last equation we have used that

$$p(x|y) = \frac{p(x, y)}{p(y)} = \frac{p(y|x)p(x)}{p(y)}.$$

Therefore, we need to find a formula for $p(y|x)$ in terms of the previous distributions.

For fixed y and z, the law of conditional probabilities provides

$$p(y) \;=\; \sum_i p(y|x_i)p(x_i) = \sum_x p(y|x)p(x)$$

$$p(y|z) \;=\; \sum_i p(y|x_i)p(x_i|z) = \sum_x p(y|x)p(x|z),$$

and hence for a given x we have the following partial derivatives:

$$\frac{\partial p(y)}{\partial p(y|x)} = p(x) \tag{12.10.28}$$

$$\frac{\partial p(y|z)}{\partial p(y|x)} = p(x|z). \tag{12.10.29}$$

When optimizing over $p(y|x)$, we need to add the constraint $\sum_y p(y|x) = 1$, for any given x. If the Lagrange multiplier for the previous constraint is $\lambda(x)$ (it is x-dependent), then we obtain the following functional with constraints:

$$\mathcal{L}(p(y|x)) = I(X,Y) - \beta I(Y,Z) - \sum_x \lambda(x)\Big(\sum_y p(y|x) - 1\Big). \tag{12.10.30}$$

For fixed values of x and y, we have the variational equation

$$\frac{\partial \mathcal{L}}{\partial p(y|x)} = \frac{\partial I(X,Y)}{\partial p(y|x)} - \beta\frac{\partial I(Y,Z)}{\partial p(y|x)} - \lambda(x). \tag{12.10.31}$$

Using that

$$
\begin{aligned}
I(X,Y) \;=\;& H(Y) - H(Y|X) \\
=\;& -\sum_y p(y)\ln p(y) + \sum_{x,y} p(x,y)\ln p(y|x) \\
=\;& -\sum_y p(y)\ln p(y) + \sum_{x,y} p(y|x)p(x)\ln p(y|x),
\end{aligned}
$$

chain rule together with (12.10.28) yield

$$
\begin{aligned}
\frac{\partial I(X,Y)}{\partial p(y|x)} \;=\;& \frac{\partial H(Y)}{\partial p(y)}\frac{p(y)}{\partial p(y|x)} - \frac{\partial H(Y|X)}{\partial p(y|x)} \\
=\;& -(1 + \ln p(y))p(x) + p(x)(1 + \ln p(y|x)) \\
=\;& p(x)\ln\frac{p(y|x)}{p(y)}.
\end{aligned}
$$

Similarly,

$$
\begin{aligned}
I(Y,Z) \;=\;& H(Y) - H(Y|Z) \\
=\;& -\sum_y p(y)\ln p(y) + \sum_{y,z} p(y|z)p(z)\ln p(y|z),
\end{aligned}
$$

and applying chain rule together with (12.10.29), we have

$$
\begin{aligned}
\frac{\partial I(Y,Z)}{\partial p(y|x)} &= -(1+\ln p(y))p(x) + \sum_z p(z)(1+\ln p(y|z))p(x|z) \\
&= -p(x) - p(x)\ln p(y) + \sum_z p(z)p(x|z) + \sum_z p(z)p(x|z)\ln p(y|z) \\
&= -p(x)\ln p(y) + \sum_z p(x,z)\ln p(y|z) \\
&= -\sum_z p(x,z)\ln p(y) + \sum_z p(x,z)\ln p(y|z) \\
&= \sum_z p(x,z)\ln\frac{p(y|z)}{p(y)} = p(x)\sum_z p(z|x)\ln\frac{p(y|z)}{p(y)} \\
&= p(x)\sum_z p(z|x)\ln\frac{p(y,z)}{p(y)p(z)} \\
&= p(x)\sum_z p(z|x)\ln\frac{p(z|y)}{p(z)}.
\end{aligned}
$$

Substituting into (12.10.31) yields

$$
\frac{\partial \mathcal{L}}{\partial p(y|x)} = p(x)\left\{ \ln\frac{p(y|x)}{p(y)} - \beta\sum_z p(z|x)\ln\frac{p(z|y)}{p(z)} - \frac{\lambda(x)}{p(x)} \right\}.
$$

The term $\beta\sum_z p(z|x)\ln\dfrac{p(z|y)}{p(z)}$ depends on both y and x. We shall write it as a sum between two Kullback-Leibler divergences as

$$
\begin{aligned}
\beta\sum_z p(z|x)\ln\frac{p(z|y)}{p(z)} &= \beta\sum_z p(z|x)\ln\left(\frac{p(z|y)}{p(z|x)}\frac{p(z|x)}{p(z)}\right) \\
&= \beta\sum_z p(z|x)\ln\frac{p(z|y)}{p(z|x)} + \beta\sum_z p(z|x)\ln\frac{p(z|x)}{p(z)} \\
&= -\beta D_{KL}\big(p(z|x)\|p(z|y)\big) + \beta\phi(x),
\end{aligned}
$$

where

$$
\phi(x) = \sum_z p(z|x)\ln\frac{p(z|x)}{p(z)} = D_{KL}\big(p(z|x)\|p(z)\big)
$$

is a function of x, which will be absorbed into the multiplier. Substituting in the previous partial derivative yields

$$
\frac{\partial \mathcal{L}}{\partial p(y|x)} = p(x)\left\{ \ln\frac{p(y|x)}{p(y)} + \beta D_{KL}\big(p(z|x)\|p(z|y)\big) - \beta\phi(x) - \frac{\lambda(x)}{p(x)} \right\}.
$$

Introducing the new multiplier

$$\tilde{\lambda}(x) = \beta\phi(x) + \frac{\lambda(x)}{p(x)},$$

we can finally write the derivative in the following simpler way:

$$\frac{\partial \mathcal{L}}{\partial p(y|x)} = p(x)\Big\{ \ln\frac{p(y|x)}{p(y)} + \beta D_{KL}\big(p(z|x)\|p(z|y)\big) - \tilde{\lambda}(x)\Big\}.$$

Therefore, the equation $\dfrac{\partial \mathcal{L}}{\partial p(y|x)} = 0$ is satisfied by the solution

$$
\begin{aligned}
p(y|x) &= p(y)e^{\tilde{\lambda}(x)}e^{-\beta D_{KL}(p(z|x)\|p(z|y))} \\
&= \frac{p(y)}{Z(\beta,x)}e^{-\beta D_{KL}(p(z|x)\|p(z|y))},
\end{aligned}
$$

where $Z(\beta,x) = e^{-\tilde{\lambda}(x)}$ is the normalization function. We skip the proof of the fact that this solution minimizes the given functional; for this the reader is directed to the paper [119].   ∎

Remark 12.10.2 The implicit system (12.10.25)–(12.10.27) can be solved using an iterative algorithm. Denoting the iteration step by k, we consider the iterative system

$$
\begin{aligned}
p_k(y|x) &= \frac{p_k(y)}{Z_k(x,\beta)}e^{-\beta D_{KL}(p_k(z|x)\|p_k(z|y))}, \\
p_{k+1}(z|y) &= \frac{1}{p_k(y)}\sum_x p_k(z|x)p_k(y|x)p_k(x), \\
p_{k+1}(y) &= \sum_x p_k(y|x)p_k(x).
\end{aligned}
$$

It can be shown that the sequence of solutions is convergent as $k \to \infty$ to the desired solution.

12.10.2   The information plane

This section will provide a geometric description for the evolution of the information through the layers of a deep feedforward neural network. This way, the shape of the associated curve will tell how the information flow behaves.

     We recall a few familiar notations. As usual, let $X = X^{(0)}$ and $Y = X^{(L)}$ be the input and output layers of a deep feedforward neural network. Denote

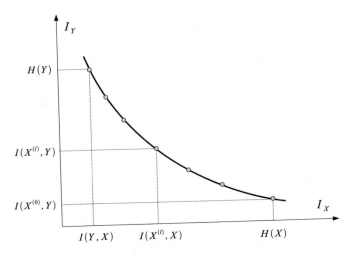

Figure 12.8: *The information path associated with a neural network in the information plane.*

by Z the target variable and by $X^{(\ell)}$ the activation of the ℓth layer of the network. Each layer, $X^{(\ell)}$, can be associated with two nonnegative numbers,

$$I_X^{(\ell)} = I(X^{(\ell)}, X), \qquad I_Y^{(\ell)} = I(X^{(\ell)}, Y),$$

which can further be considered as coordinates in a Cartesian plane, called the *information plane*. Thus, each layer of the network can be mapped into a point in the first quadrangle on this plane. We shall investigate the shape of the curve made with these points.

Data processing inequalities, see Proposition 12.7.9, provides a double sequence of inequalities

$$H(X) \geq I(X^{(1)}, X) \geq \cdots I(X^{(\ell)}, X) \geq I(X^{(\ell+1)}, X) \geq \cdots \geq I(X, Y)$$

$$I(X, Y) \leq I(X^{(1)}, Y) \leq \cdots I(X^{(\ell)}, Y) \leq I(X^{(\ell+1)}, Y) \leq \cdots \leq I(X^{(L-1)}, Y),$$

which state that the sequence $I_X^{(\ell)}$ is decreasing, while $I_Y^{(\ell)}$ is increasing with respect to ℓ

$$I_X^{(\ell)} \geq I_X^{(\ell+1)}, \qquad I_Y^{(\ell)} \leq I_Y^{(\ell+1)}.$$

Connecting the points $(I_X^{(\ell)}, I_Y^{(\ell)})$ with a monotonic continuous curve, we obtain the *information path* associated with the network, see Fig. 12.8.

It is worth noting that during the learning process, while the weights and biasses are tuned, the information path deforms in the information plane.

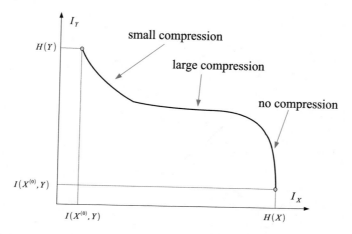

Figure 12.9: *Regions of different types of compression on the information curve.*

Another remark is that in the case of a network with compressionless layers the previous first sequence of inequalities becomes a sequence of identities, and hence the information path takes a vertical position. The reciprocal of the slope describes the compression rate along the network, see Fig. 12.9.

We note that for any feedforward network we have $I_X^{(L)} = I_Y^{(0)} = I(X, Y)$. If in addition, we also have $I_X^{(L-p)} = I_Y^{(p)}$, for all $1 \leq p \leq L-1$, then the information curve is symmetric with respect to the line $I_Y = I_X$. This curve would correspond, for instance, to a symmetric autoencoder network.

Even if the geometry of the information path provides relevant information about the network, however, it does not determine the network architecture univocally. This can be easily seen if one of the layers is modified by an invertible transformation, which changes the network architecture but leaves the mutual information invariant, see Proposition 12.7.5.

Remark 12.10.3 Bottleneck principle can be applied repeatedly, layer by layer. Applying the bottleneck principle at each layer, we would like to minimize the information $I(X^{(\ell)}, X^{(\ell+1)})$, while keeping $I(X^{(\ell+1)}, Z)$ as large as possible. This is to compress the information between the ℓth and $(\ell + 1)$th layers, while keeping enough meaningful information on Z.

12.11   Information Processing with MNIST

The efficiency of a neural network always starts with a test on the MNIST data. This is a database consisting of gray-scaled handwritten digits from

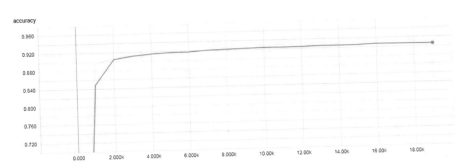

Figure 12.10: *A zero-hidden layer feedforward network using a batch size of 30, the softmax activation function, and the gradient descent method with learning rate $\lambda = 0.03$, producing a testing accuracy of 92.5%. The network uses a cost function given by the sum of squares. The diagram was generated with Tensorboard.*

0 to 9, together with a label, indicating the intended integer. Each handwritten digit is represented by a 28×28 pixel image. The labels are represented by "one-hot vectors", which are vectors of length 10 having a 1 placed on the slot with the same index as the corresponding digit, and zeros in rest.[2] MNIST data is divided into 55,000 training data, 10,000 testing data, and 5,000 validation data. A network is trained using the training data set and then is tested on the testing data, and a percentage of success is recorded, as the ratio between the correct classified test samples and total number of samples, which we shall call *accuracy*. The goal is to achieve an accuracy as close as possible to 100% on the test data, case which corresponds to a perfect classification of the test data.

Experimental work has shown that the accuracy of a neural network depends on a variety of factors, such as: number of hidden layers (network depth) and their size (network width), activation function used (logistic, ReLU, etc.), cost function (sum of square errors, cross-entropy, etc.), batch size, learning step, method of minimization (gradient descent, etc.), as well as the type of the network (fully-connected, convolutional network). Furthermore, we can state empirically some bounds for the accuracy as follows: a two-layer feedforward neural network (FNN) has a maximum of about 92.5% accuracy on MNIST data, see Fig. 12.10; a three-layer FNN cannot do better

---

[2]For instance, the labels 0 and 3 are represented by the one-hot vectors $(1, 0, 0, 0, 0, 0, 0, 0, 0, 0)$ and $(0, 0, 0, 1, 0, 0, 0, 0, 0, 0)$, respectively.

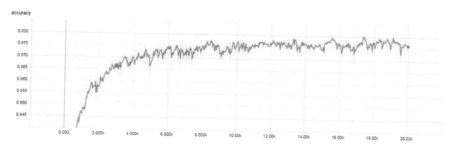

Figure 12.11: *A one-hidden layer feedforward network, with 300 neurons in the hidden layer, trained with a batch size of 40, using the ReLU and the softmax activation functions for the hidden and output layer, respectively, produces a testing accuracy of 97.6%. The learning uses ADAM method with learning rate starting at $\lambda = 0.0015$. The cost function used is the cross-entropy. The diagram was generated with Tensorboard.*

than 98% accuracy on MNIST data, see Fig. 12.11; a convolution network can exceed 98% accuracy on MNIST data.

The aim of the next section is to address this phenomenon mathematically.

12.11.1 A two-layer FNN

We shall flatten each 28×28 image into a vector, X, of length 784, following the idea of Fig. 5.2. This is considered as a random variable with values in $[0,1]^{784}$. A component with a value of $X_i = 1$ corresponds to a black pixel, while $X_i = 0$, to a white pixel. Any other value in between corresponds to a gray pixel.

The network has no hidden layers, so there are only two layers, one corresponding to the input and the other to the output.

The network output is a decision function with 10 components, which is described by the vector $Y = (Y_1, \ldots, Y_{10})$. Assume the activation function ϕ is invertible.[3] Then the output is described by the formula

---

[3]Here we assume, for instance, that the activation function is of sigmoid type. There are activation functions which are not invertible, such as ReLu, or Softmax, which is customarily used in classification. However, Softplus, which approximates ReLU, is invertible.

$$Y_j = \phi\left(\sum_{i=1}^{784} w_{ij}X_i + b^j\right), \qquad 1 \le j \le 10,$$

or, in an equivalent matrix form, $Y = \phi(W^T X + b)$. We used the notation $W = (w_{ij}) \in \mathbb{R}^{784} \times \mathbb{R}^{10}$ for the weight matrix and $b = (b^j) \in \mathbb{R}^{10}$ for the bias vector.

Since Y depends on X, the field of information defined by Y is included in the field of information defined by X, namely,

$$\mathfrak{S}(Y) \subset \mathfrak{S}(X).$$

This is a strict inequality, namely, the inverse inclusion, $\mathfrak{S}(X) \subset \mathfrak{S}(Y)$, does not hold, as we shall show next. Denoting by $T = \phi^{-1}Y - b$, we have $W^T X = T$, where W^T is the transpose of the matrix W. Variables Y and T define the same information field, $\mathfrak{S}(Y) = \mathfrak{S}(T)$, since T can be obtained from Y by an invertible transformation. The aforementioned system can be written explicitly as

$$
\begin{aligned}
w_{11}X_1 + w_{12}X_2 + \cdots + w_{1,784}X_{784} &= T_1 \\
\cdots\cdots\cdots\cdots\cdots\cdots\cdots\cdots\cdots\cdots\cdots &= \cdots \\
w_{10,1}X_1 + w_{10,2}X_2 + \cdots + w_{10,784}X_{784} &= T_{10}.
\end{aligned}
$$

Since $\text{rank}(W) \le \min\{10, 784\}$, without losing generality, we shall consider the best information transfer scenario when the rank is maximal, $\text{rank}(W) = 10$. Retaining only the independent rows and using an eventual reindexing, we write

$$
\begin{aligned}
w_{11}X_1 + w_{12}X_2 + \cdots + w_{1,10}X_{10} &= T_1 - \sum_{j=11}^{784} w_{1,j}X_j \\
\cdots\cdots\cdots\cdots\cdots\cdots\cdots\cdots\cdots &= \cdots\cdots\cdots\cdots \\
w_{10,1}X_1 + w_{10,2}X_2 + \cdots + w_{10,10}X_{10} &= T_{10} - \sum_{j=11}^{784} w_{10,j}X_j.
\end{aligned}
$$

Since the coefficient matrix is invertible, this implies that the first ten components of X can be written in terms of T (and hence, in terms of Y) and the other components of X. Consequently,

$$\mathfrak{S}(X_1, \ldots, X_{10}) \subset \mathfrak{S}(T, X_{11}, \ldots, X_{784}) = \mathfrak{S}(Y, X_{11}, \ldots, X_{784})$$

Obviously,

$$\mathfrak{S}(X_{11}, \ldots, X_{784}) \subset \mathfrak{S}(Y, X_{11}, \ldots, X_{784}).$$

Then the last two relations imply

$$\mathfrak{S}(X) = \mathfrak{S}(X_1, \ldots, X_{10}) \vee \mathfrak{S}(X_{11}, \ldots, X_{784}) \subset \mathfrak{S}(Y, X_{11}, \ldots, X_{784}).$$

$$(12.11.32)$$

On the other side, we know that

$$\mathfrak{S}(Y) \subset \mathfrak{S}(X)$$

and it is obvious that

$$\mathfrak{S}(X_{11}, \ldots, X_{784}) \subset \mathfrak{S}(X).$$

From the last two relations we infer that

$$\mathfrak{S}(Y) \vee \mathfrak{S}(X_{11}, \ldots, X_{784}) \subset \mathfrak{S}(X),$$

or, after taking the $\mathfrak{S}$ operator

$$\mathfrak{S}(Y, X_{11}, \ldots, X_{784}) \subset \mathfrak{S}(X). \qquad (12.11.33)$$

From (12.11.32) and (12.11.33) it follows that

$$\mathfrak{S}(X) = \mathfrak{S}(Y, X_{11}, \ldots, X_{784}). \qquad (12.11.34)$$

To conclude, the information casted by the output variable, Y, is strictly smaller than the input information defined by X. This corresponds to a compression of information, during which the information generated by X_{11}, ..., X_{784} is lost.

A similar approach works for the case when the matrix W has a smaller rank. In this general case, a number of $784 - \mathrm{rank}(W)$ components of X need to be included in the right term of (12.11.34).

In the following we shall provide a quantitative analysis of information through the network in the line of section 12.8.5, which is of practical interest for comparing the information processing ability of neural networks.

For the sake of simplicity, we further assume that each pixel is allowed to take only two values, black or white. The maximum entropy of the input variable X satisfies $2^{H(X)} = 784$, namely, the input contains an information of at most 9.614 bits (per image). The entropy of X, given Y, is equal to the average number of pixels corresponding to each output

$$2^{H(X|Y)} = \frac{784}{10},$$

which provides $H(X|Y) = 6.292$ bits. The mutual information of X and Y, i.e., the number of bits contained in X about Y, is given by

$$I(X,Y) = H(X) - H(X|Y) = 9.614 - 6.292 = 3.322,$$

namely, about 3.32 bits of information of each image are used toward the classification of the picture content. The number of classes provided by Y is given by

$$2^{I(X,Y)} = 2^{3.322} = 10,$$

as expected.

Writing now the mutual information as

$$I(X,Y) = H(Y) - H(Y|X),$$

and using that there is no uncertainty in Y if X is given, i.e, $H(Y|X) = 0$, (Y depends deterministically on X), it follows that $H(Y) = I(X,Y) = 3.322$ bits. This verifies the relation $2^{H(Y)} = 10$, which recovers the known fact that the number of elements of Y is 10.

To conclude, the input and output entropies are $H(X) = 9.614$ bits and $H(Y) = 3.322$ bits, respectively. The information loss due to compression is 6.292 bits (per image). If this is viewed in the light of formula (12.11.33), then it corresponds to the information lost through the components $X_{11}, \ldots, X_{784}$.

12.11.2   A three-layer FNN

Consider a three-layer FNN, with input X, output $\widetilde{Y}$, and an extra hidden layer, U, containing 100 neurons. In this case two compressions occur: one from X to U, under the ratio $784 : 100$; and another, from U to $\widetilde{Y}$, under the ratio $100 : 10$. Experiments have shown that the accuracy of the network increases from 92% (which corresponds to a zero-hidden layer network) to about 98%, when one hidden layer is used. The customary explanation of this fact is that the hidden layer is able to collect more features of the input variable, adding some extra-capacity to the network. We shall provide a mathematical formalization of this fact.

Let Y and $\widetilde{Y}$ be the outputs for a zero-hidden layer and a one-hidden layer network, respectively. The output of the one-hidden layer network is

$$\widetilde{Y} = \phi\left(\widetilde{W}^T U + \widetilde{b}\right) = \phi\left(\widetilde{W}^T \phi(W^T X + b) + \widetilde{b}\right),$$

where (W, b) and $(\widetilde{W}, \widetilde{b})$ are the systems of weights and biasses for the first-to-second layer and for the second-to-third layer, respectively. The output $\widetilde{Y}$ can be represented as a point in a space of dimension $79,510$, parametrized by

$$(\widetilde{W}, W, \widetilde{b}, b) \in \mathbb{R}^{784 \times 100} \times \mathbb{R}^{100 \times 10} \times \mathbb{R}^{100} \times \mathbb{R}^{10}.$$

Similarly, the output $Y = \phi(W^T X + b)$ of the zero-hidden layer neural network can be represented as a point in a space of dimension $7,850$, parametrized by

$$(W, b) \in \mathbb{R}^{784 \times 10} \times \mathbb{R}^{10}.$$

When the neural networks are optimized, their outcomes correspond to projections of the target variable Z onto the aforementioned spaces. Consequently, a larger dimension will produce a better approximation of the target variable Z by projections, which means the network accuracy tends to be higher in the case when we introduce a hidden layer.

12.11.3   The role of convolutional nets

The low performance of a fully-connected layer feedforward neural network used in the classification of the MNIST data is due to two kinds of information losses:

(i) One is due to the low capacity of the two-layer network. This can be fixed by adding more layers or more neurons in the hidden layer to increase the network capacity. However, there is an upper bound of about 98% for the network accuracy in this case, which cannot be exceeded, regardless of how wide the hidden layer is, or how many hidden layers are added to the network.

(ii) To gain the missing 2% we need to acknowledge another information loss in the input data, due to flattening out the image. This removes some of the 2-dimensional information, such as the relation of a pixel with its neighboring pixels. Hence, a neural network with an architecture which takes advantage of the 2-dimensional data structure is needed, and this is the convolution neural network (CNN). Chapter 16 will discuss this type of networks in more detail.

How much information is ignored when the 2-dimensional input data is flattened out into a vector in the case of the MNIST data? One way to attempt to answer this question is to asses the information provided by the surprise of misclassifying digits. This information can be assessed using the log-likelihood function.[4]

---

[4]The information contained into an event is large when the element surprise brought by that event is also large. For instance, if p is the probability to snow, then $- \ln p$ is the

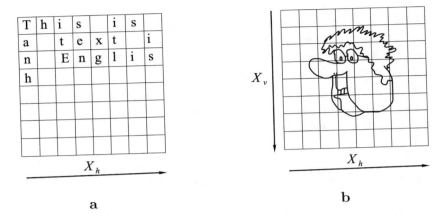

Figure 12.12: (a) *The information is given only by the horizontal component:* $H(X_h, X_v) = H(X_h)$; (b) *The information has both horizontal and vertical components:* $H(X_h, X_v) = H(X_h) + H(X_v) - I(X_h, X_v)$.

Let p denote the probability of classifying correct the MNIST data. If the accuracy is 100%, i.e., the network classifies correctly all digits, that is $p = 1$, so the log-likelihood is $-\log_2 p = 0$, i.e., there is no misclassification information in this case. If the accuracy would be of 50%, this means that the network classifies correctly one out of two digits, on average. Since $-\log_2(1/2) = 1$, there is 1 bit of information characterizing the misclassification. But if the accuracy classification is 92%, then 92 digits out of 100 are classified correctly. This corresponds to a misclassification information of $-\log_2(0.92) = 0.12$ bits. In the case of 98% accuracy, the misclassification information is $-\log_2(0.98) = 0.029$ bits. The conclusion is that after flattening the 28×28 pixel data into a vector of length 784, the most efficient FNN will still ignore about 0.029 bits per MNIST image. This loss is due to the change in topology of the figure.

In the following we shall explain this information loss using entropies. We need to introduce first the concept of *horizontal* and *vertical reading* variables. When one reads an English text, the information is extracted horizontally, line by line, from left to right. If the text is in Chinese, then the information is extracted vertically, column by column, from top to bottom.

---

information casted by this event. During the wintertime the probability to snow can be $p = 0.8$, which corresponds to an information of $-\ln 0.8 = 0.22$. During the summer the probability to snow is extremely small, let's say $p = 0.001$. This means an information of $-\ln(0.001) = 6.9$.

However, when one looks at a picture of a human face, the information in this case cannot be considered only horizontally or vertically, since the relative position each part with respect to all the other parts play a role in building the information as a whole.

We shall denote by X_h and X_v two random variables denoting the horizontal and the vertical reading of an image, respectively (For instance, when reading a doc file, we read a sequence of characters, which are considered as a random variable, X_h).

When reading an English text, the observed information is contained into the reading variable X_h and is equal to its entropy, $H(X_h)$. In this case the components X_h and X_v are independent, with $H(X_v) = 0$, i.e., no information is provided by the vertical component, see Fig. 12.12 **a**.

On the other side, when reading a Chinese text, the information is provided by $H(X_v)$. The components X_h and X_v are still independent and $H(X_h) = 0$, i.e., the horizontal component does not affect the information in this case.

Since looking at a 2-dimensional image, like an MNIST image, requires a 2-dimensional awareness for the neurons, this produces an information $H(X_h, X_v)$, which is the joint entropy of the variables (X_h, X_v), see Fig. 12.12 **b**. The components X_h and X_v are not necessarily independent in this case. The component X_h is obtained flattening out the 28×28 pixel image, by a concatenation of rows, while X_v is obtained similarly, by a concatenation of columns. The mutual information between the horizontal and vertical components is given by

$$I(X_h, X_v) = H(X_h) - H(X_h|X_v) = H(X_v) - H(X_v|X_h).$$

We note that in the case of reading a document (either in English or Chinese) we have $I(X_h, X_v) = 0$, due to components independence. This no longer holds true for the case of a 2-dimensional picture.

The total information of an image is given by the joint entropy of its components, $H(X_h, X_v)$. When the image is flattened into a vector, the information is extracted only from the horizontal component, and this is $H(X_h)$. The lost information is given by the difference between the total information, $H(X_h, X_v)$, and the partial information extracted from the flattened vector, $H(X_h)$, as in the following:

$$\mathcal{L} = H(X_h, X_v) - H(X_h). \tag{12.11.35}$$

Using that $H(X_h) + H(X_v) - H(X_h, X_v) = I(X_h, X_v)$, see equation (12.7.12), the lost information becomes

$$\begin{aligned} \mathcal{L} &= H(X_h, X_v) - H(X_h) = H(X_v) - I(X_h, X_v) \\ &= H(X_v) - \big(H(X_v) - H(X_v|X_h)\big) \\ &= H(X_v|X_h). \end{aligned}$$

Hence, the lost information is the uncertainty of the vertical component, X_v, conditioned by the horizontal component, X_h, and this is $H(X_v|X_h)$.

Remark 12.11.1 In general, the lost information depends on whether the vertical or the horizontal component entropy is subtracted from the total information $H(X_h, X_v)$. However, if the vertical and horizontal components have the same entropy, the information loss is the same in both cases.

12.12 Summary

This chapter provides some tools for assessing numerically the information contained in the layers of a neural net. Each layer output can be seen as a signal whose entropy can be evaluated. Necessary conditions for the entropy flow to be decreasing are provided.

Mutual information is used to describe quantitatively the information conveyed by one layer about another layer. The invariance property and data processing inequalities regarding the mutual information are proved and used in applications to DNNs.

The ability of a network to fit a large variety of target data is called capacity. It depends on the size of layers, depth, and number of neurons. It also represents the maximal information produced by a network under variable input information. A large capacity may lead to an overfit, so a technique that retains from the input only that part of information which is relevant to the target variable is developed.

Bottleneck information is a method that compresses the input information as much as possible such that the output contains enough meaningful information about the target set.

Examples to numerical computation of information measures in the case of the MNIST data are provided.

12.13 Exercises

Exercise 12.13.1 Consider a continuous random variable X with the density function $p(x) = \frac{1}{x(\ln x)^2} 1_{[e,\infty)}(x)$. Show that $H(X)$ is infinite.

Exercise 12.13.2 Let X be a continuous random variable with density $p(x)$. Show that:

(a) If $Var(X) < \infty$, then $H(X) < \infty$;

(b) If $p(x) \le M < \infty$, then $H(X) \ge -\ln M$.

Exercise 12.13.3 (a) Define the mutual information of X and Y, given Z as

$$I(X, Y|Z) = H(X|Z) + H(Y|Z) - H(X, Y|Z).$$

Show that $I(X, Y|Z) = D_{KL}[p(x, y, z)||p(x|z)p(y|z)]$.

(b) Show that for any three random variables X, Y, and Z, we have:

$$H(X|Z) + H(Y|Z) \leq H(X, Y|Z).$$

When is the identity satisfied?

Exercise 12.13.4 Let $b \in \mathbb{R}^m$ and A be an $n \times m$ matrix with rank m, where $m < n$. Show that the linear system $AX = b$ has at most one solution $X \in \mathbb{R}^n$.

The next few exercises refer to section 12.9.

Exercise 12.13.5 Find the input distribution $p(\mathbf{x})$ in terms of the matrix Q in the case $n = m$.

Exercise 12.13.6 Under the assumptions and notations of section 12.9, show that the capacity satisfies the following inequality:

$$C(W, b) \geq \max_j((Q^T Q)^{-1} Q^T h)_j.$$

Exercise 12.13.7 Consider a neural network obtained by the concatenation of two perceptrons. The output of the network is given by the random variable

$$Y = H(w_2 H(w_1 X + b_1) + b_2)),$$

with $X \in \{0, 1\}$. What is the capacity of this network?

Exercise 12.13.8 (a) Show that the number of parameters of the neural manifold associated with the two-layer FNN described in section 12.11.1 is $7,850$.

(b) Show that the number of parameters of the neural manifold associated with the three-layer FNN described in section 12.11.2 is $79,510$.

(c) Which of the previous networks has a larger capacity and why?

Exercise 12.13.9 How does the capacity of a network change when:

(a) An extra fully-connected layer is added to the network;

(b) Some neurons are dropped out of the network;

(c) The weights are constrained to be kept small.

The next few exercises refer to section 12.11.

Exercise 12.13.10 Assume the ranks of matrices W and $\widetilde{W}$ are maximal.
(a) Show that

$$\mathfrak{S}(X) = \mathfrak{S}(\widetilde{Y}, U_{11}, \ldots, U_{100}, X_{101}, \ldots, X_{784}).$$

(b) Verify the relation

$$\mathfrak{S}(Y, X_{11}, \ldots, X_{100}, X_{101}, \ldots, X_{784}) = \mathfrak{S}(\widetilde{Y}, U_{11}, \ldots, U_{100}, X_{101}, \ldots, X_{784}).$$

Exercise 12.13.11 With the notations of section 12.11.2 show that $H(X|U) = 2.9708$, $I(X,U) = 6.643$ and verify the inequality

$$H(X) > I(X,U) > I(X,\widetilde{Y}).$$

Exercise 12.13.12 Show that we have $H(X_h|X_v) = H(X_v|X_h)$ if and only if $H(X_h) = H(X_v)$.

Exercise 12.13.13 Consider the information loss, $\mathcal{L}$, given by (12.11.35).
(a) Find $\mathcal{L}$ given that the vertical and horizontal components are independent.
(b) Assume the vertical component depends deterministically on the horizontal component. Find $\mathcal{L}$.

An image transformation is represented as an invertible function $(X'_h, X'_v) = \Phi(X_h, X_v)$. This means that pixels change their coordinates, while keeping constant their total number. The question is how does a pixel shuffling change the entropy of an image? Since the question is too general to have an exact answer, we shall restrict the question in the next exercise to transformations of some particular type.

Exercise 12.13.14 Consider transformations of MNIST data of type

$$(X'_h, X'_v) = \big(\phi_1(X_h), \phi_2(X_v)\big),$$

with ϕ_i invertible. Show that these transformations preserve the total entropy if and only if $H(X'_h) + H(X'_v) = H(X_h) + H(X_v)$.

Exercise 12.13.15 For any $p \in (0,1)$ consider the *binary entropy function*

$$H(p) = -p \ln p - (1-p) \ln(1-p).$$

(a) Show that $H(p)$ is the entropy associated with a Bernoulli random variable.
(b) Verify the following relation between the derivative of the binary entropy and the logit function:

$$\frac{dH(p)}{dp} = -\ln\left(\frac{p}{1-p}\right).$$

Exercise 12.13.16 (decomposition into hierarchical levels) The mutual information of n random variables is defined as

$$I(X_1,\ldots,X_n) = \sum_{k=1}^{n} H(X_k) - H(X_1,\ldots,X_n).$$

(a) Show that $I(X,Y,Z) = I(X,Y) + I((X,Y),Z)$;

(b) State and prove a general statement.

Exercise 12.13.17 Consider the symmetric entropy of two discrete random variables X and Y to be given by $d(X,Y) = H(X|Y) + H(Y|X)$, and define
$D(X,Y) = \dfrac{d(X,Y)}{H(X,Y)}$.

(a) Show that $d(X,Y) = H(X,Y) - I(X,Y)$;

(b) Show that d is a distance function, i.e., it is nonnegative, symmetric, and satisfies the triangle inequality

$$d(X,Y) + d(Y,Z) \geq d(X,Z), \qquad \forall X,Y,Z;$$

(c) Verify that $D(X,Y) = 1 - \dfrac{I(X,Y)}{H(X,Y)}$;

(d) Show that D is a distance function with $D(X,X) = 0$ and $D(X,Y) \leq 1$ for any pair (X,Y).

Exercise 12.13.18 (Translation invariance and scaling equivariance) Let X be a random vector of dimension n, A be an invertible $n \times n$ matrix, and b be a constant vector.

(a) Show that $H(AX + b) = H(X) + \ln|\det A|$;

(b) What is the application to the linear neuron?

Exercise 12.13.19 Consider a matrix $W = (w_{ij}) \in M_{n \times n}$ such that $|w_{ij}| < c$, with $c \leq \frac{1}{\sqrt[n]{n!}}$. Show that $|\det W| < 1$.

Exercise 12.13.20 Let $X_1 \sim \mathcal{N}(\mu_1, \sigma_1^2)$ and $X_2 \sim \mathcal{N}(\mu_2, \sigma_2^2)$ be two normal distributed random variables, with Pearson correlation $\rho = \frac{Cov(X_1,X_2)}{\sigma_1\sigma_2}$. Show that

$$I(X_1, X_2) = -\frac{1}{2}\ln(1 - \rho^2)$$

and provide an interpretation.

Exercise 12.13.21 If X is a random variable with the cumulative distribution function F_X, then it can be shown that $U = F_X(X)$ is a random variable uniformly distributed on $[0, 1]$. Let $F_{X_1 X_2}$ denote the joint distribution function of (X_1, X_2). The copula of (X_1, X_2) is defined as the distribution function of (U_1, U_2)

$$C(u_1, u_2) = \mathbb{P}(U_1 \leq u_1, U_2 \leq u_2),$$

with $U_j = F_{X_j}(X_j)$. The density of copula is given by $c(u_1, u_2) = \dfrac{\partial^2 C(u_1, u_2)}{\partial u_1 \partial u_2}$.

(a) Prove the following relation between copula and mutual information:

$$I(X_1, X_2) = \int_0^1 \int_0^1 c(u_1, u_2) \ln c(u_1, u_2) \, du_1 du_2.$$

This formula resembles the joint entropy formula, with the probability density replaced by the copula density.

(b) Use part (a) to show that $I(X_1, X_2) = 0$ for X_1 and X_2 independent.

Exercise 12.13.22 One preprocessing technique used in neural networks is the *normalization* of the input data, X. This means transforming it into a zero mean random variable with unit standard deviation by the transformation

$$\tilde{X} = \frac{X - \mathbb{E}[X]}{\sigma_X}.$$

What is in this case the relationship between $H(X)$ and $H(\tilde{X})$?

Part IV
Geometric Theory

Chapter 13

Output Manifolds

In this chapter we shall associate a manifold with each neural network by considering the weights and biasses of a neural network as the coordinate system on the manifold. This manifold can be endowed with a Riemannian metric, which describes the intrinsic geometry of the network. Viewing a neural network in this geometric framework is useful from the following points of view. (i) The optimal weights and biasses of a network correspond to the coordinates of the orthonormal projection of the target onto the manifold.
(ii) Each learning algorithm involves a change in parameters value with respect to time and corresponds to a curve on this manifold. The most efficient learning process corresponds to the shortest curve, or geodesic, between the initial point and the projection point of the target.
(iii) The regularization problem can be treated in terms of the mean curvature and second fundamental form of the manifold into the ambient target space. Namely, since the flattest manifold produces the least overfitting to training data, regularization can be viewed as finding the output manifold with the smallest curvature.

In the next section we shall present an overview of the concept of manifold and present some results useful for deep learning. The reader interested in more introductory details on Differential Geometry is referred to [85].

13.1 Introduction to Manifolds

A *manifold* is a geometrical space which resembles, at least locally, with the numerical space $\mathbb{R}^n$. Each point in the manifold is described by a set of n parameters, which are considered as local coordinates. The number of

© Springer Nature Switzerland AG 2020
O. Calin, *Deep Learning Architectures*, Springer Series in the Data Sciences,
https://doi.org/10.1007/978-3-030-36721-3_13

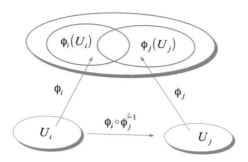

Figure 13.1: *The transitions functions $\phi_i \circ \phi_j^{-1}$ of a manifold are differentiable.*

parameters, n, is the *dimension* of the manifold. The manifold retains its own identity regardless of the parametrization. Sometimes, a manifold cannot be defined by only one parametrization and several parametrizations are needed to cover the entire space.

The manifold is called *differentiable* if the parametrizations are differentiable, i.e., the parameters are assigned to points in the manifold in a differentiable way. In the case of several parametrizations, $\phi_i : U_i \to \mathcal{M}$, their transition functions, $\phi_i \circ \phi_j^{-1}$ have to be differentiable, see Fig. 13.1. Besides this, a differentiable manifold is also required to satisfy a *regularization* condition for all parametrizations: the Jacobian of the transition functions $\phi_i \circ \phi_j^{-1}$ has a maximum rank. This condition removes any cusps, corners, or vertices on the manifold.

We note that for the scope of this book, where we need to model neural networks, we shall not use manifolds with more than one parametrization. Therefore, we do not go in too much detail about multiple parametrizations manifolds. We shall present in the following a few examples of manifolds.

Example 13.1.1 (Manifold of circles) The set of circles in the plane, $\mathcal{C}$, can be organized as a manifold using three parameters: the center coordinates, (a, b), and the circle radius, r. The parameters space is $\mathcal{U} = \mathbb{R}^2 \times (0, \infty)$ and the manifold parametrization is $\phi : \mathcal{U} \to \mathcal{C}$, where $\phi(a, b, r)$ is the circle centered at (a, b) and having radius r. In this case the manifold $\mathcal{C}$ is parametrized by only one map, ϕ. Each element of the manifold is a circle and the manifold has dimension 3.

Example 13.1.2 (Manifold of lines) Let P be a given point in the plane, with coordinates (x_0, y_0), satisfying $x_0 \neq 0$ and $y_0 \neq 0$. The family of lines in the plane, $\mathcal{L}$, passing through the point P, can be considered as a manifold in the following way. Let α and β denote, respectively, the x-intercept and the y-intercept of the lines in the previous family, see Fig. 13.2 **a**. Then $\phi : \mathbb{R} \to \mathcal{L}$, where $\phi(\alpha)$ is the line passing through P and $(\alpha, 0)$, is a parametrization for all non-horizontal lines in $\mathcal{L}$. Similarly, $\psi : \mathbb{R} \to \mathcal{L}$, where $\psi(\beta)$ is the line passing through P and $(0, \beta)$, is a parametrization for all non-vertical lines in $\mathcal{L}$. In this case the manifold $\mathcal{L}$ is one-dimensional but it is defined by two parametrizations, ϕ and ψ, since none of them can cover completely the entire manifold. The transition function $\phi \circ \psi^{-1} : \mathbb{R} \backslash \{x_0\} \to \mathbb{R} \backslash \{y_0\}$ is given by

$$\phi \circ \psi^{-1}(\alpha) = \beta = \frac{\alpha y_0}{\alpha - x_0}$$

and is bijective and differentiable. Thus, the manifold $\mathcal{L}$ becomes a differentiable manifold.

Example 13.1.3 (Surface manifold) The unit upper half-sphere

$$\mathcal{S}^+ = \{(u_1, u_2, z); (u_1, u_2) \in B(0, 1), z = (1 - u_1^2 - u_2^2)^{1/2}\}$$

is a manifold of dimension 2. Its parametrization is given by $\phi : B(0, 1) \to \mathbb{R}^3$, with $\phi(u_1, u_2) = (u_1, u_2, (1 - u_1^2 - u_2^2)^{1/2})$, see Fig. 13.2 **b**. The parameters space is the open disk $B(0, 1)$. It is worth noting that the entire unit sphere is also a 2-dimensional manifold, but in this case there are (at least) two parametrizations needed (for instance, the stereographic projections from the poles to the planes tangent to the sphere at the opposite pole).

If the coordinate z can be written as $z = f(u_1, u_2)$, with $f : \mathcal{U} \to \mathbb{R}^3$ differentiable, with $\mathcal{U}$ open set in $\mathbb{R}^2$, then the manifold is a surface, called a *Monge patch*. The surface is parametrized by two real numbers, u_1 and u_2.

Example 13.1.4 (Manifold of matrices) The set of 2×2 matrices with real entries, $\mathcal{M}_{2,2}(\mathbb{R})$, forms a 4-dimensional manifold. The parametrization is given by $\phi : \mathbb{R}^4 \to \mathcal{M}_{2,2}(\mathbb{R})$,

$$\phi(a, b, c, d) = \begin{pmatrix} a & b \\ c & d \end{pmatrix}.$$

The set of 2×2 diagonal matrices with real entries, $\mathcal{D}_{2,2}(\mathbb{R})$, forms a 2-dimensional manifold. The parametrization is $\phi : \mathbb{R}^2 \to \mathcal{M}_{2,2}(\mathbb{R})$,

$$\phi(a, d) = \begin{pmatrix} a & 0 \\ 0 & d \end{pmatrix}.$$

The matrix is parametrized by only two real numbers, a and d. In fact, $\mathcal{D}_{2,2}(\mathbb{R})$ is a *submanifold* of $\mathcal{M}_{2,2}(\mathbb{R})$, as a subset which inherits the ambient manifold structure (the coordinates).

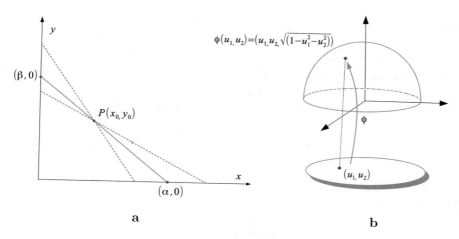

Figure 13.2: **a.** *The manifold of lines passing through the point $P(x_0, y_0)$ and their axes intercepts.* **b.** *The upper half-sphere.*

Example 13.1.5 (Manifold of densities) Consider the set of all one-dimensional Gaussian probability densities

$$\mathcal{G} = \{p_{\mu,\sigma}; \mu \in \mathbb{R}, \sigma > 0\},$$

where $p_{\mu,\sigma}(x) = \frac{1}{\sqrt{2\pi}\sigma} e^{-\frac{(x-\mu)^2}{2\sigma^2}}$, $x \in \mathbb{R}$. The set $\mathcal{G}$ becomes a 2-dimensional manifold parametrized by μ and σ.

The next example is of a special significance for the subject of this book.

Example 13.1.6 (Manifold of sigmoid neurons) Consider a sigmoid neuron with an n-dimensional input $\mathbf{x} \in \mathbb{R}^n$ and the one-dimensional output $y = \sigma(w^T\mathbf{x} + b)$, where $w \in \mathbb{R}^n$ and $b \in \mathbb{R}$ are the weights and the bias of the neuron. We take σ to be the logistic function. Then the set of outputs

$$\mathcal{S} = \{\sigma(w^T\mathbf{x} + b); w \in \mathbb{R}^n, b \in \mathbb{R}\}$$

can be regarded as an $(n+1)$-dimensional manifold, parametrized by w and b. In the following we shall verify the regularization condition by showing that the columns of the Jacobian matrix of $y = y(w, b)$ are linearly independent. Using the properties of the logistic function, we have

$$\frac{\partial y}{\partial b} = \sigma'(w^T\mathbf{x} + b) = y(1 - y)$$

$$\frac{\partial y}{\partial w_j} = \sigma'(w^T\mathbf{x} + b)x_j = y(1 - y)x_j,$$

where $\mathbf{x}^T = (x_1, \ldots, x_n)$ and $w^T = (w_1, \ldots, w_n)$. Consider the vanishing linear combination of output functions

$$\alpha_0 \frac{\partial y}{\partial b} + \sum_{j=1}^n \alpha_j \frac{\partial y}{\partial w_j} = 0, \qquad \alpha_k \in \mathbb{R}.$$

Since $y(1-y) \neq 0$, the previous relation becomes

$$\alpha_0 + \sum_{j=1}^n \alpha_j x_j = 0.$$

Since this relation holds for any $x_j \in \mathbb{R}$, it follows that $\alpha_0 = \alpha_1 = \cdots = \alpha_n = 0$. (To show this, we choose all $x_j = 0$ to get $\alpha_0 = 0$; then take $x_j = \delta_{jk}$ to obtain $\alpha_k = 0$). Therefore

$$\left\{ \frac{\partial y}{\partial b}, \frac{\partial y}{\partial w_1}, \ldots, \frac{\partial y}{\partial w_n} \right\}$$

are linearly independent. This implies that the Jacobian matrix, J_y, has rank $n+1$. We may picture the differentiable manifold $\mathcal{S}$ as an $(n+1)$-dimensional smooth surface (no corners or cusps) in the infinite-dimensional space of functions on $\mathbb{R}^n$.

Assume now that the neuron is trained to approximate the continuous target function $z = z(\mathbf{x})$. If z is a point on the manifold, $z \in \mathcal{S}$, then there are some parameters values, $w^* \in \mathbb{R}^n$ and $b^* \in \mathbb{R}$, such that we have the exact representation $y(w^*, b^*) = z$. However, in general most target functions satisfy the condition $z \notin \mathcal{S}$. In this case, we need to find by training the values

$$(w^*, b^*) = \arg \min_{w,b} \text{dist}(z, \mathcal{S}),$$

which correspond to the coordinates of the orthogonal projection of z on the surface $\mathcal{S}$. The distance is measured, for instance, in the mean square sense. Starting from the initialization (w_0, b_0), a learning algorithm should produce a sequence of approximations, $(w_n, b_n)_n$, which converges to the projection coordinates, $\lim_{n \to \infty}(w_n, b_n) = (w^*, b^*)$. If the parameters update is made continuously (implied by an infinitesimal learning rate), then we obtain a curve $c(t) = (w(t), b(t))$ joining (w_0, b_0) and (w^*, b^*). This can be lifted to the curve $\gamma(t) = y \circ c(t)$ on the manifold $\mathcal{S}$. The fastest learning algorithm corresponds to the "shortest" curve between $y(w_0, b_0)$ and $y(w^*, b^*)$. The attribute "shortest" depends on the intrinsic geometry of the manifold $\mathcal{S}$, and this topic will be discussed in the next section.

This chapter will extend the manifold ideas from this example to the general case of a neural network.

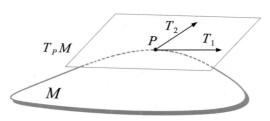

Figure 13.3: *The tangent space $T_p\mathcal{M}$.*

13.1.1   Intrinsic and Extrinsic

A manifold can be viewed from two distinct perspectives: *intrinsic* and *extrinsic*.

The intrinsic point of view is the perspective of a local observer living on the manifold. The observer's knowledge is bound to a local system of coordinates. Thus, the distance measured on the manifold, tangent vectors, their magnitudes, and angles between them belong to the intrinsic geometry of the manifold. One can picture this point of view as the perspective of an ant, which lives on the manifold, but does not have any access to the information outside of it.

The extrinsic perspective represents the knowledge about the manifold acquired by an observer, who looks at the manifold from the exterior. Geometric concepts, such as normal vector and manifold shape, are extrinsic. This can be pictured as the perspective of a satellite, which can observe the manifold from outside.

In the following we shall present a few intrinsic and extrinsic concepts of differential geometry on differentiable manifolds.

13.1.2   Tangent space

If the manifold $\mathcal{M}$ has a local parametrization $\phi : \mathcal{U} \to \mathcal{M}$, with $\mathcal{U} \subset \mathbb{R}^n$ open set, then the vector tangent to the kth coordinate curve

$$h \to \phi(x_1, \ldots, x_k + h, \ldots, x_n)$$

is given by

$$T_k(p) = \frac{\partial \phi}{\partial x_k}(p), \qquad p = \phi(x).$$

If the manifold satisfies the regularity condition that ϕ has a Jacobian matrix of maximum rank at p, then the tangent vectors $\{T_1(p), \ldots, T_n(p)\}$ are linearly independent. They form the basis of a linear space of dimension n, called the *tangent space* to $\mathcal{M}$ at p, denoted by $T_p\mathcal{M}$, see Fig. 13.3.

If $c(t)$ is a curve on the manifold $\mathcal{M}$, then in a local parametrization the velocity along $c(t)$ is a tangent vector to the manifold, which is given by

$$\dot{c}(t) = \sum_{k=1}^{n} \dot{c}^k(t) T_k(c(t)).$$

A *tangent vector field*, U, is a smooth assignment of a tangent vector $U_p \in T_p\mathcal{M}$, at each point $p \in \mathcal{M}$. In local coordinates this writes as $U_p = \sum_{k=1}^{n} U^k(p) T_k(p)$.

13.1.3   Riemannian metric

The main intrinsic concept on a manifold is the *Riemannian metric*. This is given in a local system of coordinates by a symmetric, positive definite, nondegenerate matrix, $(g_{ij})_{i,j}$, which is point dependent. Its entries represent weights used in measuring the distance between neighboring points on the manifold by a procedure similar to the Pythagorean theorem. Thus, if P and P' are two neighboring points on the manifold, having the coordinates (x_i) and (x_i'), the distance between them is measured in terms of the Riemannian metric as

$$d(P, P') = \Big(\sum_{i,j=1}^{n} g_{ij}(\Delta x)_i (\Delta x)_j \Big)^{1/2},$$

where $(\Delta x)_i = x_i - x_i'$, and $g_{ij} = g_{ij}(P)$. In the particular case of a Euclidean space, $\mathbb{R}^n$, when the coordinates are orthonormal, the metric matrix becomes the identity matrix, $g_{ij} = \delta_{ij}$. The distance is now measured by the usual Pythagorean theorem,

$$d(P, P') = \Big(\sum_{i=1}^{n} (\Delta x)_i^2 \Big)^{1/2}.$$

By a similar procedure we can measure the magnitude of a vector v tangent to the manifold as

$$\|v\|_g = \Big(\sum_{i,j=1}^{n} g_{ij} v^i v^j \Big)^{1/2},$$

where v^j denotes the jth component of the vector in a local system of coordinates. If g is the bilinear form defined on a basis $\{T_1, \ldots, T_n\}$ in $T_p\mathcal{M}$ by $g(T_i, T_j) = g_{ij}$, then the previous formula can be written as

$$\|v\|_g = g(v, v)^{1/2}.$$

The coefficients g_{ij} are historically referred to as the coefficients of the *first fundamental form g.*

On a Euclidean space this becomes the following familiar formula for the length of a vector:

$$\|v\| = \Big(\sum_{i,j=1}^{n} (v^i)^2 \Big)^{1/2} = \langle v, v \rangle^{1/2}.$$

If now $c(t)$ denotes a curve lying on the manifold and parametrized by the variable t, its velocity vector, $\dot{c}(t)$, is a tangent vector to the manifold, whose magnitude is given by

$$\|\dot{c}(t)\|_g = \Big(\sum_{i,j=1}^{n} g_{ij}(c(t))\dot{c}(t)^i\dot{c}(t)^j \Big)^{1/2} = g(\dot{c}(t), \dot{c}(t))^{1/2}.$$

Integrating the speed, $\|\dot{c}(t)\|$, with respect to time, t, we obtain the length of the curve measured with respect to the Riemannian metric

$$L(c) = \int_a^b \|\dot{c}(t)\|_g \, dt,$$

where $a \leq t \leq b$. The *Riemannian distance* between two points A and B on the manifold $\mathcal{M}$ is defined as the length of the shortest curve joining these points

$$d(A, B) = \inf_c \{L(c); c : [a, b] \to \mathcal{M}, c(a) = A, c(b) = B\}.$$

The pair $(\mathcal{M}, g)$ is called a *Riemannian manifold.* This chapter will describe the neural networks using Riemannian manifolds.

13.1.4   Geodesics

Another intrinsic concept is the notion of *geodesic.* This is the shortest curve on a manifold between two given points. If the points are close enough, there is always a geodesic between them and this is unique. In local coordinates the geodesic $c(t)$ can be described by a system of nonlinear equations as

$$\ddot{c}^k(t) + \sum_{i,j} \Gamma_{ij}^k \dot{c}^i(t)\dot{c}^j(t) = 0, \qquad 1 \leq k \leq n, \qquad (13.1.1)$$

where $\Gamma_{ij}^k = \Gamma_{ij}^k(c(t))$ are the *Christoffel symbols of second kind,* see, for instance, [85]

$$\Gamma_{ij}^k = \frac{1}{2}g^{kr}\Big(\frac{\partial g_{ir}}{\partial x_j} + \frac{\partial g_{jr}}{\partial x_i} - \frac{\partial g_{ij}}{\partial x_r} \Big), \qquad (13.1.2)$$

where $(x_1, \ldots, x_n)$ represent the local coordinates on the manifold. Since they depend on the metric coefficients g_{ij}, the Christoffel symbols are intrinsic, and hence the geodesics are too. Two simple examples are given next.

1. In the Euclidean space $(\mathbb{R}^n, \delta_{ij})$ the derivatives of the metric coefficients δ_{ij} are zero and the geodesics equations become $\ddot{c}^k(t) = 0$. This implies that the geodesics are straight lines.

2. In the case of the 2-dimensional sphere $\mathbb{S}^2$ the geodesics are arcs of great circles. The geodesics equations are more complicated to be given here, the reader being referred to a book of differential geometry.

The explicit computation of geodesics can be done on a very few particular types of manifolds and it is a relatively complex process.

13.1.5 Levi-Civita Connection

We shall start with the case of the Euclidean space. Let f be a differentiable function on $\mathbb{R}^n$ and v a vector. The *directional derivative* of f with respect to v is defined by $v(f) = \langle v, \nabla f \rangle$, where $(\nabla f)^T = (\partial_{x_1} f, \ldots, \partial_{x_n} f)$ denotes the gradient of f. The object $v(f)$ represents the rate of change of f in the direction of v.

Let $U = (U^1, \ldots, U^n) = \sum_{i=1}^n U^i e_i$ be a vector field on $\mathbb{R}^n$. The derivative of U with respect to the vector v is defined as

$$\nabla_v U = \big(v(U^1), \ldots, v(U^n)\big) = \sum_{i=1}^n v(U^i) e_i. \tag{13.1.3}$$

Now, let $V = (V^1, \ldots, V^n)$ be another vector field on $\mathbb{R}^n$. We define the *covariant derivative* of U with respect to V as

$$(\nabla_V U)_p = \nabla_v U,$$

where $v = V_p$ and the term in the right side is defined as in (13.1.3). We note that $\nabla_V U$ is a vector field on $\mathbb{R}^n$, which associates to each point $p \in \mathbb{R}^n$ the vector $(\nabla_V U)_p$.

It is not hard to show that for any differentiable function f on $\mathbb{R}^n$ and any vector fields U, V, and W on $\mathbb{R}^n$, the following properties hold:

(*i*) $\nabla_{fU} V = f \nabla_U V$,

(*ii*) $\nabla_U (V + W) = \nabla_U V + \nabla_U W$,

(*iii*) $\nabla_U (fV) = U(f) V + f \nabla_U V$.

The next property provides a compatibility between the derivation and inner product and it is a consequence of the product rule:

(*iv*) $W \langle U, V \rangle = \langle \nabla_W U, V \rangle + \langle U, \nabla_W V \rangle$,

The noncommutativity of the covariant differentiation is given by the formula

(*v*) $\nabla_U V - \nabla_V U = [U, V]$,

where $[U, V] = UV - VU$ is the *commutator* of the vector fields U and V. It can be shown by a direct computation that the commutator is always a vector field with components given by

$$[U, V]^i = U(V^i) - V(U^i) = \sum_{j=1}^{n} \Big(e_j(V^i)U^j - e_j(U^i)V^j\Big).$$

Consider now a Riemannian manifold $(\mathcal{M}, g)$. A *linear connection*, ∇, on the manifold $\mathcal{M}$ is an operator acting on vector fields of $\mathcal{M}$ satisfying the aforementioned properties (i)–(iii). There are many linear connections on a manifold. They are independent on the Riemannian structure g. However, there is only one linear connection that also satisfies properties (iv) and (v).

Theorem 13.1.7 *There is a unique linear connection on the Riemannian manifold* $(\mathcal{M}, g)$ *such that*

$$
\begin{aligned}
Wg(U, V) &= g(\nabla_W U, V) + g(U, \nabla_W V), \\
\nabla_U V - \nabla_V U &= [U, V],
\end{aligned}
$$

for all vector fields U, V, W *on* $\mathcal{M}$.

This is called the *Levi-Civita connection* on $(\mathcal{M}, g)$. Since the metric g determines uniquely this connection, the Levi-Civita connection is an intrinsic concept. For a proof of the existence and uniqueness of this connection the reader is referred, for instance, to Chapter 7 of the book [22].

It is worth noting that the Levi-Civita connection is an intrinsic concept, since it depends on the metric coefficients g_{ij} through the Christoffel symbols (13.1.2) as

$$\nabla_{T_i} T_j = \sum_k \Gamma_{ij}^k T_k,$$

where $\{T_1, \ldots, T_n\}$ is a basis of the tangent space.

13.1.6 Submanifolds

Let $\mathcal{M}$ and $\mathcal{S}$ be two manifolds, such that $\mathcal{S} \subset \mathcal{M}$ and $\mathcal{S}$ is endowed with the induced topology and differentiability structure from $\mathcal{M}$. Then any Riemannian metric g on $\mathcal{M}$ induces a Riemannian structure on $\mathcal{S}$ as

$$h(U, V) = g_{|\mathcal{S}}(U, V),$$

where U, V are vector fields on $\mathcal{S}$ and $h = g_{|\mathcal{S}}$ is the restriction of g on vector fields of $\mathcal{S}$. Then $(\mathcal{S}, h)$ becomes a Riemannian *submanifold* of $(\mathcal{M}, h)$.

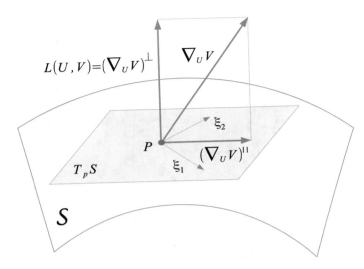

Figure 13.4: *Gauss formula* $\nabla_U V = (\nabla_U V)^{\|} + L(U, V)$.

Let n and m be the dimensions of $\mathcal{M}$ and $\mathcal{S}$, respectively. Then for any point $p \in \mathcal{S}$ the tangent space $T_p\mathcal{S}$ is a linear subspace of dimension m of the linear space $T_p\mathcal{M}$. We can consider the orthogonal split

$$T_p\mathcal{M} = T_p\mathcal{S} \oplus \mathcal{V}_p, \tag{13.1.4}$$

where $\mathcal{V}_p = \{v; g(v, u) = 0, \forall u \in T_p\mathcal{M}\}$. This means that for any vector $w \in T_p\mathcal{M}$ there are two orthogonal vectors $u \in T_p\mathcal{S}$ and $v \in \mathcal{V}_p$ such that $w = u + v$.

13.1.7 Second Fundamental Form

The shape of a submanifold with respect to the ambient manifold is described by its *second fundamental form*. Let $\mathcal{S}$ be a submanifold of the Riemannian manifold $(\mathcal{M}, g)$ and denote by ∇ the Levi-Civita connection on $(\mathcal{M}, g)$. Then for any two vector fields U and V on the submanifold $\mathcal{S}$ the vector field $\nabla_U V$ is not necessarily a vector field on $\mathcal{S}$. In general, we have according to (13.1.4) the following orthogonal decomposition:

$$(\nabla_U V)_p = (\nabla_U V)_p^{\|} + (\nabla_U V)_p^{\perp}, \tag{13.1.5}$$

with $(\nabla_U V)_p^{\|} \in T_p\mathcal{S}$ and $(\nabla_U V)_p^{\perp} \in \mathcal{V}_p$. Relation (13.1.5) is called Gauss formula, see Fig. 13.4.

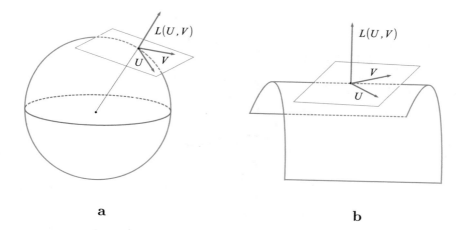

Figure 13.5: *The second fundamental form is normal to the surface and describes its shape:* **a.** *The case of a sphere.* **b.** *The case of a cylindrical surface.*

It can be proved that the operator $\overline{\nabla} = \nabla^{\|}$ is the Levi-Civita connection on the submanifold $(\mathcal{S}, g_{|\mathcal{S}})$.

The second fundamental form of the submanifold $\mathcal{S}$ with respect to $\mathcal{M}$ is defined by

$$L(U, V) = (\nabla_U V)^\perp \tag{13.1.6}$$

for any U, V vector fields on $\mathcal{S}$, see Fig. 13.5.

Proposition 13.1.8 *Let U, V, W be arbitrary vector fields on $\mathcal{S}$. The second fundamental form satisfies the following properties:*

(i) L is symmetric, $L(U, V) = L(V, U)$;

(ii) L is bilinear: $L(U + W, V) = L(U, V) + L(W, V)$;

(iii) $L(f_1 U, f_2 V) = f_1 f_2 L(U, V)$ for any two differentiable functions f_1 and f_2 on $\mathcal{S}$.

Proof: (i) Using the properties of the Levi-Civita connections ∇ and $\overline{\nabla}$ we show that L is symmetric

$$
\begin{aligned}
L(U, V) - L(V, U) &= \left(\nabla_U V - \overline{\nabla}_U V\right) - \left(\nabla_V U - \overline{\nabla}_V U\right) \\
&= \left(\nabla_U V - \nabla_V U\right) - \left(\overline{\nabla}_U V - \overline{\nabla}_V U\right) \\
&= [U, V] - [U, V] \\
&= 0.
\end{aligned}
$$

The other properties can be easily verified and are left as an exercise to the reader. ∎

It is worth noting that property (iii) states that if the vector fields U and V get scaled by the functions f_1 and f_2, then the second fundamental form gets scaled by the product $f_1 f_2$.

13.1.8 Mean Curvature Vector Field

Let $(\mathcal{S}, h)$ be a submanifold of the Riemannian manifold $(\mathcal{M}, g)$, with $h = g_{|\mathcal{S}}$, and denote the second fundamental form by L. For any fixed point $p \in \mathcal{S}$, we consider a vector basis $\{T_1, \ldots, T_m\}$ in the tangent space $T_p\mathcal{S}$ and define $L_{ij} = L(T_i, T_j)$, which is a normal vector to $\mathcal{S}$, for any $1 \leq i, j \leq m$, since $L_{ij} \in \mathcal{V}_p$. Let $h_{ij} = h(T_i, T_j)$ be the coefficients of the Riemannian metric on $\mathcal{S}$ and h^{ij} denote the coefficients of the inverse matrix $(h_{ij})^{-1}$.

The *mean curvature vector* at p is defined by the following "contraction":

$$H_p = \sum_{i,j} h^{ij}(p) L_{ij}.$$

We have $H_p \in \mathcal{V}_p$ and the mapping $p \to H_p$ defines the mean curvature vector field of $\mathcal{S}$, which is normal to $\mathcal{S}$ at each point.

The definition simplifies a little if we assume the initial chosen basis orthonormal. If $\{E_1, \ldots, E_m\}$ is an orthonormal basis in $T_p\mathcal{S}$, then $h(E_i, E_j) = \delta_{ij}$ and the mean curvature vector at p becomes

$$H_p = \sum_{i,j} \delta_{ij} L(E_i, E_j) = \sum_{i=1}^{m} L(E_i, E_i).$$

The mean curvature vector field is an extrinsic notion, since it depends on the second fundamental form.

Second fundamental form and the mean curvature measure different types of extrinsic curvatures of submanifolds. For instance, if the second fundamental form is zero, $L_{ij} = 0$, then the submanifold $\mathcal{S}$ is called *totally geodesic*. This means that any geodesic of $\mathcal{S}$ is also a geodesic of $\mathcal{M}$. To understand this concept the reader should picture the particular case of a plane included into the 3-dimensional space: any straight line in the plane is also a straight line in the space. This is compatible with the fact that the plane does not bend in the space.

If the mean curvature vector field is zero, $H = 0$, than $\mathcal{S}$ is called a *minimal submanifold* of $\mathcal{M}$. The geometric interpretation is that $\mathcal{S}$ has locally a minimal volume; that is, if the manifold is perturbed locally, then its volume increases. The concepts of second fundamental form and mean curvature will be useful for regularization purposes later.

13.2 Relation to Neural Networks

The reader might have inquired what is the relation between the differential geometry concepts presented so far and neural networks. This section briefly discusses this relation, while the later sections will present a more detailed analysis.

The role of a given neural network is to approximate a certain target function z. We assume that z is an element of a *target manifold*, $\mathcal{M}$, such as the manifold of continuous functions on $[0, 1]$. The output y of the neural network is parametrized by $\theta = (w, b)$, the weighs and biasses of the network. This way, the output y belongs to an *output manifold*, $\mathcal{S}$, which is supposed to be a submanifold of the target manifold, $\mathcal{M}$. The dimension of the submanifold $\mathcal{S}$ is equal to the number of network weights and biasses, while the dimension of $\mathcal{M}$ in this case is infinite.[1] It is worth noting that for practical applications the target is a vector $z^T = (z_1, \ldots, z_n)$, fact that implies that the manifold $\mathcal{M}$ has dimension n.

The dimension of the output manifold can be increased by adding more neurons, and hence more parameters. The larger the dimension the better the approximation. However, also the shape of the submanifold $\mathcal{S}$ plays a determinant role in trying to avoid overfitting. The second fundamental form, L, describes how does the submanifold $\mathcal{S}$ bend inside of $\mathcal{M}$. From the regularization point of view, we prefer submanifolds that bend as little as possible, so that the orthogonal projection of the target function z onto $\mathcal{S}$ is eventually unique and easy to find by the gradient descent method.

As an example, we shall consider the case of the manifold given by Example 13.1.6. The target manifold in this case can be chosen to be the space $\mathcal{M} = \mathcal{C}[0, 1]$. The manifold of outputs

$$\mathcal{S} = \{\sigma(w^T\mathbf{x} + b); w \in \mathbb{R}^n, b \in \mathbb{R}\}$$

is an $(n+1)$-dimensional submanifold of $\mathcal{M}$. At each point $y \in \mathcal{S}$ the tangent space $T_p\mathcal{S}$ is spanned by linear combinations of the functions

$$\{y(1 - y), y(1 - y)x_1, \ldots, y(1 - y)x_n\}.$$

The submanifold $\mathcal{S}$ can be seen as an $(n + 1)$-dimensional hypersurface inside the space of real-valued continuous functions $\mathcal{M} = \mathcal{C}([0, 1])$. In the case of one sigmoid neuron the target space $\mathcal{M}$ is not well approximated by

---

[1] A system of parameters for a continuous function defined on $[0, 1]$ is the set of rational numbers $Q \cap [0, 1]$.

the surface $\mathcal{S}$, since there are continuous functions $f \in \mathcal{M}$ whose distance to the surface $\mathcal{S}$ cannot be made arbitrarily small. Equivalently stated, using $\mathcal{S}$ to approximate $\mathcal{M}$ leads to an underfit.

However, increasing the number of neurons in the network leads to a larger number of parameters, and hence to a higher dimensional approximation manifold $\mathcal{S}$. The hope is that for any element of the target space, $f \in \mathcal{M}$, and any fixed $\epsilon > 0$, there is a network that produces a manifold $\mathcal{S}$ of high enough dimension such that $\text{dist}(f, \mathcal{S}) < \epsilon$, where the distance is measured by

$$\text{dist}(f, \mathcal{S}) = \inf_{s \in \mathcal{S}} \max_{x \in [0,1]} |f(x) - s(x)|.$$

The dimension of $\mathcal{S}$ can be obviously increased by adding more neurons until the desired approximation holds. However, in practice the neurons supply might be limited, fact that leads to the problem of maximizing the dimension of $\mathcal{S}$, while keeping the number of neurons constant. We shall deal with this problem in the next section.

13.3   The Parameters Space

Assume we are supplied with a fixed number, N, of computing units and we need to construct a feedforward neural network using these units as hidden neurons, such that the network acquires its maximum capacity, i.e., it has a maximum ability of approximating target functions. Specifically, we shall look for that network architecture which considers only N hidden neurons and produces a maximum dimension for the output manifold $\mathcal{S}$. This is obtained when the number of network parameters is maximized.

We shall start with an example. Assume we are provided with $N = 10$ hidden neurons and consider the following three feedforward network architectures, ordered from shallow to deep:

(i) only one hidden layer with $N = 10$ hidden neurons in that layer;

(ii) 2 hidden layers with 5 hidden neurons in each layer;

(iii) 5 hidden layers with 2 hidden neurons in each layer.

For the sake of simplicity, we assume that both the input and the output are one-dimensional.

The network given by (i) has 30 parameters: 10 weights w_j (from the input to the hidden layer); 10 biasses b_j (one for each hidden neuron); 10 weights α_j (from the hidden layer to the output). See Fig. 13.6.

The network given by (ii) has 45 parameters: 5 weights w_j (from the input to the first hidden layer); 5^2 weights $w_{ij}^{(1)}$ (from the first to the second hidden layer); 5 weights α_j (from the second hidden layer to the output); 10 biasses b_j (one for each neuron in each hidden layer). See Fig. 13.7 **a**.

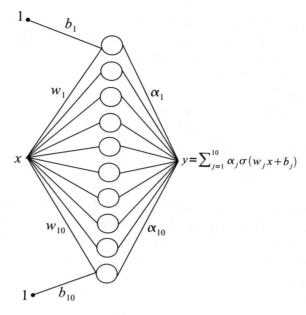

Figure 13.6: *A one-hidden layer neural network with $N = 10$ hidden neurons, which depends on 30 parameters.*

The network given by (iii) has 30 parameters: 2 weights w_j (from the input to the first hidden layer); 16 intermediate weights (4 between each two hidden layers); 2 weights α_j (from the last hidden layer to the output); 10 biasses b_j. The network is pictured in Fig. 13.7 **b**.

We conclude that the maximum number of parameters is reached in the case (ii), which is neither the shallowest, nor the deepest neural network considered. We shall show that this is a typical behavior for general feedforward neural networks architectures.

We consider now a neural network with $L - 1$ hidden layers (the layers 0 and L are reserved for the input and output layers, respectively). As usual, we denote by $d^{(\ell)}$ the number of neurons in the ℓth layer. For simplicity reasons, we choose $d^{(0)} = d^{(L)} = 1$. Since the number of hidden neurons is equal to N, we have

$$\sum_{\ell=1}^{L-1} d^{(\ell)} = N. \tag{13.3.7}$$

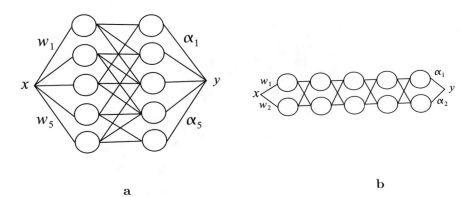

Figure 13.7: *Two neural nets with $N = 10$ neurons:* **a.** *A 2-hidden layer neural network depending on 45 parameters.* **b.** *A 5-hidden layer neural network depending on 30 parameters.*

The number of weights $w_{ij}^{(\ell)}$ between the layers $\ell - 1$ and ℓ is $d^{(\ell-1)}d^{(\ell)}$. The total number of biasses is equal to N, one for each hidden neuron. Then, the total number of parameters, including weights and biasses, is given by

$$d^{(0)}d^{(1)} + d^{(1)}d^{(2)} + \cdots + d^{(\ell-1)}d^{(\ell)} + \cdots + d^{(L-1)}d^{(L)} + N. \qquad (13.3.8)$$

The problem of finding the feedforward neural network of maximum capacity can be now formulated equivalently as:
What is the number of layers L, and the number of neurons, $d^{(\ell)}$, in each hidden layer ℓ, with $1 \leq \ell \leq L - 1$, such that the expression (13.3.8) *reaches its maximum, given the constraint* (13.3.7)?

This problem has the following geometric significance. Each product term $d^{(\ell-1)}d^{(\ell)}$ is interpreted as the area of a rectangle. Then, starting from a rectangle with dimensions $d^{(0)} \times d^{(1)}$, we continue the construction of a rectangle with dimensions $d^{(1)} \times d^{(2)}$ as in Fig. 13.8. The even sides are displayed vertically, while the odd ones are horizontal. The entire figure can be inscribed into a rectangle $\mathcal{R}$ having the width equal to $d^{(1)} + d^{(3)} + \cdots$ and the height given by $d^{(0)} + d^{(2)} + \cdots$. The constraint (13.3.7) is geometrically equivalent with the fact that the sum of the dimensions (i.e., width and height) of the rectangle $\mathcal{R}$ is constant, equal to N. The goal is to maximize the sum of the rectangles area, given the aforementioned constraint.

One approximative approach is to maximize the area of the rectangle $\mathcal{R}$ first. This occurs when the width is equal to the height, i.e., when it becomes a square. Then we try to fill in the entire square $\mathcal{R}$ by rectangles using

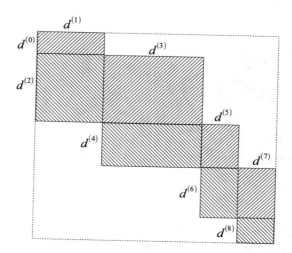

Figure 13.8: *The area significance of the sum* $d^{(0)}d^{(1)} + d^{(1)}d^{(2)} + \cdots + d^{(7)}d^{(8)}$.

the previous construction such that we fill in most of it. A combinatorial argument shows that the optimal construction occurs only when we have two hidden layers. In this case the situation looks like in Fig. 13.9 **a**. If we have more layers, let's say 3 layers, then the construction is not optimal because there is more unfilled space left, see Fig. 13.9 **b**. The reader should be able to fill in the missing details of the argument.

The problem has an exact mathematical solution if we assume from the beginning that

$$d^{(1)} = d^{(2)} = \cdots = d^{(L-1)},$$

namely, when each hidden layer has the same number of neurons. For simplicity, let $k = L - 1$ denote the number of hidden layers in the network, so each hidden layer has N/k neurons. Then the number of weights given by (13.3.8) becomes

$$f_N(k) = \frac{N}{k} + (k-1)\left(\frac{N}{k}\right)^2 + \frac{N}{k} + N.$$

This will be optimized with respect to k as in the following. We start by rewriting the expression in terms of $1/k$ as

$$
\begin{aligned}
f_N(k) &= \frac{2N}{k} + N^2 \frac{k-1}{k^2} + N \\
&= \frac{N}{k}\left(2 + N - N\frac{1}{k}\right) + N.
\end{aligned}
$$

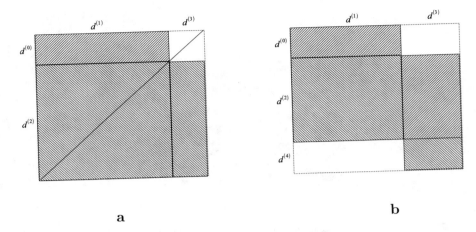

Figure 13.9: **a.** *The blank rectangle is a square of area $d^{(0)}d^{(3)} = 1$.* **b.** *The blank rectangles have an area larger than 1.*

Substituting $u = 1/k$, we obtain a quadratic function in u

$$\phi_N(u) = Nu\Big(2 + N - Nu\Big) + N = -N^2 u^2 + N(N+2)u + N.$$

The maximum of $\phi_N(u)$ is reached for

$$u = \frac{N(N+2)}{2N^2} = \frac{N+2}{2N}.$$

This corresponds to a number of hidden layers given by

$$k = \frac{2N}{N+2}.$$

Even if this number is not always an integer, for a large number of neurons N, the optimal number of hidden layers is well approximated by $k = 2$. This explains why in the case when $N = 10$, having a network with two hidden layers, each having 5 neurons, achieves the maximum capacity.

The theoretical maximum number of parameters is given by $f_N(\frac{2N}{N+2})$. In fact, this is equal to the value

$$f_N(2) = \frac{N^2}{4} + 2N. \qquad (13.3.9)$$

Therefore, the maximum dimension of the output manifold $\mathcal{S}$ grows quadratically in the number of hidden neurons N. We shall show next that this may lead sometimes to an overfit of data.

The previous formulas have been developed in the particular case of networks satisfying $d^{(0)} = d^{(L)} = 1$. For the general case, see Exercise 13.8.2.

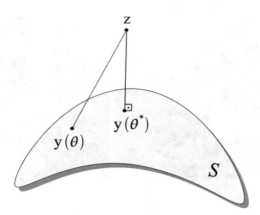

Figure 13.10: *The orthogonal projection of $\mathbf{z}$ on the manifold $\mathcal{S}$ is $\mathbf{y}(\theta^*)$.*

13.4　Optimal Parameter Values

Consider the training set $\{(x_1, z_1), (x_2, z_2), \ldots, (x_n, z_n)\}$, and let y_i be the network's one-dimensional output corresponding to the one-dimensional input x_i, with $1 \leq i \leq n$. Then $y_i = y_i(\theta)$, where $\theta \in \mathbb{R}^r$ is the parameter vector of the network. Therefore, the vector $\mathbf{y}^T = (y_1, \ldots, y_n) \in \mathbb{R}^n$ is parametrized by θ, and hence describes a manifold $\mathcal{S}$ in $\mathbb{R}^n$ of dimension r. In fact, the output manifold $\mathcal{S}$ is an r-dimensional surface in $\mathbb{R}^n$. Training the network is equivalent to finding an exact, or approximate value of θ^*, for which the distance from $\mathbf{z}^T = (z_1, \ldots, z_n)$ to $\mathcal{S}$ is minimum, namely

$$\theta^* = \arg\min_\theta \operatorname{dist}(\mathbf{z}, \mathcal{S}) = \arg\min_\theta \|\mathbf{z} - \mathbf{y}(\theta)\|,$$

where the distance is measured in the Euclidean sense. This is equivalent with the fact that $\mathbf{y}(\theta^*)$ is the orthogonal projection of $\mathbf{z}$ on $\mathcal{S}$, see Fig. 13.10.

If the network has two hidden layers, then the previous analysis provides $r = \frac{N^2}{4} + 2N$, where N is the number of hidden neurons. If N is such that $\frac{N^2}{4} + 2N \geq n$, then the network exhibits an overfit, since it memorizes the entire training set. We shall fill in some mathematical details of this fact.

Assume we have just equality, $\frac{N^2}{4} + 2N = n$. Then the submanifold $\mathcal{S}$ has the same dimension as the target space $\mathbb{R}^n$, and hence it is either the entire space, or a subset of it. Consequently, we can choose parameters θ such that the point $\mathbf{z}$ belongs to the manifold $\mathcal{S}$, fact that implies the aforementioned distance equal to zero. The details of this implication are given next. The system of n equations

$$
\begin{aligned}
y_1(\theta) &= z_1 \\
\cdots &= \cdots \\
y_n(\theta) &= z_n
\end{aligned}
$$

has n unknowns, $\theta = (\theta_1, \ldots, \theta_n)$. We can invert the system assuming the regularity hypothesis

$$
\det\left(\frac{\partial y_j}{\partial \theta_i}\right) \neq 0,
$$

which geometrically corresponds to stating that the output manifold S has a tangent plane at each point (there are no corner points or thorns on the manifold). Then the system has a unique solution, $\theta^* = \theta^*(\mathbf{y}, \mathbf{z})$. This shows that $\mathbf{z}$ is a point on the manifold S, having the corresponding parameter θ^*.

We note that the inequality $\frac{N^2}{4} + 2N > n$ cannot hold because this would mean that the submanifold has a dimension larger than the space itself. Equivalently, this means that the aforementioned system has a number of unknowns θ larger than the number of equations, n. This would imply that the weights and biasses cannot be independent, which is a contradictory statement for a neural network.

Example 13.4.1 We shall provide an example applied to the MNIST data set, which comes with $n = 55,000$ training data, $\{(\mathbf{x}_i, \mathbf{z}_i)\}$. There are some differences from the previous theory, since each input data, $\mathbf{x}_i$, is 784-dimensional (each 28×28-image is flatten out into a 784-dimensional vector), while each output, $\mathbf{z}_i$, is 10-dimensional (there are 10 digit classes). These modifications shall be taken into account when calculating the dimension of the output manifold. In the following we shall take $N = 500$ neurons and consider the following feedforward neural network architectures:

(*i*) 784-500-10: 1-hidden layer network with $N = 500$ neurons;

(*ii*) 784-250-250-10: 2-hidden layer network with 250 neurons in each layer;

(*iii*) 784-50-50-$\cdots$-50-10: 10-hidden layer network with 50 neurons in each layer;

In this example we have $d^{(L)} = 10$, because each target vector, $\mathbf{z}_i$, is 10-dimensional. Since there are 55,000 targets, this yields a space with dimension $55,000 \times 10 = 550,000$. Therefore, the output manifold S is a submanifold of the numerical space $\mathbb{R}^{550,000}$. We shall discuss next the dimensions of the output manifolds for the aforementioned architectures.

In case (*i*) the dimension of the output manifold S is

$$
r = 784 \times 500 + 500 \times 10 + 500 = 397,\!500
$$

since there are 784 weights from the input to the hidden layer, 500 biasses, and 500×10 weights from the hidden layer to the output. The accuracy in this case is about 97.2 percent and the test error is about 1.49. The small test error indicates the optimality of the network (neither overfitting nor underfitting).

In case (ii) the dimension of the output manifold is

$$r = 784 \times 250 + 250 \times 250 + 250 \times 10 + 500 = 261{,}500,$$

which is about half of the dimension of the target space. The accuracy of the network in this case is 96.5 percent, which is lower than the previous network. And the test error, which is 565, is larger than the one of the previous case.
In case (iii) the dimension of the manifold is

$$r = 784 \times 50 + 9 \times 50 \times 50 + 500 + 500 = 62{,}700,$$

which is about 17 times smaller than the dimension of the target space. This leads to an underfit of data, fact suggested by the test error, which is 1,027. The accuracy in this case is only 94.5 percent. We notice a decrease in accuracy as the network gets deeper.

These computations have been executed using 4,000 iterations using a batch size of 40. We employed the Adam optimization algorithm, which decreases the learning step as we get closer to the optimum point. The cost function used was the sum of squares errors.

Learning with information There is another way to characterize the optimal parameters, using the concept of information fields. Consider a two-layer neural network with input X and output Y, with an increasing activation function ϕ. The output is related to the input by

$$Y = \phi(W^T X + B).$$

Since ϕ is invertible, then $W^T X + B = \phi^{-1}(Y)$, and by Proposition D.5.1 we have $\mathfrak{S}(Y) = \mathfrak{S}(W^T X + B)$. Since adding constants does not change the information field, see Exercise 11.10.1, we have $\mathfrak{S}(Y) = \mathfrak{S}(W^T X)$. Hence, the output information, $\mathcal{E} = \mathfrak{S}(Y)$ is independent of the bias B but depends on the system of weights W. Since (W, B) are coordinates on the associated output manifold, we can associate with each point on this manifold the parameter-dependent sigma-field $\mathcal{E} = \mathcal{E}_{W,B}$. The information is preserved in the direction of coordinate B, i.e., it is conserved along slices $\{W = \text{constant}\}$.

In the case of a three-layer neural network, with input X and output $\tilde{Y}$, the output is related to the input by

$$\tilde{Y} = \phi\Big(W^{(2)^T}\phi(W^T X + B) + B^{(2)}\Big).$$

Similarly with the previous computation, we can show that the output information is

$$\tilde{\mathcal{E}} = \mathfrak{S}(\tilde{Y}) = \mathfrak{S}(W^{(2)^T} \phi(W^T X + B)).$$

This depends on the weights W, $W^{(2)}$, and the bias in the first layer, B, but is independent of the bias in the last layer, $B^{(2)}$.

In general, the output manifold of any feedforward neural network can be endowed with an information structure. Each point on the manifold is parametrized by some weights and biasses to which we associate an information field.

If $\mathcal{I}$, $\mathcal{E}_{W,B}$ and $\mathcal{Z}$ denote the input, output and target information fields, respectively, we assume the double inclusion holds

$$\mathcal{E}_{W,B} \subset \mathcal{Z} \subset \mathcal{I}.$$

Given the information $\mathcal{E}_{W,B}$, then the target variable, Z, is best approximated by the conditional expectation $\mathbb{E}[Z|\mathcal{E}_{W,B}]$. The optimal weights are given by

$$(W^*, B^*) = \arg\min_{W,B} \|Z - \mathbb{E}[Z|\mathcal{E}_{W,B}]\|_{L^2}.$$

The network output corresponding to the weights (W^*, B^*) is the best approximator of the target Z in the information sense.

We shall point out next the exact learning case. Assume now that, for some parameters (W, B), we have

$$\mathcal{Z} \subset \mathcal{E}_{W,B} \subset \mathcal{I}.$$

Then any target variable, Z, i.e., a random variable that is $\mathcal{Z}$-measurable, is also $\mathcal{E}_{W,B}$-measurable, case in which $Z = \mathbb{E}[Z|\mathcal{E}_{W,B}]$. This corresponds to an exact learning, since Z is completely determined by the output information of the network.

13.5   The Metric Structure

We have seen that the number of hidden neurons, N, determines the dimension of the output manifold $\mathcal{S}$. This has the maximum dimension in the case when there are two hidden layers. However, if r is smaller than the maximum value, there might be several manifolds of the same dimension r, which are associated with different feedforward network architectures.

For instance, the encoder and decoder networks represented in Fig. 13.11 **a**, **b** depend on the same number of parameters and have the same number of hidden neurons. However, their task is very different. This is explained by the fact that the architecture of the network (which here means the sequence of

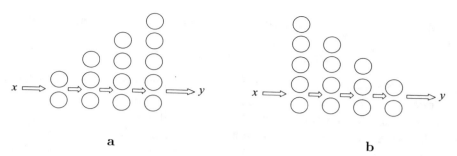

a

b

Figure 13.11: *Two symmetric networks, which depend on the same number of parameters:* **a.** *An encoder.* **b.** *A decoder. The number of parameters is 59; there are 14 biases (one for each neuron) and 45 weights.*

numbers $d^{(\ell)}$) produces different metric effects on the manifold $\mathcal{S}$. The geometric shape of $\mathcal{S}$ depends on the sensitivity of the network output y with respect to the weights and biasses. Thus, the output tends to be less sensitive to weights situated in the first layers, closer to the input and more sensitive to weights situated in the last layers, closer to the output.

We may say that a neural network is associated with the approximation manifold $\mathcal{S}$, and to train the network means to find the minimum distance from the target point $\mathbf{z}$ to the manifold $\mathcal{S}$. Hence, the geometric properties of the manifold $\mathcal{S}$, such as shape, metric structure, etc., would play an important role in the study of neural networks.

The metric structure We need to endow $\mathcal{S}$ with a Riemannian metric, which will be used to measure distances between points on $\mathcal{S}$ and angles between tangent vectors to $\mathcal{S}$.

Since $\mathcal{S}$ is a submanifold of $\mathbb{R}^n$ of dimension r, it is natural to endow $\mathcal{S}$ with the natural metric induced from the Euclidean structure of $\mathbb{R}^n$. If θ_i represent the parameters of the network (either weights or biases), then the *basic tangent vector fields* to $\mathcal{S}$ are given by partial derivatives with respect to coordinates θ_i as

$$\xi_i = \frac{\partial \mathbf{y}}{\partial \theta_i} = \left(\frac{\partial y_1}{\partial \theta_i}, \dots, \frac{\partial y_n}{\partial \theta_i} \right), \quad 1 \le i \le r.$$

The *tangent space* to $\mathcal{S}$ at $\mathbf{y}$ is the linear space, $T_{\mathbf{y}}\mathcal{S} = \text{span}\{\frac{\partial \mathbf{y}}{\partial \theta_i}; 1 \le i \le r\}$, generated by all basic tangent vectors at that point. If $\mathcal{S}$ is a smooth manifold, then $T_{\mathbf{y}}\mathcal{S}$ has a constant dimension r at each point $\mathbf{y} \in \mathcal{S}$. This condition can be stated equivalently as the maximal rank condition

$$\text{rank}\left(\frac{\partial \mathbf{y}}{\partial \theta_i} \right)_i = r,$$

which states that the basic tangent vector fields are linearly independent. This regularity condition assures that the output manifold S is smooth (no corners, cusps, etc.). A tangent vector to S at $\mathbf{y}$, $v \in T_{\mathbf{y}}S$, is defined by the linear combination

$$v = \sum_{i=1}^{r} v_i \frac{\partial \mathbf{y}}{\partial \theta_i},$$

where $v_i = v_i(\theta)$ are the components of v.

The natural metric structure of S is provided by the first fundamental form with coefficients given by

$$g_{ij}(\mathbf{y}) = \langle \xi_i, \xi_j \rangle = \left(\frac{\partial \mathbf{y}}{\partial \theta_i} \right)^T \frac{\partial \mathbf{y}}{\partial \theta_j} = \sum_{k=1}^{n} \frac{\partial y_k}{\partial \theta_i} \frac{\partial y_k}{\partial \theta_j}. \qquad (13.5.10)$$

The $r \times r$ matrix $g = (g_{ij})$ can be used to compute lengths of tangent vectors, angles between directions, lengths of curves on S, distances between points, areas of regions on S, and in general, any mathematical concept that depends on the intrinsic structure of the manifold S.

We recall that the concepts of intrinsic and extrinsic are often used in differential geometry to refer to the geometric information arisen from the local and global structures of S, respectively. For instance, measuring the angle between two curves on S can be done using the local information, namely, it can be performed by a microscopic inhabitant of the manifold, who is not allowed to leave the manifold. On the other side, the training error of the network, which is the distance from an exterior target point $\mathbf{z}$ to S, is an extrinsic concept, since it depends on the ability of a manifold inhabitant to fly above the manifold, in the exterior space, which allows him to make measurements.

Length of vectors Consider a vector, $v = \sum_{i=1}^{r} v_i \frac{\partial \mathbf{y}}{\partial \theta_i}$, tangent to the manifold S. We shall measure its length in two different ways: extrinsically, as a vector in $\mathbb{R}^n$, and intrinsically, as a tangent vector to S.

If $\{e_k; 1 \le k \le n\}$ denotes the natural orthonormal basis in $\mathbb{R}^n$, then the kth component of v in $\mathbb{R}^n$ is given by

$$\langle v, e_k \rangle = v^T e_k = \sum_{i=1}^{r} v_i \frac{\partial y_k}{\partial \theta_i}, \qquad 1 \le k \le n,$$

where $y_k = \langle \mathbf{y}, e_k \rangle$. Then the square of the *Euclidean length* of v is given by

$$\|v\|_{Eu}^2 = \sum_{k=1}^{n} \langle v, e_k \rangle^2.$$

The same length can be computed intrinsically using a scalar product with coefficients g_{ij} as in the following:

$$
\begin{aligned}
\|v\|_g^2 &= \sum_{i,j} v_i v_j g_{ij} = \sum_{i,j} v_i v_j \sum_{k=1}^n \frac{\partial y_k}{\partial \theta_i} \frac{\partial y_k}{\partial \theta_j} \\
&= \sum_{k=1}^n \Big(\sum_i v_i \frac{\partial y_k}{\partial \theta_i} \Big) \Big(\sum_j v_j \frac{\partial y_k}{\partial \theta_j} \Big) = \sum_{k=1}^n \langle v, e_k \rangle^2 \\
&= \|v\|_{Eu}^2.
\end{aligned}
$$

The fact that we obtained equal lengths in both cases was expected, since the metric g_{ij} of $\mathcal{S}$ is induced from the space $\mathbb{R}^n$ and the length is independent of the intrinsic or extrinsic approach.

Length of curves Assume the weights and biases of a neural network depend on an extra parameter s. This can be either time, or a certain hyperparameter of the network. Then $\theta_i = \theta(s)$, $1 \le i \le r$, and hence $c(s) = \mathbf{y}(\theta(s))$ represents a curve on the manifold $\mathcal{S}$. Therefore, the continuous tuning of a network hyperparameter corresponds to a curve on the manifold. If s takes values between a and b, then the length of the curve $c(s)$ is defined intrinsically by the integral

$$
L(c) = \int_a^b \|\dot{c}(s)\| \, ds = \int_a^b \sqrt{\sum_{i,j} \dot{c}_i(s) \dot{c}_j(s) g_{ij}(c(s))} \, ds,
$$

where $\dot{c}(s)$ represents the tangent vector along the curve. Chain rule provides

$$
\dot{c}(s) = \frac{d}{ds} c(s) = \frac{d}{ds} \mathbf{y}(\theta(s)) = \sum_i \frac{\partial \mathbf{y}}{\partial \theta_i} \frac{d\theta_i}{ds} = \langle \nabla_\theta \mathbf{y}, \dot{\theta}(s) \rangle. \tag{13.5.11}
$$

Geodesics Sometimes, we are interested in finding the curve of shortest length between two given points on $\mathcal{S}$. If we look for the shortest curve between two points on $\mathbb{R}^n$, this is obviously a line segment. However, in the case of the manifold $\mathcal{S}$ the characterization is more complex, the curves of shortest distance being the geodesics. One application of geodesics is to find the shortest curve between a given initial point, $\mathbf{y}(\theta^0)$, and the optimal point, $\mathbf{y}(\theta^*)$, which is the orthogonal projection of target point $\mathbf{z}$ on $\mathcal{S}$. This curve corresponds to the most efficient tuning of the network, since a parameter tuning corresponds to a curve on the manifold.

It is worth noting that the distance between two points of $\mathcal{S}$ measured in the metric of $\mathcal{S}$ is at least as large as the distance measured between the same points using the metric of the target space. These distances are equal in

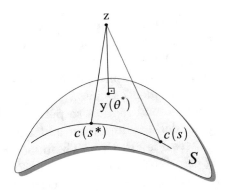

Figure 13.12: *The optimum parameter, s^*, corresponds to the closest point on the curve $c(s)$ to the target $\mathbf{z}$.*

the case when $\mathcal{S}$ is a geodesic submanifold, namely, its second fundamental form is zero, $L = 0$.

Optimal parameter value Assume that while modifying parameters $\theta(s)$ in terms of s, we notice first an improvement in accuracy followed by a decrease in accuracy. The smallest error is reached for some optimal value s^*. Geometrically, this corresponds to the point on the curve $c(s)$, which is the closest to the target point $\mathbf{z}$, see Fig. 13.12. This occurs when the vector $\overrightarrow{\mathbf{z}c(s)}$ is perpendicular to the tangent vector $\dot{c}(s)$, fact that can be written as

$$(\mathbf{z} - c(s))^T \dot{c}(s) = 0.$$

If the parameter s is modified such that the rate $\|\dot{c}(s)\|^2$ is constant,[2] then differentiating and using product rule, yields $c(h)^T \dot{c}(h) = 0$. Hence, the previous equation becomes as $\mathbf{z}^T \dot{c}(h) = 0$. Using formula (13.5.11) implies that the optimal value s^* satisfies the equation

$$\langle \nabla_\theta (\mathbf{z}^T \mathbf{y}(s^*)), \dot{\theta}(s^*) \rangle = 0.$$

13.6 Regularization

The most desired property of a neural network is to *generalize well*. This means that after optimizing the network using a training set, the network

---

[2]This is also called the arc length parameter, since it is proportional to the arc length measured along the curve $c(s)$.

should still perform with large accuracy for other unseen testing data. In order to achieve this goal, the network should be constructed such that it does not overfit the training data.

This phenomenon is easier to explain if we consider the particular case of polynomial regression. Consider 7 points, (x_i, z_i), $1 \leq i \leq 7$, in the plane and use three types of polynomial models to perform regression. The linear regression is not a good model, leading to an underfit of data, characterized by a large training error, see Fig. 13.13 **a**. The quadratic model produces a relatively small training error and is a good fit. The 7th degree interpolation polynomial produces an overfit, which is characterized by a zero training error and a large testing error. The way we should select the appropriate regression model is to be parsimonious when choosing the degree of the polynomial. In the same time, the polynomial should have a large enough degree to capture the main trend of data without overfitting it.

Similar observations apply for the case of a general neural network. The approximation polynomial in this case is replaced by the output manifold $\mathcal{S}$ and the degree of the polynomial corresponds to the dimension of $\mathcal{S}$, i.e., the number of network parameters θ_i. The principle of being parsimonious translates in this case in selecting a manifold of small dimension, namely, a network with a small number of neurons.

The setup of the problem is as follows: *Given N neurons and a training set $\{(x_i, z_i); 1 \leq i \leq n\}$, construct a neural network that learns from data, without overfitting it.*

We shall present next a few regularization techniques, that is, ways to avoid or reduce data overfitting, for a given number of neurons, N.

13.6.1 Going for a smaller dimension

Since N is given, we need to decide on the number of hidden layers and the number of neurons in each layer. We have seen that the use of only two hidden layers produce the maximum dimension for the manifold $\mathcal{S}$, so we should avoid this. We should either go shallow, with only one hidden layer, or go deep with a large enough number of hidden layers such that the dimension of $\mathcal{S}$ is sufficiently small, and the parsimony criterion holds.

13.6.2 Norm regularization

In order to reduce overfitting, smaller weights, w, should be used. The cost function, $C(w, b)$, is modified by adding an extra term involving a norm of the weights, multiplied by a positive Lagrange multiplier, λ, which describes the preference for small weights (larger λ corresponds to smaller weights).

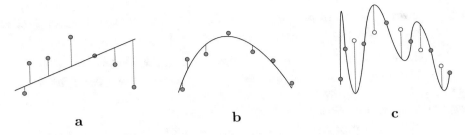

Figure 13.13: *Polynomial regression through 7 points:* (a) *Using a line leads to an underfit;* (b) *using a quadratic polynomial leads to a good fit;* (c) *the use of a 7th degree polynomial overfits the data.*

The regularized cost function becomes

$$L(w) = C(w, b) + \lambda \|w\|^2,$$

where $\|\cdot\|$ is usually either the L^1 or the L^2 norm. This type of regularization has been discussed in more detail in section 3.11. The effect of the norm regularization is to look for an optimal point on the output manifold, which is localized in a certain neighborhood.

13.6.3 Choosing the flattest manifold

We notice that there are several neural structures corresponding to a fixed dimension of the manifold $\mathcal{S}$. This can be seen, for instance, in the encoder and decoder structures given in Fig. 13.11. The question is now, *which structure is better from the regularization point of view?*

In order to decrease the testing error, we shall go for the neural network for which the manifold $\mathcal{S}$ is as *flat* as possible. By "flat" we refer to a manifold that bends as little as possible in the target space $\mathbb{R}^n$.

Flatness is an extrinsic concept that can be formalized in geometric terms by means of the *second fundamental form*. We shall start with a few examples.

Example 13.6.1 Consider a plane $\mathcal{P}$ in the space $\mathbb{R}^3$. This is flat since it does not bend. This is equivalent to observing that the normal vector to the plane is a constant vector field. The way a surface bends is described by the rate at which its normal vector changes its direction. This is called the shape operator (or the Weingarten map) of the surface and it is related to different types of curvature, see, for instance, [85].

Example 13.6.2 Another example involves a plane curve, $c(s)$, with unit tangent vector $T(s)$ and normal vector $N(s)$, where s denotes the arc length parameter. The rate of change of the normal vector is given by $N'(s) = -\kappa(s)T(s)$, where $\kappa(s)$ is the curvature of $c(s)$, which describes the bending of the curve. A zero curvature is equivalent to a zero rate of change of the normal, which corresponds to a straight curve, i.e., a line segment.

Among all manifolds of the same dimension, in order to avoid overfitting, we need to choose the one which is as flat as possible. The *second fundamental form*, L, describes how is the manifold $\mathcal{S}$ curved inside the space $\mathbb{R}^n$, as described in section 13.1.7.

We make this more explicitly in the particular case of the target space $\mathcal{M} = \mathbb{R}^n$. For any two vector fields $U = \sum_k U^k e_k$ and $V = \sum_k V^k e_k$ on $\mathbb{R}^n$, the derivation operator ∇, which defines the derivative of V with respect to U is

$$\nabla_U V = (D_U V^1, \ldots, D_U V^n),$$

where $D_U f$ denotes the directional derivative of f with respect to U. Now, if consider U and V tangent vector fields to the manifold $\mathcal{S}$, then $\nabla_U V$ can be decomposed orthogonally as

$$\nabla_U V = (\nabla_U V)^{\|} + (\nabla_U V)^{\perp},$$

where $(\nabla_U V)^{\|}$ is the projection of $\nabla_U V$ on the tangent space of $\mathcal{S}$, while $(\nabla_U V)^{\perp}$ denotes the orthogonal component of $\nabla_U V$. The normal component

$$L(U, V) = (\nabla_U V)^{\perp}$$

denotes the second fundamental form of $\mathcal{S}$ with respect to $\mathbb{R}^n$. The mapping L is symmetric and linear and can be written as

$$L(U, V) = \sum_{\alpha,\beta=1}^{r} L_{\alpha\beta} U^{\alpha} V^{\beta},$$

where the Latin superscripts describe the dependence with respect to the basic vector fields $\xi_\alpha = \dfrac{\partial \mathbf{y}(\theta)}{\partial \theta_\alpha}$, where $U = \sum_{\alpha=1}^{r} U^\alpha \xi_\alpha$. The coefficients of the second fundamental form

$$L_{\alpha\beta} = L(\xi_\alpha, \xi_\beta)$$

are vector-valued belonging to the normal space to $\mathcal{S}$, which has dimension $n - k$. If $L_{\alpha\beta} = \sum_k L_{\alpha\beta}^k$, then each component $L_{\alpha\beta}^k$ forms a symmetric square matrix of order r.

Vanishing L form A vanishing second fundamental form, $L = 0$, is equivalent to the vanishing of its coefficients, $L_{\alpha\beta} = 0$. In this case, $\mathcal{S}$ is called a *geodesic submanifold* of the Euclidean space $\mathbb{R}^n$. The equivalent characterization is that any locally length minimizing curve in $\mathcal{S}$ is a straight line segment in $\mathbb{R}^n$. The manifolds with this property are the affine subspaces of $\mathbb{R}^n$, see Exercise 13.8.6. In this case the projection of the target $\mathbf{z}$ onto $\mathcal{S}$ is unique.

The norm of the form L Since L is vector-valued, for regularization purposes we shall define and use a norm of L. For any vector U tangent to $\mathcal{S}$ at $\mathbf{y}$ we have that $L(U,U)$ is a vector in $\mathbb{R}^n$ and let $\|L(U,U)\|_{Eu}$ denote its Euclidean length. We define the norm of L by

$$\|L\| = \max\left\{\frac{\|L(U,U)\|_{Eu}}{\|U\|^2}; U \text{ tangent to } \mathcal{S}\right\}. \tag{13.6.12}$$

Here, $\|U\|$ denotes the length of U measured either in $\mathbb{R}^n$ or using the metric on $\mathcal{S}$. Using the scaling properties of L, see Proposition 13.1.8, part *(iii)*, this norm can be written equivalently as

$$\|L\| = \max_{\|U\|=1}\left\{\|L(U,U)\|_{Eu}; U \text{ tangent to } \mathcal{S}\right\}.$$

This norm is related to the eigenvalues of L as in the following. Due to symmetry, $L^k_{\alpha\beta} = L^k_{\beta\alpha}$, so that the matrix L^k has real eigenvalues. The expression

$$\max_{\|U\|=1}|L^k(U,U)| = |\lambda_k|,$$

provides the absolute value of the largest eigenvalue of L^k, see Appendix G. Therefore,

$$\|L\| = (\lambda_1^2 + \cdots + \lambda_n^2)^{1/2}.$$

From the geometric point of view each λ_k represents the curvature of the submanifold $\mathcal{S}$ into a certain normal direction, and hence, $\|L\|$ represents an extrinsic measure of the curvature of $\mathcal{S}$. By keeping $\|L\|$ small, all curvatures are kept small and by this we can control how much $\mathcal{S}$ bends inside $\mathbb{R}^n$.

The new cost function is the regularization of the square of distance using the previous norm

$$C(w,b;\mu) = \frac{1}{2}\|\mathbf{y}(w,b) - \mathbf{z}\|^2 + \mu\|L\|, \tag{13.6.13}$$

where the hyperparameter μ is a Lagrange multiplier. Thus, the regularization process is obtained as a trade-off between two effects: the minimization of the training error (the distance from target $\mathbf{z}$ to $\mathcal{S}$) and the maximization of the flatness of the manifold, see Fig. 13.14. The hyperparameter μ captures this trade-off effect: larger values of μ correspond to flatter manifolds, while smaller values of μ correspond to manifolds that pass closed by the target $\mathbf{z}$.

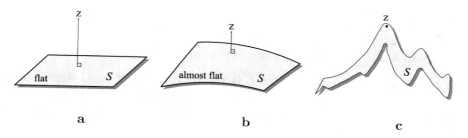

Figure 13.14: *Regularization with a manifold $\mathcal{S}$ of the same dimension but different degrees of flatness:* (a) *Using a plane leads to a large distance from the target* $\mathbf{z}$, *which underfits data;* (b) *using a trade-off between curvature and distance to* $\mathbf{z}$ *leads to a good fit;* (c) *using a largely curved manifold we can always force the target point* $\mathbf{z}$ *to belong to the manifold, case which corresponds to an overfit.*

Example 13.6.1 (Polynomial regression) This example presents the case of a polynomial regression in the context of the output manifold concept. The polynomial of degree r

$$\psi(x; \theta) = x^r + \theta_1 x^{r-1} + \theta_2 x^{r-2} + \cdots + \theta_{r-1} x + \theta_r \qquad (13.6.14)$$

is used to approximate in the least squares sense n points with coordinates given by

$$\mathcal{T} = \{(x_1, z_1), (x_2, z_2), \ldots, (x_n, z_n)\}.$$

Using the training set $\mathcal{T}$, the polynomial coefficients θ_i shall be tuned such that the sum of squares errors is minimized. We shall assume that the values x_i are distinct and that $r < n$. We shall consider the vectors $\mathbf{x}^T = (x_1, x_2, \ldots, x_n)$ and $\mathbf{z}^T = (z_1, z_2, \ldots, z_n)$ in $\mathbb{R}^n$. The input vector $\mathbf{x}$ and the parameter vector $\theta = (\theta_1, \ldots, \theta_r) \in \mathbb{R}^r$ are used to construct the manifold $\mathcal{S}$ parametrized by $\theta \to \mathbf{y}(\mathbf{x}; \theta) \in \mathbb{R}^n$ as

$$\mathbf{y}(\mathbf{x}; \theta) = (\psi(x_1; \theta), \ldots, \psi(x_n; \theta)).$$

Given the polynomial relation (13.6.14), the vector fields tangent to $\mathcal{S}$ take the form $\xi_j = \sum_{k=1}^n x_k^{r-j} e_k$. More specifically,

$$\xi_1 = \frac{\partial \mathbf{y}}{\partial \theta_1} = (x_1^{r-1}, x_2^{r-1}, \ldots, x_n^{r-1})$$

$$\xi_2 = \frac{\partial \mathbf{y}}{\partial \theta_2} = (x_1^{r-2}, x_2^{r-2}, \ldots, x_n^{r-2})$$

$$\cdots \quad \cdots \quad \cdots\cdots$$

$$\xi_{r-1} = \frac{\partial \mathbf{y}}{\partial \theta_{r-1}} = (x_1, x_2, \ldots, x_n)$$

$$\xi_r = \frac{\partial \mathbf{y}}{\partial \theta_r} = (1, 1, \ldots, 1).$$

Since x_i have distinct values, the following Vandermonde determinant does not vanish

$$\det \begin{pmatrix} 1 & 1 & \cdots & 1 \\ x_1 & x_2 & \cdots & x_r \\ \cdots & \cdots & \cdots & \cdots \\ x_1^{r-1} & x_2^{r-1} & \cdots & x_n^{r-1} \end{pmatrix} = \prod_{i<j}(x_j - x_i) \neq 0,$$

and therefore, the maximal rank condition is satisfied, namely, $\operatorname{rank}\left(\frac{\partial \mathbf{y}}{\partial \theta}\right) = r$. Thus, the vector fields $\{\xi_1, \ldots, \xi_r\}$ span the tangent space, $T_{\mathbf{y}}\mathcal{S}$, of $\mathcal{S}$ at each point $\mathbf{y} \in \mathcal{S}$. The intrinsic geometry of $\mathcal{S}$ is described by the metric tensor g having the components

$$g_{ij} = \langle \xi_i, \xi_j \rangle = \sum_{k=1}^{n} x_k^{r-i} x_k^{r-j} = \sum_{k=1}^{n} x_k^{2r-(i+j)}, \quad 1 \leq i, j \leq r.$$

It is important to note that the matrix coefficients g_{ij} (and hence the intrinsic geometry of $\mathcal{S}$) does not depend on parameters θ_k. (A similar case occurs for the Euclidean space $\mathbb{R}^n$ with the natural metric δ_{ij}). This corresponds to an *intrinsically flat* submanifold $\mathcal{S}$. Since $\partial g_{ij}/\partial \theta_k = 0$, it follows that the Christoffel symbols vanish, $\Gamma_{ij}^k = 0$. This implies that the Riemannian curvature[3] of the submanifold $\mathcal{S}$ is zero. In particular, its Levi-Civita connection vanishes on the base $\{\xi_i\}$

$$\overline{\nabla}_{\xi_i}\xi_j = \sum_{k} \Gamma_{ij}^k \theta_k = 0.$$

Consequently, the second fundamental form can be written only in terms of the Levi-Civita connection of $\mathbb{R}^n$ as

$$L(\xi_i, \xi_j) = \nabla_{\xi_i}\xi_j - \underbrace{\overline{\nabla}_{\xi_i}\xi_j}_{=0} = \nabla_{\xi_i}\xi_j = \sum_{k=1}^{n}(D_{\xi_i}\xi_j^k)e_k = 0,$$

because $D_{\xi_i}\xi_j^k = \langle \xi_i, \operatorname{grad}\xi_j^k \rangle = 0$, as the component $\xi_j^k = x_k^{r-j}$ is constant (depending on the fixed input entry x_k). Therefore, $L = 0$ and hence the submanifold $\mathcal{S}$ is also *extrinsically flat*. The cost function

$$C(\theta; \mu) = \frac{1}{2}\|\mathbf{y}(\mathbf{x}, \theta) - \mathbf{z}\|^2$$

---

[3]The Riemannian curvature of a manifold is described by the tensor

$$R_{ijk}^r = \partial_{\theta_i}\Gamma_{jk}^r - \partial_{\theta_j}\Gamma_{ik}^r + \Gamma_{ih}^r\Gamma_{jk}^h - \Gamma_{jh}^r\Gamma_{ik}^h,$$

with summation over the repeated indices.

in this case does not need the regularization term $\mu\|L\|$.

In fact, the output manifold $\mathcal{S}$ is an affine hyperplane in $\mathbb{R}^n$ of dimension r. The optimal solution is obtained by projecting the point $\mathbf{z}$ onto this hyperplane. This projection can be computed explicitly using the Moore-Penrose pseudoinverse, see section G.2 in the Appendix. We shall write the condition constraints $\psi(x_j; \theta) = z_j$, $1 \le j \le n$, as an overdetermined linear system

$$x_j^r + \theta_1 x_j^{r-1} + \cdots + \theta_{r-1} x_j + \theta_r = z_j, \qquad 1 \le j \le n.$$

This can be written in the matrix form as

$$A\theta = \beta,$$

where $\theta^T = (\theta_1, \ldots, \theta_r)$, $\beta^T = (z_1 - x_1^r, \ldots, z_n - x_n^r)$ and

$$A = \begin{pmatrix} x_1^{r-1} & \cdots & x_1 & 1 \\ \cdots & \cdots & \cdots & \cdots \\ x_n^{r-1} & \cdots & x_n & 1 \end{pmatrix}$$

The optimal parameter, θ^* can be obtained applying the pseudoinverse

$$\theta^* = A^+ \beta,$$

where A^+ is given by formula (G.2.6).

The projection of $\mathbf{z}$ onto the hyperplane $\mathcal{S}$ is given by $\mathbf{y}^* = \mathbf{y}(\mathbf{x}; \theta^*)$.

We shall discuss next the output manifold associated with only one neuron. The computation is complicated even in this simple case, while in the case of a general neural network it cannot always be performed explicitly.

Example 13.6.2 We shall consider the case of a single sigmoid neuron with the input $x \in \mathbb{R}$, output $y = \sigma(wx + b) \in \mathbb{R}$, and real parameters w, b, see Fig. 13.15. We take the activation function σ to be the logistic function. If the training set is given by

$$(\mathbf{x}, \mathbf{z}) = \{(x_1, z_1), (x_2, z_2), \ldots, (x_n, z_n)\},$$

then the manifold $\mathcal{S}$ associated with the previous neuron is defined by the map $\psi : \mathbb{R}^2 \to \mathbb{R}^n$

$$\psi(w, b) = \mathbf{y}(w, b) = (y_1, \ldots, y_n) = \left(\sigma(wx_1 + b), \ldots, \sigma(wx_n + b) \right).$$

This represents a 2-dimensional surface included in the space $\mathbb{R}^n$, endowed with the Riemannian metric induced by the Euclidean structure of $\mathbb{R}^n$. The basic tangent vector fields to $\mathcal{S}$ are given by the partial derivatives as

$$\xi_1 = \frac{\partial \mathbf{y}(w, b)}{\partial w} = \sigma'(w\mathbf{x} + b) \odot \mathbf{x} = \mathbf{y} \odot (1 - \mathbf{y}) \odot \mathbf{x}$$

$$\xi_2 = \frac{\partial \mathbf{y}(w, b)}{\partial b} = \sigma'(w\mathbf{x} + b) = \mathbf{y} \odot (1 - \mathbf{y}),$$

Output Manifolds

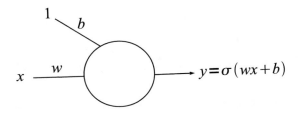

Figure 13.15: *The manifold $\mathcal{S}$ associated with a sigmoid neuron is 2-dimensional.*

where we used that $\sigma' = \sigma(1 - \sigma)$ and $\odot$ denotes the Hadamard product. The linear independence of $\{\xi_1, \xi_2\}$ is assured by the condition $\mathrm{rank}(\xi_1, \xi_2) = 2$, which is obviously satisfied. The intrinsic geometry of the output manifold $\mathcal{S}$ is described in terms of the following metric coefficients

$$
\begin{aligned}
g_{11} &= \langle \xi_1, \xi_1 \rangle = \langle \mathbf{y} \odot (1 - \mathbf{y}) \odot \mathbf{x}, \, \mathbf{y} \odot (1 - \mathbf{y}) \odot \mathbf{x} \rangle \\
&= \langle \mathbf{y} \odot \mathbf{y}, \mathbf{x} \odot \mathbf{x} \rangle - 2\langle \mathbf{y} \odot \mathbf{y} \odot \mathbf{y}, \mathbf{x} \odot \mathbf{x} \rangle + \langle \mathbf{y} \odot \mathbf{y} \odot \mathbf{x}, \mathbf{y} \odot \mathbf{y} \odot \mathbf{x} \rangle \\
g_{12} &= \langle \xi_1, \xi_2 \rangle = \langle \mathbf{y} \odot (1 - \mathbf{y}) \odot \mathbf{x}, \, \mathbf{y} \odot (1 - \mathbf{y}) \rangle \\
&= \langle \mathbf{y} \odot \mathbf{y}, \mathbf{x} \rangle - 2\langle \mathbf{y} \odot \mathbf{y}, \mathbf{y} \odot \mathbf{x} \rangle + \langle \mathbf{y} \odot \mathbf{y}, \mathbf{y} \odot \mathbf{y} \odot \mathbf{x} \rangle \\
g_{22} &= \langle \xi_2, \xi_2 \rangle = \langle \mathbf{y} \odot (1 - \mathbf{y}), \, \mathbf{y} \odot (1 - \mathbf{y}) \rangle = \langle \mathbf{y} - \mathbf{y} \odot \mathbf{y}, \, \mathbf{y} - \mathbf{y} \odot \mathbf{y} \rangle \\
&= \langle \mathbf{y}, \mathbf{y} \rangle - 2\langle \mathbf{y} \odot \mathbf{y}, \mathbf{y} \rangle + \langle \mathbf{y} \odot \mathbf{y}, \mathbf{y} \odot \mathbf{y} \rangle \\
g_{12} &= g_{21}.
\end{aligned}
$$

All the previous formulas can be expressed in terms of sums of powers, for instance,

$$
\langle \mathbf{y} \odot \mathbf{y}, \mathbf{x} \odot \mathbf{x} \rangle = \sum_k y_k^2 x_k^2, \quad \langle \mathbf{y} \odot \mathbf{y}, \mathbf{y} \rangle = \sum_k y_k^3, \quad \langle \mathbf{y} \odot \mathbf{y}, \mathbf{y} \odot \mathbf{y} \rangle = \sum_k y_k^4,
$$

where $y_k = \sigma(w x_k + b)$.

Differentiating the coefficients g_{ij}, one can potentially compute the Christoffel symbols of second type (13.1.2) and then obtain the Levi-Civita connection on $\mathcal{S}$ using a basis of tangent vectors as $\overline{\nabla}_{\xi_i} \xi_k = \sum_k \Gamma_{ij}^k(\mathbf{y}) \xi_k$. Since the computation is laborious, we shall proceed differently.

With notation $(\theta_1, \theta_2) = (w, b)$, the Levi-Civita connection on the target space $\mathbb{R}^n$ becomes $\nabla_{\xi_i} \xi_j = \frac{\partial^2 \psi}{\partial \theta_i \partial \theta_j}$. Then the Gauss decomposition (13.1.5) writes as

$$
\frac{\partial^2 \psi}{\partial \theta_i \partial \theta_j} = \left(\frac{\partial^2 \psi}{\partial \theta_i \partial \theta_j} \right)^{\|} + \left(\frac{\partial^2 \psi}{\partial \theta_i \partial \theta_j} \right)^{\perp}.
$$

The second fundamental form is given by the normal part

$$L_{ij} = \left(\frac{\partial^2 \psi}{\partial \theta_i \partial \theta_j}\right)^{\perp} = \frac{\partial^2 \psi}{\partial \theta_i \partial \theta_j} - \left(\frac{\partial^2 \psi}{\partial \theta_i \partial \theta_j}\right)^{\|}. \qquad (13.6.15)$$

Both terms in the right side are computable. We shall start with $\frac{\partial^2 \psi}{\partial \theta_i \partial \theta_j}$. Using the relation $\sigma''(x) = \sigma(x)(1 - \sigma(x))(2 - \sigma(x))$, we obtain

$$
\begin{aligned}
\frac{\partial^2 \psi}{\partial \theta_1 \partial \theta_1} &= \frac{\partial^2 \psi}{\partial^2 w} = \sigma''(w\mathbf{x} + b) \odot x \odot x \\
&= \mathbf{y} \odot (1 - \mathbf{y}) \odot (2 - \mathbf{y}) \odot x \odot x; \\
\frac{\partial^2 \psi}{\partial \theta_2 \partial \theta_2} &= \frac{\partial^2 \psi}{\partial^2 b} = \sigma''(w\mathbf{x} + b) \\
&= \mathbf{y} \odot (1 - \mathbf{y}) \odot (2 - \mathbf{y}); \\
\frac{\partial^2 \psi}{\partial \theta_1 \partial \theta_2} &= \frac{\partial^2 \psi}{\partial w \partial b} = \sigma''(w\mathbf{x} + b) \odot x \\
&= \mathbf{y} \odot (1 - \mathbf{y}) \odot (2 - \mathbf{y}) \odot x; \\
\frac{\partial^2 \psi}{\partial \theta_2 \partial \theta_1} &= \frac{\partial^2 \psi}{\partial \theta_1 \partial \theta_2}.
\end{aligned}
$$

The tangential component is a linear combination of basic vector fields

$$\left(\frac{\partial^2 \psi}{\partial \theta_i \partial \theta_j}\right)^{\|} = \alpha_{ij}^1 \xi_1 + \alpha_{ij}^2 \xi_2,$$

where the coefficients α_{ij}^k can be found explicitly as the solution of the linear system

$$
\begin{aligned}
g_{11}\alpha_{ij}^1 + g_{12}\alpha_{ij}^2 &= \langle \frac{\partial^2 \psi}{\partial \theta_i \partial \theta_j}, \xi_1 \rangle \\
g_{12}\alpha_{ij}^1 + g_{22}\alpha_{ij}^2 &= \langle \frac{\partial^2 \psi}{\partial \theta_i \partial \theta_j}, \xi_2 \rangle.
\end{aligned}
$$

This is

$$\alpha_{ij}^1 = g^{11}\langle \frac{\partial^2 \psi}{\partial \theta_i \partial \theta_j}, \xi_1 \rangle + g^{12}\langle \frac{\partial^2 \psi}{\partial \theta_i \partial \theta_j}, \xi_2 \rangle$$

$$\alpha_{ij}^2 = g^{21}\langle \frac{\partial^2 \psi}{\partial \theta_i \partial \theta_j}, \xi_1 \rangle + g^{22}\langle \frac{\partial^2 \psi}{\partial \theta_i \partial \theta_j}, \xi_2 \rangle$$

where (g^{ij}) is the inverse matrix $(g_{ij})^{-1}$. The scalar product terms can be computed easily. For instance:

$$\langle \frac{\partial^2 \psi}{\partial \theta_1 \partial \theta_1}, \xi_1 \rangle = \langle \mathbf{y} \odot (1-\mathbf{y}) \odot (2-\mathbf{y}) \odot \mathbf{x} \odot \mathbf{x}, \mathbf{y} \odot (1-\mathbf{y}) \odot \mathbf{x} \rangle$$

$$= \sum_{k=1}^{n} y_k^2 (1-y_k)^2 (2-y_k)^2 x_k^3.$$

The coefficients L_{ij} can be now computed using (13.6.15). For instance,

$$L_{11} = \frac{\partial^2 \psi}{\partial w^2} - \left(\frac{\partial^2 \psi}{\partial w^2} \right)^{\|}$$

$$= \mathbf{y} \odot (1-\mathbf{y}) \odot (2-\mathbf{y}) \odot \mathbf{x} \odot \mathbf{x} - \alpha_{11}^1 \xi_1 - \alpha_{11}^2 \xi_2$$

$$= \mathbf{y} \odot (1-\mathbf{y}) \odot [(2-\mathbf{y}) \odot \mathbf{x} \odot \mathbf{x} - \alpha_{11}^1 \mathbf{x} - \alpha_{11}^2].$$

Consider the norm

$$\|L\| = \sup_{\|U\|=1} \|L(U, U)\|,$$

which is measures the flatness of $\mathcal{S}$. The associated variational problem in this case is to minimize the regularized cost function

$$C(w, b; \mu) = \frac{1}{2} \|\mathbf{y}(w, b) - \mathbf{z}\|^2 + \mu\|L\| = \frac{1}{2} \sum_{k=1}^{n} \left(\sigma(wx_k + b) - z_k \right)^2 + \mu\|L\|.$$

13.6.4 Model averaging

A reliable technique for reducing the test error is to average the outputs of N different neural networks with the same input, which are trained separately. Each particular network makes an error ϵ_i. We shall assume the errors are independent random variables with zero mean and variance v, having the same distribution. In the virtue of the Central Limit Theorem (Theorem D.6.4 of Appendix) the average error, $\epsilon_{ave} = \frac{1}{N} \sum_{i=1}^{N} \epsilon_i$, tends to be distributed normally, with zero mean and variance v/N, for N large enough. This implies that the method of averaging performs better than each of its members.

This idea can be also formalized in the context of output manifolds. The main idea is to project the target $\mathbf{z}$ onto several output manifolds associated with some neural nets, and then consider the average of the projections as an approximation of the target $\mathbf{z}$. For the sake of simplicity, we shall discuss this technique only for the case of two networks.

Consider two feedforward neural networks of the same depth, having the same input $\mathbf{x} \in \mathbb{R}^k$, and learning the same target $\mathbf{z} \in \mathbb{R}^n$. Let $\mathbf{y} = \mathbf{y}(w, b)$ and $\widetilde{\mathbf{y}} = \widetilde{\mathbf{y}}(\widetilde{w}, \widetilde{b})$ be the outcomes of the two nets, see Fig. 13.16. They are

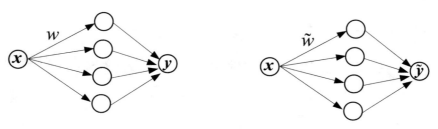

Figure 13.16: *Two neural nets with the same input,* $\mathbf{x}$, *and outputs* $\mathbf{y}(w, b)$, $\widetilde{\mathbf{y}}(\widetilde{w}, \widetilde{b})$, *learning the same target* $\mathbf{z}$.

regarded as points belonging to an output manifold each, $\mathbf{y} \in \mathcal{S}$ and $\widetilde{\mathbf{y}} \in \widetilde{\mathcal{S}}$. After training, the weights and biasses are set equal to

$$(w, b) = \arg\min \|\mathbf{z} - \mathbf{y}\|^2, \qquad (\widetilde{w}, \widetilde{b}) = \arg\min \|\mathbf{z} - \widetilde{\mathbf{y}}\|^2.$$

This means that $\mathbf{y}$ and $\widetilde{\mathbf{y}}$ are the orthogonal projections of the target $\mathbf{z}$ onto the manifolds $\mathcal{S}$ and $\widetilde{\mathcal{S}}$, respectively. The outputs average, $\frac{1}{2}(\mathbf{y} + \widetilde{\mathbf{y}})$, is an approximation of the target $\mathbf{z}$, hopefully better than each $\mathbf{y}$ and $\widetilde{\mathbf{y}}$.

However, we can do better than this by employing a convex combination. There are points on the line segment $\{\lambda \mathbf{y} + (1 - \lambda)\widetilde{\mathbf{y}}; \lambda \in [0, 1]\}$ that are closer to $\mathbf{z}$ than both $\mathbf{y}$ and $\widetilde{\mathbf{y}}$, see Fig. 13.17. The closest point is the projection of $\mathbf{z}$ onto this line. This corresponds to a better approximator of the target, which can be obtained as the output of only one network. This is the *model combination* of the previous two neural networks into only one net with the following properties, see Fig. 13.18:

(*i*) the input is $\mathbf{x}$;

(*ii*) its depth is one unit more than the given nets;

(*iii*) its ℓth layer is the union of the ℓth layers of the given nets;

(*iv*) the last layer contains only one linear neuron;

(*v*) its parameters are given by $\{w, b, \widetilde{w}, \widetilde{b}, \lambda, 1 - \lambda\}$;

(*vi*) its outcome is $\lambda \mathbf{y} + (1 - \lambda)\widetilde{\mathbf{y}}$.

The model combination can be applied to any number of neural nets. The resulting net has the output given by a convex combination of the individual network outcomes, with the coefficients chosen such that the distance between the target $\mathbf{z}$ and the affine space determined by the outcomes is minimum.

13.6.5 Dropout

Dropout is a powerful method to reduce overfitting, which works well for a large family of neural nets. The main idea is to drop or remove temporarily

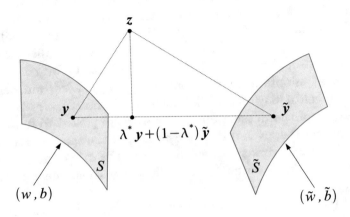

Figure 13.17: *The orthogonal projection of* $\mathbf{z}$ *on the line segment* $\mathbf{y}\widetilde{\mathbf{y}}$ *is a better approximator than both* $\mathbf{y}$ *and* $\widetilde{\mathbf{y}}$. *This is given by* $\lambda^*\mathbf{y} + (1-\lambda^*)\widetilde{\mathbf{y}}$, *where* λ^* *is obtained as* $\lambda^* = \arg\min \|\mathbf{z} - \lambda\mathbf{y} - (1-\lambda)\widetilde{\mathbf{y}}\|$.

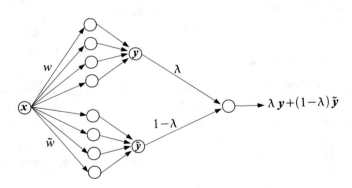

Figure 13.18: *The model combination of two nets is a net that produces a better learning than both of its parts.*

neurons (hidden, input, but not output) from the network. The choice of which neurons are removed is random. The success of this method consists in breaking the coadaptations[4] formed among neurons by the standard back-propagation algorithm. Dropout trains each neuron to be able to act without the help of other neurons. Consequently, the resulting network will generalize well to new unseen data, and hence, it will produce a smaller test error.

Sometimes, dropout is explained more plastically by comparison with a company that adopts a training policy by which a certain percent of its workers are given a day off, while the rest of the workers are trained to perform the job of the missing ones. The workers picked to have the day off are randomly selected, even if the percent is kept the same. At the end of this training period each worker knows the job of the others and, consequently, the workers are able to perform more efficiently when the company faces a new, unseen, task.

Dropout technique also resembles with the L^2-regularization in the following sense. Since the dropout idea is to train neurons to act as independently as possible, this implies an indifference of the network to any specific feature. Since features are stored into weights, it follows that the weights system has to be shrank enough, fact similar with the case of a small norm.

Dropping random neurons from a net, including their incoming and outgoing connections, is equivalent to sampling a subnetwork, which is associated with an output submanifold. Training this subnetwork is equivalent to finding the projection of the target $\mathbf{z}$ onto the associated output submanifold. Applying this process for several subnetworks produces estimations of the target by projections onto the associated submanifolds. Their average is taken as an estimator for the target.

Dropping a hidden neuron Consider a neuron in the ℓth layer of a feed-forward neural network, with $\ell \notin \{0, L\}$, i.e., the neuron belongs to a hidden layer. By dropping this neuron from the network, we understand removing the neuron together with all its weights (to and from the neuron), including also its bias. This will lead to a new neural network with the same input as the former one.

The dimension of the parameter space of the new network is with $d^{(\ell-1)} + d^{(\ell+1)} + 1$ less than the dimension of the parameter space of the former net.

---

[4]This can be easily understood, for instance, if you try to recite the alphabet in the reverse order. The brain builds coadaptations when learning the alphabet in chronological order from A to Z. The difficulty faced when trying to recite the alphabet in the reverse order shows the existence of certain coadaptations formed among neurons during the learning process.

This follows from the fact that we have removed $d^{(\ell-1)}$ incoming weights, $d^{(\ell+1)}$ outgoing weights, and 1 bias. As usual, $d^{(\ell)}$ denotes the number of neurons in the ℓth layer. Therefore, by dropping a neuron the network's output depends on less parameters, which decreases the network capacity and reduces any eventual overfit.

After training, the network output becomes the projection of the target $\mathbf{z}$ on an output manifold of a smaller dimension. It is not clear whether this projection is closer to $\mathbf{z}$ than the former approximation applied before the neuron dropout. It is also not obvious which neuron dropout produces the best approximation.

Dropping several neurons The dropout technique removes randomly a certain percentage of neurons from each layer and then takes the average of the resulting outputs, see Fig. 13.19. However, dropping too many neurons will decrease the dimension of the parameter space too much and, consequently, it will lead to an underfit.

When a certain number of neurons are dropped, the resulting associated output manifold is a submanifold of the output manifold associated with the initial network. The codimension of this submanifold[5] is given by

$$k = \sum_{\ell=1}^{L-1} n^{(\ell)}(d^{(\ell-1)} + d^{(\ell+1)} + 1),$$

where $n^{(\ell)}$ is the number of neurons dropped from the ℓth layer. If the same percent, q, is dropped from each layer, then $n^{(\ell)} = qd^{(\ell)}/100$.

After each dropout, the trained network output represents the projection of the target $\mathbf{z}$ onto the associated output submanifold. Each of these projections is approximation of the target $\mathbf{z}$. By an approach similar to the Monte Carlo method, the average of all these projections represents an approximation of $\mathbf{z}$, which diminishes the overfit and is less prone to bias than any of the $\mathbf{y}_j$.

Example 13.6.3 We shall exemplify the method using a neural network with one hidden layer, one-dimensional input and output, see Fig. 13.19. We consider N neurons in the hidden layer, and drop uniformly one neuron at a time, obtaining the following outputs

$$\mathbf{y}_j = \mathbf{y} - \lambda_j \sigma(w_j x + b_j), \quad j = 1, \ldots, N,$$

---

[5]The codimension of a submanifold $\mathcal{S}$ of a manifold $\mathcal{M}$ is the difference of their dimensions, $k = \dim \mathcal{M} - \dim \mathcal{S}$.

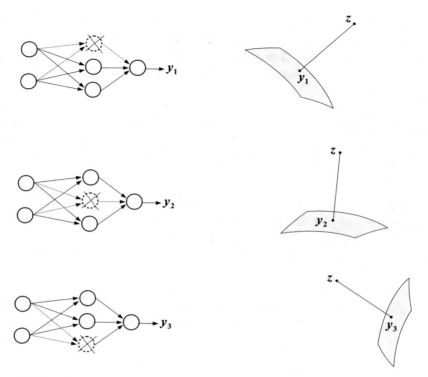

Figure 13.19: *When one neuron is dropped at a time, the network output produces projections of the target* $\mathbf{z}$ *onto the associated output manifolds. The average of projections,* $\frac{1}{3}(\mathbf{y}_1 + \mathbf{y}_2 + \mathbf{y}_3)$*, is supposed to be a better approximation of* $\mathbf{z}$ *than any of the* $\mathbf{y}_j$*.*

where $\mathbf{y} = \displaystyle\sum_{j=1}^{N} \lambda_j \sigma(w_j x + b_j)$ denotes the initial network output. Since each $\mathbf{y}_j$ is taken with probability $q_i = q = \frac{1}{N}$, then the expected network output is given by the average of the outputs as

$$\sum_{j=1}^{N} q_j \mathbf{y}_j = \frac{1}{N} \sum_{j=1}^{N} \mathbf{y}_j = \left(1 - \frac{1}{N}\right)\mathbf{y} = (1-q)\mathbf{y},$$

which is proportional to the output of the initial network, $\mathbf{y}$.

Multiplicative noise Dropout can be also seen as adding multiplicative noise to the network. Since each neuron is retained with the probability p, this means that the neuron's output remains unchanged with probability p and vanishes with probability $q = 1-p$ (i.e., it is dropped with probability q). This is equivalent to a multiplication with a Bernoulli random variable. If the ith output of the ℓth layer before dropout is $x_i^{(\ell)}$, then after dropout it becomes $\tilde{x}_i^{(\ell)} = R_i^{(\ell)} x_i^{(\ell)}$, with $R_i^{(\ell)} \sim \text{Bernoulli}(p)$, where the definition of the Bernoulli random variable can be found in the Appendix, section D.2. Therefore, the feedforward operation described by the master equation (6.2.24)

$$x_j^{(\ell)} = \phi\Big(\sum_{i=1}^{d^{(\ell-1)}} w_{ij}^{(\ell)} x_i^{(\ell-1)} - b_j^{(\ell)} \Big), \qquad 1 \le j \le d^{(\ell)}$$

in the case of dropout becomes

$$x_j^{(\ell)} = \phi\Big(\sum_{i=1}^{d^{(\ell-1)}} w_{ij}^{(\ell)} \tilde{x}_i^{(\ell-1)} - b_j^{(\ell)} \Big), \qquad 1 \le j \le d^{(\ell)},$$

where $\tilde{x}_i^{(\ell-1)} = R_i^{(\ell-1)} x_i^{(\ell-1)}$. In the equivalent matrix form, the equation (6.2.29)

$$X^{(\ell)} = \phi\Big(W^{(\ell)^T} X^{(\ell-1)} - B^{(\ell)} \Big)$$

becomes

$$X^{(\ell)} = \phi\Big(W^{(\ell)^T} R^{(\ell-1)} \odot X^{(\ell-1)} - B^{(\ell)} \Big)$$

where $R^{(\ell-1)} = (R_i^{(\ell-1)})$ is a vector of independent Bernoulli random variables and $\odot$ denotes the Hadamard product of vectors.

Remark 13.6.4 Empirical evidence has shown that the optimal retention rate for hidden layers is usually $p = 0.5$, while for the input layer is about $p = 0.8$.

The next section establishes a relation between dropout and L_2-regularization.

Linear regression with dropout This section deals with the application of dropout in the case of the classical problem of linear regression. Consider the input vector $X \in \mathbb{R}^n$ and the target $\mathbf{z} \in \mathbb{R}$. We need to learn the weights vector $\mathbf{w} \in \mathbb{R}^n$ such that $\|\mathbf{z} - X\mathbf{w}\|^2$ is minimized. Applying dropout, the new objective function becomes

$$f(\mathbf{w}) = \mathbb{E}[\|\mathbf{z} - R \odot X\mathbf{w}\|^2],$$

where $R^T = (R_1, \ldots, R_n)$ is a vector of independent Bernoulli random variables, $R_i \sim \text{Bernoulli}(p)$. Using that $\|a - b\|^2 = \|a\|^2 - 2a^T b + \|b\|^2$ and the linearity of the expectation operator, the objective function becomes

$$
\begin{aligned}
f(\mathbf{w}) &= \|\mathbf{z}\|^2 - 2\mathbf{z}^T \mathbb{E}[R] \odot X\mathbf{w} + \mathbb{E}[\|R \odot X\mathbf{w}\|^2] \\
&= \|\mathbf{z}\|^2 - 2p\mathbf{z}^T X\mathbf{w} + p^2\|X\mathbf{w}\|^2 + \mathbb{E}[\|R \odot X\mathbf{w}\|^2] - p^2\|X\mathbf{w}\|^2 \\
&= \|\mathbf{z} - pX\mathbf{w}\|^2 + \mathbb{E}[\|R \odot X\mathbf{w}\|^2] - p^2\|X\mathbf{w}\|^2 \\
&= \|\mathbf{z} - pX\mathbf{w}\|^2 + \mathbb{E}[\mathbf{w}^T (R \odot X)^T (R \odot X)\mathbf{w}] - p^2\|X\mathbf{w}\|^2 \\
&= \|\mathbf{z} - pX\mathbf{w}\|^2 + \mathbb{E}[R^2]\mathbf{w}^T X^T X\mathbf{w} - p^2\|X\mathbf{w}\|^2 \\
&= \|\mathbf{z} - pX\mathbf{w}\|^2 + p\mathbf{w}^T X^T X\mathbf{w} - p^2\|X\mathbf{w}\|^2 \\
&= \|\mathbf{z} - pX\mathbf{w}\|^2 + p(1 - p)\|X\mathbf{w}\|^2,
\end{aligned}
$$

where we have added and subtracted the term $p^2\|X\mathbf{w}\|^2$ to form a square of the norm and used that the second moment of a Bernoulli variable is p.

Absorbing the factor p into the weight $\mathbf{w}$, the objective function becomes

$$f(w) = \|\mathbf{z} - Xw\|^2 + \lambda\|Xw\|^2,$$

which is an L^2-regularization problem with the Lagrange multiplier $\lambda = \frac{1-p}{p}$ and $w = p\mathbf{w}$. When p tends to 1, all neurons are retained and λ gets small. The constant λ represents the ratio between the non-retained and the retained neurons during the dropout process. Hence, a linear regression with dropout is equivalent to a L^2-regularization problem.

Gaussian noise The idea of introducing noise into a neural network in order to reduce overfitting works also for other types of noise. Srivastava et al. [116] describes a method of adding Gaussian noise to each neuron proportional to its activation. This means that the output $X^{(\ell)}$ of a hidden neuron is perturbed by a Gaussian noise proportional with the activation, i.e., to $X^{(\ell)} + X^{(\ell)}G$, with $G \sim \mathcal{N}(0, 1)$. The perturbation can be written equivalently in a multiplicative way as $X^{(\ell)}G'$, with $G' \sim \mathcal{N}(1, 1)$. It is worth noting that this new type of dropout works at least as well as the regular dropout involving Bernoulli random variables.

In conclusion, removing neurons from each hidden layer and input layer of a neural network reduces substantially the dimension of the associated output manifold, leading to a decrease in the network capacity, and hence to a reduction of any overfitting effect. The reader can find more details in the paper [116].

Regularization by inserting noise One way to prevent neural networks from overfitting training data is to insert noise in the network during training and then average over the noise during testing. The noise can be, for instance, multiplicative or additive. In the case of a multiplicative noise, we multiply the outputs of each layer by a random variable R (Bernoulli, uniform, Gaussian, etc.). The network output, which depends now on both the input, X, and noise, R, is given by $Y = f_w(X, R)$, and becomes a random variable. At training time we find the optimal values of w for outputs of this noisy type. The optimal value depends on R as $w^* = w^*(R)$. To remove the randomness we need to average over the random variable R at the testing time as

$$y = f(x) = \mathbb{E}_R[f_{w(R)}(x, R)] = \int f_{w(r)}(x, r)p(r)\, dr, \qquad (13.6.16)$$

where $p(r)$ is the probability density of the random variable R.

Formula (13.6.16) has more theoretical value than practical, as the integral in the right side is difficult to compute exactly. In practice, we train the network for N instances of the random variable R, given by $r_1, \ldots, r_N$, and obtain the optimal values of the weights as $w_i^*, \ldots, w_N^*$. This means

$$w_i^* = \arg\min_w \|\mathbf{z} - f_w(x, r_i)\|,$$

where $\mathbf{z} = \mathbf{z}(x)$ is the target function that needs to be learned by the network. At the testing time we consider the average of all N outputs, evaluating the expectation (13.6.16) by the following Monte Carlo formula:

$$f(x) = \frac{1}{N} \sum_{i=1}^{N} f_{w_i^*}(x, r_i). \qquad (13.6.17)$$

It is worth noting that the classical dropout technique is obtained as a particular case of (13.6.17) by considering R to be a Bernoulli random variable. This means that R takes the value 1 with probability p and the value 0 with probability $1 - p$. A neuron activation multiplied by $R = 0$ is equivalent to a dropped neuron, while an activation multiplied by the value $R = 1$ is a retained neuron. Therefore, $100(1 - p)$ percent of neurons in each layer are randomly dropped, while $100p$ percent of neurons are retained.

Since multiplying a neuron activation by 0 is the same as assuming all the weights (ingoing to and outgoing from the neuron) vanishing, then (13.6.17) represents an average of outcomes of N trained subnetworks.

13.7 Summary

This chapter discusses neural networks from a geometric point of view. An output manifold is associated with each network. The local coordinates on the manifold are the weights and biasses of the network. The output manifold concept is useful for understanding several aspects such as optimal weights, learning process, overfitting and underfitting, as well as regularization techniques.

The optimal weights and biasses of a network correspond to the coordinates of the orthonormal projection of the target onto the output manifold. Each learning algorithm changes coordinates on the manifold and corresponds to a curve on it. Endowing the manifold with a Riemannian metric enables the computation of curve lengths and also defining the geodesic, which is the shortest curve between two points. The geodesic between the initial point and the projection point of the target onto the manifold corresponds to the most efficient learning algorithm.

A target point which is too distant from the output manifold indicates an underfit, while a target point which is too close, or on the manifold, represents an overfit.

Different types of regularization methods can be treated in terms of output manifolds. Going for a smaller dimension of the output manifold means to decrease the number of weights and consequently means to have less neural units into the network, which leads to a decrease in the network capacity.

Choosing the flattest output manifold produces the least overfitting to training data. Model averaging chooses by means of minimizing distances a model with a better fit than any of its component networks. Dropout technique falls into this class and can be also seen as a multiplicative noise regularization method. A relation between dropout and L_2-regularization is discusses.

13.8 Exercises

Exercise 13.8.1 A feedforward neural network of type 784-200-100-50-10 is used to classify the MNIST data. Find the dimension of the associated output manifold. (784 is the input size and 10 represents the number of digit classes).

Exercise 13.8.2 Consider a neural network with input and output sizes given by $d^{(0)}$ and $d^{(L)}$, respectively. The number of hidden neurons is denoted by N. We assume there is an equal number of neurons in each hidden layer. Show that the number of hidden layers for which the output manifold has a maximum dimension is

$$k = \frac{2N}{d^{(0)} + d^{(L)} + N}.$$

Exercise 13.8.3 A one-hidden layer feedforward neural net, 784-N-10, is used to classify the MNIST data. Find the range of the number of hidden neurons, N, for which the network overfits the training data.

Exercise 13.8.4 A two-hidden layer feedforward neural net, 784-h-h-10, is used to classify the MNIST data. Find the range of the number h, for which the network overfits the training data.

Exercise 13.8.5 Let $v, u \in T_y \mathcal{S}$ be two tangent vectors. Show that v and u are orthogonal in $\mathbb{R}^n$ if and only if $g(u, v) = \sum_{i,j} u_i v_j g_{ij} = 0$.

Exercise 13.8.6 A subset $\mathcal{A}$ of $\mathbb{R}^n$ is called an affine subspace if for any two points $A, B \in \mathcal{A}$ we have $\lambda A + (1 - \lambda)B \in \mathcal{A}$, $\forall \lambda \in \mathbb{R}$. Let L be the second fundamental form of $\mathcal{S}$ with respect to $\mathbb{R}^n$. Show that the following are equivalent:

(a) $L = 0$;

(b) Any geodesic in $\mathcal{S}$ is a straight line in $\mathbb{R}^n$;

(c) $\mathcal{S}$ is an affine subspace of $\mathbb{R}^n$.

Exercise 13.8.7 Let $\mathcal{S}$ be a submanifold of the manifold $\mathcal{M}$. Show that the following are equivalent:

(a) The second fundamental form of $\mathcal{S}$ with respect to $\mathcal{M}$ is zero, $L = 0$;

(b) Any curve, which is a geodesic in $\mathcal{S}$, is also a geodesic in $\mathcal{M}$.

Exercise 13.8.8 Let $c(s)$ be a curve on the output manifold $\mathcal{S}$, $s \in [a, b]$. Its length and energy are defined, respectively, by

$$L(c) = \int_a^b \|\dot{c}(s)\| \, ds, \qquad \mathcal{E}(c) = \frac{1}{2} \int_a^b \|\dot{c}(s)\|^2 \, ds,$$

where $\|\dot{c}\|$ represents the length of the velocity along the curve in the metric structure of $\mathcal{S}$.

(a) Show that the length and energy of a curve are invariant under curve parametrizations. This is, if $\phi : [c, d] \to [a, b]$ is a strictly increasing function, then the curve $\gamma(t) = c(\phi(t))$ and $c(s)$ have the same length and energy.

(b) Show that $L(c)^2 \le 2(b - a)\mathcal{E}(c)$. When is the identity reached?

(c) Let $c_u(s)$, $|u| \le \epsilon$, be a smooth variation of $c(s)$, with $c_0(s) = c(s)$. It can be shown that both variational equations

$$\frac{d}{du} L(c_u)_{|u=0} = 0, \qquad \frac{d}{du} \mathcal{E}(c_u)_{|u=0} = 0$$

can be written as

$$\ddot{c}^k(s) + \sum_{i,j} \Gamma_{ij}^k(c(s))\dot{c}^i(s)\dot{c}^j(s) = 0, \quad 1 \le k \le n,$$

where $c(s) = (c^1(s), \ldots, c^n(s))$. Furthermore, the previous equation represents the zero acceleration equation along the submanifold $\mathcal{S}$ and can be also written as $\nabla_{\dot{c}}\dot{c} = 0$. What is the significance of these facts?

Exercise 13.8.9 (a) Find the embedding curvature of the 2-dimensional unit sphere, $\mathbb{S}^2$, in the 3-dimensional Euclidean space, $\mathbb{R}^3$.
(b) Use part (a) to find the norm $\|L\|$. Experiment with different sphere parametrizations. What do you notice?

Exercise 13.8.10 Consider the model combination of two sigmoid neurons. Write the output of the combination and specify the dimension of the associated output manifold.

Exercise 13.8.11 List a few effects of dropping neurons from a network on the associated output manifold.

Exercise 13.8.12 For any two vector fields in $\mathbb{R}^n$

$$U = \sum_k U^k e_k, \qquad V = \sum_k V^k e_k,$$

define $\nabla_U V = \sum_k U(V^k)e_k$. Let f be a smooth function on $\mathbb{R}^n$. Show the following relations:
(a) $\nabla_{fU} V = f\nabla_U V$;
(b) $\nabla_U fV = U(f)V + f\nabla_U V$;
(c) $\nabla_U V = \nabla_V U$;
(d) $U\langle V, W \rangle = \langle \nabla_U V, W \rangle + \langle V, \nabla_U W \rangle$, where W is any other vector field.

Chapter 14

Neuromanifolds

In this chapter we shall approach the study of neural networks from the Information Geometry perspective. This applies both techniques of Differential Geometry and Probability Theory to neural networks.

The difference from the theory introduced in Chapter 13 is that here the network's input and target are probability densities of random variables and the neural network output contains some noisy perturbation. This way, the family of joint probability densities of the input and output, $p(x, y; \theta)$, becomes a statistical manifold, which is parametrized by θ; thus, the weights and biasses play the role of a coordinate system for the associated statistical manifold. The intrinsic distance between two neural networks is measured in this space using the Fisher information metric. Roughly speaking, the Fisher metric represents the amount of information about network's own weights and biasses that is contained in the training distribution. The associated statistical manifold endowed with the Fisher metric becomes a Riemannian manifold, called a neuromanifold.

In this chapter we compute explicitly the Fisher metric for several simple types of networks and present the natural gradient learning algorithm. The understanding of the Fisher metric leads to the characterization of shortest curves in the parameter space – the geodesics. This is important since each motion in the neural manifold corresponds to a learning process. The natural gradient is defined as the gradient computed with respect to the Riemannian metric induced by the Fisher information.

The natural gradient descent method is presented, as a better alternative to the usual gradient descent, which converges faster to the minimum

© Springer Nature Switzerland AG 2020
O. Calin, *Deep Learning Architectures*, Springer Series in the Data Sciences,
https://doi.org/10.1007/978-3-030-36721-3_14

of the cost function. The relation between the distance, curve length and energy in the parameter space, and the Kullback-Leibler divergence on the neuromanifold is also made.

14.1   Statistical Manifolds

First we shall recall some terminology from Information Geometry regarding statistical manifolds from the neural networks point of view. Consider the input to a neural network given by the random variable X. The output variable is $Y = f_\theta(X)$, where f_θ is the input-output mapping and $\theta = (\mathbf{w}, \mathbf{b})$ are the network parameters. The input and output distributions are denoted by $p_X(x)$ and $p_Y(y; \theta)$, respectively. The joint input-output distribution is indicated by $p(x, y; \theta)$. Here, θ has been included to suggest the density dependence on network parameters.

The target is given by the random variable Z. The joint distribution of (X, Z) is the training distribution, $p(x, z)$. Since the target is not replicated perfectly by the network output, we have $Z = Y + \epsilon(\theta)$, where $\epsilon(\theta)$ is an error term that depends on the network parameters. In the case when the mean square cost function is used as a loss function, we have to minimize the second moment of the error

$$C(\theta) = \frac{1}{2}\mathbb{E}[(Z - Y)^2] = \frac{1}{2}\mathbb{E}[\epsilon(\theta)^2].$$

Noisy neurons Information geometry is applied to neural networks in the context of noisy neurons. This means that in order to offset as much as possible of the error $\epsilon(\theta)$ we need to insert some noise in the network. The noise idea is not new. We have seen it in section 12.8.2 and in section 13.6.5, where adding multiplicative noise to the network leads to the dropout technique. Here, we consider an additive noise term, n, so the output of the network is given now by

$$Y = f_\theta(X) + n. \tag{14.1.1}$$

It is worth noting the role of the inserted noise played in regularization.

Assuming the output one-dimensional, we shall consider two cases of noise terms:

1. One possibility is to assume the noise to be a standard normal random variable, $n \sim \mathcal{N}(0, 1)$. Then, the conditional probability of the output, Y, given the input, X, is given in this case by

$$p(y|x; \theta) = \frac{1}{\sqrt{2\pi}}e^{-\frac{1}{2}(y - f_\theta(x))^2}. \tag{14.1.2}$$

This is based on the fact that $\mathbb{E}[Y|X = x] = f_\theta(x)$, and $Var(Y|X = x) = 1$.

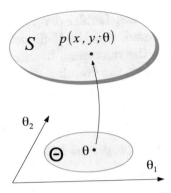

Figure 14.1: *A statistical manifold* $\mathcal{S} = \{p(x, y; \theta); \theta \in \Theta\}$.

2. Another possibility is to consider a uniform random noise between -1 and 1, $n \sim \text{Unif}[-1, 1]$. In this case the conditional probability is given by

$$p(y|x; \theta) = \begin{cases} \frac{1}{2}, & \text{if } f_\theta(x) - 1 \leq y \leq f_\theta(x) + 1 \\ 0, & \text{otherwise.} \end{cases}$$

The joint distribution of (X, Y) can be found now using the conditional probability formula

$$p(x, y; \theta) = p(x)p(y|x; \theta). \tag{14.1.3}$$

The goal is to tune the parameter θ such that $p(x, y; \theta)$ matches as much as possible the true distribution $p(x, z)$. This can be achieved geometrically as in the following.

Statistical manifold The family of density functions, $\{\theta \to p(x, y; \theta); \theta \in \Theta\}$, parametrized over θ, can be considered as a submanifold of the infinite-dimensional space of probability density functions. Here, Θ denotes the parameters space. The following regularity condition is also assumed: the functions

$$\frac{\partial}{\partial \theta_1} p(x, y; \theta), \dots, \frac{\partial}{\partial \theta_N} p(x, y; \theta) \tag{14.1.4}$$

are linearly independent, where $\theta^T = (\theta_1, \dots, \theta_N) \in \Theta$. This condition assures that the submanifold is smooth and admits a tangent space at each point $p(x, y; \theta)$. The manifold $\mathcal{S} = \{p(x, y; \theta); \theta\}$ is called a *statistical manifold*, see Fig. 14.1.

In this setup the training density, $p(x, z)$, represents a point in the space of probability densities, which in general, is not situated on the manifold $\mathcal{S}$, see Fig 14.2. By tuning the network parameter vector, θ, we try to minimize the proximity between the given density and the corresponding manifold. The optimal parameter, if exists, is given by

$$\theta^* = \arg\min_\theta d\big(p(x, z), p(x, y; \theta)\big) = \arg\min_\theta D_{KL}\big(p(x, z)\|p(x, y; \theta)\big),$$

where D_{KL} denotes the Kullback-Leibler divergence. We note that any other cost function can be considered, but the Kullback-Leibler divergence is preferred due to its relation to the maximum likelihood estimation.

If the distance is zero, then the training distribution, $p(x, z)$, belongs to the statistical manifold, $\mathcal{S}$, and the learning becomes exact, i.e., there is an exact value of the parameter θ such that $p(x, z) = p(x, y; \theta)$.

The log-likelihood function for the statistical manifold $\mathcal{S}$ is defined by

$$\ell(\theta) = \ell(x, y; \theta) = \ln p(x, y; \theta).$$

In practice, the aforementioned distance is measured as a training error over the training data $\{(x_1, z_1), \ldots, (x_n, z_n)\}$. This is given by the average of the negative log-likelihoods evaluated at (x_k, z_k) as

$$C_{\text{train}}(\theta) = -\frac{1}{n} \sum_{k=1}^{n} \ell(x_k, z_k; \theta) = -\frac{1}{n} \prod_{k=1}^{n} \ln p(x_k, z_k; \theta).$$

The optimal parameter, $\theta^* = \arg\min_\theta C_{\text{train}}(\theta)$, is the maximum likelihood estimator, $\theta^* = \theta_{MSE}$. The relation between $C_{\text{train}}(\theta)$ and the Kullback-Leibler divergence between the training and the model distributions has been pointed out in section 3.6 and is mainly based on the following argument

$$
\begin{aligned}
C_{\text{train}}(\theta) &= -\mathbb{E}^{P_{XZ}}[\ell(X, Y; \theta)] \\
&= -\mathbb{E}^{P_{XZ}}[\ln p(X, Y; \theta) - \ln p(X, Z) + \ln p(X, Z)] \\
&= \mathbb{E}^{P_{XZ}}\Big[\ln \frac{p(X, Z)}{p(X, Y; \theta)}\Big] - \mathbb{E}^{P_{XZ}} \ln[p(X, Z)] \\
&= D_{KL}(p(X, Z) \| p(X, Y; \theta)) - H(p(X, Z)).
\end{aligned}
$$

Since the Shannon entropy $H(p(X, Z))$ is independent of the parameter θ, then the optimal parameter

$$\theta^* = \arg\min_\theta C_{\text{train}}(\theta) = \arg\min_\theta D_{KL}(p(X, Z) \| p(X, Y; \theta))$$

minimizes the Kullback-Leibler divergence between the training distribution, $p(X, Z)$, and the statistical manifold $\mathcal{S} = \{p(X, y; \theta); \theta \in \Theta\}$.

The aforementioned statistical manifold, $\mathcal{S} = \{p(x, y; \theta); \theta \in \Theta\}$, can be endowed with a metric structure and will be regarded as a Riemannian manifold. We shall introduce this metric in the next section.

It is worth noting that formulas containing target values z_k have an extrinsic character, since they access information regarding an exterior point, $p(x, z)$, of the statistical manifold. Formulas involving x_k and $y_k = f_\theta(x_k)$ have an intrinsic character, and hence they belong to the local geometry of the statistical manifold $\mathcal{S}$. From this point of view, the aforementioned training error, $C_{\text{train}}(\theta)$, is an extrinsic object. The next goal is to introduce the Fisher metric on a statistical manifold, which is an intrinsic object.

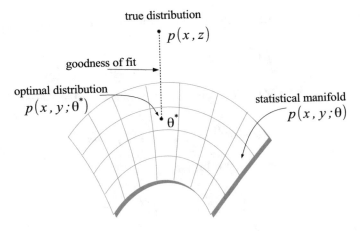

Figure 14.2: *Geometric image of the goodness of fit (loss function).*

14.2 Fisher Information

Assume that a probability density $p(x; \theta)$ depends on the real-valued parameter θ, and we would like to estimate this parameter by an unbiased estimator $\hat{\theta} = \hat{\theta}(x)$, which depends on data x. This means the average of the difference $\hat{\theta}(x) - \theta$ over all sets of data in the presence of parameter θ is zero

$$\mathbb{E}[\hat{\theta}(x) - \theta] = \int (\hat{\theta}(x) - \theta)p(x; \theta)\, dx = 0.$$

Denote, for simplicity, $p = p(x; \theta)$, and differentiate the previous relation with respect to θ using the product rule

$$0 = \frac{\partial}{\partial \theta}\mathbb{E}[\hat{\theta}(x) - \theta] = \int (\hat{\theta}(x) - \theta)\frac{\partial p}{\partial \theta}\, dx - \int p\, dx.$$

Since $\int p\, dx = 1$ and $\frac{\partial p}{\partial \theta} = \frac{\partial \ln p}{\partial \theta}p$, the previous relation implies

$$\int (\hat{\theta}(x) - \theta)\frac{\partial \ln p}{\partial \theta}p\, dx = 1.$$

A key feature is to split the density p into the product of two square roots and then rewrite the expression as

$$\int \left((\hat{\theta}(x) - \theta)\sqrt{p}\right)\left(\frac{\partial \ln p}{\partial \theta}\sqrt{p}\right)dx = 1.$$

Taking the square and using the Cauchy's integral inequality, yields

$$1 \leq \int (\hat{\theta}(x) - \theta)^2 p \, dx \int \left(\frac{\partial \ln p}{\partial \theta}\right)^2 p \, dx.$$

Each integral on the right side has the significance of an expectation. The first one,

$$\mathbb{E}[(\hat{\theta} - \theta)^2] = \int (\hat{\theta}(x) - \theta)^2 p \, dx,$$

is the mean square error and the latter,

$$I(\theta) = \mathbb{E}\left[\left(\frac{\partial \ln p}{\partial \theta}\right)^2\right] = \int \left(\frac{\partial \ln p}{\partial \theta}\right)^2 p \, dx,$$

is the *Fisher information* with respect to θ. The previous inequality can be written now as

$$\mathbb{E}[(\hat{\theta} - \theta)^2] I(\theta) \geq 1,$$

or equivalently, as

$$\mathbb{E}[(\hat{\theta} - \theta)^2] \geq \frac{1}{I(\theta)}. \tag{14.2.5}$$

This is called the *Cramér-Rao inequality*. It states that the inverse of the Fisher information is a lower bound for the mean square error.[1] Therefore, the minimum squared error estimator, $\hat{\theta}_{MSE}$, satisfies the identity

$$\mathbb{E}[(\hat{\theta}_{MSE} - \theta)^2] = \frac{1}{I(\theta)}. \tag{14.2.6}$$

Hence, when the information content is high, the error is low and vice versa. The Fisher information, $I(\theta)$, represents the derivative content of the log-likelihood function $\ell(\theta) = \ln p(x; \theta)$. The faster $\ell(\theta)$ changes with respect to θ, the larger is the information $I(\theta)$ and the smaller the mean squared error. We shall see later that estimators that realize equality in the Cramér-Rao inequality are called Fisher-efficient.

Fisher information is an assessment of the information about the unknown parameter θ contained in the random variable X, which is modeled by the family of densities $p(\theta)$. Hence, an estimator of θ has the variance larger than the inverse of the previous information and it becomes efficient when its variance is the lowest possible. If there is little information about θ contained in X, then Cramér-Rao inequality states that any estimation of θ is loose, in the sense that it has a large variance.

---

[1]Sometimes, this is stated equivalently as $Var(\hat{\theta}) \geq \frac{1}{I(\theta)}$.

We also note that the Fisher information can be written in terms of the log-likelihood function as

$$I(\theta) = \mathbb{E}\left[\left(\frac{\partial\ell(\theta)}{\partial\theta}\right)^2\right].$$

This expression serves the purpose of a density function, $p(x;\theta)$, which depends on only one parameter, θ. The multidimensional case is treated in the following.

The multivariate case A similar concept can be introduced in the case when the parameter θ is n-dimensional, $\theta = (\theta_1, \ldots, \theta_n)$. In this case we obtain the *Fisher information matrix*

$$g_{ij}(\theta) = \mathbb{E}\left[\frac{\partial\ell(\theta)}{\partial\theta_i}\frac{\partial\ell(\theta)}{\partial\theta_j}\right], \tag{14.2.7}$$

where $\ell(\theta) = \ln p(x;\theta)$, and the expectation is taken with respect to $p(x;\theta)$. Other equivalent expressions are given in the following:

Proposition 14.2.1 *The Fisher matrix can be also expressed as:*

$$g_{ij}(\theta) = -\mathbb{E}\left[\frac{\partial^2\ell(\theta)}{\partial\theta_i\partial\theta_j}\right] \tag{14.2.8}$$

$$g_{ij}(\theta) = 4\int\frac{\partial\sqrt{p(x;\theta)}}{\partial\theta_i}\frac{\partial\sqrt{p(x;\theta)}}{\partial\theta_j}\,dx. \tag{14.2.9}$$

Proof: Taking the derivative in $\int p(x;\theta)\,dx = 1$ with respect to θ_i yields

$$\int \partial_{\theta_i}p(x;\theta)\,dx = 0,$$

which is equivalent to

$$\int \partial_{\theta_i}\ln p(x;\theta)\,p(x;\theta)\,dx = 0.$$

Differentiating one more time with respect to θ_j and applying product rule, we get

$$\int \partial_{\theta_j}\partial_{\theta_i}\ln p(x;\theta)\,p(x;\theta)\,dx + \int \partial_{\theta_i}\ln p(x;\theta)\,\partial_{\theta_j}p(x;\theta)\,dx = 0$$

$$\mathbb{E}\left[\frac{\partial^2\ell(\theta)}{\partial\theta_i\partial\theta_j}\right] + \int \partial_{\theta_i}\ln p(x;\theta)\,\partial_{\theta_j}\ln p(x;\theta)\,p(x;\theta)\,dx = 0$$

$$\mathbb{E}\left[\frac{\partial^2\ell(\theta)}{\partial\theta_i\partial\theta_j}\right] + g_{ij}(\theta) = 0,$$

which implies relation (14.2.8).

In order to prove relation (14.2.9) we apply the following straightforward computation:

$$
\begin{aligned}
g_{ij}(\theta) &= \int \partial_{\theta_i} \ln p(x;\theta)\, \partial_{\theta_j} \ln p(x;\theta)\, p(x;\theta)\, dx \\
&= \int \frac{\partial_{\theta_i} p(x;\theta)}{p(x;\theta)} \frac{\partial_{\theta_j} p(x;\theta)}{p(x;\theta)}\, p(x;\theta)\, dx \\
&= 4 \int \frac{\partial_{\theta_i} p(x;\theta)}{2\sqrt{p(x;\theta)}} \frac{\partial_{\theta_j} p(x;\theta)}{2\sqrt{p(x;\theta)}}\, dx \\
&= 4 \int \frac{\partial \sqrt{p(x;\theta)}}{\partial \theta_i} \frac{\partial \sqrt{p(x;\theta)}}{\partial \theta_j}\, dx.
\end{aligned}
$$

∎

Relation (14.2.8) relates the Fisher matrix to the negative Hessian of the log-likelihood function and has the following geometric interpretation. If $\ell(x;\theta) = \ln p(x;\theta)$ is the log-likelihood function corresponding to the observation x and parameter θ, the maximum likelihood estimator of θ is

$$
\hat{\theta}_{MLE} = \arg\max_{\theta} \ell(x;\theta).
$$

Since there is a maximum at $\theta = \hat{\theta}_{MLE}$, then $\frac{\partial}{\partial \theta_i} \ell(x;\hat{\theta}_{MLE}) = 0$. Expanding about $\hat{\theta}_{MLE}$, we have

$$
\begin{aligned}
\ell(x;\theta) &= \ell(x;\hat{\theta}_{MLE}) + \frac{1}{2} \sum_{i,j} \frac{\partial^2}{\partial \theta_i \partial \theta_j} \ell(x;\hat{\theta}_{MLE})(\hat{\theta}_{MLE,i} - \theta_i)(\hat{\theta}_{MLE,j} - \theta_j) \\
&\quad + O(\|\hat{\theta}_{MLE} - \theta\|^3)
\end{aligned}
$$

The Fisher matrix, given by the expectation of the negative coefficient of the sum in the right side as in (14.2.8), measures how curved is the crest of $\ell(x;\theta)$ at $\hat{\theta}_{MLE}$.

It can be shown that under the regularity assumption (14.1.4) relation (14.2.9) can be used to show that $g_{ij}(\theta)$ is a symmetric, positive definite and nondegenerate matrix (see, for instance, Proposition 1.6.2 in [22]).

Hence, the Fisher information matrix provide the coefficients of a Riemannian metric (see section 13.1.3) on the statistical manifold $\mathcal{S} = \{\theta \to p(x;\theta)\}$, called the *Fisher metric*. This allows for computing lengths of vectors, angles, distances and areas on statistical manifolds. Sometimes, the Fisher metric $g_{ij}(\theta)$ is considered on the parameter space, Θ.

A natural question is what makes the Fisher metric distinguishable among all Riemannian metrics that can be defined on a statistical manifold? It can be shown that the Fisher metric has the following two properties, see [4] and [22]:

1. g_{ij} is invariant under reparametrizations of the sample space. This means the statistical manifolds $\mathcal{S} = \{p(x; \theta); \theta \in \Theta\}$ and $\tilde{\mathcal{S}} = \{p(h(x); \theta); \theta \in \Theta\}$, with h invertible and differentiable function, have equal metrics, $g_{ij}(\theta) = \tilde{g}_{ij}(\theta)$. This invariance property can be found in Theorem 1.6.4 of [22].

2. g_{ij} is covariant under reparametrizations. This means that if we consider a different parametrization $\xi_j = \xi(\theta_1, \ldots, \theta_N)$ depending on θ, the Fisher matrices in both parametrizations are related by the relation:

$$g_{ij}(\theta) = \sum_{k,r} g_{kr}(\xi)\Big|_{\xi=\xi(\theta)} \frac{\partial \xi^k}{\partial \theta^i} \frac{\partial \xi^r}{\partial \theta^j}.$$

For a proof of this fact, see Theorem 1.6.5 of [22].

What makes the Fisher metric so special is that it is the only metric satisfying the previous two invariance conditions. The proof of this distinguished result can be found in [28].

Cramér-Rao inequality There is a multivariate version of the inequality (14.2.5). The vector parameter is $\theta = (\theta_1, \ldots, \theta_N)^T \in \mathbb{R}^N$ and represents a coordinate system for the parameters space (Θ, g). Consider an estimator

$$\hat{\theta}(X) = (\hat{\theta}_1(X), \ldots, \hat{\theta}_N(X))^T$$

which is unbiased, $\mathbb{E}[\hat{\theta}(X)] = \theta$. Then

$$Cov(\hat{\theta}(X)) \geq g(\theta)^{-1}. \tag{14.2.10}$$

This means that the difference matrix $A_{ij} = Cov(\hat{\theta}_i(X), \hat{\theta}_j(X)) - g^{ij}(\theta)$ is positive semidefinite for all $\theta \in \Theta$, i.e., it has nonnegative eigenvalues. This inequality will be useful later when discussing about efficient estimators.

For further applications of Fisher information in science the reader is referred to [40], [41], and [39]. For applications to other type of neurons, see [121].

14.3 Neuromanifold of a Neural Net

A *neuromanifold* is a Riemannian manifold associated with a neural network as in the following. Let $y = f_\theta(x)$ be the input-output mapping of a given neural network with input and output densities $p_X(x)$ and $p_Y(y; \theta)$, and joint density $p(x, y; \theta)$, where θ is a vector parameter consisting of all the weights and biasses of the network. The statistical manifold associated with the neural network is $\mathcal{S} = \{p(x, y; \theta); \theta \in \Theta\}$. The Fisher metric can be defined on $\mathcal{S}$ by formula (14.2.7), where we consider the log-likelihood function given by $\ell(x, y; \theta) = \ln p(x, y; \theta)$. This can also be expressed as the following double integral:

$$g_{ij}(\theta) = \iint \frac{\partial \ell(x, y; \theta)}{\partial \theta_i} \frac{\partial \ell(x, y; \theta)}{\partial \theta_j} p(x, y; \theta)\, dx dy. \tag{14.3.11}$$

Since $g = g_{ij}$ is a Riemannian metric on $\mathcal{S}$, then $(\mathcal{S}, g)$ becomes a Riemannian manifold.

Definition 14.3.1 *The neuromanifold associated with the aforementioned neural network is the Riemannian manifold $(\mathcal{S}, g)$, where $\mathcal{S} = \{p(x, y; \theta); \theta \in \Theta\}$ is the statistical manifold of the joint input-output densities of the neural network, θ are the network weights and biasses, and g is the Fisher metric.*

Each joint probability density associated with a neural net, $p(x, y; \theta)$, can be regarded as a point on this manifold. The learning process, which is an adjustment of parameters, can be visualized as a curve on the neuromanifold.

It is worth noting that the metric $g_{ij}(\theta)$ is independent of the target values z_k, i.e., it is an intrinsic object. All concepts derived from the Fisher information will form the intrinsic geometry of the neuromanifold.

The next computations will be performed under the assumption that the noise in formula (14.1.1) is standard normal noise, $n \sim \mathcal{N}(0, 1)$. Sometimes, one may consider a scaled noise, $n \sim \mathcal{N}(0, s^2)$, and consider the standard deviation, s, as a hyperparameter. The use of (14.1.3) yields the following log-likelihood function

$$\begin{aligned} \ell(x, y; \theta) &= \ln p(x) + \ln p(y|x; \theta) \\ &= \ln p(x) - \ln(\sqrt{2\pi}) - \frac{1}{2}\big(y - f_\theta(x)\big)^2, \end{aligned}$$

with the partial derivative

$$\frac{\partial \ell(x, y; \theta)}{\partial \theta_k} = \big(y - f_\theta(x)\big) \frac{\partial f_\theta(x)}{\partial \theta_k}. \tag{14.3.12}$$

The sensitivity of the input-output mapping with respect to a parameter, $\frac{\partial f_\theta(x)}{\partial \theta_k}$, is specific to each type of neural network and is a measure of complexity of each network. We shall compute it in a few particular cases and then provide a general recursive formula in the case of feedforward neural nets.

14.4 Fisher Metric for One Neuron

In this section we shall provide explicit formulas for the Fisher metric in the case of a single neuron. We consider a neuron with the input given by the n-dimensional random vector $X = (X_1, \ldots, X_n)^T$, input-output mapping $f_\theta(x) = \phi(w^T x + b)$, with parameters $\theta = (w, b)$, and differentiable activation function ϕ. An application of the chain rule produces the partial derivatives

$$\frac{\partial f_\theta(x)}{\partial w_k} = x_k \phi'(w^T x + b), \qquad \frac{\partial f_\theta(x)}{\partial b} = \phi'(w^T x + b).$$

Then formula (14.3.12) provides

$$\frac{\partial \ell(x, y; \theta)}{\partial w_k} = x_k \big(y - \phi(w^T x + b)\big)\phi'(w^T x + b), \qquad 1 \leq k \leq n$$

$$\frac{\partial \ell(x, y; \theta)}{\partial b} = \big(y - \phi(w^T x + b)\big)\phi'(w^T x + b).$$

Since $\theta = (w_1, \ldots, w_n, b)$, the Fisher matrix is $(n+1)$-dimensional. Using (14.3.11) and changing the order of integration, we have

$$g_{00}(w, b) = \iint \left(\frac{\partial \ell(x, y; \theta)}{\partial b}\right)^2 p(x, y)\, dx dy$$

$$= \iint \big(y - \phi(w^T x + b)\big)^2 \phi'(w^T x + b)^2 p(x) p(y|x; \theta)\, dx dy$$

$$= \int \phi'(w^T x + b)^2 p(x) \int \big(y - \phi(w^T x + b)\big)^2 p(y|x; \theta)\, dy\, dx.$$

Substituting $p(y|x; \theta)$ from (14.1.2) and changing the variable $u = \phi(w^T x + b)$ yields

$$g_{00}(w, b) = \int \phi'(w^T x + b)^2 p(x) \bigg(\underbrace{\int u^2 \frac{1}{\sqrt{2\pi}} e^{-\frac{1}{2}u^2}\, du}_{=1}\bigg)\, dx$$

$$= \int \phi'(w^T x + b)^2 p(x)\, dx$$

$$= \mathbb{E}^{P_X}[\phi'(w^T X + b)^2],$$

where $\mathbb{E}^{P_X}$ denotes the expectation operator taken under the law of input X.

Similarly,

$$
\begin{aligned}
g_{0k}(w, b) &= \iint \frac{\partial \ell(x, y; \theta)}{\partial b} \frac{\partial \ell(x, y; \theta)}{\partial w_k} p(x, y) \, dx dy \\
&= \iint \left(y - \phi(w^T x + b)\right)^2 \phi'(w^T x + b)^2 x_k p(x) p(y|x; \theta) \, dx dy \\
&= \int \phi'(w^T x + b)^2 x_k p(x) \underbrace{\int \left(y - \phi(w^T x + b)\right)^2 p(y|x; \theta) \, dy}_{=1} \, dx \\
&= \mathbb{E}^{Px}[X_k \phi'(w^T X + b)^2].
\end{aligned}
$$

Also,

$$
\begin{aligned}
g_{jk}(w, b) &= \iint \frac{\partial \ell(x, y; \theta)}{\partial w_j} \frac{\partial \ell(x, y; \theta)}{\partial w_k} p(x, y) \, dx dy \\
&= \iint \left(y - \phi(w^T x + b)\right)^2 \phi'(w^T x + b)^2 x_j x_k p(x) p(y|x; \theta) \, dx dy \\
&= \int \phi'(w^T x + b)^2 x_j x_k p(x) \underbrace{\int \left(y - \phi(w^T x + b)\right)^2 p(y|x; \theta) \, dy}_{=1} \, dx \\
&= \mathbb{E}^{Px}[X_j X_k \phi'(w^T X + b)^2].
\end{aligned}
$$

To conclude, the last three formulas can be written in only one formula as

$$
\tilde{g}_{ij}(\tilde{w}) = \mathbb{E}^{Px}[X_i X_j \phi'(\tilde{w}^T \tilde{X})^2], \qquad 0 \le i, j \le n, \tag{14.4.13}
$$

where $\tilde{w}^T = (w^T, b)$ and $\tilde{X} = (X_0, X)$, with $X_0 = 1$. In general, formulas (14.4.13) cannot be simplified any further. However, if X_i are independent, standard normally distributed, and $b = 0$, then a closed-form solution exists for the Fisher matrix and for its inverse, see [6]. We shall develop this idea after we investigate a few particular types of neurons.

Linear neuron In this case the activation function is $\phi(x) = x$, so replacing the derivative $\phi'(x)$ by 1 in the previous formulas, yields

$$
g_{00} = 1, \qquad g_{0k}(w, b) = \mathbb{E}^{Px}[X_k], \qquad g_{jk}(w, b) = \mathbb{E}^{Px}[X_j X_k].
$$

These formulas suggest that the Fisher matrix of a linear neuron describes the auto-covariance of the input vector $\tilde{X} = (1, X)$. Furthermore, since $g_{jk}(w, b)$ do not depend on $\theta = (w, b)$, then $\frac{\partial g_{ij}}{\partial \theta_k} = 0$, which implies vanishing Christoffel symbols, $\Gamma_{ij}^k = 0$. Therefore, the neuromanifold of a linear neuron is intrinsically flat and all its intrinsic geometry is induced only by the inputs covariances.

ReLU neuron Consider the activation function $\phi(x) = ReLU(x)$, which is piecewise differentiable. We shall differentiate it in the generalized sense, as in Appendix, section F.2. Then $\phi'(x) = ReLU(x)' = H(x)$, see Exercise 8.8.1. Since $H^2(x) = H(x)$, we obtain

$$
\begin{aligned}
g_{00} &= \mathbb{E}^{Px}[ReLU'(w^T X + b)^2] = \mathbb{E}^{Px}[H(w^T X + b)^2] = \mathbb{E}^{Px}[H(w^T X + b)] \\
&= \int_{\{w^T x + b \geq 0\}} p(x)\, dx = \mathbb{P}(w^T X + b \geq 0) = \mathbb{P}(X \in \mathcal{H}_{w,b}).
\end{aligned}
$$

Therefore, the coefficient g_{00} represents the probability that the input vector X belongs to the half-space $\mathcal{H}_{w,b} = \{w^T x + b \geq 0\}$. We also note that $0 \leq g_{00} \leq 1$. Its value depends on the values of the hyperplane translation parameter b as

$$
\lim_{b \to \infty} g_{00} = 0, \qquad \lim_{b \to -\infty} g_{00} = 1.
$$

The other metric coefficients are given by

$$
\begin{aligned}
g_{0k} &= \mathbb{E}^{Px}[X_k ReLU'(w^T X + b)^2] = \mathbb{E}^{Px}[X_k H(w^T X + b)] \\
&= \int_{\{w^T x + b \geq 0\}} x_k p(x)\, dx, \\
g_{jk} &= \mathbb{E}^{Px}[X_j X_k ReLU'(w^T X + b)^2] = \mathbb{E}^{Px}[X_j X_k H(w^T X + b)] \\
&= \int_{\{w^T x + b \geq 0\}} x_j x_k p(x)\, dx.
\end{aligned}
$$

Using $0 \leq H(w^T x + b) \leq 1$, it follows that $g_{0k} \leq \mathbb{E}^{Px}[X_k]$, $g_{jk} \leq \mathbb{E}^{Px}[X_j X_k]$. In fact, we have

$$
\lim_{b \to \infty} g_{0k} = \lim_{b \to \infty} g_{jk} = 0, \qquad \lim_{b \to -\infty} g_{0k} = \mathbb{E}[X_k], \qquad \lim_{b \to -\infty} g_{jk} = \mathbb{E}[X_j X_k].
$$

Hyperbolic tangent neuron The activation function $\phi(x) = \tanh x$ satisfies $\phi'(x) = 1 - \tanh^2 x$. As usually, we denote for simplicity, $\mathbf{t}(x) = \tanh x$. The Fisher metric coefficients are given by

$$
\begin{aligned}
g_{00} &= \mathbb{E}^{Px}[\mathbf{t}'(w^T X + b)^2] = \mathbb{E}^{Px}[(1 - \mathbf{t}^2(w^T X + b))^2] \\
g_{0k} &= \mathbb{E}^{Px}[X_k \mathbf{t}'(w^T X + b)^2] = \mathbb{E}^{Px}[X_k(1 - \mathbf{t}^2(w^T X + b))^2] \\
g_{jk} &= \mathbb{E}^{Px}[X_j X_k \mathbf{t}'(w^T X + b)^2] = \mathbb{E}^{Px}[X_j X_k(1 - \mathbf{t}^2(w^T X + b))^2],
\end{aligned}
$$

for $1 \leq j, k \leq n$. It is worth to note the inequalities

$$
0 \leq g_{00} \leq 1, \qquad g_{0k} \leq \mathbb{E}^{Px}[X_k], \qquad g_{jk} \leq \mathbb{E}^{Px}[X_j X_k].
$$

14.5   The Fisher Matrix and Its Inverse

In order to compute the Fisher matrix (14.4.13) and its inverse some additional conditions have to be assumed. We shall consider $(X_1, \ldots, X_n) \sim \mathcal{N}(0, \mathbb{I}_n)$, i.e., the input is a multivariate standard normal random variable. Denote $|w|^2 = w^T w = \sum w_i^2$ and consider the functions

$$C_1(w, b) = \frac{1}{|w|^2 \sqrt{2\pi}} \int \phi'(|w|\epsilon + b)^2 e^{-\frac{1}{2}\epsilon^2}\, d\epsilon \qquad (14.5.14)$$

$$C_2(w, b) = \frac{1}{|w|^2 \sqrt{2\pi}} \int \phi'(|w|\epsilon + b)^2 \epsilon^2 e^{-\frac{1}{2}\epsilon^2}\, d\epsilon \qquad (14.5.15)$$

$$C_3(w, b) = \frac{1}{|w|^2 \sqrt{2\pi}} \int \phi'(|w|\epsilon + b)^2 \, \epsilon\, e^{-\frac{1}{2}\epsilon^2}\, d\epsilon. \qquad (14.5.16)$$

Using that $w^T X = \sum w_i X_i \sim \mathcal{N}(0, |w|^2)$, we can write $w^T X = |w|\epsilon$, with $\epsilon \sim \mathcal{N}(0, 1)$. Then we have

$$
\begin{aligned}
g_{00}(w, b) &= \mathbb{E}[\phi'(w^T X + b)^2] = \mathbb{E}[\phi'(|w|\epsilon + b)^2] \\
&= \frac{1}{\sqrt{2\pi}} \int \phi'(|w|\epsilon + b)^2 e^{-\frac{1}{2}\epsilon^2}\, d\epsilon = |w|^2 C_1(w, b). \quad (14.5.17)
\end{aligned}
$$

We compute next $g_{0k} = \mathbb{E}[X_k \phi'(w^T X + b)^2]$, $1 \le k \le n$. Writing again $w^T X = |w|\epsilon$, we have

$$
\begin{aligned}
\sum_k g_{0k} w_k &= \mathbb{E}[w^T X \phi'(w^T X + b)^2] = \mathbb{E}[|w|\epsilon \phi'(|w|\epsilon + b)^2] \\
&= \frac{|w|}{\sqrt{2\pi}} \int \epsilon \phi'(|w|\epsilon + b)^2 e^{-\frac{1}{2}\epsilon^2}\, d\epsilon \\
&= |w|^3 C_3(w, b). \qquad\qquad\qquad\qquad\qquad (14.5.18)
\end{aligned}
$$

Let v be an arbitrary unit vector, orthogonal to w. Then

$$
\begin{aligned}
\sum_k g_{0k} v_k &= \mathbb{E}[v^T X \phi'(w^T X + b)^2] = \mathbb{E}[v^T X] \mathbb{E}[\phi'(w^T X + b)^2] \\
&= \mathbb{E}[\epsilon] \mathbb{E}[\phi'(w^T X + b)^2] = 0,
\end{aligned}
$$

where we used $v^T X = |v|\epsilon = \epsilon \sim \mathcal{N}(0, 1)$, the fact that $\mathbb{E}[\epsilon] = 0$ and that $w^T X$ and $v^T X$ are independent, see Exercise 14.13.3. It follows that the vector $(g_{01}, \ldots, g_{0n})^T$ is normal to all vectors v (which are perpendicular to w). Hence, $(g_{01}, \ldots, g_{0n})^T$ has to be proportional to w, i.e., there is $\lambda \in \mathbb{R}$ such that

$$g_{0k} = \lambda w_k, \quad 1 \le k \le n. \qquad (14.5.19)$$

To determine λ, multiply by w_k and take the sum

$$\sum g_{0k}w_k = \lambda \sum w_k^2 = \lambda|w|^2,$$

from where

$$\lambda = \frac{1}{|w|^2}\sum g_{0k}w_k = \lambda \sum w_k^2 = |w|C_3(w,b),$$

where we used (14.5.18). Then (14.5.19) yields

$$g_{0k} = w_k|w|C_3(w,b), \quad 1 \le k \le n.$$

Remark 14.5.1 We make the remark that if $b = 0$ and $\phi(x) = \tanh(x)$, then $C_3(w,b) = 0$, because $\phi'(|w|\epsilon)^2$ is an even function in ϵ. Consequently, $g_{0k} = 0$, $1 \le k \le n$.

We shall show in the following that the matrix

$$g_{jk} = \mathbb{E}[X_j X_k \phi'(w^T X + b)^2], \quad 1 \le j, k \le n$$

has the explicit form

$$g_{jk} = |w|^2 C_1(w,b)\delta_{jk} + (C_2(w,b) - C_1(w,b))w_j w_k,$$

where δ_{jk} is 1 for $j = k$ and 0 otherwise. In equivalent matrix form, the matrix

$$g = \mathbb{E}[XX^T \phi'(w^T X + b)^2]$$

can be written as the following sum:

$$g = |w|^2 C_1(w,b)\mathbb{I}_n + (C_2(w,b) - C_1(w,b))ww^T. \tag{14.5.20}$$

For simplicity, shall denote the matrix in right side of (14.5.20) by h. Since both g and h are symmetric matrices, in order to show that $g = h$, we shall use Exercise 14.13.1. Therefore, it suffices to prove that $w^T gw = w^T hw$ and $v^T gv = v^T hv$, for all unit vectors v normal to w. We shall do this in two steps:

Step 1: Show $w^T gw = w^T hw$. The left side can be computed as

$$
\begin{aligned}
w^T gw &= w^T \mathbb{E}[XX^T \phi'(w^T X + b)^2]w = \mathbb{E}[(w^T X)^2 \phi'(w^T X + b)^2] \\
&= \mathbb{E}[|w|^2 \epsilon^2 \phi'(|w|\epsilon + b)^2] = \frac{|w|^2}{\sqrt{2\pi}}\int \epsilon^2 \phi'(|w|\epsilon + b)^2 e^{\frac{1}{2}\epsilon^2} d\epsilon \\
&= |w|^4 C_2(w,b). \tag{14.5.21}
\end{aligned}
$$

The right side is

$$
\begin{aligned}
w^T hw &= |w|^2 C_1(w,b)w^T w + (C_2(w,b) - C_1(w,b))w^T ww^T w \\
&= |w|^4 C_1(w,b) + (C_2(w,b) - C_1(w,b))|w|^4 \\
&= |w|^4 C_2(w,b). \tag{14.5.22}
\end{aligned}
$$

Since (14.5.21) and (14.5.22) agree, we have obtained the desired identity.

Step 2: Show $v^T g v = v^T h v$. Using relation (14.5.17) and the fact that $w^T X$ and $v^T X$ are independent, see Exercise 14.13.2, the left side becomes

$$
\begin{aligned}
v^T g v &= v^T \mathbb{E}[XX^T \phi'(w^T X + b)^2] v = \mathbb{E}[(v^T X)^2 \phi'(w^T X + b)^2] \\
&= \mathbb{E}[(v^T X)^2] \mathbb{E}[\phi'(w^T X + b)^2] = \underbrace{\mathbb{E}[\epsilon^2]}_{=1} \underbrace{\mathbb{E}[\phi'(|w|\epsilon + b)^2]}_{=g_{00}} \\
&= |w|^2 C_1(w, b).
\end{aligned}
\tag{14.5.23}
$$

For the right side we have the straightforward computation

$$
\begin{aligned}
v^T h v &= v^T \big(|w|^2 C_1(w, b)\mathbb{I}_n + (C_2(w, b) - C_1(w, b))ww^T\big) v \\
&= |w|^2 C_1(w, b) v^T v + (C_2(w, b) - C_1(w, b))(v^T w)^2 \\
&= |w|^2 C_1(w, b),
\end{aligned}
\tag{14.5.24}
$$

where we used the orthonormality conditions $v^T v = 1$ and $v^T w = 0$. Since (14.5.23) and (14.5.24) agree, we proved the desired identity.

To conclude, the Fisher matrix is given by the following $(n+1) \times (n+1)$ matrix

$$
\tilde{g} : \begin{cases}
g_{00} = |w|^2 C_1(w, b) \\
g_{0k} = g_{k0} = w_k |w| C_3(w, b), \quad 1 \le k \le n \\
g_{jk} = |w|^2 C_1(w, b)\delta_{jk} + (C_2(w, b) - C_1(w, b))w_j w_k, \quad 1 \le j, k \le n.
\end{cases}
$$

Finding the inverse of $\tilde{g}$ It is easier if we write the Fisher matrix as

$$
\tilde{g} = \begin{pmatrix}
g_{00} & g_{01} & \cdots & g_{0n} \\
g_{10} & & & \\
\vdots & & g & \\
g_{n0} & & &
\end{pmatrix}.
$$

The $(n \times n)$-block $g = g_{ij}$, $1 \le i, j \le n$, is invertible in closed form. We shall look for an inverse of a form similar to (14.5.20)

$$
g^{-1} = \rho_1 \mathbb{I}_n + \rho_2 ww^T,
$$

and determine the functions ρ_1 and ρ_2 in terms of w and b, such that $gg^{-1} = \mathbb{I}_n$. Using $ww^T ww^T = w|w|^2 w^T = |w|^2 ww^T$, an algebraic computation provides

$$
\begin{aligned}
gg^{-1} &= (|w|^2 C_1 \mathbb{I}_n + (C_2 - C_1)ww^T)(\rho_1 \mathbb{I}_n + \rho_2 ww^T) \\
&= |w|^2 C_1 \rho_1 \mathbb{I}_n + [\rho_1(C_2 - C_1) + |w|^2 \rho_2 C_2]ww^T.
\end{aligned}
$$

By coefficient identification, we ask

$$
\begin{aligned}
|w|^2 C_1 \rho_1 &= 1 \\
\rho_1(C_2 - C_1) + |w|^2 \rho_2 C_2 &= 1,
\end{aligned}
$$

with solutions

$$
\rho_1 = \frac{1}{|w|^2 C_1}, \quad \rho_2 = \frac{1}{|w|^4}\left(\frac{1}{C_2} - \frac{1}{C_1}\right).
$$

Therefore, the inverse of g is given by

$$
g^{-1} = \frac{1}{|w|^2 C_1}\mathbb{I}_n + \frac{1}{|w|^4}\left(\frac{1}{C_2} - \frac{1}{C_1}\right)ww^T. \tag{14.5.25}
$$

This result appears as Theorem 4 in [6].

When inverting the matrix $\tilde{g}$ we consider two cases:

Case 1: $b = 0$ and $\phi(x) = \tanh(x)$. In this case

$$
C_3(w, 0) = \frac{1}{|w|^2 \sqrt{2\pi}} \int \phi'(|w|\epsilon)^2 \, \epsilon \, e^{-\frac{1}{2}\epsilon^2} \, d\epsilon = 0,
$$

since $\phi'(|w|\epsilon)^2$ is an even function. It follows that $g_{0k} = g_{k0} = 0$, for $1 \le k \le n$, and the matrix can be inverted block by block as

$$
\tilde{g}^{-1} = \begin{pmatrix} \frac{1}{g_{00}} & 0 & \cdots & 0 \\ 0 & & & \\ \vdots & & g^{-1} & \\ 0 & & & \end{pmatrix},
$$

with g^{-1} given by (14.5.25).

Case 2: The general case. We shall indicate how to compute $\tilde{g}^{-1}$ in an iterative way. First, we decompose $\tilde{g}$ as a sum of two matrices

$$
\tilde{g} = \begin{pmatrix} 0 & g_{01} & \cdots & g_{0n} \\ g_{10} & & & \\ \vdots & & \mathbb{O}_n & \\ g_{n0} & & & \end{pmatrix} + \begin{pmatrix} g_{00} & 0 & \cdots & 0 \\ 0 & & & \\ \vdots & & g & \\ 0 & & & \end{pmatrix} = A_1 + A_2.
$$

The matrix A_2 is invertible, with the known inverse

$$
A_2^{-1} = \begin{pmatrix} \frac{1}{g_{00}} & 0 & \cdots & 0 \\ 0 & & & \\ \vdots & & g^{-1} & \\ 0 & & & \end{pmatrix}.
$$

The inversion of a sum of two matrices is covered in Appendix, section G. Using the expansion method, we have

$$\tilde{g}^{-1} = (A_1 + A_2)^{-1} = A_2^{-1} \sum_{k \geq 0} (-1)^k (A_1 A_2^{-1})^k. \tag{14.5.26}$$

Since the product

$$A_1 A_2^{-1} = \begin{pmatrix} 0 & \sum_{j=1}^{n} g_{0j} g^{j1} & \cdots & \sum_{j=1}^{n} g_{0j} g^{jn} \\ g_{10} & & & \\ \vdots & & \mathbb{O}_n & \\ g_{n0} & & & \end{pmatrix}$$

is a sparse matrix, its powers are not costly to compute.

Another iterative method to find the inverse is to construct the sequence $(\tilde{g}_n^{-1})_{n \geq 0}$ defined recursively by $\tilde{g}_0^{-1} = \mathbb{O}_n$, $\tilde{g}_{n+1}^{-1} = f(\tilde{g}_n^{-1})$, where $f(M) = A_2^{-1} - M A_1 A_2^{-1}$ is a contraction. The sequence $\tilde{g}_n^{-1}$ tends to the fixed point of the mapping f, which is the inverse $\tilde{g}^{-1}$.

Convergence conditions Series (14.5.26) and the sequence $\tilde{g}_n^{-1}$ converge provided some conditions are satisfied. Following section G of Appendix, the required condition is $\|A_1 A_2^{-1}\| < 1$, where the norm is the value of the largest eigenvalue. We shall show that $\|A_1 A_2^{-1}\| = g_{00}$ and that the condition $\|A_1 A_2^{-1}\| < 1$ is satisfied by some familiar classes of neurons.

Lemma 14.5.2 *Let $a, b \in \mathbb{R}^n$ be two vectors such that $a^T b > 0$. Then the eigenvalues of the matrix*

$$M = \begin{pmatrix} 0 & a_1 & \cdots & a_n \\ b_1 & & & \\ \vdots & & \mathbb{O}_n & \\ b_n & & & \end{pmatrix}$$

are $\lambda_1 = (a^T b)^{1/2}$, $\lambda_2 = -(a^T b)^{1/2}$, $\lambda_j = 0$, for all $j \geq 3$.

Proof: See Exercise 14.13.4.   ∎

We let $M = A_1 A_2^{-1}$ and show that $a^T b < 1$. We have

$$\begin{aligned}
a^T b &= g_{10} \sum_{j_1} g_{0j_1} g^{j_1 1} + \cdots + g_{n0} \sum_{j_n} g_{0j_n} g^{j_n n} \\
&= \sum_p \sum_{j_r} g_{p0} g_{0j_r} g^{j_r p} = \sum_{j_r} g_{0j_r} \sum_p g_{p0} g^{j_r p} \\
&= g_{00}.
\end{aligned}$$

If the activation function satisfies $\|\phi'\|_\infty < 1$ a.e. (namely, its steepest slope is less than 1 almost everywhere), then

$$g_{00} = \frac{1}{\sqrt{2\pi}} \int \phi'(|w|\epsilon + b)^2 \, e^{-\frac{1}{2}\epsilon^2} \, d\epsilon < \frac{1}{\sqrt{2\pi}} \int e^{-\frac{1}{2}\epsilon^2} \, d\epsilon = 1.$$

Several activations functions, such as $\tanh(\cdot)$ or logistic function $\sigma(\cdot)$, satisfy the aforementioned property. Hence, in these cases, $g_{00} < 1$ and the convergence condition is satisfied.

14.6   The Fisher Metric Structure of a Neural Net

Even if we cannot hope for explicit formulas for the Fisher metric in the case of a feedforward neural network, however, we can obtain the metric structure by an iterative method that is similar to the backpropagation method. The computation is still performed under the assumption that the noise inserted into the network is standard normal, $n \sim \mathcal{N}(0,1)$. Even if this modeling assumption seems to be limited, we consider it for the sake of simplicity. Other types of noise can be considered, but the computation will not run as smooth.

Denote $C(x, y; \theta) = \frac{1}{2}\big(f_\theta(x) - y\big)^2$. Then relation (14.3.12) writes as

$$\frac{\partial \ell(x, y; \theta)}{\partial \theta_k} = -\frac{\partial C(x, y; \theta)}{\partial \theta_k}.$$

This is equivalent to $\nabla_\theta \ell(x, y; \theta) = -\nabla_\theta C(x, y; \theta)$. If now, we regard $C(x, y; \theta)$ to be a quadratic cost function (even if in this context it has a different significance), we can compute the gradient $\nabla_\theta C(x, y; \theta)$ by the backpropagation method presented in Chapter 6. The parameter θ_k will be replaced by $w_{ij}^{(\ell)}$ and $b_j^{(\ell)}$, respectively. Following the notations and the computation from Chapter 6, we obtain

$$\frac{\partial \ell(x, y; \theta)}{\partial w_{ij}^{(\ell)}} = -\frac{\partial C(x, y; \theta)}{\partial w_{ij}^{(\ell)}} = -\frac{\partial C(x, y; \theta)}{\partial s_j^{(\ell)}} \frac{\partial s_j^{(\ell)}}{\partial w_{ij}^{(\ell)}}$$

$$= -\delta_j^{(\ell)} x_i^{(\ell-1)}$$

$$\frac{\partial \ell(x, y; \theta)}{\partial b_j^{(\ell)}} = -\frac{\partial C(x, y; \theta)}{\partial b_j^{(\ell)}} = -\frac{\partial C(x, y; \theta)}{\partial s_j^{(\ell)}} \frac{\partial s_j^{(\ell)}}{\partial b_j^{(\ell)}}$$

$$= \delta_j^{(\ell)},$$

where $\delta_j^{(\ell)}$ represents the sensitivity of $C(x, y; \theta)$ with respect to the signal $s_j^{(\ell)}$ and can be computed using the backpropagation formula (6.2.22)

$$\delta_i^{(\ell-1)} = \phi'(s_i^{(\ell-1)}) \sum_{j=1}^{d^{(\ell)}} \delta_j^{(\ell)} w_{ij}^{(\ell)}. \tag{14.6.27}$$

The delta in the last layer is computed as

$$
\begin{aligned}
\delta_j^{(L)} &= \frac{\partial C}{\partial s_j^{(L)}} = \frac{\partial}{\partial s_j^{(L)}} \left(\frac{1}{2} (\phi(s_j^{(L)}) - y)^2 \right) \\
&= \phi'(s_j^{(L)})(\phi(s_j^{(L)}) - y) \\
&= \phi'(s_j^{(L)})(f_\theta(x) - y).
\end{aligned} \tag{14.6.28}
$$

In order to overcome the writing difficulties of the expression of the Fisher matrix, we use notations $\alpha = (i, j, \ell)$, $\alpha' = (i', j', \ell')$, $X_0 = 1$, and $w_{0j}^{(\ell)} = b_j^{(\ell)}$. Then $\theta_\alpha = w_{ij}^{(\ell)}$ and we obtain the metric coefficients

$$
\begin{aligned}
g_{\alpha\alpha'} &= \mathbb{E}^{P_{XY}} \left[\frac{\partial \ell(x, y; \theta)}{\partial \theta_\alpha} \frac{\partial \ell(x, y; \theta)}{\partial \theta_{\alpha'}} \right] = \mathbb{E}^{P_{XY}} \left[\frac{\partial \ell(x, y; \theta)}{\partial w_{ij}^{(\ell)}} \frac{\partial \ell(x, y; \theta)}{\partial w_{i'j'}^{(\ell')}} \right] \\
&= \mathbb{E}^{P_{XY}} \left[X_i^{(\ell-1)} X_{i'}^{(\ell'-1)} \delta_j^{(\ell)} \delta_{j'}^{(\ell')} \right],
\end{aligned}
$$

with the indices in the following ranges

$$0 \le i, i' \le d^{(\ell-1)}, \quad 1 \le j, j' \le d^{(\ell)}, \quad 1 \le \ell \le L,$$

where L represents the depth of the network. The expressions of deltas $\delta_j^{(\ell)}$ and $\delta_{j'}^{(\ell')}$ are obtained by the backpropagation formula (14.6.27).

If $\ell = \ell' = L$, using (14.6.28) the expression of the metric coefficients becomes

$$
\begin{aligned}
g_{\alpha\alpha'} \big|_{\ell=\ell'=L} &= \mathbb{E}^{P_{XY}} \left[X_i^{(L-1)} X_{i'}^{(L-1)} \delta_j^{(L)} \delta_{j'}^{(L)} \right] \\
&= \mathbb{E}^{P_{XY}} \left[X_i^{(L-1)} X_{i'}^{(L-1)} \phi'(s_j^{(L)}) \phi'(s_{j'}^{(L)}) (f_\theta(X) - Y)^2 \right].
\end{aligned}
$$

If the activation function in the last layer is linear, $\phi(x) = x$, the expression

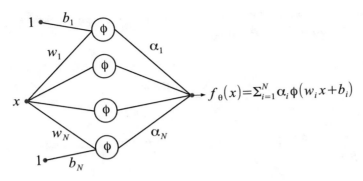

Figure 14.3: *One-hidden layer neural network with activation function ϕ and input-output mapping $f_\theta(x)$. The activation in the output neuron is linear.*

is more simple and can be computed as in the following:

$$
\begin{aligned}
g_{\alpha\alpha'}\big|_{\ell=\ell'=L} &= \mathbb{E}^{P_{XY}}\left[X_i^{(L-1)}X_{i'}^{(L-1)}(f_\theta(X)-Y)^2\right] \\
&= \iint x_i^{(L-1)}x_{i'}^{(L-1)}(f_\theta(x)-y)^2 p(x,y)\,dx\,dy \\
&= \int x_i^{(L-1)}x_{i'}^{(L-1)}p(x)\int(f_\theta(x)-y)^2 p(y|x;\theta)\,dy\,dx \\
&= \int x_i^{(L-1)}x_{i'}^{(L-1)}p(x)\,dx \\
&= \mathbb{E}^{P_X}\left[X_i^{(L-1)}X_{i'}^{(L-1)}\right],
\end{aligned}
$$

where we used that

$$
\int(f_\theta(x)-y)^2 p(y|x;\theta)\,dy = Var(n) = 1,
$$

namely, the variance of the noise $n \sim \mathcal{N}(0,1)$ is 1. The layer activations $X_i^{(L-1)}$ are computed iteratively by the forward pass formula

$$
X_i^{(\ell)} = \phi(W^{(\ell)^T}X^{(\ell-1)} - b_i^{(\ell)}).
$$

We shall compute explicitly the Fisher metric in the following concrete case.

Example 14.6.1 (Fisher metric for a one-hidden layer network) We shall consider the case of a feedforward neural network with one hidden layer and one-dimensional input and output, see Fig. 14.3. The activation function in the output neuron is linear, while in the hidden layer is denoted by ϕ. The input-output mapping is given by

$$f_\theta(x) = \sum_{i=1}^{N} \alpha_i \phi(w_i x + b_i).$$

A straightforward computation provides the partial derivatives with respect to parameters

$$\frac{\partial f_\theta(x)}{\partial \alpha_j} = \phi(w_j x + b_j)$$

$$\frac{\partial f_\theta(x)}{\partial w_i} = \alpha_i x \phi'(w_i x + b_i)$$

$$\frac{\partial f_\theta(x)}{\partial b_i} = \alpha_i \phi'(w_i x + b_i).$$

The partial derivatives of the log-likelihood functions are

$$\frac{\partial \ell}{\partial \alpha_j} = -\frac{\partial C}{\partial \alpha_j} = -\frac{1}{2}\frac{\partial}{\partial \alpha_j}(f_\theta(x) - y)^2$$

$$= -(f_\theta(x) - y)\frac{\partial f_\theta(x)}{\partial \alpha_j} = -(f_\theta(x) - y)\phi(w_j x + b_j)$$

$$\frac{\partial \ell}{\partial w_i} = -\frac{\partial C}{\partial w_i} = -(f_\theta(x) - y)\frac{\partial f_\theta(x)}{\partial w_i}$$

$$= -\alpha_i x (f_\theta(x) - y)\phi'(w_i x + b_i)$$

$$\frac{\partial \ell}{\partial b_i} = -\frac{\partial C}{\partial b_i} = -\alpha_i(f_\theta(x) - y)\phi'(w_i x + b_i).$$

Then coefficients of the Fisher matrix in directions α_j, α_k can be computed as

$$g_{\alpha_j \alpha_k} = \mathbb{E}^{P_{XY}}\left[\frac{\partial \ell}{\partial \alpha_j}\frac{\partial \ell}{\partial \alpha_k}\right] = \mathbb{E}^{P_{XY}}\left[\phi(w_j X + b_j)\phi(w_k X + b_k)(f_\theta(X) - Y)^2\right]$$

$$= \int \phi(w_j x + b_j)\phi(w_k x + b_k)p(x) \int (f_\theta(x) - y)^2 p(y|x; \theta)\, dy\, dx$$

$$= \int \phi(w_j x + b_j)\phi(w_k x + b_k)p(x)\, dx$$

$$= \mathbb{E}^{P_X}\left[\phi(w_j X + b_j)\phi(w_k X + b_k)\right].$$

Similarly,

$$
\begin{aligned}
g_{w_i w_j} &= \mathbb{E}^{P_{XY}}\left[\frac{\partial \ell}{\partial w_i}\frac{\partial \ell}{\partial w_j}\right] \\
&= \mathbb{E}^{P_{XY}}\left[\alpha_i \alpha_j X^2 \phi'(w_i X + b_i)\phi'(w_j X + b_j)(f_\theta(X) - Y)^2\right] \\
&= \alpha_i \alpha_j \int x^2 \phi'(w_i x + b_i)\phi'(w_j x + b_j)p(x)\underbrace{\int (f_\theta(x) - y)^2 p(y|x;\theta)dy}_{=1}\, dx \\
&= \alpha_i \alpha_j \int x^2 \phi'(w_i x + b_i)\phi'(w_j x + b_j)p(x)\, dx \\
&= \alpha_i \alpha_j \mathbb{E}^{P_X}\left[X^2 \phi'(w_i X + b_i)\phi'(w_j X + b_j)\right].
\end{aligned}
$$

Also, by similar manipulations we obtain

$$
\begin{aligned}
g_{b_i b_j} &= \mathbb{E}^{P_{XY}}\left[\frac{\partial \ell}{\partial b_i}\frac{\partial \ell}{\partial b_j}\right] = \alpha_i \alpha_j \mathbb{E}^{P_X}\left[\phi'(w_i X + b_i)\phi'(w_j X + b_j)\right] \\
g_{\alpha_j w_i} &= \alpha_i \mathbb{E}^{P_X}\left[X\phi(w_j X + b_j)\phi'(w_i X + b_i)\right] \\
g_{\alpha_j b_i} &= \alpha_i \mathbb{E}^{P_X}\left[\phi(w_j X + b_j)\phi'(w_i X + b_i)\right] \\
g_{w_i b_k} &= \alpha_i \alpha_k \mathbb{E}^{P_X}\left[X\phi'(w_i X + b_i)\phi'(w_k X + b_k)\right].
\end{aligned}
$$

We note that all the above coefficients depend on the input density, $p(x)$, the activation function of the neurons in the hidden layer, $\phi(x)$, as well as the parameters of the network.

14.7 The Natural Gradient

In order to minimize the cost function, $C(w, b)$, which depends on weights and biasses, the gradient descent method was employed. This involved taking a step $\eta > 0$ into the direction of the negative gradient, $-(\nabla_w C, \nabla_b C)$. This gradient is computed using the flat geometry of the parameter space Θ induced by the Euclidean metric δ_{ij}. The idea of this section is to apply the gradient descent method on the coordinate space Θ but with a gradient computed with respect to the Fisher metric. This method is desirable because it converges faster to the optimal parameter value, θ^*, since the steepest direction is not captured by the Euclidean gradient, but by its natural gradient, as it was pointed out in Amari [6]. This section will introduce this concept and present its main properties.

Let $(\mathcal{S} = \{p(x, y; \theta); \theta \in \Theta\}, g)$ be the neuromanifold associated with a given neural network. The parameter space Θ can be endowed with the metric $g(\theta)$ induced from $(\mathcal{S}, g)$ as in the following. If $\left\{\frac{\partial}{\partial \theta_i}\right\}_{1 \leq i \leq N}$ denote the coordinate vectors on Θ, then it suffices to define the metric on this basis as

$$g\left(\frac{\partial}{\partial \theta_i}, \frac{\partial}{\partial \theta_j}\right) = g_{ij}(\theta).$$

Thus, the parameter space together with $g(\theta)$ becomes the Riemann manifold $(\Theta, g(\theta))$.

Consider a smooth function defined on the parameters space, $f : \Theta \to \mathbb{R}$ (in particular, this can be any cost function). The multidirectional change of f is described by its gradient. The *Euclidean gradient* is the vector field

$$\nabla_{Eu} f = \sum_{k=1}^{N} \frac{\partial f}{\partial \theta_k} e_k = \left(\frac{\partial f}{\partial \theta_1}, \ldots, \frac{\partial f}{\partial \theta_N}\right)^T,$$

where $\{e_k\}_k$ is the natural orthonormal basis in $\mathbb{R}^N$. This type of gradient was very useful in the classical gradient descent method presented in Chapter 4. However, in the case when Θ is endowed with the Fisher metric $g(\theta)$, the gradient of f has to be adjusted accordingly.

The *natural gradient* of f is the gradient taken with respect to the Fisher metric $g(\theta)$. This can be written using the basis of coordinate vectors $\left\{\frac{\partial}{\partial \theta_k}\right\}_k$ as

$$\nabla_g f = \sum_{k=1}^{N} (\nabla_g f)^k \frac{\partial}{\partial \theta_k}, \tag{14.7.29}$$

with components given by $(\nabla_g f)^k = \sum_{j=1}^{N} g^{kj}(\theta) \frac{\partial f}{\partial \theta_j}$, where $g^{kj}(\theta)$ are the coefficients of the inverse matrix, $g^{-1}(\theta)$. An equivalent formula for the natural gradient (14.7.29) in terms of the Euclidean gradient is

$$\nabla_g f = g(\theta)^{-1} \nabla_{Eu} f. \tag{14.7.30}$$

As an application, the magnitudes of the Euclidean and natural gradients with respect to the Fisher metric are related by

$$\begin{aligned}
\|\nabla_g f\|_g^2 &= (\nabla_{Eu} f)^T g^{-1}(\theta) \nabla_{Eu} f \\
\|\nabla_{Eu} f\|_g^2 &= (\nabla_{Eu} f)^T g(\theta) \nabla_{Eu} f,
\end{aligned}$$

see Exercise 14.13.12. The multiplication by the matrix $g(\theta)^{-1}$ in formula (14.7.30) rotates and scales the Euclidean gradient $\nabla_{Eu} f$ to obtain the natural gradient $\nabla_g f$. Since the gradients $\nabla_g f$ and $\nabla_{Eu} f$ vanish at the same value of

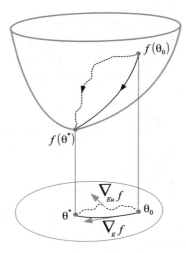

Figure 14.4: *The natural gradient descent arrives faster to the minimum than the Euclidean gradient descent method does.*

θ, see Exercise 14.13.12, it follows that both variants of the gradient descent method arrive to the same minimum $f(\theta^*)$ starting from the same initial point $f(\theta_0)$, but on different paths, see Fig. 14.4. Replacing the Euclidean gradient by the natural gradient in the gradient descent method improves the efficiency of the method. The next section deals with applications of this concept.

Remark 14.7.1 It is worth noting that the Euclidean gradient of the log-likelihood function can be used to represent the Fisher matrix $g(\theta) = \big(g_{ij}(\theta)\big)_{ij}$ as

$$g(\theta) = \mathbb{E}^{P_{XY}}\big[\nabla_{Eu}\ell(X,Y;\theta)(\nabla_{Eu}\ell(X,Y;\theta))^T\big] = \mathbb{E}^{P_{XY}}\big[\nabla_{Eu}\ell\,(\nabla_{Eu}\ell)^T\big].$$

14.8   The Natural Gradient Learning Algorithm

We have seen in Chapter 4 that the steepest direction of a cost function, $C(\theta)$, defined on a Euclidean space is given by its Euclidean gradient, $\nabla_{Eu}C$. This result is not valid in the case of a cost function, $C(\theta)$, which is defined on a curved space, such as the Riemannian manifold, (Θ, g). The steepest direction in this case is realized in the natural gradient direction, $\nabla_g C$. This section deals with the effects of using the natural Riemannian gradient in neural learning and it is based on the work of Amari et al., see [5], [6], [99], [130].

The steepest descent direction We start by considering a unit vector field, $V = \sum V^i(\theta)\frac{\partial}{\partial \theta_i}$, with $\|V\|_g = 1$, tangent to the parameters space (Θ, g) and investigate the change of the cost function, $C(\theta)$, in the direction of V. The rate of change of $C(\theta)$ with respect to V is denoted either by $V(C)$, or by $\frac{\partial C}{\partial V}$, and is equal to

$$\frac{\partial C}{\partial V} = V(C) = \sum_i V^i \frac{\partial C}{\partial \theta_i} = \langle V, \nabla_{Eu} C \rangle. \tag{14.8.31}$$

By Exercise 14.13.11, part (c), we have

$$\langle V, \nabla_{Eu} C \rangle = g(V, \nabla_g C).$$

Using Cauchy-Schwarz inequality together with equation (14.8.31) yields

$$\frac{\partial C}{\partial V} = g(V, \nabla_g C) \le \|V\|_g \|\nabla_g C\|_g = \|\nabla_g C\|_g,$$

with equality for the case when V and $\nabla_g C$ are proportional. Therefore, the rate $\frac{\partial C}{\partial V}$ is highest in the direction of $V = \nabla_g C / \|\nabla_g C\|_g$.

Therefore, the steepest descent direction of the cost function $C(\theta)$ is the negative natural gradient, which is given by

$$-\nabla_g C(\theta) = -g^{-1}(\theta)\nabla_{Eu} C(\theta).$$

For a proof variant involving Lagrange multipliers see Exercise 14.13.9.

It is worth noting that in the particular case when (Θ, g) is the Euclidean space, $(\mathbb{R}^n, \delta_{ij})$, then $g^{-1} = \mathbb{I}_n$, and hence we retrieve the direction of the Euclidean gradient.

The *natural gradient learning algorithm* introduced in [6] updates the parameter θ_n by the rule

$$\theta_{n+1} = \theta_n - \eta_n \nabla_g C(\theta_n), \tag{14.8.32}$$

where the learning rate $\eta_n \to 0$ as $n \to \infty$ in a certain way.

It was suggested in [130] that replacing $\nabla_{Eu} C$ by $\nabla_g C$ helps with the elimination of situations when the iteration is being trapped in a plateau. There are also other reasons why the natural gradient learning is more efficient than the usual gradient descent. Before getting to them, we recall first two types of learning algorithms, batch and online learning.

Batch learning In this case all training examples in a batch are used to obtain the optimal weight vector. If the training set is $\{(x_1, z_1), \ldots, (x_n, z_n)\}$, then the cost function depends on all samples as

$$C(\theta) = \frac{1}{2n} \sum_{j=1}^{n} |z_j - f_\theta(x_j)|^2.$$

If data is sampled from the same training distribution, $p_{XZ}(\theta)$, the cost can be also written as an expectation

$$C(\theta) = \frac{1}{2} E^{P_{XZ}(\theta)}[(Z - f_\theta(X))^2].$$

The regular gradient descent method in this case is described by taking steps in as

$$\theta_{n+1} = \theta_n - \eta_n \nabla C(\theta_n), \quad n = 0, 1, 2, \ldots.$$

Online learning This uses each example only once, at the observation time, assuming that the examples are given one at a time. The cost function takes the simple form

$$C(x_n, z_n, \theta_n) = \frac{1}{2}|z_n - f_{\theta_n}(x_n)|^2.$$

The gradient descent method proposed in [3] and [107] employs the rule

$$\theta_{n+1} = \theta_n - \eta_n \nabla C(x_n, y_n, \theta_n).$$

In general, the convergence of θ_n to the true minimum, θ^*, of the cost function is more accurate in the case of batch learning rather than in the case of the online learning. However, if the learning rate, η_n converges to 0 in a certain manner, and the Euclidean gradient, $\nabla C(x_n, y_n, \theta_n)$, is replaced by the natural gradient, then the online learning becomes asymptotically as efficient as the batch learning. In order to present this idea further, we shall introduce first a few notions regarding estimators.

Types of estimators Let $\mathcal{S} = \{p_\theta; \theta \in \Theta\}$ be a family of densities and $\hat{\theta} = \hat{\theta}(x_1, \ldots, x_n)$ be an estimator of the parameter θ based on data $\{x_1, \ldots, x_n\}$ sampled from the distribution p_θ. Then:

- $\hat{\theta}$ is called *unbiased* if $\mathbb{E}^{\hat{P}_X}[\hat{\theta}] = \theta$.
- $\hat{\theta}_n = \hat{\theta}(x_1, \ldots, x_n)$ is called *consistent* if $\hat{\theta}_n \to \theta$ in probability, i.e., $\lim_{n\to\infty} P(|\hat{\theta}_n - \theta| < \epsilon) = 1$, for any $\epsilon > 0$. See Appendix, section D.6.1, for the definition of convergence in probability.
- $\hat{\theta}$ is *Fisher-efficient* if it is unbiased and reaches the lower bound in the Cramér-Rao inequality

$$Cov(\hat{\theta}) \geq g^{-1}(\theta), \quad \forall \theta \in \Theta,$$

i.e., it is a minimum variance unbiased estimator.[2]

---

[2]If A and B are two square matrices, we write $A \geq B$ if $A - B$ is positive semidefinite, i.e., all its eigenvalues are nonnegative.

- $\hat{\theta}_n$ is *asymptotically Fisher-efficient* if it attains equality in the Cramér-Rao bound asymptotically, i.e.,

$$\lim_{n\to\infty} Cov(\hat{\theta}_n) = g^{-1}(\theta), \quad \forall \theta \in \Theta.$$

For instance, in a correctly specified model, a well-known result states that the maximum likelihood estimator, $\hat{\theta}_{MLE,N}$, depending on N independent samples x_j,

$$\hat{\theta}_{MLE,N} = \arg\min_{\theta} \frac{1}{N}\sum_{j=1}^{N} \ln p_\theta(x_j) = \arg\max_{\theta} \prod_{j=1}^{N} p_\theta(x_j),$$

is both consistent, ($\hat{\theta}_{MLE,N} \to \theta$, $N \to \infty$, in probability) and asymptotically efficient (Cramér-Rao lower bound is reached when the sample size, n, tends to infinity). For other examples, see Exercises 14.13.13 and 14.13.14.

In this case Cramér-Rao inequality can be written as

$$\mathbb{E}[(\hat{\theta}_{MLE,N} - \theta)(\hat{\theta}_{MLE,N} - \theta)^T] \geq \frac{1}{N}g^{-1}(\theta),$$

see Exercise 14.13.16. The fact that the maximum likelihood estimator is asymptotically Fisher-efficient can be written as

$$\lim_{N\to\infty} N\mathbb{E}[(\hat{\theta}_{MLE,N} - \theta)(\hat{\theta}_{MLE,N} - \theta)^T] = g^{-1}(\theta),$$

see also Exercise 14.13.16.

Fisher efficiency in online learning Since in online learning training examples are used only once, as they appear, the asymptotic performance of online learning should be not as good as the optimal batch procedure, when all examples are reused for several epochs. The next result states that actually the efficiency holds, provided some extra conditions are satisfied. The next result can be found in Amari [6]:

Theorem 14.8.1 *Let the cost function be the log-likelihood function, $C(x, z; \theta) = \ln p(x, z; \theta)$. Let θ^* represent the parameter true value of the distribution from which the data are sampled, i.e., $(x_n, z_n) \sim p(x, z; \theta^*)$. Then the natural gradient learning rule for the online learning*

$$\hat{\theta}_{n+1} = \hat{\theta}_n - \frac{1}{n}\nabla_g C(x_n, z_n; \hat{\theta}_n)$$

produces an estimator $\hat{\theta}_n$, which is asymptotically Fisher-efficient, i.e.,

$$\lim_{n\to\infty} n\mathbb{E}[(\hat{\theta}_n - \theta^*)(\hat{\theta}_n - \theta^*)^T] = g(\theta^*)^{-1}.$$

The proof idea is to consider the covariance matrix

$$V_n = \mathbb{E}[(\hat{\theta}_n - \theta^*)(\hat{\theta}_n - \theta^*)^T]$$

and show that it verifies the asymptotical relation

$$V_n = \frac{1}{n} g(\theta^*)^{-1} + O\left(\frac{1}{n^2}\right).$$

This is obtained by subtracting θ^* from both sides of the online learning relation

$$\hat{\theta}_{n+1} = \hat{\theta}_n - \frac{1}{n} g^{-1}(\hat{\theta}_n) \nabla_\theta \ell(x_n, z_n; \hat{\theta}_n)$$

and then taking the expectation of the square of both sides. The computation involves the linear approximation of the derivative of the log-likelihood function

$$\nabla_\theta \ell(x_n, z_n; \hat{\theta}_n) = \nabla_\theta \ell(x_n, z_n; \theta^*) + (\hat{\theta}_n - \theta^*)^T \nabla_\theta \nabla_\theta \ell(x_n, z_n; \theta^*) + O(\|\hat{\theta}_n - \theta^*\|^2),$$

as well as a few more relations

$$\mathbb{E}[\nabla_\theta \ell(x, y; \theta^*)] = 0$$
$$\mathbb{E}[\nabla_\theta \nabla_\theta \ell(x, y; \theta^*)] = -g(\theta^*)$$
$$g(\theta_n) = g(\theta^*) + O\left(\frac{1}{n}\right).$$

Adaptive implementation The natural gradient algorithm requires that the inverse of the Fisher metric, $g(\theta)^{-1}$ is known, fact that hardly occurs in a closed form. An adaptive method for directly estimating the inverse $g(\theta)^{-1}$ and applying the natural gradient online learning is given in [8]

$$\hat{g}_{n+1}^{-1} = (1 + \epsilon_n)\hat{g}_n^{-1} - \epsilon_n \hat{g}_n^{-1} \nabla_{Eu} f_n (\nabla_{Eu} f_n)^T \hat{g}_n^{-1}$$
$$\theta_{n+1} = \theta_n - \eta_n \hat{g}_n^{-1} \nabla_{Eu} \ell(x_n, z_n; \theta_n),$$

where $f_n = f_\theta(x_n)$ is the input-output mapping, $g_n = g(\theta_n)$, and $\epsilon_n > 0$ is a small learning rate.

14.9 Log-likelihood and the Metric

This section states a relation between the magnitude in the change of the log-likelihood function in terms of the Euclidean gradient of the input-output mapping.

If the parameters of a neural network are perturbed infinitesimally from θ to $\theta' = \theta + d\theta$, then the input-output mapping changes from $f_\theta(x)$ to $f_{\theta'}(x)$, where

$$f_{\theta'}(x) = f_\theta(x) + \sum_{k=1}^{N} \frac{\partial f_\theta(x)}{\partial \theta_k} d\theta_k = f_\theta(x) + \langle \nabla_{Eu} f, d\theta \rangle, \qquad (14.9.33)$$

with the infinitesimal perturbation vector $d\theta = (d\theta_1, \ldots, d\theta_N)^T$.

The square of the distance between the infinitesimally separated points θ and θ' in the parameters space Θ with respect to the metric g is given by the quadratic form

$$\|d\theta\|_g^2 = \|\theta' - \theta\|_g^2 = (d\theta)^T g(\theta) d\theta = \sum_{i,j} g_{ij}(\theta) d\theta_i d\theta_j.$$

We note the similarity with the Euclidean distance, $\|d\theta\|_{Eu}^2 = \sum_j (d\theta_j)^2$. The infinitesimal change of parameters has an effect on the change of the log-likelihood function. This is given by the next result.

Proposition 14.9.1 (*a*) *The infinitesimal change in the log-likelihood function is*

$$d\ell(x, y; \theta) = (y - f_\theta(x)) df_\theta(x);$$

(*b*) *The square of the magnitude is given by*

$$\|d\ell(x, y; \theta)\|_g^2 = (y - f_\theta(x))^2 (\nabla_{Eu} f_\theta(x))^T g(\theta) \nabla_{Eu} f_\theta(x) O(\|d\theta\|^2).$$

Proof: (*a*) For $\theta' = \theta + d\theta$, the change in the log-likelihood function $\ell(\theta) = \ell(x, y; \theta)$ is

$$
\begin{aligned}
\ell(\theta') - \ell(\theta) &= \sum_k \frac{\partial \ell(\theta)}{\partial \theta_k} d\theta_k = \sum_k (y - f_\theta(x)) \frac{\partial f_\theta(x)}{\partial \theta_k} d\theta_k \\
&= (y - f_\theta(x)) \sum_k \frac{\partial f_\theta(x)}{\partial \theta_k} d\theta_k = (y - f_\theta(x))(f_{\theta'}(x) - f_\theta(x)),
\end{aligned}
$$

where we used formulas (14.3.12) and (14.9.33). Substituting now $d\ell(x, y; \theta) = \ell(\theta') - \ell(\theta)$ and $df_\theta(x) = f_{\theta'}(x) - f_\theta(x)$, we obtain the desired formula.

(*b*) Taking the square of the magnitude in the g-metric in the relation from part (*a*), we have

$$\|d\ell(x, y; \theta)\|_g^2 = (y - f_\theta(x))^2 \|df_\theta(x)\|_g^2.$$

The second factor in the right side can be evaluated as

$$
\|df_\theta(x)\|_g^2 = (df_\theta(x))^T g(\theta) df_\theta(x) = \sum_k \frac{\partial f_\theta}{\partial \theta_k}(d\theta_k)^T g(\theta) \sum_j \frac{\partial f_\theta}{\partial \theta_j} d\theta_j
$$

$$
= \sum_{j,k} \frac{\partial f_\theta}{\partial \theta_k} \frac{\partial f_\theta}{\partial \theta_j}(d\theta_k)^T g(\theta) d\theta_j = \sum_{j,k} \frac{\partial f_\theta}{\partial \theta_k} \frac{\partial f_\theta}{\partial \theta_j} g_{jk}(\theta) \|d\theta_k\| \, \|d\theta_j\|
$$

$$
= g(\nabla_{Eu} f_\theta, \nabla_{Eu} f_\theta) O(\|d\theta\|^2) = \|\nabla_{Eu} f_\theta\|_g^2 O(\|d\theta\|^2).
$$

We have used the formula $(d\theta_k)^T g(\theta) d\theta_j = g_{jk}(\theta) O(\|d\theta\|^2)$, which follows from the linear algebra relation $e_k^T A e_j = A_{jk}$, where A is a matrix, A_{jk} the (j,k)th entry and $\{e_k\}$ an orthonormal basis; in our case, $A = g(\theta)$ and $e_k = d\theta_k/\|d\theta_k\|$. Expressing $\|\nabla_{Eu} f_\theta\|_g^2$ as in Exercise 14.13.12, part (b), we obtain the desired relation. ∎

14.10 Relation to the Kullback-Leibler Divergence

It is known that the proximity between two probability densities, $p(x, y; \theta)$ and $p(x, y; \theta')$, on the neuromanifold $\mathcal{S}$ associated to a neural net can be measured using the Kullback-Leibler divergence. This section shows the relation between this proximity and the Riemannian distance between θ and $\theta' = \theta + d\theta$ in the parameters space $(\Theta, g(\theta))$, see Fig. 14.5.

The following result will be useful shortly.

Lemma 14.10.1 *If $\ell(x, y; \theta)$ denotes the log-likelihood function, we have*

$$
\mathbb{E}^{P_{XY}(\theta)}\left[\frac{\partial}{\partial \theta_j}\ell(X, Y; \theta)\right] = 0.
$$

Proof: Using the definition of the expectation and log-likelihood function, we have

$$
\mathbb{E}^{P_{XY}(\theta)}\left[\frac{\partial}{\partial \theta_j}\ell(X, Y; \theta)\right] = \iint \frac{\partial}{\partial \theta_j}\ell(x, y; \theta)\, p(x, y; \theta)\, dx dy
$$

$$
= \iint \frac{\partial}{\partial \theta_j} p(x, y; \theta)\, dx dy = \frac{\partial}{\partial \theta_j} \underbrace{\iint p(x, y; \theta)\, dx dy}_{=1}
$$

$$
= 0.
$$

∎

The previous result can be used to write the Fisher information in the covariance form:

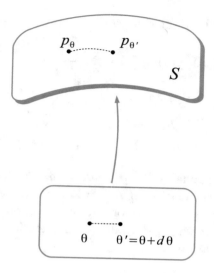

Figure 14.5: *The Riemannian distance between θ and θ' is related to the Kullback-Leibler divergence of p_θ and $p_{\theta'}$.*

Corollary 14.10.2 *The Fisher matrix is given by the covariance matrix*

$$g_{ij}(\theta) = Cov(\partial_{\theta_i}\ell, \, \partial_{\theta_j}\ell), \qquad (14.10.34)$$

where $\partial_{\theta_i}\ell = \frac{\partial}{\partial\theta_i}\ell(X, Y; \theta)$.

Proof: Using Lemma 14.10.1 and the covariance definition, we have

$$
\begin{aligned}
Cov(\partial_{\theta_i}\ell, \, \partial_{\theta_j}\ell) \; &= \; \mathbb{E}^{P_{XY}(\theta)}\left[\frac{\partial}{\partial\theta_i}\ell(X, Y; \theta)\frac{\partial}{\partial\theta_j}\ell(X, Y; \theta)\right] \\
&\quad - \underbrace{\mathbb{E}^{P_{XY}(\theta)}\left[\frac{\partial}{\partial\theta_i}\ell(X, Y; \theta)\right]}_{=0} \underbrace{\mathbb{E}^{P_{XY}(\theta)}\left[\frac{\partial}{\partial\theta_j}\ell(X, Y; \theta)\right]}_{=0} \\
&= \; g_{ij}(\theta).
\end{aligned}
$$

$\blacksquare$

Denote, for simplicity, $p_\theta = p(x, y; \theta)$ and consider $\theta' = \theta + d\theta$. The next result shows that the proximity between p_θ and $p_{\theta'}$ measured by the Kullback-Leibler divergence is half the squared Riemannian distance between θ and θ' in the space (Θ, g).

Proposition 14.10.3 *The linear and quadratic approximations of the Kullback-Leibler divergence are given by:*

(a) $D_{KL}(p_\theta||p_{\theta'}) = O(||d\theta||^2);$

(b) $D_{KL}(p_\theta||p_{\theta'}) = \frac{1}{2}||d\theta||_g^2 + O(||d\theta||^3).$

Proof: (a) Let $F_\theta : \mathbb{R}^N \to [0, \infty)$, given by $F_\theta(u) = D_{KL}(p_\theta||p_{\theta+u})$. From divergence properties, $F_\theta(0) = D_{KL}(p_\theta||p_\theta) = 0$. In order to compute the partial derivative, we consider the variation in the e_j-direction. Using the definition of the derivative as a limit, we have

$$
\begin{aligned}
\frac{\partial}{\partial u_j} F_\theta(0) &= \lim_{t \to 0} \frac{F_\theta(te_j) - F_\theta(0)}{t} = \lim_{t \to 0} \frac{D_{KL}(p_\theta||p_{\theta+te_j})}{t} \\
&= \lim_{t \to 0} \frac{1}{t} \mathbb{E}^{p_\theta}[\ell(\theta) - \ell(\theta + te_j)] = -\lim_{t \to 0} \mathbb{E}^{p_\theta}\left[\frac{\ell(\theta + te_j) - \ell(\theta)}{t}\right] \\
&= -\mathbb{E}^{p_\theta}\left[\lim_{t \to 0} \frac{\ell(\theta + te_j) - \ell(\theta)}{t}\right] = -\mathbb{E}^{p_\theta}\left[\frac{\partial}{\partial \theta_j}\ell(\theta)\right] = 0,
\end{aligned}
$$

where the last identity is provided by Lemma 14.10.1.
Since the first two terms of the right side of the linear approximation

$$
F_\theta(u) = F_\theta(0) + \sum_j \frac{\partial}{\partial u_j} F_\theta(0) du_j + O((||du||^2)
$$

are zero, we obtain $F_\theta(u) = O((||du||^2)$, which is equivalent to

$$
D_{KL}(p_\theta||p_{\theta'}) = O(||d\theta||^2).
$$

(b) Similarly, taken second partial derivatives we obtain

$$
\begin{aligned}
\frac{\partial}{\partial u_i} \frac{\partial}{\partial u_j} F_\theta(u) &= \frac{\partial}{\partial u_i} \frac{\partial}{\partial u_j} D_{KL}(p_\theta||p_{\theta+u}) \\
&= \frac{\partial}{\partial u_i} \frac{\partial}{\partial u_j} \mathbb{E}^{p_\theta}[\ell(\theta) - \ell(\theta + u)] \\
&= -\mathbb{E}^{p_\theta}\left[\frac{\partial}{\partial u_i} \frac{\partial}{\partial u_j}\ell(\theta + u)\right].
\end{aligned}
$$

Using the definition of the Fisher metric coefficients given by (14.2.8), we have

$$
\begin{aligned}
\frac{\partial}{\partial u_i} \frac{\partial}{\partial u_j} F_\theta(u)\Big|_{u=0} &= -\mathbb{E}^{p_\theta}\left[\frac{\partial}{\partial u_i} \frac{\partial}{\partial u_j}\ell(\theta + u)\right]\Big|_{u=0} \\
&= -\mathbb{E}^{p_\theta}\left[\frac{\partial}{\partial u_i} \frac{\partial}{\partial u_j}\ell(\theta)\right] = g_{ij}(\theta).
\end{aligned}
$$

We have assumed the derivatives commute with the expectation operator, fact that always hold for densities of Gaussian type.

The quadratic approximation

$$
\begin{aligned}
F_\theta(u) &= F_\theta(0) + \sum_j \frac{\partial}{\partial u_j} F_\theta(0) du_j + \frac{1}{2} \sum_{j,k} \frac{\partial}{\partial u_i} \frac{\partial}{\partial u_j} F_\theta(0)\, du_i du_j + O((\|du\|^3) \\
&= \frac{1}{2} \sum_{j,k} \frac{\partial}{\partial u_i} g_{ij}(\theta)\, du_i du_j + O(\|du\|^3)
\end{aligned}
$$

can be written as

$$
\begin{aligned}
D_{KL}(p_\theta \| p_{\theta'}) &= \frac{1}{2} \sum_{j,k} \frac{\partial}{\partial u_i} g_{ij}(\theta)\, d\theta_i d\theta_j + O(\|d\theta\|^3) \\
&= \frac{1}{2}(d\theta)^T g(\theta) d\theta + O(\|d\theta\|^3) = \frac{1}{2}\|d\theta\|_g^2 + O(\|d\theta\|^3).
\end{aligned}
$$

∎

Some concepts of Differential Geometry defined on the Riemannian manifold (Θ, g) can be expressed in terms of statistical concepts on the neuromanifold $\mathcal{S}$. We shall do this for the energy and length of a curve.

Let $\theta : [a, b] \to \Theta$ be a differentiable curve in the parameter space Θ, endowed with the Fisher metric $g(\theta)$. The *energy* of the curve is the integral of the kinetic energy density along the curve

$$
\mathcal{E}(\theta) = \frac{1}{2} \int_a^b \|\dot{\theta}(t)\|_g^2\, dt.
$$

We shall provide a quantitative characterization of the energy in terms of the Kullback-Leibler divergence of the probability density $p_{\theta(t)}$. Note that the assignment $t \to p_{\theta(t)}$ is a curve on the neuromanifold $\mathcal{S}$.

We consider an equidistant partition $a = t_0 < t_1 < \cdots < t_n = b$, with $\Delta t = t_{k+1} - t_k = (b - a)/n$, and denote $\theta_k = \theta(t_k)$. The Riemannian distance between the points θ_k and θ_{k+1}, for n large, can be expressed by Proposition 14.10.3, part (b), as

$$
\frac{1}{2}\|\theta_{k+1} - \theta_k\|_g^2 = D_{KL}(p_{\theta_k} \| p_{\theta_{k+1}}).
$$

Using this, we can evaluate the energy as

$$\mathcal{E}(\theta) = \frac{1}{2} \int_a^b \|\dot\theta(t)\|_g^2 \, dt = \lim_{n\to\infty} \sum_{k=1}^n \frac{1}{2} \frac{\|\theta_{k+1} - \theta_k\|_g^2}{(\Delta t)^2} \Delta t$$

$$= \lim_{n\to\infty} \sum_{k=1}^n \frac{1}{\Delta t} D_{KL}(p_{\theta_k}\|p_{\theta_{k+1}})$$

$$= \lim_{n\to\infty} \frac{n}{b-a} \sum_{k=1}^n D_{KL}(p_{\theta_k}\|p_{\theta_{k+1}}).$$

A similar problem regarding deformation of oval curves has been recently asked in [21].

The length of the curve $\theta(t)$ is obtained integrating the speed $\|\dot\theta(t)\|_g$ along the curve with respect to the time parameter t as

$$L(\theta) = \int_a^b \|\dot\theta(t)\| \, dt.$$

The length can be expressed in terms of Kullback-Leibler divergence as in the following

$$L(\theta) = \int_a^b \|\dot\theta(t)\| \, dt = \lim_{n\to\infty} \sum_{k=1}^n \frac{\|\theta_{k+1} - \theta_k\|_g}{\Delta t} \Delta t$$

$$= \lim_{n\to\infty} \sum_{k=1}^n \|\theta_{k+1} - \theta_k\|_g = \sqrt{2} \lim_{n\to\infty} \sum_{k=1}^n D_{KL}(p_{\theta_k}\|p_{\theta_{k+1}})^{1/2},$$

where the last identity has used Proposition 14.10.3.

14.11 Simulated Annealing Method

In the previous sections we have added a Gaussian noise, $n \sim \mathcal{N}(0,1)$, to the output of a neural network, see (14.1.1), and then we approached the problem by techniques of Information Geometry. We've seen that learning is performed by the natural gradient algorithm, which involves the inverse of the Fisher metric. In this section we shall use the previous results to make a relation with the simulated annealing method.

The regular gradient descent method applied to a deep neural network with the output $Y = f_\theta(X)$ provides in most cases, due to the high non-linearity of f_θ, only local minima of the cost function. In order to obtain a global minimum, a variant of the simulated annealing method will be used. For this we shall consider an adjustable noise, $n_T \sim \mathcal{N}(0, T^2)$, where T plays the role of temperature.

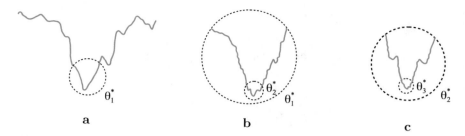

Figure 14.6: *Annealing method:* **a.** *For a large temperature, T_1, the optimal parameter θ_1^* is located in a neighborhood of the global minimum.* **b.** *Decreasing the temperature to T_2, we obtain a more accurate approximation of the global minimum given by the new optimum value, θ_2^*.* **c.** *Continuing to decrease temperature we obtain more and more accurate approximations of the global minimum.*

The heuristic idea is to start optimizing the cost function starting from a large temperature, T, and then decreasing it to zero, according to a certain schedule. If the schedule is $T_1 > T_2 > \cdots > T_N > 0$, we denote by θ_1^* the optimal parameter corresponding to temperature T_1, obtained by the natural gradient learning method. The search of the next optimal parameter value, θ_2^*, which corresponds to temperature T_2, starts from θ_1^*, see Fig. 14.6. In general, the optimal value θ_k^*, corresponding to the temperature, T_k, is obtained by the natural gradient descent, which starts the search from the initial value θ_{k-1}^*. The last optimal value, θ_N^*, corresponding to the lowest temperature, T_N, is the closest to the true global minimum of the cost function.

14.12   Summary

This chapter provides an introduction to the Informational Geometry of neural networks. Such networks are noisy and the output is characterized by a probability density parametrized by weights and biasses. Thus, each distribution can be considered as a point in a space, which becomes a Riemannian manifold when endowed with the Fisher metric. This is the neuromanifold associated with the given network.

The topics covered here deal mainly with the intrinsic geometry of a neuromanifold, which is defined by the Fisher information metric. This metric is computed explicitly for a few particular types of networks and is applied to the natural gradient learning algorithm, which is an adapted version of the gradient descent algorithm for Riemannian manifolds. Inserting noise in a neural network is like increasing temperature of a thermodynamical system. A variant of the simulated annealing method works in combination with the natural gradient descent method in order to obtain the global minimum of the cost function.

There are several important topics of Information geometry which are left out of this chapter, such as the extrinsic geometry of the neuromanifold, which describes the relative geometry of a network with respect to a larger manifold of probability densities. Topics, such as embedded curvature, dual connections, etc., can be found by the interested reader in Amari [4] or Calin et al. [22]. For more applications of Informational Geometry to Machine Learning, the reader is referred to [7].

14.13 Exercises

Exercise 14.13.1 Let $\{v_1, \ldots, v_n\}$ be an orthonormal basis in $\mathbb{R}^n$ (i.e., a set of n vectors such that $v_i^T v_j = \delta_{ij}$).

(a) If G is an $n \times n$ symmetric matrix such that $v_j^T G v_j = 0$, for all $1 \le j \le n$, show that $G = \mathbb{O}_n$ (the n-dimensional zero matrix).

(b) If A and B are two $n \times n$ symmetric matrices such that $v_j^T A v_j = v_j^T B v_j$, for all $1 \le j \le n$, show that $A = B$.

Exercise 14.13.2 A 2×2 matrix is said to be a rotation of angle ϕ if it has the form

$$R = \begin{pmatrix} u_1 & u_2 \\ v_1 & v_2 \end{pmatrix} = \begin{pmatrix} \cos \phi & \sin \phi \\ -\sin \phi & \cos \phi \end{pmatrix}.$$

We note that u and v are orthonormal vectors, and $\det R = 1$.

(a) Let $X = (X_1, X_2) \sim \mathcal{N}(0, \mathbb{I}_2)$ and consider the rotation matrix R as above. Show that $u^T X$ and $v^T X$ are independent, where $u^T = (u_1, u_2)$ and $v^T = (v_1, v_2)$.

(b) What are the distributions of $u^T X$ and $v^T X$?

(c) Show that if u and v are two orthonormal vectors in the plane, then there is $\phi \in [0, 2\pi)$ such that $u^T = (\cos \phi, \sin \phi)$ and $v^T = (-\sin \phi, \cos \phi)$.

Exercise 14.13.3 Let $X = (X_1, X_2)^T$, with X_1, X_2 independent random variables.

(a) Consider u and w orthonormal vectors in $\mathbb{R}^2$. Show that $Y_1 = u^T X$ and $Y_2 = W^T X$ are also independent.

(b) Show that part (a) holds just if u and v are only orthogonal (the magnitudes of the vectors do not matter).

Exercise 14.13.4 Prove Lemma 14.5.2

Exercise 14.13.5 (a) Find the Fisher metric of a sigmoid neuron with the activation function $\phi(x) = \sigma(x)$, where $\sigma(x)$ denotes the logistic function. (b) Show the inequalities

$$0 \le g_{00} \le \frac{1}{4^2}, \qquad g_{0k}^2 \le \frac{1}{4^4}\mathbb{E}[X_k^2], \qquad g_{ij}^2 \le \frac{1}{4^4}\mathbb{E}[X_j^2 X_k^2].$$

(c) State and prove a variant of the inequalities given in part (b) in the general case of an activation function ϕ.

Exercise 14.13.6 Find the Fisher metric coefficients for a neuron with the input $X = (X_1, \ldots, X_n)$, with X_i independently identic distributed, $X_i \sim Unif[0,1]$.

Exercise 14.13.7 Find the Fisher metric coefficients for a one-hidden layer neural network with the activation function, $\phi(x) = x$. Write the result in terms of the network parameters and the first two moments of the input variable, X.

Exercise 14.13.8 Find the Fisher metric coefficients for a one-hidden layer neural network with the activation function, $\phi(x)$ in the case when the input is $X \sim \mathcal{N}(0,1)$.

Exercise 14.13.9 Consider the loss function $L : \Theta \to \mathbb{R}$, a vector v in the tangent space $T_\theta\Theta$ with $\|v\|_g^2 = 1$, and a learning step, $\eta > 0$. A small change of the parameters in the direction of v, of magnitude η, can be written as $dw = \eta v$, so that the linear approximation becomes

$$L(\theta + d\theta) = L(\theta) + \eta \nabla_{Eu} L(\theta)^T v.$$

We need to find the direction v such that $L(\theta + d\theta)$ is minimized. For this, we consider the Lagrange functional

$$F(v, \lambda) = \nabla L(\theta)^T v - \lambda \|v\|_g^2.$$

(a) Show that the variational equations $\dfrac{\partial F}{\partial v_i} = 0$ imply $\nabla_{Eu}L(w) = 2\lambda g(\theta)v$.
(b) Show that $v = \nabla_g L(\theta)/\|\nabla_g L(\theta)\|_g$.

Exercise 14.13.10 Let $p_{X_1}(x_1; \theta)$ and $p_{X_2}(x_2; \theta)$ be the probability densities of random variables X_1 and X_2, respectively. Then

$$g(X_1, X_2; \theta) = g(X_1; \theta) + g(X_2|X_1; \theta) = g(X_2; \theta) + g(X_1|X_2; \theta),$$

where $g(X_1|X_2; \theta)$ is the Fisher information defined by the conditional probability density $p_{X_1|X_2}(x_1|x_2; \theta)$ (i.e., the amount of information about θ contained in X_1, given X_2).

Exercise 14.13.11 Let $X = \sum_{k=1}^{N} X^k \frac{\partial}{\partial \theta_k}$ be a vector field on Θ. Show that:

(a) $\langle \nabla_{Eu} f, X \rangle_{Eu} = \sum_{k=1}^{N} X^k \frac{\partial f}{\partial \theta_k}$.

(b) $g(\nabla_g f, X) = \sum_{k=1}^{N} X^k \frac{\partial f}{\partial \theta_k}$.

(c) $\langle \nabla_{Eu} f, X \rangle_{Eu} = g(\nabla_g f, X)$.

Exercise 14.13.12 Show that:

(a) $\|\nabla_g f\|_g^2 = (\nabla_{Eu} f)^T g^{-1}(\theta) \nabla_{Eu} f$;

(b) $\|\nabla_{Eu} f\|_g^2 = (\nabla_{Eu} f)^T g(\theta) \nabla_{Eu} f$;

(c) $\nabla_{Eu} f$ and $\nabla_g f$ vanish at the same points.

Exercise 14.13.13 Consider the 1-dimensional random variable $X \sim \mathcal{N}(\mu, 1)$ and $\hat{\mu}(x_1, \ldots, x_n) = \frac{1}{n} \sum_{i=1}^{n} x_i$ be an estimator for the mean μ by n independent observations of the variable X.

(a) Show that $\hat{\mu}(x_1, \ldots, x_n)$ is an unbiased estimator of the mean, μ;

(b) Find the Fisher information of X;

(c) Show that $\hat{\mu}(x_1, \ldots, x_n)$ is Fisher-efficient.

Exercise 14.13.14 Let $X \sim Pois(\lambda)$ be a Poisson-distributed discrete random variable with parameter λ, i.e.,

$$P(X = k) = \frac{\lambda^k}{k!} e^{-\lambda}, \qquad k = 0, 1, 2, \ldots.$$

Construct a Fisher-efficient estimator for the parameter λ.

Exercise 14.13.15 Consider the independent, identically distributed random variables, $X_1, \ldots, X_N \sim X$, with $X \sim \mathcal{N}(\mu, 1)$, and consider their average

$$\overline{X} = \frac{1}{N}(X_1 + \cdots + X_N).$$

Show that the information contained in $\overline{X}$ about μ is the sum of the information of each individual variable about μ, i.e., $I(\overline{X}) = N\, I(X)$, where I denotes the Fisher information of a one-dimensional random variable.

Exercise 14.13.16 (a) Let X_1 and X_2 be two independent random variables with probability densities, $p_{X_1}(x_1; \theta)$, $p_{X_2}(x_2; \theta)$, which depend on the parameter θ. The information about θ contained in X_i is given by the Fisher information $g(X_i; \theta)$. Prove that the Fisher information contained in the pair (X_1, X_2) is the sum of individual Fisher informations

$$g(X_1, X_2; \theta) = g(X_1; \theta) + g(X_2; \theta).$$

(b) State and prove a generalization to n independent random variables.

(c) Show that the inverse of the Fisher information matrix contained in N identically distributed independent random variables $X_1, \ldots, X_N \sim X$ about θ is $\frac{1}{N} g^{-1}(X; \theta)$.

(d) Use part (c) to explain why the definition of an asymptotically efficient estimator $\hat{\theta}(N)$ of θ based on N independent identically distributed random variables, $X_1, \ldots, X_N$, is given by

$$\lim_{N \to \infty} N \mathbb{E}[(\hat{\theta}(N) - \theta)(\hat{\theta}(N) - \theta)^T] = g^{-1}(\theta),$$

where $g(\theta)$ is the Fisher information matrix corresponding to one of the random variables.

Part V
Other Architectures

Chapter 15

Pooling

Pooling is a machine learning technique that provides a summary of the input, selecting some essential local features such as maxima, minima, averages, etc.

It also acts as an information contractor; in the discrete case it decreases the dimension of the input by a certain factor. Hence its usefulness in classification problems.

The idea of pooling is to consider a partition of the domain of a function and replace the function on each partition element by the "most representative" value of the function on that set. This procedure leads to a simple function. A two-dimensional variant of pooling is used in the construction of convolutional neural networks.

15.1   Approximation of Continuous Functions

This section deals with the *max, min,* and *average-pooling* techniques applied in the context of a continuous function on a compact set. For the sake of simplicity, we shall prove the results just for the case of a one-dimensional compact interval, $[a, b]$, while the reader can easily extend the results to multiple dimensions.

Max-Pooling Let $f : [a, b] \to \mathbb{R}$ be a continuous function and consider the equidistant partition of the interval $[a, b]$

$$a = x_0 < x_1 < \cdots < x_{n-1} < x_n = b.$$

The partitions size, $\frac{b-a}{n}$, is called *stride*. Denote by $M_i = \max\limits_{[x_{i-1},x_i]} f(x)$ and

consider the simple function $S_n(x) = \sum\limits_{i=1}^{n} M_i 1_{[x_{i-1},x_i)}(x)$. The process of

© Springer Nature Switzerland AG 2020
O. Calin, *Deep Learning Architectures*, Springer Series in the Data Sciences,
https://doi.org/10.1007/978-3-030-36721-3_15

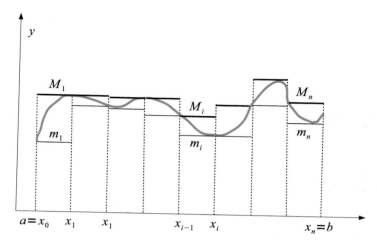

Figure 15.1: *When n increases, the difference $M_i - m_i$ decreases towards zero.*

approximating the function $f(x)$ by the simple function $S_n(x)$ is called *max-pooling*. More details can be found in Zhou and Chellappa [132].

Min-Pooling As a variant, we may consider $m_i = \min\limits_{[x_{i-1}, x_i]} f(x)$ and define the simple function $s_n(x) = \sum\limits_{i=1}^{n} m_i 1_{[x_{i-1}, x_i)}(x)$. The *min-pooling* is the process of approximating the function $f(x)$ by the step function $s_n(x)$. We note that the following double inequality holds:

$$s_n(x) \leq f(x) \leq S_n(x), \qquad \forall n \geq 1.$$

Average-Pooling Consider the average of the function f on the interval $[x_{i-1}, x_i]$, given by $\mu_i = \dfrac{1}{x_i - x_{i-1}} \displaystyle\int_{x_{i-1}}^{x_i} f(u)\, du$. Pooling the average of the function on each interval, we obtain the function $A_n(x) = \sum\limits_{i=1}^{n} \mu_i 1_{[x_{i-1}, x_i)}(x)$.

The next result states that all previous pooling functions are "good approximators" for f.

Theorem 15.1.1 *Let $f : [a, b] \to \mathbb{R}$ be a continuous function. Then all three function sequences, $(S_n)_n$, $(s_n)_n$, and $(A_n)_n$, converge uniformly to f on $[a, b]$, as $n \to \infty$. This means that $\forall \epsilon > 0$, there is $N \geq 1$ such that*

$$|S_n(x) - f(x)| < \epsilon, \quad |s_n(x) - f(x)| < \epsilon, \quad |A_n(x) - f(x)| < \epsilon,$$

$\forall x \in [a, b], \forall n \geq N.$

Proof: Construct the sequence

$$u_n(x) = S_n(x) - s_n(x) = \sum_{i=1}^{n}(M_i - m_i)1_{[x_{i-1},x_i)}(x),$$

which satisfies the following properties, see Fig. 15.1:
(*i*) $u_n(x) \geq 0$;
(*ii*) $u_{n+1}(x) \leq u_n(x)$, for any $n \geq 1$;
(*iii*) $u_n(x) \to 0$, as $n \to \infty$, for any fixed x.

Step 1. We show that $(u_n)_n$ converges to 0, uniformly on $[a, b]$.
Let $\epsilon > 0$ be arbitrary fixed. From the uniform continuity of f on $[a, b]$, there is $N \geq 1$ such that if $|x - x'| < \dfrac{b-a}{N}$, then $|f(x) - f(x')| < \epsilon$. Now, in each interval of the partition there are values $\xi_i, \xi_i' \in [x_{i-1}, x_i]$, such that $M_i = f(\xi_i)$ and $m_i = f(\xi_i')$. Since $|\xi - \xi'| < \dfrac{b-a}{N}$, then $M_i - m_i = |f(\xi_i) - f(\xi_i')| < \epsilon$. This implies

$$\sum_{i=1}^{n}(M_i - m_i)1_{[x_{i-1},x_i)}(x) < \epsilon, \qquad \forall x \in [a, b].$$

This means that $|u_n(x)| < \epsilon$, $\forall x \in [a, b]$, and $\forall n > N$, i.e., $(u_n)_n$ converges uniformly to 0.

Step 2. We show that $(S_n)_n$ converges to f, uniformly on $[a, b]$.
Since $s_n \leq f$, then the following inequality holds:

$$S_n - f \leq (S_n - s_n) + (s_n - f) \leq S_n - s_n = u_n,$$

for any $n \geq 1$. Let $\epsilon > 0$ be arbitrary fixed. Using *Step 1* together with the previous inequality, we have

$$|S_n(x) - f(x)| \leq |u_n(x)| < \epsilon, \quad n \geq 1, \forall x \in [a, b].$$

This means that $(S_n)_n$ converges uniformly to f as $n \to \infty$.

Step 3. We show that $(s_n)_n$ converges to f, uniformly on $[a, b]$.
This is similar to *Step 2*. Since $S_n \geq f$, then

$$f - s_n = (f - S_n) + (S_n - s_n) \leq S_n - s_n = u_n,$$

for any $n \geq 1$. Let $\epsilon > 0$ be arbitrary fixed. Using *Step 1* and the previous inequality, we have

$$|f(x) - s_n(x)| \leq |u_n(x)| < \epsilon, \quad n \geq 1, \forall x \in [a, b].$$

This means that $(s_n)_n$ converges uniformly to f as $n \to \infty$.

Step 4. We show that $(A_n)_n$ converges to f, uniformly on $[a, b]$.
Let $\epsilon > 0$ be arbitrary fixed. By the integral version of the Mean Value Theorem, there is a $x_i^* \in [x_{i-1}, x_i]$ such that $\mu_i = f(x_i^*)$. Therefore,

$$m_i \leq \mu_i \leq M_i.$$

Multiplying by the indicator function $1_{[x_{i-1}, x_i)}(x)$ and summing over i yields

$$s_n(x) \leq A_n(x) \leq S_n(x).$$

This implies $|A_n(x) - s_n(x)| \leq u_n(x)$ and $|S_n(x) - A_n(x)| \leq u_n$. Using *Step 1* it follows that $|A_n(x) - s_n(x)| \to 0$ and $|S_n(x) - A_n(x)| \to 0$ uniformly, as $n \to \infty$. Now, triangle inequality provides

$$
\begin{aligned}
|A_n(x) - f(x)| &= |A_n(x) - f(x) + S_n(x) - S_n(x)| \\
&\leq |S_n(x) - A_n(x)| + |S_n(x) - f(x)| \\
&\leq \frac{\epsilon}{2} + \frac{\epsilon}{2} = \epsilon,
\end{aligned}
$$

where we used *Step 2*.

∎

Remark 15.1.2 The pooling can be extended to the multidimensional case, where $f : K \to \mathbb{R}$ is a continuous function defined on a compact set $K \subset \mathbb{R}^n$. Consider the covering $K = \bigcup_{i=1}^n \bar{A}_i$ of the compact K, where $A_i \cap A_j = \emptyset$ for $i \neq j$ with A_i disjoint open sets and $\bar{A}_i$ denotes the closure of A_i.[1] We pool the maxima $M_i = \max_{\bar{A}_i} f(x)$ and consider the approximation $S_n(x) = \sum_{i=1}^n M_i 1_{A_i}(x)$. If

$$\max_{1 \leq i \leq n} \sup_{x, y \in A_i} |x - y| \to 0$$

as $n \to \infty$, then a proof similar to the previous one shows that S_n converges uniformly to f on K.

15.2 Translation Invariance

In this section we shall prove the property of *local translation invariance* for the max and min-pooling. Consider the notation T_a for the *translation operator* defined by $(T_a \circ f)(x) = f(x - a)$, for any real variable function f and $a \in \mathbb{R}$. We also denote by $\mathcal{P}(f)$ the min- or max-pooling function of f associated with a given partition.

---

[1] The closure of an open set A is the set A together with its boundary.

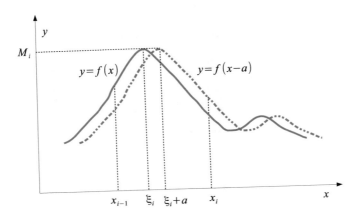

Figure 15.2: *Max-pooling for f and $T_a \circ f$.*

Proposition 15.2.1 *Let $f : \mathbb{R} \to \mathbb{R}$ be a continuous function. There is a partition of $\mathbb{R}$ such that*

$$\mathcal{P}(T_a \circ f) = \mathcal{P}(f),$$

for any small enough value of a.

Proof: We shall perform the proof in the case of max-pooling. The min-pooling can be treated similarly. The proof idea is that under small translations the maxima do not leave the partition intervals, see Fig. 15.2.

Choose a finite partition $[x_i, x_{i+1})$ $0 \le i \le N - 1$ such that the maxima, ξ_i, of the restriction $f_{|[x_i,x_{i+1})}$ are inside the open intervals (x_i, x_{i+1}). There is a $\eta > 0$ such that $x_i + \eta < \xi_i < x_{i+1} - \eta$. Then choose $a \in \mathbb{R}$ such that $|a| < \eta$.

Since the graph of $T_a \circ f$ is obtained by shifting horizontally the graph of f by an amount a, then the maxima do not leave the intervals and we have

$$M_i(f) = \max_{[x_i,x_{i+1}]} f(x) = \max_{[x_i,x_{i+1}]} f(x - a) = \max_{[x_i,x_{i+1}]} (T_a \circ f)(x) = M_i(T_a \circ f).$$

Therefore, the functions f and $T_a \circ f$ will have the same max-pooling functions. ∎

Remark 15.2.1 (i) The invariance property extends to several dimensions with only minor alterations in the proof.

(ii) The previous property provides stability of the pooling under small input variations.

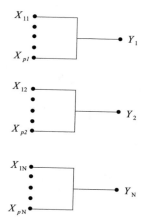

Figure 15.3: *Max-pooling layer with* $Y_j = \max\{X_{1j}, \ldots, X_{pj}\}$, $1 \leq j \leq N$.

15.3 Information Approach

Another way to look at pooling is by investigating its effect on information content. This section deals with the case of max-pooling, but similar results hold also for the average-pooling.

Consider n random variables, $X_1, X_2, \ldots, X_n$ and let $Y = \max\{X_1, \ldots, X_n\}$. Let $\mathfrak{S}(X_i)$ be the sigma-algebra generated by X_i, and

$$\mathfrak{S}(X) = \mathfrak{S}(X_1, \ldots, X_n) = \mathfrak{S}(X_i) \vee \cdots \vee \mathfrak{S}(X_n)$$

be the information field generated by all X_i. For any $b \in \mathbb{R}$ we have

$$\begin{aligned} Y^{-1}(-\infty, b] &= \{\omega; Y(\omega) \leq b\} = \{\omega; X_i(\omega) \leq b, \forall i = 1, \ldots, n\} \\ &= \bigcap_{i=1}^{n} \{\omega; X_i(\omega) \leq b\} = \bigcap_{i=1}^{n} X_i^{-1}(-\infty, b] \in \bigcap_{i=1}^{n} \mathfrak{S}(X_i). \end{aligned}$$

Consequently, $\mathfrak{S}(Y) \subset \bigcap_{i=1}^{n} \mathfrak{S}(X_i)$, that is, the information field of the maximum of n random variables is included in the information field generated by each variable. Next, we shall apply this result to neural networks.

Definition 15.3.1 *We say that the ℓth layer of a feedforward neural network is a pooling layer if*
(i) the $(\ell-1)$th layer is divided into a partition of N classes of neurons;

(ii) all neurons of the $(\ell-1)$th layer that belong to the same class are mapped into the same neuron in the ℓth layer, whose activation is their corresponding maximum value;
(iii) the number of neurons in the ℓth layer is $d^{(\ell)} = N$, where N is the number of partition classes.

Roughly stated, a pooling layer replaces each partition class of a layer by the maximum neuron value in that class. In Fig. 15.3, the $(\ell-1)$th layer contains neurons with values X_{ij}, $1 \le i \le p$, $1 \le j \le N$, divided into N classes

$$\{X_{11}, \ldots, X_{p1}\}, \{X_{12}, \ldots, X_{p2}\}, \ldots, \{X_{1N}, \ldots, X_{pN}\},$$

each class having p neurons. Each class is pooled into its maximum value

$$Y_j = \max\{X_{1j}, \ldots, X_{pj}\}, \qquad 1 \le j \le N.$$

From the previous computation, the information in each of the neurons of the pooling layer satisfies the inclusion

$$\mathfrak{S}(Y_j) \subset \bigcap_{i=1}^{p} \mathfrak{S}(X_{ij}). \tag{15.3.1}$$

The information generated by the pooling layer is given by

$$\mathfrak{S}(Y) = \mathfrak{S}(Y_1, \ldots, Y_N) = \bigvee_{j=1}^{N} \mathfrak{S}(Y_j) = \mathfrak{S}\Big[\bigcup_{j=1}^{N} \mathfrak{S}(Y_j)\Big].$$

Inclusion (15.3.1) implies

$$\mathfrak{S}(Y) \subset \mathfrak{S}\Big[\bigcup_{j=1}^{N} \bigcap_{i=1}^{p} \mathfrak{S}(X_{ij})\Big]. \tag{15.3.2}$$

Using formula (b') of Appendix section A, yields

$$\bigcup_{j=1}^{N} \bigcap_{i=1}^{p} \mathfrak{S}(X_{ij}) = \bigcup_{j=1}^{N} \bigcap_{i_r=1}^{p} \mathfrak{S}(X_{i_r j}) = \bigcap_{i_1, \ldots, i_p} \Big(\bigcup_{j=1}^{N} \mathfrak{S}(X_{i_r j})\Big).$$

Then (15.3.2), with the help of Exercise 15.6.1 part (a), becomes

$$\mathfrak{S}(Y) \subset \mathfrak{S}\Big[\bigcap_{i_1, \ldots, i_p} \Big(\bigcup_{j=1}^{N} \mathfrak{S}(X_{i_r j})\Big)\Big] \subset \bigcap_{i_1, \ldots, i_p} \mathfrak{S}\Big[\Big(\bigcup_{j=1}^{N} \mathfrak{S}(X_{i_r j})\Big)\Big],$$

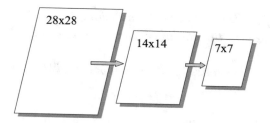

Figure 15.4: *Two max-pooling layers applied to an MNIST image.*

which can be also written as

$$\mathfrak{S}(Y) \subset \bigcap_{i_1,\ldots,i_p} \bigvee_{j=1}^{N} \mathfrak{S}(X_{i_r j}) = \bigcap_{i_1,\ldots,i_p} \mathfrak{S}(X_{i_1 1}, \ldots, X_{i_N N}).$$

This relation has the following interpretation. From the first class we pick an arbitrary neuron, say $X_{i_1 1}$. The information generated by this neuron is $\mathfrak{S}(X_{i_1 1})$. If this is done for each class, then the information generated by these arbitrarily class-picked neurons is $\mathfrak{S}(X_{i_1 1}, \ldots, X_{i_N N})$. The previous inclusion states that the information of the pooling layer, $\mathfrak{S}(Y)$, is contained in any of the information sets $\mathfrak{S}(X_{i_1 1}, \ldots, X_{i_N N})$, regardless of the arbitrary choice of neurons.

15.4   Pooling and Classification

Pooling is customarily used when the dimension of the input needs to be lowered to match the number of classification classes. For instance, in the case of the MNIST data, each input image has $28 \times 28 = 784$ pixels, while there are 10 classification classes (the numbers $0, 1, \ldots, 9$). This can be partitioned, for instance, into 14×14 squares, each having 2×2 pixels. From each 2×2 square we retain only the pixel of maximum intensity. As a consequence, we obtain a 14×14 pixel image, which is the result of the first pooling. The second pooling divides the 14×14 image into 7×7 squares, each of 2×2 pixels; again, we retain from each of these only the pixel of maximum intensity, see Fig. 15.4. This is a process by which information is thrown away in an irreversible way.

Pooling is usually used as a companion to convolution. The convolution layer filters the input signal, removing noise, while the pooling layer selects a rough summary of the filtered signal features, see Fig. 15.5. The convolution operation and convolutional networks will be discussed in more detail in the next chapter.

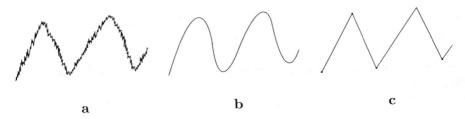

Figure 15.5: *The effects of convolution and pooling:* **a.** *Raw and noisy input signal;* **b.** *Smoothed signal obtained by convolution;* **c.** *Pooled signal obtained by selecting maxima and minima.*

15.5   Summary

Pooling is a machine learning technique by which the dimension of the input is decreased by a certain factor. It can be also considered as an information contractor by which the information is thrown away in an irreversible way. Pooling is used in classification problems together with convolution.

15.6   Exercises

Exercise 15.6.1 Let $(\mathcal{C}_i)$ be a collection of measurable sets. Show that

(a) $\mathfrak{S}\left(\bigcap_i \mathcal{C}_i\right) \subset \bigcap_i \mathfrak{S}(\mathcal{C}_i)$;

(b) $\mathfrak{S}\left(\bigcup_i \mathcal{C}_i\right) \supset \bigcup_i \mathfrak{S}(\mathcal{C}_i)$.

Exercise 15.6.2 There are $N = 2^n$ participants to a chess competition. The participants compete in pairs. At each round, the Wiener of each pair is competing against the Wiener of another pair. The final Wiener is obtained after the nth round. Explain this procedure in the light of a max-pooling process.

Exercise 15.6.3 (a) Prove Proposition 15.2.1 in the case of min-pooling. (b) Formulate and prove a version of Proposition 15.2.1 for two-dimensional functions.

In a neural network it is not recommended to place pooling layers consecutively, since their composition can be written as an only one pooling layer. The next exercise deals with a more precise statement:

Exercise 15.6.4 (a) Assume all layers of a neural network are max-pooling layers. Show that the final output of the net is the largest input.
(b) Assume all layers of a neural network are average-pooling layers. Show that the final output of the net is the average of the input.

(*c*) If all layers of a neural network are min-pooling layers, show that the final output of the net is the min of the input.

Exercise 15.6.5 In a neural network a max-pooling layer is followed by a min-pooling layer.

(*a*) Show that the network output change if we switch the order of these pooling layers.

(*b*) Does the result still hold if the min-pooling layer is replaced by an average-pooling layer?

Chapter 16

Convolutional Networks

Convolutional neural networks (CNN) are feedforward neural networks with shared weights and sparse interactions, that is, most weights are equal to zero. Given their fewer number of parameters, convolution networks are more efficient to train than any similarly sized fully-connected layer networks, with only minor negative consequences on their performance.

The excellent performance of CNNs in image processing is due to their adaptability to a 2-D grid-like topology. CNN is a biologically inspired piece of AI, whose design is based on neuroscientific principles, and successfully applied to pattern recognition, see LeCun et al. [74].

Like almost any other feedforward neural network, CNNs are trained with a version of the backpropagation algorithm and use ReLU as activation function. We shall discuss in this chapter the architecture of CNNs based on concepts of local receptive field, kernel, convolution, and feature map.

16.1 Discrete One-dimensional Signals

A *discrete one-dimensional signal* can be described by the double infinite sequence of real numbers

$$y = [\ldots, y_{-2}, y_{-1}, y_0, y_1, y_2, \ldots],$$

where y_k represents the signal amplitude measured at time t_k (Fig. 16.1). A signal y is called:
- *finite signal* if $\max_k |y_k| < \infty$, i.e., $\|y\|_\infty < \infty$;
- L^1-*finite signal* if $\sum_{k=-\infty}^{\infty} |y_k| < \infty$, i.e., $\|y\|_1 < \infty$;
- *finite energy signal* if $\sum_{k=-\infty}^{\infty} |y_k|^2 < \infty$, i.e., $\|y\|_2 < \infty$.

© Springer Nature Switzerland AG 2020

O. Calin, *Deep Learning Architectures*, Springer Series in the Data Sciences,

https://doi.org/10.1007/978-3-030-36721-3_16

517

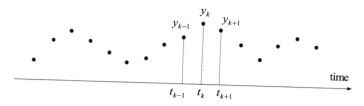

Figure 16.1: *Discrete one-dimensional signal.*

- *compact support signal* if there is $N \geq 1$ such that $y_k = 0$ for any $|k| > N$.

It is worth noting that a finite energy signal with compact support is L^1-finite. This follows from the Cauchy's inequality

$$\Big(\sum_{k=-N}^{N} |y_k| \Big)^2 \leq 2N \sum_{k=-N}^{N} |y_k|^2.$$

A signal can be processed by filtering. This involves a convolution operation between the signal and a *kernel* or *filter*, as we shall define next. A kernel (filter) is a compact support sequence of weights

$$w = [\ldots, w_{-2}, w_{-1}, w_0, w_1, w_2, \ldots].$$

The *convolved signal* between y and w is the signal

$$z = [\ldots, z_{-2}, z_{-1}, z_0, z_1, z_2, \ldots],$$

denoted by $z = y * w$, and defined by

$$z_j = \sum_{k=-\infty}^{\infty} y_{j+k} w_k. \tag{16.1.1}$$

The above infinite sum makes sense since w has only a finite number of nonzero elements. Each component of the convolved signal z is a weighted sum of the components of the initial signal y. The effect of convolution is to average out a signal using some given weights system. Equivalently, sliding the filter w, then multiplying by y and summing, produces the filtered signal z.

It is worth noting that formula (16.1.1) is known in signal processing as *cross-correlation*.

Example 16.1.1 (Moving average signal) If the kernel is given by

$$w = [\ldots, 0, 0, w_0 = \frac{1}{2}, w_1 = \frac{1}{2}, 0, 0, \ldots]$$

then $z = y * w$ is a *moving average signal*. In this case each term of the sequence is replaced by the arithmetic average of two consecutive terms, i.e.,

$$z_j = \frac{1}{2}(y_{j-1} + y_j)$$

The shifting details can be inferred from the next table:

	y_{-2}	y_{-1}	y_0	y_1	y_2	y_3	y_4	z_j
$j=0$	0	0	$1/2$	$1/2$	0	0	0	$z_0 = (y_0 + y_1)/2$
$j=1$	0	0	0	$1/2$	$1/2$	0	0	$z_1 = (y_1 + y_2)/2$
$j=2$	0	0	0	0	$1/2$	$1/2$	0	$z_2 = (y_2 + y_3)/2$

Similarly, if the filter is given by $w = [\ldots, 0, 0, \frac{1}{3}, \frac{1}{3}, \frac{1}{3}, 0, 0, \ldots]$, we obtain a moving average signal as an average of three consecutive terms

$$z_j = \frac{1}{3}(y_{j-1} + y_j + y_{j+1}).$$

16.2 Continuous One-dimensional Signals

In this case the signal amplitude is given as a continuous function of time, $y = y(t)$. The previous definitions can be adapted for the continuous case as:

- *finite signal* if $\|y\|_\infty < \infty$;
- L^1-*finite signal* if $\|y\|_1 = \int_{\mathbb{R}} |y(t)|\, dt < \infty$;
- *finite energy signal* if $\|y\|_2 = \left(\int_{\mathbb{R}} |y(t)|^2\, dt \right)^{1/2} < \infty$;
- *compact support signal* if there is $u \geq 0$ such that $y(t) = 0$ for any $|t| > u$.

The filter in this case is a continuous function with compact support, $w = w(t)$. This means $w(t) = 0$ for $|t|$ large enough. The convolved signal, $z = y * w$, is defined by

$$z(t) = (y * w)(t) = \int_{\mathbb{R}} y(u + t)w(u)\, du = \int_{\mathbb{R}} y(v)w(v - t)\, dv.$$

This formula is the continuous version of the cross-correlation between the continuous signals y and w. It is worthy to note that convolution in mathematical literature is defined using a flipped sign than in the previous formula. However, this is the way this the concept becomes useful in neural networks.

Similarly with the discrete case, the statement that any finite energy signal with compact support is L^1-finite holds also in the continuous case. Furthermore, any filtered L^1-finite signal is also L^1-finite. This follows from the following estimation

$$\|z\|_1 = \int |z(t)|\,dt = \int \left| \int_{\mathbb{R}} w(u)y(u+t)\,du \right| dt$$

$$\leq \int \int \left| w(u)y(u+t) \right| du\,dt = \int |w(u)| \int |y(u+t)|\,dt\,du$$

$$= \int |w(u)| \int |y(r)|\,dr\,du = \int |w(u)|\,\|y\|_1\,du = \|y\|_1\,\|w\|_1,$$

where we inverted the order of integration using Fubini's theorem.

16.3　Discrete Two-dimensional Signals

A *discrete two-dimensional signal* is an infinite matrix $y = [y_{ij}]_{i,j}$, where y_{ij} is the activation of the (i,j)-pixel. This way, any black and white picture can be considered as a two-dimensional signal. If the (i,j)-pixel is black than the activation is $y_{ij} = 1$; if the pixel is white, the activation is $y_{ij} = 0$; any other gray tone is a number between 0 and 1.

A signal y is called:

- L^1-*finite* if $\sum_{k=-\infty}^{\infty} \sum_{j=-\infty}^{\infty} |y_{jk}| < \infty$;
- *finite energy signal* if $\sum_{k=-\infty}^{\infty} \sum_{j=-\infty}^{\infty} |y_{jk}|^2 < \infty$;
- *compact support signal* if there is $N \geq 1$ such that $y_{jk} = 0$ for any $|j| > N$ and $|k| > N$.

A kernel in this case is a compact support signal $w = [w_{ij}]$. The signal $y = [y_{jk}]$ can be convolved with the kernel $w = [w_{ij}]$ into the signal $z = y * w$, as

$$z_{ij} = \sum_{k=-\infty}^{\infty} \sum_{r=-\infty}^{\infty} y_{i+k,j+r} w_{k,r}.$$

In the two-dimensional case the output z_{ij} is also called a *feature map*, since is supposed to contain some image features that are characteristic to the kernel w.

Remark 16.3.1 For a continuous two-dimensional signal, $y : \mathbb{R}^2 \to \mathbb{R}$, its convolution with a continuous filter $w : \mathbb{R}^2 \to \mathbb{R}$ is obtain by the integral formula

$$z(t_1, t_2) = \iint_{\mathbb{R}^2} y(u_1 + t_1, u_2 + t_2)\, w(u_1, u_2)\, du_1 du_2$$

$$= \iint_{\mathbb{R}^2} y(v_1, v_2)\, w(v_1 - t_1, v_2 - t_2)\, dv_1 dv_2.$$

The reader should have no difficulty to extend the definitions of the L^1-finite energy signals and compact support signals from the discrete case to the continuous case.

Example 16.3.2 (Moving average in 2d) Consider the feature map with a 2×2 support

$$w = \begin{pmatrix} 0 & 0 & 0 & 0 \\ 0 & 1/4 & 1/4 & 0 \\ 0 & 1/4 & 1/4 & 0 \\ 0 & 0 & 0 & 0 \end{pmatrix}.$$

The convolved signal is obtained by averaging out four neighboring activations

$$z_{ij} = \frac{1}{4}(y_{ij} + y_{i,j+1} + y_{i+1,j} + y_{i+1,j+1}).$$

In the following we shall denote the *convolution operator* by $\mathcal{C}$, so $\mathcal{C}(y)$ is the output of a convolution layer with the input y. In the previous notations we have $z = \mathcal{C}(y)$. We shall also denote by $T_{a,b}$ the translation operator in the direction of the vector (a, b) by $(T_{a,b} \circ y)_{ij} = y_{i-a,j-b}$. We note that the L^1-finiteness, finite energy and compact support properties transfer from a signal to its translation.

Proposition 16.3.3 (Equivariance to translation) *Convolution operation preserves translations, that is*

$$\mathcal{C}(T_{a,b} \circ y) = T_{a,b} \circ \mathcal{C}(y).$$

Proof: First we note that $(T_{a,b} \circ y)_{ij} = y_{i-a,j-b}$. Then for any fixed indices i and j, we have

$$\begin{aligned} \mathcal{C}(T_{a,b} \circ y)_{ij} &= \big((T_{a,b} \circ y) * w\big)_{ij} = \sum_k \sum_r (T_{a,b} \circ y)_{i+k,j+r}\, w_{kr} \\ &= \sum_k \sum_r y_{i+k-a,j+r-b}\, w_{kr} = (y * w)_{i-a,j-b} \\ &= \Big(T_{a,b}(y * w)\Big)_{ij} = \Big(T_{a,b} \circ \mathcal{C}(y)\Big)_{ij}. \end{aligned}$$

∎

The previous result states that if the input is affected by a translation, then after the convolution, the output is also affected by the same translation. Hence, since many input image features, such as corners, edges, etc., are invariant by translation, they will still be present in the output of the convolution layer.

Remark 16.3.4 Pooling is invariant to small translations of data, see section 15.2, that is $\mathcal{P}(T_{a,b} \circ y) = \mathcal{P}(y)$. This property is compatible with the translation equivariance of convolution, fact that makes pooling and convolution to be applied together. This is, if pooling is applied after convolution, we have

$$\mathcal{P} \circ \mathcal{C}(T_{a,b} \circ y) = \mathcal{P} \circ \mathcal{C}(y).$$

Conversely, if convolution is applied after pooling, then

$$\mathcal{C} \circ \mathcal{P}(T_{a,b} \circ y) = \mathcal{C} \circ \mathcal{P}(y).$$

16.4 Convolutional Layers with 1-D Input

A convolution layer resembles a fully-connected layer, the difference being that it has lots of zero weights and repeating nonzero weights. Consider a neural network with the input given by the compact supported signal $x = [x_1, x_2, \ldots, x_n]$ and let the sliding kernel be $w = [w_1, w_2]$. The lag by which the kernel slides with respect to the signal is called *stride*. In Fig. 16.2 **a** and **b** the strides are $s = 1$ and $s = 2$, respectively. In both cases, the neurons in the second layer have a sigmoid activation function.

In the case of Fig. 16.2 **a** the network output can be written in the familiar form, $Y = \sigma(WX + B)$, where $X = (x_1, \ldots, x_6)^T$, $B = (b, b, b, b)^T$, $Y = (y_1, \ldots, y_5)^T$. The system of weights can be written as the following 5×6 sparse matrix

$$W = \begin{pmatrix} w_1 & w_2 & 0 & 0 & 0 & 0 \\ 0 & w_1 & w_2 & 0 & 0 & 0 \\ 0 & 0 & w_1 & w_2 & 0 & 0 \\ 0 & 0 & 0 & w_1 & w_2 & 0 \\ 0 & 0 & 0 & 0 & w_1 & w_2 \end{pmatrix}.$$

We note the repeating weights on each row of the matrix.

Similarly, in the case of Fig. 16.2 **b** the network output can be written as $Y = \sigma(WX + B)$, where $X = (x_1, \ldots, x_6)^T$, $B = (b, b, b)^T$, $Y = (y_1, y_2, y_3)^T$, and the system of weights is given by the 3×6 matrix

$$W = \begin{pmatrix} w_1 & w_2 & 0 & 0 & 0 & 0 \\ 0 & 0 & w_1 & w_2 & 0 & 0 \\ 0 & 0 & 0 & 0 & w_1 & w_2 \end{pmatrix}.$$

Convolution with linear neurons Assume now that all neurons of a convolution neural net have a linear activation, $\phi(x) = x$. We shall show that this neural network, having several convolution layers, is equivalent to a network with only one convolution layer. Therefore, using a nonlinear activation function is essential for deep learning in convolution networks.

It suffices to show that two convolution layers are equivalent to one convolution layer. Consider the forward propagations in two consecutive layers

$$X^{(1)} = W^{(1)} X^{(0)} + B^{(1)}$$

$$X^{(2)} = W^{(2)} X^{(1)} + B^{(2)},$$

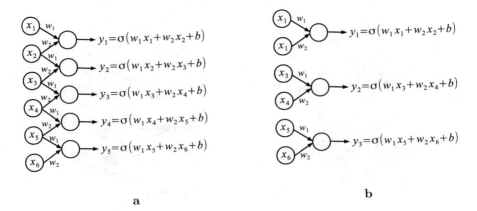

Figure 16.2: *Two convolution layers:* **a** *stride equal to 1;* **b** *stride equal to 2. All missing arrows are assigned a zero weight. All neurons have the same bias b.*

with the weight matrices of sparse type. The composition provides

$$X^{(2)} = W^{(2)}(W^{(1)}X^{(0)} + B^{(1)}) + B^{(2)} = WX^{(0)} + B,$$

with weight matrix $W = W^{(2)}W^{(1)}$ and bias vector $B = W^{(2)}B^{(1)} + B^{(2)}$. We need to show that the matrix W is also of sparse type. We shall discuss this on the previous two cases represented in Fig. 16.2 **a**, **b**.

The convolution layer represented in Fig. 16.2 **a** has stride $s = 1$ and uses only two shared weights, w_1, w_2 (the support width is 2). We consider two layers of this type as in Fig. 16.3 **a**. Then Y_1 depends on X_1, X_2, X_3; Y_2 depends on X_2, X_3, X_4, and so forth. The convolutional network is equivalent to the two-layer net given in Fig. 16.3 **b**. The new net depends on three sharing weights, ν_1, ν_2, ν_3 (the support width is 3), with the stride is 1. Both networks in Fig. 16.3 **a**, **b** satisfy the information relations

$$\mathfrak{S}(Y_1) \subset \mathfrak{S}(X_1, X_2, X_3), \quad \mathfrak{S}(Y_2) \subset \mathfrak{S}(X_2, X_3, X_4)$$

$$\mathfrak{S}(Y_3) \subset \mathfrak{S}(X_3, X_4, X_5), \quad \mathfrak{S}(Y_4) \subset \mathfrak{S}(X_4, X_5, X_6).$$

This can be stated by saying that the *receptive field* of Y_1 consists of the units X_1, X_2, X_3, unlike the case of a fully-connected layer, when the receptive field would consist of all previous neurons. Consequently, a convolution layer passes less information than a fully-connected layer.

On the other side, the convolution layer represented in Fig. 16.2 **b** has stride $s = 2$ and uses only two shared weights, w_1, w_2 (the support width is 2). We consider two layers of this type as in Fig. 16.4 **a**. The convolutional net is

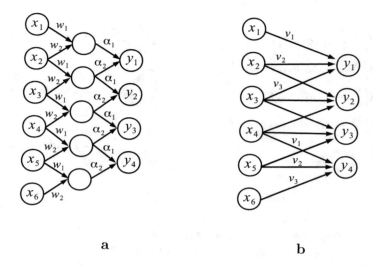

a   b

Figure 16.3: *Two equivalent convolution networks:* **a** *one-hidden layer convolutional net with stride 1 and support width 2;* **b** *two-layer convolutional net with stride 1 and support width 3.*

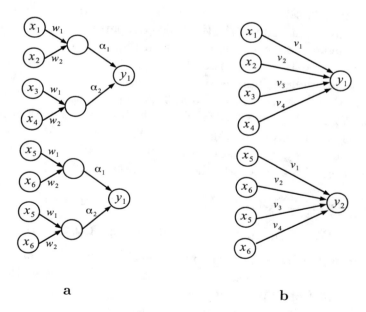

a   b

Figure 16.4: *Two equivalent convolutional networks:* **a** *one-hidden layer convolutional net with stride equal to 2;* **b** *two-layer convolutional net with stride equal to 4.*

equivalent to the two-layer net given in Fig. 16.4 **b**. The new net depends on four sharing weights, $\nu_1, \nu_2, \nu_3, \nu_4$ (the support width is 4), with the stride 1. Both networks in Fig. 16.4 **a**, **b** satisfy the information relations

$$\mathfrak{S}(Y_1) \subset \mathfrak{S}(X_1, X_2, X_3, X_4), \quad \mathfrak{S}(Y_2) \subset \mathfrak{S}(X_2, X_3, X_4, X_5).$$

In particular, the receptive field of Y_1 consists of X_1, X_2, X_3, X_4.

Remark 16.4.1 The convolution compresses the information. If s denotes the stride, then $d^{(\ell)} = d^{(\ell-1)} + s$, i.e., at each layer the number of neurons decrease by the stride number. Since s is a small number, the compression is smaller than in the case of pooling.

16.5 Convolutional Layers with 2-D Input

Convolutional networks have shown excellent performance in the case of processing 2-dimensional images. In this case each input is a colored image (in RGB format), which can be seen as a tensor of type $r \times c \times 3$, where r is the number of rows and c the number of pixel columns in the image. This is equivalent to 3 channels of dimension $r \times c$, each channel corresponding to one color. Before getting any further, we shall discuss first the convolution of the spatial slice of an image with a given kernel.

In the following we shall convolve a 2×2 convolution kernel given in Fig. 16.5 **a** with a 3×3 input image, which is illustrated by the matrix in Fig. 16.5 **b**. The convolution kernel is overlapped on the matrix and moved horizontally and vertically, in all possible positions. In each position we sum the products of the kernel entries and the matrix entries, the numbers obtained being an output. The kernel overlap starts from the top left of the image and slides by one pixel to the right. Then we continue the operation for the bottom row, as follows:

$$1 \cdot 2 - 1 \cdot 1 + 2 \cdot 4 + 1 \cdot 3 = 12$$

$$1 \cdot 1 - 1 \cdot 1 + 2 \cdot 3 + 1 \cdot 5 = 11$$

$$1 \cdot 4 - 1 \cdot 3 + 2 \cdot 7 + 1 \cdot 6 = 21$$

$$1 \cdot 3 - 1 \cdot 5 + 2 \cdot 6 + 1 \cdot 0 = 10.$$

Using the convolution operator, $*$, the previous computation writes as

$$\begin{pmatrix} 2 & 1 & 1 \\ 4 & 3 & 5 \\ 7 & 6 & 0 \end{pmatrix} * \begin{pmatrix} 1 & -1 \\ 1 & 1 \end{pmatrix} = \begin{pmatrix} 12 & 11 \\ 21 & 10 \end{pmatrix}.$$

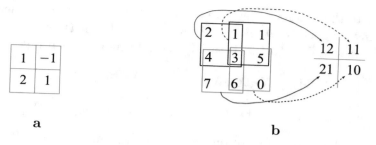

a   **b**

Figure 16.5: *The convolution operation:* **a** *A* 2×2 *kernel;* **b** *the input and output of a convolution.*

We notice that the convolution between a 2×2 kernel with a 3×3 image is an output of size 2×2. In general, if the kernel is of size $h \times k$ and the image is $H \times K$, the output has the size $(H - h + 1) \times (K - k + 1)$. For the case of an arbitrary stride, see Exercise 16.9.11.

In the previous example the spatial slice of the image has been convoluted with a 2×2 kernel. This operation is also called a *feature map*, since this procedure can detect different sorts of features, depending on the kernel used, such as horizontal or vertical edges, corners, etc., as the reader can see in the exercise section.

Each color channel can be convoluted separately with the same or a different kernel. It is possible, however, to convolute all 3-color channel image with a tensor of order three, i.e., with a sequence of 3 kernels of the same type. In this case the convolution operation is defined in a similar way. The kernel is overlapped on the image at the location $(0, 0, 0)$ and then we take the sum of the products of their entries to obtain the first output. Then the kernel is moved by one pixel from top to bottom and then from left to right to complete the operation.

For a fixed kernel, the entries of the tensor $X^{(\ell)}$ are addressed by the triple indexed entry $X^{(\ell)}_{ijk}$, with $1 \leq i \leq r^{(\ell)}$, $1 \leq j \leq c^{(\ell)}$, and $1 \leq k \leq 3$, see Fig. 16.6. One feature map in the ℓth layer, for the kth channel and corresponding to the given kernel $w^{(\ell)}$ and bias $b^{(\ell)}$ is given by

$$X^{(\ell)}_{ijk} = \phi\left(\sum_s \sum_p X^{(\ell-1)}_{i+p,j+r,k} w^{(\ell)}_{prk} + b^{(\ell)} \right),$$

where the activation function ϕ is usually taken to be a ReLU in order to avoid vanishing gradients.

Since k represents the color channel, then $1 \leq k \leq 3$. The ℓth layer of the CNN, $X^{(\ell)}$, is given by the collection of all tensors of order 3 of the form $X^{(\ell)}_{ijk}$ corresponding to all kernels $w^{(\ell)}$. If $f^{(\ell)}$ denotes the number of feature maps

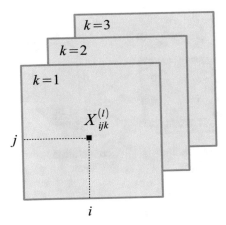

Figure 16.6: *The entries of the tensor $X^{(\ell)} \in \mathbb{R}^{r^{(\ell)} \times c^{(\ell)} \times 3}$. The entry $X_{ijk}^{(\ell)}$ represents the activation in the spatial location (i, j) situated in the kth channel.*

in the ℓth layer (the number of kernels used for that layer), $r^{(\ell)} \times c^{(\ell)}$ are the dimensions of the image in the ℓth layer (rows times columns), then we can write the output of the ℓth layer as an order 4 tensor $X^{(\ell)} \in \mathbb{R}^{r^{(\ell)} \times c^{(\ell)} \times 3 \times f^{(\ell)}}$.

Customarily, the sequences $r^{(\ell)}$ and $c^{(\ell)}$ are decreasing with respect to ℓ. This is due to the fact that processing with convolution layers tends to decrease the image dimensions with a number equal to the stride.[1] (Besides this, using pooling between convolution layers also compresses the dimension by a certain factor). The number of channels remains the same, but the number of features, $f^{(\ell)}$, increases with ℓ.

If the input layer is denoted by $X^{(0)}$, then the first hidden layer, $X^{(1)}$, contains features of the layer $X^{(0)}$. The second hidden layer, $X^{(2)}$, contains features of $X^{(1)}$, i.e., features on features of the input. In general, each layer contains features of the previous layer. At the end, a fully-connected layer is introduced to put together all the information from the last convolution layer, which contains a large number of features. Then a softmax layer can be employed if the network is meant for classification purposes, see Fig. 16.7.

---

[1]In order to avoid this dimension reduction, one can use the trick of padding with zeros.

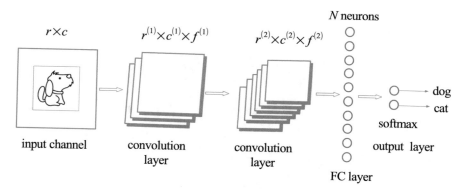

Figure 16.7: *CNN with two convolution layers, one fully-connected layer and a softmax layer (output layer).*

16.6   Geometry of CNNs

In Chapter 13 we have associated a manifold with each neural network. The network weights and biasses were coordinate systems on the associated manifold. Its dimension is equal to the number of network parameters, which can be expressed in terms of layers size by formula (13.3.8)

$$d^{(0)}d^{(1)} + d^{(1)}d^{(2)} + \cdots + d^{(\ell-1)}d^{(\ell)} + \cdots + d^{(L-1)}d^{(L)} + N. \qquad (16.6.2)$$

The fact that in a CNN the weights and biases are shared among neurons reduces substantially the number of total parameters of the network, and consequently, the dimension of the associated output manifold. This fact has regularization consequences, and hence the CNN is usually not prone to overfitting the training data.

We shall work the comparison on two examples at hand. In the case of the CNN shown in Fig. 16.3 **a**, the dimension of the associated neural net is given by $2+2+5+4 = 13$ (four weights and nine biases). If keeping the same number of neurons in the layers, this network is replaced by a fully-connected layer net, then formula (13.3.8) yields the dimension $6 \cdot 5 + 5 \cdot 4 + 5 + 4 = 59$.

As a second example, consider the CNN shown in Fig. 16.4 **a**. Then the dimension of the associated neural net is given by $2 + 2 + 4 + 2 = 10$ (four weights and six biases). The associated fully-connected layer network has an associated manifold with the dimension $6 \cdot 4 + 4 \cdot 2 + 4 + 2 = 38$. We note that in both cases the dimension of the neural manifold is substantially larger in the case of a fully-connected layer network than in the case of a similar CNN. The same phenomenon holds in general for all CNN networks.

16.7 Equivariance and Invariance

Convolutional neural networks are able to detect local patterns in an image regardless of their position. This is because CNNs ensure equivariance to translations, see Proposition 16.3.3. This means if the input image is translated by some vector, then the activation pattern in each higher layer of the network is also translated by the same vector. Therefore, a key component of the CNN success in image recognition tasks is due to their equivariance property.

The next level of abstraction is to replace the set of translations by any group of transformations of the input and explain the symmetries of the network parameters by the network equivariance with respect to the considered transformations.

Applications of group theory in neural networks can be found in Ravanbakhsh et al. [100], Kondor and Trivedi [66], Kondor [65], Cohen et al. [26, 27], and Bartók et al. [13]. In the following we shall use ideas from the aforementioned papers to briefly discuss this new emerged direction of research.

16.7.1 Groups

The next definition defines an algebraic structure that will be useful shortly.

Definition 16.7.1 *A group is a set G endowed with a composition law $G \times G \to G$ denoted multiplicatively and satisfying the following properties:*

(i) $g_1 g_2 \in G$, *for any* $g_1, g_2 \in G$;
(ii) $g_1(g_2 g_3) = (g_1 g_2)g_3$, *for any* $g_1, g_2, g_3 \in G$;
(iii) there is a unique element $e \in G$ *such that* $xe = ex = x$, *for any* $x \in G$;
(iv) for any $g \in G$ *there is* $g^{-1} \in G$ *such that* $gg^{-1} = g^{-1}g = e$.

Property (i) states that G is closed with respect to the group law, while (ii) says that the group law is associative; (iii) indicates the existence of the neutral element in the group; the existence of the inverse element is given by (iv).

If the order of elements in the group law composition does not matter, $g_1 g_2 = g_2 g_1$ for any $g_1, g_2 \in G$, then the group G is called *commutative*. Depending on the number of elements, the group G can be finite or infinite.

Any subset H of G, which forms a group under the same law as G, is called a *subgroup* and is denoted by $H \leq G$.

Example 16.7.2 The set of integers, $G = \mathbb{Z}$, endowed with the addition operation forms a commutative group. The inverse of n is $-n$ and the neutral element is $e = 0$. Similarly, the integer lattice, $G = \mathbb{Z} \times \mathbb{Z}$, with addition on components

$$(n_1, n_2) + (m_1, m_2) = (n_1 + m_1, n_2 + m_2)$$

forms also a commutative group. Its neutral element is $(0, 0)$.

The set $H = 3\mathbb{Z} = \{3m; m \in \mathbb{Z}\}$ forms a subgroup of G, while $K = \{(2i, 2j); i, j \in \mathbb{Z}\}$ forms a subgroup of $\mathbb{Z} \times \mathbb{Z}$.

Example 16.7.3 Let $v \in \mathbb{R}^3$ be a vector and define the translation $\tau_v : \mathbb{R}^3 \to \mathbb{R}^3$ by $\tau_v(x) = x + v$. The set $G = \mathcal{T}(\mathbb{R}^3) = \{\tau_v; v \in \mathbb{R}^3\}$ forms a group with respect to functions composition, called the *translations group* on $\mathbb{R}^3$. We have $\tau_v \circ \tau_u = \tau_{v+u}$ and $(\tau_v)^{-1} = \tau_{-v}$. The neutral element is $\tau_0 = Id$, the identity transformation of $\mathbb{R}^3$.

Example 16.7.4 The 2×2 matrix

$$R_\theta = \begin{pmatrix} \cos\theta & -\sin\theta \\ \sin\theta & \cos\theta \end{pmatrix}$$

represents a counterclockwise rotation of the plane $\mathbb{R}^2$ about the origin. The set of these rotations, $SO(2) = \{R_\theta; \theta \in \mathbb{R}\}$, form a group under the matrix multiplication, called the *special orthogonal group* of $\mathbb{R}^2$. We can easily check that $R_\theta R_{\theta'} = R_{\theta'} R_\theta = R_{\theta+\theta'}$ and $R_\theta^{-1} = R_{-\theta}$.

The set $H = \{R_0, R_\pi\}$, formed by the identity transform and the 180-degree rotation (or flip about the origin) forms a subgroup of $SO(2)$.

Also, the set $K = \{R_0, R_{\pi/4}, R_{\pi/2}, R_{3\pi/4}\}$ forms another subgroup of $SO(2)$, formed by the 90-degree rotations about the origin.

Example 16.7.5 The set $G = \mathbb{R}^3$ together with the composition law

$$(x_1, x_2, x_3) \circ (y_1, y_2, y_3) = (x_1 + y_1, x_2 + y_2, x_3 + y_3 + x_1 y_2)$$

forms a group, called the three-dimensional *Heisenberg group*. This is not a commutative group. The inverse of an element is given by

$$(x_1, x_2, x_3)^{-1} = (-x_1, -x_2, -x_3 + x_1 x_2).$$

The neutral element is $e = (0, 0, 0)$.

16.7.2   Actions of groups on sets

Definition 16.7.6 *Let G be a group and M be a set. An action of G on M is a mapping $\alpha : G \times M \to M$ such that:*

(i) $\alpha(gg', x) = \alpha(g, \alpha(g', x)), \quad \forall g, g' \in G$ and $x \in M$;

(ii) $\alpha(e, x) = x, \forall x \in M$.

We say that the group G acts on the set M with action α and the action of the element g on x is $\alpha(g, x)$. Thus, part (i) states that the action of the product gg' on x is the composition of action of g with the action of g' on x. Part (ii) states that the neutral element action is the identity map.

Any action of a group G on the set M, given by $\alpha : G \times M \to M$, produces a family of transformations of M as follows. For any fixed $g \in G$, let $T_g : M \to M$ be defined by $T_g x = \alpha(g, x)$. The set of these transformations, $\{T_g ; g \in G\}$, forms a group by function composition. This follows from the properties of the action, which imply $T_g T_{g'} = T_{gg'}$ and $T_e x = x$. We also have the inverse transformation $(T_g)^{-1} = T_{g^{-1}}$.

For a given element $x \in M$, the set $\mathcal{O}_x = \{T_g x ; g \in G\}$ is called the *orbit* of x. If $y, z \in \mathcal{O}_x$ are two elements in the orbit of x, then there are $g, g' \in G$ such that $y = T_g x$ and $z = T_{g'} x$. If let $u = g'g^{-1}$, then $z = T_u y$, namely, $z \in \mathcal{O}_y$. In fact, it can be shown that $\mathcal{O}_x = \mathcal{O}_y = \mathcal{O}_z$.

Definition 16.7.7 *An action $\alpha : G \times M \to M$ is called transitive if for any two elements $x, y \in M$, there is $g \in G$ such that $y = T_g x$.*

Equivalently stated, the action α is transitive if for any $x, y \in M$ we have $\mathcal{O}_x = \mathcal{O}_y$. In fact, the action α is transitive if and only if it has only one orbit, namely, $M = \mathcal{O}_x$, for all $x \in M$.

We say that M is a *homogeneous space* of G if for any $x, y \in M$ there is a $g \in G$ such that $y = \alpha(g, x)$. This is equivalent with the fact that the action has only one orbit.

Example 16.7.8 Let $M = \mathbb{R}^3$ and $G = \mathcal{T}(\mathbb{R}^3)$, the group of translations in the space $\mathbb{R}^3$ under the operation of vector addition. Then G acts on M as follows: if $x \in \mathbb{R}^3$ is a vector and $g = \tau_v$ is the translation of vector v, then we define the action of g on x by $\alpha(g, x) = \tau_v(x)$, which is the vector obtained from x by adding the vector v. More explicitly, we have

$$\alpha(g, x) = \tau_v(x) = x + v.$$

The reader can easily check the action definition properties. The action is transitive, as any element $x \in \mathbb{R}^3$ can be translated into any other element $y \in \mathbb{R}^3$.

Example 16.7.9 Consider $M = \mathbb{R}^2$ and $G = SO(2)$, the group of rotations about the origin of the two-dimensional Euclidean plane $\mathbb{R}^2$ under the operation of composition. The group G acts on the set M as follows: if $x \in \mathbb{R}^2$ is a vector and $g = R_\theta$ is the rotation of angle θ, then we define the action of g on x by $\alpha(g, x) = R_\theta x$, which is the vector obtained from x under a counterclockwise rotation of angle θ. More explicitly, we have

$$\alpha(g, x) = \begin{pmatrix} \cos\theta & -\sin\theta \\ \sin\theta & \cos\theta \end{pmatrix} \begin{pmatrix} x_1 \\ x_2 \end{pmatrix} = \begin{pmatrix} x_1\cos\theta - x_2\sin\theta \\ x_1\sin\theta + x_2\cos\theta \end{pmatrix}.$$

This action is not transitive. The orbit of an element $x \in \mathbb{R}^2$ is the circle centered at the origin and having radius $\|x\|$

$$\mathcal{O}_x = \{y \in \mathbb{R}^2; \|y\| = \|x\|\}.$$

Example 16.7.10 In this example $M = \mathbb{R}^3$ and $G = (\mathbb{R}^3, \circ)$ represents the three-dimensional Heisenberg group. Then G acts on M as

$$\alpha(g, x) = (g_1 + x_1, g_2 + x_2, g_3 + x_3 + g_1 x_2),$$

where $g = (g_1, g_2, g_3)$ and $x = (x_1, x_2, x_3)$. The easiest way to check that α is an action is to notice that

$$\alpha(g, x) = L_g x, \tag{16.7.3}$$

where L_g is the left translation on the group $(H, \circ)$, namely, $L_g x = g \circ x$. Then

$$\alpha(g', \alpha(g, x)) = \alpha(g', L_g x) = L_{g'} L_g x = L_{g'g} x = \alpha(g'g, x).$$

The second property of the action is obviously satisfied since the neutral element in the Heisenberg group is $e = (0, 0, 0)$. It is worth noting that an action of a group on itself, $\alpha : G \times G \to G$, defined by relation (16.7.3) is called *action by left multiplication*. The action of the Heisenberg group on the space $\mathbb{R}^3$ is transitive, since for any $x, y \in \mathbb{R}^3$ we have $T_g x = y$, with $g = (y_1 - x_1, y_2 - x_2, y_3 - x_3 + x_1 x_2 - y_1 x_2)$.

The last example brings up the situation when a group G acts on the set $M = G$, which is the group itself; in this case the action acts as $\alpha : G \times G \to G$. If we define $\alpha(g, x) = xg^{-1}$, then α is called the *action by left multiplication*. If consider $\alpha(g, x) = gxg^{-1}$, then α is called the *action by conjugation*.

16.7.3   Extension of actions to functions

We have seen how does an action $\alpha : G \times M \to M$ induce for each group element $g \in G$ a transformation on M, $T_g : M \to M$. Now we shall extend this transformation to functions on M. We denote by $\mathcal{F}(M) = \{f : M \to \mathbb{R}\}$ the set of real functions on M. For any element $g \in G$ we define the transform $\mathbb{T}_g : \mathcal{F}(M) \to \mathcal{F}(M)$ by $\mathbb{T}_g f = f'$, with

$$f'(T_g(x)) = f(x), \qquad x \in M.$$

This can be written equivalently as

$$(\mathbb{T}_g f)(x') = f(T_{g^{-1}}(x')), \quad \forall x' \in M.$$

We need this concept since the activations of each layer of a neural network are considered as functions. For instance, each MNIST data is considered as a function defined on a set M of dimension 28×28 with integer values between 0 and 255.

Example 16.7.11 Let $\mathbb{Z}$ represent the set of integers and consider the group $G = (\mathbb{Z} \times \mathbb{Z}, +)$ with addition on components. This acts on the lattice $M = \mathbb{Z} \times \mathbb{Z}$ by the action $\alpha : G \times M \to M$

$$\alpha\big((g_1, g_2), (x_1, x_2)\big) = (g_1 + x_1, g_2 + x_2), \quad \forall g_i, x_i \in \mathbb{Z}.$$

The induced transform on M is

$$T_{(g_1, g_2)}(x_1, x_2) = (g_1 + x_1, g_2 + x_2), \quad \forall (x_1, x_2) \in \mathbb{Z} \times \mathbb{Z}.$$

The associated extended transform on $\mathcal{F}(\mathbb{Z} \times \mathbb{Z})$ is given by

$$(\mathbb{T}_{(g_1, g_2)} f)(x_1, x_2) = f\big(T_{(g_1, g_2)^{-1}}(x_1, x_2)\big) = f(x_1 - g_1, x_2 - g_2),$$

which is the composition between the function f and the translation of vector $-(g_1, g_2)$.

Example 16.7.12 In the case of Example 16.7.9, the induced transform on $\mathbb{R}^2$ is given by

$$T_{R_\theta} x = (x_1 \cos\theta - x_2 \sin\theta, \ x_1 \sin\theta + x_2 \cos\theta),$$

which is a clockwise rotation of angle θ. The extended transform on functions is

$$(\mathbb{T}_{R_\theta} f)(x) = f\big(\mathbb{T}_{R_\theta^{-1}} x\big) = f(x_1 \cos\theta + x_2 \sin\theta, \ x_2 \cos\theta - x_1 \sin\theta).$$

16.7.4 Definition of equivariance

Consider a group G acting on two sets, M_1 and M_2, with actions

$$\alpha_1 : G \times M_1 \to M_1, \qquad \alpha_2 : G \times M_2 \to M_2.$$

Then for any $g \in G$ the actions induce the transformations

$$T_g : M_1 \to M_1, \qquad T'_g : M_2 \to M_2.$$

Consider the extended transforms to functions

$$\mathbb{T}_g : \mathcal{F}(M_1) \to \mathcal{F}(M_1), \qquad \mathbb{T}'_g : \mathcal{F}(M_2) \to \mathcal{F}(M_2).$$

A map $\Phi : \mathcal{F}(M_1) \to \mathcal{F}(M_2)$ is called *G-equivariant* if for any group element $g \in G$ we have

$$\Phi(\mathbb{T}_g f) = \mathbb{T}'_g(\Phi(f)), \qquad \forall f \in \mathcal{F}(M_1).$$

These concepts can be applied to neural networks as in the following. We consider M_1 and M_2 be the set of neurons in the input and output layers of a feedforward neural network, respectively. The activations of these layers, which are the input $x^{(0)}$ and the output $x^{(L)}$, are functions defined on M_1 and M_2, respectively. The function $\Phi = f_{w,b}$, which maps $x^{(0)}$ into $x^{(L)}$ is the input-output mapping of the network. The equivariance property of the network with respect to the group action G becomes

$$f_{w,b}\big(\mathbb{T}_g x^{(0)}\big) = \mathbb{T}'_g\big(f_{w,b}(x^{(0)})\big) = \mathbb{T}'_g\big(x^{(L)}\big), \quad \forall g \in G. \tag{16.7.4}$$

This states that the output $x^{(L)}$ is transformed in a predictable way as we transform the input $x^{(0)}$ within the family of transforms induced by the group action.

The equivariance of a network can be defined on each layer as in the following, see Kondor [66]. Let $\mathcal{N}$ be a feedforward neural network with $L+1$ layers and layer activations $x^{(0)}, x^{(1)}, \ldots, x^{(L)}$. If M_ℓ denote the set of neurons in the ℓth layer, then $x^{(\ell)} \in \mathcal{F}(M_\ell)$. Assume there is a group G, which acts on the sets $M_0, \ldots, M_L$. The induced transforms on $\mathcal{F}(M_0), \ldots, \mathcal{F}(M_L)$ are correspondingly denoted by $\mathbb{T}^{(0)}, \ldots, \mathbb{T}^{(L)}$. The neural network $\mathcal{N}$ is called a *G-equivariant feedforward network* if, when inputs are transformed by $x^{(0)} \to \mathbb{T}^{(0)}_g(x^{(0)})$, then the layer activations transform by $x^{(\ell)} \to \mathbb{T}^{(\ell)}_g(x^{(\ell)})$ for any $g \in G$.

It is worth noting that this definition holds for both cases, when the set of neurons, M_ℓ, are either discrete or continuous.

16.7.5   Convolution and equivariance

The prototype example for equivariance is given by convolutional networks, which are equivariant to the actions of the translations group. With the previous notations this means

$$(\mathbb{T}_g x^{(0)}) * w = \mathbb{T}'_g(x^{(0)} * w), \quad \forall g \in G,$$

for any filter w, see Fig. 16.8. This relation can be shown in both discrete and continuous cases. For the sake of simplicity we consider just the one-dimensional case. The discrete case follows from

$$[(\mathbb{T}_g x^{(0)}) * w]_p = \sum_i (\mathbb{T}_g x^{(0)})_{i+p} w_i = \sum_i x^{(0)}_{i+p-g} w_i$$

$$= (x^{(0)} * w)_{p-g} = [\mathbb{T}'_g(x^{(0)} * w)]_p.$$

The verification of equivariance in the continuous case is given by

$$[(\mathbb{T}_g x^{(0)}) * w](t) = \int_{\mathbb{R}} (\mathbb{T}_g x^{(0)})(u+t) w(u)\, du = \int_{\mathbb{R}} x^{(0)}(u+t-g) w(u)\, du$$

$$= (x^{(0)} * w)(t-g) = [\mathbb{T}'_g(x^{(0)} * w)](t).$$

The previous computations can be carried over to groups as in the following. First, the convolution of any two functions $f, \psi : G \to \mathbb{R}$ defined on a discrete group G is defined by

$$(f * \psi)(t) = \sum_{y \in G} f(y) \psi(t^{-1}y).$$

The verification of the G-equivariance relation is similar with the previous computation

$$(\mathbb{T}_g x^{(0)} * w)(t) = \sum_{y \in G} (\mathbb{T}_g x^{(0)})(y)\, w(t^{-1}y) = \sum_{y \in G} x^{(0)}(g^{-1}y)\, w(t^{-1}y)$$

$$= \sum_{v \in G} x^{(0)}(v)\, w(t^{-1}gv) = \sum_{v \in G} x^{(0)}(v)\, w((g^{-1}t)^{-1}v)$$

$$= \Big(\mathbb{T}_g(x^{(0)} * w)\Big)(t),$$

where we used the change of variables $v = g^{-1}y$ and the fact that the variable $v \in g^{-1}G = G$.

The equivariance theory can be extended to continuous compact groups, see [66]. The convolution definition of any two functions $f, \psi : G \to \mathbb{R}$ in this case is

$$(f * \psi)(t) = \int_G f(y) \psi(t^{-1}y)\, d\mu(y),$$

where μ is a left-translation invariant measure on G, with $\mu(G) = 1$, called the *Haar measure* on G. The verification of the equivariance is done similarly with the discrete case, replacing the sums by integrals and using the invariance property of the Haar measure

$$(\mathbb{T}_g x^{(0)} * w)(t) = \int_G (\mathbb{T}_g x^{(0)})(y) w(t^{-1}y) d\mu(y) = \int_G x^{(0)}(g^{-1}y) w(t^{-1}y) d\mu(y)$$

$$= \int_G x^{(0)}(v) w(t^{-1}gv) d\mu(v) = \int_G x^{(0)}(v) w((g^{-1}t)^{-1}v) d\mu(v)$$

$$= \Big(\mathbb{T}_g(x^{(0)} * w)\Big)(t).$$

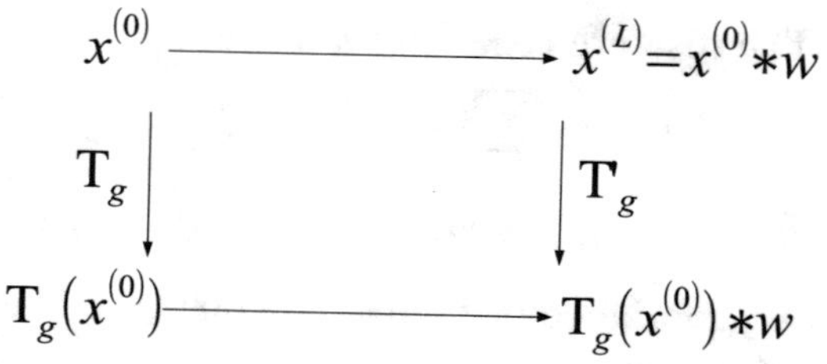

$$
\begin{array}{ccc}
x^{(0)} & \longrightarrow & x^{(L)} = x^{(0)} * w \\[2mm]
\Big\downarrow \mathbb{T}_g & & \Big\downarrow \mathbb{T}_g \\[2mm]
\mathbb{T}_g\big(x^{(0)}\big) & \longrightarrow & \mathbb{T}_g\big(x^{(0)}\big) * w
\end{array}
$$

Figure 16.8: *The equivariance can be seen as a commutative diagram.*

It is worth noting that Cohen et al. [26] applied the technique for analyzing spherical images. They adapted the definition of convolution from planar domains to the sphere $\mathbb{S}^2 = \{x \in \mathbb{R}^3; \|x\| = 1\}$ by

$$
(f * \psi)(R) = \int_{\mathbb{S}^2} f(x)\, \psi(R^{-1}x)\, dx,
$$

where $\psi, f : \mathbb{S}^2 \to \mathbb{R}$ are two spherical signals, and $R \in SO(3)$ is a rotation. They used the action of the special orthogonal group $SO(3)$ (namely, the group of 3×3 matrices that preserve distance and have determinant 1) on the sphere $\mathbb{S}^2$ to prove the rotation equivariance.

Even if a planar convolution is always equivariant to the actions of the translations group, it is not covariant with respect to the rotations group $SO(2)$, unless some additional assumptions are made. This is accomplished in the following.

Let $R \in SO(2)$ be a rotation of the plane and $f, w : \mathbb{R}^2 \to \mathbb{R}$. Then

$$
\begin{aligned}
[(\mathbb{T}_R f) * w](x) &= \sum_{y \in \mathbb{Z}^2} (\mathbb{T}_R f)(y)\, w(y - x) = \sum_{y \in \mathbb{Z}^2} f(R^{-1}y)\, w(y - x) \\
&= \sum_{u \in \mathbb{Z}^2} f(u)\, w(Ru - x) = \sum_{u \in \mathbb{Z}^2} f(u)\, w\big(R(u - R^{-1}x)\big) \\
&= \sum_{u \in \mathbb{Z}^2} f(u)\, (\mathbb{T}_{R^{-1}}w)(u - R^{-1}x) \\
&= (f * \mathbb{T}_{R^{-1}}w)(R^{-1}x) = \mathbb{T}_R(f * \mathbb{T}_{R^{-1}}w).
\end{aligned}
$$

If now we consider f to be a signal and w a filter, and further assume that w is a rotation-invariant filter, namely, $\mathbb{T}_{R^{-1}}w = w$, then the previous formula becomes

$$
[(\mathbb{T}_R f) * w](x) = \mathbb{T}_R(f * \mathbb{T}_{R^{-1}}w) = \mathbb{T}_R(f * w),
$$

which represents the covariance of the planar convolution with respect to rotation-invariant filters.

In the case of the group $G = \mathbb{Z}^2$, we consider the rotations R to be a 90, 180 and 270-degrees rotation of the lattice $\mathbb{Z}^2$ about the origin. The rotation invariance of the filter, $w(Rx) = w(x)$, in this case becomes

$$w_{i,j} = w_{-j,i} = w_{i,-j} = w_{j,-i}, \quad \forall (i,j) \in \mathbb{Z}^2.$$

16.7.6 Definition of invariance

The *invariance* is a particular case of equivariance. In this case the network output does not change when the input is transformed by the family of transformations $\{\mathbb{T}_g; g \in G\}$. In this case formula (16.7.4) becomes

$$f_{w,b}(\mathbb{T}_g x^{(0)}) = f_{w,b}(x^{(0)}) = x^{(L)}, \quad \forall g \in G.$$

This means $\mathbb{T}'_g = \text{Id}$ for all $g \in G$, where Id is the identity map of $\mathcal{F}(M_L)$. Then for any $f \in \mathcal{F}(M_L)$

$$(\mathbb{T}'_g f)(x) = f(T'_{g^{-1}}(x)) = f(x), \quad \forall x \in M_L.$$

Therefore, $T'_g(x) = x$ for all $x \in M_L$ and $g \in G$. This is equivalent to the fact that the orbit of each element in M_L is the element itself, namely, $\mathcal{O}_x = \{x\}$, for all $x \in M_L$.

A prototype example of network invariance to local translation occurs in the case of pooling, see section 15.2. We shall discuss next the pooling process in the context of groups.

Let G be a group and $f : G \to \mathbb{R}$ be a feature map.[2] Let $U \subset G$ be a subset of G, which contains the neutral element, $e \in U$, called *pooling domain*. The max pooling operator, $\mathcal{P}$, is defined by $\mathcal{P}f : G \to \mathbb{R}$

$$(\mathcal{P}f)(x) = \max_{u \in xU} f(u).$$

A distinguished case is the one when the pooling domain U is a subgroup $H \leq G$. Then the pooling domains $\{xH; x \in G\}$ form a partition of the group G, namely, if $x, y \in G$, then $xH = yH$ or $xH \cap yH = \emptyset$. Furthermore, if G is a finite group, the H and xH have the same number of elements, see Exercise 16.9.12. The partition sets $\{xH; x \in G\}$ are called *cosets* and represent the equivalence classes of the following equivalence relation on G

$$x \sim y \iff x^{-1}y \in H.$$

---

[2]If $f : M \to \mathbb{R}$ then f is an activation map, since it describes the activations of neurons in the set M. However, if $f : G \to \mathbb{R}$, then f is a feature map since it describes the features captured by the group G.

Pooling induces an application ϕ on the coset space defined by

$$\phi(xH) = (\mathcal{P}f)(x).$$

Example 16.7.13 This is a classical example of max pooling using a 2×2 pooling domain, which shifts across the two-dimensional lattice of integers. For this, let $G = \mathbb{Z} \times \mathbb{Z}$ be a group with addition on components and consider the activation $f : G \to \mathbb{R}$. Consider the pooling domain

$$U = \{(i,j); -2 \le i \le 2, \ -2 \le j \le 2\}.$$

The pooling becomes $(\mathcal{P}f)(x) = \max\limits_{u \in x+U} f(u)$, where

$$x + U = \{(i,j); -2 + x_1 \le i \le 2 + x_1, \ -2 + x_2 \le j \le 2 + x_2\},$$

is the translation of the domain U with the vector (x_1, x_2), $x_i \in \mathbb{Z}$.

If the pooling domain is a subgroup of G

$$H = \{(3n_1, 3n_2); (n_1, n_2) \in \mathbb{Z} \times \mathbb{Z}\}$$

then the maximum is taken over the shifts of x by multiples of 3 pixels, which corresponds to a stride of $s = 2$ pixels.

The max pooling invariance in the groups context is shown in the next result, [27].

Proposition 16.7.14 *Pooling commutes with the group action:*

$$\mathcal{P}(\mathbb{T}_g f)(x) = \mathbb{T}_g(\mathcal{P}f)(x), \quad \forall x, g \in G, \quad f : G \to \mathbb{R}.$$

Proof: Making the substitution $u = g^{-1}h$ and using the definition of $\mathbb{T}_g$ we have

$$
\begin{aligned}
\mathcal{P}(\mathbb{T}_g f)(x) &= \max\limits_{h \in xU}(\mathbb{T}_g f)(h) = \max\limits_{h \in xU} f(g^{-1}h) \\
&= \max\limits_{u \in g^{-1}xU} f(u) = \mathcal{P}f(g^{-1}x) \\
&= \mathbb{T}_g(\mathcal{P}f)(x).
\end{aligned}
$$

∎

Remark 16.7.15 We have considered for the sake of simplicity only activations and filters that have only one channel. The results can be extended to multiple-channel case by considering $f, w : M \to \mathbb{R}^K$, where K is the number of channels. The convolution can be defined in the discrete case as

$$(f * w)(g) = \sum_{y \in G} \sum_{k=1}^{K} f_k(y) \, w_k(g^{-1}y) = \sum_{y \in G} \langle f(y), w(g^{-1}y) \rangle$$

and in the continuous case by

$$(f * w)(g) = \int_G \sum_{k=1}^{K} f_k(y) \, w_k(g^{-1}y) \, d\mu(y) = \int_G \langle f(y), w(g^{-1}y) \rangle \, d\mu(y),$$

where μ is the Haar measure on the compact group G.

16.8 Summary

CNNs are neural networks specialized in processing image data. They have been extremely successful in practical applications such as handwritten digits recognition. A CNN contains convolution layers which use convolution with a kernel instead of a regular matrix multiplication. The kernel is usually a matrix of relatively small dimension, compared with the input image dimensions.

In a CNN layers have sparse interactions and share weighs and biasses. A local receptive field matching the dimensions of the kernel slides one (or more) pixel(s) at a time, first horizontally and then vertically, from top to bottom, until the entire input data is completely scanned. The sum of the products between the kernel entries and the activations of the receptive field produces a number, which is stored as an entry in a feature map. The convolution between the input data and each kernel produces a different feature map, which retain certain features of the data, such as corners, edges, or other simple shapes.

A certain number of kernels are used to produce feature maps. In the case of a multiple-layer CNN, the later layers consist of features on features. The convolution layers are used alternatively with pooling layers. At the end, the CNN contains one (or more) fully-connected layers and then a softmax activation for the output layer.

Convolutional neural networks are able to detect local patterns in an image regardless of their position, because CNNs are equivariant to translations. This means if the input image is translated by some vector, then the activation pattern in each higher layer of the network is also translated by the same vector. The equivariance property can be defined for any group of transformations acting on the layers of a neural network. The theory is done for both discrete groups and continuous compact groups.

16.9   Exercises

Exercise 16.9.1 Compute the following matrix convolution:

$$\begin{pmatrix} 3 & 1 & 1 \\ 4 & 2 & 0 \\ 5 & -1 & 1 \end{pmatrix} * \begin{pmatrix} -1 & 0 \\ 1 & -2 \end{pmatrix}.$$

Exercise 16.9.2 (a) Formulate and prove a variant of Proposition 16.3.3 for the case of one-dimensional inputs.

(b) Does the equivariance to translation still hold true if the input is a tensor?

Exercise 16.9.3 (Sobel operators) (a) Show that the convolution of an image with the 3×3 kernel

$$K = \begin{pmatrix} 1 & 2 & 1 \\ 0 & 0 & 0 \\ -1 & -2 & -1 \end{pmatrix}$$

emphasizes horizontal edge detection.

(b) Show that the convolution with the transpose kernel, K^T, filters the vertical edges.

Exercise 16.9.4 Show that the convolution of an image with the following 3×3 kernels blur the image:

(a) (Box blur)

$$K = \begin{pmatrix} 1/9 & 1/9 & 1/9 \\ 1/9 & 1/9 & 1/9 \\ 1/9 & 1/9 & 1/9 \end{pmatrix}$$

(b) (Gaussian blur)

$$K = \begin{pmatrix} 1/16 & 1/8 & 1/16 \\ 1/8 & 1/4 & 1/8 \\ 1/16 & 1/8 & 1/16 \end{pmatrix}$$

Exercise 16.9.5 Provide arguments supporting the fact that the convolution of an image with the 3×3 kernel

$$K = \begin{pmatrix} 0 & -1 & 0 \\ -1 & 5 & -1 \\ 0 & -1 & 0 \end{pmatrix}$$

sharpens the image.

Exercise 16.9.6 Consider the following 3×3 convolution kernels

$$K_1 = \begin{pmatrix} 1 & 0 & -1 \\ 0 & 0 & 0 \\ -1 & 0 & 1 \end{pmatrix}, \qquad K_2 = \begin{pmatrix} 0 & 1 & 0 \\ 1 & -4 & 1 \\ 0 & 1 & 0 \end{pmatrix}$$

$$K_3 = \begin{pmatrix} -1 & -1 & -1 \\ -1 & 8 & -1 \\ -1 & -1 & -1 \end{pmatrix}.$$

Show arguments supporting that the effect of convolution with these kernels is edge detection. Note that the sum of the entries in each kernel is 0. What does this mean?

Exercise 16.9.7 Consider the following convolution image processing: each pixel value in the original image is subtracted from its neighboring pixel value on the left. Write a 3×3 kernel which does this operation.

Exercise 16.9.8 Let y be a two-dimensional discrete signal. If $\mathcal{C}$ and $\mathcal{P}$ denote, respectively, the convolution and pooling operators, show the following relations:

(a) $\mathcal{P} \circ \mathcal{C}(T_{a,b} \circ y) = \mathcal{P} \circ \mathcal{C}(y)$.

(b) $\mathcal{C} \circ \mathcal{P}(T_{a,b} \circ y) = \mathcal{C} \circ \mathcal{P}(y)$.

Exercise 16.9.9 If one would like to regularize the CNN given in Fig. 16.7, in which layer should the dropout technique be used?

Exercise 16.9.10 Explain why between two networks with the same input and the same depth and width, a CNN is less prone to overfitting than a fully-connected neural network.

Exercise 16.9.11 A convolution layer compresses the information, in the sense that the output size is smaller than input size.

(i) In the case of a one-dimensional convolution layer, if N denotes the size of the input, using a filter of size F that is moved with a stride S, we obtain an output of size O, given by

$$O = \frac{N - F}{S} + 1.$$

If padding is used, denoted by P, then the formula becomes

$$O = \frac{N - F + 2P}{S} + 1.$$

(*ii*) In the case of a two-dimensional convolution layer, if the input dimension is $W_1 \times H_1$ and F is the size of a square filter, then the output has dimension $W_2 \times H_2$, with

$$W_2 = \frac{W_1 - F + 2P}{S} + 1, \qquad H_2 = \frac{H_1 - F + 2P}{S} + 1.$$

Exercise 16.9.12 Let H be a subgroup of the group G and $x \in G$. Denote by $xH = \{xh; h \in H\}$. Show that the subsets xH have the following properties:

(*i*) $x \in xH$;

(*ii*) If $x, y \in G$, then either $xH = yH$, or $xH \cap yH = \emptyset$;

(*iii*) If G is a finite group, then H and xH have the same number of elements;

(*iv*) Consider the relation "$\sim$" defined on $G \times G$, defined by $x \sim y \iff y^{-1}x \in H$. Show that "$\sim$" satisfies:

    (*a*) $x \sim x$

    (*b*) $x \sim y \Rightarrow y \sim x$;

    (*c*) $x \sim y$ and $y \sim z \Rightarrow x \sim z$.

Chapter 17

Recurrent Neural Networks

A feedforward fully-connected neural network cannot be used successfully for modeling sequences of data. A few basic reasons are the following: they cannot handle variable-length input sequences, do not share parameters, cannot track long-term dependencies, and cannot maintain information about the order of input data. Hence the need of a neural architecture that can handle successfully all the previous requirements. This is a recurrent neural network, or RNN (Rumelhart et al. [108]), which is the subject of this chapter.

17.1   States Systems

States systems are easier to understand than RNNs and will constitute a basis for later discussions on RNNs. This is the reason why we shall start by

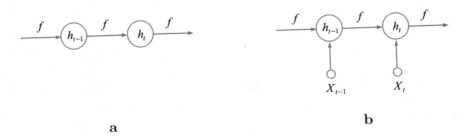

a   b

Figure 17.1: **a.** *A dynamical system isolated from exterior, given by equation (17.1.1).* **b.** *A dynamical system driven by the process X_t as given by equation (17.1.2).*

© Springer Nature Switzerland AG 2020
O. Calin, *Deep Learning Architectures*, Springer Series in the Data Sciences, https://doi.org/10.1007/978-3-030-36721-3_17

considering a dynamical system, whose state h_t at time t updates as

$$h_t = f(h_{t-1}; \theta), \qquad t \geq 1 \tag{17.1.1}$$

where the *transition function* $f : \mathbb{R}^k \times \mathbb{R}^m \to \mathbb{R}^k$ is a Borel-measurable function and $\theta \in \mathbb{R}^m$ represents a vector parameter, which is independent of t, see Fig. 17.1 **a**. We assume the states h_t are random variables (discrete or continuous) and denote by $\mathcal{H}_t = \mathfrak{S}(h_t)$ the sigma-algebra generated by h_t (see the definition in section D.5 of the Appendix). Each state h_t contains some information, $\mathcal{H}_t$, about the dynamical system. Since the state h_t is determined by h_{t-1}, Proposition D.5.1 implies $\mathcal{H}_t \subset \mathcal{H}_{t-1}$, i.e., we obtain a descending sequence of sigma-algebras. Namely, each state contains less or the same information as the previous state does.

If $f(\cdot\,; \theta)$ is invertible in the first argument, then $\mathcal{H}_t = \mathcal{H}_{t-1}$, namely, the information does not shrink. However, this condition is too restrictive, and in general the information decreases strictly as t increases. In the following we shall consider a similar example involving an application of the contraction principle, Theorem 7.7.1.

Example 17.1.1 Let's assume that f is differentiable and satisfies the inequality $\|\partial f / \partial h\| < \lambda < 1$. Then by the Mean Value Theorem the function f becomes a λ-contraction in the first argument

$$\|f(h; \theta) - f(h'; \theta)\| \leq \max_h \|\partial_h f\| \, \|h - h'\| < \lambda \|h - h'\|, \qquad \forall h, h' \in \mathbb{R}^k.$$

For any two states in the sample space, $\omega, \omega' \in \Omega$, we shall evaluate the distance between their corresponding states at time t by unfolding the recurrence as

$$\begin{aligned}
\|h_t(\omega) - h_t(\omega')\| &= \|f(h_{t-1}(\omega); \theta) - f(h_{t-1}(\omega'); \theta)\| < \lambda \|h_{t-1}(\omega) - h_{t-1}(\omega')\| \\
&< \lambda^2 \|h_{t-2}(\omega) - h_{t-2}(\omega')\| < \cdots < \lambda^t \|h_0(\omega) - h_0(\omega')\|.
\end{aligned}$$

Since the random variable h_0 is finite almost everywhere, then $\|h_0(\omega) - h_0(\omega')\|$ is finite for almost all ω and ω'. Taking $t \to \infty$ and using the Squeeze Theorem, we obtain

$$\lim_{t \to \infty} \|h_t(\omega) - h_t(\omega')\| = 0,$$

for almost all ω and ω'. This implies that $\lim_{t \to \infty} h_t = c$, constant, almost everywhere, and this convergence holds almost surely, i.e., $\mathbb{P}(\omega; h_t(\omega) \to c) = 1$. In fact, c is the fixed point of the contraction function f, i.e.,

$$f(c; \theta) = c.$$

It is obvious that c depends on the parameter θ. In this example the sequence of sigma-algebras $\mathcal{H}_t$ is strictly descending, with its inferior limit $\mathcal{H}_\infty = \bigcap_{t \geq 1} \mathcal{H}_t = \mathfrak{S}(c) = \{\varnothing, \Omega\}$, which is the trivial sigma-algebra. Hence, in the long run, the system losses all its information. Forgetting information about the past is a characteristic of dynamical systems with contractive transition function. This will constitute the cause of the vanishing gradient problem for recurrent neural networks, which will be approached later in this chapter.

Since in general the information cast by $\mathcal{H}_t$ decreases as t increases, we shall consider additional information inserted into the system at each time instance. Thus, we consider a dynamical system driven by an external signal modeled by the stochastic process X_t, $t \geq 1$. The recurrence in this case can be written as

$$h_t = f(h_{t-1}, X_t; \theta), \qquad t \geq 1, \tag{17.1.2}$$

see Fig. 17.1 **b**. Since h_t is determined now by both h_{t-1} and X_t, we obtain $\mathcal{H}_t \subset \mathfrak{S}(h_{t-1}, X_t)$. Using the definition of the joint sigma-algebra of two random variables

$$\mathfrak{S}(h_{t-1}, X_t) = \mathfrak{S}\Big(\mathfrak{S}(h_{t-1}) \cup \mathfrak{S}(X_t)\Big) = \mathfrak{S}(\mathcal{H}_{t-1} \cup \mathcal{I}_t),$$

the previous inclusion becomes

$$\mathcal{H}_t \subset \mathfrak{S}(\mathcal{H}_{t-1} \cup \mathcal{I}_t), \tag{17.1.3}$$

where $\mathcal{I}_t$ denotes the input information generated by the input variable X_t.

Proposition 17.1.2 *Consider a dynamical system with a given initial state h_0 and driven by a stochastic process X_t as in (17.1.2). Then*

$$\mathcal{H}_t \subset \mathfrak{S}(\mathcal{I}_1 \cup \cdots \cup \mathcal{I}_t). \tag{17.1.4}$$

Proof: The proof follows from a repetitive application of formula (17.1.3) and Exercise 17.10.1, which implies

$$\begin{aligned}\mathcal{H}_t \quad &\subset \quad \mathfrak{S}(\mathcal{H}_{t-1} \cup \mathcal{I}_t) \\ &\subset \quad \mathfrak{S}\Big(\mathfrak{S}(\mathcal{H}_{t-2} \cup \mathcal{I}_{t-1}) \cup \mathcal{I}_t\Big) = \mathfrak{S}\Big(\mathcal{H}_{t-2} \cup \mathcal{I}_{t-1} \cup \mathcal{I}_t\Big).\end{aligned}$$

Inductively, we obtain

$$\mathcal{H}_t \subset \mathfrak{S}\Big(\mathcal{H}_0 \cup \mathcal{I}_1 \cup \cdots \cup \mathcal{I}_t\Big).$$

Since h_0 is a given constant, then $\mathcal{H}_0 = \{\varnothing, \Omega\}$ and hence

$$\mathcal{H}_0 \cup \mathcal{I}_1 \cup \cdots \cup \mathcal{I}_t = \mathcal{I}_1 \cup \cdots \cup \mathcal{I}_t.$$

∎

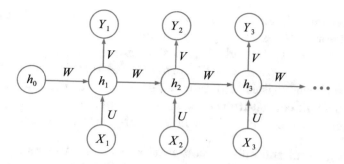

Figure 17.2: *A "many-to-many" configuration of an RNN.*

Remark 17.1.3 It is worth noting the use of dependent variables. Unfolding we have

$$h_t = f(h_{t-1}, X_t) = f(f(h_{t-2}, X_{t-1}), X_t) = g(h_{t-2}, X_{t-1}, X_t),$$

with g measurable. Since h_t is determined by h_{t-2}, X_{t-1}, and X_t, by Proposition D.5.1 we obtain $\mathfrak{S}(h_t) \subset \mathfrak{S}(h_{t-2}, X_{t-1}, X_t)$.

17.2   RNNs

A recurrent neural network can be introduced in a couple of ways.

1. First, it can be considered as a state system having the property that each state, but the initial one, provides an outcome, see Fig. 17.2. Thus, the state h_t is fed information from the previous state, h_{t-1}, and from the present input, X_t, and provides an outcome Y_t and an input to the next state, h_{t+1}. The state h_t is called the tth *hidden state* of the RNN. X_t and Y_t are the tth input and output, respectively.

2. We can also think of an RNN as a sequence of plain vanilla feedforward neural networks, $X_t \to h_t \to Y_t$, (with input layer X_t, hidden layer h_t and output layer Y_t), who feeds information from one hidden layer to the next.

The standard equations for the forward propagation in an RNN are given by

$$\begin{aligned} h_t &= \tanh(W h_{t-1} + U X_t + b) & (17.2.5) \\ Y_t &= V h_t + c. & (17.2.6) \end{aligned}$$

Thus, the hidden states take values between -1 and 1 and the output is an affine function of the current hidden state. The transition function f is a composition between the hyperbolic tangent and an affine function. The parameter $\theta = (W, U, b, c)$ consists of two matrices and two vectors. Matrices W and U, represent the hidden-to-hidden state transition and input-to-hidden state transition. The matrix V represents the transition from hidden state-to-output, while b and c are bias vectors.

17.3 Information in RNNs

We introduce some terminologies first. The input information generated by X_t is denoted by $\mathcal{I}_t$. The hidden information is generated by the hidden state, $\mathcal{H}_t = \mathfrak{S}(h_t)$, and the output information is generated by the output, $\mathcal{E}_t = \mathfrak{S}(Y_t)$. This section deals with the relation between these sigma-algebras.

The first result shows how the output information relates to the input and hidden information.

Proposition 17.3.1 *Consider an RNN with a given initial state h_0 and inputs X_t satisfying the forward pass (17.2.5)–(17.2.6). Then*

$$\mathcal{E}_t \subset \mathcal{H}_t \subset \mathfrak{S}(\mathcal{I}_1 \cup \cdots \cup \mathcal{I}_t). \tag{17.3.7}$$

Proof: From relation (17.2.6), the variable Y_t is determined by h_t, so by Proposition D.5.1 we have $\mathfrak{S}(Y_t) \subset \mathfrak{S}(h_t)$, which shows the first inclusion of (17.3.7). The second inclusion follows from Proposition 17.1.2. ∎

It is natural to consider the case when the information $\mathcal{I}_t$ inserted into the system at each time step is "new", namely the case when the input variables X_t, $t \in \{1, 2, \ldots\}$ are independent. We obtain the following history independence property.

Proposition 17.3.2 *Consider an RNN with a given initial state h_0 and independent inputs X_t satisfying the forward pass (17.2.5)–(17.2.6). Then both sigma-algebras $\mathcal{E}_{t-1}$ and $\mathcal{H}_{t-1}$ are independent from $\mathcal{I}_t$ (namely, Y_{t-1} and h_{t-1} are independent from X_t).*

Proof: Since X_t is independent of $X_1, \ldots, X_{t-1}$, then $\mathcal{I}_t$ is a body of information independent from $\{\mathcal{I}_1, \ldots \mathcal{I}_{t-1}\}$. Then their generated sigma-algebra, $\mathfrak{S}(\mathcal{I}_1 \cup \cdots \cup \mathcal{I}_{t-1})$ is independent of $\mathcal{I}_t$, too. Since by Proposition 17.3.1 we have $\mathcal{H}_{t-1} \subset \mathfrak{S}(\mathcal{I}_1 \cup \cdots \cup \mathcal{I}_{t-1})$, it follows that $\mathcal{H}_{t-1}$ is independent of $\mathcal{I}_t$. The fact that $\mathcal{E}_{t-1}$ is independent of $\mathcal{I}_t$ follows from the inclusion $\mathcal{E}_t \subset \mathcal{H}_t$, see Proposition 17.3.1. ∎

In order to asses the information using entropy, we shall assume that variables h_t, X_t, and Y_t have the same vectorial dimension, so U, V, and W

become square matrices. We shall further assume that they are also nonsingular. In this case we can solve for h_t from (17.2.6) as $h_t = V^{-1}(Y_t - c)$. This implies via Proposition D.5.1 that $\mathcal{H}_t \subset \mathcal{E}_t$. Since the inverse inclusion has been shown in Proposition 17.3.1, it follows that $\mathcal{H}_t = \mathcal{E}_t$.

Proposition 17.3.3 *Assume that variables h_t, X_t, and Y_t have the same vectorial dimension and $(X_t)_{t \geq 1}$ are independent random variables.*

(a) The conditional entropy of one hidden state with respect to the previous state satisfies

$$H(h_t | h_{t-1}) < H(X_t) + \ln |\det U|.$$

(b) The conditional entropy of an output with respect to the previous hidden state satisfies

$$H(Y_t | h_{t-1}) < H(X_t) + \ln |\det(UV)|.$$

Proof: (a) Let $a_t = W h_{t-1} + U X_t + b$, so $h_t = \tanh a_t$. We evaluate the conditional entropy using Proposition 12.1.2

$$
\begin{aligned}
H(h_t | h_{t-1}) &= H(\tanh(a_t) | h_{t-1}) \\
&= H(a_t | h_{t-1}) - \mathbb{E}^{P_{a_t | h_{t-1}}} [\ln |\det J_{\tanh^{-1}}(h_t)|]. \quad (17.3.8)
\end{aligned}
$$

We shall compute next the Jacobian. Since $a_t = \tanh^{-1}(h_t)$, and $\tanh^{-1}$ acts on components, then $a_t^j = \tanh^{-1}(h_t^j)$, so

$$J_{\tanh^{-1}}(h_t) = \left(\frac{\partial a_t^i}{\partial h_t^k} \right)_{j,k} = \left(\frac{\partial \tanh^{-1}(h_t^i)}{\partial h_t^k} \right)_{j,k} = \frac{\delta_{ik}}{1 - (h_t^i)^2}.$$

Since the Jacobian is diagonal, it follows that

$$\det J_{\tanh^{-1}}(h_t) = \prod_i \frac{1}{1 - (h_t^i)^2}.$$

Substituting back into (17.3.8) yields

$$
\begin{aligned}
H(h_t | h_{t-1}) &= H(a_t | h_{t-1}) + \mathbb{E}^{P_{a_t | h_{t-1}}} [\ln \prod_i |1 - (h_t^i)^2|] \\
&< H(a_t | h_{t-1}), \quad\quad\quad\quad\quad\quad\quad\quad\quad (17.3.9)
\end{aligned}
$$

since $1 - (h_t^i)^2 \in (0, 1)$. We evaluate next the conditional entropy $H(a_t | h_{t-1})$ as

$$
\begin{aligned}
H(a_t | h_{t-1}) &= H(W h_{t-1} + U X_t + b | h_{t-1}) \\
&= H(U X_t | h_{t-1}) = H(U X_t) = H(X_t) + \ln |\det U|.
\end{aligned}
$$

We have taken into account that X_t and h_{t-1} are independent and then used Proposition 12.1.2. Substituting back into (17.3.9) yields the desired inequality of part (a).

(b) Using $Y_t = V h_t + c$, a similar computation provides

$$\begin{aligned} H(Y_t|h_{t-1}) &= H(h_t|h_{t-1}) + \ln|\det V| \\ &< H(X_t) + \ln|\det U| + \ln|\det V|, \end{aligned}$$

where we have used the inequality from part (a). ∎

Remark 17.3.4 If the entries of matrices U and V are small enough, their absolute value of determinants is smaller than 1. Under this regularization condition we obtain the following upper bounds:

$$H(h_t|h_{t-1}) < H(X_t), \quad H(Y_t|h_{t-1}) < H(X_t).$$

This asserts that the entropy of the driving signal, X_t, is larger than the entropies of the output and hidden state, conditioned by the history of the hidden states.

We also note that in the virtue of Corollary 3.5.2 the conditional entropies $H(h_t|h_{t-1})$ are controlled by the input variance, $Var X_t$, in the sense that a small variance of X_t implies a small value for the conditional entropy $H(h_t|h_{t-1})$.

17.4   Loss Functions

Consider an RNN with T cells. We shall denote the inputs and outputs by $(X_1, \ldots, X_T)$ and $(Y_1, \ldots, Y_T)$, respectively. The target in the case of a "many-to-many" RNN, which is represented in Fig. 17.2, is given by the sequence $(Z_1, \ldots, Z_T)$. The loss function, L, for an RNN represents the "distance" between the sequences $(Y_1, \ldots, Y_T)$ and $(Z_1, \ldots, Z_T)$. This can be considered as the sum

$$L = \sum_{t=1}^{T} L_t,$$

where the individual loss function, L_t measures the proximity of Y_t with respect to Z_t.

For the individual loss function L_t we have a few choices. In the case of random variables, the individual loss can be either the mean square error

$$L_t = \mathbb{E}[|Y_t - Z_t|^2],$$

either the Kullback-Leibler divergence

$$L_t = D_{KL}\big(p_{(X_1,\ldots,X_t),Z_t} || p_{\theta;(X_1,\ldots,X_t),Z_t}\big)$$

or the cross-entropy

$$L_t = S\big(p_{(X_1,\ldots,X_t),Z_t}, p_{\theta;(X_1,\ldots,X_t),Z_t}\big) = -\mathbb{E}^{p_{(X_1,\ldots,X_t),Z_t}}[\log p_{\theta;(X_1,\ldots,X_t),Z_t}],$$

where $\theta = (U, V, W, b, c)$ is the parameter of the model.

In the case when the variables are continuous we may choose the individual loss as the square of the Euclidean distance

$$L_t = \frac{1}{2}(Y_t - Z_t)^2.$$

Regardless of the loss function considered, computing the gradient with respect to parameters, $\nabla_\theta L$, is an expensive operation and leads in many cases to gradient problems, as we shall see later.

17.5 Backpropagation Through Time

The forward pass through an RNN is given by formulas (17.2.5)–(17.2.6), which provide values for h_t, Y_t, and losses L_t. Now, in order to minimize the loss function, it suffices to apply the gradient descent method, which requires the computation of the gradient $\nabla_\theta L$. The method of computing the gradient is called the *backpropagation through time*, and it is a variant of the backpropagation method studied in earlier chapters. However, this is more complex now, since it is a composition between a backpropagation at each individual time step and a backpropagation across time.

Since the exposition of the general case contains complicated notations, which might be potentially confusing, we shall exemplify the method of backpropagation through time in the case of an RNN with only two steps, see Fig. 17.3. In this case there are two inputs X_1, X_2, two outputs Y_1, Y_2, as well as two hidden states, h_1, h_2. For the sake of simplicity we shall consider them one-dimensional continuous variables. The forward pass equations can be written as

$$
\begin{aligned}
a_1 &= W h_0 + U X_1 + b & a_2 &= W h_1 + U X_2 + b \\
h_1 &= \tanh a_1 & h_2 &= \tanh a_2 \\
Y_1 &= V h_1 + c & Y_2 &= V h_2 + c.
\end{aligned}
$$

These formulas are used to compute the loss function. For simplicity, we consider $L_t = \frac{1}{2}(Y_t - Z_t)^2$, so the loss function becomes

$$L = L_1 + L_2 = \frac{1}{2}(Y_1 - Z_1)^2 + \frac{1}{2}(Y_2 - Z_2)^2 = \frac{1}{2}\|Y - Z\|_{Eu}^2.$$

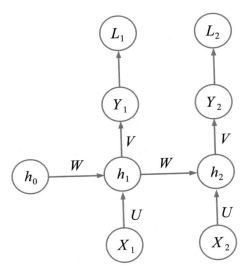

Figure 17.3: *An RNN configuration with two steps.*

We need to compute five gradients

$$\nabla_\theta L = \left(\frac{\partial L}{\partial W}, \frac{\partial L}{\partial V}, \frac{\partial L}{\partial U}, \frac{\partial L}{\partial b}, \frac{\partial L}{\partial c} \right).$$

We shall compute beforehand a few derivatives that will be useful shortly. Using the chain rule, we have

$$
\begin{aligned}
\frac{\partial h_1}{\partial W} &= \frac{\partial}{\partial W} \tanh(a_1) = \operatorname{sech}^2 a_1 \frac{\partial a_1}{\partial W} \\
&= (1 - \tanh^2 a_1) h_0 = (1 - h_1^2) h_0.
\end{aligned}
$$

A similar computation can be used to obtain

$$
\begin{aligned}
\frac{\partial h_2}{\partial W} &= (1 - h_2^2) h_1 \\
\frac{\partial h_1}{\partial b} &= 1 - h_1^2 \\
\frac{\partial h_2}{\partial b} &= 1 - h_2^2.
\end{aligned}
$$

Since the dependence of L_1 on h_1 is done only through Y_1, then chain rule yields

$$
\frac{\partial L_1}{\partial h_1} = \frac{\partial L_1}{\partial Y_1} \frac{\partial Y_1}{\partial h_1} = (Y_1 - Z_1) V.
$$

Similarly,

$$\frac{\partial L_2}{\partial h_2} = (Y_2 - Z_2)V.$$

We also have

$$
\begin{aligned}
\frac{\partial h_2}{\partial h_1} &= \frac{\partial}{\partial h_1} \tanh a_2 = \operatorname{sech}^2 a_2 \frac{\partial a_2}{\partial h_1} \\
&= (1 - \tanh^2 a_2)W = (1 - h_2^2)W.
\end{aligned}
$$

Another application of chain rule provides

$$
\begin{aligned}
\frac{\partial h_1}{\partial U} &= \frac{\partial}{\partial U} \tanh a_1 = (1 - h_1^2)\frac{\partial a_1}{\partial U} = (1 - h_1^2)X_1 \\
\frac{\partial h_2}{\partial U} &= \frac{\partial}{\partial U} \tanh a_2 = (1 - h_2^2)\frac{\partial a_2}{\partial U} = (1 - h_2^2)X_2.
\end{aligned}
$$

Now we are well prepared to compute the gradients of the loss function. We shall start with the gradient with respect to V. Since the loss L_t depends on V through Y_t, we have

$$
\begin{aligned}
\frac{\partial L}{\partial V} &= \frac{\partial L_1}{\partial V} + \frac{\partial L_2}{\partial V} \\
&= \frac{\partial L_1}{\partial Y_1}\frac{\partial Y_1}{\partial V} + \frac{\partial L_2}{\partial Y_2}\frac{\partial Y_2}{\partial V} \\
&= (Y_1 - Z_1)\frac{\partial}{\partial V}(Vh_1 + c) + (Y_2 - Z_2)\frac{\partial}{\partial V}(Vh_2 + c) \\
&= (Y_1 - Z_1)h_1 + (Y_2 - Z_2)h_2 = \sum_t (Y_t - Z_t)h_t.
\end{aligned}
$$

The derivative with respect to c is computed similarly

$$
\begin{aligned}
\frac{\partial L}{\partial c} &= \frac{\partial L_1}{\partial c} + \frac{\partial L_2}{\partial c} \\
&= \frac{\partial L_1}{\partial Y_1}\frac{\partial Y_1}{\partial c} + \frac{\partial L_2}{\partial Y_2}\frac{\partial Y_2}{\partial c} \\
&= (Y_1 - Z_1)\frac{\partial}{\partial c}(Vh_1 + c) + (Y_2 - Z_2)\frac{\partial}{\partial c}(Vh_2 + c) \\
&= (Y_1 - Z_1) + (Y_2 - Z_2) = \sum_t (Y_t - Z_t).
\end{aligned}
$$

When computing the gradient with respect to W we take into account that L_1 depends on W only through h_1, while L_2 depends on W through both h_1 and h_2 (there are two edges containing W). An application of the chain rule provides

$$
\begin{aligned}
\frac{\partial L}{\partial W} &= \frac{\partial L_1}{\partial W} + \frac{\partial L_2}{\partial W} \\
&= \frac{\partial L_1}{\partial h_1}\frac{\partial h_1}{\partial W} + \frac{\partial L_2}{\partial h_2}\frac{\partial h_2}{\partial W} + \frac{\partial L_2}{\partial h_2}\frac{\partial h_2}{\partial h_1}\frac{\partial h_1}{\partial W} \\
&= (Y_1 - Z_1)V(1 - h_1^2)h_0 + (Y_2 - Z_2)V(1 - h_2^2)h_1 \\
&\quad + (Y_2 - Z_2)V(1 - h_2^2)W(1 - h_1^2)h_0.
\end{aligned}
$$

Since there are two vertical edges involving U, a similar computation can be applied to obtain

$$
\begin{aligned}
\frac{\partial L}{\partial U} &= \frac{\partial L_1}{\partial U} + \frac{\partial L_2}{\partial U} \\
&= \frac{\partial L_1}{\partial h_1}\frac{\partial h_1}{\partial U} + \frac{\partial L_2}{\partial h_2}\frac{\partial h_2}{\partial U} + \frac{\partial L_2}{\partial h_2}\frac{\partial h_2}{\partial h_1}\frac{\partial h_1}{\partial U} \\
&= (Y_1 - Z_1)V(1 - h_1^2)X_1 + (Y_2 - Z_2)V(1 - h_2^2)X_2 \\
&\quad + (Y_2 - Z_2)V(1 - h_2^2)W(1 - h_1^2)X_1.
\end{aligned}
$$

The last gradient is computed with respect to the bias b

$$
\begin{aligned}
\frac{\partial L}{\partial b} &= \frac{\partial L_1}{\partial b} + \frac{\partial L_2}{\partial b} \\
&= \frac{\partial L_1}{\partial h_1}\frac{\partial h_1}{\partial b} + \frac{\partial L_2}{\partial h_2}\frac{\partial h_2}{\partial b} + \frac{\partial L_2}{\partial h_2}\frac{\partial h_2}{\partial h_1}\frac{\partial h_1}{\partial b} \\
&= (Y_1 - Z_1)V(1 - h_1^2) + (Y_2 - Z_2)V(1 - h_2^2) \\
&\quad + (Y_2 - Z_2)V(1 - h_2^2)W(1 - h_1^2).
\end{aligned}
$$

Similar formulas can be obtained for an RNN with more than two hidden states.

17.6 The Gradient Problems

The following difficulties in dealing with gradients in an RNN have been pointed out first in Hochreiter [55] and Bengio et al. [14, 15].

Vanishing gradient problem We notice from the previous gradient formulas that the gradients with respect to W, U, and b involve products involving matrices W, V, as well as the factors $(1 - h_2^2)$ and $(1 - h_1^2)$. Given that $h_t = \tanh a_t \in (-1, 1)$, then $1 - h_2^2 \in (0, 1)$. Therefore, a product of these type of factors have the effect of decreasing the gradient. The longer the RNN, the more products involving $(1 - h_t^2)$ will be, and then the smaller the gradient.

The factor involving the matrix W in the gradient formula comes from the derivative $\frac{\partial h_2}{\partial h_1}$. In the case of an RNN of length T there are more products involving derivatives of type $\frac{\partial h_t}{\partial h_{t-1}}$, which will have the effect of producing a power W^{T-1}. If the matrix W has eigenvalues $|\lambda_i| < 1$, then the power W^{T-1} will have eigenvalues $|\lambda_i|^{T-1} < 1$. This follows from the eigenvalues decomposition $W = MDM^t$, which implies $W^{T-1} = MD^{T-1}M^t$, where D is the diagonal form having eigenvalues along the diagonal. Since for T large we have $|\lambda_i|^{T-1} \to 0$, then $D^{T-1} \to O$, and hence W^{T-1} tends to the zero matrix when the length of the RNN increases, see Proposition G.1.2 in the Appendix.

The previous discussion can be summarized under the notion of *vanishing gradient problem*. There are several remedies for this problem, which prevent the gradient from shrinking in a larger or a smaller extent.

(i) One solution to the problem is to change the activation function. Since the factors $(1 - h_t^2)$ come from the derivative $\tanh'(a_t)$, one idea is to replace the activation function with one whose derivative is not everywhere less than 1, such as *ReLU*, which is equal to 1 just on positive activations.

(ii) Another solution is to initialize the weights to the unit matrix, $W = \mathbb{I}$. This fact will prevent the weights to shrink to zero too fast, since the eigenvalues of the powers W^p will stay closer to 1 for a larger number of iterations.

(iii) The most robust fix of the vanishing gradient problem is to employ a novel architecture involving "gated cells", such as LSTM or GRU cells. We shall deal with this type of architecture in the next section.

Exploding gradient problem We assume now the matrix W has one eigenvalue satisfying $|\lambda_i| > 1$. Since $W^p = MD^pM^t$, then D^p has the entry λ_i^p tending to infinity for p large. Therefore, some entries of W^p will tend to infinity, a fact that leads to the *gradient exploding problem*. One fix is to initialize the weights W to be the identity matrix, hoping for preventing the weights to explode too soon. Another useful technique is *gradient clipping*, which is based on rescaling back the gradients when they become too large, see Mikolov [84] and Pascanu et al. [96].

17.7   LSTM Cells

The *Long Short-Term Memory* network, LSTM, has been introduced in 1997 by Hochreiter and Schmidhuber [56]. This a type of RNN, which contains special cells that are capable of learning long-term dependencies. They use the concept of *gates* that optionally allows information passing through a sigmoid layer and a pointwise multiplication. The functionality of an LSTM involves three types of gates: *forget*, *update*, and *output*. It is based on introducing an internal cell state, C_t, and introducing an inner loop in each cell. The description of each gate is as follows.

1. The forget gate is a sigmoid layer that is used to forget the irrelevant history information. It is defined by

$$f_t = \sigma(W_f h_{t-1} + U_f X_t + b_f),$$

where W_f, U_f are matrices and b_f is a bias vector. Since σ is a logistic sigmoid function, then $f_t \in (0, 1)$. This represents the fraction of the past state that will be forgotten. It depends on the last state h_{t-1} as well as the present input, X_t.

2. The update gate selectively updates the internal cell state value, C_t, as

$$C_t = f_t C_{t-1} + i_t \tilde{C}_t,$$

where i_t is a scale factor defined by the sigmoid layer

$$i_t = \sigma(W_i h_{t-1} + U_i X_t + b_i),$$

with matrices W_i, U_i, and bias b_i, and

$$\tilde{C}_t = \tanh(W_c h_{t-1} + U_c X_t + b_c)$$

represents a candidate that could be added to the internal state, which belongs to the interval $(-1, 1)$. The product $i_t \tilde{C}_t$ represents a fraction of the candidate, which updates the internal state C_t. The term $f_t C_{t-1}$ represents how much is left after the fraction f_t was forgotten from the internal state value.

3. The output gate provides the value of the hidden state

$$h_t = o_t \tanh(C_t),$$

as the product between a factor that decides how much to output,

$$o_t = \sigma(W_o h_{t-1} + U_o X_t + b_o),$$

and a state between -1 and 1 obtained by applying tanh function to the internal state, C_t.

The use of LSTM cells in RNNs help with solving the vanishing gradient problem, since LSTM cells allow gradients to flow unchanged. However, they still suffer of the exploding gradient problem. There are other architectures, which serve similar purposes such as peephole LSTM, [42], peephole convolution LSTM, [113], GRU cells [24], etc.

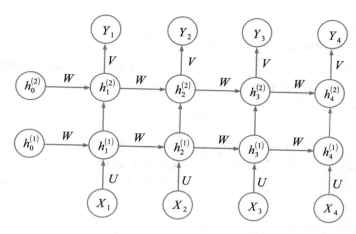

Figure 17.4: *A deep RNN with $T = 4$ and $L = 2$.*

17.8   Deep RNNs

We have seen that an RNN can be seen as a sequence of one-hidden layer parameter-sharing feedforward neural networks, whose hidden layers exchange information by a transition function. The first idea to introduce depth in an RNN appeared in Graves [50] and was shortly followed in Pascanu et al. [95].

A deep RNN is obtained by unfolding a deep parameter-sharing feedforward neural network. In this case the hidden state of the RNN, $h_t^{(\ell)}$, has two indices. The lower index, t, represents the position of the state across time, while the upper index, ℓ, provides the layer number the state belongs to, see Fig. 17.4. A deep RNN with dimensions $T \times L$ means an architecture with L horizontal layers, $1 \leq \ell \leq L$ and T vertical unfoldings, with $1 \leq t \leq T$.

Training deep RNNs is harder and more expensive than training simple RNNs. We shall provide a supporting argument for this. Consider the deep RNN with 2 layers and 4 recurrences given in Fig. 17.4. In the process of finding the gradient of the loss function with respect to W we need to compute, among other derivatives, the derivative $\partial Y_4 / \partial W$. Since Y_4 depends on W through the intermediate variables $h_t^{(\ell)}$, $1 \leq t \leq 4$, $1 \leq \ell \leq 2$, an application of the chain rule yields products involving two types of factors: $\frac{\partial h_t^{(\ell)}}{\partial h_{t-1}^{(\ell)}}$ and $\frac{\partial h_t^{(2)}}{\partial h_t^{(1)}}$. The first one introduces a multiplication by W, as we have seen in the computations of section 17.5. We look now to the second factor. If we assume the layers are related by a logistic sigmoid function as

$$h_t^{(2)} = \sigma(W_h h_t^{(1)} + b_h),$$

then using the properties of the sigmoid we obtain

$$\frac{\partial h_t^{(2)}}{\partial h_t^{(1)}} = W_h \sigma'(W_h h_t^{(1)} + b_h) = W_h \sigma(W_h h_t^{(1)} + b_h)(1 - \sigma(W_h h_t^{(1)} + b_h))$$

$$= W_h h_t^{(2)}(1 - h_t^{(2)}).$$

Since $\sigma' \leq 1/4$, this term also induces a shrinkage effect on the derivative. The deeper the RNN, the more prominent this effect will be. All the aforementioned effects lead to a more pronounced vanishing gradient problem.

17.9 Summary

RNNs are neural networks specialized in processing sequential data such as audio and video. They have been proved very successful in practical applications such as speech recognition, handwritten generation, text generation, machine translation, image captioning, unconstrained handwritten recognition, etc.

An RNN is obtained as a finite sequence of a parameter-sharing feedforward neural network of the same architecture, whose hidden layers exchange information by a transition function. RNNs train by a variant of the backpropagation method, called backpropagation through time, which is more expensive than the regular FNN training.

Plain vanilla RNNs are plagued by two problems: vanishing gradient and exploding gradient. There are several partial remedies to these problems. The exploding gradient descent can be improved by gradient clipping and initialization by unit matrix, while the vanishing gradient problem can be fixed by employing a new type of RNNs architectures that include gates cells, such as LSTM, GRU, etc.

RNNs containing several horizontal layers are called deep RNNs. They are useful in problems regarding more complicated pattern extraction from the raw sequence data.

17.10 Exercises

Exercise 17.10.1 *(a)* Let $\mathcal{G}_1$, $\mathcal{G}_2$, and $\mathcal{G}_3$ be three sigma-algebras. Show that

$$\mathfrak{S}(\mathcal{G}_1 \cup \mathcal{G}_2 \cup \mathcal{G}_3) = \mathfrak{S}\Big(\mathfrak{S}(\mathcal{G}_1 \cup \mathcal{G}_2) \cup \mathcal{G}_3\Big).$$

(b) Formulate and prove a generalization.

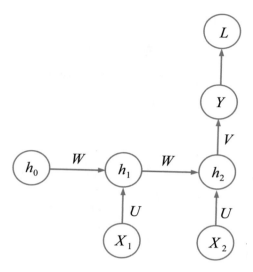

Figure 17.5: *A "2-to-1" RNN for Exercise 17.10.3.*

Exercise 17.10.2 (*a*) Consider the matrix

$$
W = \begin{pmatrix} \frac{1}{10} & \frac{2}{10} \\[2mm] \frac{3}{10} & -\frac{4}{10} \end{pmatrix}
$$

Show that $\lim_{n \to \infty} W^n = O_2$, where O_2 denotes the 2×2 zero matrix.

(*b*) Let A be a $k \times k$ symmetric matrix and denote by $\rho(A) = \max_{1 \le i \le k} |\lambda_i|$ its spectral radius, where λ_i denote the eigenvalues of A. Consider the matrix $W = \frac{1}{1+\rho(A)} A$. Prove that $\lim_{n \to \infty} W^n = O_k$.

Exercise 17.10.3 Consider an RNN with two one-dimensional inputs, X_1, X_2, and one output, Y, see Fig. 17.5. The loss function is $L = \frac{1}{2}(Y - Z)^2$, where Z denotes the target. Find the equations of the backpropagation through time algorithm in this case.

Exercise 17.10.4 Consider the RNN given in Fig. 17.5 and assume the inputs X_1, X_2 are random variables, the initial state h_0 is given, and the target Z is a random variable. Denote the output sigma-algebra by $\mathcal{E} = \mathfrak{S}(Y)$. We say that Z is learnable if $\mathfrak{S}(Z) \subset \mathcal{E}$.

Which of the following is always true:

(a) If Z is learnable than $\mathfrak{S}(Z) \subset \mathfrak{S}(\mathcal{I}_1 \cup \mathcal{I}_2)$.

(b) If $\mathfrak{S}(Z) \subset \mathfrak{S}(\mathcal{I}_1 \cup \mathcal{I}_2)$ than Z is learnable.

Exercise 17.10.5 Consider a one-dimensional dynamical system whose state updates as

$$h_n = f(h_{n-1}; \theta), \quad n \geq 1,$$

where the transition function is $f(x; \theta) = \tanh(\theta x)$, with $|\theta| < 1$. Find the hidden state of the system in the long run, $\lim_{n \to \infty} h_n$.

Exercise 17.10.6 The same question as in Exercise 17.10.5, but replacing the transition function by a logistic sigmoid, $f(x; \theta) = \sigma(\theta x)$.

Exercise 17.10.7 The same hypothesis as in Exercise 17.10.5, but replacing the transition function by a sine function, $f(x; \theta) = \sin(\theta x)$. Show that the long run behavior of the dynamical system depends on the parameter θ and initial state h_0.

Chapter 18

Classification

In classification problems a neural network has to be able to classify clusters, i.e., to assign a label with each cluster. These labels can be either natural numbers, points in the space or vectors, belonging to a label space. The classification procedure is equivalent to being able to learn a "cluster splitting function" or a decision map. The training set provides labels with each cluster. This assignment defines a decision map. The network will be able to classify the testing data by learning this decision map, i.e., to state to which cluster testing points belong to.

18.1   Equivalence Relations

Consider the hypercube, $I_n = [0,1]^n$. Any subset $\mathcal{S} \subset I_n \times I_n$ is called a *relation* on I_n. The following properties will be useful shortly:

(i) $\mathcal{S}$ is called *reflexive* if it contains the hypercube diagonal $\{(x,x); x \in I_n\}$. This states that $\forall x \in I_n$ then $(x,x) \in \mathcal{S}$.

(ii) $\mathcal{S}$ is called *symmetric* if it is symmetric with respect to the diagonal $\{(x,x); x \in I_n\}$. This means that if $(x,y) \in \mathcal{S}$, then also $(y,x) \in \mathcal{S}$.

(iii) $\mathcal{S}$ is called *transitive* if $(x,y) \in \mathcal{S}$ and $(y,z) \in \mathcal{S}$, then $(x,z) \in \mathcal{S}$. The geometric interpretation of this property is that $\mathcal{S}$ is closed to rectangles which have one of the vertices on the diagonal. More precisely, if $(x,y) \in \mathcal{S}$, then there is a unique rectangle with one vertex at this point and with another vertex at (y,y), see Fig. 18.2 **a**. Transitivity reduces to the fact that if three of the rectangle vertices, (x,y), (y,y), (y,z), belong to the set $\mathcal{S}$, then also the fourth vertex, (x,z), belongs to $\mathcal{S}$.

© Springer Nature Switzerland AG 2020
O. Calin, *Deep Learning Architectures*, Springer Series in the Data Sciences,
https://doi.org/10.1007/978-3-030-36721-3_18

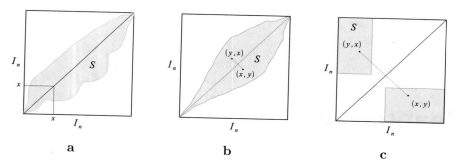

Figure 18.1: **a**. *Reflexive, nonsymmetric relation.* **b**. *Reflexive and symmetric relation.* **c**. *Symmetric nonreflexive relation.*

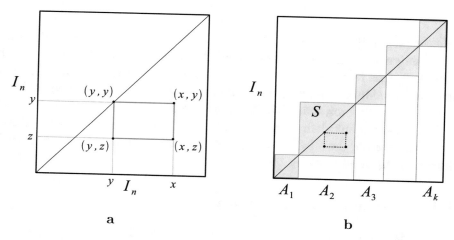

Figure 18.2: **a**. *The rectangle rule: if* $(x,y), (y,y), (y,z) \in \mathcal{S}$, *then* $(x,z) \in \mathcal{S}$. **b**. *Equivalence relation* $\mathcal{S}$, *which makes a finite partition of* I_n.

An example of transitive relation is the lattice of rational numbers in $[0,1]$, given by $\mathcal{S} = (I_n \times I_n) \cap (\mathbb{Q} \times \mathbb{Q})$, as the reader can easily check the previous rectangle property.

The previous relation properties are mutually exclusive, see Fig. 18.1 **a**, **b**, and **c**.

A relation $\mathcal{S}$ which is reflexive, symmetric, and transitive is called an *equivalence relation*. See Fig. 18.2 **b** for an example of equivalence relation. The set $\mathcal{S}$ in this case is a union of interior of squares along the diagonal.

Two points $x, y \in I_n$ are called *equivalent* under relation $\mathcal{S}$ if $(x,y) \in \mathcal{S}$. Customarily, we write $x \sim y$. All points which are equivalent to a given x is denoted by $C_x = \{y \in I_n; x \sim y\}$ and is called the *equivalence class* of x. The set of all equivalence classes, denoted by $I_n/\sim$, is called the *quotient set*.

A *partition* of the set I_n is a collection of subsets $\{A_i\}_i$ with the following properties:

(*i*) $A_i \neq \emptyset$, for all i;

(*ii*) $A_i \cap A_j = \emptyset$, for $i \neq j$;

(*iii*) $\bigcup_i A_i = I_n$.

If the collection of indices i is finite, then $\{A_i\}_i$ is a *finite partition*. They can be used to classify the points of I_n into n distinct classes.

The relation between partitions and equivalence relations is given by the following result:

Proposition 18.1.1 *Let $\sim$ be an equivalence relation on I_n. Then there is a partition $\{A_i\}_i$ of I_n such that:*

(*i*) *for each i, $\forall x, y \in A_i$ we have $x \sim y$;*

(*ii*) *$\forall x, y \in I_n$ with $x \sim y$ there is an i such that $x, y \in A_i$.*

Proof: The result can be restated by saying that any equivalence relation makes a partition of I_n and the elements of the partition are the equivalence classes of the relation. Let C_x be the equivalence class of x. We show that the collection $\{C_x\}_x$ satisfies the properties of a partition. Since $x \in C_x$, then obviously $C_x \neq \emptyset$ and $\bigcup_{x \in I_n} C_x = I_n$. It is easy to see that $x \sim y$ is equivalent to $C_x = C_y$. Assume we have an element in the intersection of two distinct classes, $z \in C_x \cap C_y$. Then $z \in C_x$ and hence $x \sim z$ and $z \in C_y$ and $z \sim y$. By transitivity $x \sim y$, which implies $C_x = C_y$, contradiction. It follows that any two distinct classes have an empty intersection. Hence, the set $\{C_x\}_x$ satisfies the properties of a partition. ∎

It is worth noting that the converse also holds true: any partition defines an equivalence relation. If $\{A_i\}_i$ is a partition, then the relation $x \sim y$ if and only if there is an i such that $x, y \in A_i$ is an equivalence relation on I_n.

The way of obtaining the partition $\{A_j\}_j$ from the equivalence relation $\mathcal{S}$ using projection can be seen in Fig. 18.2 **b**. This provides also a visualization of the equivalence relation $\mathcal{S}$ associated with a given partition: the set $\mathcal{S}$ consists in the union of rectangles aligned along diagonal constructed from the projections A_j. This set $\mathcal{S}$ contains the diagonal (is reflexive), is symmetric, and satisfies the rectangle rule (is transitive).

In the next sections we shall associate with a given partition different objects, such as entropy, decision functions, labels, decision maps, etc.

18.2 Entropy of a Partition

In this section we shall extend the notion of entropy to a partition. For this purpose, we shall consider a probability space $(\Omega, \mathcal{F}, \mu)$ and a finite measurable partition $\mathcal{A} = (A_j)_{j \leq m}$ of the set Ω, i.e., a partition with $A_j \in \mathcal{F}$. The measure μ can be used to assess numerically the sets A_j. The *entropy of the partition* $\mathcal{A}$ with respect to the probability measure μ is defined by

$$H(\mathcal{A}, \mu) = -\sum_{j=1}^{m} \mu(A_j) \ln \mu(A_j). \tag{18.2.1}$$

Since $\mu(A_j) \in (0, 1)$, the entropy is positive, $H(\mathcal{A}, \mu) > 0$. It can be shown that the entropy of the partition $\mathcal{A}$ is maximum when all the sets in the partition have the same measure $\mu(A_1) = \cdots = \mu(A_m) = \frac{1}{m}$.

Example 18.2.1 We assume that each element ω of Ω is associated with a nonnegative numerical label, such as a weight or a mass, $m(\omega)$. The probability measure in this case is

$$\mu(A) = \frac{1}{M} \sum_{x \in A} \delta_x(A),$$

where $M = \mu(\Omega)$ is the total mass of Ω and δ_x denotes the Dirac's measure sitting at x. The number $\mu(A)$ provides the proportion of mass corresponding to the set A. The entropy (18.2.1) represents the uncertainty of dividing the set Ω in parts of unequal masses.

Example 18.2.2 Let $\Omega \subset \mathbb{R}^n$ be a bounded Borel set. For any Borel set, $A \in \mathcal{B}(\Omega)$, we define the probability measure $\mu(A) = \frac{\lambda(A)}{\lambda(\Omega)}$, where λ denotes the Lebesgue measure on Ω. In this case, the entropy (18.2.1) represents the uncertainty of dividing the set Ω in subsets of unequal volumes.

Example 18.2.3 Let μ be a measure absolutely continuous with respect to the measure ν on the measurable space $(\Omega, \mathcal{F})$. By the Radon-Nikodym Theorem, see Theorem C.7 in Appendix, there is a measurable nonnegative function p such that

$$\mu(A) = \int_A p(x) \, d\nu(x),$$

for any measurable set A in Ω. If p is a density function, i.e., $\int_\Omega p(x) \, d\nu(x) = 1$, then μ becomes a probability measure. The associated entropy with the partition $\mathcal{A}$ and measure μ is

$$H(\mathcal{A}, \mu) = -\sum_{i=1}^{m} \int_{A_i} p(x) \, d\nu(x) \ln \left(\int_{A_i} p(x) \, d\nu(x) \right).$$

In the particular case when the measures are proportional, i.e., $\mu = c\nu$, with c constant, the density function is $p(x) = \frac{1}{\nu(\Omega)}$, and the previous entropy becomes

$$
\begin{aligned}
H(\mathcal{A}, \mu) &= -\sum_{i=1}^{m} \frac{1}{\nu(\Omega)} \nu(A_i) \ln\left(\frac{1}{\nu(\Omega)} \nu(A_i) \right) \\
&= -\frac{1}{\nu(\Omega)} \sum_{j=1}^{m} \nu(A_i)\left(-\ln \nu(\Omega) + \ln \nu(A_i) \right) \\
&= \ln \nu(\Omega) + \frac{1}{\nu(\Omega)} H(\mathcal{A}, \nu),
\end{aligned}
$$

which is a relation between the entropies of the partition $\mathcal{A}$ with respect to two proportional measures.

18.3 Decision Functions

Let $\{A_1, \ldots, A_k\}$ be a finite measurable partition of I_n, i.e., a partition with A_i Borel sets, $A_i \in \mathcal{B}(I_n)$, for all $i = 1, \ldots, k$. A *decision function* is a measurable function which associates an integer with each class in the partition, i.e., $f : I_n \to \mathbb{N}$, $f(x) = j$ for any $x \in A_j$. We can regard of j as being the *label* associated with the class A_j. Equivalently, $f = \sum_{i=1}^{k} j 1_{A_j}$, where 1_{A_j} is the indicator function of the set A_j. Decision functions are used to classify data into classes. Note that $A_j = f^{-1}(j)$, see Fig. 18.3 **b**.

The set $\{1, 2, \ldots, k\}$ is called the *label set* and the space that contains the label set, in this case, $\mathbb{R}$ is the *label space*. The labels are considered consecutive integers just for convenience reasons.

Example 18.3.1 (case $k = 2$) Consider two separable clusters of points in $\mathbb{R}^n$ separated by the hyperplane $\{w^T x + \theta = 0\}$. Then a classical perceptron can binary decide on each of the clusters a point belongs to, using the decision function $f(x) = 1 + H(w^T x + \theta)$. The label set is $\{1, 2\}$ and the label space is $\mathbb{R}$.

Example 18.3.2 Consider a certain attribute of the points in the hypercube I_n, such as, for instance, color. Assume there are k possible colors the points can get. Then an equivalence relation can be defined on I_n: two points are equivalent if and only if they have the same color. Let A_j be the set of points of jth color. Then the sets $\{A_j\}$ form a partition of I_n and a mapping f from I_n to the label space $\{1, \ldots, k\}$ defined by $f(A_j) = j$ is a classification rule. Sometimes the function f is called a *classifier*. We note that in this example the sets A_j are not necessary Borel-measurable.

The next result deals with the implementation of a decision function, see Cybenko [30]. Since in real life clusters are not perfectly separable, the

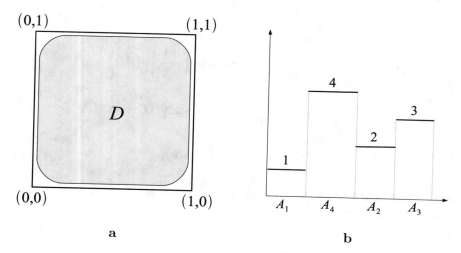

Figure 18.3: **a.** *Subset D in I_2 with the Lebesgue measure of the complement small.* **b.** *Decision function associated with a partition.*

result will take this into consideration. It states that a neural network with a single internal layer can implement any decision function such that the total Lebesgue measure, λ, of the incorrectly classified points can be made arbitrarily small, see Fig. 18.3 **a.**

Proposition 18.3.3 *Let f be a decision function associated with the measurable finite partition $\{A_i\}_i$ on I_n and let σ be a continuous sigmoidal function. For any $\epsilon > 0$, there is a finite sum $G(x) = \sum_{j=1}^{N} \alpha_j \sigma(w_j^T x + \theta_j)$, $w_j^T \in \mathbb{R}^n$, $\alpha_j, \theta_j \in \mathbb{R}$ and a set $D \subset I_n$ such that $\lambda(D) \geq 1 - \epsilon$ and*

$$|G(x) - f(x)| < \epsilon, \quad \forall x \in D.$$

Proof: By Luzin's theorem (see Appendix, section C.8), there is a continuous function $g : I_n \to \mathbb{R}$ and a set D such that $\lambda(D) > 1 - \epsilon$ and $g(x) = f(x)$ for all $x \in D$. By Theorem 9.3.8 (or Theorem 9.3.6) the sums of the previous form $G(x)$ are dense in $C(I_n)$, so for the previous $g \in C(I_n)$ we find a $G(x)$ such that $|G(x) - g(x)| < \epsilon$ for all $x \in I_n$. Therefore

$$|G(x) - f(x)| = |G(x) - g(x)| < \epsilon, \quad \forall x \in D.$$

∎

Note that this is an existence result; the actual construction of the function $G(x)$ (i.e., finding the weights w_j, α_j and thresholds θ_j) is a completely different problem.

Remark 18.3.4 To each decision function we can associate an entropy as in the following. Given a finite partition, $\mathcal{A} = (A_i)_i$, and a decision function f, we define the measure μ such that $\mu(A_i) = \frac{f(A_i)}{\sum_i f(A_i)}$ and consider the entropy $H(\mathcal{A}, \mu)$ to be the entropy associated to the partition $\mathcal{A}$ and decision function f.

18.4   One-hot-vector Decision Maps

Sometimes, it is more convenient to replace the integer labels by one-hot vectors. For instance, instead of labels 1, 2, etc., we can associate the one-hot vectors $e_1 = (1, 0, \ldots, 0)^T$, $e_2 = (0, 1, 0, \ldots, 0)^T$, etc. The label set is formed by $\{e_1, \ldots, e_k\}$, while the label space is $\mathbb{R}^n$. Thus, we arrive at the following definition.

Let $\{A_1, \ldots, A_k\}$ be a finite measurable partition of I_n, i.e., a partition with A_i Borel sets, $A_i \in \mathcal{B}(I_n)$, for all $i = 1, \ldots, k$. A *one-hot-vector decision map* is a measurable function $f : I_n \to \mathbb{R}^k$, which associates a one-hot vector with each class in the partition, i.e., $f(x) = e_j$ for any $x \in A_j$, $e_j = (0, \ldots, 1, \ldots, 0)^T$. In this case, the label associated with the class A_j is a k-dimensional unit vector, and all these label vectors form a basis in $\mathbb{R}^k$.

What is the advantage of using one-hot vectors as labels? In the case when the labels are just integers, the set I_n is mapped into the real line and this provides some localization for the testing sets around some given integers.

In the case of one-hot vectors as labels, the set I_n is mapped into a higher dimensional space, $\mathbb{R}^k$. This provides more room for the testing sets to be pooled toward linearly independent directions, leading to a better separation of classes.

Choosing the one-hot vectors e_j as labels is just a convenience. We may choose as labels any other k vectors that are linearly independent in $\mathbb{R}^k$, eventually organized as an orthonormal basis. Equivalently, instead of considering k independent vectors, we may consider the labels to be k points in $\mathbb{R}^k$, denoted by $P_1, P_2, \ldots, P_k$, whose position vectors are linearly independent.

The label space can have a smaller dimension than k, which is the number of classes in I_n. For instance, in Fig. 18.4 we consider a partition of I_n into $k = 4$ classes, and associate to each class a label point in $\mathbb{R}$ and $\mathbb{R}^2$, respectively.

The next two linear algebra results state the relation between using labels as either points or one-hot vectors.

Proposition 18.4.1 *Consider k distinct points, $P_1, \ldots, P_k$, in $\mathbb{R}^k$. Then there is a linear function $f : \mathbb{R}^k \to \mathbb{R}^k$ such that $f(e_j) = P_j$, for $j = 1, \ldots, k$.*

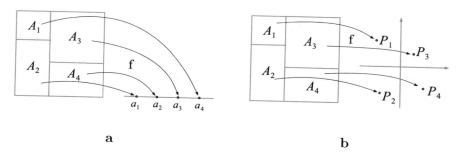

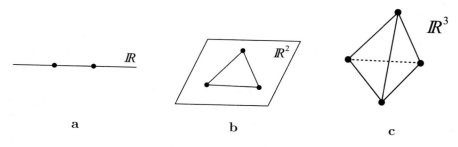

Figure 18.4: *Decision map associated with a partition:* **a.** *The label space is* $\mathbb{R}$. **b.** *The label space is* $\mathbb{R}^2$.

Figure 18.5: *Points in a general position:* **a.** *Two distinct points on a line.* **b.** *Three noncollinear point in th eplane.* **c.** *Four noncoplanar points in the 3-dimensional space.*

Proof: Let $v_j = (v_j^1, \ldots, v_j^k)^T$ be the coordinate vector of the point P_j in $\mathbb{R}^k$, so we can write $v_j = \sum_{i=1}^{k} v_j^i e_i$. Then the linear function $f(x) = Wx$, with the matrix $W_{ij} = v_j^i$ is the desired function. ∎

The converse is not necessarily true. To make this work, we need to impose an extra condition, which will be introduced next.

First, we note that through 2 distinct points pass a unique line, while through 3 noncollinear points pass a unique plane. Through 4 noncoplanar points pass a unique 3-dimensional hyperplane, and so on, see Fig. 18.5. If k points are situated in a general enough position, the dimension of the hyperplane determined by them is $k - 1$; otherwise the dimension of the hyperplane is strictly smaller than $k - 1$.

Definition 18.4.2 *The points $P_1, \ldots, P_k$ in $\mathbb{R}^k$ are said to be in a general position if there is no hyperplane of dimension less than $k - 1$ containing them.*

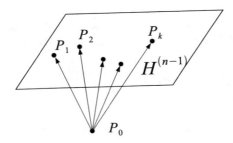

Figure 18.6: *The points $P_1, \ldots, P_k$ are in a general position if the vectors $P_0 P_j$ are linearly independent.*

Equivalently stated, the lowest dimension of the hyperplane containing the given points is $k - 1$. Another equivalence statement is given in Exercise 18.11.2. The next result is a useful characterization of this concept using vectors. It roughly states that changing the origin of the space, the position vectors of the points become linearly independent.

Proposition 18.4.3 *The points $P_1, \ldots, P_k$ in $\mathbb{R}^k$ are in a general position if and only if there is a point P_0 in $\mathbb{R}^k$ such that the vectors $\overrightarrow{P_0 P_j}$ are linearly independent in $\mathbb{R}^k$, $j \in \{1, \ldots, k\}$.*

Proof: " $\implies$ " Assume the points $P_1, \ldots, P_k$ are in a general position. Then by Exercise 18.11.3, there is a unique $(k-1)$-hyperplane, $\mathcal{H}$, which contains the points. Then choosing any point $P_0 \notin \mathcal{H}$ yields the linear independent vectors $\overrightarrow{P_0 P_1}, \ldots, \overrightarrow{P_0 P_k}$, see Fig. 18.6. To show that we form a vanishing linear combination

$$\sum_{i=1}^{k} c_i \overrightarrow{P_0 P_i} = 0,$$

and show that $c_i = 0$. Using the vector decomposition $\overrightarrow{P_0 P_i} = \overrightarrow{P_0 P_1} + \overrightarrow{P_1 P_i}$, we can write

$$\sum_{i=1}^{k} c_i \overrightarrow{P_0 P_i} = \Big(\sum_{i=1}^{k} c_i \Big) \overrightarrow{P_0 P_1} + \sum_{i=2}^{k} c_i \overrightarrow{P_1 P_i} = 0.$$

The set $\{\overrightarrow{P_1 P_2}, \ldots, \overrightarrow{P_1 P_k}\}$ forms a system of independent vectors in $\mathcal{H}$, see Exercise 18.11.3. Since $P_0 \notin \mathcal{H}$, the vector $\overrightarrow{P_1 P_0}$ is independent of the previous system, since it points outside the hyperplane $\mathcal{H}$. Therefore, the previous linear combination has zero coefficients, $c_i = 0$.

" $\impliedby$ " Let $P_0 \in \mathbb{R}^k$ such that $\{\overrightarrow{P_0 P_1}, \ldots, \overrightarrow{P_0 P_k}\}$ are linearly independent.

If the points $\{P_1, \ldots, P_k\}$ are not in a general position, then they must be contained inside of a hyperplane $\mathcal{P}$ of dimension p, with $p < k - 1$. We have

$$\mathcal{P} = \{Q \in \mathbb{R}^k;\ \overrightarrow{P_0 Q} = \sum_{j=1}^{k} c_j \overrightarrow{P_0 P_j},\ \sum_{j=1}^{k} c_j = 1\}.$$

Since $\{\overrightarrow{P_0 P_1}, \ldots, \overrightarrow{P_0 P_k}\}$ are linearly independent, the hyperplane $\mathcal{P}$ has dimension $k - 1$, which leads to a contradiction. ∎

The next result is the converse of Proposition 18.4.1.

Proposition 18.4.4 *Consider k distinct points, $P_1, \ldots, P_k \in \mathbb{R}^k$, in a general position. Then there is a linear function $f : \mathbb{R}^k \to \mathbb{R}^k$ such that $f(P_j) = e_j$, for $j = 1, \ldots, k$. The function f is invertible.*

Proof: By Proposition 18.4.3, we can choose a point P_0 such that $\overrightarrow{P_0 P_i}$ are linearly independent. These vectors actually form a basis in $\mathbb{R}^k$. Let $g : \mathbb{R}^k \to \mathbb{R}^k$ be the unique linear function such that $g(\overrightarrow{P_0 P_j}) = e_j$, $j = 1, \ldots, k$. Denote by r the function that assigns to each point P in $\mathbb{R}^k$ the vector $\overrightarrow{P_0 P}$, i.e., $r(P) = \overrightarrow{P_0 P}$. Construct the function f by the composition $f = g \circ r$. Then f is linear, as a composition of linear functions, and satisfies the property $f(P_j) = e_j$.

∎

Proposition 18.4.4 assures the equivalence between choosing labels either as one-hot vectors, e_j, or as general form points, P_j, in $\mathbb{R}^k$. We shall deal with both cases in future sections.

18.5   Linear Separability

A cluster $\mathcal{G}$ of points in $\mathbb{R}^n$ is a set of n-uples $(x_1, \ldots, x_n)$ supposed to have a certain individual identity. Two clusters in $\mathbb{R}^n$, $\mathcal{G}_1$ and $\mathcal{G}_2$, are called *linearly separable* if there is a hyperplane $\mathcal{H}$ of dimension $n - 1$, which separates the clusters. This means:

(i) The hyperplane $\mathcal{H}$ divides the space $\mathbb{R}^n$ in two half-spaces, $\mathcal{S}_1$ and $\mathcal{S}_2$.

(ii) Each cluster is contained in one of the half-spaces: $\mathcal{G}_1 \subset \mathcal{S}_1$ and $\mathcal{G}_2 \subset \mathcal{S}_2$.

If the hyperplane $\mathcal{H}$ is defined by the equation

$$h(x) = a_1 x_1 + \cdots + a_n x_n + d = 0,$$

then the separability of $\mathcal{G}_1$ and $\mathcal{G}_2$ can be written as $h(g_1)h(g_2) < 0$, for any points $g_1 \in \mathcal{G}_1$, $g_2 \in \mathcal{G}_2$. This means that h keeps constant opposite signs on each of the clusters, see Fig. 18.7 **a**.

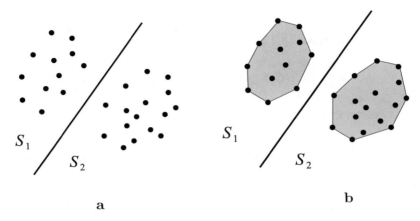

Figure 18.7: **a**. *Linear separability of two clusters.* **b**. *Linear separability of the convex hulls of two clusters.*

Example 18.5.1 Two clusters in $\mathbb{R}$, $\mathcal{G}_1$ and $\mathcal{G}_2$, are separable if there is a number α such that $(g_1 - \alpha)(g_2 - \alpha) < 0$, for all $g_1 \in \mathcal{G}_1$, $g_2 \in \mathcal{G}_2$. This means that either $g_1 < \alpha < g_2$ or $g_2 < \alpha < g_1$, for all $g_1 \in \mathcal{G}_1$, $g_2 \in \mathcal{G}_2$.

A set $K \subset \mathbb{R}^n$ is called *convex* if for any two points $A, B \in K$, the line segment AB is included in the set K. For instance, a disk, the interior of a triangle or a tetrahedron, are convex sets.

In general, a cluster is not a convex set. The *convex hull* of a cluster $\mathcal{G}$ is the set of all convex combinations

$$hull(\mathcal{G}) = \left\{ \sum_{g_i \in \mathcal{G}} \lambda_i g_i; \ \sum_{i=1}^{n} \lambda_i = 1, \lambda_i \geq 0 \right\}.$$

For instance, if a cluster has only 2 points, its hull is the closed line segment defined by the points. If the cluster contains 3 points, its cluster is the triangle (including the interior) with vertices at the given points. It can be shown that the set $hull(\mathcal{G})$ is always a convex set, which contains the cluster $\mathcal{G}$, see Exercise 18.11.5

Proposition 18.5.2 *Two clusters $\mathcal{G}_1$ and $\mathcal{G}_2$ are linearly separable if and only if $hull(\mathcal{G}_1)$ and $hull(\mathcal{G}_2)$ are linearly separable.*

Proof: " $\Longrightarrow$ " If $\mathcal{G}_1$ and $\mathcal{G}_2$ are linearly separable, then let $\mathcal{H}$ be the hyperplane that divides the space into two half-spaces, $\mathcal{S}_1$ and $\mathcal{S}_2$, which separate the clusters, i.e.,

$$\mathcal{G}_1 \subset \mathcal{S}_1, \qquad \mathcal{G}_2 \subset \mathcal{S}_2.$$

Since the half-spaces $\mathcal{S}_1$ and $\mathcal{S}_2$ are convex sets, using the convex minimality of the hull, see Exercise 18.11.5, we have

$$\mathcal{G}_1 \subset hull(\mathcal{G}_1) \subset \mathcal{S}_1, \qquad \mathcal{G}_2 \subset hull(\mathcal{G}_2) \subset \mathcal{S}_2.$$

Hence, the convex hulls $hull(\mathcal{G}_1)$ and $hull(\mathcal{G}_2)$ are separated by the hyperplane $\mathcal{H}$.

A variant of proof, starting direct from the definition, is given in the following:

Assume $\mathcal{G}_1$ and $\mathcal{G}_2$ are linearly separable and let $\mathcal{H}$ be the separation hyperplane with the equation

$$h(x) = a_1 x_1 + \cdots + a_n x_n + d = 0.$$

For any two points in the clusters' hulls

$$g^1 = \sum_{g_i^1 \in \mathcal{G}_1} \lambda_i^1 g_i^1 \in hull(\mathcal{G}_1), \quad g^2 = \sum_{g_i^2 \in \mathcal{G}_2} \lambda_i^2 g_i^2 \in hull(\mathcal{G}_2),$$

using the linearity of the function h, yields

$$\begin{aligned}
h(g^1)h(g^2) &= h\Big(\sum_{g_i^1 \in \mathcal{G}_1} \lambda_i^1 g_i^1 \Big) h\Big(\sum_{g_j^2 \in \mathcal{G}_2} \lambda_j^2 g_j^2 \Big) \\
&= \Big(\sum_{g_i^1 \in \mathcal{G}_1} \lambda_i^1 h(g_i^1) \Big) \Big(\sum_{g_j^2 \in \mathcal{G}_2} \lambda_j^2 h(g_i^1) \Big) \\
&= \sum_{g_i^1 \in \mathcal{G}_1} \sum_{g_j^2 \in \mathcal{G}_2} \lambda_i^1 \lambda_j^2 h(g_i^1)h(g_i^1) < 0,
\end{aligned}$$

because $\lambda_i^1 > 0$, $\lambda_j^2 > 0$ and we used the clusters separability condition $h(g_i^1)h(g_i^1) < 0$.

" $\Longleftarrow$ " If $hull(\mathcal{G}_1)$ and $hull(\mathcal{G}_2)$ are linearly separable, there is a hyperplane $\mathcal{H}$ that divides the space $\mathbb{R}^n$ in two half-spaces, $\mathcal{S}_1$ and $\mathcal{S}_2$, such that $hull(\mathcal{G}_1) \subset \mathcal{S}_1$ and $hull(\mathcal{G}_2) \subset \mathcal{S}_2$. Using the obvious inclusions $\mathcal{G}_1 \subset hull(\mathcal{G}_1)$ and $\mathcal{G}_2 \subset hull(\mathcal{G}_2)$ it follows that $\mathcal{G}_1 \subset \mathcal{S}_1$ and $\mathcal{G}_2 \subset \mathcal{S}_2$, and hence $\mathcal{G}_1$ and $\mathcal{G}_2$ are linearly separable. See also Fig. 18.7 **b**. ∎

Even if two clusters have distinct points, sometimes they are close enough to each other such that their convex hulls intersect. In this case the separability can be eventually achieved only by a nonlinear function.

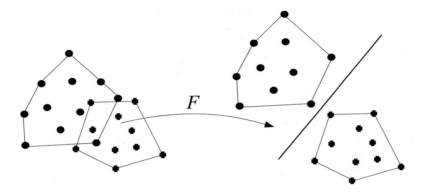

Figure 18.8: *The separation function F of two intersecting clusters is nonlinear.*

Proposition 18.5.3 *Let $\mathcal{G}_1$ and $\mathcal{G}_2$ be two clusters in $\mathbb{R}^n$ such that*

$$hull(\mathcal{G}_1) \cap hull(\mathcal{G}_2) \neq \emptyset.$$

Then there is no linear function $F : \mathbb{R}^n \to \mathbb{R}^p$ such that $F(\mathcal{G}_1)$ and $F(\mathcal{G}_2)$ are linearly separable.

Proof: By contradiction, assume there is linear function $F : \mathbb{R}^n \to \mathbb{R}^p$ such that $F(\mathcal{G}_1)$ and $F(\mathcal{G}_2)$ are linearly separable, i.e., there is a hyperplane in $\mathbb{R}^p$ defined by the equation $h(x) = \sum_{i=1}^{p} a_i x_i + d = 0$, such that

$$\Phi(g_i^1)\Phi(g_i^2) < 0$$

for all $g_i^1 \in \mathcal{G}_1$ and $g_i^2 \in \mathcal{G}_2$, where $\Phi = h \circ F$.

Consider an element in the intersection

$$g \in hull(\mathcal{G}_1) \cap hull(\mathcal{G}_2),$$

which therefore has two representations

$$g = \sum_i \lambda_i^1 g_i^1 = \sum_i \lambda_i^2 g_i^2, \qquad g_i^1 \in \mathcal{G}_1, g_i^2 \in \mathcal{G}_2.$$

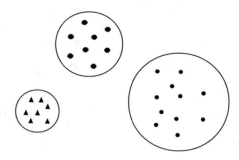

Figure 18.9: *The convex separation of three clusters.*

Using the linearity of F we obtain the following contradiction:

$$
\begin{aligned}
0 \leq \Phi(g)\Phi(g) &= \Phi\Big(\sum_i \lambda_i^1 g_i^1\Big)\Phi\Big(\sum_i \lambda_i^2 g_i^2\Big) \\
&= \Big(\sum_i \lambda_i^1 \Phi(g_i^1)\Big)\Big(\sum_j \lambda_j^2 \Phi(g_j^2)\Big) \\
&= \sum_{i,j} \lambda_j^1 \lambda_j^2 \Phi(g_i^1)\Phi(g_i^2) < 0.
\end{aligned}
$$

Therefore, there is no separation function F that is linear, see Fig. 18.8.
∎

Remark 18.5.4 (*i*) It is worth noting that there is also no separation function which is affine, i.e., of the form $F(x) = Wx + b$, with W an $n \times n$ matrix and $b \in \mathbb{R}^n$. This follows from the fact that separability is translation invariant. Consequently, a linear neuron cannot separate two clusters whose convex hulls intersect. For this job we should employ neural networks with nonlinear activation functions.

(*ii*) Two clusters in $\mathbb{R}^n$, $\mathcal{G}_1$ and $\mathcal{G}_2$ are called *F-separable* if there is an invertible bi-continuous mapping $F : \mathbb{R}^n \to \mathbb{R}^n$ such that the cluster images, $F(\mathcal{G}_1)$, $F(\mathcal{G}_2)$, are linearly separable in $\mathbb{R}^n$. A such function F is called a *homeomorphism* of $\mathbb{R}^n$. Standard results of neural network theory show that a feedforward neural network (with enough hidden layers) can learn the nonlinear continuous function F. Augmenting the network with a perceptron we can perform the final linear classification. Hence, a classification problem is reduced to the learning of the continuous nonlinear function F.

(*iii*) The role of the function F is to pull apart the clusters, so that they can be linearly classified. However, there are cases when the clusters cannot be separated by a homeomorphism of $\mathbb{R}^n$. In this case we need an extra-dimension to separate the clusters and the continuous function should be

$F : \mathbb{R}^n \to \mathbb{R}^p$, with $p > n$. For instance, $\mathcal{G}_1 = \{x \in \mathbb{R}^2; \|x\| < 1\}$ and $\mathcal{G}_2 = \{x \in \mathbb{R}^2; \|x\| > 2\}$ cannot be separated in $\mathbb{R}^2$ by pulling them apart continuously, but they can be separated in $\mathbb{R}^3$ if we pull one of them vertically.

The case of k clusters Consider a family of k clusters, $\mathfrak{G} = \{\mathcal{G}_1, \ldots, \mathcal{G}_k\}$. We say that $\mathfrak{G}$ is a *linearly separable family* if its clusters are mutually separable, i.e., any two clusters, $\mathcal{G}_i$ and $\mathcal{G}_j$, are linearly separable, for $i \neq j$.

Remark 18.5.5 (*i*) By Proposition 18.5.2, the family $\mathfrak{G} = \{\mathcal{G}_1, \ldots, \mathcal{G}_k\}$ is linearly separable if and only if $\{hull(\mathcal{G}_1), \ldots, hull(\mathcal{G}_k)\}$ is a linearly separable family of convex sets.

(*ii*) Denote by G_j the center of mass of the cluster $\mathcal{G}_j$. This means that G_j is obtained as an average of the elements of the cluster $\mathcal{G}_j$. Applying Exercise 18.11.7, we obtain that $G_j \in hull(\mathcal{G}_j)$, $j = 1, \ldots, k$.

18.6   Convex Separability

Separability can be also considered in a slightly different, but equivalent way. A family of clusters $\mathfrak{G} = \{\mathcal{G}_1, \ldots, \mathcal{G}_k\}$ in $\mathbb{R}^n$ is called *convex separable* if there are k closed balls, $B_1, \ldots, B_k$, in $\mathbb{R}^n$ such that

(i) $B_1, \ldots, B_k$ are disjoint;
(ii) $\mathcal{G}_j \subset B_j$, for all $j \in \{1, \ldots, k\}$.

In particular, two clusters in $\mathbb{R}^2$ are convex separable if they are included in two disjoint disks. For the case of three clusters, see Fig. 18.9. The following result shows the equivalence between these two types of convexity.

Proposition 18.6.1 *Two clusters $\mathcal{G}_1$, $\mathcal{G}_2$ in $\mathbb{R}^n$ are convex separable if and only if they are linearly separable.*

Proof: "$\Longrightarrow$" If $\mathcal{G}_1$ and $\mathcal{G}_2$ are convex separable in $\mathbb{R}^n$, then there are two disjoint balls, B_1, B_2 in $\mathbb{R}^n$ such that $\mathcal{G}_1 \subset B_1$ and $\mathcal{G}_2 \subset B_2$. Let $\mathcal{H}$ be an $(n-1)$-hyperplane which separates the disjoint balls B_1 and B_2 (this hyperplane can be constructed perpendicular to the centers segment, passing through a point which is exterior to both balls). Then $\mathcal{H}$ separates the clusters $\mathcal{G}_1$ and $\mathcal{G}_2$.

"$\Longleftarrow$" For the sake of simplicity, we shall consider $n = 2$; similar reasons can be carried out for higher dimensions. Assume there is a line ℓ in the plane, which separates $\mathcal{G}_1$ and $\mathcal{G}_2$ and divides the plane into two semiplanes, $\mathcal{S}_1$ and $\mathcal{S}_2$, with $\mathcal{G}_1 \subset \mathcal{S}_1$ and $\mathcal{G}_2 \subset \mathcal{S}_2$. Let M be the closest point on the line ℓ to the cluster $\mathcal{G}_1$. There a small enough $\epsilon > 0$ such that the half-lines starting at M and making angles equal to ϵ with the line ℓ do not intersect $hull(\mathcal{G}_1)$, see

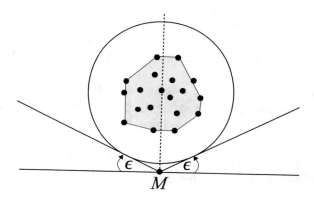

Figure 18.10: *Construction of a disk that contains the convex hull of a cluster and is contained in a given half-plane.*

Fig. 18.10. Let β be the angle bisector of the angle made by the half-lines. Then for any point $O \in \beta$, there is a unique circle centered at O, which is tangent to both half-lines. When the distance $\|OM\|$ is large enough, the cluster $\mathcal{G}_1$ lies inside the ball centered at O. This is the construction of the ball B_1. Similarly, we can construct the other ball B_2, on the other side of the line ℓ, containing the cluster $\mathcal{G}_2$.

∎

Remark 18.6.2 Even if convex separability looks to be a more general concept and seems to make more sense, however, it is the linear separability which can be tackled using neural networks.

The next result deals with the existence of a one-hot-vector decision map for a family of clusters.

Proposition 18.6.3 *Let $\mathfrak{G} = \{\mathcal{G}_1, \ldots, \mathcal{G}_k\}$ be a family of k clusters in $\mathbb{R}^k$, which are linearly separable. Then there is a decision map $F : \mathbb{R}^k \to \mathbb{R}^k$ with $F(\mathcal{G}_j) = e_j$, $j = 1, \ldots, k$.*

Proof: Since $\mathfrak{G} = \{\mathcal{G}_1, \ldots, \mathcal{G}_k\}$ is linearly separable, then it is convex separable. Therefore, there are k mutually disjoint balls, $B_1, \ldots, B_k$, in $\mathbb{R}^k$ such that $\mathcal{G}_j \subset B_j$, for all j. Consider a decision map $f : \mathbb{R}^k \to \mathbb{R}^k$ which maps the jth ball into e_j, i.e., $f(B_j) = e_j$, for $j = 1, \ldots, k$. This ends the proof.

∎

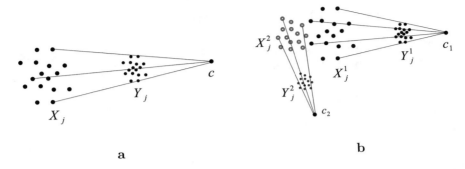

Figure 18.11: **a.** *Contraction of a cluster toward a point.* **b.** *Contraction of two clusters toward two distinct points.*

18.7   Contraction Toward a Center

Separation of clusters often involves transformations that pull apart clusters toward certain points, which are then considered as labels.

We shall start with the simplest case of a single cluster, which is pulled toward a point, called *center*. Let X_j be the position vectors of the points in a cluster $\mathcal{G}$ and C be a given center, with the position vector c. The transformation given by

$$Y_j = \lambda X_j + (1 - \lambda)c,$$

with $\lambda \in (0, 1)$, contracts the cluster $\mathcal{G}$ into a cluster situated in a proximity of the point C, see Fig. 18.11 **a**. The smaller the value of λ, the closer the image cluster is to the point C.

Consider now two clusters, $\mathcal{G}_1$ and $\mathcal{G}_2$, and two distinct centers, C_1 and C_2. Let X_j^r be the position vectors of the elements of the cluster r, with $r \in \{1, 2\}$, and c_i the position vectors of the centers C_i. The transformations

$$Y_j^1 = \lambda_1 X_j^1 + (1 - \lambda_1)c_1,$$

$$Y_j^2 = \lambda_2 X_j^2 + (1 - \lambda_2)c_2,$$

push the first cluster into the direction of the center C_1, while the second cluster toward C_2, see Fig. 18.11 **b**. The new clusters are more separate than the former.

The previous equations are written only for the cluster points. The question is whether we can extend them to a global function defined on the entire space where the clusters live in. This will be addressed in the next section.

18.8   Learning Decision Maps

Classification of clusters can be achieved by learning certain decision maps. This section deals with this topic.

18.8.1   Linear Decision Maps

Consider two clusters in $\mathbb{R}^2$

$$\mathcal{G} = \{(x_1, y_1), \ldots, (x_N, y_N)\}, \quad \widetilde{\mathcal{G}} = \{(\widetilde{x}_1, \widetilde{y}_1), \ldots, (\widetilde{x}_M, \widetilde{y}_M)\},$$

and assume that $hull(\mathcal{G}) \cap hull(\widetilde{\mathcal{G}}) = \varnothing$. By Exercise 18.11.9, the clusters $\mathcal{G}$ and $\widetilde{\mathcal{G}}$ are linearly separable. We shall investigate the cases of one and two-dimensional labels.

One-dimensional labels We associate two labels α and $\widetilde{\alpha}$ with the clusters $\mathcal{G}$ and $\widetilde{\mathcal{G}}$, respectively. The labels are two distinct real numbers. Since the clusters $\mathcal{G}$ and $\widetilde{\mathcal{G}}$ are linearly separable, it makes sense to look for a linear function $f : \mathbb{R}^2 \to \mathbb{R}$ that maps the cluster $\mathcal{G}$ in a neighborhood of α and the cluster $\widetilde{\mathcal{G}}$ in a neighborhood of $\widetilde{\alpha}$, and also hope that the midpoint $(\alpha + \widetilde{\alpha})/2$ separates the sets $f(\mathcal{G})$ and $f(\widetilde{\mathcal{G}})$.

We shall look for a function f to be the input-output function of a feedforward neural network with one-dimensional output (one neuron in the output layer) and no hidden layers, i.e., we assume it is given by

$$f_{w,b}(x, y) = w_1 x + w_2 y + b.$$

The real parameters w_i, b have to be chosen such that the image sets $f(\mathcal{G})$ and $f(\widetilde{\mathcal{G}})$ are tightly localized about α and $\widetilde{\alpha}$, respectively. This can be achieved by taking

$$(w, b) = \arg \min F(w, b),$$

where F is the following quadratic cost function that measures the distance between labels and images:

$$F(w, b) = \frac{1}{2} \sum_{i=1}^{N} \Big(f_{w,b}(x_i, y_i) - \alpha \Big)^2 + \frac{1}{2} \sum_{j=1}^{M} \Big(f_{w,b}(\widetilde{x}_j, \widetilde{y}_j) - \widetilde{\alpha} \Big)^2$$

$$= \frac{1}{2} \sum_{i=1}^{N} \Big(w_1 x_i + w_2 y_i + b - \alpha \Big)^2 + \frac{1}{2} \sum_{j=1}^{M} \Big(w_1 \widetilde{x}_j + w_2 \widetilde{y}_j + b - \widetilde{\alpha} \Big)^2.$$

The minimum is realized for the solution of the equation $\nabla F(w, b) = 0$. In the following we shall compute the gradient $\nabla F = (\partial_{w_1} F, \partial_{w_2} F, \partial_b F)$. A straightforward differentiation using chain rule yields

$$\partial_{w_1} F = \sum_{i=1}^{N} x_i(w_1 x_i + w_2 y_i + b - \alpha) + \sum_{j=1}^{M} \widetilde{x}_j(w_1 \widetilde{x}_j + w_2 \widetilde{y}_j + b - \widetilde{\alpha})$$

$$= w_1 \Big(\sum_{i=1}^{N} x_i^2 + \sum_{j=1}^{M} \widetilde{x}_j^2 \Big) + w_2 \Big(\sum_{i=1}^{N} x_i y_i + \sum_{j=1}^{M} \widetilde{x}_j \widetilde{y}_j \Big)$$

$$+ b \Big(\sum_{i=1}^{N} x_i + \sum_{j=1}^{M} \widetilde{x}_j \Big) - \alpha \sum_{i=1}^{N} x_i - \widetilde{\alpha} \sum_{j=1}^{M} \widetilde{x}_j$$

Similarly, we have

$$\partial_{w_2} F = w_1 \Big(\sum_{i=1}^{N} x_i y_i + \sum_{j=1}^{M} \widetilde{x}_j \widetilde{y}_j \Big) + w_2 \Big(\sum_{i=1}^{N} y_i^2 + \sum_{j=1}^{M} \widetilde{y}_j^2 \Big)$$

$$+ b \Big(\sum_{i=1}^{N} y_i + \sum_{j=1}^{M} \widetilde{y}_j \Big) - \alpha \sum_{i=1}^{N} y_i - \widetilde{\alpha} \sum_{j=1}^{M} \widetilde{y}_j,$$

$$\partial_b F = w_1 \Big(\sum_{i=1}^{N} x_i + \sum_{j=1}^{M} \widetilde{x}_j \Big) + w_2 \Big(\sum_{i=1}^{N} y_i + \sum_{j=1}^{M} \widetilde{y}_j \Big)$$

$$+ (M + N)b - \alpha N - \widetilde{\alpha} M.$$

Consider the following 3×3 matrices:

$$A = \begin{pmatrix} \sum x_i^2 & \sum x_i y_i & \sum x_i \\ \sum x_i y_i & \sum y_i^2 & \sum y_i \\ \sum x_i & \sum y_i & N \end{pmatrix}, \qquad \widetilde{A} = \begin{pmatrix} \sum \widetilde{x}_j^2 & \sum \widetilde{x}_j \widetilde{y}_j & \sum \widetilde{x}_j \\ \sum \widetilde{x}_j \widetilde{y}_j & \sum \widetilde{y}_j^2 & \sum \widetilde{y}_j \\ \sum \widetilde{x}_j & \sum \widetilde{y}_j & M \end{pmatrix},$$

which contain information about the clusters $\mathcal{G}$ and $\widetilde{\mathcal{G}}$, respectively, such as size, first and second moments for the x and y variables, as well as their correlation. Consider two more vectors

$$\beta = \alpha \begin{pmatrix} \sum x_i \\ \sum y_i \\ N \end{pmatrix}, \qquad \widetilde{\beta} = \widetilde{\alpha} \begin{pmatrix} \sum \widetilde{x}_j \\ \sum \widetilde{y}_j \\ M \end{pmatrix},$$

which depend on the labels and the first moments. Then the vector equation $\nabla F = 0$ can be written as the linear matrix equation

$$(A + \widetilde{A})X = \beta + \widetilde{\beta},$$

where $X^T = (w_1, w_2, b)$. The solution is given by

$$X = (A + \widetilde{A})^{-1}(\beta + \widetilde{\beta}) = \widetilde{A}^{-1}(\mathbb{I} + A\widetilde{A}^{-1})^{-1}(\beta + \widetilde{\beta}),$$

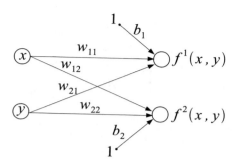

Figure 18.12: *Neural network with no hidden layers and two outputs.*

where we used the inverse of the sum formula (G.1.5) from Appendix G. The existence of the solution holds under the condition $\|A\widetilde{A}^{-1}\| < 1$, which means that the eigenvalues of A are, respectively, smaller than the eigenvalues of $\widetilde{A}$. The geometrical significance of the previous condition is roughly the following: the cluster which is the longer is also the wider. For all practical purposes, the inverse matrix $(A + \widetilde{A})^{-1}$ can be computed as indicated in the Appendix G.

It is worth noting that this method can be applied to more than two clusters, as long as their convex hulls are mutually disjoint.

The labels α and $\widetilde{\alpha}$ considered above were real numbers; for simplicity reasons, one may consider $\alpha = 1$ and $\widetilde{\alpha} = 0$. However, this is not the only good possibility.

Two-dimensional labels We shall perform in the following the linear classification of the clusters $\mathcal{G}$, $\widetilde{\mathcal{G}}$ using two-dimensional labels, such as vectors or distinct points in the plane. Assume the label of the cluster $\mathcal{G}$ is the point A in $\mathbb{R}^2$ having coordinates (a_1, a_2). Similarly, the label of the cluster $\widetilde{\mathcal{G}}$ is the point $\widetilde{A}$ with coordinates $(\widetilde{a}_1, \widetilde{a}_2)$.

As before, Proposition 18.5.2 states that the clusters $\mathcal{G}$ and $\widetilde{\mathcal{G}}$ are linearly separable. In this case, we shall look for a linear function $f : \mathbb{R}^2 \to \mathbb{R}^2$ that maps the cluster $\mathcal{G}$ in a neighborhood of the point A and the cluster $\widetilde{\mathcal{G}}$ in a neighborhood of $\widetilde{A}$, see Fig. 18.13.

The linear function $f = (f^1, f^2)$ is constructed as the input-output function of a feedforward neural network with a two-dimensional output and no hidden layers, see Fig. 18.12. This means that f is linear in each component

$$f^1_{w,b}(x, y) = w_{11}x + w_{21}y + b_1$$
$$f^2_{w,b}(x, y) = w_{12}x + w_{22}y + b_2.$$

Classification

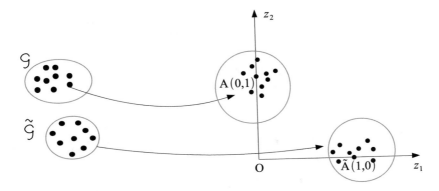

Figure 18.13: *The linearly separable clusters* $\mathcal{G}$, $\widetilde{\mathcal{G}}$ *are mapped into disjoint neighborhoods of point labels* A *and* $\widetilde{A}$.

The weights, w_{ij}, and biasses, b_j, have to be tuned such that the image sets $f(\mathcal{G})$ and $f(\widetilde{\mathcal{G}})$ are as close as possible to the points A and $\widetilde{A}$, respectively. This can be achieved by taking

$$(w, b) = \arg\min G(w, b),$$

where G is the sum of squared distances to the point labels, given by

$$G(w, b) = \frac{1}{2}\sum_{i=1}^{N} d\big(f_{w,b}(x_i, y_i), A\big)^2 + \frac{1}{2}\sum_{j=1}^{M} d\big(f_{w,b}(\widetilde{x}_j, \widetilde{y}_j), \widetilde{A}\big)^2$$

$$= \frac{1}{2}\sum_{i=1}^{N}(f^1(x_i, y_i) - a_1)^2 + \frac{1}{2}\sum_{i=1}^{N}(f^2(x_i, y_i) - a_2)^2$$

$$+ \frac{1}{2}\sum_{j=1}^{M}(f^1(\widetilde{x}_j, \widetilde{y}_j) - \widetilde{a}_1)^2 + \frac{1}{2}\sum_{j=1}^{M}(f^2(\widetilde{x}_j, \widetilde{y}_j) - \widetilde{a}_2)^2$$

$$= F_1(w_{11}, w_{21}, b_1) + F_2(w_{12}, w_{22}, b_2),$$

with

$$F_1(w_{11}, w_{21}, b_1) = \frac{1}{2}\sum_{i=1}^{N}(w_{11}x_i + w_{21}y_i + b_1 - a_1)^2$$

$$+ \frac{1}{2}\sum_{j=1}^{M}(w_{11}\widetilde{x}_j + w_{21}\widetilde{y}_j + b_1 - \widetilde{a}_1)^2$$

$$F_2(w_{12}, w_{22}, b_2) = \frac{1}{2}\sum_{i=1}^{N}(w_{12}x_i + w_{22}y_i + b_2 - a_2)^2$$

$$+ \frac{1}{2}\sum_{j=1}^{M}(w_{12}\widetilde{x}_j + w_{22}\widetilde{y}_j + b_2 - \widetilde{a}_2)^2.$$

The minimum is realized for the solution of the equation $\nabla G(w,b) = 0$, where

$$\nabla G(w,b) = (\nabla_{w_{i1}}F_1, \nabla_{b_1}F_1, \nabla_{w_{i2}}F_2, \nabla_{b_2}F_2) \in \mathbb{R}^6.$$

The system contains 6 equations and 6 unknowns (4 weights and 2 biases). Given the formulas of F_1 and F_2, it can be shown, similarly with the case of one-dimensional labels, that the system is equivalent to the following two linear systems:

$$(A + \widetilde{A})X_1 = \beta_1 + \widetilde{\beta}_1$$

$$(A + \widetilde{A})X_2 = \beta_2 + \widetilde{\beta}_2,$$

where A, $\widetilde{A}$ are defined as in the previous case, $X_j^T = (w_{1j}, w_{2j}, b_j)$, and

$$\beta_i = a_i\begin{pmatrix}\sum x_i \\ \sum y_i \\ N\end{pmatrix}, \qquad \widetilde{\beta}_i = \widetilde{a}_i\begin{pmatrix}\sum \widetilde{x}_j \\ \sum \widetilde{y}_j \\ M\end{pmatrix}.$$

Using formula (G.1.5) from Appendix G, the solution is given by

$$X_1 = \widetilde{A}^{-1}(\mathbb{I} + A\widetilde{A}^{-1})^{-1}(\beta_1 + \widetilde{\beta}_1),$$

$$X_2 = \widetilde{A}^{-1}(\mathbb{I} + A\widetilde{A}^{-1})^{-1}(\beta_2 + \widetilde{\beta}_2),$$

provided the condition $\|A\widetilde{A}^{-1}\| < 1$ holds.

The use of softmax Most classification problems use an additional layer that implements a softmax function, which was introduced in section 2.1. We shall do this in the case of the previous example, see Fig. 18.14. For the sake of simplicity, we may choose the labels to be one-hot vectors, i.e., $(a_1, a_2) = (0, 1) = e_1$ and $(\widetilde{a}_1, \widetilde{a}_2) = (1, 0) = e_2$.

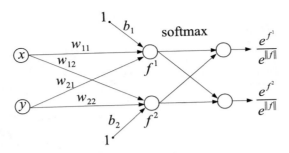

Figure 18.14: *Neural network with an extra layer implementing the softmax function.*

The outcome of the new network is $z = (z_1, z_2)$, with

$$z_i = \frac{e^{f^i}}{\|e^f\|} = \frac{e^{f^i}}{e^{f^1} + e^{f^2}}, \qquad i = 1, 2.$$

Since $z_i > 0$ and $z_1 + z_2 = 1$, it follows that z belongs to the line segment joining the points $A = (0, 1)$ and $\widetilde{A} = (1, 0)$. Then the clusters $\mathcal{G}$ and $\widetilde{\mathcal{G}}$ are mapped into the line segment $A\widetilde{A}$, with the image of $\mathcal{G}$ closer to the label point A and the image of $\widetilde{\mathcal{G}}$ closer to $\widetilde{A}$, see Fig. 18.15.

Assume the separation point of the cluster images is the middle of the segment, $(1/2, 1/2)$. Now, let (x, y) be a given point in the plane, and we have to decide to which cluster the point (x, y) belongs to. We run (x, y) through the neural network and if the outcome belongs to the upper part of the segment $A\widetilde{A}$, then the point belongs to the cluster $\mathcal{G}$. Otherwise, it belongs to $\widetilde{\mathcal{G}}$. This test can be done using either the horizontal or the vertical coordinate. For instance, using the coordinate z_1, we have

if $z_1 < 1/2$, then (x, y) belongs to $\mathcal{G}$;

if $z_1 > 1/2$, then (x, y) belongs to $\widetilde{\mathcal{G}}$.

The use of softmax is also successful for classification problems involving more than 2 clusters. For instance, in the case of 3 clusters, $\mathcal{G}_1$, $\mathcal{G}_2$, and $\mathcal{G}_3$, we associate the labels $A_1 = (1, 0, 0)$, $A_2 = (0, 1, 0)$, and $A_3 = (0, 0, 1)$ in $\mathbb{R}^3$, which forms an equilateral triangle, see Fig. 18.16. In this case the separation point is replaced by a system of separation curves obtained by taking perpendiculars from the triangle center onto its sides, see Fig. 18.17. In order to decide the decision region a point belongs to, it suffices to evaluate the distance to the vertices. For instance, if $\text{dist}(P, A_i) < 1/2$, then the point P belongs to the image of cluster $\mathcal{G}_i$.

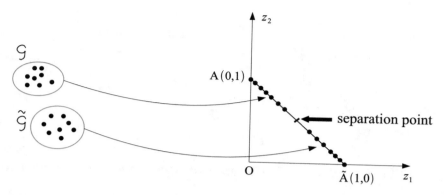

Figure 18.15: *The linearly separable clusters $\mathcal{G}$ and $\widetilde{\mathcal{G}}$ are mapped into the terminal regions of a line segment. In the interior of the segment there is a separation point.*

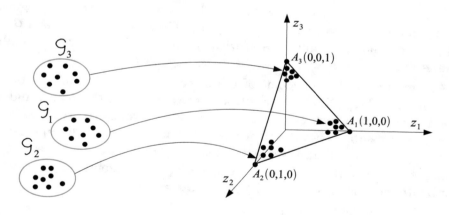

Figure 18.16: *The linearly separable clusters $\mathcal{G}_1$, $\mathcal{G}_2$ and $\mathcal{G}_3$ are mapped into the vertex regions of a triangle.*

18.8.2 Nonlinear Decision Maps

Consider two clusters in $\mathbb{R}^2$

$$\mathcal{G} = \{(x_1, y_1), \ldots, (x_N, y_N)\}, \quad \widetilde{\mathcal{G}} = \{(\widetilde{x}_1, \widetilde{y}_1), \ldots, (\widetilde{x}_M, \widetilde{y}_M)\},$$

and assume that $hull(\mathcal{G}) \cap hull(\widetilde{\mathcal{G}}) \neq \varnothing$.

We associate two real numbers, α and $\widetilde{\alpha}$, as labels with the clusters $\mathcal{G}$ and $\widetilde{\mathcal{G}}$, respectively. By Proposition 18.5.2 the clusters $\mathcal{G}$ and $\widetilde{\mathcal{G}}$ are not linearly separable. Therefore, it makes sense to look for a nonlinear function $f : \mathbb{R}^2 \to \mathbb{R}$ that maps the cluster $\mathcal{G}$ into a neighborhood of α and the cluster $\widetilde{\mathcal{G}}$ into a neighborhood of $\widetilde{\alpha}$.

The sigmoid neuron The easiest case of nonlinear function is produced by a sigmoid neuron, with the input-output function

$$f_{w,b}(x, y) = \sigma(w_1 x + w_2 y + b),$$

where $\sigma(x)$ is the logistic function. We shall assume in this case that the labels are given by $\alpha = 0$ and $\widetilde{\alpha} = 1$. Thus, the cluster $\mathcal{G}$ will be mapped toward 0, while the cluster $\widetilde{\mathcal{G}}$ toward 1. In order to make this map as localized as possible about the labels values, we need to minimize the cost function

$$F(w, b) = \frac{1}{2} \sum_{i=1}^{N} \sigma(w_1 x_i + w_2 y_i + b)^2 + \frac{1}{2} \sum_{j=1}^{M} [\sigma(w_1 \widetilde{x}_j + w_2 \widetilde{y}_j + b) - 1]^2.$$

Using the differentiation property of the logistic sigmoid, $\sigma' = \sigma(1 - \sigma)$, we obtain

$$\partial_{w_1} F = \sum_{i=1}^{N} x_i \, \sigma^2(1 - \sigma)\Big|_{w_1 x_i + w_2 y_i + b} - \sum_{j=1}^{M} \widetilde{x}_j \, \sigma(1 - \sigma)^2\Big|_{w_1 \widetilde{x}_j + w_2 \widetilde{y}_j + b}$$

$$\partial_{w_2} F = \sum_{i=1}^{N} y_i \, \sigma^2(1 - \sigma)\Big|_{w_1 x_i + w_2 y_i + b} - \sum_{j=1}^{M} \widetilde{y}_j \, \sigma(1 - \sigma)^2\Big|_{w_1 \widetilde{x}_j + w_2 \widetilde{y}_j + b}$$

$$\partial_b F = \sum_{i=1}^{N} \sigma^2(1 - \sigma)\Big|_{w_1 x_i + w_2 y_i + b} - \sum_{j=1}^{M} \sigma(1 - \sigma)^2\Big|_{w_1 \widetilde{x}_j + w_2 \widetilde{y}_j + b}.$$

Due to nonlinearity, there isn't a closed-form solution for the minimum. The way to find the minimum is by applying the gradient descent method. The approximating sequence is given iteratively by

$$
\begin{aligned}
w_1^{(n+1)} &= w_1^{(n)} - \eta \, \partial_{w_1} F(w_1^{(n)}, w_2^{(n)}, b^{(n)}) \\
w_2^{(n+1)} &= w_2^{(n)} - \eta \, \partial_{w_2} F(w_1^{(n)}, w_2^{(n)}, b^{(n)}) \\
b^{(n+1)} &= b^{(n)} - \eta \, \partial_b F(w_1^{(n)}, w_2^{(n)}, b^{(n)}),
\end{aligned}
$$

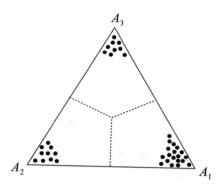

Figure 18.17: *The dashed lines separate the triangle into three decision regions.*

with the learning rate $\eta > 0$ and initial values $w_1^{(0)} = w_2^{(0)} = b^{(0)} = 0$. It is worth noting that this is a variant of the logistic regression method and works only for the classification of two clusters.

One-hidden layer network We may assume that the nonlinear function f is the input-output function of a feedforward neural network with three layers, having a one-dimensional output. The neurons in the hidden layer are assumed to have nonlinear activation functions. Then the input-output map is

$$f_{w,\lambda,b}(x, y) = \sum_{k=1}^{K} \lambda_k \sigma(w_{1k}x + w_{2k}y + b_k),$$

where K is the number of hidden neurons, w_{ij} are the weights from the input to the hidden layer, λ_k are the weights from the hidden layer to the output, and b_k are the biases of hidden neurons, see Fig. 18.18. The function $f_{w,\lambda,b}$ depends on $4K$ parameters ($2K$ weights w_{ij}, K biasses b_k, and K weights λ_k). The parameters have to be tuned such that the following cost function is minimized:

$$G(w, b, \lambda) = \frac{1}{2}\sum_{i=1}^{N}[f_{w,\lambda,b}(x_i, y_i) - \alpha]^2 + \frac{1}{2}\sum_{j=1}^{M}[f_{w,\lambda,b}(\widetilde{x}_j, \widetilde{y}_j) - \widetilde{\alpha}]^2, \quad (18.8.2)$$

where α, $\widetilde{\alpha}$ are the labels associated with the clusters $\mathcal{G}$, $\widetilde{\mathcal{G}}$, respectively. Since there is no closed-form solution for the minimum of the previous cost function, a gradient descent type algorithm is needed, see Exercise 18.11.10.

In the case of multiple clusters we may use multiple labels and a similar method as before. However, using vector labels, such as one-hot vectors, would provide a more elegant variant for tackling the classification problem.

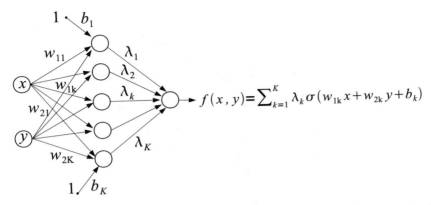

Figure 18.18: *Neural network with one-hidden layer and one output.*

In this case the label space is $\mathbb{R}^p$, where p is the number of clusters that need to be classified, see Exercise 18.11.11.

18.9 Decision Boundaries

Consider two clusters, $\mathcal{G}_1$ and $\mathcal{G}_2$, in $\mathbb{R}^2$. A *decision boundary* is a continuous curve in $\mathbb{R}^2$ that separates the clusters, see Fig. 18.19. This means that for any $g_1 \in \mathcal{G}_1$ and $g_2 \in \mathcal{G}_2$, any continuous curve from g_1 to g_2 intersects the decision boundary curve.

The next result states that the existence of a decision boundary is locally equivalent to the linear separation of the clusters.

Theorem 18.9.1 (Rectification Theorem) *Consider two clusters $\mathcal{G}_1$ and $\mathcal{G}_2$ in $\mathbb{R}^2$ and a smooth decision boundary between them. Then there is a smooth nonlinear function F defined on a neighborhood of the decision curve with values in $\mathbb{R}^2$, such that F maps the decision curve into the x-axis, the cluster $\mathcal{G}_1$ into the upper half-plane and the cluster $\mathcal{G}_2$ into the lower half-plane.*

Proof: Denote by $c(s)$, $0 \leq 0 \leq T$, the decision curve and consider its evolution through the normal flow, $\Phi_t(c(s))$, with $-\epsilon < t < \epsilon$. This is $\Phi_t(c(s)) = c(s) + tN(s)$, where N is the unit normal vector to $c(s)$, see Fig. 18.20. For small enough ϵ, the flow does not have any singularities. The cluster $\mathcal{G}_1$ corresponds to values $t > 0$ and the cluster $\mathcal{G}_2$ corresponds to $t < 0$. The set $\mathcal{U} = \{\Phi_t(c(s)); |t| < \epsilon, 0 \leq s \leq T\}$ defines a neighborhood of the decision curve. Then define the mapping $F : \mathcal{U} \to \mathbb{R}^2$ by $F(\Phi_t(c(s))) = (s, t)$. Since

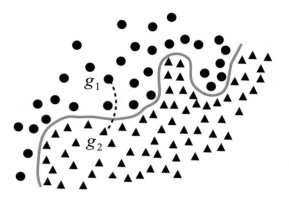

Figure 18.19: *Decision boundary curve between two clusters in* $\mathbb{R}^2$.

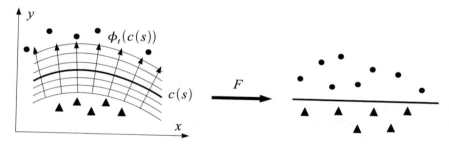

Figure 18.20: *The function F maps the decision curve into the x-axis.*

$\Phi_0(c(s)) = c(s)$, then F maps $c(s)$ into $(s, 0)$, and sends the cluster $\mathcal{G}_1 \cap \mathcal{U}$ into the upper half-plane and the cluster $\mathcal{G}_2 \cap \mathcal{U}$ into the lower half-plane.

∎

This result can be extended to higher dimensions. If the clusters $\mathcal{G}_1$ and $\mathcal{G}_2$ are in $\mathbb{R}^n$, a *decision boundary* is a $(n-1)$-hypersurface, $\mathcal{H}$, in $\mathbb{R}^n$, which separates the clusters, i.e., for any $g_1 \in \mathcal{G}_1$ and $g_2 \in \mathcal{G}_2$, any continuous curve from g_1 to g_2 intersects the hypersurface $\mathcal{H}$.

Similarly with the previous case, we can construct a system of coordinates in a neighborhood of $\mathcal{H}$. Any point P in this neighborhood can be projected onto $\mathcal{H}$ in a point P'. The coordinates associated with P are $(x_1, \ldots, x_{n-1}, x_n)$, where $(x_1, \ldots, x_{n-1})$ are the coordinates of P' on $\mathcal{H}$ and x_n is the length of the projection of P onto $\mathcal{H}$, considered with the sign given by the orientation of the hypersurface. Thus, the function $F(P) = (x_1, \ldots, x_n)$ maps $\mathcal{H}$ into the hyperplane $\{x_n = 0\}$.

The case involving more than 2 clusters is more complex, as in this case the decision boundary has a richer structure. For instance, if the clusters are situated in a plane, we have the following definition. Consider p clusters of points $\mathcal{G}_1, \mathcal{G}_2, \ldots, \mathcal{G}_p$ in $\mathbb{R}^2$. In this case, a *decision boundary* is a connected system of continuous curves in $\mathbb{R}^2$ that separates the clusters. This is, for any $g_i \in \mathcal{G}_i$ and $g_j \in \mathcal{G}_j$, with $i \neq j$, any continuous curve from g_i to g_j intersects the decision boundary.

18.10   Summary

During the classification process of clusters, labels have to be accurately assigned to each cluster. There are several types of labels that can be assigned: numbers, one-hot vectors, points, etc. The mapping assigning labels to clusters is a decision map which can be learned by neural networks.

Two clusters are linearly separable if their convex hulls are disjoint. A single perceptron can learn the associated decision function in this case.

If the convex hulls of two clusters are not disjoint, the clusters are not linearly separable, and in this case a neural network with nonlinear activation function is used to learn the associated decision function.

The use of the softmax activation function in the last layer of a neural network maps the n clusters in the corners of an $(n-1)$-dimensional polytop.

The Rectification Theorem states that the existence of a decision boundary between the clusters is locally equivalent to the linear separation of the clusters.

18.11   Exercises

Exercise 18.11.1 Let $\mathcal{S} = (I_1 \times I_1) \cap (\mathbb{Q} \times \mathbb{Q})$ be the lattice of rational numbers in $[0, 1] \times [0, 1]$.

(a) Show that $\mathcal{S}$ is a transitive relation;

(b) Is $\mathcal{S}$ an equivalence relation?

Exercise 18.11.2 Show that k points are in a general position if an only if there is a unique $(k-1)$-hyperplane containing the points.

Exercise 18.11.3 Let $P_1, \ldots, P_k$ be k points in $\mathbb{R}^k$ in a general position, and let $\mathcal{H}$ be the unique $(k-1)$-hyperplane containing them. Show that the set $\{\overrightarrow{P_1 P_2}, \ldots, \overrightarrow{P_1 P_k}\}$ forms a system of independent vectors in $\mathcal{H}$.

Exercise 18.11.4 Consider two distinct real numbers $x_1, x_2 \in \mathbb{R}$. Show that there is a unique affine function $f(x) = ax + b$ such that $f(x_1) = 1$ and $f(x_2) = 2$.

Exercise 18.11.5 Let $\mathcal{G}$ be a cluster in $\mathbb{R}^n$.

(*a*) Show that $hull(\mathcal{G})$ is a convex set;

(*b*) Prove that $hull(\mathcal{G})$ is the smallest convex set in $\mathbb{R}^n$ which contains $\mathcal{G}$;

(*c*) Verify that

$$hull(\mathcal{G}) = \bigcap_K \{K; \mathcal{G} \subset K \subset \mathbb{R}^n \text{convex}\}.$$

Exercise 18.11.6 Let $\Phi : \mathbb{R}^2 \to \mathbb{R}^2$ be a nondegenerate affine function, i.e., $\Phi(x) = Wx + b$, with $\det W \neq 0$. Consider two linearly separable clusters in $\mathbb{R}^2$, $\mathcal{G}_1$ and $\mathcal{G}_2$. Show that $\Phi(\mathcal{G}_1)$ and $\Phi(\mathcal{G}_2)$ are linearly separable. In other words, affine functions preserve linear separability.

Exercise 18.11.7 Let $\mathcal{G}$ be a finite cluster of points and denote by G its center of mass. Show that $G \in hull(\mathcal{G})$.

Exercise 18.11.8 Show that a family of clusters $\mathfrak{G} = \{\mathcal{G}_1, \ldots, \mathcal{G}_k\}$ in $\mathbb{R}^n$ is convex separable if and only if the family is mutually convex separable, i.e., any two clusters $\mathcal{G}_i$ and $\mathcal{G}_j$ are convex separable for any $i \neq j$.

Exercise 18.11.9 (*a*) If A and B are two convex sets such as $A \cap B = \emptyset$, then A and B are linearly separable;

(*b*) If $\mathcal{G}_1$ and $\mathcal{G}_2$ are two clusters such that $hull(\mathcal{G}_1) \cap hull(\mathcal{G}_2) = \emptyset$, then $\mathcal{G}_1$ and $\mathcal{G}_2$ are linearly separable.

Exercise 18.11.10 (*a*) Find the gradient $\nabla_{w,b,\lambda} G$, where G is given by (18.8.2).

(*b*) Write the gradient descent recursion for the approximation sequence of the minimum.

Exercise 18.11.11 Consider p clusters, $\mathcal{G}_1, \ldots, \mathcal{G}_p$, of points in $\mathbb{R}^2$. Write the cost function that associates the vector e_j as a label for the cluster $\mathcal{G}_j$, for all $j = 1, \ldots, p$.

Exercise 18.11.12 Prove that the entropy associated to a partition given by (18.2.1) is maximum if and only of all sets A_j have equal measure.

Chapter 19

Generative Models

So far, neural networks were useful for two main types of important problems: regression and classification problems. While pursuing regression a neural network with a one-dimensional output has been used, which is having a linear activation function in the output layer. In the case of classification problems it is useful to employ a neural network with a multidimensional output, which is having a softmax activation function in the output layer.

In this section we shall deal with another important application, which is the construction of *generative models*. If a regression problem can forecast future patterns and a classification problem can recognize typefaces, a generative model can generate examples which are very much like the training data.

19.1   The Need of Generative Models

There are many reasons for considering generative models. Their successful applications made them very popular in many areas of industry. The ability of generative models of generating examples that are very similar to training data makes them useful in situations when available data is insufficient or even missing. In many real-life situations data augmentation might be costly and time-consuming. Generative models can achieve this job at a lower cost. We shall provide a few examples.

1. For safety reasons training a neural network for driving a car, or operating a robot, cannot be done in a real-life environment. Therefore, a generative model is used to provide a simulated environment where the car is trained until the model is ready to be deployed in real life.

© Springer Nature Switzerland AG 2020
O. Calin, *Deep Learning Architectures*, Springer Series in the Data Sciences,
https://doi.org/10.1007/978-3-030-36721-3_19

2. In the case of setting a business plan, a generative model can be employed to provide a simulation of possible future business environments. These possible environments are generated by similarity with past events.

3. Another application is to price financial contracts such as derivatives. These are contracts whose price depends on the price of an underlying stock. The traditional computational technique is to use the Monte Carlo method. This assumes a large number of possible stock price simulations under which the financial contract is priced. Then an average of all the simulated contract prices is considered as the computed contract price. Generative models can be used to simulate stock markets and price financial contracts in these markets, providing an idea about the real contract price.

4. Other examples are the use of generative models in cartoons and movie industry, where real-life-like environments need to be generated, such as buildings, trees, mountains, etc. Generative models can be also used to generate music, paintings, people's faces, etc. For instance, the iGAN software was created to generate pictures starting from a rough hand sketch.

We shall discuss next the two types of generative models: the ones that provide a *density estimation* and the ones that produce *sample generations*.

19.2 Density Estimation

The task here is to create a machine that observes many samples from a given distribution, $p_{data}(x)$, and it is able to create later more samples from that distribution. We distinguish the following two cases: *parametric* and *nonparametric* distributions.

Parametric case If there is a reason to believe that the given distribution, $p_{data}(x)$, can be estimated by a parametric distribution, $p_{model}(x; \theta)$, then we just need to adjust θ such that the log-likelihood of the model distribution evaluated on the data is maximum

$$\theta^* = \arg \max_{\theta} \mathbb{E}_{x \sim p_{data}} \big[\ln p_{model}(x; \theta) \big].$$

We note that in practice the expectation is computed as an average

$$\mathbb{E}_{x \sim p_{data}} \big[\ln p_{model}(x; \theta) \big] = \frac{1}{n} \sum_{i=1}^{n} \ln p_{model}(x_i; \theta),$$

where $x_i \sim p_{data}$ is a sample extracted from the given distribution.

Another equivalent way to select the parameter θ is by requiring the Kullback-Leibler divergence between the data and the model distributions to be minimum

$$\theta^* = \arg\min_\theta D_{KL}[p_{data}||p_{model}].$$

This follows from the computation

$$\theta^* = \arg\max_\theta \mathbb{E}_{x \sim p_{data}}\left[\ln p_{model}(x;\theta)\right] = \arg\min_\theta \mathbb{E}_{x \sim p_{data}}\left[-\ln p_{model}(x;\theta)\right]$$

$$= \arg\min_\theta \left\{-\int p_{data}(x)\ln p_{model}(x)\,dx\right\} = \arg\min_\theta S(p_{data}, p_{model})$$

$$= \arg\min_\theta [S(p_{data}, p_{model}) - H(p_{data})] = \arg\min_\theta D_{KL}(p_{data}||p_{model}),$$

where we used the independence of the entropy of the given data, $H(p_{data})$, from θ and the definitions of the cross-entropy S, and Kullback-Leibler divergence.

Nonparametric case The idea, in this case, is the following: We take a simple distribution (such as a uniform or a Gaussian) and then apply a nonlinear transformation G to samples from that distribution to obtain samples from the desired distribution, $p_{data}(x)$.

We shall introduce first some terminologies and notations. The space where x takes values is denoted by $\mathcal{X}$ and represents the space we care about. The simple distribution, $p(z)$, subject to be transformed by G into $p_{data}(x)$, is called the *code distribution*; the space $\mathcal{Z}$ where z is sampled from is called the *latent space*. Now, we would like for each random selection $z \in \mathcal{Z}$ to obtain $x = G(z)$ as a sample from the space $\mathcal{X}$.

It might be useful to make the connection to the random variables terminology. The latent variables z are instances of a random variable denoted by Z taking values in the latent space $\mathcal{Z}$. The samples x from $p_{data}(x)$ are considered as instances of a random variable X on the space $\mathcal{X}$. Under these notations, we are looking for a transformation G that maps the variable Z into X. This can be achieved in a couple of ways.

1. We assume the data distribution p_{data} is one-dimensional and we shall consider a uniform coding distribution, $p_{code} \sim Unif[0,1]$. We shall show that the transformation $G : [0,1] \to \mathcal{X}$, which satisfies $G(Z) = X$ is given by $G = F_{data}^{-1}$, where $F_{data}(x) = \int_{-\infty}^x p_{data}(s)\,ds$ is the cumulative distribution function of the random variable X.

The key point here is that if $Z \sim Unif[0,1]$, then $P(Z \le z) = z$, for all $z \in [0,1]$. Now, in order to show that the random variables X and $Y = G(Z)$ have the same distribution, we use the computation

$$F_Y(x) = P(Y \le x) = P(G(Z) \le x) = P(Z \le F_{data}(x)) = F_{data}(x), \quad \forall x \in \mathcal{X}.$$

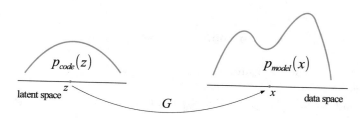

Figure 19.1: *The transformation G between the latent space and data space.*

To conclude, for generating n samples in space $\mathcal{X}$ following distribution p_{data}, we select uniformly n random instances $z_i \in [0,1]$ and consider $x_i = G(z_i)$, provided $G = F_{data}^{-1}$ is known. Then all the constructed instances satisfy $x_i \sim p_{data}$.

Since seldom in practice the distribution is one-dimensional, we shall consider next a multidimensional case.

2. In this case the coding distribution, p_{code}, is multidimensional and not necessarily uniform. The latent space is denoted by $\mathcal{Z}$ and we consider an invertible differentiable transform $G : \mathcal{Z} \to \mathcal{X}$, whose both its inverse and determinant of the Jacobian can be computed, see Fig. 19.1. Then this change of variables produces a new density on $\mathcal{X}$ given by

$$p_{model}(x) = p_X(x) = \frac{p_{code}(G^{-1}(x))}{|\det J_G(z)|}. \tag{19.2.1}$$

Example 19.2.1 As an example, we shall show how we can construct samples from the normal multivariate distribution $\mathcal{N}(\mu, \Sigma)$, where μ is a given vector in $\mathbb{R}^n$ and Σ is a symmetric, nondegenerate, and positive-definite n-dimensional matrix. We consider the affine transform $G : \mathbb{R}^n \to \mathbb{R}^n$, given by $G(z) = \mu + Az$, where $AA^T = \Sigma$. Obviously, $\det J_G(z) = \det A \neq 0$, and hence G is invertible, with the inverse

$$G^{-1}(x) = A^{-1}(x - \mu).$$

We choose the coding distribution to be the standard normal distribution with zero mean and identity covariance

$$p_{code}(z) = \frac{1}{(2\pi)^{n/2}} e^{-\frac{1}{2}\|x\|^2},$$

namely, $Z \sim \mathcal{N}(0, \mathbb{I}_n)$. Using formula (19.2.1) we obtain the following density:

$$
\begin{aligned}
p_X(x) &= \frac{1}{|\det A|} \frac{1}{(2\pi)^{n/2}} e^{-\frac{1}{2}\|A^{-1}(x-\mu)\|^2} \\
&= \frac{1}{(\det \Sigma)^{1/2}} \frac{1}{(2\pi)^{n/2}} e^{-\frac{1}{2}\langle A^{-1}(x-\mu), A^{-1}(x-\mu)\rangle^2} \\
&= \frac{1}{((2\pi)^n \det \Sigma)^{1/2}} e^{-\frac{1}{2}\langle (AA^T)^{-1}(x-\mu), (x-\mu)\rangle^2} \\
&= \frac{1}{\sqrt{(2\pi)^n \det \Sigma}} e^{-\frac{1}{2}(x-\mu)\Sigma^{-1}(x-\mu)},
\end{aligned}
$$

where we used the algebra relations $(A^{-1})^T A^{-1} = (A^T)^{-1}A^{-1} = (AA^T)^{-1}$.

It is worth noting that using this method following formula (19.2.1) has some downsides. First, the method assumes that the determinant of the Jacobian and the inverse of the transform G are computable. Another downside of this method is that the dimension of the latent space is equal to the dimension of the data space, fact that involves too many parameters. If the data space represents a picture, then using this method will involve as many latent variables as pixels in the picture! This is the reason for constructing a method which has a latent space of a much smaller dimension than the data space.

The second type of generative models is the sample generation, which generates more samples rather than finding the density function. In this group, we shall discuss the *generative adversarial networks* (GANs) and *generative moment matching networks*. We shall discuss in the next section the *adversarial* aspect first.

19.3 Adversarial Games

We consider two players who are engaged in a competitive game, fighting over the same payoff function: one wants it to be high, while the other wants it to be low. This way, the players become *adversaries*, being interested in opposite rewards. Each of them controls certain variables of the payoff function and tries to adjust them to get the maximum benefit. For certain parameter values the players might arrive at an equilibrium. This can be understood using the following example.

We may think of one player as the *seller* and of the other as the *buyer* of a certain product. Both parties have to agree upon the product price. The seller will try to push the price up, while the buyer would like the price to be as low as possible. At the end of the negotiation process the seller agrees to lower the product price enough and the seller agrees to pay sufficiently more such that the parties enter a seller-buyer agreement. This way they arrive at the *equilibrium price*, which corresponds to the Nash equilibrium point of the game.

The problem can be formulated as a *minimax problem* as in the following. Consider a payoff function, $V(x, y)$, which depends on variables x and y. The first player controls the variable x and the second player has control over the variable y. The first player wants the payoff function $V(x, y)$ to be minimized, while the second one wants $V(x, y)$ to be maximized (which is equivalent to minimize $-V(x, y)$). The problem can be formulated as

$$(x^*, y^*) = \arg \max_y \min_x V(x, y).$$

The equilibrium point can be obtained using a *simultaneous gradient descent method* in continuous time variant as in the following.

Since the first player controls the variable x and intends to minimize $V(x, y)$, it should adjust x in the direction of the negative gradient of V by a step $\eta = \Delta t$

$$x(t + \Delta t) = x(t) - \eta \frac{\partial V}{\partial x}.$$

Similarly, since the second player controls the variable y and would like to maximize $V(x, y)$, it should adjust y in the direction of the positive gradient of V by a step $\eta = \Delta t$

$$y(t + \Delta t) = y(t) + \eta \frac{\partial V}{\partial y}.$$

Taking $\Delta t \to 0$ we obtain a simultaneous gradient descent with an infinitesimal learning rate, where the learning process follows a smooth trajectory $(x(t), y(t))$ that satisfies the following continuous time differential system:

$$\frac{dx}{dt} = -\frac{\partial V}{\partial x} \tag{19.3.2}$$

$$\frac{dy}{dt} = \frac{\partial V}{\partial y}. \tag{19.3.3}$$

The initial condition is $(x(0), y(0)) = (x_0, y_0)$ and the equilibrium point, if exists, is obtained as the limit $(x^*, y^*) = \lim_{t \to \infty} (x(t), y(t))$. Since at equilibrium we have $\frac{dx^*}{dt} = \frac{dy^*}{dt} = 0$, then the equilibrium points satisfy the system

$$\frac{\partial V}{\partial x} = 0$$

$$\frac{\partial V}{\partial y} = 0.$$

It is worth noting that the change of the payoff function $V(x, y)$ along the learning curve $(x(t), y(t))$ is given by its differential, which is given by

$$dV = \frac{\partial V}{\partial x} dx + \frac{\partial V}{\partial y} dy = -\dot{x} dx + \dot{y} dy.$$

Example 19.3.1 The revenue obtained by selling q units of a certain product at the price p per unit is $V(p, q) = pq$. The seller can control the price, p, and is interested in maximizing the revenue $V(p, q)$. The buyer can control the number of units sold, q, and would like to minimize the price paid, $V(p, q)$. This becomes a minimax problem and it will be approached using the simultaneous gradient descent with an infinitesimal learning rate. The learning process follows a trajectory $(p(t), q(t))$, which satisfies the continuous-time system (19.3.2)–(19.3.3), which in this case becomes

$$\frac{dp}{dt} = q$$
$$\frac{dq}{dt} = -p.$$

Differentiating one more time with respect to t, we obtain $\ddot{p} = -p$ and $\ddot{q} = -q$, which implies that both $p(t)$ and $q(t)$ are linear combinations of $\cos t$ and $\sin t$

$$p(t) = A_1 \cos t + B_1 \sin t$$
$$q(t) = A_2 \cos t + B_2 \sin t,$$

with A_i and B_i constants. Substituting into the differential system and using the initial conditions we determine the constants and write the solution as

$$p(t) = p_0 \cos t + q_0 \sin t$$
$$q(t) = q_0 \cos t - p_0 \sin t.$$

Since this can be represented in the matrix form as $(p, q)^T = \mathcal{R}(t)(p_0, q_0)^T$, where

$$\mathcal{R}(t) = \begin{pmatrix} \cos t & \sin t \\ -\sin t & \cot t \end{pmatrix}$$

denotes the *rotation* in the plane of angle t, it follows that the solution $(p(t), q(t))$ follows a circle trajectory centered at the origin and having the radius $r = \sqrt{p_0^2 + q_0^2}$. This solution does not approach any equilibrium point as long as $r \neq 0$. The only equilibrium point is obtained at the origin, $(p^*, q^*) = (0, 0)$, and this occurs in the case $p_0 = q_0 = 0$.

19.4 Generative Adversarial Networks

Generative Adversarial Networks (GANs) have been introduced in 2014 by Goodfellow et al. [47] and are considered nowadays to be one of the most powerful generative models.

GANs represent a noncooperative game played by two neural networks: the *discriminator* and the *generator*. These networks act as adversaries, having opposite rewards, the worst-case input for one network being produced by the other network. During this competitive game each network forces the other network to improve.

Generator network The generator is a network whose input is some random noise selected from a latent space and the output is an image x, which is supposed to resemble the images in the data space. The generator output is $x = G(z; \theta^{(g)})$, where $\theta^{(g)}$ are the generator network parameters and z is a latent vector variable in a latent space $\mathcal{Z}$. We should represent by Z the random variable with instance z taking values in the latent space $\mathcal{Z}$ and by $p_{code}(z)$ its probability density. We denote by X the output random variable, namely $X = G(Z; \theta^{(g)})$, and denote its density by $p_{model}(x; \theta^{(g)})$. The generator function G must be differentiable, with the dimension of Z less than the dimension of X.

Discriminator network The discriminator network works as a classifier. This means that the discriminator is a network that is being given an input x (for instance, an image) and produces a number $D(x; \theta^{(d)})$ between 0 and 1, where $\theta^{(d)}$ are the discriminator network parameters. This number can be considered as the probability that the input x is regarded as a genuine training data, with a value of 0 assigned to the case when the discriminator rejects completely the input x as belonging to the training data. If the input is an image, the discriminator can be considered as a convolutional network and can be trained using the gradient descent. The discriminator and generator network functions can be seen in Fig. 19.2.

The generator's job is to full the discriminator, making it believe that its output is a real training data; this means the generator wants $D(G(z))$ close to 1. On the other side, the discriminator's goal is to prove the generator wrong, namely, it would like to output a value $D(G(z))$ close to 0, for all $z \in \mathcal{Z}$.

The training process functions on cycles. In the beginning the generator is not that smart, producing some random noise. Over time, the generator will improve, producing images more and more similar to the images in the training dataset. In the same time, the discriminator is trained on fake and real images, using a batch of images produced by the generator and a batch of images selected from the training data set. As the discriminator becomes

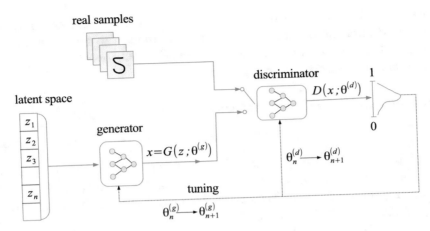

Figure 19.2: *The anatomy of a GAN: latent variables are fed into the generator producing "fake" data. Then repeatedly sample two minibatches of data, one from the real data and another from the generated data and let the discriminator decide whether it is fake or genuine. Using this decision, we can update the networks parameters to improve the GANs functionality.*

more skillful in its job, the generator will tend to produce images that resemble more and more the ones in the data space. If the training is successful, the generator will produce at the end images exactly indistinguishable from the real training images. Also, in the end the discriminator will not be able to distinguish whether the output is a fake or real, providing a probability output of 0.5. At this point the discriminator becomes useless and can be neglected, while keeping only the generator network.

Example 19.4.1 Consider a GAN which is supposed to generate prime numbers. We assume that the training data is the set of the first n prime numbers, $\mathcal{X} = \{p_1, p_2, \ldots, p_n\}$. If the generator network produces a number x, the discriminator network will test whether this number is prime by checking whether the division between x and p_i is even. The discriminator produces $D(x) = 0$ if there is a prime number p_i that is a divisor of x, and $D(x) = 1$, otherwise. In the beginning the generator might produce, for instance, $x = p_1 p_2$ and the discriminator will easily classify it as a nonprime number. But over time, the generator will learn that it has to produce a number larger than all the prime numbers provided, which is not a multiple of them, such as $x = p_1 p_2 \ldots p_n + 1$. In this case the generator will produce $D(x) = 1$, i.e, will classify the number x as prime. The generator can use this information, producing next time the output $x = p_1 \ldots p_n (p_1 p_2 \ldots p_n + 1) + 1$, and so forth.

We note that several discriminator functions $D(x)$ can be used. If we denote by $div(x)$ the number of divisors of x among the set of known prime numbers, then we can also define $D(x) = e^{2-div(x)}$, or $D(x) = \frac{2}{div(x)}$ and consider them as discriminator functions.

Loss function The discriminator tries to maximize the payoff

$$V(G, D) = \mathbb{E}_{x \sim p_{data}}[\ln D(x)] + \mathbb{E}_{x \sim p_{model}}[\ln(1 - D(x))], \qquad (19.4.4)$$

Maximizing the first term assures for a correct classification of true data, while maximizing the second term ensures for the correct classification of the model generated data. This follows from the fact that the discriminator's output $D(x)$ tends to be close to 1 for true data x, while $1 - D(x)$ tends to be close to 1 for generated data x.

Example 19.4.2 The simplest example of discriminator network is a sigmoid neuron. If the input is a vector $x^T = (x_1, \ldots, x_n)$, then the discriminator produces the probability $D(x; w, b) = \sigma(w^T x + b) = \frac{1}{1+e^{-(w^T x+b)}}$, where w and b denote the weights and the bias, respectively. In the following we shall evaluate the payoff function (19.4.4). Using the properties of logarithmic and softplus functions, we have

$$\ln D(x) = -\ln(1 + e^{-(w^T x+b)}) = -sp(-(w^T x + b)) = x - sp(w^T x + b),$$

and then

$$\mathbb{E}_{x \sim p_{data}}[\ln D(x)] = \mu_X - \mathbb{E}_{x \sim p_{data}}[sp(w^T x + b)],$$

where μ_X denotes the mean of generator's output. We also have

$$1 - \ln D(x) = 1 - \sigma(w^T x + b) = \sigma(-w^T x - b) = \frac{1}{e^{w^T x+b}},$$

so that

$$\ln(1 - D(x)) = -\ln(1 + e^{w^T x+b}) = -sp(w^T x + b).$$

Taking expectation, we obtain

$$\mathbb{E}_{x \sim p_{model}}[\ln(1 - D(x))] = -\mathbb{E}_{x \sim p_{model}}[sp(w^T x + b)].$$

Therefore, the payoff function becomes

$$\begin{aligned}
V(G, D) &= \mu_X - \mathbb{E}_{x \sim p_{data}}[sp(w^T x + b)] - \mathbb{E}_{x \sim p_{model}}[sp(w^T x + b)] \\
&= \mu_X - \int \big(p_{data}(x) + p_{model}(x)\big) sp(w^T x + b)\, dx \\
&= \mu_X - \mathbb{E}_{x \sim p_m}[sp(w^T x + b)],
\end{aligned}$$

where $p_m(x) = \frac{1}{2}\big(p_{data}(x) + p_{model}(x)\big)$.

The generator's payoff is $-V(G, D)$ (and hence the name of a zero-sum game), namely the generator intends to maximize $-V(G, D)$, or equivalently, to minimize $V(G, D)$.

Since each network attempts to maximize its own payoff, this can be set now as the following minimax problem:

$$G^* = \arg\min_G \max_D V(G, D),$$

where G^* denotes the optimum generator.

For the time being we shall consider the generator fixed and work on the inner maximum loop. For a given generator, G, the optimum discriminator function, D_G^* is given by

$$D_G^* = \arg\max_D V(G, D). \tag{19.4.5}$$

The next result provides the value of this discriminator.

Proposition 19.4.3 *For a fixed generator, G, the optimal discriminator (19.4.5) is given by*

$$D_G^*(x) = \frac{p_{data}(x)}{p_{data}(x) + p_{model}(x)}. \tag{19.4.6}$$

Proof: Considering distributions p_{data} and p_{model} defined on the space $\mathcal{X}$, the payoff function can be written as the following integral:

$$V(G, D) = \int_{\mathcal{X}} \Big[p_{data}(x) \ln D(x) + p_{model}(x) \ln(1 - D(x)) \Big]\, dx.$$

Since G is fixed, we consider the payoff as a functional of $D(x)$ as

$$F(D) = \int_{\mathcal{X}} L(D(x))\, dx,$$

with the Lagrangian function

$$L(G) = p_{data}(x) \ln D(x) + p_{model}(x) \ln(1 - D(x)).$$

Critical points satisfy the variational equation

$$\frac{dL(D)}{dD} = 0.$$

This becomes

$$\frac{p_{data}(x)}{D(x)} - \frac{p_{model}(x)}{1 - D(x)} = 0.$$

Solving for D we obtain the solution

$$D_G^*(x) = \frac{p_{data}(x)}{p_{data}(x) + p_{model}(x)}.$$

In order to show that this critical point corresponds to a maximum, we consider the second variation and show that it has a negative value

$$\frac{d^2 L(D)}{dD^2} = -\frac{p_{data}(x)}{D(x)^2} - \frac{p_{model}(x)}{(1 - D(x))^2}.$$

Hence, $D_G^*(x)$ corresponds to a maximum of $V(G, D)$ for a given G. ∎

Remark 19.4.4 (i) We note that the optimum $D_G^*(x)$ depends on G through the density $p_{model}(x)$, which is the density of $G(z)$, with $z \sim p_{code}$. (ii) This is a theoretical result of existence and uniqueness of the optimum. However, the result is not practical if p_{data} is not given.

The generator optimum, G^*, is now given by

$$G^* = \arg\min_G V(G, D_G^*).$$

In order to find it, we first need to evaluate the maximum value of the payoff $V(G, D)$ at its optimum point.

Proposition 19.4.5 *If D_{JS} denotes the Jensen-Shannon divergence given by (3.7.4), then the maximum value of the payoff function is*

$$V(G, D_G^*) = 2D_{JS}(p_{model}(x) \| p_{data}(x)) - \ln 4. \qquad (19.4.7)$$

Substituting the value given by formula (19.4.6) into the payoff we obtain

$$\begin{aligned}
V(G, D_G^*) &= \int_{\mathcal{X}} \left[p_{data}(x) \ln D_G^*(x) + p_{model}(x) \ln(1 - D_G^*(x)) \right] dx \\
&= \int_{\mathcal{X}} \Big[p_{data}(x) \ln \frac{p_{data}(x)}{p_{data}(x) + p_{model}(x)} \\
&\quad + p_{model}(x) \ln \frac{p_{model}(x)}{p_{data}(x) + p_{model}(x)} \Big] dx.
\end{aligned}$$

In the following, we shall add and then subtract $\ln 2\, p_{data}(x)$ and $\ln 2\, p_{model}(x)$ and use some algebraic manipulations to split it into three integrals

$$\begin{aligned}
V(G, D_G^*) &= \int_{\mathcal{X}} \Big[(\ln 2 - \ln 2) p_{data}(x) + p_{data}(x) \ln \frac{p_{data}(x)}{p_{data}(x) + p_{model}(x)} \\
&\quad + (\ln 2 - \ln 2) p_{model}(x) + p_{model}(x) \ln \frac{p_{model}(x)}{p_{data}(x) + p_{model}(x)} \Big] dx \\
&= I_1 + I_2 + I_3,
\end{aligned}$$

where

$$I_1 = -\int_{\mathcal{X}} \ln 2 \big(p_{data}(x) + p_{model}(x)\big)\, dx$$

$$I_2 = \int_{\mathcal{X}} p_{data}(x)\left[\ln 2 + \ln \frac{p_{model}(x)}{p_{data}(x) + p_{model}(x)}\right] dx$$

$$I_3 = \int_{\mathcal{X}} p_{model}(x)\left[\ln 2 + \ln \frac{p_{model}(x)}{p_{data}(x) + p_{model}(x)}\right] dx.$$

In the following we evaluate each integral. Using that p_{data} and p_{model} are probability density functions, we have

$$I_1 = -\ln 2 \Big[\underbrace{\int_{\mathcal{X}} p_{data}(x)\, dx}_{=1} + \underbrace{\int_{\mathcal{X}} p_{model}(x)\, dx}_{=1}\Big] = -2\ln 2 = -\ln 4.$$

Using properties of logarithmic function, we evaluate the second integral as a Kullback-Leibler divergence

$$\begin{aligned}
I_2 &= \int_{\mathcal{X}} p_{data}(x) \ln \frac{2 p_{model}(x)}{p_{data}(x) + p_{model}(x)}\, dx \\
&= \int_{\mathcal{X}} p_{data}(x) \ln \frac{p_{model}(x)}{\frac{1}{2}(p_{data}(x) + p_{model}(x))}\, dx \\
&= D_{KL}\Big(p_{data}(x) \,\|\, \frac{p_{data}(x) + p_{model}(x)}{2}\Big).
\end{aligned}$$

The third integral can be evaluated in a similar way

$$\begin{aligned}
I_3 &= \int_{\mathcal{X}} p_{model}(x) \ln \frac{2 p_{model}(x)}{p_{data}(x) + p_{model}(x)}\, dx \\
&= \int_{\mathcal{X}} p_{model}(x) \ln \frac{p_{model}(x)}{\frac{1}{2}(p_{data}(x) + p_{model}(x))}\, dx \\
&= D_{KL}\Big(p_{model}(x) \,\|\, \frac{p_{data}(x) + p_{model}(x)}{2}\Big).
\end{aligned}$$

Using the definition of the Jensen-Shannon divergence (3.7.4), we obtain

$$\begin{aligned}
V(G, D_G^*) &= I_1 + I_2 + I_3 \\
&= -\ln 4 + D_{KL}\Big(p_{data}(x) \,\|\, \frac{p_{data}(x) + p_{model}(x)}{2}\Big) \\
&\quad + D_{KL}\Big(p_{model}(x) \,\|\, \frac{p_{data}(x) + p_{model}(x)}{2}\Big) \\
&= -\ln 4 + 2 D_{JS}(p_{model}(x) \| p_{data}(x)).
\end{aligned}$$

The next result shows an expected fact of GANs, namely, that at the equilibrium convergence the model distribution approximates the data distribution (Goodfellow et al. [47]).

Proposition 19.4.6 *The global minimum*

$$G^* = \arg \min_G V(G, D_G^*)$$

is achieved if and only if $p_{model} = p_{data}$. At this minimum point the value is $-\ln 4$.

Proof: Using relation (19.4.7) we have

$$
\begin{aligned}
G^* &= \arg \min_G V(G, D_G^*) = \arg \min_G \left(2D_{JS}(p_{model}(x)\|p_{data}(x)) - \ln 4 \right) \\
&= \arg \min_G \left(2D_{JS}(p_{model}(x)\|p_{data}(x)) \right).
\end{aligned}
$$

By Proposition 3.7.1 the right-side expression reaches the global minimum value of zero if and only if $p_{model}(x) = p_{data}(x)$. In this case we obviously have $\min_G V(G, D_G^*) = V(G^*, D_G^*) = -\ln 4$. ∎

The discriminant output at equilibrium is obtained substituting $p_{model} = p_{data}$ into (19.4.6)

$$D^*(x) = \frac{p_{data}(x)}{p_{data}(x) + p_{model}(x)} = \frac{1}{2}.$$

The fact that the discriminator's output values are 0.5 means that it cannot distinguish whether the generated data x are fake or true. At the equilibrium the maximum value of the payoff function is

$$
\begin{aligned}
V(G^*, D^*) &= \int_{\mathcal{X}} \left[p_{data}(x) \ln D^*(x) + p_{model}(x) \ln(1 - D^*(x)) \right] dx \\
&= \int_{\mathcal{X}} \left[p_{data}(x) \ln \frac{1}{2} + p_{model}(x) \ln \frac{1}{2} \right] dx \\
&= \ln \frac{1}{2} \int_{\mathcal{X}} \left[p_{data}(x) + p_{model}(x) \right] dx \\
&= 2 \ln \frac{1}{2} = -\ln 4,
\end{aligned}
$$

which recovers the result of the previous proposition.

In general, the dimension of the latent space $\mathcal{Z}$ is smaller than the dimension of the data space $\mathcal{X}$. In the case when they are equal and the generator transformation $G : \mathcal{Z} \to \mathcal{X}$ is differentiable and invertible, the model and code distributions are related by

$$p_{model}(x) = p_{code}(G^{-1}(x)) \det J_{G^{-1}(x)}.$$

Moreover, the second term of the payoff $V(G, D)$ can be written as an expectation over the coding density as

$$\mathbb{E}_{x \sim p_{model}}[\ln(1 - D(x))] = \mathbb{E}_{z \sim p_{code}}[\ln(1 - D(G(z)))].$$

The computation involves the change of variables $z = G^{-1}(x)$ as in the following:

$$
\begin{aligned}
\mathbb{E}_{x \sim p_{model}}[\ln(1 - D(x))] &= \int_{\mathcal{X}} \ln(1 - D(x))\, p_{model}(x)\, dx \\
&= \int_{\mathcal{X}} \ln(1 - D(x))\, p_{code}(G^{-1}(x)) \det J_{G^{-1}(x)}\, dx \\
&= \int_{\mathcal{Z}} \ln(1 - D(G(z)))\, p_{code}(z)\, dz \\
&= \mathbb{E}_{z \sim p_{code}}[\ln(1 - D(G(z)))].
\end{aligned}
$$

Remark 19.4.7 The minimax problem $\min_{G} \max_{D} V(G, D)$ is not the same as $\max_{D} \min_{G} V(G, D)$. The former is the correct version, while the latter does not work well since the generator might place all mass on most likely points, fulling the discriminator. This leads to a mode collapse, which is a setting where the GAN produces the same output.

Training procedure A GAN is trained by an application of a simultaneous stochastic gradient descent method. We repeatedly sample two minibatches of data, one from the training set and another from generated samples. Then we run the gradient descend on both players crossed simultaneously. This will lead to one update for each player in the direction, or the opposite direction of the gradient as follows:

$$
\begin{aligned}
\theta_{n+1}^{(d)} &= \theta_n^{(d)} - \eta \nabla_{\theta^{(d)}} V(\theta_n^{(d)}) \\
\theta_{n+1}^{(g)} &= \theta_n^{(g)} + \eta \nabla_{\theta^{(g)}} V(\theta_n^{(g)}),
\end{aligned}
$$

where $\eta > 0$ is the learning rate and V is the payoff defined by (19.4.4). The gradients can be computed as in the following. Using that $D(x) = D(x, \theta^{(d)})$ and that p_{model} and p_{data} are independent of $\theta^{(d)}$, we have

$$
\begin{aligned}
\nabla_{\theta^{(d)}} V &= \nabla_{\theta^{(d)}} \mathbb{E}_{x \sim p_{data}}[\ln D(x, \theta^{(d)})] + \nabla_{\theta^{(d)}} \mathbb{E}_{x \sim p_{model}}[\ln(1 - D(x, \theta^{(d)}))] \\
&= \mathbb{E}_{x \sim p_{data}}\left[\frac{1}{D(x)} \frac{\partial D(x)}{\partial \theta^{(d)}}\right] - \mathbb{E}_{x \sim p_{model}}\left[\frac{1}{1 - D(x)} \frac{\partial D(x)}{\partial \theta^{(d)}}\right].
\end{aligned}
$$

If $D(x) = \sigma(a(x, \theta^{(d)})$, where σ is the logistic sigmoid, then using

$$\frac{\partial D(x)}{\partial \theta^{(d)}} = \sigma'(a(x, \theta^{(d)})) \frac{\partial a(x)}{\partial \theta^{(d)}} = D(x)(1 - D(x)) \frac{\partial a(x)}{\partial \theta^{(d)}},$$

the gradient computation can be continued as

$$
\begin{aligned}
\nabla_{\theta^{(d)}} V &= \mathbb{E}_{x \sim p_{data}}\left[(1 - D(x))\frac{\partial a(x)}{\partial \theta^{(d)}}\right] - \mathbb{E}_{x \sim p_{data}}\left[D(x)\frac{\partial a(x)}{\partial \theta^{(d)}}\right] \\
&= \frac{\partial}{\partial \theta^{(d)}}\mathbb{E}_{x \sim p_{data}}[a(x)] - 2\mathbb{E}_{x \sim \frac{1}{2}(p_{data}+p_{model})}\left[D(x)\frac{\partial a(x)}{\partial \theta^{(d)}}\right].
\end{aligned}
$$

It is worth noting that the first expectation at the equilibrium point is equal to the Kullback-Leibler divergence $D_{KL}(p_{data}\|p_{model})$, see Exercise 19.7.5. We have assumed the probability densities p_{data} and p_{model} decrease to zero at infinity fast enough such that the derivation operator commutes with the expectation operators.

The gradient $\nabla_{\theta^{(g)}} V$ is computed similarly, see Exercise 19.7.7.

Even if GANs are considered the most successful generative models, there are still some unsolved problems, which are active areas of research nowadays. These include: vanishing gradients, mode collapse, solving the problems dealing with counting, perspective, and global structure.

19.5 Generative Moment Matching Networks

Generative moment matching networks have been introduced by Li et al. [76] in 2015. In this case the generator is trained by *moment matching*, a technique by which the generator's outcome has moments as close as possible to the corresponding moments of the training data.

This idea is based on the result stating that if two random variables have equal moments then their distributions are the same. If the random variables are of a particular type, less moments would be needed. In particular, two Gaussian random variables with the same mean and variance are the same.

This generative model is a generator network, such as a convolutional network, which is provided a sample z from a uniform (or gaussian) distribution and produces an output x, such as an image of a face or digit, for example. If the network parameters are denoted by θ, the outcome variable X can be written in terms of the input variable Z as $X = G(Z; \theta)$, with $Z \sim Unif[0, 1]$, for example. The density of X will be denoted by p_θ.

We assume the corresponding random variable describing the training data (such as a set of human faces or MNIST data digits) is denoted by Y and has density q. The network parameter θ has to be tuned such that the distributions of X and Y are as close as possible, and this would be done my matching the first k moments. For this, we choose the function $\phi(x) = (x, x^2, \ldots, x^k)^T$ and consider the parameter θ that minimizes the maximum mean discrepancy defined by (3.8.5)

$$\theta^* = \arg\min_\theta d_{MMD}(p_\theta, q) = \arg\min_\theta \|\mu_\phi(G(Z;\theta)) - \mu_\phi(Y)\|_{Eu}.$$

In practice we pick n random inputs $\{z_1, \ldots, z_n\}$ that yield an output sample $\{x_1, \ldots, x_n\}$ and choose another random sample $\{y_1, \ldots, y_m\}$ from the training data. In order to match the moments of the distributions underlying these two samples we need to consider the parameter

$$\theta^* = \arg\min_\theta \left\{ \frac{1}{n^2} \sum_{i,j} K(x_i, x_j) + \frac{1}{m^2} \sum_{i,j} K(y_i, y_j) - \frac{2}{mn} \sum_{i,j} K(x_i, y_j) \right\},$$

where $K(x,y) = \phi(x)^T \phi(y)$ and we used formula (3.8.6). The generator network is trained using the backpropagation method.

19.6 Summary

We have shown how generative networks can be used either to produce samples from a given probability distribution, or to generate examples that resemble data in the training set. The idea of a generative model is to provide a random seed to the model and to obtain an outcome that resembles the training data or a sample from the data distribution.

A central role was dedicated to GANs, a generative model whose architecture consists of two networks, a generator and a discriminator network, which play a competitive game. This noncooperative game is usually expressed as the relation between a painting counterfeiter (the generator) and the expert investigator (the discriminator). The counterfeiter tries to full the expert, while the expert intends to identify the forgery as fake. The expert trains on fake and real paintings. In the long run they force each other to improve: the counterfeiter does such a good job such that the expert gets confused about whether a painting is fake or genuine. In the most common setup the game is a zero-sum game, or a minimax problem. The equilibrium point is a saddle point of the payoff.

Another model discussed is the generative moment matching network. This consists of a generator network whose parameters are tuned such that the outcome moments match the corresponding moments of the training distribution.

19.7 Exercises

Exercise 19.7.1 We apply the simultaneous gradient descent method for the minimax problem with payoff $V(x,y) = xy$ in the discrete case when the learning rate is $\eta > 0$.

(i) Write the system update.

(ii) Write the system state (x_n, y_n) in terms of (x_0, y_0).

(iii) Is the sequence (x_n, y_n) convergent?

Exercise 19.7.2 Consider a competitive game between two players, who fight over the linear payoff function $V(x, y) = ax + by$, with $a^2 + b^2 \neq 0$. The first player controls variable x and wants to minimize $V(x, y)$, while the second player controls variable y and wants to maximize the payoff function.

(i) Write the differential system of the learning dynamics given by the simultaneous gradient decent with infinitesimal learning rate and solve for the learning trajectory.

(ii) Show that there is no equilibrium solution.

Exercise 19.7.3 Consider a competitive game between two players with the payoff function $V(x, y) = \frac{1}{2}(x^2 - y^2)$. The player who controls the x variable intends to minimize the payoff, while the player who controls the y variables wants to maximize it. Show that this game has a unique equilibrium solution. Consider both continuous and discrete learning.

Exercise 19.7.4 We recall that the expansion or contraction of a vector field $U = (U_1, U_2)$ in the plane is given by its divergence $\text{div} U = \partial_x U_1 + \partial_y U_2$. The velocity vector field along the learning flow $(x(t), y(t))$ is defined by $U(x, y) = (\dot{x}(t), \dot{y}(t))$.

(i) Show that the velocity vector of the learning flow associated with the payoff function $V(x, t) = x^2 + y^2$ has a zero divergence.

(ii) Prove that the velocity vector of the learning flow has a zero divergence if and only if the associated payoff function is of the form $V(x, t) = F(x + y) + G(x - y)$, with F and G two arbitrary twice differentiable functions.

(iii) Show that if the divergence of the learning flow is zero, then there are no equilibrium points along that flow.

Exercise 19.7.5 (i) Consider a GAN with a fixed generator, G. Show that the optimum discriminator satisfies $D_G^*(x) = \sigma(a(x))$, where

$$e^{a(x)} = \frac{p_{data}(x)}{p_{model}(x)}$$

and σ is the logistic function.

(ii) With the $a(x)$ given by (i), show that $\mathbb{E}_{x \sim p_{data}}[a(x)] = D_{KL}(p_{data} \| p_{model})$.

Exercise 19.7.6 Consider a GAN with the model distribution $p_{model}(x, \theta^{(g)})$. We assume the parameter vector $\theta^{(g)}$ is obtained by the maximum likelihood estimation, namely

$$\theta^{(g)} = \theta_{MLE} = \arg\max_\theta \mathbb{E}_{x \sim p_{data}} [\ln p_{model}(x; \theta)].$$

Find a function $f(x)$ such that $\theta^* = \theta_{MLE}$, where

$$\theta^* = \arg\max_\theta \mathbb{E}_{x \sim p_{model}} [f(x)].$$

Exercise 19.7.7 Consider the GAN payoff given by (19.4.4).

(i) Show that

$$\nabla_{\theta^{(g)}} V = \int_{\mathcal{X}} \ln(1 - D(x)) \nabla_{\theta^{(g)}} p_{model}(x; \theta^{(g)}) \, dx.$$

(ii) Assume the generator function $G : \mathcal{Z} \to \mathcal{X}$, $x = G(z, \theta^{(g)})$ is invertible and differentiable. Show that

$$\nabla_{\theta^{(g)}} V = -\mathbb{E}_{z \sim p_{code}} \left[\frac{1}{1 - D(G(z))} \frac{\partial D(x)}{\partial x} \bigg|_{x=G(z)} \frac{\partial G(z)}{\partial \theta^{(g)}} \right].$$

(iii) Assume $D(x) = \sigma(a(x))$, where σ is the logistic function. Show that the gradient formula in part (ii) becomes

$$\nabla_{\theta^{(g)}} V = -\mathbb{E}_{z \sim p_{code}} \left[D(G(z)) \frac{da(x)}{dx} \bigg|_{x=G(z)} \frac{\partial G(z)}{\partial \theta^{(g)}} \right].$$

Exercise 19.7.8 Consider a zero-sum game where one player controls the variable y and wants to minimize the cross-entropy payoff

$$V(y, \hat{y}) = y \ln \hat{y} + (1 - y) \ln(1 - \hat{y}),$$

while the other player controls the variable $\hat{y}$ and intends to minimize the payoff $-V(y, \hat{y})$. Both variables take values in $(0, 1)$. Find the equilibrium point of the game.

Exercise 19.7.9 In order to avoid the vanishing gradient problem the following choice of the generator payoff has been used:

$$J^{(G)} = \mathbb{E}_{z \sim p_{code}} [\ln D(G(z, \theta^{(g)}))].$$

In this case the generator maximizes the log probability of the discriminator being mistaken. Find the gradient $\nabla_{\theta^{(g)}} J^{(G)}$. Assume the discriminator is optimal.

Chapter 20

Stochastic Networks

This chapter deals with stochastic networks, such as Hopfield networks and Boltzmann machines. A Hopfield network is an ensemble of perceptrons which interact with each other, until a certain cost function which is defined in terms of the interaction between the neurons (called energy function) is minimized. They are useful in solving combinatorial optimization problems. A Boltzmann machine is like a Hopfield network for which the perceptrons have been replaced by binary stochastic neurons. This way, a Boltzmann machine can be seen as the noisy version of a Hopfield network. They are used to avoid Hopfield networks to get stuck in local minima of the energy function. We also present the equilibrium distribution of a Boltzmann machine, its entropy, and its associated Fisher information metric.

20.1 Stochastic Neurons

We have seen that a deterministic perceptron is a computing unit, which accepts n inputs, $x_1, \ldots, x_n$, and provides the one-dimensional output

$$y = H(w^T \mathbf{x} + b) = H\Big(\sum_{i=1}^{n} w_i x_i + b\Big) = \begin{cases} 1, & \text{if } w^T \mathbf{x} + b > 0 \\ 0, & \text{otherwise.} \end{cases}$$

A *binary stochastic neuron* is a unit which accepts n inputs, $x_1, \ldots, x_n$, and provides the output Y, given by the binary random variable $Y \in \{0, 1\}$, having the distribution

$$
\begin{aligned}
P(Y = 1|\mathbf{x}) &= \sigma(w^T \mathbf{x} + b), \\
P(Y = 0|\mathbf{x}) &= 1 - \sigma(w^T \mathbf{x} + b) = \sigma(-w^T \mathbf{x} - b),
\end{aligned}
$$

© Springer Nature Switzerland AG 2020
O. Calin, *Deep Learning Architectures*, Springer Series in the Data Sciences,
https://doi.org/10.1007/978-3-030-36721-3_20

where σ is the logistic function. The conditional by $\mathbf{x}$ can be omitted if the input is deterministic.

When the input satisfies $w^T\mathbf{x} + b > 0$, the deterministic perceptron provides the value 1 (with probability 1), while the stochastic neuron provides the value 1 with probability $\sigma(w^T\mathbf{x} + b)$.

Fisher metric The information of a stochastic neuron about its weights is assessed by its Fisher matrix. This will be explicitly computed in the following. For any $1 \leq i, j \leq n$, we have

$$
\begin{aligned}
g_{ij} &= \mathbb{E}[\partial_{w_i}\ell \, \partial_{w_j}\ell] \\
&= P(Y = 1|\mathbf{x})\partial_{w_i}\ln P(Y = 1|\mathbf{x})\partial_{w_j}\ln P(Y = 1|\mathbf{x}) \\
&\quad + P(Y = 0|\mathbf{x})\partial_{w_i}\ln P(Y = 0|\mathbf{x})\partial_{w_j}\ln P(Y = 0|\mathbf{x}) \\
&= \sigma(w^T\mathbf{x} + b)\partial_{w_i}\ln\sigma(w^T\mathbf{x} + b)\partial_{w_j}\ln\sigma(w^T\mathbf{x} + b) \\
&\quad + (1 - \sigma(w^T\mathbf{x} + b))\partial_{w_i}\ln(1 - \sigma(w^T\mathbf{x} + b))\partial_{w_j}\ln(1 - \sigma(w^T\mathbf{x} + b)).
\end{aligned}
$$

Using the differentiation rule $(\ln f(x))' = f'(x)/f(x)$ and the sigmoid property $\sigma' = \sigma(1 - \sigma)$, we obtain

$$
\begin{aligned}
g_{ij} &= \sigma(w^T\mathbf{x} + b)(1 - \sigma(w^T\mathbf{x} + b))^2 x_i x_j + \sigma(w^T\mathbf{x} + b)^2(1 - \sigma(w^T\mathbf{x} + b))x_i x_j \\
&= \sigma(w^T\mathbf{x} + b)(1 - \sigma(w^T\mathbf{x} + b))x_i x_j.
\end{aligned}
$$

The formula can be extended if write $\widetilde{\mathbf{x}} = (1, \mathbf{x})^T$ and $\widetilde{w} = (b, w)$ as

$$
\widetilde{g}_{ij} = \sigma(w^T\mathbf{x} + b)(1 - \sigma(w^T\mathbf{x} + b))\widetilde{x}_i\widetilde{x}_j,
$$

or, in matrix form, as $\widetilde{g} = \sigma(w^T\mathbf{x} + b)(1 - \sigma(w^T\mathbf{x} + b))\widetilde{\mathbf{x}}\widetilde{\mathbf{x}}^T$. Since any two columns of the matrix $\mathbf{x}\mathbf{x}^T$ are proportional, we have $\det\widetilde{g} = 0$. Also, Trace $\widetilde{g} = (1 + \|x\|_2^2)\sigma(w^T\mathbf{x} + b)(1 - \sigma(w^T\mathbf{x} + b)) > 0$.

The information density in direction $\widetilde{w} = (b, w)$ is given by

$$
\begin{aligned}
\widetilde{w}^T\widetilde{g}\widetilde{w} &= \sigma(w^T\mathbf{x} + b)(1 - \sigma(w^T\mathbf{x} + b))\widetilde{w}^T\widetilde{\mathbf{x}}\widetilde{\mathbf{x}}^T\widetilde{w} \\
&= \sigma(w^T\mathbf{x} + b)(1 - \sigma(w^T\mathbf{x} + b))(\widetilde{\mathbf{x}}^T\widetilde{w})^2 \\
&= (w^T\mathbf{x} + b)^2\sigma(w^T\mathbf{x} + b)(1 - \sigma(w^T\mathbf{x} + b)).
\end{aligned}
$$

This implies a zero density along the hyperplane $\{\mathbf{x} \in \mathbb{R}^n; w^T\mathbf{x} + b = 0\}$. Also, the information density tends to zero when the sigmoid saturates, i.e., for $w^T\mathbf{x} + b \to \pm\infty$.

Maximum likelihood A stochastic neuron learns by maximizing the likelihood. For this, we consider the random variable $U = 2Y - 1$. Then $U = -1$ for $Y = 0$ and $U = 1$ for $Y = 1$. Moreover,

$$
\begin{aligned}
P(U = 1) &= P(Y = 1) = \sigma(w^T\mathbf{x} + b) = \sigma((w^T\mathbf{x} + b)U) \\
P(U = -1) &= P(Y = 0) = 1 - \sigma(w^T\mathbf{x} + b) = \sigma(-w^T\mathbf{x} - b) = \sigma((w^T\mathbf{x} + b)U).
\end{aligned}
$$

The advantage of using the new variable U is that the probability has the same expression in both cases. We transform the training set $\{(\mathbf{x}_1, z_1), \ldots, (\mathbf{x}_n, z_n)\}$ into $\{(\mathbf{x}_1, t_1), \ldots, (\mathbf{x}_n, t_n)\}$, where $t_j = 2z_j - 1$. The optimal parameters are given by

$$
\begin{aligned}
(w^*, b^*) &= \arg\max_{w,b} \prod_{j=1}^{n} \sigma((w^T \mathbf{x}_j + b)t_j) = \arg\max_{w,b} \ln \prod_{j=1}^{n} \sigma((w^T \mathbf{x}_j + b)t_j) \\
&= \arg\max_{w,b} \frac{1}{n} \sum_{j=1}^{n} \ln \sigma((w^T \mathbf{x}_j + b)t_j).
\end{aligned}
$$

The cost function

$$
C(w, b) = \frac{1}{n} \sum_{j=1}^{n} \ln \sigma((w^T \mathbf{x}_j + b)t_j)
$$

can be maximized by the *gradient ascent* method. This means taking a step of size η in the gradient direction at each parameter update

$$
\begin{aligned}
w^{(m+1)} &= w^{(m)} + \eta \nabla_w C(w, b) \\
b^{(m+1)} &= b^{(m)} + \eta \nabla_w C(w, b),
\end{aligned}
$$

with the gradient computed using the chain rule and properties of the sigmoid:

$$
\begin{aligned}
\frac{\partial C}{\partial w_k} &= \frac{1}{n} \sum_{j=1}^{n} \frac{x_j^k t_j \sigma'((w^T \mathbf{x}_j + b)t_j)}{\sigma((w^T \mathbf{x}_j + b)t_j)} = \frac{1}{n} \sum_{j=1}^{n} x_j^k t_j \sigma(-(w^T \mathbf{x}_j + b)t_j) \\
\frac{\partial C}{\partial b} &= \frac{1}{n} \sum_{j=1}^{n} t_j \sigma(-(w^T \mathbf{x}_j + b)t_j).
\end{aligned}
$$

Simulated annealing method A binary stochastic neuron converges in a certain sense to a deterministic perceptron. First, we make the following modification in the outcome distribution of the stochastic neuron

$$
\begin{aligned}
P(Y = 1 | \mathbf{x}) &= \sigma_c(w^T \mathbf{x} + b), \\
P(Y = 0 | \mathbf{x}) &= 1 - \sigma_c(w^T \mathbf{x} + b) = \sigma_c(-w^T \mathbf{x} - b),
\end{aligned}
$$

where $\sigma_c(x) = \sigma(cx) = \dfrac{1}{1 + e^{-cx}}$, with $c \geq 0$. The associated cost function is

$$
C(w, b) = \frac{1}{n} \sum_{j=1}^{n} \ln \sigma_c((w^T \mathbf{x}_j + b)t_j).
$$

For $c \to 0$ we obtain equiprobable states, i.e.,

$$P(Y = 1|\mathbf{x}) = \frac{1}{2}, \quad P(Y = 0|\mathbf{x}) = \frac{1}{2},$$

regardless of the input value, $\mathbf{x}$. This corresponds to the maximal noisy case.

When c increases unboundedly, the function $\sigma_c(x)$ tends almost everywhere (but the origin) to the Heaviside step function $H(x)$. For $c \to \infty$, we obtain the distribution

$$
\begin{aligned}
P(Y = 1|\mathbf{x}) &= H(w^T\mathbf{x} + b) \\
P(Y = 0|\mathbf{x}) &= 1 - H(w^T\mathbf{x} + b).
\end{aligned}
$$

This is equivalent to the fact that $Y = 1$ if $w^T\mathbf{x}+b > 0$ and $Y = 0$, otherwise which corresponds to a deterministic perceptron.

We note that from the annealing point of view, the constant c is regarded as the inverse of temperature. Increasing the value of c means to decrease the temperature in order to obtain a global minimum of the perceptron cost function.

20.2   Boltzmann Distribution

The thermodynamic equilibrium of a system of particles at a given temperature is described by a probability distribution, called the *Boltzmann distribution*. Given its importance in the study of stochastic neural networks, we shall provide a detailed presentation of this distribution in the following.

Consider a thermodynamic system with N states and a random variable $\mathbf{x}$ that describes the state of the system. The possible values of the state variable, $\mathbf{x}$, are $x_1, \ldots, x_N$, which are taken with probabilities $p_j = P(\mathbf{x} = x_j)$. Furthermore, we assume that each state, x_j, is associated with an energy level of the system, E_j, which is a positive real number. Equivalently stated, the system takes the state of energy E_j with probability p_j.

The state of the system changes due to particle interactions. By the second law of Thermodynamics[1] the change of the states should be such that the total entropy of the system increases. In the absence of any other constraints, the system tends in the long run to the uniform distribution, which is known to be the distribution with the largest entropy. This is realized for the uniform distribution $p_1 = \cdots = p_N = \frac{1}{N}$.

---

[1]This states that the entropy of an isolated thermodynamic system increases.

However, the particle interactions are assumed to take place at a given prescribed temperature, T. The temperature is proportional to the average energy of the system, i.e.,

$$T \sim \sum_{j=1}^{N} p_j E_j.$$

Therefore, it suffices to search for the distribution of maximum entropy, which is subject to the constraint $\sum_{j=1}^{N} p_j E_j = k$, with k positive constant. This corresponds to the state of the system with a maximum uncertainty at a given temperature k.

In order to find the distribution with the largest entropy, $H(p) = -\sum_{j=1}^{N} p_j \ln p_j$, subject to constraints

$$\sum_{j=1} p_j = 1 \tag{20.2.1}$$

$$\sum_{j=1} p_j E_j = k, \tag{20.2.2}$$

we construct the following function:

$$F(p_1, \ldots, p_N, \lambda_1, \lambda_2) = -\sum_{j=1}^{N} p_j \ln p_j + \lambda_1 \Big(\sum_{j=1} p_j E_j - k \Big) + \lambda_2 \Big(\sum_{j=1} p_j - 1 \Big),$$

where $\lambda_1, \lambda_2 \in \mathbb{R}$ are Lagrange multipliers. Since the probability vector belongs to a compact set, $(p_1, \ldots, p_n) \in [0, 1] \times \cdots \times [0, 1]$, and F is a continuous function, a well-known result states the existence of the maximum of F on the aforementioned hypercube.[2] Since each state is taken with a positive probability, it makes sense to search for the maxima in the interior of the domain. Therefore, we can use the variational equations

$$\frac{\partial F}{\partial p_j} = -(1 + \ln p_j) + \lambda_1 E_j + \lambda_2 = 0, \qquad 1 \le j \le N,$$

which imply $p_j = c e^{\lambda_1 E_j}$, with $c = e^{\lambda_2 - 1}$. This is a truly maximum since the Hessian matrix

$$\frac{\partial^2 F}{\partial p_i \partial p_j} = \begin{pmatrix} -\frac{1}{p_1} & 0 & 0 \\ \vdots & \ddots & \vdots \\ 0 & 0 & -\frac{1}{p_N} \end{pmatrix}$$

---

[2] The theorem states that a continuous function defined on a compact set is bounded and reaches its minima and maxima.

is negative definite. To show the uniqueness of this distribution it suffices to state the uniqueness of the Lagrange multipliers λ_1, λ_2. We shall show there is only one pair of numbers (λ_1, c) that satisfy the system of constraints (20.2.1)–(20.2.2). This can be written equivalently as the following nonlinear system:

$$
\begin{aligned}
G_1(c, \lambda_1) &= 1 \\
G_2(c, \lambda_1) &= k
\end{aligned}
$$

where $G_1(c, \lambda_1) = \sum_{j=1}^{N} c e^{\lambda_1 E_j}$ and $G_2(c, \lambda_1) = \sum_{j=1}^{N} c e^{\lambda_1 E_j} E_j$. By the Inverse Functions Theorem, see Theorem F.1 in Appendix, the system has a unique solution provided its Jacobian is nonzero. The Jacobian can be evaluated as

$$
\begin{aligned}
J_G &= \begin{vmatrix} \frac{\partial G_1}{\partial c} & \frac{\partial G_1}{\partial \lambda_1} \\ \frac{\partial G_2}{\partial c} & \frac{\partial G_2}{\partial \lambda_1} \end{vmatrix} = \begin{vmatrix} \sum e^{\lambda_1 E_j} & \sum c e^{\lambda_1 E_j} E_j \\ \sum e^{\lambda_1 E_j} E_j & \sum c e^{\lambda_1 E_j} E_j^2 \end{vmatrix} = \begin{vmatrix} \frac{1}{c} & k \\ \frac{k}{c} & \sum c e^{\lambda_1 E_j} E_j^2 \end{vmatrix} \\
&= \frac{1}{c} \Big[\sum_{j=1}^{N} p_j E_j^2 - \Big(\sum_{j=1}^{N} p_j E_j \Big)^2 \Big] > 0,
\end{aligned}
$$

which is a consequence of Cauchy's inequality

$$
\Big(\sum_{j=1}^{N} p_j E_j \Big)^2 = \Big(\sum_{j=1}^{N} \sqrt{p_j} (\sqrt{p_j} E_j) \Big)^2 < \sum_{j=1}^{N} p_j \sum_{j=1}^{N} p_j E_j^2 = \sum_{j=1}^{N} p_j E_j^2.
$$

The partition function If we define the *partition function* $Z = \sum_{j=1}^{N} e^{\lambda_1 E_j}$, the solution can be also written as

$$
p_j = \frac{e^{\lambda_1 E_j}}{Z}, \quad 1 \leq j \leq N. \tag{20.2.3}
$$

The partition function, Z, can be computed as in the following. Differentiating the function $Z(\lambda_1)$ with respect to λ_1, we obtain

$$
\frac{d}{d\lambda_1} Z(\lambda_1) = \sum_{j=1}^{N} e^{\lambda_1 E_j} E_j = \sum_{j=1}^{N} Z p_j E_k = kZ,
$$

where we used that $\sum_{j=1}^{N} p_j E_k = k$. Therefore, $Z(\lambda_1)$ is the solution of the differential equation

$$
\begin{aligned}
\frac{d}{d\lambda_1} Z(\lambda_1) &= kZ \\
Z(0) &= N,
\end{aligned}
$$

and hence, $Z = Ne^{k\lambda_1}$. Solving for $\lambda_1 = -\frac{\ln(N/Z)}{k}$ and substituting into (20.2.3) yields

$$p_j = \frac{e^{-\frac{\ln(N/Z)}{k} E_j}}{Z} = e^{-\frac{E_j}{T}},$$

where the temperature T is a physical measure proportional to the average energy of the system, k. In the literature this is called either the Boltzmann distribution[3] or the *Gibbs distribution* [72] and it was introduced by Boltzmann, [112]. This is the probability distribution that characterizes a system in thermal equilibrium with N different states and associated energies $E_1, E_2, \ldots, E_N$. The system takes states of lower energy with higher probability and the probability decreases exponentially with the increase in the energy levels.

20.3   Boltzmann Machines

A *Boltzmann machine* is a neural network made out of mutually connected binary stochastic neurons (see section 20.1) with symmetric weights, $w_{ij} = w_{ji}$, and $w_{ii} = 0$, which make stochastic decisions about whether to be on or off. A Boltzmann machine with n neurons defines a probability distribution over the state space $\mathcal{X} = \{0, 1\}^n$, which is parametrized by an energy function, which describes the interactions within the model. More precisely, we have:

Definition 20.3.1 *A Boltzmann machine is a set of n noisy neurons with states $x_1, \ldots, x_n$, which form a network with symmetric weights. The state of the ith neuron is updated stochastically according to the rule*

$$x_i = \begin{cases} 1, & \text{with probability } p_i; \\ 0, & \text{with probability } 1 - p_i, \end{cases}$$

where

$$p_i = \sigma_{1/T}\left(\sum_j w_{ji} x_j + b_i\right) = \frac{1}{1 + e^{-(\sum_j w_{ji} x_j + b_i)/T}}. \tag{20.3.4}$$

The constant $T > 0$ denotes the temperature and $w_{ij} = w_{ji}$ are the weights between the ith and jth neurons, with self-recurrent connection $w_{ii} = 0$. The constants b_j denote the bias of the jth neuron.

---

[3]The version used in Physics includes a Boltzmann constant, $\kappa > 0$, which we considered equal to 1. The distribution actually writes as $p(\mathbf{x}) = \frac{1}{Z} e^{-E(\mathbf{x})/(\kappa T)}$.

Moreover, the probability that the ith neuron remains inactive is

$$1 - p_i = \sigma_{1/T}\left(-\sum_j w_{ji}x_j + b_i\right) = \frac{1}{1 + e^{(\sum_j w_{ji}x_j + b_i)/T}},$$

where we used the complementarity property of the sigmoid function, $\sigma(x) + \sigma(-x) = 1$.

The *state of a Boltzmann machine* is described by the vector $\mathbf{x} = (x_1, \ldots, x_n)^T$. Since each neuron activates binary, $x_i \in \{0, 1\}$, it follows that there are $N = 2^n$ states that a Boltzmann machine can take.

Boltzmann machine as a thermodynamical system We shall introduce first an energy function depending on the network state. The *signal potential* of the ith neuron is defined by the action of all other neurons on itself, including its own bias, as $u_i = -\left(\frac{1}{2}\sum_j w_{ji}x_j + b_i\right)$. The minus sign is included for convenience (to end up with a distribution resembling Boltzmann's distribution) and the factor $\frac{1}{2}$ is included because half of the weight w_{ij} counts for the ith neuron and the other half counts for the jth neuron. The activation (either 0 or 1) of all other neurons contribute additively to the potential through the synaptic strength given by the weight w_{ij}. The energy of the ith neuron is the product between its activation, x_i, and the signal potential, $\epsilon_i = u_i x_i$. The total energy of the network state $\mathbf{x} = (x_1, \ldots, x_n)^T$ is the sum of all individual neuron energies in that state

$$\begin{aligned}
E(\mathbf{x}) &= \sum_i \epsilon_i = \sum_i u_i x_i \\
&= -\frac{1}{2}\sum_{i,j} w_{ji}x_j x_i - \sum_i b_i x_i = -\sum_{i<j} w_{ji}x_j x_i - \sum_k b_i x_k.
\end{aligned}$$

In matrix form we can write $E(\mathbf{x}) = -\frac{1}{2}\mathbf{x}^T w \mathbf{x} - \mathbf{x}^T b$. Hence, the associated energy function is quadratic with respect to the neuron activations x_i.

Consider now a thermodynamical system with $N = 2^n$ states, $\mathbf{x}_1, \ldots, \mathbf{x}_N$. Each state corresponds to an energy level $E_j = E(\mathbf{x}_j)$, where E is the quadratic energy function previously defined.

The Boltzmann machine is assumed to have a certain state at time zero. Through the stochastic update rule the entropy of the system increases until the thermodynamical equilibrium is achieved. The corresponding equilibrium distribution is the Boltzmann distribution $p_j = \frac{e^{-E_j/T}}{Z}$, $1 \leq j \leq n$, and this is achieved regardless of the initial state of the machine.

There are many ways one can endow a network with an energy function. However, the quadratic function introduced before seems to be the natural choice, since it is compatible with a transition probability from Thermodynamics, which states that the probability of transition from the state $\mathbf{x}_i$ with energy E_i to the state $\mathbf{x}_j$ with energy E_j is given by $p_{ij} = \frac{1}{1+e^{(E_j - E_i)/T}}$.

Assume the network updates from the state $\mathbf{x} = (x_1, \ldots, x_k, \ldots, x_n)$ with energy $E = E(\mathbf{x})$ to the state $\mathbf{x}' = (x_1, \ldots, x'_k, \ldots, x_n)$ with energy $E' = E(\mathbf{x}')$. By Exercise 20.10.1 the difference between the energy levels is

$$E' - E = -\Big(\sum_{i=1}^{n} w_{ki} x_i + b_k\Big)(x'_k - x_k).$$

Assume the state $x_k = 0$ updates to $x'_k = 1$. Then the difference becomes $E' - E = -\Big(\sum_{i=1}^{n} w_{ki} x_i + b_k\Big)$ and the update formula (20.3.4) can be written as

$$p = \frac{1}{1 + e^{-(\sum_{i=1}^{n} w_{ki} x_i + b_k)/T}} = \frac{1}{1 + e^{(E' - E)/T}},$$

which agrees with the aforementioned transition probability formula.

If the state $x_k = 1$ updates to $x'_k = 0$, then the difference becomes $E' - E = \sum_{i=1}^{n} w_{ki} x_i + b_k$ and we obtain the transition probability again

$$1 - p = \frac{1}{1 + e^{(\sum_{i=1}^{n} w_{ki} x_i + b_k)/T}} = \frac{1}{1 + e^{(E' - E)/T}}.$$

To facilitate understanding, we shall consider an example.

Example 20.3.2 Consider a network with three neurons, having symmetric connections and $w_{ii} = 0$. Their signal potentials are given by

$$\begin{aligned}
u_1 &= -(w_{12} x_2 + w_{13} x_3 + b_1) \\
u_2 &= -(w_{12} x_1 + w_{23} x_3 + b_2) \\
u_3 &= -(w_{13} x_1 + w_{23} x_2 + b_3).
\end{aligned}$$

Then the energy of the network is given by

$$E(\mathbf{x}) = -(w_{12} x_1 x_2 + w_{13} x_1 x_3 + w_{23} x_2 x_3 + b_1 x_1 + b_2 x_2 + b_3 x_3).$$

There are $N = 2^3 = 8$ states of the system

$$\mathcal{X} = \{(0,0,0), (1,0,0), (0,1,0), (0,0,1), (0,1,1), (1,0,1), (1,1,0), (1,1,1)\}$$

corresponding to the following energy levels:

$$\begin{aligned}
E_1 &= E(0,0,0) = 0 & E_5 &= E(0,1,1) = -w_{23} - b_2 - b_3 \\
E_2 &= E(1,0,0) = -b_1 & E_6 &= E(1,0,1) = -w_{13} - b_1 - b_3 \\
E_3 &= E(0,1,0) = -b_2 & E_7 &= E(1,1,0) = -w_{12} - b_1 - b_2 \\
E_4 &= E(0,0,1) = -b_3 & E_8 &= E(1,1,1) = -w_{12} - w_{13} - w_{23} - b_1 - b_2 - b_3.
\end{aligned}$$

The Boltzmann distribution, $p_j = \frac{e^{-E_j}}{Z}$, for the case $T = 1$, is given by

$$p = \frac{1}{Z}(1, e^{b_1}, e^{b_2}, e^{b_3}, e^{w_{23}+b_2+b_3}, e^{w_{13}+b_1+b_3}, e^{w_{12}+b_1+b_2}, e^{w_{12}+w_{13}+w_{23}+b_1+b_2+b_3}),$$

where Z is a normalization factor. We note that all energy levels (with the exception of the first one) depend on the network weights and biasses. Hence, any change in the network parameters leads to a change in the energy levels, and therefore to a change of the Boltzmann distribution.

One might be interested to learn an arbitrary distribution q on the states space $\mathcal{X}$ using a Boltzmann distribution by adjusting the network parameters. In the present case this cannot be done exactly, since we need to learn 7 unknowns, $q_1, \ldots, q_7$ (since q_8 depends on the others) using only 6 parameters, $b_1, b_2, b_3, w_{12}, w_{13}$, and w_{23}. The Boltzmann learning algorithm will provide a way of learning q in an approximate way by minimizing a Kullback-Leibler divergence.

However, there are cases when the learning is exact. We shall deal with this in the next example.

Example 20.3.3 (exact learning) Consider a Boltzmann machine with $n = 2$ neurons. The $N = 2^2$ states are given by

$$\mathcal{X} = \{(0,0), (1,0), (0,1), (1,1)\}.$$

The associated energy is

$$E(\mathbf{x}) = -w_{12}x_1x_2 - b_1x_1 - b_2x_2,$$

and the energy levels on each state are given by

$$E(0,0) = 0, \quad E(1,0) = -b_1, \quad E(0,1) = -b_2, \quad E(1,1) = -w_{12} - b_1 - b_2.$$

The thermal equilibrium at $T = 1$ is realized for the Boltzmann distribution

$$p = \frac{1}{Z}(1, e^{b_1}, e^{b_2}, e^{w_{12}+b_1+b_2}).$$

We consider now an arbitrary distribution q on $\mathcal{X}$

$$q_1 = q(0,0), \quad q_1 = q(1,0), \quad q_2 = q(0,1), \quad q_3 = q(1,1)$$

and we shall find the exact values of parameters w, b such that $p = q$, i.e., the Boltzmann machine learns q exactly. We identify parameters as

$$p_1 = \frac{1}{Z} = q_1 \quad \Rightarrow \quad Z = 1/q_1$$

$$p_2 = \frac{e^{b_1}}{Z} = q_2 \quad \Rightarrow \quad b_1 = \ln \frac{q_2}{q_1}$$

$$p_3 = \frac{e^{b_2}}{Z} = q_3 \quad \Rightarrow \quad b_2 = \ln \frac{q_3}{q_1}$$

$$p_4 = \frac{e^{w_{12}+b_1+b_2}}{Z} = q_4 \quad \Rightarrow \quad w_{12} = \ln \frac{q_1 q_4}{q_2 q_3}.$$

Given the distribution q, we can write an exact expression for the Boltzmann distribution, which learns exactly q, as

$$
\begin{aligned}
p(\mathbf{x}) &= \frac{e^{-E(\mathbf{x})}}{Z} = q_1 e^{w_{12}x_1 x_2 + b_1 x_1 + b_2 x_2} = q_1 (e^{w_{12}})^{x_1 x_2} (e^{b_1})^{x_1} (e^{b_2})^{x_2} \\
&= q_1 \left(\frac{q_1 q_4}{q_2 q_3}\right)^{x_1 x_2} \left(\frac{q_2}{q_1}\right)^{x_1} \left(\frac{q_3}{q_1}\right)^{x_2} \\
&= q_1^{(1-x_1)(1-x_2)} q_2^{x_1(1-x_2)} q_3^{x_2(1-x_1)} q_4^{x_1 x_2}.
\end{aligned}
$$

This exact learning of a distribution q works just for $n = 2$ neurons, since in this case the number of equations is equal to the number of variables.

In the general case the number of unknown variables in distribution p is $n(n+1)/2$ (we used symmetry conditions $w_{ij} = w_{ji}$, $w_{ii} = 0$, and also included b_k), while the number of equations is $2^n - 1$ (we subtracted 1 to account for the linear relationship $\sum q_i = 1$). For $n > 2$ we always have $n(n+1)/2 < 2^n - 1$, with equality only for the case $n = 2$.

20.4　Boltzmann Learning

Boltzmann machines can function as approximators of distributions defined on the state space $\mathcal{X}$, see [2]. In order for a Boltzmann machine to learn a given distribution q defined on the state space $\mathcal{X}$, we need to choose among all distributions p generated by the machine the distribution p^*, which is the closest to q in the sense of the Kullback-Leibler divergence, i.e.,

$$p^* = \arg\min_p D_{KL}(q\|p) = \arg\min_p \sum_{\mathbf{x}\in X} q(\mathbf{x}) \ln \frac{q(\mathbf{x})}{p(\mathbf{x})}.$$

Since the entropy $H(q) = \sum_{x\in\mathcal{X}} q(\mathbf{x}) \ln q(\mathbf{x})$ is independent of p, the previous search is equivalent to

$$p^* = \arg\max_p \sum_{\mathbf{x}\in X} q(\mathbf{x}) \ln p(\mathbf{x}).$$

The idea is that the Boltzmann distribution, $p(\mathbf{x}) = \frac{e^{-E(\mathbf{x})}}{Z(\mathbf{x})}$, changes when the machine parameters change. The parameters are updated following a gradient ascent method. Using chain rule, we find first the derivative of the log-likelihood function

$$
\begin{aligned}
\frac{\partial}{\partial w_{ij}} \ln p(\mathbf{x}) &= -\frac{\partial}{\partial w_{ij}} E(\mathbf{x}) - \frac{\partial}{\partial w_{ij}} \ln Z(\mathbf{x}) \\
&= \frac{\partial}{\partial w_{ij}} (\frac{1}{2}\mathbf{x}^T w \mathbf{x} + \mathbf{x}^T b) - \frac{1}{Z(\mathbf{x})} \frac{\partial}{\partial w_{ij}} Z(\mathbf{x}) \\
&= x_i x_j - \frac{1}{Z(\mathbf{x})} \sum_{\mathbf{x} \in \mathcal{X}} \frac{\partial}{\partial w_{ij}} e^{-E(\mathbf{x})} \\
&= x_i x_j - \sum_{\mathbf{x} \in \mathcal{X}} p(\mathbf{x}) x_i x_j \\
&= x_i x_j - \mathbb{E}^p[x_i x_j].
\end{aligned}
$$

Similarly, we have

$$
\begin{aligned}
\frac{\partial}{\partial b_j} \ln p(\mathbf{x}) &= -\frac{\partial}{\partial b_j} E(\mathbf{x}) - \frac{\partial}{\partial b_j} \ln Z(\mathbf{x}) \\
&= x_j - \sum_{\mathbf{x} \in \mathcal{X}} p(\mathbf{x}) x_j \\
&= x_j - \mathbb{E}^p[x_j].
\end{aligned}
$$

The gradient components of the cost function $C(w, b) = \sum_{\mathbf{x} \in X} q(\mathbf{x}) \ln p(\mathbf{x})$ are computed now as

$$
\begin{aligned}
\frac{\partial}{\partial w_{ij}} C(w, b) &= \sum_{\mathbf{x} \in X} q(\mathbf{x}) \frac{\partial}{\partial w_{ij}} \ln p(\mathbf{x}) \\
&= \sum_{\mathbf{x} \in X} q(\mathbf{x}) x_i x_j - \sum_{\mathbf{x} \in X} q(\mathbf{x}) \mathbb{E}^p[x_i x_j] \\
&= \mathbb{E}^q[x_i x_j] - \mathbb{E}^p[x_i x_j].
\end{aligned}
$$

$$
\begin{aligned}
\frac{\partial}{\partial b_j} C(w, b) &= \sum_{\mathbf{x} \in X} q(\mathbf{x}) \frac{\partial}{\partial b_j} \ln p(\mathbf{x}) \\
&= \sum_{\mathbf{x} \in X} q(\mathbf{x}) x_j - \sum_{\mathbf{x} \in X} q(\mathbf{x}) \mathbb{E}^p[x_j] \\
&= \mathbb{E}^q[x_j] - \mathbb{E}^p[x_j].
\end{aligned}
$$

The learning follows the adjustment rule

$$\Delta w_{ij} = \eta \frac{\partial}{\partial w_{ij}} C(w, b)$$

$$\Delta b_j = \eta \frac{\partial}{\partial b_j} C(w, b),$$

which recovers the learning rule obtained in [2]

$$\Delta w_{ij} = \eta \Big(\mathbb{E}^q[x_i x_j] - \mathbb{E}^p[x_i x_j] \Big)$$

$$\Delta b_j = \eta \Big(\mathbb{E}^q[x_j] - \mathbb{E}^p[x_j] \Big),$$

where $\eta > 0$ is the learning rate. It is worth noting that the learning rule has two phases: in the first phase the connection weight w_{ij} is increased by the average activation of x_i and x_j under the given distribution, q; in the second phase, the connection weight w_{ij} is decreased by the average activation of x_i and x_j under the Boltzmann distribution, p.

Remark 20.4.1 Changes in weights and biasses lead to perturbations in the Boltzmann distribution. Its sensitivity with respect to parameters is given by

$$
\begin{aligned}
dp(\mathbf{x}) &= p(\mathbf{x}) \, d \ln p(\mathbf{x}) \\
&= p(\mathbf{x}) \sum_{i,j} \frac{\partial \ln p(\mathbf{x})}{\partial w_{ij}} \, dw_{ij} + p(\mathbf{x}) \sum_j \frac{\partial \ln p(\mathbf{x})}{\partial b_j} \, db_j \\
&= p(\mathbf{x}) \sum_{i,j} \Big(x_i x_j - \mathbb{E}^p[x_i x_j] \Big) \, dw_{ij} + p(\mathbf{x}) \sum_j \Big(x_j - \mathbb{E}^p[x_j] \Big) \, db_j.
\end{aligned}
$$

20.5 Computing the Boltzmann Distribution

The Boltzmann distribution is an equilibrium distribution, i.e., regardless of the initial state choice, the machine settles in the long run to the same distribution. In this section we shall directly compute the Boltzmann distribution for a Boltzmann machine having two neurons, using a limiting procedure. Starting from an arbitrary initial state, $\mathbf{x}^0 = (x_1^0, x_2^0)$, adjusting the state according to the updating rule (20.3.4) of a Boltzmann machine, we obtain a sequence of states $\mathbf{x}^n = (x_1^n, x_2^n)$, whose distribution will converge to the Boltzmann distribution. Thus, it suffices to compute the distribution of the sequence of states $\mathbf{x}^n$ and then take $n \to \infty$ to obtain the equilibrium distribution. This procedure does not use the results introduced in section 20.2.

In this model we have 3 parameters, b_1, b_2, and $w = w_{12} = w_{21}$. Let $a_n = P(x_2^n = 1)$. In a first stage we shall find a recurrence relation for the sequence a_n and also find its limit.

The state of the second neuron at the $(n+1)$th step, x_2^{n+1}, depends on the state of the first neuron at the $(n+1)$th step, x_1^{n+1}, as in the following:

$$x_2^{n+1} = \begin{cases} 1, & \text{with probability } \sigma(wx_1^{n+1} + b_2) \\ 0, & \text{with probability } 1 - \sigma(wx_1^{n+1} + b_2). \end{cases}$$

Using the probability chain rule[4] we can express the probability of the second neuron state in terms of the first one using conditional probabilities as

$$\begin{aligned} P(x_2^{n+1} = 1) & = P(x_2^{n+1} = 1 | x_1^{n+1} = 1)P(x_1^{n+1} = 1) \\ & \quad + P(x_2^{n+1} = 1 | x_1^{n+1} = 0)P(x_1^{n+1} = 0) \\ & = \sigma(w + b_2)P(x_1^{n+1} = 1) + \sigma(b_2)P(x_1^{n+1} = 0) \quad (20.5.5) \end{aligned}$$

Now, the state x_1^{n+1} depends on x_2^n as

$$x_1^{n+1} = \begin{cases} 1, & \text{with probability } \sigma(wx_2^n + b_1) \\ 0, & \text{with probability } 1 - \sigma(wx_2^n + b_1). \end{cases}$$

The probability chain rule yields

$$\begin{aligned} P(x_1^{n+1} = 1) & = P(x_1^{n+1} = 1 | x_2^n = 1)P(x_2^n = 1) \\ & \quad + P(x_1^{n+1} = 1 | x_2^n = 0)P(x_2^n = 0) \\ & = \sigma(w + b_1)P(x_2^n = 1) + \sigma(b_1)P(x_2^n = 0). \quad (20.5.6) \end{aligned}$$

Substituting (20.5.6) into (20.5.5) yields the first-order recurrence

$$a_{n+1} = \alpha a_n + \beta, \quad (20.5.7)$$

with

$$\begin{aligned} \alpha & = \big(\sigma(w + b_1) - \sigma(b_1)\big)\big(\sigma(w + b_2) - \sigma(b_2)\big) \\ \beta & = \sigma(w + b_2)\sigma(b_1) + \sigma(b_2)\sigma(-b_1). \end{aligned}$$

Since σ is increasing, it follows that $\alpha > 0$ for $w \neq 0$. Using the Mean Value Theorem we can estimate the following upper bound:

$$\alpha = \sigma'(c_1)w\sigma'(c_2)w \leq \|\sigma'\|_\infty w^2 = \frac{w^2}{16},$$

where we used that the largest slope of σ is $1/4$. Since $\sigma \in (0,1)$, we have $0 < \beta < 2$.

---

[4]For any two events A and B we have $P(A) = P(A,B) + P(A,B^c) = P(A|B)P(B) + P(A|B^c)P(B^c)$.

The recurrence (20.5.7) can be solved inductively as

$$
\begin{aligned}
a_{n+1} &= \alpha^{n+1} a_0 + \beta(1 + \alpha + \cdots + \alpha^n) \\
&= \alpha^{n+1} a_0 + \beta \frac{1 - \alpha^{n+1}}{1 - \alpha}.
\end{aligned}
$$

Assume $|w| < 4$ (which implies the stability condition). Then $0 < \alpha < 1$, and hence, $\alpha^n \to 0$, as $n \to \infty$. Therefore $a_n \to \frac{\beta}{1-\alpha}$, or

$$
\lim_{n \to \infty} P(x_2^n = 1) = \frac{\beta}{1 - \alpha}.
$$

The next goal is to find the following equilibrium distribution:

$$
\begin{aligned}
p(0,0) &= \lim_{n \to \infty} P(x_1^n = 0, x_2^n = 0) \\
p(0,1) &= \lim_{n \to \infty} P(x_1^n = 0, x_2^n = 1) \\
p(1,0) &= \lim_{n \to \infty} P(x_1^n = 1, x_2^n = 0) \\
p(1,1) &= \lim_{n \to \infty} P(x_1^n = 1, x_2^n = 1).
\end{aligned}
$$

Taking the limit $n \to \infty$ in the following conditional probability relations

$$
\begin{aligned}
P(x_1^{n+1} = 0, x_2^n = 0) &= P(x_1^{n+1} = 0 | x_2^n = 0) P(x_2^n = 0) = \sigma(-b_1) P(x_2^n = 0) \\
P(x_1^{n+1} = 0, x_2^n = 1) &= P(x_1^{n+1} = 0 | x_2^n = 1) P(x_2^n = 1) = \sigma(-w - b_1) P(x_2^n = 1) \\
P(x_1^{n+1} = 1, x_2^n = 0) &= P(x_1^{n+1} = 1 | x_2^n = 0) P(x_2^n = 0) = \sigma(b_1) P(x_2^n = 0) \\
P(x_1^{n+1} = 1, x_2^n = 1) &= P(x_1^{n+1} = 1 | x_2^n = 1) P(x_2^n = 1) = \sigma(w + b_1) P(x_2^n = 1)
\end{aligned}
$$

provides the equilibrium distribution

$$
\begin{aligned}
p(0,0) &= \sigma(-b_1)\left(1 - \frac{\beta}{1 - \alpha}\right) \\
p(0,1) &= \sigma(-w - b_1)\frac{\beta}{1 - \alpha} \\
p(1,0) &= \sigma(b_1)\left(1 - \frac{\beta}{1 - \alpha}\right) \\
p(1,1) &= \sigma(w + b_1)\frac{\beta}{1 - \alpha}.
\end{aligned}
$$

We shall present next two ways of assessing information on a Boltzmann manifold, using the entropy and evaluating the Fisher information.

20.6   Entropy of Boltzmann Distribution

The Boltzmann distribution is the distribution with the largest entropy on the state space $\mathcal{X}$, given a fixed value of the average energy $\mathbb{E}^p[E(\mathbf{x})] = k$. The entropy can be computed as in the following:

$$
\begin{aligned}
H(p) &= -\sum_{\mathbf{x} \in \mathcal{X}} p(\mathbf{x}) \ln p(\mathbf{x}) = -\sum_{\mathbf{x} \in \mathcal{X}} p(\mathbf{x}) \ln \frac{e^{-\frac{E(\mathbf{x})}{T}}}{Z} \\
&= \frac{1}{T} \sum_{\mathbf{x} \in \mathcal{X}} p(\mathbf{x}) E(\mathbf{x}) + \sum_{\mathbf{x} \in \mathcal{X}} p(\mathbf{x}) \ln Z \\
&= \frac{1}{T} \mathbb{E}^p[E(\mathbf{x})] + \ln Z = \frac{k}{T} + \ln Z.
\end{aligned}
$$

Since in Physics the entropy is determined up to an additive constant, the constant term, $\ln Z$ can be neglected. Then the entropy becomes the quotient between the average system energy, $\mathbb{E}^p[E(\mathbf{x})]$, and the temperature, T.

Let p_{unif} denote the uniform distribution on $\mathcal{X}$, i.e., $p_{unif}(\mathbf{x}) = \frac{1}{N}$. Its entropy

$$
H(p_{unif}) = -\sum_{\mathbf{x}} \frac{1}{N} \ln \frac{1}{N} = \ln N,
$$

is the largest among all entropies of distributions on $\mathcal{X}$. Since

$$
\begin{aligned}
D_{KL}(p \| p_{unif}) &= \sum_{\mathbf{x}} p(\mathbf{x}) \ln \frac{p(\mathbf{x})}{p_{unif}(\mathbf{x})} = -H(p) - \sum_{\mathbf{x}} p(\mathbf{x}) \ln \frac{1}{N} \\
&= \ln N - H(p),
\end{aligned}
$$

it follows that the difference between the largest possible entropy and the entropy of the Boltzmann distribution is given by

$$
H(p_{unif}) - H(p) = D_{KL}(p \| p_{unif}) > 0.
$$

The left term represents the reduction of the maximum entropy due to the constraint on the average energy, $\mathbb{E}^p[E(\mathbf{x})] = k$. The right term shows that this is given by the Kullback-Leibler divergence.

20.7   Fisher Information

Any Boltzmann machine defines a probability distribution of the form $p(\mathbf{x}) = \dfrac{e^{-E(\mathbf{x})/T}}{Z}$, $\mathbf{x} \in \mathcal{X}$. Conversely, any distribution of this form defines uniquely a Boltzmann machine. Thus, the family of Boltzmann machines can be identified with the family of Boltzmann distributions, and hence can be parametrized by w_{ij} and b_k. The Riemannian metric on the associated manifold having coordinates (w_{ij}, b_k) is given by the Fisher metric. Using the computation from section 20.4

$$
\begin{aligned}
\partial_{w_{ij}} \ln p(\mathbf{x}) &= x_i x_j - \mathbb{E}^p[x_i x_j] \\
\partial_{w_{kl}} \ln p(\mathbf{x}) &= x_k x_l - \mathbb{E}^p[x_k x_l] \\
\partial_{b_k} \ln p(\mathbf{x}) &= x_k - \mathbb{E}^p[x_k] \\
\partial_{b_l} \ln p(\mathbf{x}) &= x_l - \mathbb{E}^p[x_l],
\end{aligned}
$$

the linearity of the expectation operator yields

$$
\begin{aligned}
g_{ij,kl}(w,b) &= \mathbb{E}^p[\partial_{w_{ij}} \ln p(\mathbf{x})\, \partial_{w_{kl}} \ln p(\mathbf{x})] \\
&= \mathbb{E}^p[x_i x_j x_k x_l] - \mathbb{E}^p[x_i x_j]\mathbb{E}^p[x_k x_l] \\
&= Cov(x_i x_j, x_k x_l). \\
g_{k,r}(w,b) &= \mathbb{E}^p[\partial_{b_k} \ln p(\mathbf{x})\, \partial_{b_r} \ln p(\mathbf{x})] \\
&= \mathbb{E}^p[x_k x_r] - \mathbb{E}^p[x_k]\mathbb{E}^p[x_r] \\
&= Cov(x_k, x_r).
\end{aligned}
$$

We note that the Fisher information depends on the correlations of neuron activations x_j and is independent of the weights and biases, i.e., of the neural manifold coordinates.[5] This implies that the derivatives of the metric coefficients vanish

$$
\frac{\partial g_{ij,kl}(w,b)}{\partial w_{ik}} = \frac{\partial g_{ij,kl}(w,b)}{\partial b_r} = \frac{\partial g_{k,r}(w,b)}{\partial w_{ij}} = \frac{\partial g_{k,r}(w,b)}{\partial b_j} = 0.
$$

Consequently, all Christoffel symbols (13.1.2) vanish. Then the associated manifold is intrinsically flat (the Riemannian curvature tensor is zero). The geodesics equations (13.1.1) on this manifold become $\ddot{c}^\alpha(t) = 0$, which means that the geodesic components $c^\alpha(t)$ are affine functions in t. Since the distance between the initial point and the optimal point on the manifold cannot be smaller than the length of the geodesic, the previous relation provides a lower bound on how fast learning can be accomplished on this manifold. This situation is similar to the geometry of the neuromanifold associated with a linear neuron described in section 14.4.

---

[5]Something similar happens with the Euclidean metric in an Euclidean space, $\mathbb{R}^n$. These types of metrics are called translation invariant.

For efficiency reasons, the Fisher matrix can be used in combination with the natural gradient learning algorithm introduced in section 14.8. The updating rule in this case becomes

$$
\begin{aligned}
\Delta w_{ij} &= \eta \sum_{k,l} g^{ij,kl} \frac{\partial}{\partial w_{kl}} C(w,b) = \eta \sum_{k,l} g^{ij,kl} \Big(\mathbb{E}^q[x_k x_l] - \mathbb{E}^p[x_k x_l] \Big) \\
&= \eta \Big(\mathbb{E}^q \Big[\sum_{k,l} g^{ij,kl} x_k x_l \Big] - \mathbb{E}^p \Big[\sum_{k,l} g^{ij,kl} x_k x_l \Big] \Big) \\
\Delta b_k &= \eta \sum_{l} g^{kl} \frac{\partial}{\partial b_l} C(w,b) = \eta \sum_{l} g^{kl} \Big(\mathbb{E}^q[x_l] - \mathbb{E}^p[x_l] \Big) \\
&= \eta \Big(\mathbb{E}^q \Big[\sum_{l} g^{kl} x_l \Big] - \mathbb{E}^p \Big[\sum_{l} g^{kl} x_l \Big] \Big).
\end{aligned}
$$

20.8   Applications of Boltzmann Machines

Boltzmann machines are mainly used for solving combinatorial optimization problems and carrying out learning tasks.

1. Distributions approximator As we have seen in section 20.4, a Boltzmann machine has the capability of learning any discrete distribution, $q(\mathbf{x})$, on the state space $\mathcal{X}$ using a distribution of the exponential form

$$
p(\mathbf{x}) = \frac{e^{-(\mathbf{x}^T w x + b^T \mathbf{x})/T}}{Z}. \tag{20.8.8}
$$

In order to refine this result, we approximate the distribution $q(\mathbf{x})$ by a convex combination of distributions of type (20.8.8)

$$
q(\mathbf{x}) \approx \sum_{i=1}^{m} \alpha_i q_i(\mathbf{x}),
$$

with $q_i(\mathbf{x}) = \frac{e^{-(\mathbf{x}^T w(i) x + b(i) \mathbf{x})/T}}{Z(i)}$. The Boltzmann machine $B(i)$ defined by coordinates $(w(i), b(i))$ learns the distribution $q_i(\mathbf{x})$. Consider now a neural network which combines the Boltzmann machines $B(i)$. Its output, $y(\mathbf{x}) = \sum_{i=1}^{m} \alpha_i q_i(\mathbf{x})$ is an approximation of the distribution $q(\mathbf{x})$.

Remark 20.8.1 (i) It has been shown that Boltzmann machines with hidden units are universal approximators of probability mass functions over discrete variables, see [103].

(ii) A similar approximation result holds for the continuous case. The class of combinations of exponentials is well known to be dense in the set of distributions on $(0, \infty)$. Interested readers are referred to [35].

2. Simulated annealing method A Boltzmann machine is a network of mutually connected binary stochastic neurons. When the temperature parameter $T \searrow 0$, each stochastic neuron becomes a regular perceptron, see section 20.1. Hence, the Boltzmann machine tends to a neural network made out of perceptrons, which is called a *Hopfield network*. We shall briefly present this type of network in the following.

Hopfield networks These type of networks had been introduced by the physicist John Hopfield[6] in 1982, see [57]. The physical importance consists of the fact that a Hopfield network is isomorphic to the Ising model of magnetism at zero temperature, see [61].

A Hopfield network consists of n perceptrons[7] that preserve their individual states until they are selected for a new update, which is made at random. The perceptrons are totally coupled, that is, each neuron is connected with all the others, except itself.[8] The weights between the ith and the jth neuron, w_{ij}, are symmetric, and can be modeled by a symmetric matrix w with zero diagonal entries. An example of a Hopfield network with $n = 6$ neurons is shown in Fig. 20.1 **a**.

The network starts with an initial state, $\mathbf{x}^0 = (x_1^0, \dots, x_n^0) \in \{0, 1\}^n$, and the updates occur one at a time (i.e., they are asynchronous). Assume the jth neuron is chosen for an update. The effect of all the other neurons on the jth neuron, including its bias, is $\sum_{i \neq j} w_{ij} x_i + b_j$. The value of the neuron is updated to the value of the Heaviside function $H\left(\sum_{i \neq j} w_{ij} x_i + b_j\right)$, which either 0 or 1.

The energy associated with a Hopfield network evaluated at the state $\mathbf{x}$ is the same as the one in case of a Boltzmann machine

$$
\begin{aligned}
E(\mathbf{x}) &= -\frac{1}{2}\mathbf{x}^T w \mathbf{x} - \mathbf{x}^T b \\
&= -\frac{1}{2}\sum_{i,j} w_{ji} x_j x_i - \sum_i b_i x_i = -\sum_{i<j} w_{ji} x_j x_i - \sum_k b_i x_k.
\end{aligned}
$$

The task of a Hopfield network is to minimize the aforementioned energy by the updating procedure. The convergence of the network to a stable state, which corresponds to a local minimum of the energy function, is shown in the following. Consider the update from the state $\mathbf{x} = (x_1, \dots, x_k, \dots, x_n)$ to the new state $\mathbf{x}' = (x_1, \dots, x_k', \dots, x_n)$. By Exercise 20.10.1 the energy

---

[6]Professor at Princeton University since 1964.

[7]In our case the perceptrons take values 1 and 0, while in Hopfield's approach they take values 1 and -1, because his model was derived from a physical model where particles have a spin that is either "up" or "down".

[8]This avoids a permanent feedback of its own state value.

difference is

$$E(\mathbf{x}') - E(\mathbf{x}) = -\left(\sum_{i=1}^{n} w_{ki}x_i + b_k\right)(x_k' - x_k).$$

We consider two cases:

(i) If the kth neuron updates its value from $x_k = 0$ to $x_k' = 1$, then we have $H\left(\sum_{i=1}^{n} w_{ki}x_i + b_k\right) = 1$, which implies $\sum_{i=1}^{n} w_{ki}x_i + b_k > 0$. In this case

$$E(\mathbf{x}') - E(\mathbf{x}) = -\left(\sum_{i=1}^{n} w_{ki}x_i + b_k\right)(1 - 0) < 0.$$

(ii) If the kth neuron updates its value from $x_k = 1$ to $x_k' = 0$, then we have $H\left(\sum_{i=1}^{n} w_{ki}x_i + b_k\right) = 0$, so $\sum_{i=1}^{n} w_{ki}x_i + b_k < 0$. Then

$$E(\mathbf{x}') - E(\mathbf{x}) = -\left(\sum_{i=1}^{n} w_{ki}x_i + b_k\right)(0 - 1) < 0.$$

Then, every time a neuron state is updated, the total energy decreases. Since there are only a finite number of states that can be taken, 2^n, it follows that at some point the energy cannot be reduced any further, which corresponds to a state of minimum energy.

Sometimes, we arrive at a minimum that is not an absolute minimum. In order to avoid getting stuck in local minima of the energy function, we add noise to the system. This is done by transforming the Hopfield network into a Boltzmann machine and then use the simulated annealing method with $T \searrow 0$ to approach the global minimum of the energy function.

Example 20.8.2 (The n rooks problem) We now get back to the problem introduced in section 1.3. We make it here a little more general by asking to place n rooks in an $n \times n$ chess board so that no one endangers another. This optimization problem can be solved using a Hopfield network. First, the objective function, which needs to get minimized, is the sum

$$E(x_{11}, \ldots, x_{nn}) = \sum_{j=1}^{n}\left(\sum_{i=1}^{n} x_{ij} - 1\right)^2 + \sum_{i=1}^{n}\left(\sum_{j=1}^{n} x_{ij} - 1\right)^2,$$

where x_{ij} is the state of the (i, j)th square. The state is 1, if there is a rook placed at that spot and 0, otherwise. The minimum value of $E(x_{11}, \ldots, x_{nn})$ is zero and this is accomplished when there is only one rook on each row and on each column. Since $x_{ij} \in \{0, 1\}$, then $x_{ij}^2 = x_{ij}$. We shall show that some algebraic manipulation reduces the function $E(x_{11}, \ldots, x_{nn})$ to the energy of a Hopfield network. We start expanding the square

$$
\begin{aligned}
\Big(\sum_{i=1}^{n} x_{ij} - 1 \Big)^2 &= \Big(\sum_{i=1}^{n} x_{ij} \Big)^2 - 2 \sum_{i=1}^{n} x_{ij} + 1 \\
&= \sum_{i=1}^{n} x_{ij}^2 + 2 \sum_{k \neq i} x_{ij} x_{kj} - 2 \sum_{i=1}^{n} x_{ij} + 1 \\
&= 2 \sum_{k \neq i} x_{ij} x_{kj} - \sum_{i=1}^{n} x_{ij} + 1.
\end{aligned}
$$

The first term of E is evaluated as

$$
F_1 = \sum_{j=1}^{n} \Big(\sum_{i=1}^{n} x_{ij} - 1 \Big)^2 = 2 \sum_{j=1}^{n} \sum_{k \neq i} x_{ij} x_{kj} - \sum_{i,j=1}^{n} x_{ij} + n
$$

Similarly, the second term of E becomes

$$
F_2 = \sum_{i=1}^{n} \Big(\sum_{j=1}^{n} x_{ij} - 1 \Big)^2 = 2 \sum_{i=1}^{n} \sum_{k \neq j} x_{ik} x_{ij} - \sum_{i,j=1}^{n} x_{ij} + n.
$$

Then

$$
\begin{aligned}
E(x_{11}, \ldots, x_{nn}) &= F_1(x_{11}, \ldots, x_{nn}) + F_2(x_{11}, \ldots, x_{nn}) \\
&= 2 \Big\{ \sum_{j=1}^{n} \sum_{k \neq i} x_{ij} x_{kj} + \sum_{i=1}^{n} \sum_{k \neq j} x_{ik} x_{ij} \Big\} - 2 \sum_{i,j=1}^{n} x_{ij} + 2n \\
&= -\frac{1}{2} \sum_{\alpha, \beta} w_{\alpha, \beta} x_\alpha x_\beta - \sum_\alpha b_\alpha x_\alpha + 2n,
\end{aligned}
$$

where $b_\alpha = 2$ and

$$
w_{\alpha, \beta} = \begin{cases} -4, & \text{if } \alpha \text{ and } \beta \text{ are placed on the same row or column;} \\ 0, & \text{otherwise.} \end{cases}
$$

Neglecting the irrelevant constant $2n$, the rest of the expression is the energy of a Hopfield network, see Fig. 20.1 **b**. The bias of each unit is 2 and the connection weights are nonzero only for horizontal and vertical connected neurons.

Each of the n^2 squares of the board corresponds to a perceptron, and hence the network can take 2^{n^2} states. The state of a neuron is 1 if there is a rook placed in that square and 0, otherwise. Each rook's configuration,

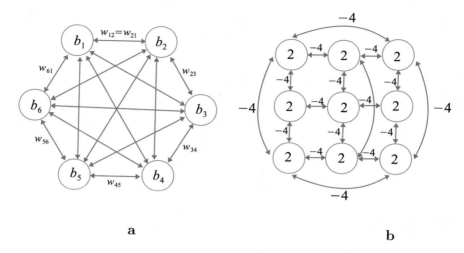

a   b

Figure 20.1: **a** *A Hopfield network with* $n = 6$ *neurons;* **b**. *A Hopfield network associated with a* 3×3 *chess board has nonzero connection weights only for vertically or horizontally connected units.*

$\mathbf{x} = (x_{11}, \ldots, x_{nn})$ corresponds to a state of the Hopfield network. The stable state of the network minimizes the energy, and hence provides a solution for the rooks' problem. The learning algorithm consists in choosing a square at random and making an update. It is worth noting that regardless of the initial state of the Hopfield network the final configuration of the problem will contain only n active perceptrons, i.e., there are n rooks that solve the problem.

Example 20.8.3 The people of a community have to vote for or against a certain candidate leader. We shall model the community behavior as a Hopfield network. The state of a member who votes for the candidate is 1. If he votes against, the state is 0. The mutual influence between the ith and jth members is given by the weight w_{ij}. Each individual member has his/her own threshold of belief, denoted by $-b_i$. If the input belief influence on the ith member from the other members is larger than his individual threshold of belief, i.e., if $\sum_j w_{ij} x_j > -b_i$, then the ith individual will vote for the candidate, i.e., will have associated the state 1. Otherwise, if $\sum_j w_{ij} x_j < -b_i$, he will vote against, i.e., has the value 0. In both cases, the state of the ith individual is given by the output of a a perceptron $H(\sum_j w_{ij} x_j + b_i)$. Thus, each member can be considered as a perceptron, and the community as a Hopfield network.

In the long run the state of the network, $\mathbf{x}$, which is a sequence of 0s and 1s, is maximizing the quadratic function $f(\mathbf{x}) = \frac{1}{2}\mathbf{x}^T w\mathbf{x} + x^T b$. Hence, if one knows the mutual influence between the members, w_{ij}, and the individual thresholds of belief, b_i, one can find the state of the community. Therefore, the candidate will get a number of $\sum_i x_i$ favorable votes. If $\sum_i x_i > 1 + n/2$, then he wins the election. Here, n is the size of the community.

20.9 Summary

A binary stochastic neuron is a neuron whose output is a random variable taking values in $\{0, 1\}$. The probability of taking the value 1 is given by a sigmoid function applied to the input of the neuron. A Boltzmann machine is a neural network that consists of an ensemble of symmetrically connected stochastic neurons in complex interaction with all the other neurons in the net. An energy function is introduced to harness the complexity of the model. The stable state of a Boltzmann machine is described by an equilibrium distribution parametrized by the energy function, called Boltzmann distribution. When the temperature parameter tends to 0, the Boltzmann machine becomes a Hopfield network, that is, a network made out of perceptrons. Hopfield networks are useful for solving complex combinatorial problems. To avoid falling into a local minimum, a Boltzmann machine is used instead in combination with the simulated annealing method for obtaining a global minimum of the energy function.

20.10 Exercises

Exercise 20.10.1 Let $\mathbf{x} = (x_1, \ldots, x_k, \ldots, x_n)$ and $\mathbf{x}' = (x_1, \ldots, x'_k, \ldots, x_n)$ be two states of a Boltzmann machine, with corresponding energies $E = E(\mathbf{x})$ and $E' = E(\mathbf{x}')$. Prove that

$$E' - E = -\left(\sum_{i=1}^{n} w_{ki} x_i + b_k \right)(x'_k - x_k).$$

Exercise 20.10.2 Consider all notations introduced in Example 20.3.2. If $q = (q_1, \ldots, q_8)$ is a distribution on the state space $\mathcal{X}$ show that the Boltzmann machine can learn exactly the distribution q if and only if

$$q_8 = \frac{q_5 q_6 q_7}{q_2 q_3 q_4}.$$

Exercise 20.10.3 The *state transition matrix* $P_T = p_{ij}$ of a Boltzmann machine with n neurons is a $N \times N$ matrix, with $N = 2^n$, whose elements p_{ij} represent the probability of the transition from the state j to the state i in a single step at the temperature T. This is defined by

$$p_{ij} = \frac{1}{1 + e^{(E_i - E_j)/T}}, \text{ if } i \neq j$$

$$p_{ii} = 1 - \sum_{j \neq i}^{N} \frac{1}{1 + e^{(E_j - E_i)/T}}.$$

(a) Prove that the Boltzmann distribution $p = \frac{1}{Z}(e^{-E_1/T}, \ldots, e^{-E_N/T})^t$ is a fixed point for P_T, i.e., $P_T p = p$ (or equivalently, p is an eigenvector of P_T with eigenvalue equal to 1).

(b) Show that the largest eigenvalue of P_T is equal to 1.

(c) Prove that for any initial state q_0, the sequence $(q_n)_n$, defined recursively by $q_{n+1} = P_T q_n$, converges in $\mathbb{R}^N$ to the Boltzmann distribution, p. What is the physical significance of this?

Exercise 20.10.4 Find the explicit formula for the Fisher information matrix in the case of a Boltzmann machine with 2 neurons.

Exercise 20.10.5 Let w_{ij}, b_k be coordinates on a Boltzmann machine and consider the linear operator[9]

$$A_{w,b} = \frac{1}{2} \sum_{i,j} w_{ij} \frac{\partial}{\partial w_{ij}} + \sum_k \frac{\partial}{\partial b_k}.$$

(a) Show that

$$A_{w,b} \, p(\mathbf{x}) = \Big(\mathbb{E}^p[E(\mathbf{x})] - E(\mathbf{x}) \Big) p(\mathbf{x}).$$

(b) Consider the smooth deformation $p_t(\mathbf{x})$ of $p(\mathbf{x})$ induced by the exponential transformation of coordinates $w_{ij}(t) = w_{ij} e^{\alpha t}$, $b_k(t) = b_k e^{\alpha t}$, with $\alpha > 0$. Show that

$$\frac{\partial}{\partial t} p_t(\mathbf{x}) = \alpha \Big(\mathbb{E}^p[E(\mathbf{x})] - E(\mathbf{x}) \Big) p_t(\mathbf{x}).$$

(c) Consider the following evolution equation of a Boltzmann distribution

$$\begin{aligned} \frac{\partial}{\partial t} p_t(\mathbf{x}) &= \alpha A_{w,b} \, p(\mathbf{x}), \qquad \alpha > 0 \\ p_0(\mathbf{x}) &= p(\mathbf{x}), \end{aligned}$$

which corresponds to a curve $(w(t), b(t))$ in the parameters space. Find the components $w_{ij}(t)$, $b_k(t)$ of this curve.

Exercise 20.10.6 8 distinct rooks are placed on a 8×8 chess board at random. Find the probability that all rooks are safe from one another.

---

[9]This can be considered as a vector field on the neural manifold.

Exercise 20.10.7 (Restricted Boltzmann machines, [114]) We consider the more general situation when the neurons of a Boltzmann machine are divided into two groups, visible neurons, $\mathbf{v}$, and hidden neurons, $\mathbf{h}$, and there are no connections between the units of the same group (hence the name of "restricted"). Thus, the state of the machine is $\mathbf{x} = (\mathbf{v}, \mathbf{h}) \in \mathcal{V} \times \mathcal{H} = \mathcal{X}$. The energy is defined by

$$E(\mathbf{v}, \mathbf{h}) = -\frac{1}{2}\mathbf{v}^T w \mathbf{h} - b^T \mathbf{v} - c^T \mathbf{h}$$

and the joint probability of the visible and hidden neurons by

$$p(\mathbf{v}, \mathbf{h}) = \frac{1}{Z}e^{-E(\mathbf{v},\mathbf{h})},$$

where Z is the partition function.

(a) Show that the hidden states, $h_j \in \mathcal{H}$, are conditional independent given the visible states;

(b) Find the conditional probability $p(h_j|\mathbf{v})$ for $h_j \in \{0, 1\}$;

(c) Compute the conditional log-likelihood function $\ell(\mathbf{h}|\mathbf{v}) = \ln p(\mathbf{h}|\mathbf{v})$ and its partial derivatives $\partial_{b_k}\ell(\mathbf{h}|\mathbf{v})$, $\partial_{c_k}\ell(\mathbf{h}|\mathbf{v})$, $\partial_{w_{ij}}\ell(\mathbf{h}|\mathbf{v})$;

(d) The information contained in the hidden states about the parameters, $\theta = (w, b, c)$, given the visible states is described by the Fisher information metric $g_{ij}(\theta|\mathbf{v}) = E^{p_{\mathbf{h}|\mathbf{v}}}[\partial_{\theta_i}\ell(\mathbf{h}|\mathbf{v}) \, \partial_{\theta_j}\ell(\mathbf{h}|\mathbf{v})]$. Compute $g_{ij}(\theta|\mathbf{v})$;

(e) Let $q(\mathbf{h}|\mathbf{v})$ be a given conditional probability distribution. Using a computation similar with the one done in section 20.4 provide a learning rule for the weights and biasses such that $D_{KL}(q(\mathbf{h}|\mathbf{v})\|p(\mathbf{h}|\mathbf{v}))$ is minimized.

Hints and Solutions

We have selected a number of exercises for which we provide hints or full solutions. The reader is encouraged to try the other exercises based on the expertise acquired from the chapter examples and other similar solved exercises.

Chapter 1

Exercise 1.9.1 (a) The problem can be modeled by a neuron as shown in Fig. 1 **a**. If $x < b$, then the factory does not produce anything, so $y = 0$. If $x \geq b$, then the revenue is $y = k(x - b)$, where k is a positive constant related to the cost of production. Consider the following activation function, see Fig. 1 **b**:

$$\varphi(x) = \begin{cases} 0, & \text{if } x < 0 \\ kx, & \text{otherwise.} \end{cases}$$

Then the revenue can be modeled as the composition

$$y = \varphi(x - b) = \varphi\left(\sum_{i=1}^{n} c_i x_i - b\right).$$

(b) The emerging learning problem is the following: *Given the amounts x_i, what are the values of the road capacities c_i to meet or be very close to a given revenue value y?* One of the error functions subject to be minimized in this case is $\frac{1}{2}(y - \varphi(x - b))^2$.

Exercise 1.9.2 (a) The problem is modeled by a neuron as shown in Fig. 2. The output is given by

$$y = \begin{cases} 0, & \text{if } x \leq M \\ k(x - M), & \text{if } x > M. \end{cases}$$

© Springer Nature Switzerland AG 2020

O. Calin, *Deep Learning Architectures*, Springer Series in the Data Sciences,

https://doi.org/10.1007/978-3-030-36721-3

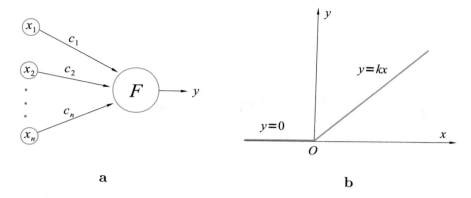

Figure 1: **a.** *Suppliers send products at capacities c_j to the factory F which has a revenue y.* **b.** *The revenue function y is piecewise linear.*

If consider the activation function

$$\varphi(x) = \begin{cases} 0, & \text{if } x \le 0 \\ kx, & \text{if } x > 0, \end{cases}$$

which can be seen in Fig. 1 **b**, then the output becomes

$$y = \varphi(x - M) = \varphi(x_1 w_1 + \cdots + x_n w_n - M).$$

(*b*) The learning problem can be stated now as: *Adjust the investing rates w_i such that an a priori planned profit z on the fund is obtained at a prescribed time t.*

Even if y and z might never be equal, an accepted answer is given as a solution of the variational problem

$$\mathbf{w} = \arg\min \frac{1}{2}(z - y)^2 = \arg\min \frac{1}{2}(z - \varphi(\mathbf{w}^T \mathbf{x} - M))^2,$$

where $\mathbf{w}^T = (w_1, \ldots, w_n)$.

Exercise 1.9.3 (*a*) We have $C(a) = \frac{1}{2} \int_0^1 (ax - f(x))^2 \, dx$ and $C'(a) = a \int_0^1 x^2 \, dx - \int_0^1 x f(x) \, dx$, and $C''(a) = \int_0^1 x^2 \, dx > 0$. Then $a = 3 \int_0^1 x f(x) \, dx$ and $b = f(0)$.

(*b*) Let $C(a, b) = \frac{1}{2} \int_0^1 \int_0^1 (ax + by - f(x, y))^2 \, dx dy$. Then the equation $(\frac{\partial C}{\partial a}, \frac{\partial C}{\partial b}) = (0, 0)$ becomes

$$\frac{1}{3}a + \frac{1}{4}b = \int_0^1 \int_0^1 x f(x, y) \, dx dy$$

$$\frac{1}{4}a + \frac{1}{3}b = \int_0^1 \int_0^1 y f(x, y) \, dx dy,$$

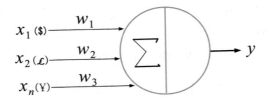

Figure 2: *Deposits at rates w_j of amounts x_j in certain currencies into a given fund.*

which is a linear system with a unique solution a and b. The last coefficient is $c = f(0,0)$.

Exercise 1.9.4 (a) Using Cauchy's inequality, we have

$$\det \rho_{ij} = \begin{vmatrix} \int_K x^2 & \int_K xy \\ \int_K xy & \int_K y^2 \end{vmatrix} = \left(\int_K x^2 \right) \left(\int_K y^2 \right) - \left(\int_K xy \right)^2 > 0.$$

The inequality is strict because the functions cannot be proportional.
(b) In this case we can compute explicitly

$$\rho_{ii} = \int_{[0,1]^n} x_i^2 \, dx_1 \cdots dx_n = \int_0^1 x_i^2 \, dx_i = \frac{1}{3}$$

$$\rho_{ij} = \int_{[0,1]^n} x_i x_j \, dx_1 \cdots dx_n = \int_0^1 x_i \, dx_i \int_0^1 x_j \, dx_j = \frac{1}{4}, \ i \neq j.$$

Chapter 2

Exercise 2.5.1 (a) The function $f : (0,1) \to \mathbb{R}$, $f(t) = -t^2 + t$ is positive and has a maximum at $t = 1/2$, which is equal to $f(1/2) = 1/4$. Then using the sigmoid property, we have $\sigma' = \sigma(1 - \sigma) = f(\sigma)$. Therefore $0 < \sigma' \leq 1/4$.
(b) Since we have

$$\sigma_c'(x) = \frac{d}{dx}\sigma(cx) = c\sigma'(cx) = c\sigma(cx)(1 - \sigma(cx)) = cf(\sigma(cx)),$$

it follows that $0 \leq \sigma_c'(x) \leq \frac{c}{4}$.

Exercise 2.5.2 (a) We have

$$2H(x) - 1 = 2 \begin{cases} 0, & \text{if } x < 0 \\ 1, & \text{if } x \geq 0 \end{cases} - 1 = \begin{cases} -1, & \text{if } x < 0 \\ 1, & \text{if } x \geq 0 \end{cases} = S(x).$$

(b) From part (a), solving for $H(x)$, yields $H(x) = \frac{1}{2}(S(x) + 1)$. Then $ReLU(x) = xH(x) = \frac{1}{2}x(S(x) + 1)$.

Exercise 2.5.3 (a) Using chain rule, we have

$$
\begin{aligned}
sp'(x) &= \Big(\ln(1 + e^x)\Big)' = \frac{(1 + e^x)'}{1 + e^x} \\
&= \frac{e^x}{e^x(e^{-x} + 1)} = \frac{1}{1 + e^{-x}} = \sigma(x).
\end{aligned}
$$

(b) Since $sp'(x) = \sigma(x) > 0$, the function $sp(x)$ is increasing. Its inverse is obtained as

$$
sp(x) = y \iff \ln(1 + e^x) = y \iff 1 + e^x = e^y
$$

$$
x = \ln(e^y - 1) \iff sp^{-1}(y) = \ln(e^y - 1).
$$

(c) Differentiating in the formula $sp(x) - sp(-x) = x$ we obtain $sp'(x) + sp'(-x) = 1$. Using part (a) we get $\sigma(x) + \sigma(-x) = 1$.

Exercise 2.5.4 An algebraic computation provides

$$
2\sigma(2x) - 1 = \frac{2}{1 + e^{-2x}} - 1 = \frac{2e^{2x}}{e^{2x} - 1} = \frac{e^{2x} - 1}{e^{2x} + 1} = \tanh(x).
$$

Exercise 2.5.5 (a) Note that $so(-x) = -so(x)$, i.e., the so function is odd. Let first $x > 0$. Then $so(x) = \frac{x}{1+x} = 1 - \frac{1}{1+x}$ is increasing since $\frac{1}{1+x}$ is decreasing. Therefore, $so(x)$ is increasing on $(0, \infty)$.

Let $x_2 < x_1 < 0$. Using that $so(x)$ is an odd function and the fact that so is increasing on $(0, +\infty)$, we have

$$
so(x_1) - so(x_2) = so(-x_2) - so(-x_1) > 0,
$$

which implies that so is increasing on $(-\infty, 0)$.

(b) $so(x)$ is continuous since $so(0-) = so(0+) = so(0) = 0$. Also, we have $so(\infty+) = 1$, $so(-\infty) = -1$. Therefore, so applies $\mathbb{R}$ one-to-one, onto $(-1, 1)$. The inverse is computed on branches. If $y \in (0, 1)$ we have $so(x) = y \iff \frac{x}{1+x} = y \iff so^{-1}(y) = \frac{y}{1-y}$. If $y \in (-1, 0)$ we have $so(x) = y \iff \frac{x}{1-x} = y \iff so^{-1}(y) = \frac{y}{1+y}$. The last two expressions imply $so^{-1}(y) = \frac{y}{1-|y|}$.

(c) Starting from $|x + y| \le |x| + |y|$ we apply that so is increasing

$$
\begin{aligned}
so(|x + y|) &\le so(|x| + |y|) = \frac{|x| + |y|}{1 + |x| + |y|} \\
&= \frac{|x|}{1 + |x| + |y|} + \frac{|y|}{1 + |x| + |y|} \\
&\le \frac{|x|}{1 + |x|} + \frac{|y|}{1 + |y|} = so(|x|) + so(|y|).
\end{aligned}
$$

Exercise 2.5.6 An algebraic manipulation provides

$$softmax(y + \mathbf{c})_i = \frac{e^{y_i+c}}{\sum_j e^{y_j+c}} = \frac{e^c \cdot e^{y_i}}{e^c \cdot \sum_j e^{y_j}} = \frac{e^{y_i}}{\sum_j e^{y_j}} = softmax(y).$$

Exercise 2.5.7 (a) $\sum_i \rho(y)_i = \frac{\sum y_i^2}{\|y\|_2^2} = 1.$ (b) $\rho(\lambda y)_i = \frac{\lambda^2 y_i^2}{\lambda^2 \|y\|_2^2} = \rho(y)_i.$

Exercise 2.5.8 The function can be written as a sum of two compact supported functions, $\varphi(x) = \varphi_1(x) + \varphi_2(x)$, with

$$\varphi_1(x) = \frac{1}{2}\left(1 + \cos(x + \frac{3\pi}{2})\right) 1_{[-\frac{\pi}{2}, \frac{\pi}{2}]}(x)$$
$$\varphi_2(x) = 1_{(\frac{\pi}{2}, \infty)}(x).$$

The function $\varphi_1(x)$ is increasing on $[-\frac{\pi}{2}, \frac{\pi}{2}]$ and constant in rest, while the function $\varphi_2(x)$ is equal to 1 for $x \geq 1$ and equal to 0 in rest. It is obvious that $\varphi(x) = 0$ for any $x \leq -\frac{\pi}{2}$ and $\varphi(x) = 1$ for any $x \geq \frac{\pi}{2}$. The graph of the function is given in Fig. 2.11 **b**.

Exercise 2.5.9 (a) It follows from the fact that a squashing function is a sigmoidal function which is nonincreasing.
(b) We choose any sigmoidal function which is decreasing on some interval. For instance, the function

$$\varphi(x) = x\sigma(x) 1_{[x_0, \infty)}(x),$$

where x_0 is the unique positive solution of $\sigma(x) = 1/x$, is nonincreasing and satisfies $\varphi(-\infty) = 0$ and $\varphi(\infty) = 1$.

Chapter 3

Exercise 3.15.1 It follows from the linearity property of the integral as well as from the properties of the logarithmic function.

Exercise 3.15.2 Using that $\ln x \leq x - 1$, we obtain

$$
\begin{aligned}
S(p, q) &= -\int p(x) \ln q(x) \, dx \geq -\int p(x)(q(x) - 1) \, dx \\
&= -\int p(x)q(x) \, dx + \int p(x) \, dx \\
&= 1 - \int p(x)q(x) \, dx.
\end{aligned}
$$

Exercise 3.15.3 Each of the terms are nonnegative, $D_{KL}(p\|q) \geq 0$, $D_{KL}(q\|p) \geq 0$, reaching their minimum, which is zero, simultaneously, for $p = q$.

Exercise 3.15.4 (*a*) A computation shows

$$
\begin{aligned}
D_{KL}(p_1\|p_2) &= \int_0^\infty p_1(x)\ln\frac{p_1(x)}{p_2(x)}\,dx \\
&= \int_0^\infty \ln\frac{\xi^1}{\xi^2}p_1(x)\,dx + (\xi^2 - \xi^1)\int_0^\infty xp_1(x)\,dx \\
&= \ln\frac{\xi^1}{\xi^2} + (\xi^2 - \xi^1)\frac{1}{\xi^1} \\
&= \frac{\xi^2}{\xi_1} - \ln\frac{\xi^2}{\xi_1} - 1.
\end{aligned}
$$

(*b*) Let $f(x) = x - \ln x - 1$. Since $f\left(\frac{\xi^2}{\xi^1}\right) \neq f\left(\frac{\xi^1}{\xi^2}\right)$, then $D_{KL}(p_1\|p_2) \neq D_{KL}(p_2\|p_1)$.

Exercise 3.15.5 Let $p_i = P(X = x_i)$, $1 \leq i \leq n$. Since $p_i \in [0,1]$, then $-\ln p_i \geq 0$, so $H(X) = -\sum_i p_i \ln p_i \geq 0$.

Exercise 3.15.6 Since $H(X) \geq 0$, see **Exercise** 3.15.5, then

$$
D_{KL}(p\|q) = S(p,q) - H(p) \geq S(p,q).
$$

Exercise 3.15.7 Since Z is $\mathcal{E}$-measurable, $\mathbb{E}[Z|\mathcal{E}] = Z$. Then the error is

$$
\|Z - \mathbb{E}[Z|\mathcal{E}]\| = \|Z - Z\| = 0.
$$

This corresponds to an exact learning.

Exercise 3.15.8 The mapping $(w,b) \to f_{w,b}(\mathbf{x})$ corresponds to a hyperplane in $\mathbb{R}^n$. The optimal parameters, (w^*, b^*), correspond to the coordinates of the orthogonal projection of the target $\mathbf{z}$ onto the hyperplane. By geometric reasons, this projection is unique. The normal equations are linear and, hence, they can be solved explicitly for w^* and b^*.

Exercise 3.15.9 By L'Hospital's rule, we have

$$
\begin{aligned}
\lim_{\alpha \to 1} H_\alpha(p) &= \lim_{\alpha \to 1}\frac{1}{1-\alpha}\ln\int p^\alpha(x)\,dx = \lim_{t \to 0}\frac{\ln\int p^{1-t}(x)\,dx}{t} \\
&= \lim_{t \to 0}\frac{d}{dt}\ln\int p^{1-t}(x)\,dx = \lim_{t \to 0}\frac{\frac{d}{dt}\int p^{1-t}(x)\,dx}{\int p^{1-t}(x)\,dx} \\
&= \lim_{t \to 0}\frac{-\int p^{1-t}(x)\ln p(x)\,dx}{\int p^{1-t}(x)\,dx} = \frac{-\int p(x)\ln p(x)\,dx}{\int p(x)\,dx} \\
&= -\int p(x)\ln p(x)\,dx = H(p).
\end{aligned}
$$

Exercise 3.15.10 (*a*) It is a straightforward computation involving changes of variable and completing the square

$$\phi_\sigma * \phi_\sigma(v) = \int \phi_\sigma(t)\phi_\sigma(t-v)\,dt = \frac{1}{2\pi\sigma^2} \int e^{-\frac{t^2}{2\sigma^2}} e^{-\frac{(t-v)^2}{2\sigma^2}}\,dt$$

$$= \frac{1}{2\pi\sigma^2} \int e^{-\frac{(t^2-tv+v^2/2)}{\sigma^2}}\,dt = \frac{1}{2\pi\sigma^2} e^{-\frac{v^2}{4\sigma^2}} \int e^{-\frac{(t-v^2/2)^2}{\sigma^2}}\,dt$$

$$= \frac{1}{2\pi\sigma} e^{-\frac{v^2}{4\sigma^2}} \int e^{-u^2}\,du = \frac{1}{2\sqrt{\pi}\sigma} e^{-\frac{v^2}{4\sigma^2}} = \frac{1}{\sqrt{2\pi}\sigma'} e^{-\frac{v^2}{2\sigma'^2}} = \phi_{\sigma'}(v),$$

for $\sigma' = \sigma\sqrt{2}$.
(*b*) A similar computation shows $\phi_\sigma * \phi_{\sigma'} = \phi_s$, with $s = \sqrt{\sigma^2 + \sigma'^2}$.

Exercise 3.15.11 (*a*) If $D_{CS}(p,q) = 0$ then $\int pq = \sqrt{\int p^2 \int q^2}$. Since this means equality in Schwartz inequality, the functions must be proportional. Then, there is a constant λ such that $p(x) = \lambda q(x)$. Since $\int p = \int q = 1$, it follows that $\lambda = 1$, and hence $p = q$. The converse is obvious.
(*b*) By Schwartz inequality $\int pq \leq \sqrt{\int p^2 \int q^2}$, so the argument of ln belongs to $(0,1]$, where the logarithmic function is negative.
(*c*) Obvious. (*d*) It follows from the properties of the logarithmic function and definition of Renyi entropy.

Exercise 3.15.12 Use the fact that $|\tanh| \leq |x|$. This follows from $\tanh' x = 1 - \tanh^2 x \leq 1$.

Chapter 4

Exercise 4.17.1 (*a*) Compute the Laplacian

$$\Delta f(x) = \frac{\partial^2 f}{\partial x_1^2} + \frac{\partial^2 f}{\partial x_2^2} = e^{x_1} \sin x_2 - e^{x_1} \sin x_2 = 0.$$

(*b*) Since $\nabla f(x) = (e^{x_1} \sin x_2, e^{x_1} \cos x_2)$, we have

$$\|\nabla f\| = e^{x_1} \|(\sin x_2, \cos x_2)\| = e^{x_1}.$$

(*c*) $\nabla f(x) = 0 \iff \|\nabla f\| = 0 \iff e^{x_1} = 0$, which does not have solutions.
(*d*) Since f is harmonic (or, because $\nabla f \neq 0$), the function f reaches its minima and maxima on the boundary of $[0,1] \times [0, \frac{\pi}{2}]$. Since the functions e^{x_1}, $x_1 \in [0,1]$ and $\sin x_2$, $x_2 \in [0, \pi/2]$ are both increasing, it follows that the maximum of $f(x)$ is reached for $(x_1, x_2) = (1, \pi/2)$ and the minimum is reached for $(x_1, x_2) = (0,0)$. Furthermore, $\min f(x) = f(0,0) = 0$ and $\max f(x) = f(1, \pi/2) = e$.

Exercise 4.17.2 (a) $\nabla Q(\mathbf{x}) = A\mathbf{x} - b$.

(b) The iteration given by the gradient descent is

$$\begin{aligned}
\mathbf{x}^{n+1} &= \mathbf{x}^n - \eta \nabla Q(\mathbf{x}^n) \\
&= \mathbf{x}^n - \eta(A\mathbf{x}^n - b) = (\mathbb{I} - A)\mathbf{x}^n + b.
\end{aligned}$$

(c) The Hessian is $H_Q = \frac{1}{2}A$.

(d) The iteration given by Newton's formula is

$$\begin{aligned}
\mathbf{x}^{n+1} &= \mathbf{x}^n - H_Q^{-1}(\mathbf{x}^n)\nabla Q(\mathbf{x}^n) \\
&= \frac{1}{2}\mathbf{x}^n\mathbf{x}^n - \frac{1}{2}A^{-1}(A\mathbf{x}^n - b) \\
&= \frac{1}{2}\mathbf{x}^n + \frac{1}{2}A^{-1}b.
\end{aligned}$$

Assuming $\mathbf{x}^* = \lim_{n\to\infty} \mathbf{x}^n$, taking the limit in the previous relation yields

$$\mathbf{x}^* = \frac{1}{2}\mathbf{x}^* + \frac{1}{2}A^{-1}b,$$

which implies $\mathbf{x}^* = A^{-1}b$, which was expected. The existence of the limit follows from the inductive iteration

$$\begin{aligned}
\mathbf{x}^{n+1} &= \frac{1}{2}\mathbf{x}^n + \frac{1}{2}A^{-1}b \\
&= \frac{1}{2}\left(\frac{1}{2}\mathbf{x}^{n-1} + \frac{1}{2}A^{-1}b\right) + \frac{1}{2}A^{-1}b \\
&= \cdots\cdots\cdots\cdots\cdots\cdots \\
&= \frac{1}{2^{n+1}}\mathbf{x}^0 + \left(\frac{1}{2^{n+1}} + \cdots + \frac{1}{2}\right)A^{-1}b.
\end{aligned}$$

The first term tends to zero and the second tends to $A^{-1}b$.

Exercise 4.17.3 (a) The cost function can be written as

$$\begin{aligned}
C(\mathbf{x}) &= \frac{1}{2}\|A\mathbf{x} - b\|^2 = \frac{1}{2}\langle A\mathbf{x} - b, A\mathbf{x} - b\rangle \\
&= \frac{1}{2}\langle A^T A\mathbf{x}, \mathbf{x}\rangle - \langle A\mathbf{x}, b\rangle + \frac{1}{2}\|b\|^2.
\end{aligned}$$

The gradient is $\nabla C(\mathbf{x}) = A^T A\mathbf{x} - Ab$ and the Hessian is given by $H_C(\mathbf{x}) = A^T A$. Since $\det A \neq 0$ and the matrix $A^T A$ is symmetric, H_C has real nonzero eigenvalues, which are the squares of the eigenvalues of A. Therefore, H_C is positive definite.

(b) We have

$$\begin{aligned}
\mathbf{x}^{n+1} &= \mathbf{x}^n - \eta \nabla C(\mathbf{x}^n) = \mathbf{x}^n - \eta A^T A\mathbf{x}^n - Ab \\
&= (I - \eta A^T A)\mathbf{x}^n - Ab.
\end{aligned}$$

Exercise 4.17.4 (*a*) Inductively, we have

$$
\begin{aligned}
a_{n+1} &\le \mu a_n + K \\
&\le \mu(\mu a_{n-1} + K) + K \\
&= \mu^2 a_{n-1} + K(\mu + 1)
\end{aligned}
$$

$$
\cdots\cdots\cdots\cdots\cdots\cdots\cdots
$$

$$
\begin{aligned}
&\le \mu^{n+1} a_0 + K(\mu^n + \mu^{n-1} + \cdots + \mu + 1) \\
&= \mu^{n+1} a_0 + K\frac{1 - \mu^{n+1}}{1 - \mu} < a_0 + \frac{K}{1 - \mu}.
\end{aligned}
$$

(*b*) Taking the norm in Equation (4.4.17) yields

$$
\|v^{n+1}\| \le \mu\|v^n\| + \eta\|\nabla f(x^n)\|.
$$

Consider the sequence $a_n = \|v^n\|$. Then $a_{n+1} \le \mu a_n + K$, with $K = \eta M$. By part (*a*), the sequence $(a_n)_n$ is bounded.

Exercise 4.17.5 (*a*) It follows from Fubini's Theorem.
(*b*) Using part (*a*) applied to $|f|$ and $|g|$, we obtain

$$
\begin{aligned}
\|f * g\|_1 &= \int |f * g|(x)\, dx = \int \left| \int f(y)g(x - y)\, dy \right| dx \\
&\le \int \int |f(y)|\,|g(x - y)|\, dy\, dx \\
&= \int (|f| * |g|)(x)\, dx = \int |f|(x)\, dx \int |g|(x)\, dx \\
&= \|f\|_1 \|g\|_1.
\end{aligned}
$$

(*c*) It follows from part (*b*) by taking $g = G_\sigma$ and using $\|G_\sigma\|_1 = \int_{\mathbb{R}} G_\sigma(x)dx = 1$. More generally, for any $1 \le p \le \infty$ we have $\|f_\sigma\|_p \le \|f\|_p \|G_\sigma\|_1 = \|f\|_p$, i.e., if $f \in L^p$ then $f_\sigma \in L^p$. In particular, for $p = 2$, we obtain that by filtering a finite energy signal we also obtain a finite energy signal.

Exercise 4.17.6 It is a straightforward computation as in the following:

$$
\begin{aligned}
(G_{\sigma_1} * G_{\sigma_2})(x) &= \int G_{\sigma_1}(u) G_{\sigma_2}(x - u)\, du \\
&= \frac{1}{2\pi\sigma_1\sigma_2} \int e^{-\frac{1}{2}[(\frac{u}{\sigma_1})^2 + (\frac{x-u}{\sigma_2})^2]}\, du.
\end{aligned}
$$

After completing the square, the exponent can be written as

$$
\left(\frac{u}{\sigma_1}\right)^2 + \left(\frac{x - u}{\sigma_2}\right)^2 = \frac{\sigma_1^2 + \sigma_2^2}{\sigma_1^2 \sigma_2^2}\left(u - \frac{\sigma_1^2}{\sigma_1^2 + \sigma_2^2}x\right)^2 + \frac{x^2}{\sigma_1^2 + \sigma_2^2}.
$$

Changing the variable and evaluating the Gaussian integral, we have

$$(G_{\sigma_1} * G_{\sigma_2})(x) = \frac{1}{2\pi\sigma_1\sigma_2} \int e^{-\frac{\sigma_1^2+\sigma_2^2}{2\sigma_1^2\sigma_2^2}\left(u-\frac{\sigma_1^2}{\sigma_1^2+\sigma_2^2}x\right)^2} e^{-\frac{x^2}{2(\sigma_1^2+\sigma_2)^2}} \, du$$

$$= \frac{1}{2\pi\sigma_1\sigma_2} e^{-\frac{x^2}{2(\sigma_1^2+\sigma_2^2)}} \int e^{-\frac{\sigma_1^2+\sigma_2^2}{\sigma_1^2\sigma_2^2}\frac{v^2}{2}} \, dv$$

$$= \frac{1}{\sqrt{2\pi}\sqrt{\sigma_1^2+\sigma_2^2}} e^{-\frac{x^2}{2(\sigma_1^2+\sigma_2^2)}} = G_{\sqrt{\sigma_1^2+\sigma_2^2}}(x).$$

There is a proof variant using Fourier transforms as in the following. If $\mathcal{F}(f)(\xi) = \int f(x)e^{-2\pi i x\xi} \, dx$ denotes the Fourier transform of f, using the properties

$$\mathcal{F}(f * g)(\xi) = \mathcal{F}(f)(\xi)\,\mathcal{F}(g)(\xi)$$

$$\mathcal{F}(e^{-ax^2}) = \sqrt{\frac{\pi}{a}} e^{-\frac{(\pi a)^2}{a}},$$

we have

$$\mathcal{F}(G_{\sigma_1} * G_{\sigma_2})(\xi) = \mathcal{F}(G_{\sigma_1})(\xi)\,\mathcal{F}(G_{\sigma_2})(\xi) = e^{-(2\pi\sigma_1\xi)^2} e^{-(2\pi\sigma_2\xi)^2}$$

$$= e^{-2^2\pi^2(\sigma_1^2+\sigma_2^2)\xi^2} = \mathcal{F}(G_\sigma),$$

with $\sigma = \sqrt{\sigma_1^2+\sigma_2^2}$. Applying the inverse transform, $\mathcal{F}^{-1}$ yields the desired result.

Exercise 4.17.7 We consider the variational problem with constraint

$$L = \frac{1}{2}(\sigma_1^2 + \cdots + \sigma_n^2) - \lambda(\sigma_1 + \cdots + \sigma_n - s),$$

with λ Lagrange multiplier. The minimum satisfies

$$\frac{\partial L}{\partial \sigma_j} = \sigma_j - \lambda = 0 \Longrightarrow \sigma_j = \lambda, \quad \forall 1 \le j \le n.$$

Chapter 5

Exercise 5.10.1 (*a*) The mapping is $(0,0) \to 0$, $(0,1) \to 0$, $(1,0) \to 1$, $(1,1) \to 0$. The point $(1,0)$ is separated from the other points by the line $y = x - 1/2$. Then the perceptron with the output $y = H(x - y - 1/2)$ learns the previous assignment. We have $w_1 = 1, w_2 = -1, b = -1/2$.
(*b*) Similarly with (*a*).

(*c*) The output of the 1-dimensional perceptron is $y = H(-x + 1/2)$. The weight is $w = -1$ and the bias is $b = 1/2$. Since for x Boolean variable we have $\neg x = 1 - x$, the output of the associated linear neuron is $y = wx + b$, with weight and bias given by $w = -1$ and $b = 1$.

(*d*) We can visualize the learning of $x_1 \wedge x_2 \wedge x_3$ as a separation of the cube corner $(1, 1, 1)$ from the other corners. This is done by the plane $x_1 + x_2 + x_3 = 5/2$. The perceptron which learns $x_1 \wedge x_2 \wedge x_3$ has weights $w_1 = w_2 = w_3 = 1$, bias $b = -5/2$, and the output $y = H(x_1 + x_2 + x_3 - 5/2)$. The perceptron which learns $x_1 \vee x_2 \vee x_3$ has the output $y = H(x_1 + x_2 + x_3 - 1/2)$.

Exercise 5.10.2 Show that the sets A and B can be separated by two distinct lines and then find a line in between with rational weights.

Exercise 5.10.3 (*a*) The matrix $A = \mathbb{E}[XX^T]$ is diagonal, with $A_{ij} = \sigma_j^2 \delta_{ij}$. The optimal weights are $\mathbf{w}^* = A^{-1}\mathbf{b}$, with $w_j^* = b_j/\sigma_j^2 = \mathbb{E}[ZX_j]/\sigma_j^2$.
(*b*) Since Z and X_j are independent, $\mathbb{E}[ZX_j] = \mathbb{E}[Z]\mathbb{E}[X_j] = 0$. Then the cost function is $\xi(\mathbf{w}) = c + \mathbf{w}^T A\mathbf{w}$, with $c = \mathbb{E}[Z^2]$. If A is positive definite, its minimum is reached for $\mathbf{w} = 0$. A zero mean target, independent of the input, is learned with zero weights; by this choice the input does not matter.
(*c*) Newton's iteration is

$$
\begin{aligned}
\mathbf{w}^{(j+1)} &= \mathbf{w}^{(j)} - H^{-1}\nabla\xi(\mathbf{w}^{(j)}) \\
&= \mathbf{w}^{(j)} - 2A^{-1}(2A\mathbf{w}^{(j)} - 2\mathbf{b}) \\
&= -3\mathbf{w}^{(j)} + 4A^{-1}\mathbf{b}.
\end{aligned}
$$

Exercise 5.10.4 In linear regression there are provided with N vectors in $\mathbb{R}^n$, $\mathbf{x}_j$, corresponding to N numbers, z_j, $1 \le j \le N$. We look for $n + 1$ parameters, $w_0, w_1, \ldots, w_n$ such that the sum of squared errors

$$
C(\mathbf{w}) = \frac{1}{N}\sum_{j=1}^{N}(w_1\mathbf{x}_j^1 + \cdots + w_n\mathbf{x}_j^n - w_0 - z_j)^2
$$

is minimum. Let X and Z be the input and the target variables of a linear neuron. They are provided in practice by N measurements, $\mathbf{x}_j = X(\omega_j)$ and $z_j = Z(\omega_j)$, for some N state of the world $\omega_j \in \Omega$, $1 \le j \le N$. Also, let $X_0 = -1$.

The training set is given by $\mathcal{T} = \{(\mathbf{x}_j, z_j); 1 \le 1 \le N\}$. The cost function can be written using the empirical evaluation of the expectation as

$$
C(\mathbf{w}) = \mathbb{E}[(X^T\mathbf{w} - Z)^2] = \frac{1}{N}\sum_{j=1}^{N}(\mathbf{x}_j^T\mathbf{w} - z_j)^2.
$$

Hence, both the linear regression and the linear neuron minimize the same cost function.

Exercise 5.10.5 Assume that $x_0 \in (0, 1)$. Then

$$y = H\Big(\int_0^1 x \, d\delta_{x_0}(x) \Big) = H(x_0) = 1,$$

since $x_0 > 0$.
Assume $x_0 \notin (0, 1)$. Then

$$
\begin{aligned}
y &= H\Big(\int_0^1 x \, d\delta_{x_0}(x) \Big) = H\Big(\int_{\mathbb{R}} x 1_{[0,1]}(x) \, d\delta_{x_0}(x) \Big) \\
&= H(x_0 1_{[0,1]}(x_0)) = 1.
\end{aligned}
$$

Exercise 5.10.6 We have

$$\langle w_1, \mathbf{x}_{i_0} \rangle = \langle w_0 + \mathbf{x}_{i_0}, \mathbf{x}_{i_0} \rangle = \langle w_0, \mathbf{x}_{i_0} \rangle + \|\mathbf{x}_{i_0}\|^2 = \langle w_0, \mathbf{x}_{i_0} \rangle + 1 > 0,$$

since $\langle w_0, \mathbf{x}_{i_0} \rangle \le \|w_0\| \, \|\mathbf{x}_{i_0}\| = 1$.
Since all points P_j belong to a half-circle, let w^* be the unit normal vector to the diameter of the circle, which separates the points from the rest of the circle. Denote by ρ_m the angle between w^* and w_m and let $\delta = \min\{\cos \angle(w^*, \mathbf{x}_i); 1 \le i \le n\}$. Then

$$
\begin{aligned}
\langle w^*, w_m \rangle &= \langle w^*, w_{m-1} + \mathbf{x}_{i_{m-1}} \rangle = \langle w^*, w_{m-1} \rangle + \langle w^*, \mathbf{x}_{i_{m-1}} \rangle \\
&\ge \langle w^*, w_{m-1} \rangle + \delta.
\end{aligned}
$$

Iterating the inequality, we obtain $\langle w^*, w_m \rangle \ge \langle w^*, w_0 \rangle + m\delta$. Then

$$\cos \rho_m = \frac{\langle w^*, w_m \rangle}{\|w^*\| \, \|w_m\|} = \langle w^*, w_m \rangle \ge \langle w^*, w_0 \rangle + m\delta.$$

The inequality

$$1 \ge \cos \rho_m \ge \langle w^*, w_0 \rangle + m\delta$$

is contradictory, since the right term increases unbounded as $m \to \infty$, while the left side is bounded. Hence, the process should end after finitely many steps. If N is the maximum number of iterations, then $N\delta \sim 1$ which provides the estimation $N \sim 1/\delta$.

Exercise 5.10.7 The perceptron learning algorithm is valid for the case of points in a half-plane. The reason is that if a diameter separates the points defined by the unitary vectors $\mathbf{x}_i$, then the same diameter separates any other points whose vectors have the same direction as the previous ones. This follows from the equivalence $\langle w, \mathbf{x}_i \rangle > 0 \iff \langle w, \lambda \mathbf{x}_i \rangle > 0$ for all $\lambda > 0$.

Exercise 5.10.9 Consider the mean square error computed at step k, $C(w_k) = \frac{1}{2}(z_k - w_k^T x_k)^2$. The gradient is $\nabla_{w_k} C = -(z_k - w_k^T x_k)x_k$. The gradient descent rule is $w_{k+1} = w_k - \eta \nabla_{w_k} C = w_k + \eta \epsilon_k x_k$. If let $c = |x_k|$ and define $\alpha = \eta c^2$ we obtain $w_{k+1} = w_k + \alpha \frac{\epsilon_k x_k}{|x_k|^2}$, which is the α-LMS update rule.

Chapter 6

Exercise 6.6.2 (a) Consider the unit cube with one vertex at the origin and with three sides in the direction of the axes of $\mathbb{R}^3$. Since no single plane can separate the cube vertices $(0,0,0)$ and $(1,1,1)$, it follows that a single perceptron cannot learn the mapping. (b) Try a one-hidden layer network.

Exercise 6.6.3
$$y = H\big(-2H(x_1 + x_2 + 1/2) - 3/2\big) - H\big(H(x_1 + x_2 + 1/2) + 5/2\big) - 1.$$

Exercise 6.6.4 (a) By chain rule $\frac{\partial y}{\partial w} = x\sigma'(wx + b)$ and $\frac{\partial y}{\partial b} = \sigma'(wx + b)$. The gradient is $\nabla C = (\frac{\partial C}{\partial w}, \frac{\partial C}{\partial b})$, with

$$\begin{aligned}
\frac{\partial C}{\partial w} &= (y - z)\frac{\partial C}{\partial w} \\
&= (\sigma(wx + b) - z)x\sigma'(wx + b) \\
\frac{\partial C}{\partial b} &= (y - z)\frac{\partial C}{\partial b} \\
&= (\sigma(wx + b) - z)\sigma'(wx + b).
\end{aligned}$$

Using that $\sigma' \leq 1/4$ and $\sigma < 1$, we get

$$\begin{aligned}
\|\nabla C\| &= \sqrt{\Big(\frac{\partial C}{\partial w}\Big)^2 + \Big(\frac{\partial C}{\partial b}\Big)^2} \\
&= |\sigma(wx + b) - z|\,\sigma'(wx + b)\sqrt{1 + x^2} < \frac{1}{4}\sqrt{1 + x^2}\,(1 + |z|).
\end{aligned}$$

(b) For a step $\eta > 0$, we have
$$\begin{aligned}
w_{n+1} &= w_n - \eta x\sigma'(w_n x + b_n)(\sigma(w_n x + b_n) - z) \\
b_{n+1} &= b_n - \eta \sigma'(w_n x + b_n)(\sigma(w_n x + b_n) - z).
\end{aligned}$$

Exercise 6.6.5 (a) Let $X^{(0)} = X$. Taking the norm in the sensitivity relation
$$dY = \phi'(s^{(2)})W^{(2)^T}\phi'(s^{(1)})W^{(1)^T}\,dX^{(0)},$$

we obtain
$$\begin{aligned}
\|dY\| &= |\phi'(s^{(2)})|\|W^{(2)^T}\|\,|\phi'(s^{(1)})|\,\|W^{(1)^T}\|\,\|dX^{(0)}\| \\
&\leq \|\phi'\|^2\|W^{(1)}\|\,\|W^{(2)}\|\,\|dX^{(0)}\|
\end{aligned}$$

It suffices to choose $\eta = \frac{\epsilon}{\|\phi'\|^2 \|W^{(1)}\| \|W^{(2)}\|}$.

(b) If the input X is noisy, then small variations dX occur. For noise removal purposes, these variations should have the least possible effect on the output variation dY. This can be achieved in two ways: keeping the norms of the weight matrices small or choosing an activation function with ϕ' small. The weights regularization for noise removal can be added as a constraint, such as $\|W^{(1)}\|^2 + \|W^{(2)}\|^2 < 1$, to the cost function.

(c) We need to choose the activation function with the smallest derivative norm $\|\phi'\|$, i.e., the activation function with the slowest saturation. Then use that tangent hyperbolic saturates faster than logistic sigmoid.

Exercise 6.6.6 (a) The signals and outcomes of the layers are given by $s^{(1)} = w_1 x - b_1$, $x^{(1)} = \sigma(s^{(1)}) = \sigma(w_1 x - b_1)$, $s^{(2)} = w_2 \sigma(s^{(1)}) - b_2$, and $y = x^{(2)} = \sigma(s^{(2)})$. Using chain rule

$$\delta^{(2)} = \frac{\partial C}{\partial s^{(2)}} = \frac{1}{2}(y - z)\frac{\partial y}{\partial s^{(2)}} = \frac{1}{2}(y - z)\sigma'(s^{(2)})$$

$$\delta^{(1)} = \frac{\partial C}{\partial s^{(1)}} = \frac{\partial C}{\partial s^{(2)}}\frac{\partial s^{(2)}}{\partial s^{(1)}} = \delta^{(2)} w_2 \sigma'(s^{(1)}).$$

(b) The components of the gradient, ∇C, are given by

$$\frac{\partial C}{\partial w_2} = \frac{\partial C}{\partial s^{(2)}}\frac{\partial s^{(2)}}{\partial w_2} = \delta^{(2)}\sigma(s^{(1)})$$

$$\frac{\partial C}{\partial w_1} = \frac{\partial C}{\partial s^{(1)}}\frac{\partial s^{(1)}}{\partial w_1} = \delta^{(1)} x$$

$$\frac{\partial C}{\partial b_2} = \frac{\partial C}{\partial s^{(2)}}\frac{\partial s^{(2)}}{\partial b_2} = -\delta^{(2)}$$

$$\frac{\partial C}{\partial b_1} = \frac{\partial C}{\partial s^{(1)}}\frac{\partial s^{(1)}}{\partial b_1} = -\delta^{(1)}.$$

Exercise 6.6.7 (a) We shall use backwards induction. Note that $W_{ij}^{(\ell)}$ and $\delta_i^{(L)}$ are independent. Then assume that $W_{ij}^{(\ell)}$ and $\delta_i^{(k)}$ are independent for any $\ell + 1 \leq k \leq L$. We need to show that $W_{ij}^{(\ell)}$ and $\delta_i^{(\ell)}$ are independent. Since $\phi'(x) = 1$, we can express the deltas of the ℓth layer in terms of the deltas in the $(\ell + 1)$th layer as

$$\delta_i^{(\ell)} = \sum_j \delta_j^{(\ell+1)} W_{ij}^{(\ell+1)}.$$

Now, $W_{ij}^{(\ell)}$ is independent of $\delta_j^{(\ell+1)}$ by the induction hypothesis and independent of $W_{ij}^{(\ell+1)}$ as weights in distinct layers. Therefore, $W_{ij}^{(\ell)}$ is independent

of the combination $\sum_j \delta_j^{(\ell+1)} W_{ij}^{(\ell+1)}$, and hence independent of $\delta_i^{(\ell)}$.
(b) By induction, similar with part (a).

Exercise 6.6.8 Using the approximate formula $Var f(X) \approx f'(\mathbb{E}[X])^2 Var(X)$
for $f(x) = \sigma(wx + b)$, we obtain

$$VarY \approx \sigma'(w\mathbb{E}[X] + b)^2 Var(wX + b) = \sigma'(b)^2 w^2 Var(X) = \sigma'(b)^2 w^2.$$

When the bias takes a large positive or negative value, the sigmoid saturates
and its derivative becomes small; consequently $Var(Y)$ decreases.

Exercise 6.6.9 Using an idea similar with the one used in the previous problem, we have

$$
\begin{aligned}
Var(Y) &= \sum_i \alpha_i^2 Var \sigma(w_i X + b_i) \approx \sum_i \alpha_i^2 \sigma'(b_i)^2 w_i^2 Var \sigma(X) \\
&= \sum_i \alpha_i^2 \sigma'(b_i)^2 w_i^2.
\end{aligned}
$$

Exercise 6.6.10 Since $p(x) = 1/(b - a)$, then

$$H(p) = -\int_a^b p(x) \ln p(x)\, dx = -\frac{1}{b-a} \int_a^b \ln \frac{1}{b-a}\, dx = \ln(b-a).$$

Exercise 6.6.11 Let $p(x) = \frac{1}{\sigma\sqrt{2\pi}} e^{-\frac{(x-\mu)^2}{2\sigma^2}}$. Using relations $\int_{\mathbb{R}} p(x)\, dx = 1$ and
$\int_{\mathbb{R}} (x - \mu)^2 p(x)\, dx = \sigma^2$, we have

$$
\begin{aligned}
H(p) &= -\int_{\mathbb{R}} p(x) \ln p(x)\, dx \\
&= \frac{1}{2} \ln(2\pi) \int_{\mathbb{R}} p(x)\, dx + \ln \sigma \int_{\mathbb{R}} p(x)\, dx + \frac{1}{2\sigma^2} \int_{\mathbb{R}} (x - \mu)^2 p(x)\, dx \\
&= \frac{1}{2} \ln(2\pi) + \ln \sigma + \frac{1}{2} = \ln(\sigma\sqrt{2\pi e}).
\end{aligned}
$$

Chapter 7

Exercise 7.11.1 It follows from $|\tanh x| \le 1$.

Exercise 7.11.2 (a) Let $f \in \mathcal{F}$, with $f(x) = ax + b$. Then modulus inequalities
yield $|f(x)| \le |a||x| + |b| \le |a| + |b| < 1$, for all $x \in [0, 1]$. The family $\mathcal{F}$ is
uniformly bounded with $M = 1$.
Fix $\epsilon > 0$ and choose $\eta = \epsilon/|a|$. Then for any $x, x' \in [0, 1]$, with $|x - x'| < \eta$,
we have $|f(x) - f(x')| = |ax + b - ax' - b| = |a||x - x'| < \epsilon$. Hence, the family
$\mathcal{F}$ is equicontinuous.

(*b*) Applying the Arzela-Ascoli Theorem, we obtain: Among all possible outcomes of a linear neuron (with 1-dimensional input and output), $y = ax + b$, which satisfy the regularization constraints, $|a| + |b| < 1$, there is a sequence $y_n = a_n x + b_n$ which converges uniformly to an affine function, $f(x) = ax + b$. On short, the linear neuron has the capability of learning some affine function.

Exercise 7.11.3 By the Fundamental Theorem of Calculus, we have $f(x_2) - f(x_1) = \int_{x_1}^{x_2} f'(x)\, dx$. Then Cauchy's inequality provides

$$|f(x_2) - f(x_1)|^2 = \left| \int_{x_1}^{x_2} f'(x)\, dx \right|^2 \leq \left(\int_{x_1}^{x_2} 1\, dx \right) \left(\int_{x_1}^{x_2} f'(x)^2\, dx \right)$$
$$\leq M |x_2 - x_1|.$$

Then for any given $\epsilon > 0$, choose $\eta = \epsilon^2 / M$ and apply the definition of equicontinuity.

Exercise 7.11.4 Fix $\epsilon = 1$. Then for any $x_0 \in D$, there is $\eta > 0$ such that for $x \in (x_0 - \eta, x_0 + \eta)$ we have

$$|f(x)| \leq |f(x) - f(x_0)| + |f(x_0)| \leq 1 + M, \quad \forall f \in \mathcal{F}.$$

Note that $\bigcup_{x_0 \in D} (x_0 - \eta, x_0 + \eta) = (a, b)$, any $x \in (a, b)$ belongs to a neighborhood of the type $(x_0 - \eta, x_0 + \eta)$.

Exercise 7.11.5 Let $\phi \in \mathcal{F}$. Fix $\epsilon > 0$. Then by equicontinuity of $\mathcal{F}$, there is $\eta_k > 0$ such that $|f_j(x) - f_j(x')| < \epsilon / k$ for $|x - x'| < \eta_k$, $1 \leq j \leq k$. Choose $\eta = \min_{1 \leq j \leq k} \eta_j$. We have

$$|\phi(x) - \phi(x')| \leq \sum_{j=1}^{k} |w_j|\, |f_j(x) - f_j(x')| \leq \sum_{j=1}^{k} |f_j(x) - f_j(x')| < \epsilon,$$

for $|x - x'| < \eta$.

Exercise 7.11.6 It follows from Dini's Theorem applied to the decreasing positive sequence of functions $f_n = |f(x) - G_n(x)|$.

Exercise 7.11.7 Periodic functions with period T can be considered as functions on the circle of radius $R = 2\pi/T$, which is a compact set, denoted here by K. Therefore, $f \in C(K)$. The set of all functions F, which can be written as trigonometric sums, form a subalgebra of $C(K)$, which separates points and contains constants. An application of the Stone-Weierstrass Theorem produces the desired result.

Exercise 7.11.8 The main ingredient in the proof of Proposition 7.7.2 is that $\|\sigma'\| \leq 1/4$. Given that most sigmoid functions are symmetric with respect to the origin, we have $\|\sigma'\| = \sigma'(0)$. Therefore, the result holds for all increasing differentiable sigmoid functions satisfying $\sigma'(0) < \lambda < 1$.

Exercise 7.11.9 By Arzela-Ascoli Theorem, we may assume, eventually passing to a subsequence, that f_j converges uniformly to f on $[a, b]$. Then f_j^2 converges uniformly to f^2 on $[a, b]$. Then $\int_a^b f_j^2(x)\, dx \to \int_a^b f^2(x)\, dx$. Using hypothesis, it follows that $\int_a^b f^2(x)\, dx = 0$. Since $f^2 \geq 0$, it follows that $f = 0$. That is $\lim_{j \to \infty} f_j = 0$, uniformly. We also note that the proof could have been done by the contradiction method, without using Arzela-Ascoli Theorem.

Exercise 7.11.10 (a) Let $\epsilon > 0$ be arbitrary fixed. Using the uniform continuity of the function $K(s, t)$ (continuous function on a compact set), for any $\rho > 0$, there is $\eta_\rho > 0$ such that if $|s' - s| < \eta_\rho$ then $|K(s', t) - K(s, t)| < \rho$. Since ρ is arbitrary, we choose it such that $\rho < \epsilon / \sqrt{M(d - c)}$.
For any $g \in \mathcal{F}_M$, applying Cauchy's inequality, we have

$$
\begin{aligned}
|g(s') - g(s)| &= \left| \int_c^d \Big(K(s', t) - K(s, t) \Big) h(t)\, dt \right| \\
&\leq \left(\int_c^d |K(s', t) - K(s, t)|^2\, dt \right)^{1/2} \left(\int_c^d h(t)^2\, dt \right)^{1/2} \\
&\leq \rho \sqrt{d - c} \sqrt{M} < \epsilon.
\end{aligned}
$$

Hence, we have $|g(s') - g(s)| < \epsilon$, whenever $|s' - s| < \eta$, with ϵ and η independent of g.
(b) Using Cauchy's inequality, for any $g \in \mathcal{F}_M$, we have

$$
\begin{aligned}
|g(s)| &= \left| \int_c^d K(s, t) h(t)\, dt \right| \leq \int_c^d |K(s, t)|\, |h(t)|\, dt \\
&\leq \left(\int_c^d K^2(s, t)\, dt \right)^{1/2} \left(\int_c^d h(t)^2\, dt \right)^{1/2} \\
&\leq \left(\int_c^d K^2(s, t)\, dt \right)^{1/2} M^{1/2}.
\end{aligned}
$$

Being continuous on the compact $[a, b]$, the function $s \to \int_c^d K^2(s, t)\, dt$ is bounded. It follows that all functions g have a common upper bound.

Exercise 7.11.11 (a) It's just a verification of the algebra axioms. (b) It follows from the Stone-Weierstrass Theorem of approximation. (c) Any continuous function on $[a, b]$ can be learned uniformly by a sequence of polynomials in e^x.

Chapter 8

Exercise 8.8.1 (a) For any $\phi \in C_0^\infty$ we have

$$-\int H(x - x_0)\phi'(x)\,dx = -\int_{x_0}^\infty \phi'(x)\,dx = -(\phi(\infty) - \phi(x_0)) = \phi(x_0)$$
$$= \int \phi\delta(x - x_0)\,dx.$$

(b) For any $\phi \in C_0^\infty$

$$-\int ReLU(x)\phi'(x)\,dx = -\int_0^\infty x\phi'(x)\,dx = -\left(x\phi(x)\Big|_0^\infty - \int_0^\infty \phi(x)\,dx\right)$$
$$= \int_0^\infty \phi(x)\,dx = \int H(x)\phi(x)\,dx.$$

The others are similar.

Exercise 8.8.2 Shifting indices and parsing the sums, we have

$$\sum_{i=0}^{N-1} \alpha_i 1_{[x_i, x_{i+1})}(x) = \sum_{i=0}^{N-1} \alpha_i[H(x - x_i) - H(x - x_{i+1})]$$
$$= \sum_{i=0}^{N-1} \alpha_i H(x - x_i) - \sum_{j=1}^{N} \alpha_{j-1} H(x - x_j)$$
$$= \alpha_0 H(x - x_0) + \sum_{i=1}^{N-1} \alpha_i H(x - x_i)$$
$$- \sum_{i=1}^{N-1} \alpha_{i-1} H(x - x_i) - \alpha_{N-1} H(x - x_N)$$
$$= \alpha_0 H(x - x_0) + \sum_{i=1}^{N-1} (\alpha_i - \alpha_{i-1}) H(x - x_i) - \alpha_{N-1} H(x - x_N).$$

Equating against the sum $\sum_{i=0}^{N} c_i H(x - x_i)$, we obtain the following coefficients:
$c_0 = \alpha_0$, $c_i = \alpha_i - \alpha_{i-1}$, and $c_N = \alpha_{N-1}$.

Exercise 8.8.4 Consider a decreasing sequence of positive numbers, $\epsilon_n \searrow 0$, and consider the function $f_n(x) = |g(x) - g_{\epsilon_n}(x)|$. Given the construction of $g_{\epsilon_n}(x)$, we have $0 \leq f_{n+1}(x) \leq f_n(x)$ for any x. By Dini's Theorem, the sequence of functions f_n converges uniformly to 0, i.e., g_{ϵ_n} converges, uniformly, to $g(x)$.

Exercise 8.8.6 Changing variables and using L'Hospital's rule, we have

$$
\begin{aligned}
\lim_{\alpha \searrow 0} \varphi_\alpha(x) &= \lim_{\alpha \searrow 0} \alpha \ln(1 + e^{x/\alpha}) = \lim_{\alpha \searrow 0} \frac{\ln(1 + e^{x/\alpha})}{\frac{1}{\alpha}} \\
&= \lim_{t \nearrow \infty} \frac{\ln(1 + e^{tx})}{t} = \lim_{t \nearrow \infty} \frac{(1 + e^{tx})'}{1 + e^{tx}} \\
&= \lim_{t \nearrow \infty} \frac{x e^{tx}}{1 + e^{tx}} = \lim_{t \nearrow \infty} \frac{x}{1 + e^{-tx}} \\
&= \lim_{t \nearrow \infty} x \sigma_t(x) = \begin{cases} x, & \text{if } x \geq 0 \\ 0, & \text{if } x < 0, \end{cases}
\end{aligned}
$$

which leads to the first part of the result. If let $t = 1/\alpha > 0$ then

$$
\varphi_\alpha(x) = \alpha \ln(1 + e^{x/\alpha}) = \frac{\ln(1 + e^{xt})}{t},
$$

which, for α small (t large), behaves as $\frac{x}{1 + e^{-xt}}$, which is increasing in t for $x > 0$.

Chapter 9

Exercise 9.8.1 In this case the target space is the metric space $(\mathbb{R}, d)$, with the distance function $d(x, y) = |x - y|$. The approximation space is $\mathbb{Q}$. Any real number can be learned by rational numbers. Actually, one of the constructions of the real number's field is based on this property.

Exercise 9.8.2 The target space is the space of continuous functions $C[a, b]$, endowed with the distance $d(f, g) = \sup_{[a,b]} |f(x) - g(x)|$. The approximation space is the set of all polynomials $\mathcal{P}[a, b]$. Any continuous function can be learned by polynomial functions on a compact interval.

Exercise 9.8.3 Let $\mathcal{T} = \{\alpha x_1 + \lambda x_0; \alpha, \lambda \in \mathbb{R}\}$ be the space generated by the noncollinear vectors x_0 and x_1. Define the functional $L : \mathcal{T} \to \mathbb{R}$, $L(t) = \lambda$, where $t = \alpha x_1 + \lambda x_0$. It is obvious that L is linear. By a procedure similar with the one used in the proof of Lemma 9.3.1 we can show that L is also bounded, with the norm $\|L\| < 1/\delta$, where δ is the distance between x_0 and the support line of x_1. By Hahn-Banach Theorem the functional L can be extended to a linear bounded functional on $\mathcal{X}$ keeping the same bound. We can easily verify that $L(x_0) = 1$ and $L(x_1) = 0$.

Exercise 9.8.4 Applying the construction used in **Exercise 9.8.3** symmetrically for x_0 and then for x_1, we obtain two linear bounded functionals L_1 and L_2 on $\mathcal{X}$ such that

$$
L_1(x_0) = 0, \quad L_1(x_1) = 1, \quad L_2(x_0) = 1, \quad L_2(x_1) = 0.
$$

Consider the average functional $L = \frac{1}{2}(L_1 + L_2)$, which is linear and bounded. We have $L(x_0) = L(x_1) = 1/2$. Moreover, its norm satisfies

$$\|L\| = \frac{1}{2}\|L_1 + L_2\| \leq \frac{1}{2}(\|L_1\| + \|L_2\|) \leq \frac{1}{2}\left(\frac{1}{\delta_0} + \frac{1}{\delta_1}\right) = \frac{\delta_0 + \delta_1}{2\delta_0\delta_1}.$$

By a similar procedure we can prove the following more general statement: given the linearly independent set of vectors $\{x_0, x_1, \ldots, x_N\}$ in $\mathcal{X}$, there is a bounded linear functional L on $\mathcal{X}$, such that $L(x_0) = L(x_1) = \cdots = L(x_N)$.

Exercise 9.8.5 Consider in **Exercise** 9.8.4 the space $\mathcal{X} = C[a, b]$ and the independent vectors $x_0 = \sin t$, $x_1 = \cos t$. Then there is a bounded linear functional $L : C[a, b] \to \mathbb{R}$ such that $L(\sin t) = L(\cos t)$. By the representation theorem, Theorem E.5.6, exists a unique finite signed Borel measure μ on $[a, b]$, such that

$$L(f) = \int_a^b f(t)\, d\mu(t), \qquad \forall f \in C[a, b].$$

Therefore, the identity $L(\sin t) = L(\cos t)$ becomes

$$\int_a^b \sin t\, d\mu(t) = \int_a^b \sin t\, d\mu(t).$$

Moreover, since $\|L\| = |\mu|([a, b])$. An upper bound for $\|L\|$ is provided in **Exercise** 9.8.4.

Exercise 9.8.6 (*a*) Note that $L(P) = P(1)$, so L is a linear functional. Since

$$|L(P)| = |P(1)| \leq \sup_{[0,1]} |P(x)| = \|P\|_\infty,$$

it follows that $\|L\| \leq 1$.

(*b*) The functional L can be extended linearly to $C[0, 1]$ by Hahn-Banach Theorem, keeping the bound $\|L\| \leq 1$. Since now L is a bounded linear functional on $C[0, 1]$, the representation theorem, Theorem E.5.6, provides the measure μ such that $L(f) = \int_0^1 f(x)\, d\mu(x)$, for all $f \in C[0, 1]$. In particular, for $f = P$ we obtain $L(P) = \int_0^1 P(x)\, d\mu(x)$, or equivalently,

$$\int_0^1 P(x)\, d\mu(x) = a_0 + a_1 + \cdots + a_n, \qquad \forall P \in \mathcal{P}([0, 1]).$$

Exercise 9.8.7 (*b*) The function $\phi(x) = e^{-x^2} \in L^1(\mathbb{R})$ and $\int_{\mathbb{R}} e^{-x^2}\, dx \neq 0$. By point 2. of Remark 9.3.19 the function $\phi(x)$ is discriminatory in L^1-sense.

Exercise 9.8.8 (*a*) The output is

$$y = \sum_{i=1}^{N_2} \alpha_i \sigma\left(\sum_{j=1}^{N_1} w_{ji}\sigma(\lambda_j x + b_j) + \beta_i\right), \tag{1}$$

with the following notations: b_j are the biasses of neurons in the first hidden layer, β_i denote the biasses of the neurons in the second hidden layer, λ_j are the weights from the input to the first layer, w_{ji} are the weights between the hidden layers, and α_i are the weights between the second hidden layer and the output.

(b) Adding, subtracting, and multiplying by scalars expressions of type (1), we obtain something of the same type, with different parameters, eventually some equal to zero.

Exercise 9.8.9 (a) Take the continuous function to be affine, $f(x) = w^T x + b$. (b) Let $\mathcal{U}$ denote the set of continuous functions of type (1), i.e., the set of all outputs of two-hidden layer FNNs. We show that $\mathcal{U}$ is dense in $C(I_n)$. By contradiction, if $\mathcal{U}$ is not dense, by Lemma 9.3.2 there is a nonzero bounded linear functional L on $C(I_n)$ such that $L_{|\mathcal{U}} = 0$. By the representation theorem, Theorem E.5.6, there is a signed measure μ on I_n such that

$$\int_{I_n} \sum_{i=1}^{N_2} \alpha_i \sigma \Big(\sum_{j=1}^{N_1} w_{ji} \sigma(\lambda_j x + b_j) + \beta_i \Big) \, d\mu(x) = 0,$$

for all values of the parameters. In particular,

$$\int_{I_n} \sigma \Big(\sum_{j=1}^{N_1} w_{ji} \sigma(\lambda_j x + b_j) + \beta_i \Big) \, d\mu(x) = 0.$$

Since $f(x) = \sum_{j=1}^{N_1} w_{ji} \sigma(\lambda_j x + b_j) + \beta_i$ is continuous and σ is strong discriminatory, it follows that $\mu = 0$, fact that contradicts $\|L\| = |\mu|(I_n)$.

Chapter 10

Exercise 10.11.1 (a) The network output is of the form

$$G(x) = \sum_{i=1}^{2} \alpha_i H(\mathbf{w}^T \mathbf{x} + b_i) = \sum_{i=1}^{2} \alpha_i H(w_{1i} x_1 + w_{2i} x_2 + b_i).$$

We make the simplificative assumptions $w_{11} = w_{12} = w_1$ and $w_{21} = w_{22} = w_2$. Then we are left with 6 parameters, which satisfy 4 equations:

$$
\begin{aligned}
0 &= G(0,0) = \alpha_1 H(b_1) + \alpha_2 H(b_2) \\
1 &= G(0,1) = \alpha_1 H(w_2 + b_1) + \alpha_2 H(w_2 + b_2) \\
1 &= G(1,0) = \alpha_1 H(w_1 + b_1) + \alpha_2 H(w_1 + b_2) \\
0 &= G(1,1) = \alpha_1 H(w_1 + w_2 + w_2 + b_1) + \alpha_2 H(w_1 + b_2).
\end{aligned}
$$

One possible solution is $\alpha_1 = 1$, $\alpha_2 = -1$, $w_1 = -0.5$, $w_2 = -2$, $b_1 = 2.25$, $b_2 = 0.25$. Hence, an output which learns the XOR function is

$$G(x_1, x_2) = H(-0.5x_1 - 2x_2 + 2.25) - H(-0.5x_1 - 2x_2 + 0.25).$$

Exercise 10.11.3 Obviously, $\psi(x) = e^{-x^2} \in L^1(\mathbb{R})$, with the Fourier transform $\Psi(\xi) = \widehat{\psi}(\xi) = \sqrt{\pi}e^{-\xi^2/4}$, so $\Psi(1) = \sqrt{\pi}/e^{1/4}$. Then Irie-Miyake's formula becomes

$$f(\mathbf{x}) = \frac{e^{1/4}}{(2\pi)^n \sqrt{\pi}} \int_{\mathbb{R}^{n+1}} e^{i\omega_0 - (\mathbf{x}^t \mathbf{w} - w_0)^2} F(\mathbf{w}) \, d\mathbf{w}.$$

Exercise 10.11.4 Consider the function

$$f(x) = \begin{cases} e^{-\frac{1}{x-\frac{1}{2}}}, & \text{if } \frac{1}{2} < x \leq 1 \\ 0, & \text{if } 0 \leq x \leq \frac{1}{2}. \end{cases}$$

It can be shown that f is continuous on $[0, 1]$ but it is not analytic. (If f would be analytic on $(0, 1)$, then, since $f_{|(0,1/2)} = 0$, then by the identity theorem of analytic functions we get $f = 0$ on $(0, 1)$, contradiction.) Since the logistic sigmoid is analytic, then the outcome of the network is also analytic, while f is not.

Exercise 10.11.5 Use that condition (ii) becomes the convergence of the series $\displaystyle\sum_{i \geq 1} \frac{g_i^2}{\lambda_i^{2(n+1)}}$.

Exercise 10.11.6 Integrating in relation $K^{(n)}(t, t) = \sum_{i \geq 1} \lambda_i^n e_i^2(t)$ yields

$$\int_0^1 K^{(n)}(t, t) \, dt = \sum_{i \geq 1} \lambda_i^n \int_0^1 e_i^2(t) \, dt = \sum_{i \geq 1} \lambda_i^n,$$

and use that the integral on the left side is finite.

Exercise 10.11.7 (a) Consider intervals of length 2 centered at 1, 3, 5, and 7. Then choose the simple function

$$c(x) = 1_{[0,2)}(x) + 3.1_{[2,4)}(x) + 2.1_{[4,6)}(x) + 1_{[7,8)}(x),$$

which learns the data. (b) Since each indicator function can be written as a difference of step functions, for instance, $1_{[2,4)}(x) = H(x - 2) - H(x - 4)$, we obtain

$$c(x) = H(x - 0) + 2H(x - 2) - H(x - 4) - H(x - 6) - H(x - 8),$$

and choose this function as $G(x)$.

Exercise 10.11.8 It follows from the composition associativity property.

Chapter 11

Exercise 11.10.1 (a) Since $\{\omega; c \le t\} = \begin{cases} \varnothing, & \text{if } c > t \\ \Omega, & \text{if } c \le t \end{cases}$, the sigma-field generated by c is the sigma-field generated by the sets $\varnothing$ and Ω, which is the trivial algebra $\{\varnothing, \Omega\}$.

(b) It is shown by double inclusion as in the following: Since

$$\{\omega; c + X(\omega) \le t\} = \{\omega; X(\omega) \le t - c\} \in \mathfrak{S}(X),$$

then $\mathfrak{S}(c + X) \subset \mathfrak{S}(X)$. Similarly, since

$$\{\omega; X(\omega) \le t\} = \{\omega; c + X(\omega) \le t + c\} \in \mathfrak{S}(c + X),$$

it follows that $\mathfrak{S}(X) \subset \mathfrak{S}(c + X)$.

(c) Assume $c > 0$. Use that $\{\omega; cX(\omega) \le t\} = \{\omega; X(\omega) \le t/c\} \in \mathfrak{S}(X)$ and hence $\mathfrak{S}(cX) \subset \mathfrak{S}(X)$. The converse inclusion can be shown similarly.

Exercise 11.10.2 (a) Since the value of Y_n is determined by $X_1, \ldots, X_n$, then $Y_n = F(X_1, \ldots, X_n)$, and hence $\mathfrak{S}(Y_n) \subset \mathfrak{S}(X_1, \ldots, X_n)$. The strict inclusion follows from part (b).

(b) Since $X_n = Y_n - Y_{n-1}$, then $\mathfrak{S}(X_n) \subset \mathfrak{S}(Y_{n-1}, Y_n)$. We have

$$\begin{aligned} \mathfrak{S}(X_1, \ldots, X_n) &= \mathfrak{S}\Big(S(X_1) \cup \cdots \cup \mathfrak{S}(X_n)\Big) \\ &\subset \mathfrak{S}\Big(\mathfrak{S}(Y_0, Y_1) \cup \cdots \cup \mathfrak{S}(Y_{n-1}, Y_n)\Big) \\ &\subset \mathfrak{S}(Y_1, \ldots, Y_n), \end{aligned}$$

which proves the converse inclusion.

(c) The formula for Z_n can be shown inductively. The identity of sigma-fields follows from part (b), which implies

$$\mathfrak{S}(Z_1, \ldots, Z_n) = \mathfrak{S}(Y_1, \ldots, Y_n) = \mathfrak{S}(X_1, \ldots, X_n).$$

Exercise 11.10.3 We draw the Venn diagram with sets A and B intersecting. We obtain 4 regions: $R_1 = A \backslash B$, $C_2 = A \cap B$, $R_3 = (A \cup B)^c$, and $R_4 = B \backslash A$. The desired sigma-algebra is generated by the partition $\{R_1, R_2, R_3, R_4\}$, i.e., its elements are obtained by taking unions of subsets. There are 2^4 elements. Using de Morgan's relations we obtain the sets given in the exercise. For instance, $R_1 \cup R_4 = (A \backslash B) \cup (B \backslash A) = (A \cap B^c) \cup (A^c \cap B)$.

Exercise 11.10.4 We draw the Venn diagram with sets A, B, and C intersecting. We obtain a partition of Ω into 8 distinct regions. The $\mathfrak{S}$-field will have $2^8 = 256$ elements, too many to be written down explicitly.

Exercise 11.10.5 (a) We have

$$\mathcal{F}_{E_1,E_2} = \{\emptyset, E_1, E_2, E_1 \cup E_2, E_1 \cap E_2, (E_1 \cup E_2)^c, (E_1 \cap E_2)^c, \Omega\}.$$

It suffices to recover the sets E_1 and E_2. This is done by

$$E_1 = \Big(\bigcup_{j=2^N} E_j \Big)^c, \qquad E_2 = \Big(\bigcup_{j=1, j \neq 2}^{N} E_j \Big)^c.$$

(b) Since $\mathcal{F}_{E_1} = \{\emptyset, E_1, E_1^c, \Omega\}$, and E_1 can be recovered from the other sets as $E_1 = \bigcup_{i=2}^{N} E_i$, it follows that $\mathcal{F}_{E_1}$ is a recoverable body of information. Using that $\mathcal{F}_{E_1} \subset \mathcal{F}_{E_1,E_2}$, it follows that $\mathcal{F}_{E_1}$ is not maximal.

Exercise 11.10.6 This is an application of Theorem C.1.3. It will be proved by double inclusion. Let $\mathcal{P} = \{\bigcap_{i=1}^{k} \{\omega; X_i(\omega) \leq x_i\}, k \geq 1, x_i \in \mathbb{R}\}$, which is a p-system (closed to intersections). For $k = 1$, we have $\{\omega; X_i(\omega) \leq x_i\} \in \mathcal{P}$, which implies $\mathfrak{S}(\{\omega; X_i(\omega) \leq x_i\}) \in \mathfrak{S}(\mathcal{P})$, or equivalently $\mathfrak{S}(X_i) \subset \mathfrak{S}(\mathcal{P})$. Taking unions yields $\bigcup_i \mathfrak{S}(X_i) \subset \mathfrak{S}(\mathcal{P})$, and taking sigma-algebras we get $\mathfrak{S}\Big(\bigcup_i \mathfrak{S}(X_i)\Big) \subset \mathfrak{S}(\mathcal{P})$, which means the inclusion $\mathfrak{S}(X) \subset \mathfrak{S}(\mathcal{P})$. Next, we show the opposite inclusion. Since $\mathcal{P}$ is included in $\bigcap_i \mathfrak{S}(X_i)$, we also have $\mathcal{P} \subset S\Big(\bigcup_{i=1}^{n} \mathfrak{S}(X_i)\Big) = \mathcal{D}$. As a p-system, $\mathcal{P}$, included into a d-system, $\mathcal{D}$, by Theorem C.1.3 from Appendix, we obtain

$$\mathfrak{S}(\mathcal{P}) \subset S\Big(\bigcup_{i=1}^{n} \mathfrak{S}(X_i) \Big) = \mathfrak{S}(X),$$

which proves the opposite inclusion.

Exercise 11.10.7 (a) Since $\tilde{Y}$ is determined by Y, we have $\tilde{Y} = f(Y)$. Then, by Proposition D.5.1 of Appendix, we have $\mathfrak{S}(\tilde{Y}) \subset \mathfrak{S}(Y)$, or $\tilde{\mathcal{E}} \subset \mathcal{E}$. Then $\mathcal{I} \backslash \mathcal{E} \subset \mathcal{I} \backslash \tilde{\mathcal{E}}$. Taking the generating sigma-algebra on both sides yields $\mathcal{L} \subset \tilde{\mathcal{L}}$. (b) If a neuron is dropped from the ℓth layer, a verbatim argument shows that $\mathcal{L}^{(\ell)} \subset \tilde{\mathcal{L}}^{(\ell)}$. To conclude, dropping out neurons leads to information loss in the layer where the dropping occurs.

Exercise 11.10.8 (a) In this case Y depends on $\tilde{Y}$, so $S(Y) \subset S(\tilde{Y})$. A similar proof with the one use in **Exercise** 11.10.7 yields the inclusion $\tilde{\mathcal{L}} \subset \mathcal{L}$.

Exercise 11.10.9 In case **a** one unit in the hidden layer would suffice to classify the points in each of the half-planes. In case **b** we need three units in the hidden layer, each unit corresponding to a side of the triangle. Each unit learns a half-plane and a triangle can be written as the intersection of three half-planes. Similarly, in case **c** we need 4 neurons. Using the activation function $H(x)$ leads to sharp corners for the triangle and rectangle, while using a sigmoid function leads to roundup corners.

Exercise 11.10.10 It suffices to show that both parts are equal to the sigma-field $\mathfrak{S}(\mathcal{F} \cup \mathcal{G} \cup \mathcal{H})$. This will be done by double inclusion.

We show first $(\mathcal{F} \vee \mathcal{G}) \vee \mathcal{H} \subset \mathfrak{S}(\mathcal{F} \cup \mathcal{G} \cup \mathcal{H})$. Since $\mathcal{H} \subset \mathfrak{S}(\mathcal{F} \cup \mathcal{G} \cup \mathcal{H})$ and $\mathcal{F} \cup \mathcal{G} \subset \mathfrak{S}(\mathcal{F} \cup \mathcal{G} \cup \mathcal{H})$, it follows that $\mathfrak{S}(\mathcal{F} \cup \mathcal{G}) \subset \mathfrak{S}(\mathcal{F} \cup \mathcal{G} \cup \mathcal{H})$, and then $\mathfrak{S}(\mathcal{F} \cup \mathcal{G}) \cup \mathcal{H} \subset \mathfrak{S}(\mathcal{F} \cup \mathcal{G} \cup \mathcal{H})$. Taking the sigma operator on both sides yields $\mathfrak{S}(\mathfrak{S}(\mathcal{F} \cup \mathcal{G}) \cup \mathcal{H}) \subset \mathfrak{S}(\mathcal{F} \cup \mathcal{G} \cup \mathcal{H})$, which is $(\mathcal{F} \vee \mathcal{G}) \vee \mathcal{H} \subset \mathfrak{S}(\mathcal{F} \cup \mathcal{G} \cup \mathcal{H})$.

We show now $\mathfrak{S}(\mathcal{F} \cup \mathcal{G} \cup \mathcal{H}) \subset (\mathcal{F} \vee \mathcal{G}) \vee \mathcal{H}$. Starting from $\mathcal{F} \cup \mathcal{G} \subset \mathfrak{S}(\mathcal{F} \cup \mathcal{G})$, we have $\mathcal{F} \cup \mathcal{G} \cup \mathcal{H} \subset \mathfrak{S}(\mathcal{F} \cup \mathcal{G}) \cup \mathcal{H}$. Taking the sigma operator we obtain $\mathfrak{S}(\mathcal{F} \cup \mathcal{G} \cup \mathcal{H}) \subset \mathfrak{S}(\mathfrak{S}(\mathcal{F} \cup \mathcal{G}) \cup \mathcal{H})$, which is $\mathfrak{S}(\mathcal{F} \cup \mathcal{G} \cup \mathcal{H}) \subset (\mathcal{F} \vee \mathcal{G}) \vee \mathcal{H}$. Hence, $\mathfrak{S}(\mathcal{F} \cup \mathcal{G} \cup \mathcal{H}) = (\mathcal{F} \vee \mathcal{G}) \vee \mathcal{H}$. Similarly, we can show $\mathfrak{S}(\mathcal{F} \cup \mathcal{G} \cup \mathcal{H}) = \mathcal{F} \vee (\mathcal{G} \vee \mathcal{H})$.

Exercise 11.10.11 Assume the neurons are of sigmoid type. We have

$$
\begin{aligned}
X_1^{(1)} &= \sigma(w_{11} X_1^{(0)} + b_1) \\
X_2^{(1)} &= \sigma(w_{12} X_1^{(0)} + w_{22} X_2^{(0)} + b_2) \\
X_3^{(1)} &= \sigma(w_{23} X_2^{(0)} + b_3).
\end{aligned}
$$

Since $X^{(1)}$ depends on $X^{(0)}$, we obtain $\mathcal{I}^{(1)} \subset \mathcal{I}^{(0)}$. We can also solve for $X^{(0)}$ in terms of $X^{(1)}$, which implies $\mathcal{I}^{(0)} \subset \mathcal{I}^{(1)}$.

Exercise 11.10.12 Similar to the solution of **Exercise** 11.10.11.

Exercise 11.10.13 If $\mathcal{I} = \mathcal{E}$ then obviously $\mathcal{I} \backslash \mathcal{E} = \emptyset$, so $\mathcal{L} = \mathfrak{S}(\mathcal{I} \backslash \mathcal{E}) = \mathfrak{S}(\{\emptyset\}) = \{\emptyset, \Omega\}$. Conversely, if $\mathcal{L} = \{\emptyset, \Omega\}$, then $\mathfrak{S}(\mathcal{I} \backslash \mathcal{E}) = \{\emptyset, \Omega\}$, so either $\mathcal{I} \backslash \mathcal{E} = \{\emptyset\}$, or $\mathcal{I} \backslash \mathcal{E} = \{\Omega\}$, or $\mathcal{I} \backslash \mathcal{E} = \{\emptyset, \Omega\}$. The first case implies $\mathcal{I} = \mathcal{E}$. The other two cases lead to contradiction, since Ω is subtracted with $\mathcal{E}$ and cannot belong to $\mathcal{I} \backslash \mathcal{E}$.

Exercise 11.10.14 Assume, by contradiction, that $\mathcal{H} \subseteq \mathcal{E} \subseteq \mathcal{F}$. Using Proposition 11.4.5 part (a), the first inclusion yields that $\mathcal{I} = \mathcal{L}$. Similarly, Proposition 11.4.5 part (b) applied to the second inclusion implies $\mathcal{I} \neq \mathcal{L}$. The two statements obtained are contradictory.

Exercise 11.10.15 It is a consequence of Exercise 11.10.14.

Exercise 11.10.16 Let $\mathcal{U} \subset \mathcal{F}$ be a subfield of a recoverable information field. Then $\mathcal{I} \backslash \mathcal{F} \subset \mathcal{I} \backslash \mathcal{U}$. Taking the $\mathfrak{S}$-operator yields $\mathcal{I} = \mathfrak{S}(\mathcal{I} \backslash \mathcal{F}) \subset \mathfrak{S}(\mathcal{I} \backslash \mathcal{U})$. It follows that the inclusion is in fact equality.

Chapter 12

Exercise 12.13.1 Use the inequality

$$H(X) = \int_e^\infty \frac{1}{x(\ln x)^2}(\ln x + 2\ln\ln x)\,dx \geq \int_e^\infty \frac{1}{x\ln x}\,dx = +\infty.$$

Exercise 12.13.2 (a) Let $\mu = \mathcal{E}[X]$, $\sigma^2 = Var(X)$ and consider the normal density $q(x) = \frac{1}{\sigma\sqrt{2\pi}}e^{-\frac{(x-\mu)^2}{2\sigma^2}}$. Then use the inequality

$$H(X) = -\int_\mathbb{R} p(x)\ln p(x)\,dx \leq -\int_\mathbb{R} p(x)\ln q(x)\,dx = \frac{1}{2}\ln(2\pi e\sigma^2).$$

(b) It follows from

$$-H(X) = \int_\mathbb{R} p(x)\ln p(x)\,dx \leq \ln M\int_\mathbb{R} p(x)\,dx = \ln M.$$

Exercise 12.13.3 (a) We have

$$I(X,Y|Z) = \int p(x,y,z)\ln p(x|z)dxdydz + \int p(x,y,z)\ln p(y|z)dxdydz$$
$$- \int p(x,y,z)\ln p(x,y|z)\,dxdydz$$
$$= \int p(x,y,z)\ln\frac{p(x,y,z)}{p(x|z)p(y|z)}\,dxdydz$$
$$= D_{KL}[p(x,y,z)||p(x|z)p(y|z)].$$

(b) Use that the Kullback-Leibler divergence is nonnegative and the fact that

$$D_{KL}[p(x,y,z)||p(x|z)p(y|z)] = 0$$

if and only if $p(x,y,z) = p(x|z)p(y|z)$, i.e., when X and Y are independent, given Z.

Exercise 12.13.4 It suffices to prove that the system cannot have more than two distinct solutions. Assume there are two distinct solutions, X_1 and X_2, and consider their difference $X_0 = X_1 - X_2 \neq 0$. Then $AX_0 = AX_1 - AX_2 = b - b = 0$, so X_0 is a nonzero solution for the homogeneous system $AX = 0$. For any $\alpha \in \mathbb{R}$, $X = \alpha X_0$ is also a solution of the homogeneous system, since $AX = \lambda AX_0 = 0$; hence the space of solutions

$$\mathcal{S} = \{X \in \mathbb{R}^n; AX = 0\}$$

is at least one-dimensional. On the other hand, a well-known result (based on the isomorphism theorem) states that $\dim \mathcal{S} = m - rank(A) = m - m = 0$, which leads to a contradiction. It follows that $X_0 = 0$, i.e., $X_1 = X_2$.

Exercise 12.13.5 If $m = n$ then Q is invertible, then $p(\mathbf{x}) = Q^{-1}p(\mathbf{y}) = Q^{-1}softmax(Q^{-1}h)$.

Exercise 12.13.6 Using that the natural logarithmic function is increasing, we have $\ln(\sum_j e^{u_j}) \geq \ln(e^{u_j}) = u_j$. Hence, $\ln(\sum_j e^{u_j}) \geq \max_j u_j$. Then choose $u_j = ((Q^T Q)^{-1} Q^T h)_j$.

Exercise 12.13.8 For (a) and (b) apply the formula for neural manifold dimension. (c) The one at (b) has a larger capacity because has more parameters.

Exercise 12.13.9 (a) increases, (b) decreases, (c) decreases.

Exercise 12.13.10 (a) Using an analog of relation (12.11.34) for the one-hidden layer network, applied to the first and second layers, and then to the second and third layers, we have, respectively,

$$\mathcal{S}(X) = \mathcal{S}(U, X_{101}, \ldots, X_{784}), \qquad \mathcal{S}(U) = \mathcal{S}(\widetilde{Y}, U_{11}, \ldots, U_{100}).$$

Concatenating, yields

$$\mathcal{S}(X) = \mathcal{S}(\widetilde{Y}, U_{11}, \ldots, U_{100}, X_{101}, \ldots, X_{784}). \tag{2}$$

(b) From relation (12.11.34) applied to the zero-hidden layer network, with input X and output Y, we have

$$\mathcal{S}(X) = \mathcal{S}(Y, X_{11}, \ldots, X_{784}). \tag{3}$$

Comparing formulas (3) and (2), the transitivity implies

$$\mathcal{S}(Y, X_{11}, \ldots, X_{100}, X_{101}, \ldots, X_{784}) = \mathcal{S}(\widetilde{Y}, U_{11}, \ldots, U_{100}, X_{101}, \ldots, X_{784}). \tag{4}$$

Exercise 12.13.11 The maximum entropy of the input variable X is about 9.614 bits per image. The entropy of X, given U, is equal to the average number of pixels of X corresponding to each entry of U

$$2^{H(X|U)} = \frac{784}{100},$$

which implies $H(X|U) = 2.9708$ bits. Then the mutual information of X and U is given by

$$I(X, U) = H(X) - H(X|U) = 9.614 - 2.9708 = 6.643,$$

namely, about 6.643 bits of information of each image are used toward each entry of the hidden layer. From the previous analysis of the zero-hidden layer network we have $I(X, \widetilde{Y}) = 3.322$. These lead to the inequalities

$$H(X) > I(X, U) > I(X, \widetilde{Y}).$$

Exercise 12.13.12 Using that $H(X_h|X_v) = H(X_h, X_v) - H(X_h)$ and $H(X_v|X_h) = H(X_v, X_h) - H(X_v)$, it follows $H(X_h|X_v) = H(X_v|X_h)$ if and only if $H(X_h) = H(X_v)$.

Exercise 12.13.14 From the invariance property, see Proposition 12.7.5, we have $I(X'_h, X'_v) = I(X_h, X_v)$. Using formula $I(X, Y) = H(X) + H(Y) - H(X, Y)$, it follows that $H(X'_h, X'_v) = H(X_h, X_v)$ if and only if

$$H(X'_h) + H(X'_v) = H(X_h) + H(X_v).$$

Exercise 12.13.16 (*a*) By the definition of mutual information, we have

$$\begin{aligned}
I(X, Y, Z) &= H(X) + H(Y) + H(Z) - H(X, Y, Z) \\
&= [H(X) + H(Y) - H(X, Y)] + [H(X, Y) + H(Z) - H(X, Y, Z)] \\
&= I(X, Y) + I((X, Y), Z).
\end{aligned}$$

(*b*) It can be shown similarly that for any $1 \le k \le n$ we have

$$I(X_1, \ldots, X_n) = I(X_1, \ldots, X_k) + I((X_1, \ldots, X_k), (X_{k+1}, \ldots, X_n)).$$

Exercise 12.13.19 It follows from the definition of the determinant:

$$|\det W| = \left| \sum \epsilon_{i_1 \cdots i_n} w_{1i_1} \cdots w_{ni_n} \right| \le \sum |w_{1i_1}| \cdots |w_{ni_n}| \le n! \, c^n.$$

Exercise 12.13.20 Since

$$I(X_1, X_2) = H(X_1) + H(X_2) - H(X_1, X_2)$$

we need to compute the entropy terms from the right side. By **Exercise** 6.6.11 we have

$$\begin{aligned}
H(X_1) &= \frac{1}{2} \ln(2\pi) + \ln \sigma_1 + \frac{1}{2} \\
H(X_2) &= \frac{1}{2} \ln(2\pi) + \ln \sigma_2 + \frac{1}{2}.
\end{aligned}$$

Using that the bivariate distribution of (X_1, X_2) is given by

$$f(x_1, x_2) = \frac{1}{2\pi\sigma_1\sigma_2\sqrt{1-\rho^2}} e^{\left\{ -\frac{1}{1-\rho^2} \left[\frac{(x_1-\mu_1)^2}{2\sigma_1^2} - \rho\frac{(x_1-\mu_1)(x_2-\mu_2)}{\sigma_1\sigma_2} + \frac{(x_2-\mu_2)^2}{2\sigma_2^2} \right] \right\}}$$

the joint entropy term can be computed as

$$
\begin{aligned}
H(X_1, X_2) &= -\iint f(x_1, x_2) \ln f(x_1, x_2)\, dx_1 dx_2 \\
&= \ln(2\pi\sigma_1\sigma_2\sqrt{1-\rho^2}) \iint f(x_1, x_2)\, dx_1 dx_2 \\
&\quad + \frac{1}{1-\rho^2}\Big[\iint f(x_1, x_2)\frac{(x_1-\mu_1)^2}{2\sigma_1^2}\, dx_1 dx_2 \\
&\quad -\rho \iint f(x_1, x_2)\frac{(x_1-\mu_1)(x_2-\mu_2)}{\sigma_1\sigma_2}\, dx_1 dx_2 \\
&\quad + \iint f(x_1, x_2)\frac{(x_2-\mu_2)^2}{2\sigma_2^2}\, dx_1 dx_2 \Big] \\
&= \ln(2\pi\sigma_1\sigma_2\sqrt{1-\rho^2}) + \frac{1}{1-\rho^2}\Big[\frac{1}{2} - \frac{\rho}{\sigma_1\sigma_2}Cov(X_1, X_2) + \frac{1}{2}\Big] \\
&= 1 + \ln(2\pi\sigma_1\sigma_2\sqrt{1-\rho^2}).
\end{aligned}
$$

Therefore, after cancellations, we obtain

$$
\begin{aligned}
I(X_1, X_2) &= H(X_1) + H(X_2) - H(X_1, X_2) \\
&= -\ln\sqrt{1-\rho^2} = -\frac{1}{2}\ln(1-\rho^2).
\end{aligned}
$$

The mutual information depends explicitly on the correlation coefficient, ρ. Therefore, at least in the case of normal distributions, correlation is an expression of the mutual information between the variables.

Exercise 12.13.21 (a) Applying the invariance property given by Proposition 12.7.5, we have

$$
\begin{aligned}
I(X_1, X_2) &= I(F_{X_1}(X_1), F_{X_2}(X_2)) = I(U_1, U_2) \\
&= H(U_1) + H(U_2) - H(U_1, U_2) \\
&= \int_0^1 \int_0^1 c(u_1, u_2) \ln c(u_1, u_2)\, du_1 du_2,
\end{aligned}
$$

where we used that $H(U_i) = 0$, since U_i are uniformly distributed on $[0, 1]$. (b) If X_1 and X_2 are independent, then the copula is $C(u_1, u_2) = u_1 u_2$ and its density is $c(u_1, u_2) = 1$. Therefore, using (a), we obtain $I(X_1, X_2) = 0$.

Chapter 13

Exercise 13.8.1 The dimension of the associated neural manifold is

$$
r = 784 \times 200 + 200 \times 100 + 100 \times 50 + 50 \times 10 + 350 = 182,300.
$$

Exercise 13.8.2 The dimension of the neural manifold is

$$
\begin{aligned}
f_N(k) &= d^{(0)}\frac{N}{k} + (k-1)\left(\frac{N}{k}\right)^2 + \frac{N}{k}d^{(L)} + N \\
&= (d^{(0)} + d^{(L)})\frac{N}{k} + \frac{k-1}{k^2}N^2 + N.
\end{aligned}
$$

If let $u = 1/k$ the above formula is written as

$$
\begin{aligned}
\phi_N(u) &= (d^{(0)} + d^{(L)})Nu + (u - u^2)N^2 + N \\
&= -N^2u^2 + N(d^{(0)} + d^{(L)} + N)u + N.
\end{aligned}
$$

This is a quadratic function in u, which achieves its maximum for

$$
u = \frac{d^{(0)} + d^{(L)} + N}{2N}.
$$

For N large in comparison with $d^{(0)} + d^{(L)}$ the number of layers is $k = 2$.

Exercise 13.8.3 The dimension of the neural manifold is $r = 784N + 10N = 794N$. Solving the inequation $794N \geq 550000$, we get $N \geq 693$. Hence, for N larger than 700 the networks exhibit overfitting effects.

Exercise 13.8.4 Solve the inequality $h^2 + 796h \geq 550,000$.

Exercise 13.8.5 Assume first that u and v are arbitrary vectors, so they can be decomposed as

$$
u = \sum_{i=1}^{r} u_i\epsilon_i + u^N N, \qquad v = \sum_{j=1}^{r} v_j\epsilon_j + v^N N,
$$

where $\{\epsilon_1, \ldots, \epsilon_r\}$ is a basis in the tangent space $T_{\mathbf{y}}\mathcal{S}$ and N is the normal unit to $\mathcal{S}$ at $\mathbf{y}$. Their Euclidean inner product is

$$
\begin{aligned}
\langle u, v \rangle &= \left\langle \sum_{i=1}^{r} u_i\epsilon_i, \sum_{j=1}^{r} v_j\epsilon_j \right\rangle + u^N v^N \\
&= \sum u_i v_j \langle \epsilon_i, \epsilon_j \rangle + u^N v^N \\
&= \sum u_i v_j g_{ij} + u^N v^N.
\end{aligned}
$$

If the vectors u and v are tangent to $\mathcal{S}$, their normal parts are zero, so $u^N = v^N = 0$, and hence $\langle u, v \rangle = \sum u_i v_j g_{ij}$. Then use that u and v are orthogonal iff $\langle u, v \rangle = 0$.

Exercise 13.8.6 An affine subspace of $\mathbb{R}^n$ is a k-plane in $\mathbb{R}^n$, with $1 \leq k \leq n$. Any geodesic in this k-plane is a straight line. Also $L = 0$, because when a vector field U is differentiated with respect to a vector field V, both U and V being in the given k-plane, the directional derivative, $\nabla_U V$, belongs to the same k-plane, so $L(U, V) = (\nabla_U V)^\perp = 0$.

Exercise 13.8.7 For any two vector fields, U and V, tangent to S, we have the orthogonal decomposition $\nabla_U V = (\nabla_U V)^{\|} + (\nabla_U V)^\perp$. For any curve $c(s)$ in S, take $U = V = \dot{c}(s)$ and obtain the kinematic equation

$$\nabla_{\dot{c}(s)} \dot{c}(s) = (\nabla_{\dot{c}(s)} \dot{c}(s))^{\|} + (\nabla_{\dot{c}(s)} \dot{c}(s))^\perp = D_{\dot{c}(s)} \dot{c}(s) + L(\dot{c}(s), \dot{c}(s)),$$

where ∇ and D are the directional derivatives on $\mathcal{M}$ and S, respectively, and L is the second fundamental form of S with respect to $\mathcal{M}$. We use that a geodesic is a curve with zero acceleration.
$(a) \implies (b)$ Assume $L = 0$. Then $\nabla_{\dot{c}(s)} \dot{c}(s) = D_{\dot{c}(s)} \dot{c}(s)$. If $c(s)$ is a geodesic in S, then $D_{\dot{c}(s)} \dot{c}(s) = 0$. Then $\nabla_{\dot{c}(s)} \dot{c}(s) = 0$, so $c(s)$ is a geodesic in $\mathcal{M}$.
$(b) \implies (a)$ If $D_{\dot{c}(s)} \dot{c}(s) = 0$, then $\nabla_{\dot{c}(s)} \dot{c}(s) = 0$, and hence $L(\dot{c}, \dot{c}) = 0$, for any geodesic $c(s)$ in S. Since for any given vector $v \in T_p S$, there is a geodesic with initial velocity $\dot{c}(0) = v$, it follows that $L(v, v) = 0$, for all $v \in T_p S$, with p arbitrary in S. Then use the polarization formula

$$L(v, w) = \frac{1}{2}[L(v + w, v + w) - L(v, v) - L(w, w)]$$

to get $L = 0$.

Exercise 13.8.8 (a) It follows from the change of variables formula in an integral. By chain rule, $\gamma'(t) = \dot{c}(s)\phi'(t)$, so taking the norm, and using that ϕ is increasing, we get $\|\gamma'(t)\| = \|\dot{c}(s)\|\phi'(t)$. Then

$$\begin{aligned} L(\gamma) &= \int_c^d \|\gamma'(t)\| \, dt = \int_c^d \|\dot{c}(\phi(t))\| \underbrace{\phi'(t) \, dt}_{=ds} \\ &= \int_a^b \|\dot{c}(s)\| \, ds = L(c). \end{aligned}$$

Similar computation for the energy.
(b) Apply the integral version of Cauchy's inequality

$$\left(\int_a^b f(x)g(x) \, dx \right)^2 \leq \int_a^b f(x)^2 \, dx \int_a^b g(x)^2 \, dx$$

for $f = \|\dot{c}\|$ and $g = 1$. The identity is reached for constant speed curves, that is, $\|\dot{c}(s)\| = $ constant.
(c) Locally, length minimizing curves and energy minimizing curve are equivalent. A geodesic is also an energy minimizing curve; along a geodesic the acceleration is zero, and hence the magnitude of the velocity is constant.

Exercise 13.8.9 (*a*) Let $(r(t), z(t))$ be a plane curve with $r(t) > 0$. If the curve is rotated about the z-axis, a surface of revolution is obtained. This is parametrized by

$$\phi(t, s) = \big(r(t) \cos s, \, r(t) \sin s, \, z(t)\big), \qquad 0 \leq s \leq 2\pi,$$

where t measures the position on the given curve and s measures the rotation angle. It can be shown (see the book [85], for instance) that the coefficients of the second fundamental form of a revolution surface are given by

$$L_{ij} = \frac{1}{\sqrt{\dot{r}^2 + \dot{z}^2}} \begin{pmatrix} \dot{r}\ddot{z} - \dot{z}\ddot{r} & 0 \\ 0 & r\dot{z} \end{pmatrix}.$$

We shall apply this to two parametrizations of the unit sphere.
(*i*) Choose $r(t) = \cos t$, $z(t) = \sin t$, $-\pi/2 < t < \pi/2$ and obtain the sphere parametrization

$$\phi(t, s) = \big(\cos t \cos s, \, \cos t \sin s, \, \sin t\big), \qquad 0 \leq s \leq 2\pi, \, -\pi/2 < t < \pi/2.$$

Using the previous formula yields $L_{ij} = \begin{pmatrix} 1 & 0 \\ 0 & \cos^2 t \end{pmatrix}$. The norm of L is its largest eigenvalue, so $\|L\| = 1$.
(*ii*) Let $r(t) = \sqrt{1 - t^2}$, $z(t) = t$, $-1 < t < 1$. The parametrization is

$$\phi(t, s) = \big(\sqrt{1 - t^2} \cos s, \, \sqrt{1 - t^2} \sin s, \, t\big), \qquad 0 \leq s \leq 2\pi, \, -1 < t < 1.$$

A computation shows that $L_{ij} = \begin{pmatrix} \frac{1}{(1-t^2)^2} & 0 \\ 0 & 1 \end{pmatrix}$. The norm is, again, $\|L\| = 1$.

We make the remark that even if the coefficients L_{ij} are dependent on the parametrization, the norm $\|L\|$ is not.

Exercise 13.8.10 The outputs of the sigmoid neurons are $y_1 = \sigma(w_1 x + b_1)$ and $y_2 = \sigma(w_2 x + b_2)$. Their combination has the output

$$y = \lambda y_1 + (1 - \lambda) y_2 = \lambda \sigma(w_1 x + b_1) + (1 - \lambda)\sigma(w_2 x + b_2).$$

Since $(w_1, w_2, b_1, b_2, \lambda) \in \mathbb{R}^5$, the associated neural manifold has dimension 5.

Exercise 13.8.11 When a neuron is dropped, all in the incoming and outgoing weights as well as its bias are equal to zero. Hence, the dimension of the neural manifold decreases. Having fewer neurons in the network produces an outcome which is less nonlinear. Hence, the neural manifold tends to have a smaller embedded curvature. Both reduce overfitting and if too many neurons are dropped, it may lead to an underfit.

Exercise 13.8.12 All formulas are applications of the product rule and symmetry of the second derivative.

Chapter 14

Exercise 14.13.1 (a) Consider the expansion $e_i = \sum_{j=1}^{n} \alpha_{ij} v_j$. Then

$$
G_{ik} \;=\; e_i^T G e_k = \Big(\sum_{j=1}^{n} \alpha_{ij} v_j \Big)^T G \Big(\sum_{r=1}^{n} \alpha_{kr} v_r \Big) = \sum_{j,r} \alpha_{ij} \alpha_{kr} \underbrace{v_j^T G v_r}_{=0} = 0,
$$

so $G = \mathbb{O}_n$. (b) It follows from using linearity, part (a) and considering $G = A - B$.

Exercise 14.13.2 Let $Y_1 = u^T X$ and $Y_2 = v^T X$. The joint cumulative distribution function is

$$
\begin{aligned}
F_{Y_1 Y_2}(a,b) \;&=\; P(Y_1 \le a, Y_2 \le b) = P(u^T X \le a, v^T X \le b) \\
&=\; P(u_1 X_1 + u_2 X_2 \le a, v_1 X_1 + v_2 X_2 \le b) \\
&=\; \iint_{\{u^T x \le a\} \cap \{v^T x \le b\}} p_{X_1 X_2}(x_1, x_2)\, dx_1 dx_2.
\end{aligned}
$$

We compute the ratio

$$
\frac{F_{Y_1 Y_2}(a + \Delta a, b + \Delta b) - F_{Y_1 Y_2}(a,b)}{\Delta a \Delta b} \;=\; \frac{1}{\Delta a \Delta b} \iint_{D_{ab}} p_{X_1 X_2}(x_1 x_2)\, dx_1 dx_2,
$$

where $D_{ab} = \{a \le u^T x \le a + \Delta a\} \cap \{b \le v^T x \le b + \Delta b\}$ is a rectangular region with side directions parallel to u and v. Using that $R^{-1}(D_{ab}) = [a, a + \Delta a, b + \Delta b$, changing variables, using Fubini's Theorem and the fact that X is rotational invariant, we obtain

$$
\begin{aligned}
&\frac{1}{\Delta a \Delta b} \iint_{D_{ab}} p_{X_1 X_2}(x_1 x_2)\, dx_1 dx_2 \\
=\; &\frac{1}{\Delta a \Delta b} \iint_{[a, a+\Delta a] \times [b, b+\Delta b]} p_{X_1 X_2}(R(x_1, x_2)) \underbrace{\lfloor \det R \rfloor}_{=1} dx_1 dx_2 \\
=\; &\frac{1}{\Delta a} \int_a^{a+\Delta a} p_{X_1}(x_1)\, dx_1 \; \frac{1}{\Delta b} \int_b^{a+\Delta b} p_{X_2}(x_1)\, dx_1.
\end{aligned}
$$

Taking the limits $\Delta a \to 0$ and $\Delta b \to 0$, we obtain

$$
\frac{\partial^2}{\partial a \partial b} F_{Y_1 Y_2}(a,b) = p_{X_1}(a) p_{X_2}(b)
$$

or $f_{Y_1Y_2}(a,b) = f_1(a)f_2(b)$, which implies that Y_1 and Y_2 are independent.

(b) $Y_1 = u^T X$ is normally distributed with zero mean and $Var(Y_1) = |u| = 1$. Hence, $u^T X, v^T X \sim \mathcal{N}(0,1)$.

(c) If follows from the trigonometric circle. We take $\phi = \arg u$ and since u and v are orthogonal, $\arg v = \phi + \pi/2$. Then $v = (\cos(\phi + \pi/2), \sin(\phi + \pi/2)) = (-\sin\phi, \cos\phi)$.

Exercise 14.13.3 (a) We have

$$Y = \begin{pmatrix} Y_1 \\ Y_2 \end{pmatrix} = \begin{pmatrix} u_1 & u_2 \\ v_1 & v_2 \end{pmatrix} \begin{pmatrix} X_1 \\ X_2 \end{pmatrix} = MX.$$

By Exercise 14.13.2 M is a rotation matrix and it preserves independence.

(b) If $Y_1 = u^T X$ and $Y_2 = v^T X$ are independent, then $Y_1' = \alpha u^T X$ and $Y_2' = \beta v^T X$ are also independent, for nonzero constants α, β.

Exercise 14.13.4 We expand the determinant $D(\lambda) = \det(M - \lambda \mathbb{I}_n)$ over the first row and then expand each of the $(n-1)$ minors over the first column to obtain

$$D(\lambda) = -\lambda \begin{vmatrix} -\lambda & 0 & \cdots & 0 \\ \vdots & \ddots & & \vdots \\ 0 & \cdots & -\lambda & 0 \\ 0 & \cdots & 0 & -\lambda \end{vmatrix} - a_1 \begin{vmatrix} b_1 & 0 & \cdots & 0 \\ b_2 & -\lambda & \cdots & 0 \\ \vdots & & \ddots & \vdots \\ b_n & \cdots & 0 & -\lambda \end{vmatrix}$$

$$+ a_2 \begin{vmatrix} b_1 & -\lambda & \cdots & 0 \\ b_2 & 0 & \cdots & 0 \\ \vdots & & & \ddots \vdots \\ b_n & \cdots & 0 & -\lambda \end{vmatrix} + \cdots + (-1)^{n+1} a_n \begin{vmatrix} b_1 & -\lambda & \cdots & 0 \\ b_2 & 0 & -\lambda & \cdots & 0 \\ \vdots & & & \ddots & \vdots \\ b_n & \cdots & & 0 & -\lambda \end{vmatrix}$$

$$= -\lambda(-\lambda)^n - a_1 b_1 (-\lambda)^{n-1} - a_2 b_2 (-\lambda)^{n-1} - \cdots + (-1)^{2n+1} a_n b_n (-\lambda)^{n-1}$$

$$= (-\lambda)^{n-1} [\lambda^2 - \sum a_i b_i] = (-\lambda)^{n-1} [\lambda^2 - a^T b].$$

Solving the equation $D(\lambda) = 0$, we obtain the desired solutions.

Exercise 14.13.5 (a) Use $\sigma'(x) = \sigma(x)(1 - \sigma(x))$. (c) Let $\|\phi'\| = \sup_x |\phi'(x)|$. Then

$$0 \le g_{00} = \mathbb{E}^{P_X}[\phi'(w^T X + b)^2] \le \|\phi'\|^2 \mathbb{E}^{P_X}[1] = \|\phi'\|^2$$

$$g_{0k} = \int x_k \phi'(w^T x + b)^2 \, p(x) \, dx$$

$$= \int x_k \sqrt{p(x)} \, \phi'(w^T x + b)^2 \sqrt{p(x)} \, dx.$$

Then Cauchy-Schwartz inequality implies

$$g_{0k}^2 \leq \int \left(x_k \sqrt{p(x)}\right)^2 dx \int \left(\phi'(w^T x + b)^2 \sqrt{p(x)}\right)^2 dx$$

$$= \int x_k^2 p(x)\, dx \int \phi'(w^T x + b)^2 p(x)\, dx \leq \|\phi'\|^2 \int x_k^2 p(x)\, dx$$

$$= \|\phi'\|^2 \mathbb{E}[X_k^2].$$

Similarly, we have

$$g_{jk}^2 \leq \int \left(x_j x_k \sqrt{p(x)}\right)^2 dx \int \left(\phi'(w^T x + b)^2 \sqrt{p(x)}\right)^2 dx$$

$$= \int x_j^2 x_k^2 p(x)\, dx \int \phi'(w^T x + b)^2 p(x)\, dx \leq \|\phi'\|^2 \int x_j^2 x_k^2 p(x)\, dx$$

$$= \|\phi'\|^2 \mathbb{E}[X_j^2 X_k^2].$$

To obtain part (*b*) use that $\|\sigma'\| = \frac{1}{4}$.

Exercise 14.13.6

$$g_{00} = \mathbb{E}[\phi'(w^T X + b)^2] = \int_{I_n} \phi'(w^T X + b)^2 dx_1 \cdots dx_n$$

$$= \frac{1}{w_1 \cdots w_n} \int_0^{w_1} \cdots \int_0^{w_n} \phi'(u_1 + \cdots + u_n + b)^2\, du_1 \cdots du_n.$$

Similar computation applies to other coefficients.

Exercise 14.13.8

$$g_{\alpha_j \alpha_k} = \mathbb{E}[\phi(w_j X + b_j)\phi(w_k X + b_k)] = \frac{1}{\sqrt{2\pi}} \int \phi(w_j x + b_j)\phi(w_k + b_k) e^{-\frac{1}{2}x^2}\, dx.$$

Similar for other coefficients.

Exercise 14.13.10 Taking the logarithm in the relation

$$p_{X_1, X_2}(x_1, x_2; \theta) = p_{X_1}(x_1; \theta) p_{X_2|X_1}(x_2|x_1; \theta)$$

we get $\ell_{X_1, X_2}(\theta) = \ell_{X_1}(\theta) + \ell_{X_2|X_1}(\theta)$. Differentiating, then multiplying by $p_{X_1, X_2}(x_1, x_2; \theta)$ and integrating in x_1 and x_2, we obtain

$$\mathbb{E}^{P_{X_1 X_2}}[\partial_{\theta_i \theta_j} \ell_{X_1, X_2}(\theta)] = \mathbb{E}^{P_{X_1}}[\partial_{\theta_i \theta_j} \ell_{X_1}(\theta)] + \mathbb{E}^{P_{X_2|X_1}}[\partial_{\theta_i \theta_j} \ell_{X_2|X_1}(\theta)],$$

where we used $\int p_{X_2|X_1}(x_2|x_1; \theta)\, dx_2 = 1$. This means

$$g_{ij}(X_1, X_2; \theta) = g_{ij}(X_1; \theta) + g_{ij}(X_2|X_1; \theta).$$

Exercise 14.13.11 (*a*) It follows from the definition of the Euclidean scalar product.

(*b*) using the definition of the gradient, we have

$$
\begin{aligned}
g(\nabla_g f, X) &= \sum_{i,j} g_{ij}(\nabla_g f)^i X^j = \sum_{i,j,k} g_{ij} g^{ik} \frac{\partial f}{\partial \theta_k} X^j \\
&= \sum_{j,k} \delta_j^k \frac{\partial f}{\partial \theta_k} X^j = \sum_{k=1}^{N} X^k \frac{\partial f}{\partial \theta_k}
\end{aligned}
$$

(*c*) It follows from (*a*) and (*b*).

Exercise 14.13.12 (*a*) We have

$$
\begin{aligned}
\|\nabla_g f\|_g^2 &= g(\nabla_g f, \nabla_g f) = \sum_{i,j} g_{ij}(\nabla_g f)^i (\nabla_g f)^j \\
&= \sum_{i,j,k} g_{ij} g^{ik} \frac{\partial f}{\partial \theta_k} g^{jp} \frac{\partial f}{\partial \theta_p} = \sum_{k,p} g^{kp} \frac{\partial f}{\partial \theta_k} \frac{\partial f}{\partial \theta_p} \\
&= (\nabla_{Eu} f)^T g^{-1}(\theta) \nabla_{Eu} f.
\end{aligned}
$$

(*b*) Similarly,

$$
\|\nabla_{Eu} f\|_g^2 = \sum_{i,j} g_{ij}(\nabla_{Eu} f)^i (\nabla_{Eu} f)^j = (\nabla_{Eu} f)^T g(\theta) \nabla_{Eu} f.
$$

(*c*) Let $(\nabla_{Eu} f)(p) = 0$. By part (*a*) we obtain $\|(\nabla_g f)(p)\|_g^2 = 0$ and since g is nondegenerate, it follows that $(\nabla_g f)(p) = 0$. Conversely, assume now that $(\nabla_g f)(p) = 0$. Using formula $\frac{\partial f}{\partial \theta_p} = \sum_j g_{jp}(\nabla_g f)^j$ it follows that $\frac{\partial f}{\partial \theta_p}(p) = 0$, and hence $(\nabla_{Eu} f)(p) = 0$.

Exercise 14.13.13 (*a*) $\mathbb{E}[\hat{\mu}] = \frac{1}{n} \sum_{j=1}^{n} \mathbb{E}[X_j] = \frac{n\mu}{n} = \mu$.

(*b*) Since $p_\mu(x) = \frac{1}{\sqrt{2\pi}} e^{-\frac{(x-\mu)^2}{2}}$, then $\partial_\mu^2 \ln p_\theta = -1$, so $I(\mu) = -\mathbb{E}[\partial_\mu^2 \ln p_\theta] = 1$.

(*c*) The Fisher information contained in n independent random variables, $X_1, \dots, X_n$, is the sum

$$
I(X_1, \dots, X_n; \theta) = \sum_{j=1}^{n} I(X_i; \theta) = n.
$$

Then $Var(\hat{\mu}) = \frac{1}{n^2} \sum_j Var(X_i) = \frac{1}{n^2} n Var(X_1) = \frac{1}{n} = \frac{1}{I(X_1,\dots,X_n;\theta)}$, i.e., the Cramér-Rao bound is achieved.

Exercise 14.13.14 Since we know that $\lambda = \mathbb{E}[X]$, it is natural to consider the estimator $\hat{\lambda} = \frac{1}{n} \sum_{j=1}^{n} X_j$, with $X_j \sim Pois(\lambda)$ independent random variables. The Fisher information contained in n independent Poisson distributed random variables, $X_1, \ldots, X_n$, is

$$
\begin{aligned}
I(X_1, \ldots, X_n; \lambda) &= \sum_{j=1}^{n} I(X_i; \lambda) = nI(X; \lambda) = -n\mathbb{E}[\partial_\lambda^2 \ln p_\lambda(X)] \\
&= -n \sum_{k \geq 0} \frac{\lambda^k e^{-\lambda}}{k!} \partial_\lambda^2 (k \ln \lambda - \lambda) = ne^{-\lambda} \sum_{k \geq 1} \frac{\lambda^{k-1}}{(k-1)!} \\
&= \frac{n}{\lambda} e^{-\lambda} e^\lambda = \frac{n}{\lambda}.
\end{aligned}
$$

The estimator $\hat{\lambda}$ is unbiased, with variance

$$
Var(\hat{\lambda}) = \frac{1}{n^2} nVar(X_1) = \frac{\lambda}{n} = \frac{1}{I(X_1, \ldots, X_n; \lambda)}.
$$

Exercise 14.13.15 Since $\overline{X} \sim \mathcal{N}(\mu, \frac{1}{N})$, then $p_{\overline{X}}(u; \mu) = \frac{\sqrt{N}}{\sqrt{2\pi}} e^{-\frac{N}{2}(u-\mu)^2}$, so the log-likelihood function becomes

$$
\ell_{\overline{X}}(\mu) = \ln \frac{\sqrt{N}}{\sqrt{2\pi}} - \frac{N}{2}(u-\mu)^2.
$$

The square of the score function is $\left(\partial_\mu \ell_{\overline{X}}(\mu)\right)^2 = N^2(\overline{X} - \mu)^2$. Therefore, the Fisher information induced by $\overline{X}$ is

$$
\begin{aligned}
I(\overline{X}) &= \mathbb{E}^{P_{\overline{X}}}[(\partial_\mu \ell_{\overline{X}}(\mu))^2] = \mathbb{E}^{P_{\overline{X}}}[N^2(\overline{X} - \mu)^2] \\
&= N^2 Var(\overline{X}) = N^2 \frac{1}{N} = N.
\end{aligned}
$$

From Exercise 14.13.13, part (b), we have $I(X) = 1$. Hence, $I(\overline{X}) = NI(X)$.

Exercise 14.13.16 (a) Since $p_{X_1, X_2}(x_1, x_2; \theta) = p_{X_1}(x_1; \theta)p_{X_2}(x_2; \theta)$, taking the logarithmic function, we obtain $\ell_{X_1, X_2}(\theta) = \ell_{X_1}(\theta) + \ell_{X_2}(\theta)$. Differentiate to get

$$
\partial_{\theta_i \theta_j} \ell_{X_1, X_2}(\theta) = \partial_{\theta_i \theta_j} \ell_{X_1}(\theta) + \partial_{\theta_i \theta_j} \ell_{X_2}(\theta).
$$

Then multiplying by $p_{X_1, X_2}(x_1, x_2; \theta)$ and integrating with respect to x_1 and x_2, using $\int p_{X_1, X_2}(x_1, x_2; \theta)\, dx_1 = p_{X_2}(x_2; \theta)$ and $\int p_{X_1, X_2}(x_1, x_2; \theta)\, dx_2 = p_{X_1}(x_1; \theta)$, leads to

$$
\mathbb{E}^{P_{X_1 X_2}}[\partial_{\theta_i \theta_j} \ell_{X_1, X_2}(\theta)] = \mathbb{E}^{P_{X_1}}[\partial_{\theta_i \theta_j} \ell_{X_1}(\theta)] + \mathbb{E}^{P_{X_2}}[\partial_{\theta_i \theta_j} \ell_{X_2}(\theta)]
$$

which is equivalent to

$$
g_{ij}(X_1, X_2; \theta) = g_{ij}(X_1; \theta) + g_{ij}(X_2; \theta).
$$

(b) If $X_1, \ldots, X_N$ are independent random variables, then

$$g(X_1, \ldots, X_N; \theta) = \sum_{j=1}^{N} g(X_j; \theta).$$

(c) If $X_1, \ldots, X_N \sim X$ are i.i.d., by part (b) we have

$$g(X_1, \ldots, X_N; \theta) = \sum_{j=1}^{N} g(X_j; \theta) = N g(X; \theta).$$

Therefore, the inverse is

$$g(X_1, \ldots, X_N; \theta)^{-1} = \frac{1}{N} g(X; \theta)^{-1}.$$

(d) $\hat{\theta}(N)$ is an asymptotically efficient estimator for θ if for $N \to \infty$

$$Cov(\hat{\theta}(N)) = \mathbb{E}\big[(\hat{\theta}(N) - \theta)(\hat{\theta}(N) - \theta)^T\big] \sim g(X_1, \ldots, X_N; \theta)^{-1} = \frac{1}{N} g(X; \theta)^{-1},$$

Multiplying by N and taking the limit, we obtain

$$\lim_{N \to \infty} N \mathbb{E}[(\hat{\theta}(N) - \theta)(\hat{\theta}(N) - \theta)^T] = g^{-1}(\theta).$$

Chapter 15

Exercise 15.6.1 (a) For any fixed index k, we have $\bigcap_i \mathcal{C}_i \subset \mathcal{C}_k$. Using the monotonicity property of $\mathfrak{S}$ yields $\mathfrak{S}\big(\bigcap_i \mathcal{C}_i\big) \subset \mathfrak{S}(\mathcal{C}_k)$. Since k is arbitrary, it follows that $\mathfrak{S}\big(\bigcap_i \mathcal{C}_i\big) \subset \bigcap_k \mathfrak{S}(\mathcal{C}_k)$, which is the desired relation.

(b) The proof is similar, starting from $\bigcup_i \mathcal{C}_i \supset \mathcal{C}_k$.

Exercise 15.6.2 The set of participants form the input data, X, into a neural network. The participants left after each round represent the layers of the network. There are $n - 1$ hidden layers and each of them implements a max-pooling method by which the number of units decreases to half.

Exercise 15.6.3 For the sake of simplicity, we shall assume that the input has been partitioned into two classes, $\{a_1, a_2, a_3\}$, $\{b_1, b_2, b_3\}$, with three elements each.

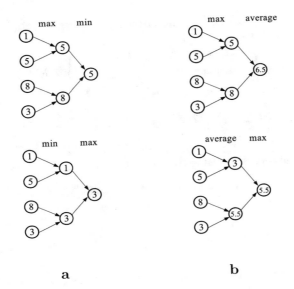

Figure 3: *For Exercise 15.6.5:* **a.** *Switching a min-pooling with a max-pooling layer provides different outputs.* **b.** *Switching an average-pooling with a max-pooling layer provides different outputs.*

(a) It follows from

$$\max\{\max\{a_1, a_2, a_3\}, \max\{b_1, b_2, b_3\}, \} = \max\{a_1, a_2, a_3, b_1, b_2, b_3\}.$$

(c) The average of the averages counts as only one average

$$\frac{1}{2}\left(\frac{a_1 + a_2 + a_3}{3} + \frac{b_1 + b_2 + b_3}{3}\right) = \frac{1}{6}(a_1 + a_2 + a_3 + b_1 + b_2 + b_3).$$

(c) Use the relation

$$\min\{\min\{a_1, a_2, a_3\}, \min\{b_1, b_2, b_3\}, \} = \min\{a_1, a_2, a_3, b_1, b_2, b_3\}.$$

Exercise 15.6.5 It follows from the counterexamples provided in Fig. 3 **a**, **b**.

Chapter 16

Exercise 16.9.1 Slide the kernel and convolve to obtain $\begin{pmatrix} -3 & 1 \\ 3 & -5 \end{pmatrix}$.

Exercise 16.9.2 (a) Define $(T_a \circ y)_i = y_{i-a}$. Then the equivariance property in one dimension becomes $\mathcal{C}(T_a \circ y) = T_a \circ \mathcal{C}(y)$. The proof is very similar to the one in two dimensions. (b) Yes.

Exercise 16.9.3 (a) Performing the convolution, we obtain

$$\begin{pmatrix} a_{i-1,j-1} & a_{i-1,j} & a_{i-1,j+1} \\ a_{i,j-1} & a_{i,j} & a_{i,j+1} \\ a_{i+1,j-1} & a_{i+1,j} & a_{i+1,j+1} \end{pmatrix} * K$$

$$= (a_{i-1,j-1} - a_{i+1,j-1}) + 2(a_{i-1,j} - a_{i+1,j}) + (a_{i-1,j+1} - a_{i+1,j+1}),$$

the effect being to subtract rows $(i-1)$ and $(i+1)$.
(b) It is similar, the effect being that of subtracting columns.

Exercise 16.9.4 (a) The convolution with this kernel produces a moving average in 2-D, which yields an uniform blur given by

$$\frac{1}{9}(a_{11} + a_{12} + a_{13} + a_{21} + a_{22} + a_{23} + a_{31} + a_{32} + a_{33}).$$

(b) This is a blur which emphasizes more the center, then the sides middles and then the corners. The sum of all weights is equal to 1.

Exercise 16.9.5 Performing the convolution, we obtain

$$\begin{pmatrix} a_{i-1,j-1} & a_{i-1,j} & a_{i-1,j+1} \\ a_{i,j-1} & a_{i,j} & a_{i,j+1} \\ a_{i+1,j-1} & a_{i+1,j} & a_{i+1,j+1} \end{pmatrix} * \begin{pmatrix} 0 & -1 & 0 \\ -1 & 5 & -1 \\ 0 & -1 & 0 \end{pmatrix}$$

$$= 5a_{ij} - (a_{i-1,j} + a_{i,j-1} + a_{i,j+1} + a_{i+1,j}).$$

The contrast results from the difference between the 5 times the central pixel activation and the vertical and horizontal neighboring pixel activations.

Exercise 16.9.7 The kernel is $K = \begin{pmatrix} 0 & 0 & 0 \\ 1 & -1 & 0 \\ 0 & 0 & 0 \end{pmatrix}$. The convolution with K of a 3×3 matrix is $a_{i,j-1} - a_{i,j}$. We can also consider the equivalent kernel $K = \begin{pmatrix} 0 & 0 & 0 \\ 0 & 1 & -1 \\ 0 & 0 & 0 \end{pmatrix}$.

Exercise 16.9.8 (a) Using the equivariance of convolution and invariance of pooling to translation, we have

$$\mathcal{P} \circ \mathcal{C}(T_{a,b} \circ y) = \mathcal{P} \circ T_{a,b}(\mathcal{C} \circ y) = \mathcal{P} \circ \mathcal{C}(y).$$

(b) Similar properties used in reverse order.

Exercise 16.9.9 Just in the FC layer. The other layers are already sparse enough and a dropout would not provide substantial improvements.

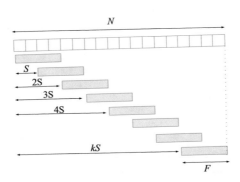

Figure 4: *For Exercise 16.9.11.*

Exercise 16.9.10 Being sparse, a CNN has fewer weights and hence a smaller capacity than a fully-connected neural network.

Exercise 16.9.11 (i) We have $N - F = kS$, where k is the number of steps the kernel is moved from one end to the other. The size of the output is one more than k, i.e., $O = k + 1$. Solving for k yields $O = \frac{N-F}{S} + 1$. In the case of padding, we add $2P$ to the dimension N (one P for each side) and obtain the second desired formula. (ii) We apply the same procedure as in part (i). See Fig. 4.

Exercise 16.9.12 (i) Since H is a subgroup of G, we have $e \in H$, and then $x = xe \in xH$. (ii) Assume $xH \cap yH = \emptyset$ and we shall show that $xH = yH$ using double inclusion. There is $z \in xH \cap yH$. So, there are $h_1, h_2 \in H$ such that $z = xh_1 = yh_2$. This implies $x = yh_2 h_1^{-1}$. Since $h_2 h_1^{-1} \in H$, it follows that $x \in yH$, and from here we get $xH \subseteq yH$. Similarly, $y = xh_1 h_2^{-1} \in xH$, and then $yH \subseteq xH$. (iii) The function $\varphi : H \to xH$ defined by $\varphi(h) = xh$ is a bijection, and hence H and xH have the same number of elements. (iv) (a) Since $x^{-1}x = e \in H$, then $x \sim x$. (b) If $x \sim y$, then $y^{-1}x \in H$, and its inverse $(y^{-1}x)^{-1} \in H$, which becomes $x^{-1}y \in H$, i.e., $y \sim x$. (c) If $x \sim y$ and $y \sim z$ then $y^{-1}x \in H$ and $z^{-1}y \in H$. Since H is closed to multiplication, we have $z^{-1}x = (z^{-1}y)(y^{-1}x) \in H$, which means $x \sim z$.

Chapter 17

Exercise 17.10.1 (a) It can be shown by double inclusion. The inclusion "$\subset$" follows from applying the $\mathfrak{S}$-operator to the inclusion

$$\mathcal{G}_1 \cup \mathcal{G}_2 \cup \mathcal{G}_3 \subset \mathfrak{S}(\mathcal{G}_1 \cup \mathcal{G}_2) \cup \mathcal{G}_3.$$

The inclusion "$\supset$" follows from the following. First, since $\mathcal{G}_3 \subset \mathcal{G}_1 \cup \mathcal{G}_2 \cup \mathcal{G}_3$, then $\mathcal{G}_3 \subset \mathfrak{S}(\mathcal{G}_1 \cup \mathcal{G}_2 \cup \mathcal{G}_3)$. Similarly, from $\mathcal{G}_1 \cup \mathcal{G}_2 \subset \mathcal{G}_1 \cup \mathcal{G}_2 \cup \mathcal{G}_3$ it follows

$$\mathfrak{S}(\mathcal{G}_1 \cup \mathcal{G}_2) \subset \mathfrak{S}(\mathcal{G}_1 \cup \mathcal{G}_2 \cup \mathcal{G}_3).$$

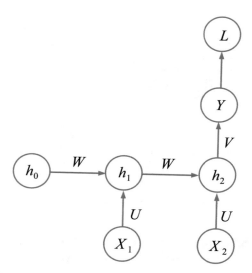

Figure 5: *A "2-to-1" RNN for Exercise* 17.10.3.

From the last two inclusions we get

$$\mathfrak{S}(\mathcal{G}_1 \cup \mathcal{G}_2) \cup \mathcal{G}_3 \subset \mathfrak{S}(\mathcal{G}_1 \cup \mathcal{G}_2 \cup \mathcal{G}_3).$$

Taking the $\mathfrak{S}$-operator on both sides we obtain the desired inclusion.
(b) We can show inductively that

$$\mathfrak{S}(\mathcal{G}_1 \cup \cdots \cup \mathcal{G}_n) = \mathfrak{S}\Big(\mathfrak{S}(\mathcal{G}_1 \cup \cdots \cup \mathcal{G}_{n-1}) \cup \mathcal{G}_n\Big).$$

$$\mathfrak{S}(\mathcal{G}_1 \cup \cdots \cup \mathcal{G}_n) = \mathfrak{S}\Big(\mathfrak{S}(\mathcal{G}_1 \cup \cdots \cup \mathcal{G}_{n-k}) \cup \mathcal{G}_{n-k+1} \cup \cdots \cup \mathcal{G}_n\Big).$$

Exercise 17.10.2 (a) The eigenvalues of W are $\lambda_1 = 1/5$ and $\lambda_2 = -1/2$. Since both are less than 1, apply Proposition G.1.2. (b) Since $\rho(W) < 1$, apply Proposition G.1.2.

Exercise 17.10.3 Hint: Follow the example presented in section 17.5 for Fig. 5.

Exercise 17.10.4 From $Y = Vh_2 + c$, we get $\mathcal{E} \subset \mathcal{H}_2$. From

$$h_2 = \tanh(Wh_1 + UX_2 + b)$$
$$h_1 = \tanh(Wh_0 + UX_1 + b)$$

we can show that $\mathcal{H}_2 \subset \mathfrak{S}(\mathcal{I}_1, \mathcal{I}_2)$. Then $\mathcal{E} \subset \mathfrak{S}(\mathcal{I}_1, \mathcal{I}_2)$. So, if $\mathfrak{S}(Z) \subset \mathcal{E}$ then $\mathfrak{S}(Z) \subset \mathfrak{S}(\mathcal{I}_1 \cup \mathcal{I}_2)$. Hence, (a) holds true.

Exercise 17.10.5 The transition function f is a contraction. Then $\lim_{n \to \infty} h_n$ is the fixed point of f, i.e., $\tanh(\theta c) = c$. It follows that $c = 0$.

Exercise 17.10.6 Since $f'(x; \theta) = \theta \sigma'(\theta x) = \theta \sigma(\theta x)(1 - \sigma'(\theta x)) \le \theta/4 < 1$, the transition function f is still a contraction. The fixed point satisfies $\sigma(\theta c) = c$. It is unique but cannot be determined in closed form.

Exercise 17.10.7 If θ is large the graphs $y = \sin(\theta x)$ and $y = x$ intersect more than once, each representing a fixed point. Depending on h_0, the system might settle to one of those points.

Chapter 18

Exercise 18.11.1 (a) The rectangle rule is easily satisfied, since if three corners of a rectangle have rational coordinates, the fourth corner has the same property.
(b) No, it's not. Since $(1/\sqrt{2}, 1/\sqrt{2}) \notin (I_1 \times I_1) \cap (\mathbb{Q} \times \mathbb{Q})$, $\mathcal{S}$ is not reflexive.

Exercise 18.11.2 "$\Longrightarrow$" By contradiction, assume there are two distinct $(k-1)$-hyperplanes containing the points. Then the points belong to the intersection of the hyperplanes, which will be a hyperplane of dimension strictly less than $k-1$, which is in contradiction with the fact that the points are in a general position. Hence, there is a unique $(k-1)$-hyperplane containing the points.
"$\Longleftarrow$" By contradiction, assume the points are not in a general position, so they belong to a hyperplane of dimension p, which is strictly smaller than $k-1$. This p-hyperplane is always inside the intersection of two distinct $(k-1)$-hyperplanes. Then the points belong to two distinct $(k-1)$-hyperplanes, contradiction. It follows that the points are in a general position.

Exercise 18.11.3 By contradiction, assume the vectors are not linearly independent. If consider the span $\mathcal{S} = span\{\overrightarrow{P_1 P_2}, \ldots, \overrightarrow{P_1 P_k}\}$, then $\dim \mathcal{S} = s < k - 1$. Then the points P_i belong to a hyperplane of dimension s, which contradicts the fact that they are in a general position.

Exercise 18.11.4 Conditions $f(x_1) = 1$ and $f(x_2) = 2$ can be written as the linear system

$$
\begin{aligned}
a x_1 + b &= 1 \\
a x_2 + b &= 2.
\end{aligned}
$$

As long as $x_1 \ne x_2$, the system has a unique solution (a, b).

Exercise 18.11.5 (a) We need to show that for any two points $A, B \in hull(\mathcal{G})$ the convex combination $tA + (1-t)B \in hull(\mathcal{G})$. We have $A = \sum \alpha_i P_i$ and

$B = \sum \beta_i P_i$, with $\sum \alpha_i = \sum \beta_i = 1$. Then

$$
\begin{aligned}
tA + (1-t)B &= t\sum \alpha_i P_i + (1-t)\sum \beta_i P_i \\
&= \sum (t\alpha_i + (1-t)\beta_i)P_i = \sum r_i P_i.
\end{aligned}
$$

This is a convex combination of P_i because

$$
\sum r_i = t\sum \alpha_i + (1-t)\sum \beta_i = t + (1-t) = 1,
$$

and hence $tA + (1-t)B \in hull(\mathcal{G})$.

(b) By contradiction, we assume there is a convex set K in $\mathbb{R}^n$ such that $\mathcal{G} \subset K \subset hull(\mathcal{G})$. Pick a point $Q \in hull(\mathcal{G})\backslash K$. Then Q is a convex combination of the points in $\mathcal{G}$, and hence belongs to K, contradiction.

(c) Use that intersection of convex sets is always convex.

Exercise 18.11.6 Let $r(t) = r_0 + tv$ be the line separator of $\mathcal{G}_1$ and $\mathcal{G}_2$. The affine transform Φ transforms this line into the line $\rho(t) = \Phi(r(t)) = \rho_0 + tu$, with $\rho_0 = wr_0 + b$ and $u = Wv$. Note that $u \neq 0$ since $\det W \neq 0$. We shall show that $\rho(t)$ is a separator of $\Phi(\mathcal{G}_1)$ and $\Phi(\mathcal{G}_2)$. By contradiction, we assume that $\rho(t)$ does not separate $\Phi(\mathcal{G}_1)$ and $\Phi(\mathcal{G}_2)$. Therefore, there are some $g_i \in \mathcal{G}_i$ such that $\rho(t)$ does not separate $\Phi(g_1)$ and $\Phi(g_2)$. Consequently, the line segment $\Phi(g_1)\Phi(g_2)$ does not intersect the line $\rho(t)$. Since $r(t)$ separates g_1 and g_2, the line segment $g_1 g_2$ intersects the line $r(t)$ at some point p. Then $\Phi(p)$ is the intersection between the line segment $\Phi(g_1)\Phi(g_2)$ and line $\rho(t)$, which is a contradiction.

It is worth noting that the statement still holds as long as the direction vector of the separation line is not in the kernel of the matrix W, i.e., $Wv \neq 0$.

Exercise 18.11.7 Let $\mathcal{G} = \{A_1, A_2, \ldots, A_m\}$. Then $\overrightarrow{OG} = \frac{1}{m}\sum_{j=1}^{k}\overrightarrow{OA_k} \in hull(\mathcal{G})$ as a convex combination of position vectors of the cluster elements.

Exercise 18.11.9 (a) Let $p \in A$ and $q \in B$ such that $\|p_0 q_0\| = \inf_{p \in A, q \in B}\|pq\|$. Choose the separation line to be perpendicular on the midpoint of the segment $p_0 q_0$. (b) it follows from part (a).

Chapter 19

Exercise 19.7.1 (i) In this case the state updates satisfy the system

$$
\begin{aligned}
x_{n+1} &= x_n - \eta y_n \\
y_{n+1} &= y_n + \eta x_n,
\end{aligned}
$$

with $\eta > 0$ learning rate. This can be written in the matrix form

$$
\begin{pmatrix} x_{n+1} \\ y_{n+1} \end{pmatrix} = \begin{pmatrix} 1 & -\eta \\ \eta & 1 \end{pmatrix}\begin{pmatrix} x_n \\ y_n \end{pmatrix}.
$$

(ii) Therefore

$$\begin{pmatrix} x_n \\ y_n \end{pmatrix} = \begin{pmatrix} 1 & -\eta \\ \eta & 1 \end{pmatrix}^n \begin{pmatrix} x_0 \\ y_0 \end{pmatrix}.$$

(iii) The matrix $A = \begin{pmatrix} 1 & -\eta \\ \eta & 1 \end{pmatrix}$ has eigenvalues equal to $1 \pm i\eta$ and has the spectral radius $\rho(A^T A) = \sqrt{1 + \eta^2}$ strictly larger than 1. Therefore, the sequence (x_n, y_n) will not converge, unless $x_0 = y_0 = 0$.

Exercise 19.7.2 (i) The system is $x'(t) = -a$, $y'(t) = b$, with solution $x(t) = -at + x_0$, $y(t) = bt + y_0$. (ii) The solution is a line in the (x, y)-plane which does not approach any equilibrium point as $t \to \infty$.

Exercise 19.7.3 Continuous case: The associated differential system is $x' = -x$, $y' = -y$, with solution $x(t) = x_0 e^{-t}$, $y(t) = y_0 e^{-t}$. Regardless of the initial value (x_0, y_0), the limit is $\lim_{t \to \infty}(x(t), y(t)) = (0, 0)$. The origin is an equilibrium point. Discrete case: We have $x_{n+1} = x_n - \eta x_n = (1 - \eta)x_n$ and $y_{n+1} = y_n - \eta y_n = (1 - \eta)y_n$, which implies $x_n = (1 - \eta)^n x_0$ and $y_n = (1 - \eta)^n y_0$. For any x_0, y_0 and $\eta \in (0, 1)$ we obtain $\lim_{n \to \infty}(x_n, y_n) = (0, 0)$.

Exercise 19.7.4 (i) $\text{div} U = \partial_x U_1 + \partial_y U_2 = -\partial_x^2 V + \partial_y^2 V = 0$. ($ii$) $\text{div} U = 0$ if and only if $\partial_x^2 V - \partial_y^2 V = 0$. With the change of variables $u = x + y$ and $v = x - y$ this becomes $\partial_u \partial_v V = 0$. This has the solution with separate variables $V = F(u) + G(v) = F(x + y) + G(x - y)$, with F and G arbitrary smooth functions. (iii) If divergence is zero, the flow is incompressible and hence does not converge toward any point in the long range.

Exercise 19.7.5 (i) We have

$$\begin{aligned} D_G^*(x) &= \frac{p_{data}(x)}{p_{data}(x) + p_{model}(x)} = \frac{1}{1 + \left(\frac{p_{data}(x)}{p_{model}(x)}\right)^{-1}} = \frac{1}{1 + e^{-\ln \frac{p_{data}(x)}{p_{model}(x)}}} \\ &= \sigma\left(\ln \frac{p_{data}(x)}{p_{model}(x)}\right) = \sigma(a(x)), \end{aligned}$$

from where by using the injectivity of σ and then taking an exponential we obtain $e^{a(x)} = \frac{p_{data}(x)}{p_{model}(x)}$. ($ii$) It follows from the definition of the Kullback-Leibler divergence

$$\mathbb{E}_{x \sim p_{data}}[a(x)] = \int p_{data}(x) \ln \frac{p_{data}(x)}{p_{model}(x)} \, dx = D_{KL}(p_{data} \| p_{model}).$$

Exercise 19.7.6 We write the condition that both expectations have the critical point at the maximum likelihood estimation

$$\partial_\theta \mathbb{E}_{x \sim p_{data}}[\ln p_{model}(x; \theta)]\big|_{\theta = \theta_{MLE}} = 0$$

$$\partial_\theta \mathbb{E}_{x \sim p_{model}}[f(x)]\big|_{\theta = \theta_{MLE}} = 0.$$

They are equivalent to

$$\int p_{data}(x) \, \partial_\theta \ln p_{model}(x; \theta)\big|_{\theta = \theta_{MLE}} dx = 0$$

$$\int f(x) \, \partial_\theta p_{model}(x; \theta)\big|_{\theta = \theta_{MLE}} dx = 0.$$

One possible function $f(x)$ satisfying this property is obtained equating the integrands. We obtain

$$f(x) = \frac{p_{data}(x)\partial_\theta \ln p_{model}(x; \theta)}{\partial_\theta p_{model}(x; \theta)}\bigg|_{\theta = \theta_{MLE}} = \frac{p_{data}(x)}{p_{model}(x; \theta_{MLE})}.$$

Exercise 19.7.7 (i) Since the first expectation term of (19.4.4) is independent of $\theta^{(g)}$, we differentiate just the second term and commute the expectation and the gradient operators.

(ii) We apply the gradient to the expectation $\mathbb{E}_{z \sim p_{code}}[\ln(1 - D(G(z, \theta^{(g)})))]$ and use the chain rule

$$\nabla_{\theta^{(g)}} \ln(1 - D(G(z, \theta^{(g)}))) = \frac{1}{1 - D(G(z))} \frac{\partial D(x)}{\partial x}\bigg|_{x = G(z)} \frac{\partial G(z)}{\partial \theta^{(g)}}.$$

(iii) Use that $\frac{\partial D(x)}{\partial x} = D(x)(1 - D(x))\frac{da(x)}{dx}$.

Exercise 19.7.8 The equation $\frac{\partial L}{\partial y} = \ln \frac{\hat{y}}{1-\hat{y}} = 0$ yields $\hat{y} = \frac{1}{2}$, while the equation $\frac{\partial L}{\partial \hat{y}} = \frac{y - \hat{y}}{\hat{y}(1-\hat{y})} = 0$ implies $y = \hat{y}$. The equilibrium point is $(y^*, \hat{y}^*) = (\frac{1}{2}, \frac{1}{2})$.

Exercise 19.7.9 Using chain rule yields

$$\nabla_{\theta^{(g)}} J^{(G)} = \mathbb{E}_{z \sim p_{code}}[\nabla_{\theta^{(g)}} \ln D(G(z, \theta^{(g)}))]$$

$$= \mathbb{E}_{z \sim p_{code}}\left[\frac{1}{D(G(z, \theta^{(g)}))} \frac{\partial D(x)}{\partial x}\bigg|_{x = G(z)} \frac{\partial G}{\partial \theta^{(g)}} \right].$$

Assuming now that the discriminator is optimized, then by Exercise 19.7.5 we have $D(x) = D_G^*(x) = \sigma(a(x))$, with $a(x) = \ln \frac{p_{data}(x)}{p_{model}(x)}$. Then using chain rule

$$\frac{\partial D_G^*(x)}{\partial x} = \sigma'(a(x))\frac{da(x)}{dx} = \sigma(a(x))(1 - \sigma(a(x)))\frac{da(x)}{dx}$$

$$= D_G^*(x)(1 - D_G^*(x))\frac{da(x)}{dx}.$$

Substituting, in the previous expression, yields

$$\nabla_{\theta^{(g)}} J^{(G)} = \mathbb{E}_{z \sim p_{code}} \left[(1 - D(G(z, \theta^{(g)}))) \frac{da(x)}{dx} \frac{\partial G}{\partial \theta^{(g)}} \right],$$

with $\dfrac{da(x)}{dx} = \dfrac{\partial_x p_{data}(x)}{p_{data}(x)} - \dfrac{\partial_x p_{model}(x)}{p_{model}(x)}.$

Chapter 20

Exercise 20.10.1 Keeping the contribution of the kth neuron separately, we have

$$
\begin{aligned}
E' - E &= E(\mathbf{x}') - E(\mathbf{x}) \\
&= -\frac{1}{2} \sum_{i \neq k} \sum_{j \neq k} w_{ij} x_i x_j - \sum_{i \neq k} b_i x_i - \sum_{i=1}^{n} w_{ki} x_i x'_k - b_k x'_k \\
&\quad + \frac{1}{2} \sum_{i \neq k} \sum_{j \neq k} w_{ij} x_i x_j + \sum_{i \neq k} b_i x_i + \sum_{i=1}^{n} w_{ki} x_i x_k + b_k x_k \\
&= \sum_{i=1}^{n} w_{ki} x_i (x_k - x'_k) + b_k (x_k - x'_k) \\
&= -\left(\sum_{i=1}^{n} w_{ki} x_i + b_k \right)(x'_k - x_k).
\end{aligned}
$$

Exercise 20.10.2 By identification of the first 7 components, we have

$$
\begin{aligned}
q_1 &= \frac{1}{Z} \\
q_2 &= e^{b_1} q_1 \Rightarrow b_1 = \ln \frac{q_2}{q_1} \\
q_3 &= e^{b_2} q_1 \Rightarrow b_2 = \ln \frac{q_3}{q_1} \\
q_4 &= e^{b_3} q_1 \Rightarrow b_3 = \ln \frac{q_4}{q_1} \\
q_5 &= e^{w_{23} + b_2 + b_3} q_1 \Rightarrow w_{23} = \ln \frac{q_5 q_1}{q_3 q_4} \\
q_6 &= e^{w_{13} + b_1 + b_3} q_1 \Rightarrow w_{13} = \ln \frac{q_6 q_1}{q_2 q_4} \\
q_7 &= e^{w_{12} + b_1 + b_2} q_1 \Rightarrow w_{12} = \ln \frac{q_7 q_1}{q_2 q_3}.
\end{aligned}
$$

We identify the last component and substitute the values obtained before

$$q_8 = e^{w_{12}} e^{w_{13}} e^{w_{23}} e^{b_1} e^{b_2} e^{b_3} q_1 = \frac{q_5 q_6 q_7}{q_2 q_3 q_4}.$$

Exercise 20.10.3 (*a*) It follows from a straightforward computation. For instance, for the first component, we have

$$
\begin{aligned}
\sum_{j=1}^{N} p_{1j}\, p_j &= \frac{1}{Z}\Big\{ p_{11} e^{-E_1/T} + \sum_{j=2}^{N} p_{1j} e^{E_j/T} \Big\} \\
&= \frac{1}{Z}\Big\{ \Big(1 - \sum_{i=2}^{N} \frac{1}{1 + e^{(E_i - E_1)/T}}\Big) e^{-E_1/T} + \sum_{j=2}^{N} \frac{e^{-E_j/T}}{1 + e^{(E_1 - E_j)/T}} \Big\} \\
&= \frac{1}{Z}\Big\{ e^{-E_1/T} - \sum_{i=2}^{N} \frac{1}{e^{E_1/T} + e^{E_i/T}} + \sum_{j=2}^{N} \frac{1}{e^{E_j/T} + e^{E_1/T}} \Big\} \\
&= \frac{1}{Z} e^{-E_1/T} = p_1.
\end{aligned}
$$

(*b*) In the next computation we shall use that all entries of P_T are nonnegative and the sum of the entries on each row is equal to 1. Using the L^1 norm, $\|v\|_1 = \sum_j |v_j|$, we have

$$
\begin{aligned}
\|P_T v\|_1 &= |\sum_j p_{1j} v_j| + \cdots + |\sum_j p_{nj} v_j| \\
&\leq \sum_j p_{1j} |v_j| + \cdots + \sum_j p_{nj} |v_j| = \sum_i \sum_j p_{ij} |v_j| \\
&= \sum_i \Big(\underbrace{\sum_j p_{ij}}_{=1} \Big) |v_j| = \sum_j |v_j| = \|v\|_1.
\end{aligned}
$$

If λ is an eigenvalue corresponding to the eigenvector v, then $P_T v = \lambda v$ and taking the norm yields $\|P_T v\|_1 = |\lambda| \|v\|_1$. Then using the previous computation we obtain $|\lambda| = \frac{\|P_T v\|_1}{\|v\|_1} \leq 1$.

(*c*) First, we show that q_n is convergent. Iterating, we obtain $q_n = P_T^n q_0$. Since $P_T^n = (MDM^{-1})^n = MD^n M^{-1}$, then $q_n = MD^n M^{-1} q_0$. Since the diagonal matrix D contains only entries with absolute value less than or equal to 1, the limit $L = \lim_{n \to \infty} D^n$ is a sparse matrix having only one entry equal to 1. Then the following limit exists:

$$
q^* = \lim_{n \to \infty} q_n = \lim_{n \to \infty} MD^n M^{-1} q_0 = MLM^{-1} q_0.
$$

On the other side, we apply the limit in the relation $q_{n+1} = P_T q_n$ and obtain $q^* = P_T q^*$. This implies that q^* and p are proportional (we used, without proof, that the dimension of the eigenspace with eigenvalue 1 is equal to 1). Since the sum of their elements is 1, it follows that $q^* = p$. The physical significance is that for any initial state, in the long run, the Boltzmann machine settles to an equilibrium distribution.

Exercise 20.10.4 The Boltzmann distribution is $p = \frac{1}{Z}(1, e^{b_1}, e^{b_2}, e^{b_1+b_2+w})$. The Fisher metric involves the computations of the type

$$
\begin{aligned}
\mathbb{E}^p[x_1 x_2] &= \frac{1}{Z}[p(0,0) \cdot 0 \cdot 0 + p(1,0) \cdot 1 \cdot 0 + p(0,1) \cdot 0 \cdot 1 + p(1,1) \cdot 1 \cdot 1] \\
&= \frac{1}{Z}p(1,1) = \frac{1}{Z}e^{b_1+b_2+w}.
\end{aligned}
$$

The others can be easily computed by the reader.

Exercise 20.10.5 (a) From $\frac{\partial}{\partial w_{ij}} \ln p(\mathbf{x}) = \frac{1}{p(\mathbf{x})} \frac{\partial}{\partial w_{ij}} p(\mathbf{x})$ and $\frac{\partial}{\partial w_{ij}} \ln p(\mathbf{x}) = x_i x_j - \mathbb{E}^p[x_i x_j]$ yields

$$
\frac{\partial}{\partial w_{ij}} p(\mathbf{x}) = \left(x_i x_j - \mathbb{E}^p[x_i x_j] \right) p(\mathbf{x}).
$$

Then

$$
\begin{aligned}
\sum_{i,j} w_{ij} \frac{\partial}{\partial w_{ij}} p(\mathbf{x}) &= \left(\sum_{i,j} w_{ij} x_i x_j - \mathbb{E}^p[w_{ij} x_i x_j] \right) p(\mathbf{x}) \\
&= \left(\mathbf{x}^T w \mathbf{x} - \mathbb{E}^p[\mathbf{x}^T w \mathbf{x}] \right) p(\mathbf{x}).
\end{aligned}
$$

Similarly,

$$
\sum_k b_k \frac{\partial}{\partial b_k} p(\mathbf{x}) = \left(\mathbf{x}^T b - \mathbb{E}^p[\mathbf{x}^T b] \right) p(\mathbf{x})
$$

and hence

$$
\begin{aligned}
A_{w,b} \, p(\mathbf{x}) &= \left(\frac{1}{2} \mathbf{x}^T w \mathbf{x} + \mathbf{x}^T b \right) p(\mathbf{x}) - \mathbb{E}^p\left[\frac{1}{2} \mathbf{x}^T w \mathbf{x} + \mathbf{x}^T b \right] p(\mathbf{x}) \\
&= \left(\mathbb{E}^p[E(\mathbf{x})] - E(\mathbf{x}) \right) p(\mathbf{x}).
\end{aligned}
$$

(b) Since $w'_{ij}(t) = \alpha w_{ij}(t)$ and $b'_k(t) = \alpha b_k(t)$, we have

$$
\begin{aligned}
\partial_t E(x) &= -\partial_t \left[\frac{1}{2} \mathbf{x}^T w(t) \mathbf{x} + \mathbf{x}^T b(t) \right] = -\left[\frac{1}{2} \mathbf{x}^T w'(t) \mathbf{x} + \mathbf{x}^T b(t) \right] \\
&= -\alpha \left[\frac{1}{2} \mathbf{x}^T w(t) \mathbf{x} + \mathbf{x}^T b(t) \right] = \alpha E(\mathbf{x}).
\end{aligned}
$$

Differentiating in $\ln p_t(\mathbf{x}) = -E(\mathbf{x}) - \ln Z(t)$, we get

$$
\begin{aligned}
\partial_t \ln p_t(\mathbf{x}) &= -\partial_t E(\mathbf{x}) - \frac{1}{Z} \partial_t Z = -\alpha E(\mathbf{x}) - \frac{1}{Z(t)} \sum_x \partial_t e^{-E(\mathbf{x})} \\
&= -\alpha E(\mathbf{x}) + \sum_x p(\mathbf{x}) \partial_t E(\mathbf{x}) = -\alpha E(\mathbf{x}) + \alpha \sum_x p(\mathbf{x}) E(\mathbf{x}) \\
&= \alpha \left(\mathbb{E}^p[E(\mathbf{x})] - E(\mathbf{x}) \right).
\end{aligned}
$$

Using $\partial_t p_t(\mathbf{x}) = \partial_t \ln p_t(\mathbf{x}) \, p_t(\mathbf{x})$, we obtain the desired formula.

(c) From (a) and (b) it follows that $w_{ij}(t) = w_{ij} e^{\alpha t}$, $b_k(t) = b_k e^{\alpha t}$ is a solution of the problem. The (local) uniqueness follows from the solution formula $p_t(\mathbf{x}) = e^{t A_{w,b}} p(\mathbf{x})$.

Exercise 20.10.6 Since rooks attack only horizontally and vertically, it must be only one rook on each row and on each column. We start placing a rook in the first row. There are 8 possibilities to do that. In the next row you place another rook. Since it can't be in the same column as the first rook, there are 7 possibilities left. On the next row there are 6 possible places, and so on, until, in the last row there is only one possible position to place the last rook. The total number of possibilities is the product $8 \cdot 7 \cdots 2 \cdot 1$. Hence, there are 8! possible ways to place the rooks without them threatening one another. These are favorable choices. The number of all possible choices is the number of possibilities in which 8 squares can be selected out of 64, that is, the binomial coefficient $\binom{64}{8}$. To get the desired probability, we need to divide the number of favorable choices to the number of all possible ways to place the rooks, $p = \frac{8!}{\binom{64}{8}}$.

Exercise 20.10.7 (a) Using the conditional probability formula, we have

$$
\begin{aligned}
p(h,v) &= \frac{p(v,h)}{p(v)} = \frac{p(v,h)}{\sum_h p(v,h)} = \frac{\frac{1}{Z} e^{-E(v,h)}}{\frac{1}{Z} \sum_h e^{-E(v,h)}} = \frac{e^{-E(v,h)}}{\sum_h e^{-E(v,h)}} \\
&= \frac{e^{v^T wh} e^{b^T v} e^{c^T h}}{\sum_h e^{v^T wh} e^{b^T v} e^{c^T h}} = \frac{e^{v^T wh} e^{c^T h}}{\sum_h e^{v^T wh} e^{c^T h}} = \frac{e^{v^T wh} e^{c^T h}}{Z'} \\
&= \frac{1}{Z'} e^{\sum_j \sum_i v_i w_{ij} h_j} \, e^{\sum_j c_j h_j} = \frac{1}{Z'} \prod_j e^{\sum_i v_i w_{ij} h_j + c_j h_j} \\
&= \prod_j \frac{1}{Z'_j} e^{(\sum_i v_i w_{ij} + c_j) h_j} = \prod_j p(h_j|v),
\end{aligned}
$$

with the partition function $Z'_k = e^{(\sum_i v_i w_{ij} + c_j) \cdot 0} + e^{(\sum_i v_i w_{ij} + c_j) \cdot 1} = 1 + e^{(\sum_i v_i w_{ij} + c_j)}$. Since $p(h,v) = \prod_j p(h_j|v)$, it follows that h_j are conditional independent, given the visible states.

(b) Using the previous formula, we have

$$
p(h_j|v) = \frac{1}{Z'_j} e^{(\sum_i v_i w_{ij} + c_j) h_j} = \frac{e^{(\sum_i v_i w_{ij} + c_j) h_j}}{1 + e^{(\sum_i v_i w_{ij} + c_j)}},
$$

which implies

$$p(h_j = 1|v) = \frac{e^{(\sum_i v_i w_{ij} + c_j)}}{1 + e^{(\sum_i v_i w_{ij} + c_j)}} = \frac{1}{1 + e^{-(\sum_i v_i w_{ij} + c_j)}} = \sigma\left(\sum_i v_i w_{ij} + c_j\right),$$

$$p(h_j = 0|v) = 1 - \sigma\left(\sum_i v_i w_{ij} + c_j\right) = \sigma\left(-\sum_i v_i w_{ij} - c_j\right).$$

(c) The log-likelihood function and its partial derivatives are given by

$$\ell(h|v) = \ln p(h|v) = \ln\frac{e^{v^T w h + c^T h}}{Z'} = v^T w h + c^T h - \ln Z'$$

$$\partial_{b_k}\ell(h|v) = 0$$

$$\partial_{c_k}\ell(h|v) = h_k - \frac{1}{Z'}\partial_{c_k}Z' = h_k - \frac{1}{Z'}\partial_{c_k}\left(\sum_h e^{v^T w h + c^T h}\right)$$

$$= h_k - \frac{1}{Z'}\sum_h e^{v^T w h + c^T h} h_k = h_k - \sum_h p(h|v)h_k$$

$$= h_k - \mathbb{E}^{p_{v|h}}[h_k];$$

$$\partial_{w_{ij}}\ell(h|v) = v_i h_j + v_j h_i - \frac{1}{Z'}\partial_{w_{ij}}Z'$$

$$= v_i h_j + v_j h_i - \frac{1}{Z'}\sum_h e^{v^T w h + c^T h}(v_i h_j + v_j h_i)$$

$$= v_i h_j + v_j h_i - \mathbb{E}^{p_{v|h}}[v_i h_j + v_j h_i].$$

(d) The following entries of the Fisher matrix are zero:

$$g_{b_i b_j} = 0, \quad g_{b_i c_j} = 0, \quad g_{b_i w_{ij}} = 0.$$

The others are given by conditional correlations of the hidden units

$$g_{c_i c_j} = Cor(h_i, h_j|v)$$

$$g_{w_{ij} c_{kl}} = Cor(v_i h_j + v_j h_i, v_k h_l v_l h_k|v).$$

(e) To minimize the Kullback-Leibler divergence $D_{KL}(q(\mathbf{h}|\mathbf{v})||p(\mathbf{h}|\mathbf{v}))$ is equivalent to maximizing the cost function $C = \sum_h q(h|v)\ln p(h|v)$. By a similar computation with the Boltzmann learning we obtain

$$\Delta w_{ij} = \eta\left(\mathbb{E}^{q_{h|v}}[h_i h_j] - \mathbb{E}^{p_{h|v}}[h_i h_j]\right)$$

$$\Delta c_k = \eta\left(\mathbb{E}^{q_{h|v}}[h_k] - \mathbb{E}^{p_{h|v}}[h_k]\right)$$

$$\Delta b = 0.$$

Appendix

In order to keep this book as self-contained as possible, we have included in this appendix a brief presentation of notions that have been used throughout the book. They are basic notions of Measure Theory, Probability Theory, Linear Algebra, and Functional Analysis.

However, this appendix is not designed to be a substitute for a complete course in the aforementioned areas; it is supposed just to supply enough information for the reader in order to be able to follow the book smoothly.

© Springer Nature Switzerland AG 2020
O. Calin, *Deep Learning Architectures*, Springer Series in the Data Sciences,
https://doi.org/10.1007/978-3-030-36721-3

Appendix A

Set Theory

Let $(A_i)_{i \in I}$ be a family of sets. Their *union* and *intersection* will be denoted, respectively, by

$$\bigcup_{i \in I} A_i = \{x; \exists i \in I, x \in A_i\}, \qquad \bigcap_{i \in I} A_i = \{x; \forall i \in I, x \in A_i\}.$$

If $(A_i)_{i \in I}$ and $(B_j)_{j \in J}$ are two family of sets, then:

(a) $\left(\bigcup_i A_i \right) \cap \left(\bigcup_j B_j \right) = \bigcup_{i,j} (A_i \cap B_j)$

(b) $\left(\bigcap_i A_i \right) \cup \left(\bigcap_j B_j \right) = \bigcap_{i,j} (A_i \cup B_j)$

These relations can be generalized to any number of family of sets, $(A_{i_1}^1)_{i_1 \in I_1}, (A_{i_2}^2)_{i_2 \in I_2}, \ldots, (A_{i_p}^p)_{i_p \in I_p}$, as

(a') $\displaystyle\bigcap_{r=1}^{p} \left(\bigcup_{i_r} A_{i_r}^r \right) = \bigcup_{i_1,\ldots,i_p} \left(\bigcap_{r=1}^{p} A_{i_r}^r \right)$

(b') $\displaystyle\bigcup_{r=1}^{p} \left(\bigcap_{i_r} A_{i_r}^r \right) = \bigcap_{i_1,\ldots,i_p} \left(\bigcup_{r=1}^{p} A_{i_r}^r \right).$

The following properties of functions are useful when dealing with sigma-fields. They can be proved by double inclusion. In the following $\mathcal{X}$ and $\mathcal{Y}$ denote two sets.

Proposition A.0.1 *Let $f : \mathcal{X} \to \mathcal{Y}$ be a function. Then:*

(a) $f\left(\bigcup_i A_i \right) = \bigcup_i f(A_i), \quad \forall A_i \subset \mathcal{X};$

© Springer Nature Switzerland AG 2020
O. Calin, *Deep Learning Architectures*, Springer Series in the Data Sciences,
https://doi.org/10.1007/978-3-030-36721-3

(b) $f\left(\bigcap_i A_i\right) = \bigcap_i f(A_i), \; \forall A_i \subset \mathcal{X};$

(c) $f^{-1}\left(\bigcup_i B_i\right) = \bigcup_i f^{-1}(B_i), \; \forall B_i \subset \mathcal{Y};$

(d) $f^{-1}\left(\bigcap_i B_i\right) = \bigcap_i f^{-1}(B_i), \; \forall B_i \subset \mathcal{Y};$

(e) $f^{-1}(B^c) = \left(f^{-1}(B)\right)^c, \; \forall B \subset \mathcal{Y}.$

A set $A \subset \mathbb{R}^n$ is called *bounded* if it can be included into a ball, i.e., there is an $r > 0$ such that $A \subset B(0, r)$. Equivalently, $\|x\| \leq r$, for all $x \in A$.

A set $A \subset \mathbb{R}^n$ is called *closed*, if A contains the limit of any convergent sequence $(x_n) \subset A$, i.e., $\lim_{n \to \infty} x_n \in A$.

A subset K of $\mathbb{R}^n$ is called *compact* if it is bounded and closed. Equivalently, K is compact if for any sequence $(x_n) \subset K$ we can extract a convergent subsequence $(x_{n_k})_k$. The prototype examples for a compact set in $\mathbb{R}^n$ in this book are the hypercube, $K = I_n = [0, 1] \times \cdots \times [0, 1]$, and the n-dimensional closed ball, $K = B(x_0, r) = \{x; \|x - x_0\| \leq r\}$.

Proposition A.0.2 (Cantor's lemma) *An intersection of a descending sequence of compact sets in $\mathbb{R}^n$ is nonempty.*

For example, if $K_n = [-1/n, 1/n]$, then $K_{n+1} \subset K_n \subset \mathbb{R}$ and $\bigcap_{k \geq 1} K_n = \{0\}$.

A binary relation "$\leq$" on a nonempty set A is called an *order relation* if for any $a, b, c \in A$ we have

    (*i*) $a \leq a$ (reflexivity);
    (*ii*) $a \leq b$ and $b \leq a$ then $a = b$ (antisymmetry);
    (*iii*) $a \leq b$ and $b \leq c$ then $a \leq c$ (transitivity).

A set A on which an order relation "$\leq$" has been defined is called an *ordered set*, and is denoted by $(A, \leq)$. If for any two elements $a, b \in A$ we have either $a \leq b$ or $b \leq a$, then $(A, \leq)$ is called a *totally ordered* set. An element $m \in A$ is called *maximal* if $m \leq x$ implies $m = x$. Consider a subset $B \subset A$. An element $a \in A$ is called an *upper bound* for B if $x \leq a$ for all $x \in B$. An ordered set $(A, \leq)$ is called *inductive* if any subset $B \subset A$ has an upper bound.

Lemma A.0.3 (Zorn) *Any nonempty ordered set which is inductive has at least one maximal element.*

Appendix B

Tensors

Let $I_1, \ldots, I_n \subset \mathbb{N}$ be n subsets of the set of natural numbers, $\mathbb{N}$, and consider the Cartesian product

$$I_1 \times \cdots \times I_n = \{(i_1, \ldots, i_n); \; i_k \in I_k, 1 \leq k \leq n\}.$$

A set of objects indexed over the multi-index $(i_1, \ldots, i_n) \in I_1 \times \cdots \times I_n$ is a *tensor* of order n. Typically, we denote it by $T_{i_1,\ldots,i_n}$.

It is worth noting, without getting into details, that the tensor concept comes from Differential Geometry and Relativity Theory, where was successfully used to describe manifolds curvature, tangent vector fields, or certain physical measures, such as velocity, energy-momentum, force, density of mass, etc.

Many objects in neural networks, such as inputs, weights, biasses, intermediate representations and outputs, are described by tensors. For instance, a vector $x \in \mathbb{R}^d$, given by $x = (x_1, \ldots, x_d)$, is a tensor of order 1 and type d. A matrix, $A = (A_{ij}) \in \mathbb{R}^{d \times r}$, is a tensor of order 2 and type $d \times r$ (d rows and r columns), see Fig. 1 **a**. A tensor of order 3, say $t \in \mathbb{R}^{d \times r \times s}$, can be viewed as a vector of length s of matrices of type $d \times r$. A generic entry of this tensor is represented by t_{ijk}, see Fig. 1 b. A scalar value can be seen as a tensor of order zero.

Example B.0.1 A color image can be represented as a tensor of order 3 of type $n \times m \times 3$, where n is the number of pixel lines, m is the number of columns and 3 stands for the number of the color channels in the RGB format.

© Springer Nature Switzerland AG 2020
O. Calin, *Deep Learning Architectures*, Springer Series in the Data Sciences,
https://doi.org/10.1007/978-3-030-36721-3

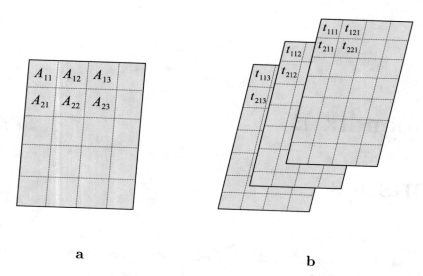

a   b

Figure 1: **a.** *An order 2 tensor* $A \in \mathbb{R}^{5 \times 4}$. **b.** *An order 3 tensor* $t \in \mathbb{R}^{7 \times 4 \times 3}$.

Appendix C

Measure Theory

The reader interested in a more detailed exposition of Measure Theory should consult reference [51].

C.1 Information and $\mathfrak{S}$-algebras

The concept of $\mathfrak{S}$-algebra is useful to describe an information structure and to define later measurable functions and measures. To make the understanding easier, the concept of sigma-algebra will be introduced as the information stored in a set of mental concepts.

We assume the brain is a set of N neurons, each neuron being regarded as a device that can be either on or off. This leads to a total of 2^N possible brain states. Any subset of this set corresponds to a representation of a mental concept. The set of all possible mental concepts will define a sigma-algebra as in the following.

Assume that one looks at an object. Then his/her mind takes a certain state associated with that specific object. For instance, looking at an "apple" and at a "bottle", separately, the mind will produce two mental concepts denoted by A and B, respectively. Then, our day-to-day experience says that the mind can understand both "apple and bottle" as a mental concept represented by the intersection $A \cap B$, which contains the common features of these two objects, such as color, shape, size, etc.

The mind can also understand the compound mental concept of "apple or bottle" as the union concept $A \cup B$, which contains all features of both apple or bottle. The mind can understand that the apple is not a bottle, and in general, if an apple is presented, then it understands all objects that are not

© Springer Nature Switzerland AG 2020
O. Calin, *Deep Learning Architectures*, Springer Series in the Data Sciences,
https://doi.org/10.1007/978-3-030-36721-3

an apple. This is done by a concept denoted by $\overline{A}$, called the complementary of the set A.

The mental concepts A and B denote two pieces of information the mind gets from the outside world. Then it has the ability to compile them into new pieces of information, such as $A \cap B$, $A \cup B$, $\overline{A}$ and $\overline{B}$.

If all possible mental concepts are denoted by $\mathcal{E}$, then the previous statements can be written as

(*i*) $A \cup B \in \mathcal{E}$, $\forall A, B \in \mathcal{E}$

(*ii*) $A \cap B \in \mathcal{E}$, $\forall A, B \in \mathcal{E}$

(*iii*) $\overline{A} \in \mathcal{E}$, $\forall A \in \mathcal{E}$.

Relation (*i*) can be generalized to n sets as in the following:

(*iv*) For any $A_1, \ldots, A_n \in \mathcal{E}$, then $\bigcup_{i=1}^{n} A_i \in \mathcal{E}$.

The structure of $\mathcal{E}$ defined by $(i) - (iv)$ is called an *algebra structure*. We note that (ii) is a consequence of the other two conditions, fact that follows from de Morgan relation $\overline{A \cap B} = \overline{A} \cup \overline{B}$.

Assume now the mind has the capability of picking up an infinite countable sequence of information and store it also as information, i.e., $\mathcal{E}$ is closed with respect to countable unions:

(*v*) For any $A_1, \ldots, A_n, \ldots \in \mathcal{E}$, then $\bigcup_{i \geq 1} A_i \in \mathcal{E}$.

In the case when $\mathcal{E}$ satisfies also condition (v), the structure is called a $\mathfrak{S}$-*algebra*. This will be the fundamental structure of modeling information.

Remark C.1.1 Using de Morgan's relation, it follows that a $\mathfrak{S}$-*algebra* is closed to countable unions, i.e., $\bigcap_{i \geq 1} A_i \in \mathcal{E}$ for any $A_1, \ldots, A_n, \ldots \in \mathcal{E}$.

Example C.1.1 Let E be a finite set. The smallest $\mathfrak{S}$-algebra on E is $\mathcal{E} = \{\emptyset, E\}$, while the largest is the set of parts $\mathcal{E} = 2^E = \{\mathcal{P}; \mathcal{P} \subseteq E\}$.

Example C.1.2 Let $\mathcal{C}$ be a set of parts of E, i.e., $\mathcal{C} \subset 2^E$. Then the smallest $\mathfrak{S}$-algebra on E that contains $\mathcal{C}$ is given by the intersection of all $\mathfrak{S}$-algebras $\mathcal{E}_\alpha$ that contain $\mathcal{C}$

$$\mathfrak{S}(\mathcal{C}) = \bigcap_\alpha \mathcal{E}_\alpha.$$

Here, we note that an intersection of $\mathfrak{S}$-algebras is also a $\mathfrak{S}$-algebra. $\mathfrak{S}(\mathcal{C})$ is the information structure generated by the collection of sets $\mathcal{C}$. It can be shown that it has the following properties:

(*i*) $\mathcal{C} \subset \mathcal{D} \Longrightarrow \mathfrak{S}(\mathcal{C}) \subset \mathfrak{S}(\mathcal{D})$;

(*ii*) $\mathcal{C} \subset \mathfrak{S}(\mathcal{D}) \Longrightarrow \mathfrak{S}(\mathcal{C}) \subset \mathfrak{S}(\mathcal{D})$;

(*iii*) If $\mathcal{C} \subset \mathfrak{S}(\mathcal{D})$ and $\mathcal{D} \subset \mathfrak{S}(\mathcal{C})$, then $\mathfrak{S}(\mathcal{C}) = \mathfrak{S}(\mathcal{D})$;

(*iv*) $\mathcal{C} \subset \mathcal{D} \subset \mathfrak{S}(\mathcal{C}) \Longrightarrow \mathfrak{S}(\mathcal{C}) = \mathfrak{S}(\mathcal{D})$.

Example C.1.3 Let E be a topological space. The Borel $\mathfrak{S}$-algebra $\mathcal{B}_E$ is the $\mathfrak{S}$-algebra generated by all open sets of E. In particular, if $E = \mathbb{R}^n$, then $\mathcal{B}_{\mathbb{R}^n}$ is the $\mathfrak{S}$-algebra generated by all open balls of $\mathbb{R}^n$.

Example C.1.4 The Borel $\mathfrak{S}$-algebra $\mathcal{B}_{\mathbb{R}}$ is the $\mathfrak{S}$-algebra generated by all the following collections: $\{(-\infty, x); x \in \mathbb{R}\}$, $\{(-\infty, x]; x \in \mathbb{R}\}$, $\{(x, y); x, y \in \mathbb{R}\}$, $\{(x, y]; x, y \in \mathbb{R}\}$, $\{(x, \infty); x \in \mathbb{R}\}$.

Definition C.1.2 *A collection $\mathcal{P}$ of subsets of Ω is called a p-system if it is closed under intersections, i.e.,*

$$A, B \in \mathcal{P} \Longrightarrow A \cap B \in \mathcal{P}.$$

A collection $\mathcal{D}$ of subsets of Ω is called a d-system on Ω if:
(i) $\Omega \in \mathcal{D}$;
(ii) $A, B \in \mathcal{D}$ *and* $B \subset A \Longrightarrow A \backslash B \in \mathcal{D}$;
(iii) $(A_n)_n \subset \mathcal{D}$ *and* $A_n \nearrow A \Longrightarrow A \in \mathcal{D}$.

Theorem C.1.3 (Dynkin) *If a d-system contains a p-system, then it contains also the sigma-algebra generated by that p-system, i.e.,*

$$\mathcal{P} \subset \mathcal{D} \Longrightarrow \mathfrak{S}(\mathcal{P}) \subset \mathcal{D}$$

A *measurable space* is a pair $(E, \mathcal{E})$, where E is a set and $\mathcal{E}$ is a set of parts on E, which is a $\mathfrak{S}$-algebra.

For instance, the configurations of the brain can be thought of as a measurable space. In this case, E is the set of brain synapses (states of the brain). The brain stores information into configurations (collections of states of the brain). A configuration of the brain is a subset of synapses set E that gets activated. The set of brain configurations, $\mathcal{E}$, forms a $\mathfrak{S}$-algebra, and hence the pair $(E, \mathcal{E})$ becomes a measurable space.

C.2 Measurable Functions

The rough idea of measuring seems to be simple: it is a procedure that assigns a number with each set. The set can be, for instance, a set of points on a line or in the plane, or a set of people with a certain characteristic, etc. However, questions like *How many people have exactly 200 lb?* are not well posed. This is because in order to measure, we need an interval to which the number should belong to. There might be many people whose weights are between 199.9 lb and 200.01 lb., and are regarded as having just 200 lb.

We might face a similar problem when writing a computer program. Let L and D denote the length and the diameter of a circle. The program:

$x = L/D$

if $(x = \pi)$

print ("it is a circle")

won't run, because the quotient will never be exactly equal to π. The correct version should include an error tolerance:

$x = L/D$

if $(abs(x - \pi) < 0.0001)$

print ("it is probably a circle")

Therefore, the correct way to measure is to assign a lower and an upper bound for the measured result. Here is where the information defined by open intervals, which is a Bored $\mathfrak{S}$-algebra, will come into the play.

Let $f : E \to \mathbb{R}$ be a function, where E denotes the synapses set of the brain. The function f is called *measurable* if regresses any open interval into an a priori given brain configuration. This means $f^{-1}(a, b) \in \mathcal{E}$, for all $a, b \in \mathbb{R}$, i.e., $f^{-1}(a, b)$ is one of the brain configurations, where we used the notation $f^{-1}(a, b) = \{x \in E; f(x) \in (a, b)\}$. We can also write this, equivalently, as $f^{-1}(\mathcal{B}_{\mathbb{R}}) \subset \mathcal{E}$. To indicate explicitly that the measurability is considered with respect to the $\mathfrak{S}$-algebra $\mathcal{B}_{\mathbb{R}}$, the function f is called sometimes *Borel-measurable*. Measurable functions can be used by the brain to get aware of the exterior world.

If $(E, \mathcal{E})$ and $(F, \mathcal{F})$ are two measurable spaces, a function $f : E \to F$ is called measurable if $f^{-1}(B) \in \mathcal{E}$ for all $B \in \mathcal{F}$. Or, equivalently, $f^{-1}(\mathcal{F}) \subset \mathcal{E}$. If $(E, \mathcal{E})$ is the measurable space associated with Ann's brain and $(F, \mathcal{F})$ is associated with Bob's brain, then the fact that the function f is measurable means that "anything Bob can think, Ann can understand." More precisely, any configuration state B in Bob's brain can be regressed by f into a configuration present in Ann's brain.

Example C.2.1 Let $A \in \mathcal{E}$ be a set and consider the *indicator function* of A

$$1_A(x) = \begin{cases} 1, & \text{if } x \in A \\ 0, & \text{if } x \notin A. \end{cases}$$

Then 1_A is measurable (with respect to $\mathcal{E}$).

Example C.2.2 A function $f : E \to F$ is called *simple* if it is a linear combination of indicator functions

$$f(x) = \sum_{j=1}^{n} a_j 1_{A_j}(x), \quad a_j \in \mathbb{R}, A_j \in \mathcal{E}.$$

Any simple function is measurable.

Given some measurable functions, one can use them to construct more measurable functions, as in the following:

1. If f and g are measurable, then $f \pm g$, $f \cdot g$, $\min(f,g)$, $\max(f,g)$ are measurable.

2. If $(f_n)_n$ is a sequence of measurable functions, then $\inf f_n$, $\sup f_n$, $\liminf f_n$, $\limsup f_n$, and $\lim f_n$ (if exists) are all measurable.

3. If f is measurable, then $f^+ = \max(f,0)$ and $f^- = -\min(f,0)$ are measurable.

It can be shown that any measurable function is the limit of a sequence of simple functions. If the function is bounded, the same bound applies to the simple functions.

C.3   Measures

A *measure* is a way to assess a body of information. This can be done using a mapping $\mu : \mathcal{E} \to \mathbb{R}_+ \cup \{\infty\}$ with the following properties:

(*i*) $\mu(\emptyset) = 0$;

(*ii*) $\mu(\bigcup_{n \geq 1} A_n) = \sum_{n \geq 1} \mu(A_n)$, for any disjoint sets A_j in $\mathcal{E}$ (countable additivity).

If $(E, \mathcal{F})$ is a measurable space associated with the states and configurations of a brain, respectively, then a measure μ is an evaluation system of beliefs; each brain configuration $A \in \mathcal{E}$ is associated with an intensity $\mu(A)$. The triplet $(E, \mathcal{E}, \mu)$ is called a *measure space.*

Example C.3.1 (Dirac measure) Let $x \in E$ be a fixed point. Then

$$\delta_x(A) = \begin{cases} 1, & \text{if } x \in A \\ 0, & \text{if } x \notin A, \end{cases} \quad \forall A \in \mathcal{E}$$

is a measure on $\mathcal{E}$.

Example C.3.2 (Counting measure) Let $D \subset E$ be fixed and finite. Define

$$\mu(A) = card(A \cap D) = \sum_{x \in D} \delta_x(A), \quad \forall A \in \mathcal{E}.$$

μ is a measure on $\mathcal{E}$. It counts the number of elements of A which are in D.

Example C.3.3 (Discrete measure) Let $D \subset E$ be fixed and discrete. Define $m(x)$ be the mass of x, with $m(x) \geq 0$, for all $x \in D$. Consider

$$\mu(A) = mass(A) = \sum_{x \in D} m(x)\delta_x(A), \quad \forall A \in \mathcal{E}.$$

μ is a measure on $\mathcal{E}$, which assess the mass of A.

Example C.3.4 (Lebesgue measure) Consider $E = \mathbb{R}$ and the $\mathfrak{S}$-algebra $\mathcal{E} = \mathcal{B}_{\mathbb{R}}$. μ is defined for open intervals as their lengths, $\mu(a, b) = |b - a|$. It can be shown that there is a unique measure that extends μ to $\mathcal{B}_{\mathbb{R}}$, called the *Lebesgue measure* on $\mathbb{R}$. In a similar way, replacing lengths by volumes and open intervals by open hypercubes, one can define the Lebesgue measure on $\mathbb{R}^n$.

Example C.3.5 (Borel measure) Let $\mathcal{B}_{\mathbb{R}^n}$ be the $\mathfrak{S}$-algebra generated by all the open sets of $\mathbb{R}^n$. A *Borel measure* is a measure $\mu : \mathcal{B}_{\mathbb{R}^n} \to \mathbb{R}$. In the case $n = 1$, μ becomes a Borel measure on the real line. If μ is a finite measure, we associate to it the function $F(x) = \mu(-\infty, x]$, called the cumulative distribution function, which is monotone increasing function, satisfying $\mu(a, b] = F(b) - F(a)$. For instance, the Lebesgue measure is a Borel measure.

Example C.3.6 (Baire measure) Let $K \subseteq \mathbb{R}^n$ and denote by $C^0(K)$ the set of all continuous real-valued functions with compact support (which vanish outside of a compact subset of K). The class of *Baire sets*, $\mathcal{B}$, is defined to be the $\mathfrak{S}$-algebra generated by $\{x; f(x) \geq a\}$, with $f \in C^0(K)$. A *Baire measure* is a measure defined on $\mathcal{B}$, such that $\mu(C) < \infty$, for all compact subsets $C \subset K$. It is worth noting that for $K \subseteq \mathbb{R}^n$ the class of Baire sets is the same as the class of Borel sets. In particular, any finite Borel measure is a Baire measure.

Proposition C.3.1 (Properties of measures) *Let $(E, \mathcal{E}, \mu)$ be a measure space. The following hold:*

(i) finite additivity:

$$A \cap B = \emptyset \Longrightarrow \mu(A \cup B) = \mu(A) + \mu(B), \quad \forall A, B \in \mathcal{E};$$

(ii) monotonicity:

$$A \subset B \Longrightarrow \mu(A) \leq \mu(B), \quad \forall A, B \in \mathcal{E};$$

(iii) sequential continuity:

$$A_n \nearrow A \Longrightarrow \mu(A_n) \nearrow \mu(A), \quad n \to \infty;$$

(iv) Boole's inequality:

$$\mu\left(\bigcup_n A_n\right) \leq \sum_n \mu(A_n), \quad \forall A_n \in \mathcal{E}.$$

If μ and λ are measures on $(E, \mathcal{E})$, then $\mu + \lambda$, $c\mu$ and $c_1\mu + c_2\lambda$ are also measures, where $c, c_i \in \mathbb{R}_+$.

Let $(E, \mathcal{E}, \mu)$ be a measure space. Then μ is called:
- *finite measure* if $\mu(E) < \infty$;
- *probability measure* if $\mu(E) = 1$;
- $\mathfrak{S}$-*finite measure* if $\mu(E_n) < \infty$, where $(E_n)_n$ is partition of E, with $E_n \in \mathcal{E}$;
- Σ-*finite measure* if $\mu = \sum_n \mu_n$, with μ_n finite measure.

For instance, the Lebesgue measure is $\mathfrak{S}$-finite but it is not finite.

A set $M \in \mathcal{E}$ is called *negligible* if $\mu(M) = 0$. Two measurable functions $f, g : E \to \mathbb{R}$ are equal almost everywhere, i.e., $f = g$ a.e., if there is a negligible set M such that $f(x) = g(x)$ for all $x \in E \backslash M$.

C.4   Integration in Measure

Let $(E, \mathcal{E}, \mu)$ be a measure space and $f : E \to \mathbb{R}$ be a measurable function. The object

$$\mu(f) = \int_E f(x)\mu(dx) = \int_E f \, d\mu$$

represents an evaluation of f though the system of beliefs μ and it is called the *integral* of f with respect to the measure μ. This is defined by the following sequence of steps:

(i) if f is simple and positive, i.e., if $f = \sum_{i=1}^n w_i 1_{A_i}$, then define

$$\mu(f) = \sum_{i=1}^n w_i \mu(A_i).$$

(ii) If f is measurable and positive, then there is a sequence $(f_n)_n$ of simple and positive functions with $f_n \nearrow f$. In this case, define $\mu(f) = \lim_n \mu(f_n)$.

(iii) If f is measurable, then let $f = f^+ - f^-$ and define

$$\mu(f) = \mu(f^+) - \mu(f^-).$$

If $\mu(f) < \infty$, then the measurable function f is called *integrable*. The non-negativity, linearity, and monotonicity properties of the integral are, respectively, given by
1. $\mu(f) \geq 0$ for $f : E \to \mathbb{R}_+$;
2. $\mu(af + bg) = a\mu(f) + b\mu(g)$, for all $a, b \in \mathbb{R}$;
3. If $f \leq g$ then $\mu(f) \leq \mu(g)$.

Example C.4.1 Let δ_x be the Dirac measure sitting at x. The integral of the measurable function f with respect to the Dirac measure is $\delta_x(f) = f(x)$.

Example C.4.2 The integral of the measurable function f with respect to the discrete measure $\mu = \sum_{x \in D} m(x) \delta_x$ is given by $\mu(f) = \sum_{x \in D} m(x) f(x)$.

Example C.4.3 Let $E = \mathbb{R}$, $\mathcal{E} = \mathcal{B}_{\mathbb{R}}$ and μ be the Lebesgue measure on $\mathbb{R}$. In this case $\mu(f) = \int_E f(x)\, dx$ is called the Lebesgue integral of f on E.

The integral of f over a set $A \in \mathcal{E}$ is defined by

$$\int_A f\, d\mu = \int_E f\, 1_A\, d\mu = \mu(f\, 1_A).$$

In particular, we have

$$\int_A d\mu = \mu(1_A) = \mu(A), \quad \forall A \in \mathcal{E}.$$

We provide next three key tools for interchanging integrals and limits.

Theorem C.4.1 (Monotone Convergence Theorem) *Let $(f_n)_n$ be a sequence of positive and measurable functions on E such that $f_n \nearrow f$. Then*

$$\lim_{n \to \infty} \int f_n\, d\mu = \int f\, d\mu.$$

Theorem C.4.2 (Dominated Convergence Theorem) *Let $(f_n)_n$ be a sequence of measurable functions with $|f_n| \leq g$, with g integrable on E. If $\lim_n f_n$ exists, then*

$$\lim_{n \to \infty} \int f_n\, d\mu = \int f\, d\mu.$$

Theorem C.4.3 (Bounded Convergence Theorem) *Let $(f_n)_n$ be a bounded sequence of measurable functions on E with $\mu(E) < \infty$. If $\lim_{n \to \infty} f_n$ exists, then*

$$\lim_{n \to \infty} \int f_n\, d\mu = \int f\, d\mu.$$

We make the remark that two measurable functions equal almost everywhere have equal integrals, i.e., the integral is insensitive at changes over negligible sets.

Appendix

C.5   Image Measures

Let $(F, \mathcal{F})$ and $(E, \mathcal{E})$ be two measurable spaces and $h : F \to E$ be a measurable function. Any measure ν on $(F, \mathcal{F})$ induces a measure μ on $(E, \mathcal{E})$ by

$$\mu(B) = \nu(h^{-1}(B)), \qquad \forall B \in \mathcal{E}.$$

The measure $\mu = \nu \circ h^{-1}$ is called the *image measure* of ν under h. It is worth noting that if $f : E \to \mathbb{R}$ is measurable, then the above relation becomes the following change of variable formula:

$$\int_E f(y) \, d\mu(y) = \int_F f(h(x)) \, d\nu(x),$$

provided the integrals exist.

Remark C.5.1 If $(F, \mathcal{F})$ and $(E, \mathcal{E})$ represent the pairs (states, configurations) for Ann's and Bob's brains, respectively, then the fact that h is measurable means that "anything Bob can think, Ann can understand." The measure μ is a system of beliefs for Bob, which is induced by the system of beliefs ν of Ann.

C.6   Indefinite Integrals

Let $(E, \mathcal{E}, \mu)$ be a measure space and $p : E \to \mathbb{R}_+$ measurable. Then

$$\nu(A) = \int_A p(x) \, d\mu(x), \quad A \in \mathcal{E}$$

is a measure on $(E, \mathcal{E})$, called the *indefinite integral* of p with respect to μ. It can be shown that for any measurable positive function $f : E \to \mathbb{R}_+$ we have

$$\int_E f(x) \, d\nu(x) = \int_E f(x) p(x) \, d\mu(x).$$

This can be informally written as $p(x) d\mu(x) = d\nu(x)$.

C.7   Radon-Nikodym Theorem

Let μ, ν be two measures on $(E, \mathcal{E})$ such that

$$\mu(A) = 0 \Rightarrow \nu(A) = 0, \qquad A \in \mathcal{E}.$$

Then ν is called *absolutely continuous* with respect to μ. The following result states the existence of a density function p.

Theorem C.7.1 *Let μ be $\mathfrak{S}$-finite and ν be absolute continuous with respect to μ. Then there is a measurable function $p : E \to \mathbb{R}_+$ such that*

$$\int_E f(x)\, d\nu(x) = \int_E f(x)p(x)\, d\mu(x),$$

for all measurable functions $f : E \to \mathbb{R}_+$.

The previous integral can be also written informally as $p(x)d\mu(x) = d\nu(x)$. The density function $p = \frac{d\nu}{d\mu}$ is called the Radon-Nikodym derivative.

Remark C.7.2 This deals with an informal understanding of the theorem. If the measures μ, ν are considered as two systems of evaluation, the fact that ν is absolute continuous with respect to μ means that the system of evaluation ν is less rigorous than the system μ. This means that the mistakes that are negligible in the system ν, i.e., $\mu(A) = 0$, also pass undetected by the system μ, i.e. $\nu(A) = 0$. Under this hypothesis, Radon-Nikodym theorem states that the more rigorous system can be scaled into the less rigorous system by the relation $d\nu = pd\mu$. The density function p becomes a scaling function between the systems of evaluation.

C.8   Egorov and Luzin's Theorems

Egorov's theorem establishes a relation between the almost everywhere convergence and uniform convergence.

Let $(E, \mathcal{E}, \mu)$ be a measure space and $f_n : E \to \mathbb{R} \cup \{\infty\}$ be a sequence of extended real value functions. We say that f_n *converges a.e.* to a limit f if $f_n(x) \to f(x)$, as $n \to \infty$, for all $x \in E \backslash N$, with N a μ-negligible set.

We say that the sequence (f_n) *converges uniformly* to f on A if $\exists n_0 > 1$ such that

$$|f_n(x) - f(x)| < \epsilon, \qquad \forall n > n_0, \forall x \in A.$$

Theorem C.8.1 (Egorov) *Let E be a measurable set of finite measure, and f_n a sequence of a.e. finite value measurable functions that converge a.e. on E to a finite measurable function f. Then for any $\epsilon > 0$, there is a measurable subset F of E such that $\mu(F) < \epsilon$ and (f_n) converges to f uniformly on $E \backslash F$.*

Loosely stated, any a.e. pointwise convergent sequence of measurable functions is uniformly convergent on nearly all its domain.

A consequence of this result is Luzin's Theorem, which states that a measurable function is nearly continuous on its domain, and hence a continuous function can be approximated by measurable functions.

Appendix

A measure μ on $(\mathbb{R}^n, \mathcal{B}_{\mathbb{R}^n})$ is called *regular* if:

$$\mu(A) = \inf\{\mu(U); A \subset U, U \text{open in } \mathbb{R}^n\} = \sup\{\mu(V); V \subset A, V \text{open in } \mathbb{R}^n\}.$$

This means that the measure structure is related with the topological structure of the space.

Theorem C.8.2 *If $f : I_n = [0,1]^n \to \mathbb{R}$ is a measurable function, then for any $\epsilon > 0$ there is a compact set K in I_n such that $\mu(I_n \backslash K) < \epsilon$ and f is continuous on K, where μ is a regular Borel measure.*

C.9   Signed Measures

The difference of two measures is not in general a measure, since is not necessarily nonnegative. In the following we shall deal with this concept.

A *signed measure* on the measurable space $(E, \mathcal{E})$ is a map

$$\nu : \mathcal{E} \to \mathbb{R} \cup \{\pm\infty\}$$

such that

(i) ν assumes at most one of the values $-\infty$, $+\infty$;

(ii) $\nu(\emptyset) = 0$;

(iii) $\nu\left(\bigcup_{i \geq 1} A_i\right) = \sum_{i \geq 1} \nu(A_i)$, for any sequence (A_i) of disjoint sets in $\mathcal{E}$ (i.e., ν is countable additive).

Example C.9.1 Any measure is a signed measure, while the reverse does not hold true in general.

Example C.9.2 Any difference of two measures, $\mu = \nu_1 - \nu_2$ is a signed measure. The reverse of this statement also holds true, as we shall see shortly.

Let ν be a signed measure. A set G in $\mathcal{E}$ is called a *positive set* with respect to ν if any measurable subset of G has a nonnegative measure, i.e.,

$$\nu(G \cap A) \geq 0, \quad \forall A \in \mathcal{E}.$$

A set F in $\mathcal{E}$ is called a *negative set* with respect to ν if any measurable subset of F has a nonpositive measure, i.e.,

$$\nu(F \cap B) \leq 0, \quad \forall B \in \mathcal{E}.$$

A set which is simultaneously positive and negative with respect to a signed measure is called a *null set*. We note that any null set has measure zero, while the converse is in general false.

Proposition C.9.3 (Hahn Decomposition Theorem) *Let ν be a signed measure on the measurable space $(E, \mathcal{E})$. Then there is a partition of E into a positive set A and a negative set B, i.e., $E = A \cup B$, $A \cap B = \emptyset$.*

It is worth noting that Hahn decomposition is not unique. However, for any two distinct decompositions $\{A_1, B_1\}$ and $\{A_2, B_2\}$ of E it can be shown that

$$\nu(F \cap A_1) = \nu(F \cap A_2), \quad \nu(F \cap B_1) = \nu(F \cap B_2), \quad \forall F \in \mathcal{E}.$$

This suggests to define for any Hahn decomposition $\{A, B\}$ of E two measures, ν^+, ν^-, such that

$$\nu^+(F) = \nu(F \cap A), \quad \nu^-(F) = -\nu(F \cap B), \quad \forall F \in \mathcal{E}.$$

We note that $\nu = \nu^+ - \nu^-$. We also have $\nu^+(B) = \nu^-(A) = 0$, with $\{A, B\}$ measurable partition of E. A pair of measures ν^+, ν^- with this property is called *mutually singular*.

Proposition C.9.4 (Jordan Measure Decomposition) *Let ν be a signed measure on the measurable space $(E, \mathcal{E})$. Then there are two measures ν^+ and ν^- on $(E, \mathcal{E})$ such that $\nu = \nu^+ - \nu^-$. Furthermore, if ν^+ and ν^- are mutually singular measures, the decomposition is unique.*

The measures ν^+ and ν^- are called the *positive part* and *negative part* of ν, respectively. Since ν takes only one of the values $\pm \infty$, then one of the parts has to be a finite measure. The sum of measures $|\nu| = \nu^+ + \nu^-$ is a measure, called the *absolute value* of ν.

The *integration* of a measurable function f with respect to a signed measure ν is defined as

$$\int f \, d\nu = \int f \, d\nu^+ - \int f \, d\nu^-,$$

provided f is integrable with respect to $|\nu|$. Furthermore, if $|f| < C$, then

$$\left| \int_E f \, d\nu \right| \leq C |\nu|(E).$$

Example C.9.1 Let g be an integrable function on the measure space $(E, \mathcal{E}, \nu)$. The measure

$$\nu(F) = \int_F g(x) \, d\nu(x), \qquad \forall F \in \mathcal{E}$$

is a finite signed measure. The decomposition $g = g^+ - g^-$ yields $\nu = \nu^+ - \nu^-$, where

$$\nu^+(F) = \int_F g^+(x) \, d\nu(x), \quad \nu^-(F) = \int_F g^-(x) \, d\nu(x).$$

Remark C.9.5 If μ is a signed finite measure on the measurable space $(\Omega, \mathcal{F})$, define its total variation as

$$\|\mu\|_{TV} = \sup \sum_{i=1}^{n} |\mu(A_i)|$$

over all finite partitions with disjoint sets, $\Omega = \bigcup_{i=1}^{n} A_i$. This set of measures with the norm $\|\cdot\|_{TV}$ forms a Banach space.

Appendix D

Probability Theory

The reader interested in probability can consult for details the book [23].

D.1   General definitions

A *probability space* is a measure space $(\Omega, \mathcal{H}, \mathbb{P})$, where $\mathbb{P}$ is a probability measure, i.e., a measure with $\mathbb{P}(\Omega) = 1$. The set Ω is the sample space, sometimes regarded also as the states of the world; it represents the outcomes set of an experiment. The $\mathfrak{S}$-algebra $\mathcal{H}$ is the history information; each set $H \in \mathcal{H}$ is called an *event*. The probability measure $\mathbb{P}$ evaluates the chance of occurrence of events. For each $H \in \mathcal{H}$, the number $\mathbb{P}(H)$ is the probability that H occurs.

A *random variable* is a mapping $X : \Omega \to \mathbb{R}$ which is $(\mathcal{H}, \mathcal{B}_{\mathbb{R}})$-measurable, i.e., $f^{-1}(\mathcal{B}_{\mathbb{R}}) \subset \mathcal{H}$. This means that for each experiment outcome $\omega \in \Omega$, X assigns a number $X(\omega)$.

The image of the probability measure $\mathbb{P}$ through X, $\mu = \mathbb{P} \circ X^{-1}$, is a measure on $(\mathbb{R}, \mathcal{B}_{\mathbb{R}})$, called the *distribution* of the random variable X. More precisely, we have

$$\mu(A) = \mathbb{P}(X \in A) = \mathbb{P}(\omega; X(\omega) \in A) = \mathbb{P}(X^{-1}(A)).$$

The measure μ describes how is X distributed. The function

$$F(x) = \mu(-\infty, x] = \mathbb{P}(X \leq x)$$

is called the *distribution function* of X.

© Springer Nature Switzerland AG 2020
O. Calin, *Deep Learning Architectures*, Springer Series in the Data Sciences,
https://doi.org/10.1007/978-3-030-36721-3

Let $(\Omega, \mathcal{H}, \mathbb{P})$ be a probability space and $(E, \mathcal{E})$, $(F, \mathcal{F})$ be two measurable spaces. If $X : \Omega \to E$ and $f : E \to F$ are measurable, consider their composition $Y = f \circ X$. If μ is the distribution of X, then $\nu = \mu \circ f^{-1}$ is the distribution of Y. We can write

$$
\begin{aligned}
\nu(B) &= \mathbb{P}(Y \in B) = \mathbb{P}(f \circ X \in B) = \mathbb{P}(X \in f^{-1}(B)) \\
&= \mu(f^{-1}(B)) = (\mu \circ f^{-1})(B), \quad \forall B \in \mathcal{F}.
\end{aligned}
$$

D.2 Examples

Bernoulli Distribution A random variable X is said to be Bernoulli distributed with parameter $p \in [0, 1]$, if $X \in \{0, 1\}$, with $P(X = 1) = p$ and $P(X = 0) = 1 - p$. The mean and variance are given, respectively, by

$$
\mathbb{E}[X] = 1 \cdot P(X = 1) + 0 \cdot P(X = 0) = p,
$$

$$
Var(X) = \mathbb{E}[X^2] - \mathbb{E}[X]^2 = 1^2 \cdot P(X = 1) + 0^2 \cdot P(X = 0) - p^2 = p - p^2 = p(1-p).
$$

We shall write $X \sim$ Bernoulli(p).

Normal Distribution A random variable X is called normal distributed with parameters μ and σ if $P(X < x) = \int_{-\infty}^{x} f(x)\, dx$, with

$$
f(x) = \frac{1}{\sqrt{2\pi}\sigma} e^{-\frac{(x-\mu)^2}{2\sigma^2}}, \qquad x \in \mathbb{R}.
$$

The mean and variance are given by

$$
\mathbb{E}[X] = \mu, \quad Var(X) = \sigma^2.
$$

We shall use the notation $X \sim \mathcal{N}(\mu, \sigma^2)$.

D.3 Expectation

The *expectation* of the random variable X is the assessment of X though the probability measure $\mathbb{P}$ as

$$
\mathbb{E}[X] = \mathbb{P}(X) = \int_\omega X(\omega)\, d\mathbb{P}(\omega).
$$

If μ is the distribution measure of X then the change of variable formula provides

$$
\mathbb{E}[f(X)] = \int_\Omega f(X(\omega))\, d\mathbb{P}(\omega) = \int_\mathbb{R} f(y)\, d\mathbb{P}(X^{-1}y) = \int_\mathbb{R} f(y)\, d\mu(y)
$$

In particular, if μ is absolute continuous with respect to the Lebesgue measure dy on $\mathbb{R}$, then there is a nonnegative measurable density function $p(y)$ such that $d\mu(y) = p(y)dy$. Consequently, the previous formula becomes

$$\mathbb{E}[f(X)] = \int_{\mathbb{R}} f(y)p(y)dy.$$

The expectation operator is nonnegative, monotonic, and linear operator, i.e.,

(i) $X \geq 0 \Longrightarrow \mathbb{E}[X] \geq 0$;
(ii) $X \geq Y \Longrightarrow \mathbb{E}[X] \leq \mathbb{E}[Y]$;
(iii) $\mathbb{E}[aX + bY] = a\mathbb{E}[X] + b\mathbb{E}[Y]$ for $a, b \in \mathbb{R}$.

D.4 Variance

Let $\mu = \mathbb{E}[X]$ be the mean of X. The *variance* of the random variable X is defined as

$$Var(X) = \mathbb{E}[(X - \mu)^2].$$

The variance is a measure of deviation from the mean in the mean square sense. If let $p(x)$ denote the probability density of X, physically, the variance represents the inertia momentum of the curve $y = p(x)$ about the vertical axis $x = \mu$. This is a measure of easiness of revolution of the graph $y = p(x)$ about a vertical axis passing through its mass center, μ.

In general, the variance is neither additive nor multiplicative. However in some particular cases it is, as we shall see next. The *covariance* of two random variables X and Y is defined by $Cov(X, Y) = \mathbb{E}[XY] - \mathbb{E}[X]\mathbb{E}[Y]$. The variance of the sum is given by

$$Var(X + Y) = Var(X) + 2Cov(X, Y) + Var(Y).$$

If X and Y are independent, then $Var(X+Y) = Var(X) + Var(Y)$. The variance is also homogeneous of order 2, i.e., $Var(cX) = c^2 var(X)$, for all $c \in \mathbb{R}$.

We also have the following exact expression for the variance of a product of independent random variables, see [49].

Lemma D.4.1 (Goodman's formula) *If X and Y are two independent random variables, then*

$$Var(XY) = \mathbb{E}[X]^2 Var(Y) + \mathbb{E}[Y]^2 Var(X) + Var(X)Var(Y).$$

In particular, if $\mathbb{E}[X] = \mathbb{E}[Y] = 0$, then

$$Var(XY) = Var(X)Var(Y).$$

Proof: Denote the right side by

$$R = \mathbb{E}[X]^2 Var(Y) + \mathbb{E}[Y]^2 Var(X) + Var(X)Var(Y).$$

The definition of variance and some algebraic manipulations together with the independence property provides

$$
\begin{aligned}
R &= Var(Y)\big[\mathbb{E}[X]^2 + Var(X)\big] + \mathbb{E}[Y]^2 Var(X) \\
&= Var(Y)\mathbb{E}[X^2] + \mathbb{E}[Y]^2 Var(X) \\
&= \big[\mathbb{E}[Y^2] - \mathbb{E}[Y]^2\big]\mathbb{E}[X^2] + \mathbb{E}[Y]^2\big[\mathbb{E}[X^2] - \mathbb{E}[X]^2\big] \\
&= \mathbb{E}[Y^2]\mathbb{E}[X^2] - \mathbb{E}[Y]^2\mathbb{E}[X]^2 = \mathbb{E}[X^2 Y^2] - \mathbb{E}[XY]^2 \\
&= Var(XY),
\end{aligned}
$$

which recovers the left side of the desired expression. ∎

Variance approximation Let $m = \mathbb{E}[X]$ and f be a differentiable function. Linear approximation about $x = m$ provides

$$
f(x) = f(m) + f'(m)(x - m) + o(x - m)^2.
$$

Then replace the variable x by the random variable X to obtain

$$
f(X) \approx f(m) + f'(m)(X - m).
$$

Taking the variance on both sides and using its properties we arrive at the following approximation formula:

$$
Var(f(X)) \approx f'(m)^2 Var(X), \tag{D.4.1}
$$

provided that f is twice differentiable and that the mean m and variance of X are finite. Therefore, a small variance of X produces a small variance of $f(X)$, provided f' is bounded.

D.5 Information generated by random variables

Let $X : \Omega \to \mathbb{R}$ be a random variable. The information field generated by X is the $\mathfrak{S}$-algebra

$$
\mathfrak{S}(X) = X^{-1}(\mathcal{B}_{\mathbb{R}}).
$$

Let $X : \Omega \to \mathbb{R}$ be a random variable and $f : \mathbb{R} \to \mathbb{R}$ be a measurable function. Consider the random variable $Y = f(X)$. Then Y is $\mathfrak{S}(X)$-measurable, i.e., $\mathfrak{S}(Y) \subset \mathfrak{S}(X)$. Equivalently stated, the information generated by X determines the information generated by Y. The converse statement holds also true. The proof of the following result can be found in Cinlar [23], Proposition 4.4, p.76.

Proposition D.5.1 *Consider two random vector variables $X, Y : \Omega \to \mathbb{R}$. Then $\mathfrak{S}(Y) \subset \mathfrak{S}(X)$ if and only if Y is determined by X, i.e., there is a measurable function f such that $Y = f(X)$.*

This can be stated also by saying that Y is determined by X if and only if the information generated by X is finer than the information generated by Y.

It is worth noting that the previous function f is constructed by a limiting procedure. The idea is given in the following: for a fixed n we consider the measurable set

$$A_{m,n} = Y^{-1}\left[\frac{m}{2^m}, \frac{m+1}{2^n}\right] \in \mathfrak{S}(Y) \subset \mathfrak{S}(X), \quad m = 0, \pm 1, \pm 2, \dots$$

so that $A_{m,n} = X^{-1}(B_{m,n})$, with $B_{m,n}$ measurable. Construct the simple function $f_n(x) = \sum_m \frac{m}{2^n} 1_{B_{m,n}}(x)$. It can be shown that

$$f_n(X) \leq Y \leq f_n(X) + \frac{1}{2^n}.$$

Then we choose $f = \lim_{n\to\infty} f_n$, which is measurable, as a limit of simple functions.

Example D.5.1 Let $\omega = (\omega_1, \omega_2, \omega_3) \in \Omega$ and define the random variables $X_i, Y_i : \Omega \to \mathbb{R}$, $i = 1, 2, 3$ by

$$X_1(\omega) = \omega_1, \quad X_2(\omega) = \omega_2, \quad X_3(\omega) = \omega_3,$$

$$Y_1(\omega) = \omega_1 - \omega_2, \quad Y_2(\omega) = \omega_1 + \omega_2, \quad Y_3(\omega) = \omega_1 + \omega_2 + \omega_3.$$

Then $\mathfrak{S}(X_1, X_2) = \mathfrak{S}(Y_1, Y_2)$ and $\mathfrak{S}(X_1, X_2, X_3) = \mathfrak{S}(Y_1, Y_2, Y_3)$, but $\mathfrak{S}(X_2, X_3) \neq \mathfrak{S}(Y_2, Y_3)$, since Y_3 cannot be written in terms of X_2 and X_3.

A *stochastic process* is a family of random variable $(X_t)_{t\in T}$ indexed over the continuous or discrete parameter t. The information generated by the stochastic process $(X_t)_{t\in T}$ is the smallest $\mathfrak{S}$-algebra with respect to which each random variable X_t is measurable. This can be written as

$$\mathcal{G} = \mathfrak{S}(X_t; t \in T) = \mathfrak{S}\left(\bigcup_{t\in T} \mathfrak{S} X_t\right) = \bigvee_{t\in T} \mathfrak{S}(X_t).$$

D.5.1   Filtrations

Let $(\Omega, \mathcal{H}, \mathbb{P})$ be a probability space. The $\mathfrak{S}$-algebra $\mathcal{H}$ can be interpreted as the entire history of the states of the world Ω. The information available until time t is denoted by $\mathcal{F}_t$. We note that the information grows in time, i.e., if $s < t$, then $\mathcal{F}_s \subset \mathcal{F}_t$. An increasing flow of information on $\mathcal{H}$, $(\mathcal{F}_t)_{t\in T}$, is called a *filtration*.

Each stochastic process, $(X_t)_{t\in T}$, defines a natural filtration

$$\mathcal{F}_t = \mathfrak{S}(X_s; s \leq t),$$

which is the history of the process up to each time instance t. In this case each random variable X_t is $\mathcal{F}_t$-measurable. Stochastic processes with this property are called *adapted* to the filtration.

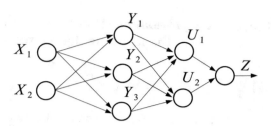

Figure 1: *In a feedforward neural network the information flow satisfies the sequence of inclusions $\mathcal{F}_Z \subset \mathcal{F}_U \subset \mathcal{F}_Y \subset \mathcal{F}_X$, which forms a filtration.*

Example D.5.2 Consider the feedforward neural network given by Fig. 1, which has two hidden layers, $Y = (Y_1, Y_1, Y_3)^T$, $U = (U_1, U_2)^T$. Its input is given by the random variable $X = (X_1, X_2)^T$ and the output is given by Z. The input information is the information generated by X, given by the sigma-algebra $\mathcal{F}_X = \mathfrak{S}(X_1, X_2)$. The information in the first and second hidden layers is given by $\mathcal{F}_Y = \mathfrak{S}(Y_1, Y_2, Y_3)$ and $\mathcal{F}_U = \mathfrak{S}(U_1, U_2)$, respectively. The output information is $\mathcal{F}_Z = \mathfrak{S}(Z)$. Since Y_j are determined by X_i, it follows that Y_j are $\mathcal{F}_X$-measurable, fact that can be written also as $\mathcal{F}_Y \subset \mathcal{F}_X$. Similarly, we have the following sequence of inclusions:

$$\mathcal{F}_Z \subset \mathcal{F}_U \subset \mathcal{F}_Y \subset \mathcal{F}_X.$$

This natural filtration describes the information flow through the network.

D.5.2   Conditional expectations

Let X be a random variable on the probability space $(\Omega, \mathcal{H}, \mathbb{P})$ and consider some partial information $\mathcal{F} \subset \mathcal{H}$. The random variable determined by $\mathcal{F}$, which is "the best approximator" of X is called the *conditional expectation* of X given $\mathcal{F}$. This is defined as the variable $\widetilde{X}$ that satisfies the properties:

(i) $\widetilde{X}$ is $\mathcal{F}$-measurable;

(ii) $\displaystyle\int_A X \, d\mathbb{P} = \int_A \widetilde{X} \, d\mathbb{P}, \ \forall A \in \mathcal{F}.$

It can be shown that in the case of a square integrable random variable X, we have

$$\|X - \widetilde{X}\|^2 \leq \|X - Z\|^2, \qquad \forall Z \in \mathcal{S}_{\mathcal{F}},$$

where $\mathcal{S}_{\mathcal{F}} = \{f \in L^2(\Omega); f \text{ is } \mathcal{F}-\text{measurable}\}$. This means that $\widetilde{X}$ is the orthogonal projection of X onto $\mathcal{S}_{\mathcal{F}}$, i.e., it is the best approximator of X in the mean square sense with elements from $\mathcal{S}_{\mathcal{F}}$. In the previous relation the norm is the one induced by expectation of the square, $\|X\|^2 = \mathbb{E}[X^2]$.

D.6 Types of Convergence

In the following $(\Omega, \mathcal{H}, \mathbb{P})$ will denote a probability space and $X_n : \Omega \to \mathbb{R}$ will be a sequence of random variables.

D.6.1 Convergence in probability

The sequence X_n *converges in probability* to X if $\lim_{n \to \infty} \mathbb{P}(|X_n - X| < \epsilon) = 1$. This type of convergence has the following interpretation. If X denotes the center of a target of radius ϵ and X_n the location of the n-th shoot, then $\{|X_n - X| < \epsilon\}$ represents the event that the n-th shoot hits the target. Convergence in probability means that in the long run the chance that the shoots X_n will hit any target centered at X with arbitrarily fixed radius ϵ approaches 1.

D.6.2 Almost sure convergence

X_n *converges almost surely* to X if $\lim_{n \to \infty} \mathbb{P}(\omega; X_n(\omega) \to X(\omega)) = 1$. This means that for almost any state $\omega \in \Omega$, the sequence $X_n(\omega)$ converges to $X(\omega)$ as a sequence of real numbers, as $n \to \infty$.

The following results provide necessary conditions for a.s. convergence.

Proposition D.6.1 (Borel-Cantelli Lemma I) *Assume*

$$\sum_{n \geq 1} P(|X_n - X| > \epsilon) < \infty, \qquad \forall \epsilon > 0.$$

Then $X_n \to X$ almost surely.

Proposition D.6.2 (Borel-Cantelli Lemma II) *Suppose there is a sequence* (ϵ_n) *decreasing to 0 such that*

$$\sum_{n \geq 1} P(|X_n - X| > \epsilon_n) < \infty.$$

Then $X_n \to X$ almost surely.

It is worth noting that the almost sure convergence implies convergence in probability.

D.6.3 L^p-convergence

The sequence X_n converges to X in L^p-sense if $X_n, X \in L^2(\Omega)$ and

$$\mathbb{E}(|X_n - X|^p) \to 0, \quad n \to \infty.$$

It is worth to note that Markov's inequality

$$P(|X_n - X| > \epsilon) \leq \frac{\mathbb{E}(|X_n - X|^p)}{\epsilon^p}$$

implies that the L^p-convergence implies the convergence in probability.

If $p = 2$, the convergence in L^2 is also called the *mean square convergence*.

A classical result involving all previous types of convergence is the Law of Large Numbers:

Theorem D.6.3 *Consider $X_1, X_2, X_3 \cdots$ independent, identically distributed random variables, with mean $\mathbb{E}[X_j] = a$ and variance $Var[X_j] = b$, both finite. Let $\overline{X}_n = \frac{1}{n}(X_1 + \cdots + X_n)$. Then $\overline{X}_n$ converges to a in mean square, almost surely and in probability.*

D.6.4   Weak convergence

Consider the random variables X_n, X, with distribution measures μ_n, μ, respectively, and denote

$$\mathbb{C}_b = \{f : \mathbb{R} \to \mathbb{R}; f \text{continuous and bounded}\}.$$

The sequence X_n *converges to X in distribution* if

$$\mathbb{E}[f \circ X_n] \to \mathbb{E}[f \circ X], \ n \to \infty \qquad \forall f \in \mathbb{C}_b.$$

The measures μ_n *converge weakly* to μ if

$$\mu_n f \to \mu f, \quad n \to \infty \qquad \forall f \in \mathbb{C}_b.$$

The relation

$$\mu(f) = \int f(x) \, d\mu(x) = \int_\Omega f \circ X \, d\mathbb{P} = \mathbb{E}[f \circ X]$$

shows that the convergence in distribution for random variables corresponds to the weak convergence of the associated distribution measures.

Another equivalent formulation is the following. If $\varphi_X(t) = \mathbb{E}[e^{itX}]$ denotes the *characteristic function* of the random variable X, then X_n converges to X in distribution if and only if $\varphi_{X_n}(t) \to \varphi_X(t)$ for all $t \in \mathbb{R}$. If μ denotes the distribution measure of X, then $\varphi_X(t) = \int_{\mathbb{R}} e^{itx} \, d\mu(x)$. This is called the *Fourier transform* of μ. Furthermore, if μ is absolute continuous with respect to dx, then X has a probability density $p(x)$, i.e., $d\mu(x) = p(x) \, dx$ and then $\varphi_X(t) = \int_{\mathbb{R}} e^{itx} p(x) \, dx$. This is called the Fourier transform of $p(x)$.

A useful property is the injectivity of this transform, i.e., if $\varphi_X = 0$ then $p = 0$. A heuristic explanation is based on the concept of frequency. More precisely, the Fourier transform of a time-domain signal $p(x)$ (i.e., $p(x)$ is the amplitude of a signal at time x) is a signal in the frequency domain, given by $\varphi_X(t)$ (i.e., $\varphi_X(t)$ is the amplitude of the signal at frequency t). Now, if the Fourier transform vanishes, i.e., $\varphi_X(t) = 0$ for each t, this implies the absence of a signal regardless of its frequency. This must be the zero signal, $p = 0$.

It is worthy to note that all previous mentioned types of convergence imply the convergence in distribution. The following classical result uses this type of convergence.

Theorem D.6.4 (Central Limit Theorem) *Let $X_1, X_2, X_3, \ldots$ be independent, identically distributed random variables, with mean $\mathbb{E}[X_j] = a$ and variance $Var(X_j) = b$, both finite. Denote $Z_n = \frac{S_n - na}{\sqrt{nb}}$, with $S_n = X_1 + \cdots + X_n$. Then Z_n tends in distribution to a standard normal variable, $\xi \sim N(0,1)$. This can be also stated as the convergence of the distribution function*

$$\lim_{n \to \infty} \mathbb{P}\left(Z_n \leq x\right) = \int_{-\infty}^{x} \frac{1}{\sqrt{2\pi}} e^{-u^2/2} \, du.$$

D.7 Log-Likelihood Function

This section provides some functional equation for the logarithmic, exponential, and linear functions.

The existence of a function which transforms products of numbers into sums is of uttermost importance for being able to define the concept of information.

Let s be an event with probability $P(s)$. The information casted by s is large when the probability $P(s)$ is small, i.e., when the event is a surprise. Then the information contained in s is given by a function of its probability, $g(P(s))$, where g is a positive decreasing function, with $g(0+) = +\infty$ and $g(1-) = 0$. This means that events with zero probability have infinite information, while events which happens with probability 1, practically, contain no information. Furthermore, if s_1 and s_2 are two independent events with probabilities of realization $P(s_1) = \pi_1$ and $P(s_2) = \pi_2$, respectively, by heuristic reasoning, the information produced by both events has to be the sum of individual information produced by each event, i.e.,

$$g(P(s_1 \cap s_2)) = g(\pi_1 \pi_2) = g(\pi_1) + g(\pi_2).$$

The next proposition shows that the only function satisfying these properties is $g(x) = -\ln x$, where the negative sign was included for positivity.

Proposition D.7.1 *Any differentiable function $f : (0, \infty) \to \mathbb{R}$ satisfying*

$$f(xy) = f(x) + f(y), \qquad \forall x, y \in (0, \infty)$$

is of the form $f(x) = c \ln x$, with c real constant.

Proof: Let $y = 1 + \epsilon$, with $\epsilon > 0$ small. Then

$$f(x + x\epsilon) = f(x) + f(1 + \epsilon). \qquad (D.7.2)$$

Using the continuity of f, taking the limit

$$\lim_{\epsilon \to 0} f(x + x\epsilon) = f(x) + \lim_{\epsilon \to 0} f(1 + \epsilon)$$

yields $f(1) = 0$. This implies

$$\lim_{\epsilon \to 0} \frac{f(1 + \epsilon)}{\epsilon} = \lim_{\epsilon \to 0} \frac{f(1 + \epsilon) - f(1)}{\epsilon} = f'(1).$$

Equation (D.7.2) can be written equivalently as

$$\frac{f(x + x\epsilon) - f(x)}{x\epsilon} = \frac{f(1 + \epsilon)}{x\epsilon}.$$

Taking the limit with $\epsilon \to 0$ transforms the above relation into a differential equation

$$f'(x) = \frac{1}{x} f'(1).$$

Let $c = f'(1)$. Integrating in $f'(x) = \frac{c}{x}$ yields the solution $f(x) = c \ln x + K$, with K constant. Substituting in the initial functional equation we obtain $K = 0$.
∎

Proposition D.7.2 *Any differentiable function $f : \mathbb{R} \to \mathbb{R}$ satisfying*

$$f(x + y) = f(x) + f(y), \qquad \forall x, y \in \mathbb{R}$$

is of the form $f(x) = cx$, with c real constant.

Proof: If let $x = y = 0$, the equation becomes $f(0) = 2f(0)$, and hence $f(0) = 0$. If take $y = \epsilon$, then $f(x + \epsilon) - f(x) = f(\epsilon)$. Dividing by ϵ and taking the limit, we have

$$\lim_{\epsilon \to 0} \frac{f(x + \epsilon) - f(x)}{\epsilon} = \lim_{\epsilon \to 0} \frac{f(\epsilon) - f(0)}{\epsilon},$$

which is written as $f'(x) = f'(0)$. On integrating, yields $f(x) = cx + b$, with $c = f'(0)$ and b real constant. Substituting in the initial equation provides $b = 0$. Hence, smooth additive functions are linear.
∎

Proposition D.7.3 *Any differentiable function* $f : \mathbb{R} \to (0, \infty)$ *satisfying*

$$f(x + y) = f(x)f(y), \qquad \forall x, y \in \mathbb{R}$$

is of the form $f(x) = e^{cx}$, *with c real constant.*

Proof: Applying the logarithm function to the given equation, we obtain $g(x + y) = g(x) + g(y)$, with $g(x) = \ln(f(x))$. Then Proposition D.7.1 yields $g(x) = cx$, with $c \in \mathbb{R}$. Therefore, $f(x) = e^{g(x)} = e^{cx}$. ∎

Remark D.7.4 It can be proved that Propositions D.7.2 and D.7.3 hold for the more restrictive hypothesis that f is just continuous.

D.8   Brownian Motion

Section 4.13.2 uses the notion of Brownian motion. For a gentle introduction to this subject the reader can consult [20]. For more specialized topics, the reader can consult [37].

Definition D.8.1 *A Brownian motion is a stochastic process* W_t, $t \geq 0$, *which satisfies:*
(i) $W_0 = 0$ *(the process starts at the origin);*
(ii) if $0 \leq u < t < s$, *then* $W_s - W_t$ *and* $W_t - W_u$ *are independent (the process has independent increments);*
(iii) $t \to W_t$ *is continuous;*
(iv) the increments are normally distributed, with $W_t - W_s \sim \mathcal{N}(0, |t - s|)$.

Note also that $\mathbb{E}[W_t] = 0$, $\mathbb{E}[W_t^2] = t$ and $Cov(W_t, W_s) = \min\{s, t\}$.

Ito's formula If X_t is a stochastic process satisfying

$$dX_t = b(X_t)dt + \sigma(X_t)dW_t,$$

with b and σ measurable functions, and $F_t = f(X_t)$, with f differentiable, then

$$dF_t = [b(X_t)f'(X_t) + \frac{1}{2}\sigma(X_t)^2 f''(X_t)]dt + \sigma f'(X_t)\, dW_t.$$

Some sort of converse is given by **Dynkin's formula**:
Consider the Ito diffusion

$$dX_t = b(X_t)dt + \sigma(X_t)dW_t, \qquad X_0 = x.$$

Then for any $f \in C_0^2(\mathbb{R}^n)$ we have

$$\mathbb{E}^x[f(X_t)] = f(x) + \mathbb{E}^x\left[\int_0^t \mathcal{A}f(X_s)\, ds\right],$$

where the conditional expectation is

$$\mathbb{E}^x[f(X_t)] = \mathbb{E}[f(X_t)|X_0 = x]$$

and A is the infinitesimal generator of X_t. This means

$$\mathcal{A}f(x) = \lim_{t \searrow 0} \frac{\mathbb{E}^x[f(X_t)] - f(x)}{t}.$$

Appendix E

Functional Analysis

This section presents the bare bones of the functional analysis results needed for the purposes of this book. The reader interested in more details is referred to Rudin's book [106].

E.1 Banach spaces

This section deals with a mathematical object endowed with both topological and algebraic structure. Let $(\mathcal{X}, +, \cdot)$ be a linear vector space, where "+" denotes the addition of elements of $\mathcal{X}$ and "·" is the multiplications with real scalars.

A *norm* on $\mathcal{X}$ is a real-valued function $\| \ \| : \mathcal{X} \to \mathbb{R}$ satisfying the following properties:

(i) $\|x\| \geq 0$, with $\|x\| = 0 \Longleftrightarrow x = 0$;

(ii) $\|\alpha x\| = |\alpha| \, \|x\|, \forall \alpha \in \mathbb{R}$;

(iii) $\|x + y\| \leq \|x\| + \|y\|, \forall x, y \in \mathcal{X}$.

The pair $(\mathcal{X}, \| \ \|)$ is called a *normed space*. The norm induces the metric $d(x, y) = \|x - y\|$, and thus (V, d) becomes a *metric space*. A *Banach space* is a normed vector space which is complete in this metric.

It is worth noting the sequential continuity of the norm: if $x_n \to x$, then $\|x_n\| \to \|x\|$, for $n \to \infty$.

A few examples of Banach spaces are given in the following:

1. The n-dimensional real vector space $\mathbb{R}^n$ together with the Euclidean norm, $\|x\| = (x_1^2 + \cdots + x_n^2)^{1/2}$, forms a Banach space.

© Springer Nature Switzerland AG 2020
O. Calin, *Deep Learning Architectures*, Springer Series in the Data Sciences,
https://doi.org/10.1007/978-3-030-36721-3

2. Let $K \subset \mathbb{R}^n$ be a compact set. The space $C(K)$ of all continuous real-valued functions on K with the norm $\|f\| = \max_{x \in K} |f(x)|$ forms a Banach space.

3. Let $p \geq 1$ and consider $L^p[0,1] = \{f; \int_0^1 |f|p < \infty\}$, where f represents a class of all measurable functions which are equal a.e. This is a vector space with respect to addition of functions and multiplication by real numbers. The norm is given by $\|f\| = \|f\|_p = \left(\int_0^1 |f|^p\right)^{1/p}$. The fact that $\|f\| = 0$ implies $f = 0$ a.e. is consistent with the definition of the space as a space of classes of functions equal a.e. $L^p[0,1]$ is a Banach space by the Riesz-Fisher theorem.

4. Consider $L^\infty[0,1]$ to be the space of a.e. bounded measurable functions on $[0,1]$. This is a vector space with respect to addition of functions and multiplication by real numbers. The norm is given by $\|f\| = \|f\|_\infty = \inf_{g = f \text{ a.e.}} \sup g$.

E.2   Linear Operators

The reader interested in this topic is referred to the comprehensive book [36]. Let $\mathcal{X}$ and $\mathcal{Y}$ be two vector spaces. A mapping $T : \mathcal{X} \to \mathcal{Y}$ is called a *linear operator* if

$$T(a_1 x_1 + a_2 x_2) = a_1 T(x_1) + a_2 T(x_2), \qquad \forall a_i \in \mathbb{R}, \forall x_i \in \mathcal{X}.$$

Assume now that $\mathcal{X}$ and $\mathcal{Y}$ are normed vector spaces. The linear operator T is called *bounded* if there is a constant $M > 0$ such that

$$\|Tx\| \leq M\|x\|, \qquad \forall x \in \mathcal{X}.$$

The smallest such M is called the *norm of the operator*, and it is denoted by $\|T\|$. We also have the equivalent definitions

$$\|T\| = \sup_{x \in V \setminus \{0\}} \frac{\|Tx\|}{\|x\|} = \sup_{\|x\|=1} \|Tx\| = \sup_{\|x\| \leq 1} \|Tx\|.$$

Since

$$\|Tx_1 - Tx_2\| \leq M\|x_1 - x_2\|, \qquad \forall x_1, x_2 \in \mathcal{X},$$

it follows that a bounded linear operator is uniformly continuous, and hence continuous. Conversely, if the linear operator T is continuous at only one point, then it is bounded.

It can be shown that the space of bounded linear operators $T : \mathcal{X} \to \mathcal{Y}$, with $\mathcal{Y}$ Banach space is also a Banach space.

If the space $\mathcal{Y} = \mathbb{R}$, the linear operator T is called a *linear functional*. In particular, the space of bounded linear functionals forms a Banach space.

E.3   Hahn-Banach Theorem

A *convex functional* on the vector space V is a function $p : \mathcal{X} \to \mathbb{R}$ such that

(*i*) $p(x + y) \le p(x) + p(y)$, i.e., p is subadditive;

(*ii*) $p(\alpha x) = \alpha p(x)$ for each $\alpha \ge 0$, i.e., p is positive homogeneous.

Example E.3.1 Let $\mathcal{X} = \mathbb{R}^n$ and consider $p(x) = \max\limits_{1 \le i \le n} |x_i|$, where $x = (x_1, \ldots, x_n)$. Then $p(x)$ is a convex functional on $\mathbb{R}^n$.

If $(\mathcal{X}, +, \cdot)$ is a linear space, a subset $\mathcal{X}_0 \subset \mathcal{X}$ is a linear subspace if $\mathcal{X}_0$ is closed with respect to the endowed operations from $\mathcal{X}$. Consequently, $\mathcal{X}_0$ becomes a linear space with respect to these operations.

The following result deals with the extension of linear functionals from a subspace to the whole space, such that certain properties are preserved.

Theorem E.3.1 (Hahn-Banach) *Let $\mathcal{X}$ be a linear real vector space, $\mathcal{X}_0$ a linear subspace, p a linear convex functional on $\mathcal{X}$, and $f : \mathcal{X}_0 \to \mathbb{R}$ a linear functional such that $f(x) \le p(x)$ for all $x \in \mathcal{X}_0$.*
Then there is a linear functional $F : \mathcal{X} \to \mathbb{R}$ such that

(*i*) $F_{\mathcal{X}_0} = f$ *(the restriction of F to $\mathcal{X}_0$ is f).*

(*ii*) $F(x) \le p(x)$ *for all $x \in \mathcal{X}$.*

We include next a few applications of the Hahn-Banach theorem.

1. Let $p : \mathcal{X} \to [0, +\infty)$ be a nonnegative, convex functional and $x_0 \in \mathcal{X}$ be a fixed element. Then there is a linear functional F on $\mathcal{X}$ such that $F(x_0) = p(x_0)$ and $F(x) \le p(x)$ for all x in $\mathcal{X}$.

2. Let x_0 be an element in the normed space $\mathcal{X}$. Then there is a bounded linear functional F on $\mathcal{X}$ such that $F(x_0) = \|F\| \, \|x_0\|$.

3. Let S be a linear subspace of the normed linear space $\mathcal{X}$ and y an element of $\mathcal{X}$ whose distance to S is at least δ, i.e.,

$$\|y - s\| \ge \delta, \qquad \forall s \in S.$$

Then there is a bounded linear functional f on $\mathcal{X}$ with $\|f\| \le 1$, $f(y) = \delta$, and such that $f(s) = 0$ for all $s \in S$.

E.4   Hilbert Spaces

A *Hilbert* space H is a Banach space endowed with a function $(\, , \,) : H \times H \to \mathbb{R}$ satisfying the following:

(*i*) $(x, x) = \|x\|^2$;

(*ii*) $(x, y) = (y, x)$;

(*iii*) $(c_1 x_1 + c_2 x_2, y) = c_1(x_1, y) + c_2(x_2, y)$, $c_i \in \mathbb{R}$, $x_i, y \in H$.

Example E.4.1 $H = \mathbb{R}^n$, with $(x, y) = \sum_{i=1}^{n} x_i y_i$.

Example E.4.2 $H = L^2[0, 1]$, with $(x, y) = \int_0^1 x(t) y(t)\, dt$.

Cauchy's inequality states that $|(x, y)| \leq \|x\| \, \|y\|$, so the linear functional $g(x) = (x, y)$ is bounded, and hence continuous from H to $\mathbb{R}$. Consequently, if $x_n \to x$ in H, then $(x_n, y) \to (x, y)$, as $n \to \infty$.

Two elements $x, y \in H$ are called *orthogonal* if $(x, y) = 0$. A set $\mathcal{U}$ is called an *orthogonal system* if any two distinct elements of $\mathcal{U}$ are orthogonal.

Example E.4.3 The set $\{1, \cos t, \sin t, \ldots, \cos nt, \sin nt, \ldots\}$ is an orthogonal system for $H = L^2[-\pi, \pi]$.

The orthogonal system $\mathcal{U}$ is called *orthonormal* if $\|x\| = 1$ for all $x \in \mathcal{U}$.

Example E.4.4 The following set

$$\left\{ \frac{1}{2\pi}, \frac{1}{\sqrt{\pi}} \cos t, \frac{1}{\sqrt{\pi}} \sin t, \ldots, \frac{1}{\sqrt{\pi}} \cos nt, \frac{1}{\sqrt{\pi}} \sin nt, \ldots \right\}$$

is an orthogonal system for $H = L^2[-\pi, \pi]$.

Let $\{x_1, x_2, \ldots\}$ be a countable orthonormal system in H. The *Fourier coefficient* of an element $x \in H$ with respect to the previous system is given by $c_k = (x, x_k)$. Then *Bessel inequality* states that

$$\sum_{k \geq 1} c_k^2 \leq \|x\|^2.$$

A linear subspace $\mathcal{X}_0$ of a Hilbert space is called *closed* if it contains the limit of any convergent sequence (x_n) in $\mathcal{X}_0$. In this case, for any element x, the number $d(x, \mathcal{X}_0) = \inf\{\|x - y\|; y \in \mathcal{X}_0\}$ is called the *distance* from x to the subspace $\mathcal{X}_0$.

Theorem E.4.1 *Let $\mathcal{X}_0$ be a closed linear subspace of the Hilbert space $\mathcal{X}$. Then for any element x in $\mathcal{X}$ there is an element $x_0 \in \mathcal{X}_0$ such that $\|x - x_0\|$ equals the distance from x to $\mathcal{X}_0$.*

The element x_0 is called the projection of x onto the subspace $\mathcal{X}_0$.

E.5 Representation Theorems

The present section provides representations of linear functionals on different spaces, which are needed for showing the universal approximator property of neural networks. Most of them are usually quoted in literature as "Riesz's theorems of representation." The reader can find the proofs, for instance, in the books of Halmos [51] or Royden [104].

The next result is a representation of bounded linear functionals on Hilbert spaces.

Theorem E.5.1 (Riesz) *Let $f : H \to \mathbb{R}$ be a bounded linear functional on the Hilbert space H endowed with the inner product $(\ ,\ )$. Then there is a unique element $y \in H$ such that $f(x) = (x, y)$ for all $x \in H$. Furthermore, $\|f\| = \|x\|$.*

In particular, if $F : L^2[0, 1] \to \mathbb{R}$ is a bounded linear functional, there is an unique $g \in L^2[0, 1]$ such that $F(f) = \int_0^1 f(t)g(t)\,dt$.

Even if L^p is not a Hilbert space for $p \neq 2$ (it is a complete space though), this result can still be extended to a representation result on the spaces L^p.

Theorem E.5.2 (Riesz) *Let F be a bounded linear functional on $L^p[0, 1]$, with $1 < p < \infty$. Then there is a unique function $g \in L^q[0, 1]$, with $\frac{1}{p} + \frac{1}{q} = 1$, such that $F(f) = \int_0^1 f(t)g(t)\,dt$. We also have $\|F\| = \|g\|_q$.*

The previous result holds also for the case $p = 1$ with some small modifications. Let $L^\infty[0, 1]$ be the space of measurable and a.e. bounded functions, which becomes a Banach space with the supremum norm, $\|\ \|_\infty$, see point 4 of Section E.1.

Theorem E.5.3 *Let F be a bounded linear functional on $L^1[0, 1]$. Then there is a unique function $g \in L^\infty[0, 1]$, such that*

$$F(f) = \int_0^1 f(t)g(t)\,dt, \quad \forall f \in L^1[0, 1].$$

We also have $\|F\| = \|g\|_\infty$.

A real-valued function g defined on the interval $[a, b]$ has *bounded variation* if for any division

$$a = x_0 < x_1 < \cdots < x_n = b,$$

the sum

$$\sum_{k=0}^{n-1} |g(x_{k+1}) - g(x_k)|$$

is smaller than a given constant. The superior limit of these sums over all possible divisions of $[a, b]$ is called the *total variation* of g and is denoted by $V_a^b(g)$.

Example E.5.1 An increasing function g on $[a, b]$ is of bounded variation, with $\bigvee_a^b(g) = g(b) - g(a)$.

The following result is an analog of Arzela-Ascoli theorem for noncontinuous functions and can be used for extracting a convergent sequence of functions from a given set.

Theorem E.5.4 (Helly) *Let $\mathcal{K}$ be an infinite set of functions from $[a, b]$ to $\mathbb{R}$ such that:*

(i) $\mathcal{K}$ is uniformly bounded, i.e., $\exists C > 0$ such that $\sup\limits_{x \in [a,b]} |f(x)| < C$, for all $f \in \mathcal{K}$;

(ii) $\exists V > 0$ such that $\bigvee_a^b(f) \leq V$, for all $f \in \mathcal{K}$.
Then we can choose a sequence $(f_n)_n$ of functions in $\mathcal{K}$ that is convergent at each point $x \in [a, b]$.

Recall that $C[0, 1]$ is the space of real-valued continuous functions defined on $[0, 1]$, and it becomes a Banach space with respect to the norm $\|\ \|_\infty$. The simplest continuous linear functional on $C([0, 1])$ is $F(f) = f(t_0)$, which assigns to f its value at a fixed point t_0. The next result states that the general form of these type of functionals are obtained by a Stieltjes combination of the aforementioned type of particular functions.

Theorem E.5.5 *Let F be a continuous linear functional on $C([0, 1])$. There is a function $g : [0, 1] \to \mathbb{R}$ with bounded variation, such that*

$$F(f) = \int_0^1 f(t)\, dg(t), \qquad \forall f \in C([0, 1]).$$

We also have $\|F\| = \bigvee_0^1(f)$.

We recall that the Stieltjes integral $\int_0^1 f(t)\, dg(t)$ is defined as the limit of the Riemann-type sums

$$\sum_{k=0}^{m-1} f(x_k)[g(x_{k+1}) - g(x_k)]$$

as the norm of the division $0 = x_0 < x_1 < \cdots < x_m = 1$ tends to zero.

The next result is a generalization of the previous one. Let K denote a compact set in $\mathbb{R}^n$, and denote by $C(K)$ the set of real-valued continuous functions on K.

Theorem E.5.6 *Let F be a bounded linear functional on $C(K)$. Then exists a unique finite signed Borel measure μ on K, such that*

$$F(f) = \int_K f(x)\,d\mu(x), \qquad \forall f \in C(K).$$

Moreover, $\|F\| = |\mu|(K)$.

The next result replaces the boundness condition of the functional by positivity. In this case the signed measure becomes a measure.

Theorem E.5.7 *Let L be a positive linear functional on $C(K)$. Then exists a unique finite Borel measure μ on K, such that*

$$L(f) = \int_K f(x)\,d\mu(x), \qquad \forall f \in C(K).$$

E.6   Fixed Point Theorem

Let (M, d) be a metric space and $T : M \to M$ be a mapping of M into itself. T is called a *contraction* if

$$d(T(x), T(x')) \leq \lambda d(x, x'), \qquad \forall x, x' \in M$$

for some positive constant λ less than 1. If the metric is induced by a norm on M, i.e., if $d(x, x') = \|x - x'\|$, the contraction condition can be written as $\|T(x) - T(x')\| \leq \lambda \|x - x'\|$.

A sequence (x_n) of points in a metric space (M, d) is called a *Cauchy sequence* if for any $\epsilon > 0$, there is an $N > 1$ such that $d(x_n, x_m) < \epsilon$, for all $n, m > N$.

The metric space (M, d) is called *complete* if any Cauchy sequence (x_n) in M is convergent, i.e., there is $x^* \in M$ such that for all $\epsilon > 0$, there is $N > 1$ such that $d(x_n, x^*) < \epsilon$, for all $n \geq N$.

Example E.6.1 The space $\mathbb{R}^n$ with the Euclidean distance forms a complete metric space.

Example E.6.2 The space of linear operators $\{L; L : \mathbb{R}^n \to \mathbb{R}^n\}$ endowed with the metric induced by the norm $\|L\| = \sup_{x \neq 0} \frac{\|Lx\|}{\|x\|}$ is a complete metric space.

Example E.6.3 The space $C[a, b] = \{f : [a, b] \to \mathbb{R}; f \text{ continuous}\}$ is a complete metric space with the metric $d(f, g) = \sup_{x \in [a,b]} |f(x) - g(x)|$.

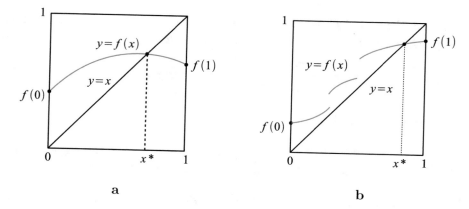

Figure 1: **a.** *A fixed point of a continuous function* $f : [0, 1] \to [0, 1]$. **b.** *A fixed point of an increasing function* $f : [0, 1] \to [0, 1]$.

A point $x^* \in M$ is called a *fixed point* for the application $T : M \to M$ if $T(x^*) = x^*$.

Example E.6.4 Any continuous function $f : [0, 1] \to [0, 1]$ has at least one fixed point. This follows geometrically from the fact that any continuous curve joining two arbitrary points situated on opposite sides of a square intersects the diagonal $y = x$, see Fig. 1 **a**. It is worth noting that Knaster proved that the fixed point property holds also in the case when the function f is monotonically increasing instead of continuous, see Fig. 1 **b**.

Theorem E.6.5 *A contraction T of the complete metric space (M, d) into itself has a unique fixed point.*

For any point $x_0 \in M$, the sequence (x_n) defined by $x_{n+1} = T(x_n)$ converges toward the fixed point, $x_n \to x^*$. Moreover, we have the estimation

$$d(x_n, x^*) \leq \frac{\lambda^n}{1 - \lambda} d(x_0, x_1).$$

Appendix F

Real Analysis

F.1 Inverse Function Theorem

The Inverse Function Theorem is a result that states the local invertibility of a continuous differentiable function from $\mathbb{R}^n$ to $\mathbb{R}^n$, which satisfies a nonzero Jacobian condition at a point. We recall that the Jacobian of the function $F = (F_1, \dots, F_n)$ is the $n \times n$ matrix of partial derivatives

$$J_F(x) = \left(\frac{\partial F_i(x)}{\partial x_j} \right)_{i,j}.$$

Theorem F.1.1 *Let $f : \mathbb{R}^n \to \mathbb{R}^n$ be a continuously differentiable function and $p \in \mathbb{R}^n$ be a point such that $\det J_F(p) \neq 0$. Then there are two open sets $\mathcal{U}$ and $\mathcal{V}$ that contain points p and $q = F(p)$, respectively, such that $F_{|\mathcal{U}} : \mathcal{U} \to \mathcal{V}$ is invertible, with the inverse continuously differentiable. The Jacobian of the inverse is given by $J_{F^{-1}}(q) = [J_F(p)]^{-1}$.*

The theorem can be reformulated in mathematical jargon by stating that a continuously differentiable function with a nonsingular Jacobian at a point is a local diffeomorphism around that point.

Another way to formulate the theorem is in terms of nonlinear systems of equations. Consider the system of n equations with n unknowns

$$
\begin{aligned}
F_1(x_1, \dots, x_n) &= y_1 \\
\cdots \quad \cdots \quad \cdots &= \cdots \\
F_n(x_1, \dots, x_n) &= y_n,
\end{aligned}
$$

© Springer Nature Switzerland AG 2020
O. Calin, *Deep Learning Architectures*, Springer Series in the Data Sciences,
https://doi.org/10.1007/978-3-030-36721-3

and assume there is a point $x^0 \in \mathbb{R}^n$ such that $\det(\frac{\partial F_i}{\partial x_j})(x^0) \neq 0$. Then there exists two open sets $\mathcal{U}$ and $\mathcal{V}$ about x^0 and $y^0 = F(x^0)$ such that the system can be solved with unique solution as long as $x \in \mathcal{U}$ and $y \in \mathcal{V}$. This means there are n continuous differentiable functions $G_i : \mathcal{V} \to \mathbb{R}$ such that

$$
\begin{aligned}
x_1 &= G_1(y_1, \ldots, y_n) \\
\cdots &= \cdots \quad \cdots \quad \cdots \\
x_n &= G_n(y_1, \ldots, y_n)
\end{aligned}
$$

for any $y = (y_1, \ldots, y_n) \in \mathcal{V}$.

We note that this is an existential result, which does not construct explicitly the solution of the system. However, in the particular case when the function is linear, i.e., $F(x) = Ax$, with A nonsingular square matrix, the linear system $Ax = y$ has a unique solution $x = A^{-1}y$. The solution in this case is global, since the Jacobian $J_F(x) = A$ has a nonzero determinant everywhere.

F.2　Differentiation in generalized sense

How do we differentiate a function that is not differentiable everywhere? If we have a piecewise differentiable function, differentiating it piecewise does not always provide the right result since we can't find the derivative at the contact points. The "derivative" of a non-differentiable function might exist sometimes in the so-called *generalized sense*. Let $f : \mathbb{R} \to \mathbb{R}$ be a function. We say that g is the derivative of function f in the generalized sense if

$$
\int_{\mathbb{R}} g(x)\varphi(x)\, dx = -\int_{\mathbb{R}} f(x)\varphi'(x)\, dx, \tag{F.2.1}
$$

for any compact supported smooth function φ. We note that the generalized differentiation is an extension of classical differentiation, since the previous relation becomes the familiar integration by parts formula

$$
\int_{\mathbb{R}} f'(x)\varphi(x)\, dx = -\int_{\mathbb{R}} f(x)\varphi'(x)\, dx.
$$

Example F.2.1 Let $f(x) = H(x)$ be the Heaviside step function. Then its derivative is given by Dirac's function, $f'(x) = \delta(x)$. We shall check it next by computing both sides of relation (F.2.1):

$$
\int_{\mathbb{R}} \delta(x)\varphi(x)\, dx = \int_{\mathbb{R}} \varphi(x)\delta(dx) = \varphi(0)
$$

$$
-\int_{\mathbb{R}} f(x)\varphi'(x)\, dx = -\int_{\mathbb{R}} H(x)\varphi'(x)\, dx = -\int_0^\infty \varphi'(x)\, dx = \varphi(0) - \varphi(\infty) = \varphi(0).
$$

Similarly, the derivative of $H(x - a)$ is $\delta_a(x)$, where $\delta_a(x) = \delta(x - a)$.

Example F.2.2 The derivative of $ReLU(x)$ is the Heaviside function, $ReLU'(x) = H(x)$. We can check it using relation (F.2.1):

$$
\begin{aligned}
\int_{\mathbb{R}} ReLU'(x)\varphi(x)\,dx &= -\int_{\mathbb{R}} ReLU(x)\varphi'(x)\,dx = -\int_{0}^{\infty} x\varphi'(x)\,dx \\
&= \int_{0}^{\infty} x'\varphi(x)\,dx = \int_{0}^{\infty} \varphi(x)\,dx = \int_{\mathbb{R}} H(x)\varphi(x)\,dx,
\end{aligned}
$$

for any $\varphi \in C_0^{\infty}(\mathbb{R})$. Hence, $ReLU'(x) = H(x)$, in the generalized sense.

F.3 Convergence of sequences of functions

Let $f_n : \mathbb{R} \to \mathbb{R}$ be a sequence of functions. Then f_n can approximate a function f in several ways.

1. The sequence of functions $(f_n)_n$ is *pointwise convergent* to f if $f_n(x)$ converges to $f(x)$ for any $x \in \mathbb{R}$.

2. Let $f \in L^2(\mathbb{R})$. The sequence of functions $(f_n)_n$ is *L^2-convergent* to f if $\|f_n - f\|_2 \to 0$, as $n \to \infty$. This means

$$
\lim_{n \to \infty} \int_{\mathbb{R}} |f_n(x) - f(x)|^2\,dx = 0.
$$

3. The sequence $(f_n)_n$ is *weakly convergent* to f if

$$
\lim_{n \to \infty} \int_{\mathbb{R}} f_n(x)\varphi(x)\,dx = \int_{D} f(x)\varphi(x)\,dx, \qquad \forall \varphi \in C_0^{\infty}(\mathbb{R}).
$$

It is worth noting that the L^2-convergence implies both the pointwise convergence and the weak convergence. The latter follows from an application of Cauchy's inequality

$$
\begin{aligned}
\left| \int_{\mathbb{R}} f_n\varphi - \int_{\mathbb{R}} f\varphi \right| &= \left| \int_{\mathbb{R}} (f_n - f)\varphi \right| \le \int_{|R} |f_n - f|\,|\varphi| \\
&= \|f_n - f\|_2\,\|\varphi\|_2
\end{aligned}
$$

and the Squeeze Theorem.

Appendix G

Linear Algebra

G.1   Eigenvalues, Norm, and Inverse Matrix

Consider a matrix with n rows and m columns, $A \in \mathcal{M}_{n,m}$, given by $A = (a_{ij})$. The *transpose* of the matrix A, denoted by A^T, is given by $A^T = (a_{ji})$ and satisfies $(A^T)^T = A$, $(AB)^T = B^T A^T$. A matrix is called *symmetric* if $A = A^T$. A necessary condition for symmetry is that A has to be a square matrix, i.e., $n = m$. If $AA^T = \mathbb{I}$ (the identity matrix), then the matrix is called *orthogonal*.

A number λ (real or complex) is called an *eigenvalue* of the square matrix A if there is a nonzero vector $\mathbf{x}$ such that $A\mathbf{x} = \lambda\mathbf{x}$. The vector $\mathbf{x}$ is called an *eigenvector* of the matrix A. The eigenvalues are the solutions of the polynomial equation $\det(A - \lambda\mathbb{I}) = 0$.

Proposition G.1.1 *Let A be a symmetric matrix. Then A has n real eigenvalues, not necessarily distinct, $\lambda_1, \ldots, \lambda_n$, and n eigenvectors, $\mathbf{x}_1, \ldots, \mathbf{x}_n$, which form an orthonormal basis in $\mathbb{R}^n$.*

A matrix of type $n \times 1$ is called a *vector*. If w and b are two vectors, then $(w^T b) = (wb^T) = \langle w, b \rangle$, where $\langle \, , \rangle$ is the Euclidean scalar product on $\mathbb{R}^n$.

One can define several norms on $\mathbb{R}^n$. Let $\mathbf{x}^T = (x_1, \ldots, x_n)$ be a vector. Then

$$\|\mathbf{x}\|_1 = \sum_{i=1}^{n} |x_i|, \qquad \|\mathbf{x}\|_2 = \Big(\sum_{i=1}^{n} x_i^2 \Big)^{1/2}, \qquad \|\mathbf{x}\|_\infty = \max_{1 \le i \le n} |x_i|$$

© Springer Nature Switzerland AG 2020
O. Calin, *Deep Learning Architectures*, Springer Series in the Data Sciences,
https://doi.org/10.1007/978-3-030-36721-3

are three norms that are used in this book. It can be shown that there are two constants $C_1, C_2 > 0$ such that

$$C_1\|\mathbf{x}\|_1 \leq \|\mathbf{x}\|_2 \leq C_2\|\mathbf{x}\|_\infty, \qquad \forall \mathbf{x} \in \mathbb{R}^n.$$

Geometrically, this means that any $\|\cdot\|_2$-ball can be simultaneously included in a $\|\cdot\|_\infty$-ball and also includes a $\|\cdot\|_1$-ball.

Each of the previous norms induces a norm for the squared matrix A. Inspired by the norm of a linear operator, we define

$$\|A\| = \sup_{\mathbf{x} \neq 0} \frac{\|Ax\|}{\|\mathbf{x}\|}, \tag{G.1.1}$$

where $\|\cdot\|$ is any of the aforementioned norms on $\mathbb{R}^n$. We note the inequalities $\|Ax\| \leq \|A\|\|\mathbf{x}\|$ and $\|AB\| \leq \|A\|\,\|B\|$, for any other square matrix B. The norms induced by the previous three norms are

$$\|A\|_1 = \max_{1 \leq j \leq n} \sum_{i=1}^n |a_{ij}|, \quad \|A\|_\infty = \max_{1 \leq i \leq n} \sum_{j=1}^n |a_{ij}|, \quad \|A\|_2 = \sqrt{\rho(A^T A)},$$

where ρ denotes the spectral radius of a matrix, given by $\rho(A) = \max |\lambda_j|$, where λ_j represent the eigenvalues of A. The matrix $B = A^T A$, being symmetric, by Proposition G.1.1 it has real eigenvalues. Since A and A^T have equal eigenvalues, the eigenvalues of $A^T A$ are λ_i^2, and hence $A^T A$ is positive definite. Therefore, $\rho(A^T A) = \max \lambda_i^2$, and then $\|A\|_2 = \max |\lambda_i|$ is the largest absolute value of eigenvalues of matrix A.

It is worth noting that if A is symmetric, then $\|A\|_2 = \rho(A)$. Also, $\|A\|_2$ is the smallest among all norms generated by formula (G.1.1), i.e., $\|A\|_2 \leq \|A\|$. This can be shown like in the following. Let λ be the eigenvalue with the largest absolute value and $\mathbf{x}$ an eigenvector of length one. Then

$$\|A\|_2 = |\lambda| = \|\lambda\mathbf{x}\| = \|A\mathbf{x}\| \leq \|A\|\,\|\mathbf{x}\| \leq \|A\|.$$

We also have $\|A\|_2 \geq \frac{1}{n}|Tr(A)|$ and $\|A\|_2 \geq |\det A|^{1/n}$.

Proposition G.1.2 *Let A be a square matrix.*
(i) The power matrix A^m converges to the zero matrix as $m \to \infty$ if and only if $\rho(A) < 1$.
(ii) If $\rho(A) < 1$ then $\mathbb{I} - A$ is invertible and we have

$$(\mathbb{I} - A)^{-1} = \mathbb{I} + A + A^2 + \cdots A^m + \cdots \tag{G.1.2}$$

(iii) If the geometric series (G.1.2) converges, then $\rho(A) < 1$.

Corollary G.1.3 *Let A be a square matrix and $\|\cdot\|$ be a norm. If $\|A\| < 1$, then $\mathbb{I} - A$ is invertible with the inverse given by (G.1.2) and*

$$\|(\mathbb{I} - A)^{-1}\| < \frac{1}{1 - \|A\|}.$$

Appendix

In the following we shall deal with a procedure of finding the inverse of a sum of two square matrices. We shall explain first the idea in the 1-dimensional case. Assume $a_1, a_2 \in \mathbb{R}\backslash\{0\}$ are two real numbers with $a_1 + a_2 \neq 0$. A simple algebraic computation yields

$$
\begin{aligned}
\frac{1}{a_1 + a_2} &= \frac{1}{a_2} - \frac{1}{a_1 + a_2}\frac{a_1}{a_2} & \text{(G.1.3)} \\
&= \frac{1}{a_1} - \frac{1}{a_1 + a_2}\frac{a_2}{a_1}.
\end{aligned}
$$

We construct two linear functions $f, g : \mathbb{R} \to \mathbb{R}$ given by $f(x) = -\dfrac{a_1}{a_2}x + \dfrac{1}{a_2}$ and $g(x) = -\dfrac{a_2}{a_1}x + \dfrac{1}{a_1}$, and consider two cases:

1. If $|a_1| < \lambda|a_2|$, with $0 < \lambda < 1$, then

$$
|f(x) - f(x')| = \left|\frac{a_1}{a_2}\right||x - x'| < \lambda|x - x'|.
$$

Therefore, f is a contraction of the complete metric space $(\mathbb{R}, |\ |)$ into itself, and hence it has a unique fixed point, see Theorem E.6.5. The fixed point, x^*, satisfies $f(x^*) = x^*$ and is given by $x^* = \frac{1}{a_1+a_2}$. Its approximation sequence is (x_n), defined by the recurrence $x_{n+1} = f(x_n)$, and $x_0 = 0$. The error estimation is given by $|x_n - x^*| < \frac{\lambda^n}{1-\lambda}|x_1 - x_0| = \frac{\lambda^n}{(1-\lambda)}\frac{1}{|a_2|}$.

2. If $|a_2| < \lambda|a_1|$, with $0 < \lambda < 1$, then $|g(x) - g(x')| < \lambda|x - x'|$, so g is a contraction of $\mathbb{R}$, and hence it has a unique fixed point, which is $\frac{1}{a_1+a_2}$. The details are similar with the first case.

Consider now two invertible $n \times n$ matrices, A_1, A_2. We claim that

$$
(A_1 + A_2)^{-1} = A_2^{-1} - (A_1 + A_2)^{-1}A_1 A_2^{-1}, \qquad \text{(G.1.4)}
$$

which is the analog relation of (G.1.3) for matrices. This can be shown by a mere multiplication on both sides by $(A_1 + A_2)$ as follows:

$$
\begin{aligned}
\mathbb{I} &= (A_1 + A_2)A_2^{-1} - A_1 A_2^{-1} \Longleftrightarrow \\
\mathbb{I} &= A_1 A_2^{-1} + \mathbb{I} - A_1 A_2^{-1}.
\end{aligned}
$$

Assuming $\|A_1 A_2^{-1}\| < 1$, it follows that $\mathbb{I} + A_1 A_2^{-1}$ is invertible, see Corollary G.1.3. Then solving for $(A_1 + A_2)^{-1}$ from relation (G.1.4) yields

$$
(A_1 + A_2)^{-1} = A_2^{-1}(\mathbb{I} + A_1 A_2^{-1})^{-1}. \qquad \text{(G.1.5)}
$$

This closed-form formula for the inverse of a sum of two matrices can't be used in practice. For computational reasons we consider the following two ways:

1. Consider the map $f : \mathcal{M}_{n \times n} \to \mathcal{M}_{n \times n}$, $f(M) = A_2^{-1} - MA_1A_2^{-1}$. Since

$$\|f(M) - f(M')\| = \|(M' - M)A_1A_2^{-1}\| \leq \|M' - M\|\|A_1A_2^{-1}\| < \lambda\|M - M'\|,$$

then f is a contraction of $\mathcal{M}_{n \times n}$ into itself. The space $\mathcal{M}_{n \times n}$ is complete, since any matrix is associated with a linear operator, and the space of linear operators on $\mathbb{R}^n$ is complete. By the fixed point theorem, the mapping f has a unique fixed point, M^*, i.e., $f(M^*) = M^*$. From (G.1.4) it follows that $M^* = (A_1 + A_2)^{-1}$. This inverse can be approximated by the sequence of matrices (M_n), given by $M_{n+1} = f(M_n)$, $M_0 = \mathbb{O}$. The error is estimated by

$$\|M_n - M^*\| < \frac{\lambda^n}{1 - \lambda}\|M_1 - M_0\| = \frac{\lambda^n}{(1 - \lambda)}\|A_2^{-1}\|.$$

2. Another way to approximate $(A_1 + A_2)^{-1}$ is to expand (G.1.4) in a series, see Proposition G.1.2

$$(A_1 + A_2)^{-1} = A_2^{-1}(\mathbb{I} + A_1A_2^{-1})^{-1} = A_2^{-1}\sum_{k \geq 0}(-1)^k(A_1A_2^{-1})^k.$$

The previous computation was conducted under the condition $\|A_1A_2^{-1}\| < 1$. This implies $\rho(A_1A_2^{-1}) < 1$, or $\rho(A_1) < \rho(A_2)$, or $\lambda_i(A_1) < \lambda_i(A_2)$, for all $i \in 1, \dots, n$, i.e., the matrix A_1 has smaller eigenvalues that A_2, respectively.

It is worth noting that due to symmetry reasons, the roles of A_1 and A_2 can be inverted, and similar formulas can be obtained if A_1 is assumed invertible.

The proof of the following result is just by mere multiplication.

Lemma G.1.4 (Matrix inversion lemma) *Let $A, B \in \mathcal{M}_{m \times n}$ be positive definite matrices, and $C \in \mathcal{M}_{m \times n}$, $D \in \mathcal{M}_{n \times n}$ positive definite. If*

$$A = B^{-1} + CD^{-1}C^T$$

then

$$A^{-1} = B - BC(D + C^TBC)^{-1}C^TB.$$

G.2   Moore-Penrose Pseudoinverse

A linear system is called *overdetermined* if it has more equations than unknowns. Typically, this type of systems do not have any solutions. The *Moore-Penrose pseudoinverse* method provides an approximate solution, [89, 97], which, for all practical purposes, is good in some sense specified at the end of this section.

We start by considering a linear system in matrix form, $AX = b$, where A is an $m \times n$ matrix, with $m > n$ (more rows than columns), X is an n-dimensional unknown vector, and b is an m-dimensional given vector. Since A is not a square matrix, the inverse A^{-1} does not make sense in this case. However, there are good chances that the square matrix $A^T A$ is invertible.[1] For instance, if A has full rank, $\text{rank} A = n$, then $\text{rank} A^T A = \text{rank} A = n$, so the $n \times n$ matrix $A^T A$ has maximal rank, and hence $\det A^T A \neq 0$, i.e., $A^T A$ is invertible.

Then multiplying the equation by the transpose matrix, A^T, to the left we obtain $A^T A X = A^T b$. Assuming that $A^T A$ is invertible, we obtain the solution $X = (A^T A)^{-1} A^T b$. The pseudoinverse of A is defined by the $n \times m$ matrix

$$A^+ = (A^T A)^{-1} A^T. \tag{G.2.6}$$

In this case the Moore-Penrose pseudoinverse solution of the overdetermined system $AX = b$ is given by $X = A^+ b$.

In the case when A is invertible we have $A^+ = A^{-1}$, i.e., the pseudoinverse is a generalization of the inverse of a matrix.

It is worth noting that if the matrix A has more columns than rows, i.e, $n > m$, then $A^T A$ does not have an inverse, since $\det A^T A = 0$. This follows from the rank evaluation of the n-dimensional matrix $A^T A$

$$\text{rank} A^T A = \text{rank} A \leq \min\{n, m\} = m < n.$$

In this case the pseudoinverse A^+, even if it always exists, cannot be expressed by the explicit formula (G.2.6).

Geometric Significance Consider the linear mapping $F : \mathbb{R}^n \to \mathbb{R}^m$, $F(X) = AX$, with $n < m$. The matrix A is assumed to have full rank, $\text{rank} A = n$. In this case the range of F is the following linear subspace of $\mathbb{R}^m$:

$$\mathcal{R} = \{AX; X \in \mathbb{R}^n\},$$

of dimension $\dim \mathcal{R} = \text{rank} A = n$.

Now, given a vector $b \in \mathbb{R}^m$, not necessary contained in the space $\mathcal{R}$, we try to approximately solve the linear system $AX = b$, using the *minimum norm solution*. This is a vector $X^* \in \mathbb{R}^n$, which minimizes the L^2-norm of the difference $AX - b$, i.e.,

$$X^* = \arg \min_{X \in \mathbb{R}^n} \|AX - b\|_2. \tag{G.2.7}$$

---

[1] This follows from the fact that the matrices A satisfying the algebraic equation $\det A^T A = 0$ form a negligible set in the set of $m \times n$ matrices.

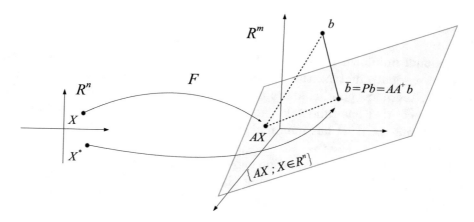

Figure 1: *The geometric interpretation of the pseudoinverse solution $X^* = A^+b$.*

Geometrically, this means that AX^* is the point in the space $\mathcal{R}$, which is closest to b, i.e., it is the orthogonal projection of b onto $\mathcal{R}$, see Fig. 1.

Let $\bar{b}$ denote the orthogonal projection of b onto the space $\mathcal{R}$, and consider the linear system $AX = \bar{b}$. Since $\bar{b} \in \mathcal{R}$, the system does have solutions. The uniqueness follows from the maximum rank condition of matrix A. Therefore, there is a unique vector $X^* \in \mathbb{R}^n$ such that $AX^* = \bar{b}$. This is the solution claimed by the equation (G.2.7). Equivalently, this means

$$\|AX - b\|_2 \geq \|AX^* - b\|_2 = \|\bar{b} - b\|_2, \qquad \forall X \in \mathbb{R}^n.$$

Next we focus on the expression of the projection $\bar{b}$. It can be shown that $\bar{b} = AA^+b$. This follows from the fact that the linear operator $P : \mathbb{R}^m \to \mathbb{R}^m$ given by $P = AA^+ = A(A^TA)^{-1}A^T$ is the orthogonal projector of $\mathbb{R}^m$ onto the subspace $\mathcal{R}$. This result is implied by the following three properties, which can be checked by a straightforward computation: $P^2 = P$, $P^T = P$, and $PA = A$; the former means that P is a projector, the latter means that P is an orthonormal projector, and the last means that the space $\mathcal{R}$ is invariant by P. Since $X^* = A^+b$ verifies the equation $AX = \bar{b}$, namely,

$$AX^* = A(A^+b) = (AA^+)b = Pb = \bar{b},$$

then X^* represents the pseudoinverse solution of the system $AX = b$.

One first application of the Moore-Penrose pseudoinverse is finding the line of best fit for m given points in the plane of coordinates (x_1, y_1), (x_2, y_2), $\ldots$, (x_m, y_m). If the line has the equation $y = ax + b$, we shall

Appendix

write the following overdetermined system of m equations:

$$ax_1 + b = y_1$$
$$\ldots\ldots = \cdots$$
$$ax_m + b = y_m,$$

which can be written in the equivalent matrix form

$$\underbrace{\begin{pmatrix} x_1 & 1 \\ x_2 & 1 \\ \vdots & \vdots \\ x_m & 1 \end{pmatrix}}_{A} (a\ b) = \underbrace{\begin{pmatrix} y_1 \\ y_2 \\ \vdots \\ y_m \end{pmatrix}}_{=Y}$$

We note that in this case $n = 2$, because there are only two parameters to determine. A straightforward computation shows

$$A^T A = \begin{pmatrix} \|x\|^2 & \sum x_i \\ \sum x_i & n \end{pmatrix}, \quad (A^T A)^{-1} = \frac{1}{n\|x\|^2 - (\sum x_i)^2} \begin{pmatrix} n & -\sum x_i \\ -\sum x_i & \|x\|^2 \end{pmatrix}.$$

Since

$$A^T Y = \begin{pmatrix} \sum x_i y_i \\ \sum y_i \end{pmatrix},$$

the pseudoinverse solution is

$$\begin{pmatrix} a \\ b \end{pmatrix} = A^+ Y = (A^T A)^{-1} A^T Y = \frac{1}{n\|x\|^2 - (\sum x_i)^2} \begin{pmatrix} n\sum x_i y_i - \sum x_i \sum y_i \\ \|x\|^2 \sum y_i - \sum x_i \sum x_i y_i \end{pmatrix}.$$

This provides the familiar expressions for the regression line coefficients

$$a = \frac{n\sum x_i y_i - \sum x_i \sum y_i}{n\sum x_i^2 - (\sum x_i)^2}, \quad b = \frac{\sum x_i^2 \sum y_i - \sum x_i \sum x_i y_i}{n\sum x_i^2 - (\sum x_i)^2}.$$

It is worth noting that a similar approach can be applied to polynomial regression.

Proposition G.2.1 *Let A be an $m \times n$ matrix of rank n. Then*

(i) $A^T A$ *is positive definite and invertible;*

(ii) $\displaystyle\lim_{t\to\infty} e^{-A^T A t} = \mathbb{O}_n.$

Proof: (i) Since for any $x \in \mathbb{R}^n$ we have

$$\langle A^T A x, x \rangle = x^T A^T A x = \|Ax\|^2 \geq 0,$$

the matrix $A^T A$ is positive definite. Using the properties of matrix rank, $rank(A^T A) = rank(A) = n$, so the matrix $A^T A$ has a maximum rank and hence it is invertible.

(ii) Part (i) can be formulated by stating that the matrix $A^T A$ has positive, nonzero eigenvalues, $\alpha_j > 0$, $1 \leq j \leq n$. Let M be an invertible $n \times n$ matrix which diagonalizes $A^T A$, namely, $A^T A = M Diag(\alpha_j) M^{-1}$. Then $(A^T A)^k = M Diag(\alpha_j^k) M^{-1}$, and hence

$$
\begin{aligned}
e^{-A^T A t} &= \sum_{k \geq 0} (-1)^k (A^T A)^k \frac{t^k}{k!} = M \sum_{k \geq 0} (-1)^k (Diag(\alpha_j))^k \frac{t^k}{k!} M^{-1} \\
&= M \sum_{k \geq 0} (-1)^k Diag(\alpha_j^k) \frac{t^k}{k!} M^{-1} = M Diag\Big(\sum_{k \geq 0} (-1)^k \alpha_j^k \frac{t^k}{k!} \Big) M^{-1} \\
&= M Diag(e^{-\alpha_j t}) M^{-1}.
\end{aligned}
$$

Using $\lim_{t \to \infty} e^{-\alpha_j t} = 0$, it follows that $\lim_{t \to \infty} e^{-A^T A t} = \mathbb{O}_n$.

∎

Bibliography

[1] E. Aarts, J. Korst, *Simulated Annealing and Boltzmann Machines* (John Wiley, Chichester, UK, 1989)

[2] D.H. Ackley, G.E. Hinton, T.J. Sejnowski, A learning algorithm for boltzmann machines. Cogn. Sci. **9**, 147–169 (1985)

[3] S. Amari, Theory of adaptive pattern classifiers. IEEE Trans. Comput. **EC-16**(3), 299–307 (1967)

[4] S. Amari, *Differential-Geometrical Methods in Statistics*. Lecture Notes in Statistics, vol. 28 (Springer, Berlin, 1985)

[5] S. Amari, Information geometry of the EM and em algorithms for neural networks. Neural Netw. **8**(9), 1379–1408 (1995)

[6] S. Amari, Natural gradient works efficiently in learning. Neural Comput. **10**(2), 251–276 (1998)

[7] S. Amari, *Information Geometry and Its Applications*. Applied Mathematical Sciences Book, vol. 194, 1st edn. (Springer, New York, 2016)

[8] S. Amari, H. Park, F. Fukumizu, Adaptive method of realizing natural gradient learning for multilayer perceptrons. Neural Comput. **12**, 1399–1409 (2000)

[9] V.I. Arnold, On functions of three variables. Dokl. Akad. Nauk SSSR **114**, 953–965 (1957)

[10] R. Arora, A. Basu, P. Mianjy, A. Mukherjee, Understanding deep neural networks with rectified linear units. ICLR (2018)

© Springer Nature Switzerland AG 2020
O. Calin, *Deep Learning Architectures*, Springer Series in the Data Sciences,
https://doi.org/10.1007/978-3-030-36721-3

[11] R.B. Ash, *Information Theory* (Dover Publications, New York, 1990)

[12] A. Barron, Universal approximation bounds for superpositions of a sigmoidal function. IEEE Trans. Inf. Theory **39**, (1993)

[13] G. Bartok, C. Szepesvári, S. Zilles, Models of active learning in grouped-structured state spaces. Inf. Comput. **208**(4), 364–384 (2010)

[14] Y. Bengio, P. Frasconi, P. Simard, The problem of learning long-term dependencies in recurrent networks, in *IEEE International Conference on Neural Networks*, San Francisco (IEEE Press, 1993), pp. 1183–1195

[15] Y. Bengio, P. Frasconi, P. Simard, Learning long-term dependencies with gradient descent is difficult. IEEE Trans. Neural Netw. (1994)

[16] J. Bergstra, G. Desjardins, P. Lamblin, Y. Bengio, Quadratic polynomials learn better image features. Technical Report 1337 (Département d'Informatique et de Recherche Opérationnelle, Université de Montréal, 2009)

[17] H. Bohr, Zur theorie der fastperiodischen funktionen i. Acta Math. **45**, 29–127 (1925)

[18] A.E. Bryson, A gradient method for optimizing multi-stage allocation processes, in *Proceedings of the Harvard University Symposium on Digital Computers and Their Applications*, April 1961

[19] P.C. Bush, T.J. Sejnowski, *The Cortical Neuron* (Oxford University Press, Oxford, 1995)

[20] O. Calin, *An Informal Introduction to Stochastic Calculus with Applications* (World Scientific, Singapore, 2015)

[21] O. Calin, Entropy maximizing curves. Rev. Roum. Math. Pures Appl. **63**(2), 91–106 (2018)

[22] O. Calin, C. Udriște, *Geometric Modelling in Probability and Statistics* (Springer, New York, 2014)

[23] E. Çinlar, *Probability and Stochastics*. Graduate Texts in Mathematics, vol. 261 (Springer, New York, 2011)

[24] K. Cho, B. van Merrienboer, C. Gulcehre, D. Bahdanau, F. Bougares, H. Schwenk, Y. Bengio. Learning phrase representations using rnn encoder-decoder for statistical machine translation. IEEE Trans. Neural Netw. (2014), arXiv:1406.1078

[25] M.A. Cohen, S. Grossberg, Absolute stability of global pattern information and parallel memory storage by competitive neural networks. IEEE Trans. Syst. Man Cybern. **SMC-13**, 815–826 (1983)

[26] T.S. Cohen, M. Geiger, J. Köhler, M. Welling, Spherical CNNS. ICRL (2018), https://openreview.net/pdf?id=Hkbd5xZRb

[27] T.S. Cohen, M. Welling, Group equivariant convolutional networks (2016), https://arxiv.org/abs/1602.07576

[28] J.M. Corcuera, F. Giummolé, A characterization of monotone and regular divergences. Ann. Inst. Stat. Math. **50**(3), 433–450 (1998)

[29] R. Courant, D. Hilbert, *Methods of Mathematical Physics*, 2nd edn. (Interscience Publishers, New York, 1955)

[30] G. Cybenko, Approximation by superposition of a sigmoidal function. Math. Control Signals Syst. **2**, 303–314 (1989)

[31] K. Diederik, J. Ba, Adam: a method for stochastic optimization (2014), arXiv:1412.6980

[32] R.J. Douglas, C. Koch, K.A. Martin, H.H. Suarez, Recurrent excitation in neocortical circuits. Science **269**(5226), 981–985 (1995). https://doi.org/10.1126/science.7638624

[33] S.E. Dreyfus, The numerical solutions of variational problems. J. Math. Anal. Appl. **5**, 30–45 (1962)

[34] J. Duchi, E. Hazan, Y. Singer, Adaptive subgradient methods for online learning and stochastic optimization. JMLR **12**, 2121–2159 (2011)

[35] D. Dufresne, Fitting combinations of exponentials to probability distributions. Appl. Stoch. Model Bus. Ind. **23**(1), (2006). https://doi.org/10.1002/asmb.635

[36] N. Dunford, J.T. Schwartz, *Linear Operators*. Pure and Applied Mathematics, vol. 1 (Interscience Publishers, New York, 1957)

[37] E.B. Dynkin, *Markov Processes I, II* (Springer, Berlin, 1965)

[38] A. Einstein, *Investigations on the Theory of Brownian Movement* (Dover Publications, Mineola, 1956) translated by A.D. Cowper

[39] B.R. Frieden, *Science from Fisher Information*, 2nd edn. (Cambridge University Press, Cambridge, 2004)

[40] B.R. Frieden, Extreme physical information as a principle of universal stability, in *Information Theory and Statistical Learning*, ed. by F. Emmert-Streib, M. Dehmer (Springer, Boston, 2009)

[41] B.R. Frieden, B.H. Soffer, Lagrangians of physics and the game of fisher-information transfer. Phys. Rev. E **52**, 2274–2286 (1995)

[42] F.A. Gers, J. Schmidhuber, LSTM recurrent networks learn simple context free and context sensitive languages. IEEE Trans. Neural Netw. **12**, 1333–1340 (2001)

[43] X. Glorot, Y. Bengio, Understanding the difficulty of training deep feedforward neural networks, in *AISTATS'2010* (2010)

[44] X. Glorot, Y. Bengio, Undertanding the difficulty of training deep feedforward neural networks, in *Proceedings of the 13th International Conference on Artificial Intelligence and Statistics 2010*, Chia Laguna resort, Sardinia, Italy. JMLR, vol. 9 (2010)

[45] X. Glorot, A. Borders, Y. Bengio, Deep sparse rectifier neural networks, in *Proceedings of the 14th International Conference on Artificial Intelligence and Statistics 2011*, Fort Lauderdale, FL, USA (2011)

[46] I. Goodfellow, Y. Bengio, A. Courville, *Deep Learning* (MIT Press, Cambridge, 2016), http://www.deeplearningbook.org

[47] I.J. Goodfellow, J. Pouget-Abadie, M. Mirza, B. Xu, D. Warde-Farley, S. Ozair, A. Courville, Y. Bengio, Generative adversarial networks, in *NIPS* (2014)

[48] I.J. Goodfellow, D. Warde-Farley, M. Mirza, A. Courville, Y. Bengio, Maxout networks, in *ICML'13*, ed. by S. Dasgupta, D. McAllester (2013), pp. 1319–1327

[49] L. Goodman, On the exact variance of products. J. Am. Stat. Assoc. **55**(292), 708–713 (1960). https://doi.org/10.2307/2281592., JSTOR 2281592

[50] A. Graves, A. Mohamed, G. Hinton, Speech recognition with deep recurrent neural networks, in *ICASSP* (2013), pp. 6645–6649

[51] P.R. Halmos, *Measure Theory*. The University Series in Higher Mathematics, 7th edn. (Van Nostrand Company, Princeton, 1961)

[52] B. Hanin, Universal function approximation by deep neural nets with bounded width and relu activations (2017), arXiv:1708.02691

Bibliography

[53] T. Hastie, R. Tibshirani, J. Friedman, *The Elements of Statistical Learning*, 2nd edn. (Springer, New York, 2017)

[54] D. Hilbert, *Grundzuge einer allgemeinen theorie der linearen integralgleichungen i*. Gott. Nachrichten, math.-phys. K1 (1904), pp. 49–91

[55] S. Hochreiter, Untersuchungen zu dynamischen neuronalen netzen, Diploma thesis, Technische Universität München, 1991

[56] S. Hochreiter, J. Schmidhuber, Long short-term memory. Neural Comput. **9**, 1735–1780 (1997)

[57] J. Hopfield, Neural networks and physical systems with emergent collective computational abilities. Proc. Natl. Acad. Sci. **79**, 2554–2558 (1982)

[58] J.J. Hopfield, Neurons with graded response have collective computational properties like those of two-state neurons. Proc. Natl. Acad. Sci. **81**, 3088–3092 (1984)

[59] K. Hornik, M. Stinchcombe, H. White, Multilayer feed-forward networks are universal approximators. Neural Netw. **2**, 359–366 (1989)

[60] B. Irie, S. Miyake, Capabilities of three-layered perceptrons, in *IEEE International Conference on Neural Networks*, vol. 1 (1988), pp. 641–648

[61] E. Ising, Beitrag zur theorie des ferromagnetismus. Z. für Phys. **31**, 253 (1925)

[62] H.J. Kelley, Gradient theory of optimal flight paths. ARS J. **30**(10), 947–954 (1960). https://doi.org/10.2514/8.5282

[63] S. Kirkpatrick, C. Gelatt, M. Vecchi, Optimization by simulated annealing. Science **220**, 671–680 (1983)

[64] A.N. Kolmogorov, On the representation of continuous functions of many variables by superposition of continuous functions in one variable and addition. Dokl. Akad. Nauk. SSSR **144**, 679–681 (1957). American Mathematical Society Translation, **28**, 55–59 (1963)

[65] R. Kondor, *Group Theoretical Methods in Machine Learning* (Columbia University, New York, 2008)

[66] R. Kondor, S. Trivedi, On the generalization of equivariance and convolution in neural networks to the action of compact groups (2018), https://arxiv.org/abs/1802.03690

[67] B. Kosko, Bidirectional associative memories. IEEE Trans. Syst. Man Cybern. **18**, 49–60 (1988)

[68] A. Krizhevsky, I. Sutskever, G.E. Hinton, ImageNet classification with deep convolutional neural networks, in *NIPS'2012* (2012)

[69] S. Kullback, R.A. Leibler, On information and sufficiency. Ann. Math. Stat. **22**, 79 (1951)

[70] S. Kullback, R.A. Leibler, *Information Theory and Statistics* (Wiley, New York, 1959)

[71] S. Kullback, R.A. Leibler, Letter to the editor: the Kullback-Leibler distance. Am. Stat. **41**(4), (1987)

[72] L.D. Landau, E.M. Lifshitz, *Statistical Physics. Course of Theoretical Physics*, vol. 5, translated by J.B. Sykes, M.J. Kearsley (Pergamon Press, Oxford, 1980)

[73] Y. LeCun, Modéles connexionists de l'apprentissage, Ph.D. thesis, Université de Paris VI, 1987

[74] Y. LeCun, L. Bottou, Y. Bengio, P. Haffner, Gradient-based learning applied to document recognition, in *Proceedings of the IEEE*, November 1998

[75] Y. LeCun, K. Kavukcuoglu, C. Farabet, Convolutional networks and applications in vision, in *Proceedings of 2010 IEEE International Symposium Circuits and Systems (ISCAS)*, pp. 253–256

[76] Y. Li, K. Swersky, R.S. Zemel, Generative moment matching networks. CoRR (2015), arXiv:abs/1502.02761

[77] S. Linnainmaa, The representation of the cumulative rounding error of an algorithm as a Taylor expansion of the local rounding errors, Master's Thesis (in Finnish), University of Helsinki, pp. 6–7, 1970

[78] S. Linnainmaa, Taylor expansion of the accumulated rounding error. BIT Numer. Math. **16**(2), 146–160 (1976). https://doi.org/10.1007/bf01931367

[79] Z. Lu, H. Pu, , F. Wang, Z. Hu, L. Wang, The expressive power of neural networks: a view from the width, in *Neural Information Processing Systems* (2017), pp. 6231–6239

[80] M.E. Hoff Jr., Learning phenomena in networks of adaptive circuits. Ph.D. thesis, Tech Rep. 1554-1, Stanford Electron. Labs., Standford, CA, July 1962

[81] D.J.C. MacKay, *Information Theory, Inference, and Learning Algorithms* (Cambridge University Press, Cambridge, 2003)

[82] W.S. McCulloch, W. Pitts, A logical calculus of idea immanent in nervous activity. Bull. Math. Biophys. **5**, 115–133 (1943)

[83] J. Mercer, Functions of positive and negative type and their connection with the theory of integral equations. Philos. Trans. R. Soc. Lond. Ser. A **209**, 415–446 (1909)

[84] T. Mikolov, Statistical language models based on neural networks. Ph.D. thesis, Brno University of Technology, 2012

[85] R.S. Millman, G.D. Parker, *Elements of Differential Geometry* (Prentice-Hall, Englewoods Cliffs, 1977)

[86] M. Minsky, Neural nets and the brain: model problem. Dissertation, Princeton University, Princeton, 1954

[87] M.L. Minsky, S.A. Papert, *Perceptrons* (MIT Press, Cambridge, 1969)

[88] M. Mohri, A. Rostamizadeh, A. Talwalkar, *Foundations of Machine Learning*, 2nd edn. (MIT Press, Boston, 2018)

[89] E.H. Moore, On the reciprocal of the general algebraic matrix. Bull. Am. Math. Soc. **26**(9), 394–95 (1920). https://doi.org/10.1090/S0002-9904-1920-03322-7

[90] V. Nair, G.E. Hinton, Rectified linear units improve restricted Boltzmann machines, in *Proceedings of the 27th International Conference on Machine Learning 2010* (2010)

[91] Y.A. Nesterov, A method of solving a convex programming problem with convergence rate $o(1\backslash\sqrt{k})$. Sov. Math. Dokl. **27**, 372–376 (1983)

[92] M. Nielsen, *Neural Networks and Deep Learning* (2017), http://www.neuralnetworksanddeeplearning.com

[93] D.B. Parker, *Learning-Logic* (MIT, Cambridge, 1985)

[94] E. Parzen, On the estimation of a probability density function and its mode. Ann. Math. Stat. **32**, 1065–1076 (1962)

[95] P. Pascanu, Ç. Gülçehre, K. Cho, Y. Bengio, How to construct deep recurrent neural networks, in *ICLR* (2014)

[96] R. Pascanu, T. Mikolov, Y. Bengio, On the difficulty of training recurrent neural networks, in *ICML*. Neural Computation (2013)

[97] R. Penrose, A generalized inverse for matrices. Proc. Camb. Philos. Soc. **51**(3), 406–413 (1955). https://doi.org/10.1017/S0305004100030401

[98] B.T. Polyak, Some methods of speeding up the convergence of iteration methods. USSR Comput. Math. Math. Phys. **4**(5), 1–17 (1964)

[99] M. Rattray, D. Saad, S. Amari, Natural gradient descent for one-line learning. Phys. Rev. Lett. **81**, 5461–5465 (1998)

[100] S. Ravanbakhsh, J. Schneider, B. Póczos, Equivariance through parameter-sharing, in *Proceedings of International Conference on Machine learning (ICML)* (2016), https://arxiv.org/pdf/1702.08389.pdf

[101] R. Rojas, *Neural Networks a Systemic Introduction* (Springer, Berlin, 1996)

[102] F. Rosenblatt, The perceptron: a probabilistic model for information storage and organization in the brain. Psychol. Rev. **65**, 386–408 (1958). Reprinted in: [Anderson and Rosenfeld 1988]

[103] N. Le Roux, Y. Bengio, Deep belief networks are compact universal approximators. Neural Comput. **22**(8), 2192–2207 (2010)

[104] H.L. Royden, *Real Analysis*, 6th edn. (The Macmillan Company, New York, 1966)

[105] H.L. Royden, P.M. Fitzpatrick, *Real Analysis* (Prentice Hall, 2010)

[106] W. Rudin, *Functional Analysis* (International Series in Pure and Applied Mathematics (McGraw-Hill, New York, 1991)

[107] D. Rumelhart, G.E. Hinton, J.R. Williams, *Learning internal representations, in Parallel Distributed Processing: Explorations in the Microstructure of Cognition, Foundations* (MIT Press, Cambridge, 1986)

[108] D.E. Rumelhart, G.E. Hinton, R.J. Williams, Learning representations by back-propagating errors. Nature **323**(6088), 533–536 (1986). https://doi.org/10.1038/323533a0. Bibcode:1986Natur.323..533R

[109] J. Schmidhuber, Deep learning in neural networks: an overview. J. Math. Anal. Appl. (2014), https://arxiv.org/pdf/1404.7828.pdf

[110] S. Shalev-Shwartz, S. Ben-David, *Understanding Machine Learning: From Theory to Algorithms* (Cambridge University Press, Cambridge, 2014)

[111] C. Shannon, A mathematical theory of communication. Bell Syst. Tech. J. **379–423**, 623–656 (1948)

[112] K. Sharp, F. Matschinsky, Translation of ludwig boltzmann's paper "on the relationship between the second fundamental theorem of the mechanical theory of heat and probability calculations regarding the conditions for thermal equilibrium". Entropy **17**, 1971–2009 (2015). https://doi.org/10.3390/e17041971

[113] Xingjian Shi, Z. Chen, H. Wang, D.-Y. Yeung, W.K. Wong, W.C. Woo, Convolutional LSTM network: a machine learning approach for precipitation nowcasting, in *Proceedings of the 28th International Conference on Neural Information Processing Systems* (2015), pp. 802–810

[114] P. Smolensky, Information processing in dynamical systems: foundations of harmony theory, in *Parallel Distributed Processing*, vol. 1, ed. by D.E. Rumelhart, J.L. McClelland (MIT Press, Cambridge, 1986), pp. 194–281

[115] D. Sprecher, On the structure of continuous functions of several variables. Trans. Am. Math. Soc. **115**, 340–355 (1964)

[116] N. Srivastava, G. Hinton, A. Krizhevsky, I. Sutskever, R. Salakhutdinov, Dropout: a simple way to prevent neural networks from overfitting. J. Mach. Learn. Res. **15**, 1929–1958 (2014)

[117] K. Steinbuch, *Automat und Mensch: Kybernetische Tatsachen und Hypothesen* (Springer, Berlin, 1965)

[118] T. Tieleman, G. Hinton, Lecture 6.5—rmsprop, coursera: neural networks for machine learning. Technical report (2012)

[119] N. Tishby, F.C. Pereira, W. Bialek, The information bottleneck method, in *The 37th Annual Allerton Conference on Communication, Control, and Computing* (1999), pp. 368–377

[120] D. Wackerly, W. Meddenhall, R. Scheaffer, *Mathematical Statistics with Applications*, 7th edn. (Brooks/Cole Cengage Learning, 2008)

[121] D. Wagenaar, *Information Geometry for Neural Networks* (Centre for Neural Networks; King's College London, 1998)

[122] S. Wang, X. Sun, Generalization of hinging hyperplanes. IEEE Trans. Inf. Theory **51**(12), 4425–4431 (2005)

[123] P.J. Werbos, Beyond regression: New tools for prediction and analysis in the behavioral sciences, Harvard University, 1975

[124] P.J. Werbos, Applications of advances in nonlinear sensitivity analysis, in *Proceedings of the 10th IFIP Conference, 31.8–4.9 NYC* (1981), pp. 762–770

[125] B. Widrow, An adaptive "adaline" neuron using chemical "memistors". Tehnical Report 1553-2 (Office of Naval Research Contract, October 1960)

[126] B. Widrow, Generalization and information storage in networks of adaline neurons, in *Self-Organizing Systems*, ed. by M. Yovitz, G. Jacobi, G. Goldstein (Spartan Books, Washington, 1962), pp. 435–461

[127] B. Widrow, M.A. Lehr, 30 years of adaptive neural networks: perceptron, madaline ad backpropagation. Proc. IEEE **78**(9), 1415–1442 (1990)

[128] N. Wiener, Tauberian theorems. Ann. Math. **33**(1), 1–100 (1932)

[129] H.R. Wilson, J.D. Cowan, Excitatory and inhibitory interactions in localized populations of model neurons. Biophys. J. **12**, 115–143 (1972)

[130] H.H. Yang, S. Amari, Complexity issues in natural gradient descent method for training multilayer perceptrons. Neural Comput. **10**, 2137–2157 (1998)

[131] R.W. Yeung, *First Course in Information Theory* (Kluwer, Dordrecht, 2002)

[132] Y.T. Zhou, R. Chellappa, Computation of optical flow using a neural network, in *IEEE 1988 International Conference on Neural Networks* (1988), pp. 71–78

Index

Symbols

F-separable, 574
$\Sigma\Pi$-class network, 260
Σ-class network, 260
Σ-finite measure, 701
d-dense, 253
d-system, 697
d_μ-convergence, 323
p-system, 697
ϵ-close neural nets, 220

A

absolutely continuous, 703
absolutely convergent, 304
abstract neuron, 133
action, 530
activation function, 21, 217, 342
AdaGrad, 100
Adaline, 157
Adam, 103
AdaMax, 104
adapted, 713
adaptive implementation, 493
algebra, 209
almost periodic, 224
analytic function, 260, 300
AND, 136
approximation sequence, 176
approximation space, 253
arctangent function, 29
area, 433

B

backpropagation, 173, 177, 180, 237, 483, 550
Baire measure, 257, 700
Banach space, 721
basin of attraction, 78, 126
batch, 188
Bernoulli
distribution, 710
random variable, 459
Bessel inequality, 303, 724
bias, 5
binary classifier, 148
bipolar step function, 22
body-cell, 9
Bohr condition, 224
Boltzmann
constant, 107
distribution, 109, 614
learning, 621
machine, 611, 617
probability, 107

Arzela-Ascoli Theorem, 203, 205, 207, 208, 726
ascending sequence, 272
autocorrelation, 154
autoencoder, 360
average
of the function, 109
average-pooling, 507
axon, 9

© Springer Nature Switzerland AG 2020
O. Calin, *Deep Learning Architectures*, Springer Series in the Data Sciences,
https://doi.org/10.1007/978-3-030-36721-3

Index

Printed in the United States
by Baker & Taylor Publisher Services